Environmental Toxicology

This volume collects selected topical entries from the *Encyclopedia of Sustainability Science and Technology* (ESST). ESST addresses the grand challenges for science and engineering today. It provides unprecedented, peer-reviewed coverage of sustainability science and technology with contributions from nearly 1,000 of the world's leading scientists and engineers, who write on more than 600 separate topics in 38 sections. ESST establishes a foundation for the research, engineering, and economics supporting the many sustainability and policy evaluations being performed in institutions worldwide.

Edward A. Laws
Editor

Environmental Toxicology

Selected Entries from the Encyclopedia of Sustainability Science and Technology

Springer

Editor
Edward A. Laws
School of the Coast and Environment
Louisiana State University
Baton Rouge, LA, USA

This book consists of selections from the Encyclopedia of Sustainability Science and Technology edited by Robert A. Meyers, originally published by Springer Science +Business Media New York in 2012. The contributions "Harmful Algal Blooms" and "Sentinel Species in Oceans and Human Health" originally appeared as part of the Oceans and Human Health section, edited by Darrell Jay Grimes. The contributions "Solar Radiation and Human Health," "Ultraviolet Radiation: Distribution and Variability," and "UV Effects on Living Organisms" originally appeared as part of the Solar Radiation section, edited by Christian A. Gueymard.

ISBN 978-1-4614-5763-3 ISBN 978-1-4614-5764-0 (eBook)
DOI 10.1007/978-1-4614-5764-0
Springer New York Heidelberg Dordrecht London

Library of Congress Control Number: 2012954388

Printed on acid-free paper

Springer is part of Springer Science+Business Media (www.springer.com)

Contents

Chapter 1
Environmental Toxicology, Introduction

Edward A. Laws

Toxicology is the quantitative study of the effects of harmful substances or stressful conditions on organisms. This rather broad field is broken down into three major divisions: economic, forensic, and environmental toxicology. Economic toxicology is concerned with the deliberate use of toxic chemicals to produce harmful effects on target organisms such as bacteria, parasites, and insects. Forensic toxicology is concerned with the medical and legal aspects of the adverse effects of harmful chemicals and stressful conditions on humans. Environmental toxicology, the subject of this chapter, is concerned with the incidental exposure of plants and animals, including humans, to pollutant chemicals and unnatural environmental stresses. On the following pages the status and challenges of this multidisciplinary field of science is discussed within the context of (1) ecological risk assessment, (2) monitoring, (3) mechanisms, (4) fate and transport, (5) prevention, and (6) correctives.

Ecological Risk Assessment

Carroquino et al. (Environmental Toxicology: Children at Risk) discuss the numerous factors that make children at greater risk from exposure to toxic substances than adults. They point out that children have higher exposures relative to body weight than adults because they drink more water (seven times more per kilogram), eat more food (three to four times as much per kilogram for children between the ages

---

This chapter, which has been modified slightly for the purposes of this volume, was originally published as part of the Encyclopedia of Sustainability Science and Technology edited by Robert A. Meyers. DOI:10.1007/978-1-4419-0851-3

E.A. Laws (✉)
School of the Coast and Environment, Louisiana State University, 3141 Energy, Coast and Environment Building, 70803 Baton Rouge, LA, USA
e-mail: edlaws@lsu.edu

E.A. Laws (ed.), *Environmental Toxicology: Selected Entries from the Encyclopedia of Sustainability Science and Technology*, DOI 10.1007/978-1-4614-5764-0_1,

of 1 and 5), and breathe more air (up to twice as much per kilogram for a child less than 1 year old). Furthermore, during the time that their central nervous systems are developing, children are susceptible to permanent neurological damage from exposure to neurotoxins such as methyl mercury, lead, and ionizing radiation. And if endocrine disruptors send false signals to developing reproductive organs, there is a high probability that the resulting dysfunction will be permanent and irreversible. Babies in utero are especially sensitive to the use of addictive substances such as drugs and alcohol by their mothers, and studies have shown that maternal smoking during pregnancy increases the risk of pregnancy loss, stillbirth, and infant mortality.

Willett and Foran (Ecological and Health Risks at Low Doses) discuss what is known and, more frequently, what is not known about the mechanisms associated with human health effects caused by exposure to low doses of xenobiotics. Traditional sigmoidal dose–response models lead to the conclusion that below a certain threshold dose, there are no adverse health effects. Application of such models has led to the use of so-called no observed adverse effect levels (NOAEL), which in turn have become the basis for setting acceptable daily intakes of substances known to cause adverse effects at higher doses.

In most cases, health effects at low doses are extrapolated from experimental results at much higher doses using a linear dose–response model. But in a number of cases, low-dose treatments with a toxicant have been shown to induce a beneficial response, a phenomenon called hormesis. And pretreatment of animals with metals, specifically cadmium, copper, mercury and zinc, is protective for subsequent exposures. The mechanism underlying the acclimation to metals is synthesis of thionein, a protein that normally only occurs in trace amounts in certain tissues (blood, gills, liver, kidney, and intestine) and can effectively sequester toxic metals if the dose is not too great. But in humans, at least, there is a growing consensus that any amount of lead in the body can be damaging, and especially so in children, in whom adverse effects of lead exposure are typically associated with brain damage. Exposure to neurotoxic organophosphorus pesticides such as chlorpyrifos and to endocrine disruptors such as Bisphenol A are additional examples of risks that are poorly quantified because of inadequate understanding of the mechanisms underlying effects at low doses.

Bain (Ecological Risk Assessment and Animal Models) discusses the many issues associated with assessing the risk to plants and animals caused by damage/modification of the environment by human activities. As noted by Bain, this is a multistep process, culminating ultimately in a phase characterized as risk management. The first few steps involve problem formulation and analysis, and these are the focus of the chapter. Much of the information upon which current environmental standards are based has come from either acute (short-term) or chronic (long-term) exposure of organisms to stress and observing the effects on survival, growth and development, and reproduction. Since the number of organisms potentially impacted by toxic substances and stresses is very large, care must be taken to carry out these bioassays with organisms that are in some sense representative of those found in the natural environment, including in particular some of the more sensitive species and life stages. Since such bioassays in most cases involve exposure to only one stress, an alternative, particularly when concern involves a combination of stresses and toxic substances, is to expose test organisms to, for example, contaminated soil or water and observe the impacts on survival, growth, and/or reproduction. In this way, the interaction of a combination of stresses is taken into account.

Alternatives to such bioassays include use of so-called bioaccumulation factors (BAFs) to estimate the effective degree of exposure based on the concentration(s) in the organism. In the last several decades, an approach to quantifying the overall health of a community of organisms has been the use of so-called biotic indices and rapid bioassessment protocols. Such surveys compare populations and community compositions of macroinvertebrates, fish, or periphyton (communities of organism and organic matter attached to surfaces) between a reference site and the site of interest. Finally, and most recently, molecular methods, so-called "omics" (genomics, proteomics, transcriptomics, and metabolomics), have begun to be used, with the expectation that their use will provide a more sensitive assay for stress than the proverbial canary in the coal mine. While this expectation is very likely true, the interpretation of results from omics studies with respect to issues such as risk management is an evolving art.

Medema (Microbial Risk Assessment of Pathogens in Water) discusses the theoretical and practical issues associated with using quantitative microbial risk assessment (QMRA) to estimate the risk to public health from the presence of pathogens in drinking water derived from surface sources. This is an approach that has become popular in recent years because of the absence of relevant epidemiological data that might otherwise be used to estimate risks associated with drinking contaminated water. The use of QMRA has benefitted from the existence of a database that can be used to relate dose to the risk of infection by the protozoan pathogen *Cryptosporidium parvum*, a database, incidentally, derived from studies with human volunteers. An important point noted by Medema is that in many cases, the risk estimates are dominated by relatively infrequent so-called hazardous events, such as floods, when the water distribution system is compromised and/or the source water is seriously polluted. The water distribution system in New Orleans following Hurricane Katrina is a case in point.

The strategy in QMRA is to use mathematical and statistical models to follow drinking water from its source to a treatment system, through a distribution system, and finally to the consumer and to estimate the probability that a person drinking water from his/her tap would be infected by a particular pathogen. The process is an iterative one, and Medema notes that one of the important products of the modeling exercise is identification of knowledge gaps.

In addition to its role in estimating risks due to consumption of contaminated water, QMRA has been used to identify the risks associated with the use of reclaimed wastewater, for example, to irrigate golf courses and crops destined for human consumption, and for recreational water use. As in the case of drinking water, the strategy is to develop a probabilistic model of the public health risk associated with such practices. Persons and agencies responsible for risk management use the output of QMRA to make decisions about appropriate levels of drinking water treatment and to establish appropriate policies for water reuse and recycling.

Monitoring

Soelberg and Furlong (Biosensors and Bioassays for Ecological Risk Monitoring and Assessment) describe a portable device for monitoring pollutants based on surface plasmon resonance (SPR) technology. This is a rather esoteric technology but simplistically takes advantage of the change in refractive index (RI) of a solution adjacent to a gold surface when the solution contains a target compound. The gold surface is coated with so-called recognition elements that provide recognition element (e.g., antibody) attachment sites for the target compound. The technology is designed to detect targets dissolved or suspended in a liquid medium, but the chapter includes a discussion of methodologies for collecting/transferring substances/compounds from air or soil samples or from the surface of solid objects.

Details of the detection process depend on the size of the target, which may range from small molecules to single-celled organisms such as viruses and bacteria. Binding of small molecules to the sensor surface usually does not produce a change in RI large enough to be of practical use, but the detection of such small molecules can be achieved using a competition/inhibition assay involving another larger analyte. For larger molecules such as proteins, binding of the target may effect a change in RI sufficient for detection, but the signal can be amplified by addition of a second recognition element that binds to another site on the target molecule. Whole cells, which are generally too large to be directly detected by SPR technology, may, nevertheless, be detected indirectly as a result of their binding to an appropriate antibody, whose concentration is assayed by SPR in the presence and absence of the cells of interest. Because of the small size of the sensor chips, it is quite possible to design a flow-through system that can detect multiple analytes in sequential fashion. Future directions may involve the further miniaturization of the technology via illumination of an array of recognition element spots in a way that would allow detection of hundreds or perhaps thousands of analytes simultaneously.

Keum et al. (Biomarkers and Metabolomics, Evidence of Stress) review the information and understanding of the effects of stress on organisms that has been derived from a study of metabolomics. This is an emerging area of omics research that lies downstream of genomics and proteomics. The rationale for identifying metabolic fingerprints associated with particular kinds of stress is to provide an early warning mechanism that could be used to trigger corrective action before serious damage has been done. To date, much of the metabolic work has involved single-celled microorganisms and plants and their responses to stresses such as nutrient deprivation, lack of water, pesticides, and/or salt stress. One of the more interesting metabolomic studies has been the investigation of the response of plants to attack by herbivorous insects and pathogens. Evolutionary arguments would lead one to expect that these responses provide some degree of defense for the plants, and elucidation of the relevant upstream portion(s) of the genome has enabled genetic engineers to produce crops resistant to attack by important plant pests. As noted by the authors, the metabolomes of mammals are more complex than those of plants and microorganisms, but ultimately, lessons learned from the study of relatively simple organisms will likely facilitate more informed investigations of the human metabolome, the etiology of various forms of cancer, and the metabolic pathways of preclinical drugs being obvious topics of interest.

Boehm and Soller (Recreational Water Risk: Pathogens and Fecal Indicators) discuss the problems associated with determining the risk of infection associated with recreational water use. Although there is a long list of pathogens that could potentially infect swimmers, more than 95% of all non-food-borne illnesses in the United States are caused by only eight: the viruses norovirus, rotavirus, and adenovirus; the bacteria *Campylobacter*, *Salmonella*, and pathogenic *E. coli*; and the protozoans *Cryptosporidium* spp. and *Giardia lamblia*. Because the concentrations of such pathogens in recreational waters are typically low, monitoring of recreational waters has traditionally relied on the concentrations of so-called indicator organism, which are found in relatively high concentrations in human feces but are not themselves pathogenic. Problems associated with the use of fecal indicator bacteria include the variable ratios of indicators to pathogens and the fact that epidemiological studies have shown that there is a human health risk associated with swimming in waters where there is no apparent source of human fecal contamination. In such cases, the risks of infection are presumably associated with animal excreta, including birds, pigs, and cattle.

A complement and possibly alternative to traditional recreational water quality monitoring techniques is quantitative microbial risk assessment (QMRA), a health risk modeling approach that translates microbial exposures into infection or illness risk estimates. The input to such models includes information on the die-off rates of pathogens in the environment, assumptions about the amount of water swallowed by recreational swimmers, and the infectivity of the various pathogens. Comparisons of the output of QMRA models with the incidence of relevant infections have provided valuable insights concerning the likely contribution of land runoff and animal feces to the risks associated with recreational water use. Historically, the pathogen that has received the most attention in QMRA simulations is the rotavirus because of its high infectivity and the availability of dose–response information. More recent work has focused on the norovirus following publication of relevant dose–response information in 2008.

Mechanisms

Wilson (Environmental Toxicology: Carcinogenesis) provides an excellent overview of the numerous and diverse mechanisms responsible for causing cancer. He points out that nine different conditions must be satisfied before a malignancy can develop and that to satisfy each of these conditions, one or more mutations are needed to alter the intracellular signaling pathways and/or response to the tissue microenvironment. The human body in fact has numerous defenses against cancer, including inter alia, DNA repair systems with phenomenally high fidelity, detoxifying enzymes for reactive species, glutathione to quench reactive electrophiles, and, in the case of mammalian and human cells, the ability to become terminally senescent or undergo self-suicide (apoptosis) upon self-recognition of excess DNA damage or perturbation of cell growth controls. However, these defenses are not infallible, and a variety of mutations generally classified as (1) single-base mutations, (2) chromosomal aberrations, (3) insertions and deletions (indels), and (4) epigenetic mutations can eventually lead to cancer, the probability being directly proportional to the dose and duration of carcinogen exposure.

With a few exceptions (e.g., mesothelioma and exposure to asbestos), most cancers have multiple etiological causes. Therefore, the development of a particular form of cancer does not, in most cases, implicate a mechanism. On the other hand, the mechanisms associated with some forms of stress are rather well established. Chronic alcohol consumption, for example, is associated with oral cavity, pharynx, larynx, esophageal, liver, and colorectal cancer, and breast cancer in women, and suspected to be involved in pancreatic and lung cancer. Alcohol (i.e., ethanol), however, is not itself genotoxic, but acetaldehyde, to which ethanol is converted in the liver, is mutagenic. Chronic alcohol use is associated with three carcinogenic pathways: DNA damage and mutation, inflammatory processes that promote proliferation, and vitamin deficiencies that perturb DNA epigenetic patterns. Exposure to ultraviolet radiation is associated with skin cancer, the most common cancer in the United States. Ultraviolet radiation in the wavelength range 290–320 nm causes DNA mutations leading to signature cyclobutane pyrimidine (6–4) pyrimidone photoproducts and pyrimidine dimers, which, if not repaired, result in signature C (cytosine) to T (thymine) and CC to TT mutations, respectively. Tobacco smoke, probably the most important anthropogenic carcinogen of the twentieth century, contains at least 300 known or suspected carcinogens and is believed to be an important etiological factor in many cases of oral cancers, esophageal, bladder, pancreatic, kidney, and possibly breast and cervical cancers. Many of the carcinogens in tobacco smoke alkylate or arylate DNA, with consequent production of promutagenic lesions, or produce DNA adducts that can lead to primarily G (guanine) to T (thymine) transversions, which are the most common single-base mutations in lung tumors. Despite the remarkable number of carcinogens in tobacco smoke, many years of smoking ($\sim$50 pack years in the case of lung cancer) are generally required before a tumor develops, a testimony, as noted by Wilson, "to the resilience of the human body to the bombardment of such a powerful combination of the carcinogens and toxic agents present in tobacco smoke."

Jones (Environmental Toxicology: Oxidative Stress) points out that understanding of the mechanisms that underlie oxidative stress has changed significantly in recent years. Thirty years ago, oxidative stress was understood to reflect a predominance of prooxidants (agents that initiate radical reactions, oxidize biologic components, or interfere with normal reductive and antioxidant functions) over antioxidants, leading to macromolecular damage. If oxidative stress were as simple as that, supplementation with antioxidants would be expected to provide protection but in fact has been found to be associated with little or no health benefits in humans. Furthermore, the discovery of NADPH oxidases that generate O_2^- and H_2O_2 as signaling molecules suggested that oxidative stress could involve disruption of redox signaling and control mechanisms. Living cells have a variety of mechanisms that almost completely prevent the sort of radical chain reactions associated with the more traditional concept of oxidative stress. Although such chain reactions may be relevant in cases of acute exposure, non-radical oxidants such as H_2O_2, lipid hydroperoxides, quinones, disulfides, and reactive nitrogen species are now believed to be far more important from the standpoint of chronic toxicity. The major mechanisms of non-radical oxidative stress involve primarily effects on thiols and selenols in proteins and secondarily mutagenic damage to DNA. Non-radical oxidants can interfere with reversible oxidation–reduction reactions of thiols or selenols (e.g., the thiol in cysteine, the thioether in methionine, and the selenol in selenocysteine) by causing abnormal oxidation or irreversible modification. These changes alter physiological function in receptor signaling, transcriptional regulation, cell proliferation, angiogenesis, and apoptosis. The hypothesized mechanisms presume that (1) all biological systems contain redox elements that function in cell signaling, (2) organization and coordination of the redox activity of these elements occurs through redox circuits dependent upon common control nodes, (3) the redox-sensitive elements are spatially and kinetically insulated, and (4) oxidative stress is a disruption of the function of these redox circuits caused by specific reaction with the redox-sensitive thiol/selenol elements, altered pathways of electron transfer, or interruption of the gating mechanisms controlling the flux through these pathways. Because biological systems are effectively engineered to conceal stress effects, subtoxic exposures can accumulate without overt signs of adverse effects. A redox proteomics/gene expression approach may therefore provide a means for systems analysis to detect evidence of oxidative stress before overt symptoms become apparent.

Laws (Toxic Chemical Risks) note that most exposure to toxic chemicals is associated with lifestyle decisions and not, as might naively be expected, because a person lives downwind or downstream from a source of pollution. In the United States, by far, the biggest source of exposure is tobacco smoke, which, as noted by Wilson (vide supra), contains several hundred known or suspected human carcinogens. The list of toxic substances includes ammonia, arsenic, benzene, benzo[a]pyrene, cadmium, carbon monoxide, nicotine, and several compounds classified as tobacco-specific N-nitrosamines. The death toll from cigarette smoking accounts for about 20% of all deaths in the United States. Fortunately, the percentage of adults who smoke in the United States has been steadily declining from 42% in 1965 to 21% in 2008–2009.

After tobacco smoke, the most significant source of exposure to a toxic chemical in the United States is alcohol (ethanol) consumption. Health problems associated with chronic alcohol consumption include liver damage, a variety of reproductive system disorders, and damage to the digestive tract. Alcohol abuse accounts for about 100,000 deaths per year in the United States. The specific causes of death include drunken driving, cirrhosis of the liver, cancer, and stroke. In addition to these direct effects on human health, alcohol abuse is associated with half of all homicides and 40% of assaults. Alcohol consumption can be addictive, and about 8% of Americans abuse alcohol or are alcoholic.

Drug misuse/abuse follows alcohol in terms of the frequency with which people are exposed in the United States. The number of deaths associated with unintentional drug overdoses has been rising more or less exponentially during the last 30 years and is now approaching 30,000 per year. The biggest single source of exposure is abuse of prescription analgesics (painkillers). The painkillers in question are opioids, which puts them in the same drug category with drugs such as heroin and cocaine. They effectively block feelings of pain by binding to opioid receptors in the brain, spinal cord, and gastrointestinal tract, but they also produce feelings of euphoria, which accounts for their abuse and compulsive use. When used as prescribed, opioid analgesics seldom result in clinical addiction. Lethal effects are associated with the fact that unprescribed opioid use can lead to a suppression of breathing that may ultimately lead to respiratory failure.

Cocaine and heroin are both illegal opioid drugs. Cocaine has virtually no beneficial application, and there is no approved medication to treat cocaine addiction. Heroin abuse is associated with a variety of health effects related to the tendency of addicts to inject the drug with a needle that in some cases is far from sterile. Thirty-six percent of AIDS cases and an estimated 70–80% of the new hepatitis C cases in the United States each year are attributable to injection drug use. Fortunately, there are a variety of effective treatments available for heroin addiction.

Obesity has become a major public health problem in the United States and reflects an old adage about toxic substances: the dose makes the poison. Food is essential for living organisms, but consumption of too much food can lead to serious health problems, including, in the case of humans, morbidity from hypertension, abnormal amounts of fat in the blood, type II diabetes, coronary heart disease, stroke, gallbladder disease, osteoarthritis, sleep apnea, respiratory problems, endometrial, and social stigmatization and discrimination. At the present time, roughly 100 million adults in the United States are overweight or obese. The solution is to consume fewer calories, but for many people, this is a difficult goal to achieve on a sustainable basis. Behavior therapy may be necessary to effect the necessary changes in lifestyle.

Of the problems with little or no lifestyle connection, lead intoxication appears to be the most pervasive. Several decades ago, the use of leaded gasoline was a significant source of exposure, but this practice was fortunately phased out in the United States in the 1970s. Today, the major lead-related issue is the fact that many residences were painted with lead-based paint prior to 1978, when the use of such paint was banned by the Environmental Protection Agency. Unfortunately, many people live in residences that were painted prior to 1978, and at the present time, it is estimated that 250,000 children in the U.S. have blood lead levels exceeding 100 μg/L, a concentration that the Centers for Disease Control feel should not be exceeded. Effects of lead intoxication are variable and depend on the degree of exposure. In children, the biggest concern is effects on the brain. The apparently obvious solution is to remove the lead-based paint in old homes, but doing so in a safe way requires special training and appropriate equipment.

Many of the other incidents of exposure to toxic chemicals involve accidents in the home. Among the most frequent involve ill-advised mixing of cleaning compounds, the most common examples being bleach mixed with either ammonia, vinegar, or toilet bowl cleaners. Chlorine gas, a highly toxic and corrosive gas, is often one of the products of such mixtures, and in some cases, the chemical reactions are very exothermic and can easily result in explosions.

Fate and Transport

Macauda and Hiscox (Airborne Toxic Chemicals) use three case studies to illustrate issues related to the fate and transport of air toxics (atmospheric pollutants). Perchloroethylene (PERC) is the most common chemical associated with dry cleaning. It has been linked to liver and kidney tumors in rats and is considered to be a carcinogen by the EPA, which has ruled that the use of PERC in dry cleaning must be completely phased out by 2020. Persons living or working in close proximity to dry cleaning establishments are at greatest risk from exposure to PERC. In 2005, the EPA ruled that large dry cleaning establishments use state-of-the-art recovery systems to ensure that PERC is not emitted to the air, and that smaller dry cleaners use generally available control technologies to control emissions. Dry cleaners that operate in apartment buildings are particularly troublesome because residents are obviously living in close proximity to a source of PERC. For such establishments, the EPA has mandated the replacement of so-called transfer machines that require PERC-soaked clothes to be moved from washing to drying machines. These must be replaced by machines that both wash and dry.

Gasoline combustion creates air pollution problems on a broader scale geographically. The most troublesome issues relate to incomplete combustion of gasoline, which can result in the release of numerous pollutants. The pollutant of greatest concern from a public health standpoint is benzene, which has been linked to leukemia. Several approaches have been taken to mitigating the problems associated with incomplete gasoline combustion. First, many states now require gasoline pumps to be equipped with vapor recovery nozzles in order to trap gases that evaporate and would otherwise escape during refueling. Second, gasoline has been reformulated, primarily by addition of ethanol, to burn more completely. This reformulation has significantly reduced carbon monoxide emissions, a cause of smog, but at the same time has led, in the case of California, to a 54% reduction in benzene emissions. A longer-term solution may be electric cars, which could be recharged at night when electric power demand is relatively low. However, such a transformative change would require an unprecedented degree of cooperation between the electric power and automobile industries.

Mercury emissions are an even longer-range problem, with emissions in Asia linked to mercury levels in North America. Although historically, industrial use of mercury has led to some very serious localized human health problems (e.g., Minamata Disease), on a global scale, coal combustion accounts for about 60% of atmospheric emissions. Inhalation of mercury vapor can lead to serious health effects because the mercury reaches the brain, where it can cause irreparable neurological damage. Ironically, the primary post-combustion technologies that reduce mercury emissions do not specifically target mercury but instead are designed to remove oxides of sulfur and nitrogen. In the USA, at least these stack gas–scrubbing technologies reduce mercury emissions by about 33%. In recent years, the USA has considered controlling the remaining mercury emissions from power plants using a cap-and-trade approach, but this idea was struck down by a Court of Appeals ruling. It now appears that mercury emissions from power plants will be regulated by the requirement for use of Maximum Achievable Control Technology.

Oliver and Heathwaite (Invisible Threats: Transfer of Pathogens and Nutrients Through and Across Agricultural Soils) discuss the tendency of agricultural soils to serve as conduits for pathogens and nutrients derived from agricultural activities. The nutrients of primary concern are nitrogen and phosphorus. The pathogens include a somewhat longer list or organisms (*E. coli* O157, *Salmonella* spp., *Campylobacter jejuni*, *Listeria monocytogenes*, *Cryptosporidium parvum*, and *Giardia intestinalis*) that are shed by livestock and are, coincidentally, pathogenic to humans. The ability of soils to sequester N is very much influenced by the fact that soil particles typically have a negative surface charge. Thus, the ammonium ion, NH_4^+, is effectively retained in soils, while nitrate, NO_3^-, is not. Phosphate, although negatively charged, is typically retained in soils because of its tendency to bind with ferric iron (Fe^{3+}), aluminum (Al^{3+}), or calcium (Ca^{2+}). The survival and movement of pathogens is very much influenced by the water content of the soil and the existence of conduits (e.g., macropores) for transport within the soil matrix. Under appropriate conditions, pathogens may survive within soils for literally months before being flushed out by seepage from storm events. Strategies such as no-till agriculture that are intended to minimize the use of pesticides and reduce nutrient runoff may actually exacerbate pathogen export by maintaining the continuity of macropores. In some cases, relatively simple strategies can be used to minimize pathogen problems, a case in point being the application of manure in a broadcast slurry to effect a more rapid destruction of associated bacteria through UV radiation and desiccation. Because of the potentially long time that pathogens and particularly nutrients may be sequestered in soils before being released, a considerable lag in time may exist between the implementation of management practices and discernible changes in the water quality of drainage systems. Awareness of this time lag is important to the informed assessment of management practices.

Bienfang et al. (Bioaccumulation/Biomagnifications in Food Chains) discuss two somewhat different case studies that illustrate problems associated with the bioaccumulation and/or biomagnifications of toxic substances in food chains. Ciguatera fish poisoning (CFP) is the most common food-borne disease related to the consumption of marine finfish, with ~50,000 reported cases each year, but the actual number may be much larger. The toxin responsible for causing CFP comes in several dozen congeners. The various forms of ciguatoxin (CTX) are believed to be metabolites of gambiertoxins produced by various species of the dinoflagellate genus *Gambierdiscus.* Because CTX is toxic at very low concentrations, and in part because there are so many congeners, analytical detection of CTX has been problematic. In fact, virtually all reported cases of CFP are based on symptomology rather than detection of CTX. Short-term effects of CFP are a variety of gastrointestinal disturbances. The long-term and more troublesome effects are of a neurological nature.

As a high-molecular-weight lipid, CTX is a prime candidate for biomagnifications, but because of the aforementioned analytical detection issues, direct evidence of biomagnification of CTX is lacking. The indirect evidence is the fact that the greatest number of reported cases of CFP and the most severe reported symptoms typically involve consumption of carnivorous as opposed to herbivorous fish.

Mercury is a well-known neurotoxin that is found in the environment in several different forms, by far the most troublesome of which is methyl mercury. In contrast to other forms of mercury that are rather efficiently eliminated from the human body, methyl mercury is, at least initially, retained with about 95% efficiency following consumption and is only slowly excreted at a rate of $\sim$1% per day. The damaging effects of methyl mercury are associated with its ability to cross the blood-brain and placental barriers. As a result, exposure to methyl mercury can cause serious and permanent neurological damage. Fetuses appear to be particularly susceptible, as women evidencing no overt symptoms of mercury intoxication have been known to give birth to babies hopelessly brain damaged from the effects of methyl mercury.

Though methyl mercury has been reported at high concentrations in some fish populations, the documented cases of methyl mercury intoxication have all involved anthropogenic discharges. Natural concentrations of methyl mercury in fish are, with very few exceptions, invariably accompanied by even higher concentrations of selenium, the only exceptions being pilot whales and mako sharks. Evidence that has accumulated during the last 50 years has shown that mercury and selenium effectively sequester one another. The absence of any apparent mercury-related adverse health effects associated with the consumption of virtually any species of marine fish reflects the fact that any mercury in the fish is rendered inert by the selenium in the same fish.

Prevention

Kennedy and Tierney (Xenobiotic Protection/Resistance Mechanisms in Organisms) point out that the toxic effects of xenobiotics can be mitigated either through acclimation or adaptation, the former involving intragenerational tolerance and the latter involving intergenerational resistance. Acclimation may involve nothing more complicated than avoidance, in some cases instinctive and in other cases the result of experiential learning. Adaptation involves selection for individuals in a population that for some reason are less sensitive/more resistant to a xenobiotic than other members of the population. Generally speaking, adaptation involves either (1) toxicokinetically derived mechanisms, which alter the way in which organisms absorb, biotransform, and excrete chemicals. and/or (2) toxicodynamically derived mechanisms, in which target sites are modified to reduce sensitivity. An important consideration in the case of toxicokinetic mechanisms is the fact that lipophilic chemicals such as PCBs and chlorinated hydrocarbons are not easily excreted because they partition back into cells and tissues from excretory media (e.g., urine, bile). Water-soluble chemicals are not reabsorbed because lipid membranes of cells lining excretory routes act as barriers to their reuptake. Thus, in the case of lipophilic toxins, toxicokinetic modes of adaptation involve transformation of the toxin into a more polar, water-soluble compound. This is generally accomplished through a series of reactions characterized as Phase I and Phase II. In Phase I reactions, a nucleophilic functional group is introduced or exposed in the parent molecule, and in Phase II, the Phase I metabolite or parent molecule already containing a functional group is conjugated to an endogenous molecule to form a water-soluble product. Cytochrome P-450 mixed-function oxidases, for example, are a superfamily of heme enzymes found in all living species that mediate Phase I transformations. Phase II involves conjugation reactions that produce substrates that are secreted with great efficiency. Glutathione conjugation mediated by glutathione S-transferases is a good example of a Phase II reaction.

Toxicodynamically derived resistance refers to alterations in xenobiotic-receptor interactions, which can be effected by alterations in target site, increased or decreased concentrations of target molecules, or circumvention of target function. Structural changes in the target site can significantly reduce binding of the xenobiotic but, as noted by the authors, can lead to biological effects from reduced functionality of the target in its normal physiological roles. So-called knockdown resistance, i.e., insensitivity of Na^+ channels to insecticide (e.g., DDT and pyrethroids) inhibition, may occur by one or another of several mechanisms, including a reduction in the number of Na^+ channels, changes in the fluidity of nerve membranes, or alterations in the binding characteristics between Na^+ channels and insecticides. In the case of citrus-scale mites, an interesting adaptation to HCN, which interferes with cytochrome oxidase in the electron transport chain, is the use of an HCN-insensitive flavoprotein as an alternate electron carrier in the place of cytochrome oxidase. Application of molecular tools such as toxicogenomics, bioinformatics, and metabolomics has the potential to greatly enhance understanding of the mechanisms by which organisms, including humans, acclimate and adapt to xenobiotics.

Consumption of fish and shellfish is associated with a number of well-documented beneficial health effects, but Meujo and Hamann (Science, Policy, and Risk Management: Case of Seafood Safety) point out that there are also health risks associated with fish and shellfish consumption. The risks reflect the fact that fish and shellfish may contain dangerous concentrations of pathogenic bacteria and viruses, parasites, toxic metals, pesticides, drugs, and a variety of toxins. In the United States, the Food and Drug Administration (FDA) is primarily responsible for ensuring that seafood is safe to eat, and the strategies taken by the FDA to fulfill this mission reflect the diverse sources of the contaminants of concern, practical issues associated with analytical detection and surveillance, and the extent to which postharvest processing can be used to reduce contaminant concentrations to an acceptable level.

Bacterial and viral pathogens can be very effectively eliminated by a process called high hydrostatic pressure (HHP) treatment. However, neither fish nor shellfish survive HHP, and that fact is a deal-killer for oysters destined for the raw oyster market. Two promising alternatives to HHP are X-ray treatment and supercritical CO_2 exposure. Both methods appear capable of killing the relevant pathogens (e.g., *Vibrio parahaemolyticus*) without killing the oysters.

Neurotoxins produced by a variety of microalgae can accumulate in fish and shellfish to levels that cause serious health effects on consumers. The most troublesome of these natural neurotoxins is ciguatoxin, but there is a long list of others, including brevitoxin, okadaic acid, and domoic acid. An additional neurotoxin, scombrotoxin, is not produced by microalgae but instead is a product of a catalytic reaction involving the conversion of histidine into histamine. One of the problematic issues with these neurotoxins is the fact that they are heat stabile, so cooking has little or no effect on the toxicity of seafood containing these compounds. From the standpoint of prevention, what is badly needed in the case of these neurotoxins is a simple and inexpensive test to determine whether seafood is safe to eat. In the case of brevitoxin, an antibody–antigen assay using ELISA (enzyme-linked immunosorbent assay) has recently been reported. The development of similar high-specificity probes for the other troublesome neurotoxins would clearly be desirable.

Correctives

Portier (Bioremediation and Mitigation) describes several case studies in which heterotrophic microorganisms have been used to clean up organic waste in soil and water. The metabolic pathways involved in the biological degradation of various classes of organic compounds are reasonably well understood, and the ease/speed with which microbes can decompose different kinds of organic wastes have been the subject of numerous investigations. Normal alkanes, for example, are oxidized to primary alcohols, which are then further oxidized to carboxylic acids. The carboxylic acids are then converted via beta-oxidation enzymes to acetic acid and a simpler alcohol. Polycyclic aromatic hydrocarbons (PAHs) are generally considered to be the most toxic components of petroleum. Simple PAHs are initially converted via dioxygenase enzymes to catechols and substituted catechols. The strategy in bioremediation is to identify a community of microorganisms with metabolic pathways that enable them to effect such transformations and, insofar as possible, to create environmental conditions that encourage their catabolism of toxic substrates.

The examples cited by Portier illustrate some of the diverse strategies that may be employed to effect bioremediation. The first example concerns the application of adapted microbial consortia to remediate diesel-contaminated soils at an active rail yard. Total petroleum hydrocarbon concentrations in the soil were reduced by almost a factor of 20 over a 6-month timeframe. In the second example, an immobilized packed bed reactor filled with support media was used to provide a habitat for adapted microbes with enzyme systems capable of degrading wood preservative contaminants in groundwater. The contaminated groundwater was pumped through the reactors and then used to irrigate surface vegetation and/or percolate back into the ground. The site in question is now zoned recreational, and groundwater is close to meeting federal maximum contaminant level water quality criteria. The third example involves in situ remediation of groundwater via use of soil seeding bioreactors called bioplugs or bioconduits. Through manipulation of subsurface hydrology, the contaminated groundwater is routed through the bioreactors. Surplus microbial biomass is allowed to escape into the surrounding soil, where the microbes infiltrate contaminated areas and further reduce the concentrations of toxic chemical constituents. Although bioremediation is not an option for all sorts of pollutant (e.g., heavy metals), the strategy has proven to be remarkably successful in cleaning up biodegradable wastes in terms of both cost and sustainability of treatment effects.

Jewett and Wascom (CERCLA, Sustainability and Public and Environmental Health) discuss another kind of corrective strategy: legal action. The particular legal subject is the Comprehensive Environmental Response, Compensation, and Liability Act (CERCLA), one of the last pieces of legislation signed into effect during the administration of President Jimmy Carter. CERCLA was intended to provide a mechanism to clean up "facilities" contaminated by inappropriate disposal of hazardous substances, with the word facility very broadly defined. Hazardous substances are also broadly defined, with one notable exception: petroleum and petroleum products, natural gas, and synthetic gas used for fuel. This is the so-called Petroleum Exception.

Cleanup activities carried out under the "Superfund" component of CERCLA were originally financed by taxes imposed on the chemical and petroleum industries and an environmental tax on corporations. More recently, Superfund has been supported by allocations from the federal general fund and through additional legislation, such as the American Recovery and Reinvestment Act.

Thirty years after its passage, the report card on CERCLA is mixed. There have been some very noteworthy success stories. Of the more than 1,600 sites at one time or another on the National Priority List (NPL), 340 have been delisted, i.e., cleaned up. The case study discussed by Jewett and Wascom, the Agricultural Street Landfill in New Orleans, is a provocative example given the apparently sweeping powers the EPA has to investigate and correct problems at facilities contaminated by inappropriate disposal of hazardous substances. One certainly has to ask why a municipality would elect to locate residential housing and an elementary school on top of a landfill known, during its operative days, as Dante's Inferno. A hazard ranking score (HRS) of 28.5 is sufficient to place a facility on the NPL. The HRS for the Agricultural Street Landfill was 50.0. Why were the results of tests conducted at the site in the 1970s never made public? And why, when the presence of lead, mercury, and arsenic was detected in the soil, was a clay barrier not constructed (as recommended) to contain the contamination prior to the construction of an elementary school?

These are troublesome questions, but there is another side to the coin. The Small Business Liability Relief and Brownfields Revitalization Act has provided a defense for parties who have generated, transported, or arranged for transport of only small amounts of hazardous material and for so-called innocent landowners in the case of facilities acquired through inheritance or bequest. Such revisions have reshaped CERCLA to be more effective and fair, but, as noted by Jewett and Wascom, by no means supplant CERCLA. In summary, CERCLA is a piece of legislation with a very ambitious goal, and as time goes by may well be modified to make it more effective and fair. The need for such legislation is clearly apparent now and is unlikely to go away in the future.

To complete the coverage of this volume, we are pleased to have contributions on Harmful Algal Blooms; Sentinel Species in Oceans and Human Health; Solar Radiation and Human Health; Ultraviolet Radiation: Distribution and Variability; and UV Effects on Living Organisms.

Chapter 2
Airborne Toxic Chemicals

April Hiscox and Mark Macauda

Glossary

Air toxic	Substances that are known or suspected to cause cancer or other serious health effects. Also known as Hazardous Air Pollutants (HAPS).
Anthropogenic source	A source of air toxics created by human beings.
Area source	A single source of pollutant that emits less that 10 t per year of one air toxic, or less than 25 t per year of any combination of air toxics.
MACT (Maximum Achievable Control Technology)	Standard that dictates emission limits of a source is set by the best performing 12% of similar sources, if more than 30 similar sources exist nationally. If less than 30 exist, the best five are used to set the standard.
Major source	A single source of pollutant that emits 10 t per year or more of one air toxic, or 25 t per year or more of any combination of air toxics.
Mobile source	A source of air toxics that moves (such as a car, truck, airplane, or boat).

This chapter was originally published as part of the Encyclopedia of Sustainability Science and Technology edited by Robert A. Meyers. DOI:10.1007/978-1-4419-0851-3

A. Hiscox (✉)
Department of Geography, The University of South Carolina, 709 Bull Street, Columbia, SC 29208, USA
e-mail: hiscox@sc.edu

M. Macauda (✉)
Health Promotion, Education, and Behavior, University of South Carolina, 800 Sumter Street, Columbia, SC 29208, USA
e-mail: macauda@mailbox.sc.edu

E.A. Laws (ed.), *Environmental Toxicology: Selected Entries from the Encyclopedia of Sustainability Science and Technology*, DOI 10.1007/978-1-4614-5764-0_2,

Definition of the Subject and Its Importance

"Air toxics" is a term that is often used colloquially, but for the purposes of this article it will be used following the specific definition set forth by the US Environmental Protection Agency (EPA). Toxic (or Hazardous) air pollutants are those pollutants that are known or suspected of causing cancer or other serious health effects [1]. Air toxics are defined by the Clean Air Act, which explains what pollutants qualify, how different sources are categorized, and how they are to be regulated [2]. The Clean Air Act (CAA) is the most comprehensive modern legislation concerning air quality in the USA. The original 1970 Act defines "Hazardous Air Pollutants" (HAPS) or air toxics as those substances known or suspected of causing cancer, birth defects, or other adverse health problems. It charges the EPA to "Significantly reduce emissions of the most potent air pollutants" [2]. Initially, the Clean Air Act listed 189 chemicals as toxics; during the revision of 1990 caprolactam was eliminated based on new scientific evidence, leaving 188 [1].

Introduction

It can be said that air pollution is as old as civilization itself. With the discovery of fire, humans started emitting substances into the atmosphere. Ancient Rome had issues with pollution of land, water, and air, including emissions from copper and lead production that were greater than Europe during the nineteenth century [3]. In 1954, smog descended on London, primarily from coal burning, for 4 days and killed 4,000 people [4–6]. Air pollution has been known to affect evolution of species; the most famous example being the peppered moths of England, who evolved a darker color to match soot covered trees near industrial operations [7].

Despite its long history and fundamental importance to environmental health and quality of life, air quality is one of the harder aspects of environmental health to grasp, and even harder to control. Unlike water or soil, air cannot be picked up and held. How it travels from place to place is still the subject of intense scientific research. Human understanding of the air, how pollutants enter into it and travel, and how they ultimately affect human and animal health is still being actively investigated. This chapter will summarize the fundamentals of air toxics, a subset of air pollutants, from both an environmental and human health perspective. The following presents a synopsis of the current state of knowledge and discusses future directions for research in this area. The article will start off with a brief summary of air toxics, where they come from and where they go. Readers are encouraged to consult the numerous air pollution textbooks listed in the references for an in-depth treatment of these topics. Following this background material, three specific toxics, benzene, mercury, and perchloroethylene and their roles in modern living will be discussed. The article will finish with a discussion of future directions for better understanding of the role of air toxics from a sustainability standpoint.

Toxics in the Air

Where Do They Come From?

Any introductory textbook on the topic of air pollution starts with the fundamental knowledge that regulated pollutants are from both biogenic and anthropogenic sources. This information is typically presented in the context of the six major criteria pollutants, which constitute the EPA regulations for ambient air quality. Toxics, however, represent a different set of pollutants and by definition it is not necessarily their ambient concentration that is of main concern. It is important to recognize that while some toxics can be emitted from natural occurrences such as volcanic eruptions and forest fires, by and large, toxics come from anthropogenic sources.

Air toxics can be emitted into the atmosphere in several ways. Attrition occurs through the mechanical wearing of physical objects. Activities such as grinding, polishing, sanding, drilling, and spraying can lead to pollution by attrition. This can be found in activities such as drilling for oil [8], or in a local auto body shop that specializes in sanding and painting cars. Vaporization occurs when a liquid converts into its gaseous form. This can happen under temperature and pressure, or because the liquid is volatile (readily evaporates at normal temperatures). Gasoline is one such volatile liquid that readily evaporates [8]. One of the byproducts of gasoline evaporation is benzene, which is discussed in detail below. The third manner in which many toxics become airborne is through combustion. Combustion occurs when a substance is combined with oxygen in a chemical reaction that creates energy. If a fuel is perfectly combustible, meaning that all of it gets used in the combustion process, the two outputs are water vapor and carbon dioxide. Both of which, while not toxic to humans are greenhouse gasses that can contribute to global warming through the trapping of heat energy from the sun. Most fuels, however, are imperfectly combustible, meaning that they do not burn completely. This leaves byproducts such as benzene, toluene, and formaldehyde [8].

For a full understanding of the sources of air toxics and regulation, one must first be familiar with the source categories. The main concern for air toxics regulators are routine emissions: Those that are produced as a byproduct of a process, rather than accidental "one time" emissions [1]. Three major source categories exist. *Mobile sources* include any mechanical object that moves (cars, planes, trains, marine, farming equipment, etc.). Over 50% of air toxic released into the atmosphere come from mobile source emissions [9]. In addition to emitting greenhouse gasses, auto emissions also include several air toxics such as formaldehyde, acetaldehyde, 1,3-butadiene, and particulate matter from diesel engines. These are all considered possible carcinogens by the EPA. The EPA estimates that over half of all cancers caused by outdoor emissions are caused by motor vehicle emissions [9].

While motor vehicles contribute about half of hazardous air toxics, air toxics are also released by stationary sources, which are those sources that do not move. Stationary sources can fall into one of two categories: *Major sources* are a single (or point) source that emits 10 t per year of any one listed toxic air pollutant or 25 t of a mixture of any listed air toxic pollutants. *Area sources* are those single sources that emit less than 10 t of one, or 25 t of more than one air pollutant. Area sources may not contribute a large portion of pollution on their own, but coupled with other small sources, such as one might find in a city, area sources can significantly impact air quality. As the population becomes denser, the impact of these small sources on overall air quality becomes greater; since there are more sources of toxics per unit area (such as combustion sources from cooking and heating). The urban poor are especially affected since they tend to live in more highly populated areas, as is the case in Hong Kong, where those more financially well off can live farther from the populous city center [10].

It should be noted that though sources that emit toxics into the atmosphere are of primary concern, a HAP may pollute an indoor area as well, such as coal burning for cooking or heating. Indoor sources present a concern as they can be contained in smaller spaces, and therefore expose those occupying that space (such as industry workers) to high concentrations of toxics. Indoor sources pose high levels of risk in developing areas like parts of China, where coal is used for home heating [11]. The full list of HAPs contains a wide array of compounds ranging from industrial chemicals to agricultural pesticides, which can be present in the air in particulate (solid), gaseous, or liquid (aerosol) form.

Where Do They Go?

Once a pollutant becomes airborne, it can have many potential pathways. It can remain in the air as is, become a component of a chemical reaction and transform, and/or it can be transported short or long distances and then follow another pathway. The complexities of each of these mechanisms depend on numerous factors ranging from the scale of long-term climates to the scale of short-term turbulence occurring within seconds of release. As long as the toxic is in the air, it can affect the global climate or an individual human in a single breath. The complexities of atmospheric interactions of chemicals can complicate efforts to reduce toxics; Each chemical that is emitted into the atmosphere can react differently with other chemicals, to the point that the reduction of one type of pollutant (such as NO_x) can actually increase the production of other pollutants (such as Ozone and aldehydes), depending on solar radiation, climate, and season [12].

The fate of pollutants can be affected by a number of complex physical transformations, including nucleation and coagulation (by which particles grow in size) or deposition through settling or precipitation. Toxics that leave the air and are deposited on the ground can continue to do environmental and health harm by contaminating land and water (including drinking water). Humans can then be exposed by drinking contaminated water, eating contaminated food, or by coming in contact with contaminated soils [8]. Toxics that have settled or precipitated out can also find their way into the food chain. As larger animals eat smaller animals that are contaminated, toxins can accumulate in biologic tissues, resulting in greater concentrations as one rises higher in the food chain. Human consumption of animals higher in the food chain can result in significant doses of toxin. The classic example of this is mercury contamination of tuna fish, which accumulate the toxin through feeding on smaller aquatic life forms [13].

The interactions between air toxics and the atmosphere is multilayered and dynamic, factors such as time of day, season, wind directions, and temperature can affect what happens to a toxic after it is released into the air. Concentrations of mercury, for example, increase in Korea during winter and at night because of increased coal burning. In contrast, in some areas of China, gaseous mercury concentrations increase in summer because high solar radiation transforms mercury trapped in soil into a gaseous form [11]. Some studies have shown that air toxics tend to have periods of low ambient concentrations that are followed by sharp spikes in output. Logue, Huff-Hartz et al. found that over 50% of the toxics measured by their high time resolved methods occurred during spikes in emissions of short duration; and that different areas (such high traffic areas, areas next to industrial sites, and general city buildings) can have very different release profiles, even if they are in close proximity (in this case 13 km) from one another [14]. Further examples of the importance of microclimates include Raymer et al. who showed that exposures to aldehydes (which are toxics formed from combustion engines, cigarette smoking, oil frying) have been shown to vary widely in different microenvironments, being higher in restaurants but lower in gas stations in the same city [15]. Similarly, areas of high automobile traffic have been shown to equate with high levels of air toxics [16]. With timing and specific location being an important part of the exposure equation, individual human movement patterns are an important part of one's risk for exposure to air toxics [16]. Two individuals living in the same city, or even the same block, can have different exposures depending on when they are present in that block relative to the time of day that an exposure might be taking place.

Though two individuals living in the same city might be exposed to different levels of toxics, they are both likely to be exposed to worse air quality than their rural counterparts. Cities, where most anthropogenic sources are centered, present unique air quality challenges that go beyond the mere high concentration of toxic pollutants due to population. The physical layout of a city affects the air that moves within and around it. The buildings in large cities create a larger surface area to collect heat in the daytime, which they reradiate at night. Warm pollutant filled air concentrates in locations with high numbers of tall buildings (usually at a city's center). That air rises and spreads out over the city, cooling as it moves. As it reaches the city's edge it is drawn back in to fill the void left by the rising warm air in the center. The result is a convection current that circulates pollutants within the city [8]. Though particular microclimates within a city will have their own levels of air toxics based upon what is occurring locally [15], the overall air quality in a city can be worse than it is in less populated areas because of the properties of the physical landscape.

Regulation and Monitoring

As one might imagine, addressing the potential health implications of almost 200 different compounds, all with different sources, means of transport, fates and health effects presents a complex problem. During attempts to regulate air toxics, the EPA has tried several strategies. From 1970 to 1990, the EPA attempted to set standards and regulate each of the 189 toxic air chemicals based on the individual health risks that were posed by each one. The strategy was to identify all of the pollutants that could cause "serious and irreversible illness and death" and reduce the emissions of each to a point that provided a margin of safety to the public. Issues arose with this approach however, as the EPA attempted to create policy based on incomplete scientific evidence; pinpointing the level of reduction needed to avoid health effects proved to be easier said than done! How risk was to be assessed and the level of acceptable risk that should be incurred by the public for each of the pollutants created an inefficient and slow system that only saw regulation of seven pollutants in 20 years [1].

In 1990, with the revision of the Clean Air Act, congress charged the EPA to implement a new system of regulation using a technology-based approach called "Maximum Achievable Control Technology" or MACT. The MACT standard dictates that the emissions limits of a certain toxin be set by the average emissions of the best performing 12% of similar sources, if there are more than 30 sources nationally that are in the same category. If there are less than 30, the average emissions of the best five are used as a standard [1]. Since air toxics may still be harmful even at the emissions levels of the best emitters, the EPA is able to assess how well current technologies reduce risks and has the power to implement additional standards to deal with any remaining risk posed by generation of toxic pollutants. The EPA must explore the remaining health risks posed by a pollutant 8 years after issuing the MACT standard [1].

The 1990 amendments to the Clean Air Act also eliminated caprolactam as a HAP. This shows that the understanding of pollution and regulation is always changing and that the links between regulations, reductions of pollutants, and the health effects of pollutants are still being actively explored [17–20].

In order to evaluate toxic levels and create appropriate regulations, there must be an understanding of how much toxic is being emitted and how it behaves in the environment. Understanding how air toxics behave, the concentration level at any one time, and how and when an individual might be exposed is an evolving science. Both direct sampling of air quality and modeling of air toxics based upon preexisting data (or a combination of both) are used. The Environmental Protection Agency does not monitor the entire USA in order to gain an understanding of the concentration of air toxics; rather they receive reports from industries who account for their own emissions, which are compiled in the Toxic Release Inventory Report [21]. These emissions data are combined with data such as the rate of toxic release, the location, the height of the release, the nature of the pollutants with respect to decay and longevity, wind speed, and wind direction. This information is then broken down by census track and used in models such as the Assessment System for Population Exposure Nationwide (ASPEN) in order to estimate the level of pollutants in a given area of USA [18, 20, 22]. The success and accuracy of a model depends upon the assumptions made and the inputs used, and different modeling techniques can produce different results for the same area [16].

Direct sampling of toxics can also take different forms. Some techniques, for example, collect an average concentration over a period of time, such as 24 h. Others are able to sample at more regular intervals, allowing a profile of different concentrations during different times of the day [14]. In addition, there may be several different sampling technologies for one pollutant, such as is the case with mercury [11]. *When* sampling occurs is also important: Concentrations can vary both by season, and by time of day. Sampling can help to test the validity of models, and help to determine what factors can make a model more accurate [18, 23].

The difficulties in monitoring and sampling air toxics have led to recent controversy. In 2009, USA Today and scientists from University of Maryland and Johns Hopkins tested the air outside 95 schools nationwide and found that seven schools had high enough levels of toxics such as benzene and chromium to elevate risks of cancer, and 57 schools had levels that were higher than their respective state guidelines. The article created a public outcry, and in response to the article, Louisiana and Pennsylvania conducted their own short-term monitoring and found that levels were not high enough to pose a health threat. This led to questions about how areas are monitored, the duration of monitoring, and whether the health threats existed or not [24, 25].

Monitoring and Regulation of hazardous materials is complex. The ways in which air toxics behave in the atmosphere are not always completely understood, and monitoring and modeling air toxic behavior is subject to several variables that can lead to different results. This can lead to controversy about the level of exposure (and thus risk) of the general public. In order to regulate such a complex issue, the EPA has taken a technological approach, based upon the level of control possible, which allows them to establish regulations despite incomplete knowledge about distribution and health effects.

The Toxic Cycle: Specific Examples

Mobile Sources: A Balancing Act

As mentioned above, mobile sources, most notably automobiles, represent a major contributor to air toxic emissions in the USA. Of high concern is benzene. Benzene is a colorless liquid with a sweet odor. It evaporates rapidly and is used in plastics, resins, and synthetic fibers. It also naturally occurs in crude oil and is present in gasoline when it is refined. The benzene that exists in gasoline can become airborne when gasoline evaporates or vaporizes. Most benzene, however, comes from incomplete combustion of other naturally occurring compounds in gasoline, namely, toluene and xylene [9]. Inhaling very high levels of benzene can cause death. At slightly lower levels, it can cause drowsiness, confusion, and increased heart rate. Most seriously, it has been linked to leukemia of the blood called AML, which is a byproduct of benzene's effect on blood cells and bone marrow [9, 26]. Children are of particular concern with respect to air toxics because their bodies are still growing, and air toxics can affect them developmentally [18, 20]. In a study by Whitworth et al., census tracks with the highest levels of air benzene based on ASPEN also had the highest levels of leukemia in children. Other studies have found that concentrations of benzene and 1,3-butadiene exceed EPA health benchmarks at hundreds of locations across the USA [18], making benzene a serious health threat.

In order to eliminate air toxics, including benzene, from gasoline, the 1990 revision of the Clean Air Act mandated that highly polluted cities use reformulated gasoline; which is required to be less likely to vaporize, and have lower levels of benzene and aromatics. In addition, in many states, gasoline pumps are required to have vapor recovery nozzles, in order to trap gasoline vapor that may evaporate during refueling [9]. Reformulation of gasoline to eliminate benzene has been successful. California saw reductions in benzene emissions of 54% in the mid-1990 s, corresponding with the introduction of reformulated gasoline [17].

Another requirement of reformulated gasoline is that it burns more efficiently. This last requirement is accomplished by adding compounds that oxygenate the gasoline, which allow gasoline to burn more completely and efficiently. Two such compounds are ethanol, and MTBE or methyl tertiary butyl ether. The goal of the increased efficiency is to reduce the production of carbon monoxide, a cause of smog and to conserve oil, which is a nonrenewable resource. Thus it serves several purposes [27].

Studies of the effectiveness and side effects of ethanol and MTBE gasoline have shown that the addition of these compounds does reduce CO emissions, but increases the concentrations of formaldehyde and ethyl-aldehyde [27]. Formaldehyde is one of the 188 EPA air toxics [28]. In addition, MTBE itself has been named as a health concern, causing irritation and nervous system effects [26]. Health concerns caused California to phase out MTBE in favor of ethanol as an oxygen booster [17]. The story of MTBE makes an important point about the efforts to lessen the impact of the industrial way of life on the environment, so that mankind sustains itself into the future. Interactions between chemicals and the atmosphere are complex, and adding something new to the equation can have unforeseen consequences, even if the original aim was to reduce the levels of a harmful substance. MTBE was introduced to help reduce smog, but it also increases levels of some air toxics, and has health concerns of its own.

The phase-in of reformulated gasoline highlights another issue in regulation; the economic, and by extension, political trade-offs inherent in regulating an industry: The phase-in of reformulated gasoline in California has led to higher gasoline prices for consumers, as well as disadvantaged smaller refiners who have had more trouble in meeting the regulations in relation to larger refiners [29]. Thus the regulation of air toxics can, and does, become a political and economic issue.

The most transformative solution to the problems of air toxics from gasoline is to simply greatly reduce or eliminate the use of gasoline altogether. In 2010, cars that run on both electric and gasoline power (i.e., hybrids) are gaining greater market share, and automotive manufacturers, both large and small, are designing and constructing vehicles that run exclusively on electric power. A complete transition from gasoline to electric power creates its own obstacles, however. Battery technology is one of them. Traditional lead acid batteries are inexpensive, but are heavy and cannot hold enough power for electric vehicles to be practical. Nickel-metal hydride batteries have more power but can become permanently damaged if over discharged. Lithium ion batteries, a newer technology, can store a good deal of power and are not affected by discharge, but some use cobalt, which is highly toxic [30]. Thus the ideal battery technology is yet to be developed, and existing technologies have inherent trade-offs between efficiency, cost, and environmental impact.

Technology development is not the only obstacle to changing the automotive landscape: The use of Electric cars could be a benefit to energy companies, allowing them to gain revenue as cars charge at night (when little electricity is being used elsewhere). The integration of electric vehicles, however, would necessitate a fundamental change in power infrastructure, which would take time and serious amounts of coordination between industries, such as the power generation industry and the automobile industry, two industries that have never worked together very closely [30]. This brings up a very important issue when talking about sustainability. The challenges and obstacles to sustainable practices are not just limited to developing technology, but are also problems of coordination, commitment and political will, and balancing the trade-offs inherent in changing from one system to another.

Perchloroethylene: Keeping Things Clean

A common site in cities (and suburbs) in the USA is a local, small neighborhood dry cleaner. In fact, there are about 27,000 free standing dry cleaners within the USA [31]. Many people, when entering a dry cleaners, will notice a particular odor; both in the establishment, and on their newly cleaned clothes. That odor is Perchloroethylene or PERC. It historically has been the most common chemical associated with dry cleaning. PERC has been linked to liver and kidney tumors in rats and is considered by the EPA to be a carcinogen. In high concentrations (such as that associated with occupational exposure) it is associated with dizziness, confusion, nausea, difficulty in speaking, unconsciousness, and death [26]. Exposure to PERC can happen from inhaling the chemical directly or ingesting contaminated water. Once in the body, it can be stored in fat cells, and releases itself slowly into the blood stream [26].

The case of PERC provides a good example of how an air toxic is regulated, the different categories of emitters, and how technology provides the basis for guidelines. The 1990 revision of the Clean Air Act required regulation based upon available technology (MACT). The first rules for PERC dry cleaners were laid out in 1991, revised and finalized in 1993 and updated in 2005–2006 as technology improved and health effects were better understood. In regulating the industry, the EPA divided dry cleaners into three groups. The first were major sources (those that produce more than 10 t of PERC per year), the second were smaller area sources that were free standing or in shopping plazas, the third (which was a category that was not originally differentiated in the 1993 guidelines, but believed to cause a greater health threat) were dry cleaners that exist as part of residential buildings [31].

In 2005, the EPA estimated that nine million people lived within 6 miles of a major source of PERC [32]. The main strategy for the large cleaners (the major sources) is the requirement to use state-of-the-art recovery systems that do not permit PERC to be released into the air, but rather trap and filter it. They are also required to implement advanced leak detection practices to ensure that systems do not leak PERC into the atmosphere accidentally [31].

The guidelines for smaller area sources are based upon the standard of "Generally Available Control Technologies" or GACT. GACT is slightly less stringent than MACT and mandates that firms that emit HAPs use control technologies that are commercially available and the most appropriate considering the economic impact on regulated entities and the ability for entities to comply [33].

As per the GACT standard, new free-standing operations are required to build facilities with closed loop systems (that do not release PERC into the air) and implement advanced leak detection practices. Dry cleaners that already exist have to eliminate what are called transfer machines, those that require PERC soaked clothes to be moved to a second machine for drying, in favor of machines that wash and dry together [32].

Dry cleaning operations that are part of apartment buildings pose a greater risk to the public because people are living in close proximity to PERC emissions; thus more stringent regulations were proposed. According to the rule, all existing dry cleaners have to eliminate transfer machines, and are not able to replace PERC machines when the old ones wear out. PERC machines will have to be phased out completely by 2020 [31].

These regulations are not without costs. For example, implementation of these regulations has been estimated to cost the smaller dry cleaners 7.3 million dollars collectively. Though greater efficiency might save 2.7 million annually, these changes represent an overall net cost to the industry [32].

Dry cleaning of clothes is a service millions of Americans use without giving much thought to the health effects or environmental impacts of the service. Even though a small neighborhood dry cleaner might not cause a large environmental problem on its own; a densely populated area, such as a city, might have tens or hundreds of small establishments. Taken together, they can have a large environmental impact. The regulations are foremost to protect health, and those operations that cause a greater immediate threat (such as residential dry cleaners) are regulated more strictly.

The regulations also attempt to take into account the ability of emitters to follow the regulations in a way that is not detrimental to their business, hence the GACT guidelines for smaller area emitters. Regulations are often a give and take.

Mercury: A Global Toxin

Air toxics are truly a global problem, and a good example of that is the harm posed by human exposure to mercury. Mercury is an element and is naturally occurring in the environment. It can be emitted through geothermal and volcanic activity. The main concern for human health, however, comes from human-made (or anthropogenic) sources of mercury emissions. In adults, mercury exposure can damage the brain, heart, kidneys, lungs, and the immune system. Brain effects can lead to personality changes, shyness, tremors, and hearing and sight problems. A bigger risk comes for children. Fetuses and young children exposed to mercury can develop neurological and developmental disorders that can lead to mental retardation, blindness, seizures, inability to speak, and lack of coordination. It can also cause damage to kidney, liver, and digestive systems [34]. Mercury can be passed from mother to infant in breast milk or from mother to fetus directly during gestation.

Mercury poisoning goes far back into human history: Mercury poisoning can result from exposure during metalworking and gold extraction, and there is evidence of increased mercury concentrations globally during times of high levels of gold mining activity, including the ancient Roman and Incan empires [35]. Small-scale gold working is still a major source of mercury pollution in places such as Africa and Asia [36, 37]. Mining and metal working are not the only industries that historically exposed one to mercury: The term "Mad as a Hatter" comes from mercury poisoning among hatters, who used the element in the creation of hat felt from animal pelts and suffered cognitive effects [38].

Currently, humans introduce mercury into their environment through metal mining, waste incineration, refining, and manufacturing. The largest source of mercury comes from the combustion of coal for electricity generation and heating. In fact 60% of global mercury production in 2000 was due to coal burning [39]. Mercury can exist in several forms: Elemental mercury (Hg^0), divalent mercury (Hg^{2+}), and particulate mercury (Hg(p)). Elemental mercury is the most common form. It can exist in the atmosphere for extended periods of time, a year or more, which means that it can travel large distances, up to 1,000 km from its source. In fact, it has been argued that elemental mercury production in Asia contributes to mercury levels in North America [36]. Mercury occurs naturally in coal. The amount of mercury depends upon the type of coal and its origin. When coal is burned for fuel, elemental mercury, particulate mercury, and divalent mercury can be released. Elemental mercury can oxidize into Hg^{2+} and deposit in waterways and land. Hg^{2+} that is deposited in waterways can become transformed by certain bacteria into methyl mercury (CH_3Mg^+). Though these basics are understood, the details about how mercury cycles through the environment are still under investigation [40].

Methyl mercury is highly toxic to humans and animals, and can accumulate in animal tissue. Through this accumulation (called bioaccumulation) humans can receive large doses of mercury from ingesting several smaller animals that have been contaminated. Consumption of toxic fish is the primary risk of mercury globally. In the USA alone, 48 out of the 50 states had to advise residents to avoid consuming fish from certain bodies of water within their borders [39].

Mercury from coal-fired power plants is a particular problem since rapid development, and the associated need for power generation, has increased coal usage worldwide. Poland, for example, derives 93% of its power from coal and has no legal limitations on mercury emissions, making it the fourth highest emitter in Europe [41]. Asia, because of rapid development, is a source of mercury emissions from all sources, with a large portion coming from manufacturing [36] and coal burning [40].

The primary post-combustion technologies that reduce mercury emissions are not specifically for mercury control, but rather for controls of particulate matter, SO_2 and NO_x [40, 42]. Combinations of control technologies can be successful for mercury reduction. South Korea, for example, has seen a 68% reduction in mercury emissions by using a combination of technologies intended for control of SO_2 and NO_x [43]. Though reductions based on current technologies are possible, the lack of specific controls indicates that the science of mercury control is lacking. Even developed countries, such as the USA, do not have a clear direction in dealing with this hazard.

The path of regulation of mercury in the USA has taken a few interesting turns. In the USA, power plants are the largest emitters of mercury, and the Clean Air Act revisions of 1990 charged EPA to study coal burning in the USA. They found that about a third of the mercury found in coal was being removed by current filters (the ones primarily for SO_2 and NO_x), while the other two thirds were being released into the atmosphere. The EPA was in the process of developing new standards for mercury emissions under MACT until 2004, when they changed course and proposed a cap and trade system instead. In the cap and trade system, the mercury emissions would be limited nationally, and power plants would possess credits for each unit of mercury produced. Plants that produced *less* mercury than permitted could then sell their excess units to another power plant. That plant could then produce more. The cap and trade system has come under fire, however, because though it would reduce emission of mercury nationally, it could create a situation where mercury emissions are high locally, affecting residents of an area disproportionately. It was ultimately struck down by the USA court of appeals, who ruled that the EPA did not follow the proper procedures for delisting mercury as an air toxic; therefore, it was still subject to the MACT standards as laid out in section 112 of the Clean Air Act. Currently the US government is likely to reinstate MACT standards for mercury, but until then, mercury control will be left to the states, many of whom adopted stricter standards than the federal ones [44].

Mercury is a highly toxic substance, and its toxic effects have affected humans for millennia. Mercury produced in one corner of the world can affect another, which can make mercury contamination a political issue. The case of the change in US regulations highlights issues of scale and social justice: The cap and trade system may reduce the levels of mercury nationally, but it may do so at the expense of some local populations. That raises questions of whom regulations benefit (such as more well-off people at the expense of poorer who may live near power generation). It also illustrates the importance of scale when thinking about pollution and regulation: Nationally the levels of mercury might look acceptable, but when the resolution is increased and examined locally some places may have unacceptably high levels. This ties back to the issue mentioned above about how an air toxic is measured, when and where; different levels of measurement can have different results, and the action taken based on those measurements can affect health and outcomes.

Human Health

Many of the health effects of different compounds that are considered air toxics are well established, and can be quite dramatic. In the early 1950s, residents of Minamata Bay, Japan began suffering from neurological illness. Adults began exhibiting symptoms such as blurred vision and difficulty in hearing, reduced ability to smell and taste, clumsiness, difficulty in walking, and mood changes. Children and newborns began suffering from impairments in chewing, swallowing, speech, and movement. All of these individuals were suffering from brain damage due to acute poisoning with methyl mercury. Contamination came from a local acetaldehyde plant (acetaldehyde is used in the making of plastics). The residents became ill from eating fish and shellfish that had stored mercury in their bodies through biomagnification. Estimates put the number affected since the 1950s at 200,000 and these acute cases of illness have been unequivocally linked to mercury [45].

The results of air toxics in lower doses (the type one might encounter living next to a dry cleaner, or living in a heavy traffic prone area) is much more difficult to understand and to illustrate. Knowing that chemicals like benzene and PERC have carcinogenic effects in laboratory animals and during heavy occupational exposures is not the same as claiming that the average residents of a city are at greater risk for cancer because they live in a high traffic area. The relationship between toxics and illness is not always clear-cut. Benzene is one of the best understood of the air toxic carcinogens, and the effects have been well explored in animal research as well as in humans who are exposed to benzene in occupational concentrations. Less is known, however, about the relationship between ambient concentrations and cancer, especially in children. Whitworth et al., for example, used EPA models of benzene levels and census track to show that census tracks that have the highest levels of benzene also have higher incidence of childhood leukemia [20]. On the other hand, Reynolds et al. found that traffic density (which is a predictor of benzene and butadiene levels) did not correlate with childhood leukemia rates [46]. Thus even though the health effects of benzene are well established, the effects of environmental concentrations on populations are more difficult to pinpoint. With respect to mercury, Lewandowski et al. note that correlations have been found between mercury levels and autism in 2002, for school districts in Texas, but that these correlations were not found for 2005 and 2006, calling into question the original correlations [47]. Understanding the health effects is an important issue since policy and regulation decisions are made, at least in part (remember the EPA can create stricter regulations based on health effects) upon the impacts on human health. Understanding how these effects unfold in real-world situations is crucial to assessing local impacts of pollutants, the benefits of regulation, and the need for new cleaner technologies.

The health effects of air toxics cannot be completely decoupled from general problems of air pollution. Often, sources of pollution, such as power plants and automobiles, release more than one harmful substance. In addition, multiple sources will cluster in certain areas, such as cities (many cities have power plants, drycleaners, and cars!). In areas that are developing quickly and do not have strict environmental controls, such as parts of Africa, individuals may be exposed to mercury, lead, pesticides, contaminated water, and various air toxics simultaneously [48]. Air toxics and other forms of pollution are also part of larger issues of health and illness: Factors such as several environmental pollutants, nutrition, levels of stress, and poverty interact and can be considered to affect health synergistically, with each problem causing the others to become more dangerous and vice versa [49, 50]. The impacts of toxics are complex and intertwined with the fates of other pollutants and linked to other aspects of human health and well-being. While it is important to understand how a particular toxin reacts, and how it may affect health, it is important not to lose sight of the larger system and the interactions that play a role in determining who and what is ultimately impacted.

Future Directions

The regulation of air toxic chemicals is an evolving science. Not all aspects of transport and fate are completely understood (mercury is a good example of this). The activities that produce air toxics are part of the activities of daily life, and are difficult to terminate. People all around the world still need to get from place to place, and power must be generated. These needs have to be balanced with the impact from the consequences of these actions. The immediate strategy is to make current technologies (coal burning, gasoline automobiles, dry cleaning establishments) as clean as they can be. This presents challenges since the way in which toxics behave in the environment is not always completely understood (as is the case with mercury), regulations have costs to consumers and producers, and monitoring for toxics is still an underdeveloped science. In some cases, such as the push for cleaner burning fuels, attempting to solve one problem (higher efficiency) leads to another (higher levels of volatile aromatics such as benzene). Some toxic-generating activities may be supplanted with newer technologies in future. This presents challenges as well. Changing from gasoline to electric powered cars will necessitate a change in the infrastructure that supports the automobile, and will create the need for collaboration between very different industries. Alternative sources of power, such as nuclear, have their own drawbacks in risks and wastes generated. Even alternatives such as wind have opponents who are concerned about the space needed to house wind generation and the impact on the aesthetics and enjoyment of land [51]. There are political issues at play as well. Much of the reason for the increase in coal burning is that it is a relatively inexpensive means of power generation for countries that are expanding rapidly. Many in those countries argue that they cannot afford to implement stringent environmental control technologies at the expense of the development that they need. The picture becomes more complex with toxics such as mercury, which can travel half way across the globe, so that the energy choices made by one country may affect another.

The way forward is likely a mixture of better monitoring and regulation of current technologies, as well as implementation of newer technologies that have less environmental impact. Many of these will likely exist in tandem (as is seen today with gasoline cars and gasoline electric "hybrid" cars). In order to create a way forward that is sustainable, solutions must be found that can benefit most people. The concerns of developed countries, developing countries, industry, and citizens must be considered, with the overall health and well-being of humans everywhere being the foremost concern. Cooperation between corporations and governments, as well as long-term thinking will be key to meeting these challenges.

Bibliography

Primary Literature

1. US EPA (2000) Taking toxics out of the air. Brochure of the US EPA Office of Air Quality Planning and Standards, Research Triangle Park, NC. Accessed 11/223/2010, Available from: http://www.epa.gov/air/oaqps/takingtoxics/p1.html
2. EPA (2007) The Plain English Guide to the Clean Air Act. 2008, US EPA Office of Air Quality Planning and Standards, Research Triangle Park, NC. Publication No. EPA-456/K-07-001
3. Keys D (2003) How Rome polluted the world. Geogr Campion Interact Publishing 75(12):45–48
4. Bell ML, Davis DL (2001) Reassessment of the lethal London fog of 1952: novel indicators of acute and chronic consequences of acute exposure to air pollution. Environ Health Perspect Suppl 109:389
5. BeruBe K et al (2005) London smogs: why did they kill? Proc R Microsc Soc 40(3):171–183
6. Parris TM (2006) Smog Season. Environment 48(4):3–3
7. Aronson SM (2009) The ignoble fate of the peppered moth. In: Medicine & Health Rhode Island, pp 291–291

8. Corman R (1978) Air pollution primer. The American Lung Association, New York
9. US EPA (1994) Environmental fact sheet: air toxics from motor vehicles. In: Environmental Protection Agency (ed) Ann Arbor.
10. Stern RE (2003) Hong Kong Haze. Asian Surv 43(5):780–800
11. Fang G-C, Wu Y-S, Chang T-H (2009) Comparison of atmospheric mercury (Hg) among Korea, Japan, China and Taiwan during 2000–2008. J Hazard Mater 162(2/3):607–615
12. Luecken DJ, Cimorelli AJ (2008) Codependencies of Reactive Air Toxic and Criteria Pollutants on Emission Reductions. J Air Waste Manage Assoc 1995 58(5):693–701
13. McSwane D (2009) Food safety, in environmental health: from global to local. In: Frumpkin H (ed). Jossey-Bass, San Franscisco
14. Logue JM et al (2009) High time-resolved measurements of organic air toxics in different source regimes. Atmos Environ 43(39):6205–6217
15. Raymer JH et al (2009) Microenvironmental characteristics important for personal exposures to aldehydes in Sacramento, CA, and Milwaukee, WI. Atmos Environ 43(25):3910–3917
16. Oñzkaynak H et al (2008) Modeling population exposures to outdoor sources of hazardous air pollutants. J Expo Sci Environ Epidemiol 18(1):45–58
17. Harley RA et al (2006) Effects of Reformulated Gasoline and Motor Vehicle Fleet Turnover on Emissions and Ambient Concentrations of Benzene. Environ Sci Technol 40(16):5084–5088
18. McCarthy MC et al (2009) Characterization of the Chronic Risk and Hazard of Hazardous Air Pollutants in the United States Using Ambient Monitoring Data. Environ Health Perspect 117(5):790–796
19. Raun LH, Marks EM, Ensor KB (2009) Detecting improvement in ambient air toxics: An application to ambient benzene measurements in Houston, Texas. Atmos Environ 43(20): 3259–3266
20. Whitworth KW, Symanski E, Coker AL (2008) Childhood Lymphohematopoietic Cancer Incidence and Hazardous Air Pollutants in Southeast Texas, 1995–2004. Environ Health Perspect 116(11):1576–1580
21. US EPA (2009) Toxics Release Inventory (TRI) Program Fact Sheet. 2009 9/20/2009 [cited 2010 1/19]; Available from: http://www.epa.gov/triinter/triprogram/tri_program_fact_sheet.htm
22. US EPA (1996) The Aspen model. Technology Transfer Network 1996 National-Scale Air Toxics Assessment 2009 6/6/2007 [cited 2010 1/19]; Available from: http://www.epa.gov/ttn/uatw/nata/aspen.html
23. Venkatram A et al (2009) Modeling the impacts of traffic emissions on air toxics concentrations near roadways. Atmos Environ 43(20):3191–3199
24. Morrison B,Heath B (2009) States' tests of air questioned. *USA Today*
25. Heath B, Morrison B (2009) EPA study: 2.2 M live in areas where air poses cancer risk. *USA Today*
26. ATSDR (1997) American Toxic Substances and Disease Registry Toxfaqs: Tetrachloroethylene. US DHHS PHS ATSDR CAS#127-18-4
27. Graham LA, Belisle SL, Baas C-L (2008) Emissions from light duty gasoline vehicles operating on low blend ethanol gasoline and E85. Atmos Environ 42(19):4498–4516
28. US EPA (2006) Master list of compounds emitted by mobile sources. Environmental Protection Agency (ed) Washington, DC
29. Brown JL (2009) The Market Effects of Going Green: Evidence from California's Wholesale Reformulated Gasoline Market. Energy J 30(3):115–127
30. Tollefson J (2008) Car industry: Charging up the future. Nature 456(7221):436–440
31. EPA (2009) Fact sheet-final amendments to air standards for perchloroethylene dry cleaners. EPA, Editor
32. US EPA (2006) Proposed amendments to air toxics standards for perchloroethylene dry cleaners. Environmental Protection Agency, Editor
33. US EPA (1993) Air pollutants, hazardous; national emission standards: perchlorethylene emissions from dry cleaning facilities. Edited by Environmental Protection Agency, Research Triangle Park, NC, p 49354

34. Department of Health and Human Services (1999) Agency for toxic substances and disease registry, *Mercury*, Department of Health and Human Services, Atlanta
35. Sun L et al (2006) A 2000-year record of mercury and ancient civilizations in seal hairs from King George Island, West Antarctica. Sci Total Environ 368(1):236–247
36. Li P et al (2009) Mercury pollution in Asia: a review of the contaminated sites. J Hazard Mater 168(2–3):591–601
37. Tillett T (2009) The Price of Progress. Environ Health Perspect 117(6):A257–A257
38. Levin P (2007) From mad hatters to dental amalgams: heavy metals: toxicity and testing. MLO Med Lab Obs 39(12):20
39. Selin NE (2009) Global Biogeochemical Cycling of Mercury: A Review. Annu Rev Environ Resour 34:43–63
40. Streets DG, Zhang Q, Wu Y (2009) Projections of Global Mercury Emissions in 2050. Environ Sci Technol 43(8):2983–2988
41. Glodek A, Pacyna JM (2009) Mercury emission from coal-fired power plants in Poland. Atmos Environ 43(35):5668–5673
42. EPA (2002) National Risk Management Research Laboratory Research, Triangle Park, NC. Control of mercury emissions from coal-fired elecric utililty boilers: interim report including errata dated 3-21-02
43. Pudasainee D, Kim JH, Seo YC (2009) Mercury emission trend influenced by stringent air pollutants regulation for coal-fired power plants in Korea. Atmos Environ 43(39):6254–6259
44. Milford JB, Pienciak A (2009) After the Clean Air Mercury Rule: Prospects for Reducing Mercury Emissions from Coal-Fired Power Plants. Environ Sci Technol 43(8):2669–2673
45. Ekino S et al (2007) Minamata disease revisited: An update on the acute and chronic manifestations of methyl mercury poisoning. J Neurol Sci 262(1–2):131–144
46. Reynolds P et al (2002) Traffic patterns and childhood cancer incidence rates in California, United States. Cancer Causes Control 13(7):665–673
47. Lewandowski TA et al (2009) An Evaluation of Surrogate Chemical Exposure Measures and Autism Prevalence in Texas. J Toxicol Env Health A 72(24):1592–1603
48. Nweke OC, Sanders WH III (2009) Modern Environmental Health Hazards: A Public Health Issue of Increasing Significance in Africa. Environ Health Perspect 117(6):863–870
49. Singer M, Clair S (2003) Syndemics and Public Health: Reconceptualizing Disease in Bio-Social Context. Med Anthropol Quart 17(4):423–441
50. Goodman AH, Leatherman TL (eds) (1998) Building a new biocultural synthesis. University of Michigan Press, Ann Arbor, pp 3–42
51. Johansson M, Laike T (2007) Intention to respond to local wind turbines: the role of attitudes and visual perception. Wind Energy 10(5):435–451

Books and Reviews

Ganesan K et al (1996) Air toxics: problems and solutions. Gordon & Breach, Amsterdam, p 342

Griffin RD (2007) Principles of air quality management, 2nd edn. Taylor and Francis, London, 334 pp

Hanna SR et al (1982) Handbook on atmospheric diffusion. Office of Energy Research, US Department of Energy, 102 pp

Harrop O (2002) Air quality assessment and management: a practical guide. Taylor & Francis, London, 504 pp

Heinsohn RJ, Kabel RL (1999) Sources and control of air pollution. Pearson Education, Upper Saddle River, p 696

Jacobson MZ (2000) Atmospheric pollution: history, science and regulation. Cambridge University Press, New York, p 399

Spicer CW et al (2002) Hazardous air pollutant handbook: measurements, properties, and fate in ambient air. Lewis, Boca Raton, p 240

Turner DB (1994) Workbook of atmospheric dispersion estimates: an introduction to dispersion modeling. Lewis, Boca Raton

USEPA (2004) Air toxics risk assessment reference Library, edited by Office of Air Quality Planning and Standards, EPA, Research Triangle Park, NC

Vallero D (2008) Fundamentals of air pollution, 4th edn. Academic Press, San Diego, 942 p

Chapter 3
Bioaccumulation/Biomagnifications in Food Chains

Paul K. Bienfang, Henry Trapido-Rosenthal, and Edward A. Laws

Glossary

Bioaccumulation	An increase in the concentration of a substance obtained from the abiotic environment in one or more tissues of an organism. Bioaccumulation occurs within a given trophic level.
Bioconcentration	An increase in the concentration of a substance from the abiotic and biotic environment in one or more tissues of an organism. Bioconcentration occurs within a given trophic level.
Biomagnification	An increase in the concentration of a substance obtained from organisms at lower trophic levels by an organism at a higher trophic level.
Biosynthesis	Many molecules that can be bioaccumulated or biomagnified by organisms at higher trophic levels enter the food web by being synthesized by organisms at lower trophic levels.

This chapter was originally published as part of the Encyclopedia of Sustainability Science and Technology edited by Robert A. Meyers. DOI:10.1007/978-1-4419-0851-3

P.K. Bienfang (✉)
Oceanography Department, School of Ocean and Earth Science and Technology (SOEST), University of Hawaii, 1000 Pope Road, Honolulu, HI 96822, USA
e-mail: bienfang@soest.hawaii.edu

H. Trapido-Rosenthal
Center for Marine Microbial Ecology and Diversity, School of Ocean and Earth Science and Technology, University of Hawaii, 1680 East-West Road, Honolulu, HI 96822, USA
e-mail: rosenthl@hawaii.edu

E.A. Laws
Louisiana State University, School of the Coast and Environment, Baton Rouge, LA, USA
e-mail: edlaws@lsu.edu

E.A. Laws (ed.), *Environmental Toxicology: Selected Entries from the Encyclopedia of Sustainability Science and Technology*, DOI 10.1007/978-1-4614-5764-0_3,

Food chain	The simplest possible representation of producer–consumer relationships in an ecosystem.
Food web	The representation of the network of producer–consumer–degrader relationships in an ecosystem.
Depuration	The metabolic alteration and excretion of a substance, usually a xenobiotic molecule.
Trophic level	The position in a food chain or food web occupied by a producer or consumer of food.

Definition of the Subject

In a meta-analytical review of biomagnification in marine ecosystems, Gray [1] noted that more than half of the papers purporting to study this topic were using the term incorrectly [1, p. 46]. He then went on to propose careful definitions of the ways in which xenobiotics can gain entry into organisms. As noted above in the glossary, bioaccumulation, bioconcentration, biomagnification, and biosynthesis of substances represent four different routes of entry. Bioaccumulation, bioconcentration, and biosynthesis typically account for the presence of toxic molecules in organisms at low trophic levels, with transfer of compounds from environment to organism via cell membranes in small organisms with high surface-to-volume ratios, and via gill membranes in many larger aquatic species. Accumulation of xenobiotics via biomagnification is more likely to be a phenomenon that occurs in larger animals at higher trophic levels. The suite of processes that leads to the accumulation of toxic compounds in organisms has been an important element of ecological research for decades and an issue of great practical concern relative to anthropogenic impacts on biological communities [2] as well as to the health and well-being of human populations [3].

Introduction

The Fundamental Ecological Context and Simple Food Chain Theory

A brief review of food chain theory will aid in understanding the processes that can lead to magnification of a non- or poorly metabolizable natural toxin or a non-biodegradable pollutant in a food chain. It is within this context of basic ecological principles that we can best view the movement, concentration, and/or transformation of such compounds that result in the degree of magnification that we find expressed in nature.

All animals need food to survive. Food may be burned (i.e., respired) to provide energy, incorporated into the animal's biomass as proteins, fats, and/or carbohydrates to provide essential structural or metabolic needs, or directed toward reproduction. Food used for these purposes takes the form of organic compounds that are collectively composed of (primarily) carbon, hydrogen, oxygen, nitrogen, phosphorus, and (secondarily) literally dozens of other elements. Organic compounds are produced from inorganic molecules such as carbon dioxide and water by a special category of organisms called autotrophs, or primary producers. Plants are the autotrophs with which most persons are familiar. They use the energy in sunlight to effect the transformation of inorganic compounds into organic matter and hence are called photoautotrophs. Less familiar to many people are the autotrophic bacteria that use chemical energy released from the mediation of oxidation-- reduction reactions to synthesize organic matter from inorganic compounds. Such bacteria are referred to as chemoautotrophs. All living organisms that lack the ability to convert inorganic compounds into organic matter are called heterotrophs and depend either directly or indirectly on primary producers as a source of food.

The production of biomass by heterotrophs involves the conversion of one form of organic matter into another, a process called secondary production. Heterotrophs that eat plants are called herbivores. Heterotrophs that eat other heterotrophs are carnivores. Heterotrophs that eat herbivores are primary carnivores, and heterotrophs that eat primary carnivores are secondary carnivores. Not all heterotrophs are so easily classified. For example, some animals have rather cosmopolitan feeding habits and are called omnivores. Other heterotrophs feed primarily on nonliving organic matter (e.g., dead animals or the waste products of other heterotrophs) and are called detritivores. Regardless of whether the consumer is a herbivore, carnivore, omnivore, or detritivore, it is possible to make some broadly relevant generalizations about what happens to organic matter once it has been synthesized by autotrophs.

The successive transfers of food from prey to predator make up what is called a food chain. Each component of such a food chain is called a trophic level. Autotrophs are logically assigned to the first trophic level in such food chains, herbivores to the second trophic level, primary carnivores to the third trophic level, and so forth (Fig. 3.1).

In most natural systems only a small percentage of consumed food is passed along from one trophic level to the next highest trophic level. This percentage is referred to as an ecological efficiency and is typically no more than 20%. This means that the rate at which food is ingested by a trophic level is at least five times greater than the rate at which food is passed on to the next trophic level. Most of the consumed food is either respired or excreted. Because ecological efficiencies are low, the flux of food from one trophic level to the next steadily decreases as one moves up a food chain, and the steady decrease in the flux of food to higher and higher trophic levels usually results in decreases in the biomass of organisms at successively higher trophic levels. A caveat to this last generalization is associated with the fact that organisms at successively higher trophic level tend to become progressively larger, i.e., predators tend to be larger than their prey. There are exceptions to this generalization, mainly in the cases of predators that hunt in packs

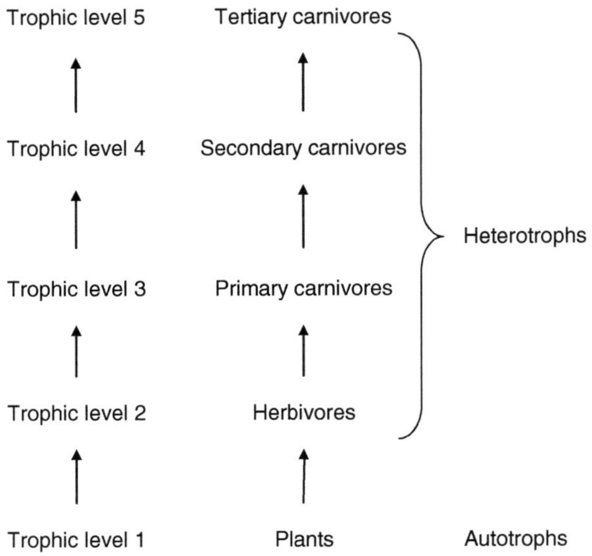

Fig. 3.1 Diagram of a simple grazing food chain with plants as the first trophic level. *Arrows* indicate transfers of biomass

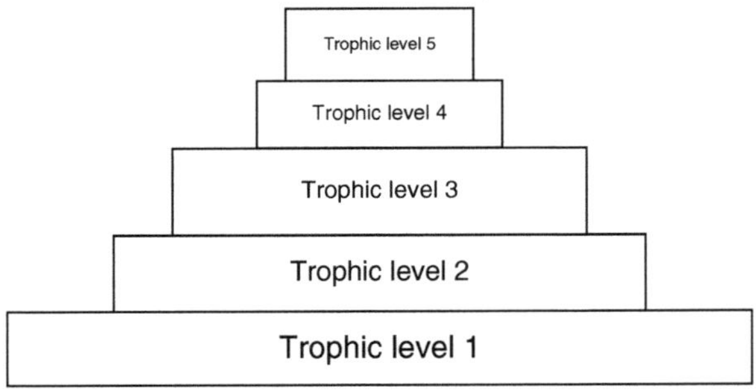

Fig. 3.2 An ecological pyramid. The decrease of biomass on successively higher trophic levels reflects the combined effects of low ecological efficiency and decreasing food requirements per unit biomass on higher trophic levels. Length of horizontal bars is proportional to biomass

or groups, such as wolves or killer whales. Because large animals generally require less food per unit biomass to sustain themselves than small animals, the amount of biomass on successively higher trophic levels does not usually decline as much as one might expect based on the magnitude of ecological efficiencies, but in general one finds a decrease in total biomass at successively higher trophic levels. In other words, although the animals tend to be larger, there are a lot fewer of them. If the total biomass on successive trophic levels is represented by a series of horizontal bars, and if those bars are stacked on top of one another, the resultant figure has the appearance of a pyramid. That pyramid is called an ecological pyramid (Fig. 3.2).

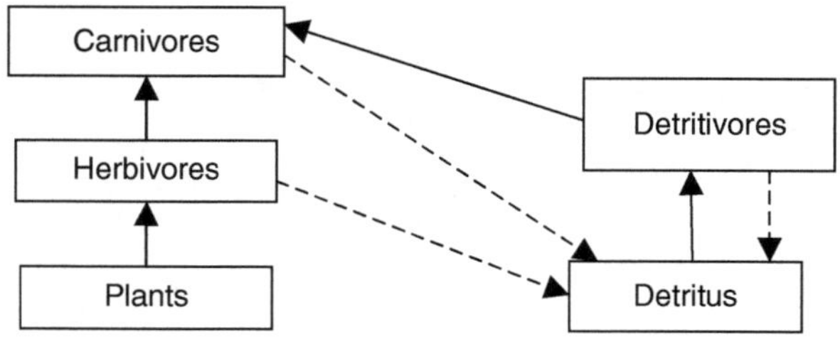

Fig. 3.3 Interactions between the grazing food chain (*left*) and detritus food chain (*right*). *Solid arrows* indicate grazing, i.e., consumption of prey (food) by a predator. *Dashed arrows* indicate excretion and/or death

The food chain consisting of plants, herbivores, and carnivores is often called the grazing food chain. Complementing the grazing food chain is another whose first trophic level consists of nonliving organic matter, i.e., dead remains of animals and plants and organic waste products excreted by heterotrophs. This nonliving organic matter is called detritus, and the associated food chain is called the detritus food chain. Simplistically, it consists of detritus, detritivores, and carnivores, and at the last step, the detritus food chain merges with the grazing food chain (Fig. 3.3). In a healthy ecosystem these two food chains work together to consume almost all the organic matter produced by the autotrophs. It is very important that the organic wastes produced by heterotrophs are consumed by detritivores, because such waste products are almost invariably toxic to the organisms that produce them. One food chain's waste is the other food chain's food, and of course, the carbon dioxide produced by the respiration of all heterotrophs is needed by all autotrophs for primary production. Thus the production and consumption of organic matter in a healthy community of autotrophs and heterotrophs is characterized by a great deal of recycling.

A variation on the concept of food chains is the depiction of feeding relationships as food webs (Fig. 3.4), in which the feeding relationships of (typically) individual species are represented in a two-dimensional diagram. Since a given predator may feed on more than one and perhaps many different prey species, such a diagram tends to have a very complex appearance and has the general characteristics of a web as opposed to a chain. However, careful examination of the feeding relationships depicted by a food web often reveals a pattern consistent with the food chain model. And the principles that govern the fate of consumed food apply regardless of whether feeding relationships are depicted by a food web or food chain. For purposes of this chapter, we will rely on the food chain model to rationalize/explain issues that arise from biomagnification, and we will use the interactions between the grazing and detritus food chains to illustrate how recycling can lead to results that are somewhat counterintuitive from the standpoint of simple food chain theory.

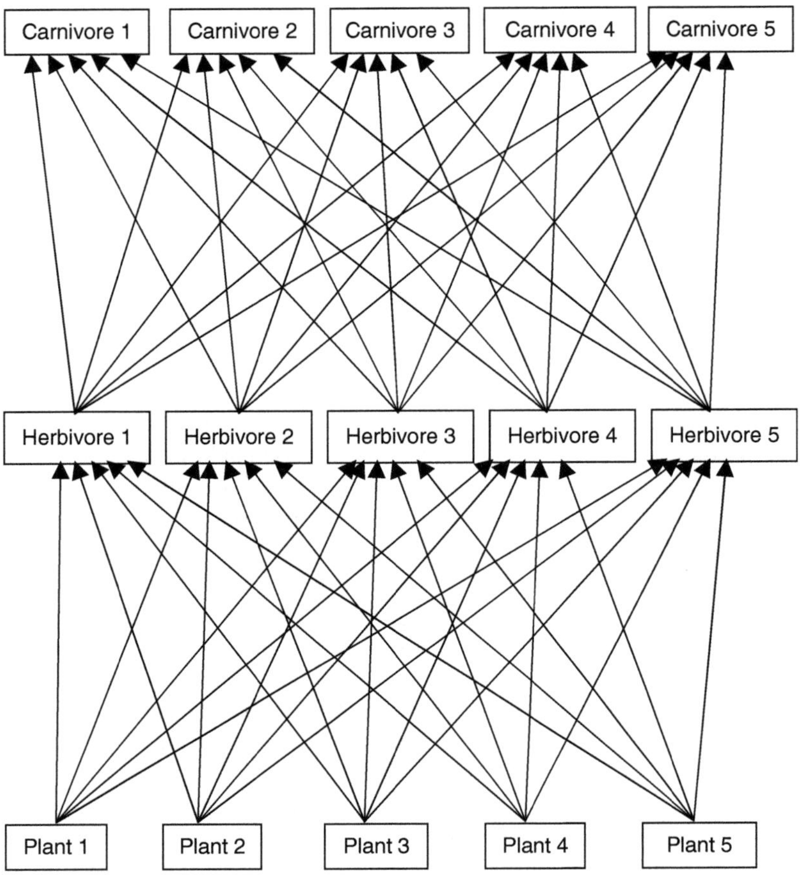

Fig. 3.4 Feeding relationships in a hypothetical food chain in which each of five plant species is grazed by each of five herbivore species, which in turn is grazed by each of five carnivore species. The arrows depicting the feeding relationships have the appearance of a web, and such diagrams are characterized as food webs. In this particular case, however, the flow of organic matter could be just as well represented by a simple food chain consisting of plants, herbivores, and carnivores, as on the left side of Fig. 3.3

Food Chain Magnification Principles

Given the rather low ecological efficiencies that characterize most food chains, it is not hard to imagine how the concentration of a relatively refractory substance might be magnified from one trophic level to the next. Assume for the sake of argument that xenobiotic substance X cannot be respired. This would be the case, for example, if X happened to be a metal such as lead or cadmium. Assume further that for whatever reason X is not effectively excreted. This happens to be the case for methylmercury. For humans, about 95% of ingested methylmercury is, at least initially, retained by the body. Assume as a worst case scenario that X is neither

Table 3.1 DDT residues (DDT + DDD + DDE) in water and organisms taken from a Long Island Salt Marsh [2]

Organism	DDT residues (ppm)
Water	0.00005
Plankton	0.04
Silverside minnow	0.23
Sheephead minnow	0.94
Pickerel (predatory fish)	1.33
Needlefish (predatory fish)	2.07
Heron (feeds on small animals)	3.57
Tern (feeds on small animals)	3.91
Herring gull (scavenger)	6.00
Fish hawk (osprey) egg	13.8
Merganser (fish-eating duck)	22.8
Cormorant (feeds on larger fish)	26.4

respired nor excreted and instead is retained with 100% efficiency in the body of the consumer organism. If the ecological efficiency in the food chain is 20%, then the concentration of X in the food consumed by one trophic level will be magnified by a factor of five in the food consumed by the next trophic level. In a grazing food chain consisting of five trophic levels, beginning with plants and ending with tertiary carnivores, the concentration of X in the tertiary carnivores should be $5^4 = 625$ times higher than the concentration of X in the plants. The fact that the concentrations of ostensibly refractory xenobiotics are sometimes found to be very high in the tissues of top-level carnivores has naturally led to speculation that exactly this sort of mechanism, i.e., biomagnification or food chain magnification, is responsible.

To illustrate this point, Table 3.1 shows concentrations of residues of the pesticide DDT in the water and in various organisms taken from a Long Island, New York, salt marsh [2]. The data illustrate how very low concentrations of toxic substances in the water and/or in organisms at low trophic levels may increase dramatically through the food chain. The residue concentrations increase steadily from the plankton to small fish, to larger and larger fish, and finally to fish-eating birds. The total concentration factor from plankton to fish-eating birds is about 600, which suggests that the concentration factor between successive trophic levels may be about a factor of 5 (see above).

Agreement between theory and observation does not, however, prove that the theory is correct. There are other processes that can affect the concentrations of xenobiotics in the tissues of animals. Fish, for example, must constantly pump water over their gills to breathe. It is certainly possible that direct transfers of xenobiotics from water to fish may occur through the gills. This exchange may lead to high concentrations in the tissues of the fish by a mechanism (bioaccumulation) that has nothing to do with food chains. In a classic study, Hamelink [4] demonstrated that the concentrations of DDT residues in the tissues of fish in experimental ecosystems were unaffected by whether the DDT residues entered the fish directly from the water or through a combination of entry via the water and food chain. The implication was that the food chain route was, at least in that case, insignificant compared to direct uptake from the water. And in contrast to the Woodwell et al. [2] results, Harvey et al. [5] found no evidence for biomagnifications of DDT residues in Atlantic Ocean food chains. In fact, the DDT residues were highest in the plankton and were ten times lower in flying fish, which feed largely on plankton. During the 1960s, when DDT was being used extensively in the United States and many other countries, sharks were found to contain high concentrations of DDT residues, but barracuda, which are also top-level carnivores, contained about 100 times lower concentrations than sharks [6]. The implication of these and other studies is that high concentrations of xenobiotics in top-level carnivores is not *prima facie* evidence of biomagnification, and among organisms occupying ostensibly similar positions in a food chain xenobiotic concentrations may vary widely as a result of differences in the extent to which organisms bioconcentrate xenobiotics directly from the environment.

An additional concern is the interaction between the grazing and detritus food chains, which is overlooked in the simple paradigm of food chain magnification. The consequences of this interaction are illustrated by a study of Isaacs [7, 8] that concerned concentrations of cesium (Cs) and potassium (K) in the muscle tissues of fish from the Salton Sea and Gulf of California. Briefly, the Cs/K ratio had been found to increase by a factor of three between successive trophic levels in the grazing food chain of the Salton Sea. However, the Cs/K ratios in similar fish from the Gulf of California were independent of the trophic level of the fish and about 16 times higher than the Cs/K ratios in the algae. Isaacs [7, 8] developed a mathematically complex unstructured food web to account for these differences. Laws [9] showed that the same results could be explained by taking into account the recycling that occurs between the grazing and detritus food chains (Fig. 3.3).

The conclusion is that a great many factors can influence the concentrations of xenobiotics in the tissues of plants and animals. Biomagnification, or food chain magnification, is one of those factors. But interactions between the grazing and detritus food chain can lead to results that are not predicted when the detritus food chain is ignored, and bioaccumulation in at least some cases can be more important than biomagnification in determining xenobiotic concentrations within organisms. With this introduction, we consider in detail two human health issues potentially impacted by biomagnifications, ciguatera fish poisoning, and methylmercury intoxication.

Ciguatera Fish Poisoning

History

At the 14th Conference of the International Society for the Study of Harmful Algae, Fraga [10] described what is believed to have been the first reported case of ciguatera fish poisoning, an incident that occurred in 1525 in the Gulf of Guinea. An entry in an old manuscript described a case that happened in a Spanish fleet led by Andrés de Urdaneta. The fleet was en route to the Pacific through the Strait of Magellan. The journal entry translates as follows:

> On this island, a very beautiful fish called barracuda was caught by the flagship, and the Captain General invited some of the captains and officers of the King. All who ate the barracuda fell ill from diarrhea and were unconscious, so we thought they had died; however our Creator wanted everyone to be saved.

The fish responsible for this incident, a barracuda (top-level carnivore), is a fish not infrequently implicated in ciguatera fish poisoning. This case was subsequently reported by other early chroniclers, demonstrating how important the incident was.

Five years later, Peter Martyr, the early historian of the West Indies, wrote that when people eat certain fish, they "are attacked by divers strange maladies" [11]. Interestingly, Martyr ascribed this poisoning to movement of a toxin through what we would now call a food chain, from an origin in a plant, to fish, to humans.

In 1601, Pedro Fernandéz de Quirós, a Portuguese navigator in the service of Spain, reported that his entire crew was poisoned after eating "pargo," fish that may well have been snapper or grouper, that they caught while anchored off the island of Espiritu Santo in what is now Vanuatu [12]. The most descriptive of the early accounts of ciguatera, however, comes from the famous British explorer Captain James Cook. In his journal entry for Sunday the 24th of July, 1774 [13], Cook wrote:

> "The Night before we came out of Port two Red fish about the size of large Bream and not unlike them were caught with hook and line of which most of the officers and some of the petty officers dined the next day. In the evening everyone who had eat of these fish were seiz'd with violant pains in the head and Limbs, so as to be unable to stand, together with a kind of Scorching heat all over the Skin, there remained no doubt but that it was occasioned by the fish being of a Poisoness nature and communicated its bad effects to everyone who had the ill luck to eat of it even to the Dogs and Hogs, one of the latter died in about Sixteen hours after and a young dog soon after shared the same fate. These must be the same sort of fish as Quiros mentions under the name of *Pargos*, which Poisoned the Crews of his Ships, so that it was some time before they recovered. We had reason to be thankfull in not having caught more of them for if we had, we should have been in the Same Situation."

Another of Cook's officers identified the suspect fish as "Groopers"; it is likely that the fish Cook's men ate on the 23rd of July, 1774 were either snapper or grouper.

Nature and Extent of the Problem

Ciguatera fish poisoning (CFP) is a worldwide health problem that represents an example of biomagnification of a naturally produced marine toxin. CFP is a non-bacterial, food-borne disease associated with the consumption of seafood that originates from coral reef environments in tropical and subtropical regions of the world. Humans contract the disease by consuming reef fish that have accumulated ciguatoxins as a result of feeding in coral reef habitats. Public health institutions throughout the world rank CFP as the most common food-borne disease related to the consumption of marine finfish [14].

In the introduction to this entry, the distribution of concentrations of DDT residues in organisms representing various trophic levels in a Long Island salt marsh were shown to be positively correlated with trophic level. The biological world frequently does not present itself as being so orderly. The relationships between concentrations of naturally occurring marine toxins and trophic levels are considerably more difficult to discern given the variability caused by the complexity of marine food web interactions [7–9]. The next few paragraphs will explore the scope of the CFP problem, including the organisms responsible for toxin synthesis, natural pathways of the toxin through food webs, and biological and clinical evidence related to ciguatera fish poisoning.

CFP represents an example of biomagnification of a marine toxin that is produced by natural processes and, while presenting substantial difficulties for quantification and study, affects humans like no other toxin from the marine realm. Early work by Randall [14], Banner et al. [15, 16], and Banner and Helfrich [17] indicated that ciguatoxin (CTX) was derived from food consumed by fish and showed that toxicity could be transferred to non-toxic fish via consumption of toxic fish. These contributions evolved into the food web concept for CFP that is the paradigm to this day.

CFP has affected coastal populations and travelers in tropical and subtropical regions throughout the world for centuries. From old ship logs noted above, through WWII chronicles when CFP was a serious problem for military troops stationed in Pacific island locales, to the present day, CFP has been a serious public health problem. Currently, there are at least 50,000 reported cases of CFP cases per year [18, 19], but due to the high degree of misdiagnosis and underreporting, it is estimated that the actual frequency of CFP cases is closer to 500,000 per year [20]. It is estimated that >50% of the populations of small islands in the Caribbean and South Pacific have suffered from CFP. Interested readers are directed to reviews by Lewis [21], Lange [22, 23], and Fleming et al. [24] for additional details on the degree of CFP incidence in populations inhabiting various locales.

Symptomology

CFP produces gastrointestinal, neurological, and cardiovascular symptoms. These normally develop within 12–24 h of eating contaminated fish. Gastrointestinal effects may disappear within 4 days. Normally gastrointestinal symptoms are followed by neurological symptoms. The gastrointestinal symptoms commonly include diarrhea, abdominal pain, nausea, and vomiting. The neurological symptoms that ensue may include numbness and tingling in the appendages (e.g., hands and feet), dizziness, altered hot/cold perception, muscle aches, low heart rates, and low blood pressure. The key pathognomonic symptom is the neurological malady of reversal of hot/cold sensation, i.e., hot feels cold and cold feels hot. Such conditions are caused by the disruption of the "vernacular" that neurons use to communicate with one another, i.e., the ion signals involving sodium and potassium. The suite of symptoms associated with CFP is caused by ciguatoxin's ability to increase Na^+ permeability through the Na^+ channels of excitable cells (e.g., neurons and muscle cells) that are open at normal resting membrane potentials. This enhanced excitability of the membrane in turn affects $Na^+ - Ca^{+2}$ exchange and mobilizes intracellular Ca^{+2}. The primary receptor site of the CTX action is the 5th domain of the Na^+ channel, where it causes increased sodium ion permeability and depolarization of the resting membrane. The prolonged depolarization of nerve cells is understood to cause the suite of sensory discomfort symptoms associated with CFP [25]. CFP symptoms may persist in some form for weeks, months, or even years [26]. Generally, the gastrointestinal symptoms and general weakness last about 1 week, but the neurosensory manifestations (e.g., muscle aches, tingling extremities, and thermal reversals) commonly represent the most prolonged discomfort. It is interesting that CFP intoxication does not confer any immunity in its victims; on the contrary, CFP intoxication results in a heightened sensitivity to ciguatoxin in the victim.

Fortunately, death from CFP is rare (i.e., <0.1%), and when it has occurred it was generally the result of respiratory failure due to cardiovascular shock induced by severe dehydration. Death is frequently associated with the consumption of internal organs of the fish (e.g., liver, brains, gonads, viscera, etc.) in addition to the muscle tissue that is usually consumed. Given the exceptional potency of ciguatoxin (CTX), the low death rate merits some discussion. The two most plausible explanations for the low frequency of fatalities associated with CFP are (a) the large size differential between the "average" human and the average serving of CTX-contaminated fish, and (b) the likelihood that ciguatoxin cannot biomagnify indefinitely in nature. To understand the consumer meal size differential, consider a person weighing between 125 and 200 lbs who consumes an 8-oz portion of contaminated fish; that serving represents 0.25–0.4% of the biomass of the consumer. Since reef organisms do not attain a size of 125–200 lbs, the final "dilution" of the toxin at the point of the human is large with respect to the dilutions that are experienced at various junctures within the marine reef ecosystem. The second explanation is related to the effects that the CTX toxin has on the various animals within the marine reef ecosystem. The ingestion of the ciguatoxin via the consumption of a contaminated organism (lower on the trophic structure) has been shown to cause behavioral changes in animals [27]. If, as has been suggested, such impaired behaviors may make those infected animals more susceptible to death, it would present a natural truncation of the CTX magnification food chain. This, together with the roughly 300-fold dilution factor at the serving-human juncture, may be the reason that humans rarely ingest levels of CTX of sufficient magnitude to cause a fatality.

Environmental and Biological Controls

The distribution of ciguatoxic fish is frequently stated as being within the 35°N and 35°S latitudinal band. Patterns of CFP incidence are well known for the Pacific, Indian, and Atlantic oceans, as well as the Caribbean and Mediterranean Seas [28]. Nonetheless, recent scientific inquires have suggested an even more expansive geographic scale for ciguatera incidence [29]. It has also been suggested that the latitudinal range for CFP is expending as a consequence of warming sea surface temperatures associated with climate changes [30]. The relevance of the expanding latitudinal range lies in the fact that the inclusion of higher latitude regions will include some areas with high population densities, which suggests the potential for increased incidence rates. Several informative reviews on CFP may be found in Lehane and Lewis [31], Bienfang et al. [31], Dickey [26, 27], Dickey and Plakas [32], and references cited therein.

In contrast to the orderly increases of DDT residue concentrations in organisms at various trophic levels presented in the introduction to this entry, the biomagnification scenario for CTX is complex. The difference for CTX is nested in the fact that the concentrations and potency of this naturally produced marine toxin are modified through the complexity of marine food web interactions. Reasons for the apparent complexity of CTX biomagnification include the following:

1. Precursors of the CTX toxin, gambiertoxins, are produced by a variety of dinoflagellates belonging to the genus *Gambierdiscus*. These species vary significantly in terms of their tendency to produce toxins and the potency of the toxin they produce.
2. Metabolites of gambiertoxins, the CTX toxins themselves, appear in a variety of chemical forms (i.e., congeners). Additionally, the toxins are potent at very low dosages and are odorless, tasteless, and highly resistant to destruction/detoxification.
3. Spatial and temporal environmental variability appear to influence the production and flux of CTX into fish populations.
4. The complexity of the marine food web leading to the distribution of CTX among fishes harvested by man presents multiple vectors that may lead to human maladies in a given situation.
5. Humans display variable sensitivities to and symptomologies from CTX intoxication.

A genus of single-celled algae called *Gambierdiscus* (division Phyrrophyta, commonly known as dinoflagellates) produces the gambiertoxins that are the natural precursors of CTX. This genus of dinoflagellates does not form conspicuous blooms that color the water like some other dinoflagellates that are known to produce "red tide" aggregations. The *Gambierdiscus* species normally grow epiphytically on various macroalgae in coral reef ecosystems. The need of these sessile, benthic macroalgae for sufficient sunlight for photosynthesis accounts for the limitation of ciguatera to organisms that inhabit and derive most of their nutrition from the relatively shallow coral reef environments found in warm, tropical areas. *Gambierdiscus toxicus* was originally designated as a new genus and species by Adachi and Fukuyo [28]. Since then, at least ten new species have been added to this genus [33–36]. It has long been suspected that the various *Gambierdiscus* species differ significantly in their degrees of toxin production [33, 34, 37–41]. These dinoflagellates are consumed by herbivorous fish, beginning the process of bioaccumulation and biomagnification through the reef food web. The herbivores are consumed by carnivores, which in turn are consumed by humans.

The Toxin

CTX is a polar, lipid-soluble, highly oxygenated polyether molecule; it consists of 13–14 rings that are fused by ether linkages into a ladder-like structure [42–48]. The seminal toxins that are produced by cells of *Gambierdiscus* are frequently designated as gambiertoxins. These are less polar than the ciguatoxins isolated from fish. These less polar precursors are considerably less potent than the CTX molecule found in high trophic level carnivores. Because the different *Gambierdiscus* species/strains often co-occur at various relative abundances and can vary with respect to their individual toxicities, the resulting variability of toxin flux into the food web has confounded monitoring efforts. Ciguatoxins are piscine metabolites of the gambiertoxins that are assimilated and metabolized through multiple trophic levels of the marine food web. Ciguatoxins in fish may comprise an assemblage of principal ciguatoxins and numerous closely related structural isomers and congeners. Toxins isolated from different regions and/or organisms have been shown to exist in a number of forms that have different molecular weights, chemical structures, and toxicities [42, 48–53]. The recognition of these congeners necessitated new nomenclature, and in the more recent literature an I, P, or C prefix refers to the ocean of origin (i.e., Indian, Pacific, or Caribbean), and a number following refers to a specific congener/form of CTX. Thus, P-CTX-1 would refer to CTX congener 1 isolated from the Pacific Ocean. There are currently understood to be 29 congeners of P-CTX, and 12 of C-CTX. More robust analytical methods have revealed important new details regarding the chemical structure of CTX and its congeners.

Ciguatoxin is one of the most potent natural toxins known; the intake of picogram (i.e., 10^{-12} g) amounts of toxin will cause clinical symptomologies in humans. Such potency at minuscule concentrations has presented imposing challenges to the analytical techniques that may be applied for measurement of CTX in fish tissues. The need for sophisticated chemistry presents exceptional challenges for the quantitative analysis of CTX at concentrations that produce clinical symptomologies in humans. The equipment required for analyses of complex organic compounds such as CTX is expensive to purchase and maintain, and requires highly talented analytical chemists to operate. This requirement for CTX detection has been addressed by a variety of approaches over the decades. Early work involved a number of non-ciguatera–specific bioassay procedures using mouse neuroblastoma cells and various invertebrates and vertebrates; these were followed by radioimmunoassay, enzyme immunoassay, and immunobead assays [54]. More recent analytical developments have focused on chromatographic methods involving high performance liquid chromatography (HPLC) combined with mass spectrometry (MS) [27, 55–57]. These robust analytical techniques are very expensive and require sophisticated infrastructural support. An important consequence of this range in analytical procedures used for CTX analyses has been a lack of uniformity, sensitivity, and specificity of analyses used by investigators working in various regions over the years.

Variability of Incidence

Within areas that are known to be prone to CFP incidence, the distribution of toxic fish can be highly variable on small scales both temporally and spatially. This is thought to be due to ecological factors that influence the degree of dietary intact of CTX by fishes, the degree of CTX biotransformation that may occur, the age/growth rates of the fish, and rates of CTX excretion (which are normally slow, and on the order of months). As a consequence, ciguatoxicity has been found to vary seasonally, from species to species, and from location to location within fairly proximate areas. A species of fish in one area may be relatively free of ciguatera but found in close proximity to an area where that same species is highly ciguatoxic. Such heterogeneous distributions of ciguatoxic fish have been described for numerous locations throughout the Caribbean Sea, Indian, and Pacific Oceans, and have been attributed variously to the combined forces of the territorial nature of carnivorous reef fish, differences in the abundance and/or toxicity of *Gambierdiscus* spp., or variations in feeding patterns between locations.

An example was described by Lewis [58] for the mid-Pacific location of Kiribati, where southern (Tarawa) and western (Maraki) reefs were known for high CFP incidence, but not other reefs in the vicinity. Dierking [59] showed similar spatial heterogeneity in the distribution of ciguatoxic *Cephalopholis argus* throughout coastal areas on the islands of Oahu and Hawaii, Hawaii. The biomass of *C. argus*, introduced to Hawaii from French Polynesia in the 1950s, now surpasses the biomass of all other reef predators combined, and is one of the principal species associated with CFP in this area. Dierking's [59] findings showed substantial (8% versus 24%) differences in the prevalence of toxic *C. argus* between the islands of Oahu and Hawaii, large variations in toxicity within sites, no clear patterns between sites on either island, and very weak correlations of toxicity and fish size. Similar lack of correlation of either incidence or severity of ciguatoxicity with fish size, and high degrees of spatial heterogeneity in the incidence of ciguatoxicity in *C. argus* has been shown by Bienfang and DeFelice (unpublished data; www.fish4science.com).

The complexity of the marine food web leading to ciguatoxic fishes that are consumed by man has challenged marine scientists who have studied the ecosystem looking for vectors that contribute to this human malady. There is evidence that individual *Gambierdiscus* strains differ considerably in their inherent ciguatoxin production capabilities [34, 39, 56, 60]. Laboratory studies have suggested that individual *Gambierdiscus* strains are adapted to particular environmental regimes (e.g., different relationships between light intensity and growth rate, toxicity and light intensity, temperature, and/or salinity [60–63]. Changes in the dietary intake of ciguatoxins within food webs leading to fish have been described as being influenced by a variety of stochastic changes in predator–prey relationships within areas. Despite a low correlation between changes in ciguatera incidence rates and catastrophic events, it has also been repeatedly speculated that anthropogenic disruptions (e.g., shipwrecks, dredging) and/or natural environmental disturbances (e.g., hurricanes, cyclones, coastal inundations) may lead to the availability of new surfaces for colonization by macrophytic algae and/or the associated epiphytic *Gambierdiscus.*

The apparent differences in sensitivity of individuals to ciguatoxic fish and/or associated variability in the severity of symptomologies from CTX intoxication among afflicted victims, together with a high (e.g., 90%) degree of underreporting has also confounded elucidation of processes associated with ciguatera prevalence. Although symptomologies that are consistent with an interference with sodium channel activity are characteristically associated with CTX [27, 57], other biochemical aberrations have also been implicated [56, 64, 65]. Also, it is relatively infrequent that remains of a meal that apparently caused CFP in a human have been available for credible analysis of ciguatoxin. When possible, analyses have shown ciguatoxicity in humans at concentrations in fish ranging from 0.1 to greater than 1 part per billion P-CTX-1.

Evidence of Biomagnification

Despite all these vagaries in the databases associated with CFP, the role of biomagnifications is suggested by the fact that high-level carnivores are by far the fish most frequently associated with the presence of CTX. The data pertaining to CFP incidence in fishes comes from both clinical records and sampling studies that broadly examined various reef species for ciguatoxicity irrespective of associated incidence of reported symptomologies by humans. Not surprisingly, data from the clinical records have had considerably more influence on our perspective regarding the piscivorous vectors leading to CFP. This comes from the standpoints of both their numerical abundance and their role in initiating the broad-scale studies of ciguatoxicity across fish species inhabiting reef ecosystems in various geographic locations. CTX is capable of bioaccumulating through the food chain through herbivorous and carnivorous (i.e., predatory) reef fish.

The fish most commonly involved in CFP incidents within various geographic locations have been listed/prioritized in several summaries [54–56, 59, 66–71]. Such listings show that the fish species that are most commonly implicated with ciguatoxicosis are primarily carnivorous. The prominent carnivorous taxa include the barracudas (*Sphyraena* spp.), groupers (*Epinephelus* spp.), jacks (*Caranx* spp.), snappers (*Lutjanus* spp.), mackerels (*Scomberomorus* spp., especially the Spanish mackerel), kingfish (*Seriola* spp.), and moray eels (e.g., *Gymnothorax* spp.).

Not all fish implicated in CFP are carnivores. Non-carnivorous taxa include the detritivorous/omnivorous surgeonfish (*Ctenochaetus* spp.), parrotfish (*Scarus* spp.), and wrasses (e.g., *Cheilinus* spp.). In the Pacific, the detritivorous grazer, *Ctenochaetus striatus* (a surgeonfish), is frequently ciguatoxic and is thought to be a key vector amplifying CTX up the food chain. High levels of CTX and other *Gambierdiscus*-associated toxins in biodetritus are believed to account for the frequency of ciguatoxicity in *C. striatus*, and occasionally, mullets (e.g., *Mugil* spp.) because both have detritivorous grazing behaviors [72]. The mouthparts of the *C. striatus* are especially well adapted for feeding on phytodetritus and certain filamentous macroalgae known to commonly be host to the epiphytic *Gambierdiscus* spp. Surgeonfish and parrotfish tend to be dominant families by weight on many coral reefs, and among the most common prey of larger piscivores. Interestingly, carnivorous species that are grown in tropical aquaculture settings have been shown not to display ciguatoxicity, despite the presence of *Gambierdiscus* spp. growing on the confining vessels used in the culture process [73]. This lack of ciguatoxicity due to bioaccumulation has been attributed to the fact that the majority of the nutrition of such cultured carnivores is provided through the provision of prepared feeds supplied to the culture systems, rather than (normal) grazing upon a heterogeneous suite of natural feedstuffs.

The perceived variety in CFP symptomologies between fishes from the Caribbean and Pacific has variously been attributed to differences in the gambiertoxins produced by different predominant *Gambierdiscus* species in these locales, and/or the portions of the fish traditionally consumed. Helfrich et al. [74] estimated that the relative toxicity of fish liver tissue was 50 times greater than that in the muscle tissue. Vernoux et al. [75] similarly showed that the relative toxicity of the liver and viscera tissues (e.g., brain, gonads, heart) was considerably higher than that of the muscle tissue. This has been attributed to the preferential sequestering of the lipid-soluble CTX molecule in the most lipid-rich portions of the fish. Such differences in the portions of a given fish species that are actually ingested by an assortment of populations from various geographic and/or ethnic backgrounds confound the analysis of the degree of trophic-level influence independent from other, potentially aliasing factors.

The high trophic level barracuda (*Sphyraena* spp.) has been shown to accumulate and retain CTX in its muscle tissue for extended periods of time, and toxicity was shown to be inducible. Banner et al. [16] and Helfrich and Banner [76] fed *Lutjanus sp.* ciguatoxic fish and showed the induction of toxicity within 6 months, and the subsequent retention of potency for 30 months following the cessation of feeding ciguatoxic fish. There exist chemistry results suggesting that CTX precursors from dinoflagellates may be oxidatively biotransformed to numerous congeners within herbivorous and carnivorous fish [53]. Biomagnification effects also appear to be evidenced in the human symptomologies caused by the consumption of herbivorous versus carnivorous fish. Bagnis and Legrand [77] and Kodama and Hokama [78] reported that the ingestion of ciguateric herbivorous fish caused mostly gastrological and mild neurological symptoms, while the consumption of carnivorous fish was associated with a broader and more severe suite of symptoms. This was attributed to differences in the ciguatoxin congeners produced by various (predominant) *Gambierdiscus* species in these locales, inevitably leading to an assortment of compounds present in fish representing a range of trophic positions.

There are many obstacles to a clearer definition of the prevalence of ciguatoxic fish and/or the distribution of ciguatoxic fish within a species or trophic structure. These impediments include vagary due to the temporally sporadic and spatially patchy distribution of both the causative agent (*Gambierdiscus* spp.) and the feeding patterns of the herbivorous and carnivorous fish species that lead to human CFP intoxications. The primary evidence for the biomagnification of CTX up the trophic structure is found in both the higher frequencies of CTX incidence in carnivores and the severity of symptomologies that have been observed in association with human consumption of carnivorous fish versus herbivorous fish.

Methylmercury

Hg: Sources and Cycling

Mercury (Hg) is a heavy metal. In nature, it exists in three chemical forms. These include the pure, elemental form (i.e., the form that in times past was used in thermometers), inorganic mercury compounds (i.e., mercury salts), and organic mercury (mercury incorporated into organic molecules). If you are old enough to remember gasoline at less than a dollar per gallon, you probably remember mercury as that strange, heavy, silver liquid that you rolled around on the counter top or in your hand. You are also likely to associate it with thermometers. If you are younger than that, you are most likely to think of mercury as a pollutant.

Methylmercury in the form of compounds such as methyl mercuric chloride, CH_3HgCl, and dimethyl mercury, $(CH_3)_2Hg$, is the most common form of mercury in the food people eat and is primarily responsible for the adverse human health effects attributed to mercury. Methylmercury is highly toxic and is readily bioaccumulated and biomagnified in freshwater invertebrates, marine invertebrates, freshwater fish, marine fish, marine mammals, and terrestrial mammals (including humans). The majority of methylmercury in freshwater and marine aquatic systems is derived from the conversion of other forms of mercury; rarely is methylmercury directly discharged into aquatic systems [79].

Mercury distribution within the environment (i.e., atmospheric, terrestrial, and aquatic) is controlled by complex biogeochemical cycles and ecological interactions. Sources of environmental mercury input may be divided into natural and anthropogenic categories: The relative magnitude of these categories is roughly equal [80, 81].

Natural contributions include such sources as volcanic eruptions, forest fires, weathering of mercury-bearing rocks/geological deposits, and aquatic volatilization [82]. Volcanic eruptions are enormous contributors on a per-event basis, but rare in terms of event frequency. Forest fires are intermediate contributors on a per-event basis but represent a higher frequency of event than volcanic eruptions. The weathering of geological deposits represents a slow but continuous contributory process. Volatilization, sometimes referred to as recycling, occurs when mercury is released back into the atmosphere as a gas from freshwater and marine aquatic systems. An example would be the breakdown of methyl mercury by sunlight [83].

The various natural inputs of mercury to the environmental enter primarily to the atmosphere. Once in the atmosphere, mercury can circulate for years, during which time it becomes widely distributed throughout the planet [84]. Mercury deposition from the atmosphere to the land or water can occur in either aqueous or dry forms. Aqueous forms include rain, snow, sleet, hail, and fog. Dry forms include gas or particulates. The principal atmospheric contributor is rain. Atmospheric deposition may enter aquatic systems directly as rainfall, or indirectly as a result of either surface runoff from terrestrial systems (i.e., in streams and/or rivers) or percolation followed by subsequent input from groundwater [83].

The anthropogenic sources of mercury are those inputs that result from the activities of humans. These anthropogenic inputs can be divided into four categories: (1) area sources, (2) combustion processes, (3) manufacturing activities associated with metals, alkali, and cement, and (4) other industrial processes. Landfills, dental preparations (e.g., mercury fillings), and laboratory operations are examples of area sources. Examples of combustion processes include coal-burning power plants and/or the burning of any other fossil fuel, medical waste incineration, and municipal waste combustion. Other industrial processes include such things as discharges from mining operations, hydroelectric plants, and pulp and paper production.

The primary anthropogenic contributors of mercury to the environment include coal combustion, chlor-alkali production (i.e., production of chlorine and sodium hydroxide), waste incineration, and metal processing. The relative proportions of mercury emissions to the atmosphere, exclusive of the burning of biomass, are illustrated by data from the year 2000 [80]. Relative proportions of mercury emissions to the atmosphere were 65% from stationary combustion, 11% from gold production [85], 6.8% from non-ferrous metal production, 6.4% from cement production, 3.0% from waste disposal (municipal and hazardous waste, crematoria and sewage sludge incineration), 1.4% from pig iron and steel production, 1.1% from mercury production (mainly for batteries), and 2.0% from other sources.

Methylmercury is the most toxic form of mercury. It is formed by the methylation of inorganic mercury [79]; methylation means a single carbon methyl group (i.e., CH_3^-) is transferred from an organic compound and becomes bound to an inorganic mercury ion (i.e., Hg^{2+}) [86]. Methylation can occur by either biotic or abiotic means, with the former being the principal mechanism [83]. Examples of such biotic methylation include catalysis by microorganisms that live in freshwater systems (lakes, rivers, wetlands, sediments, soils) and/or the open ocean [87]. To a lesser extent, mercury methylation sometimes can occur via natural processes such as photochemical reactions; these are termed "abiotic processes" [83]. Methylmercury rather than inorganic mercury is bioconcentrated because organisms at various levels in the food chain retain it more efficiently. Inorganic mercury enters bacteria via a specialized transport protein that takes the mercury across the lipid membrane of the bacterial cell wall; this uptake is a key first step in both the methylation and bioaccumulation of mercury [83]. In contrast to other forms of mercury, which are not reactive and thus diffuse out of the cell as fast as they enter, methylmercury is reactive. This is the key to its accumulation in bacterioplankton and phytoplankton, which initiates the bioaccumulation process throughout the feedweb.

Bacteria that contain methylmercury may be passed to higher trophic levels when they are consumed by bacterivores or particle feeders. Alternatively, the bacteria may excrete the methylmercury into the water, where it can become adsorbed to planktonic particles and/or nonliving particulate material suspended in ocean systems. Similarly, methylmercury can be adsorbed by marsh grasses or other microscopic or macroscopic aquatic plants in freshwater systems. Through this mechanism mercury can enter the foodweb when these plants are subsequently consumed by herbivores or detritivores.

Methylmercury may be biomagnified in both marine and freshwater aquatic food webs, i.e., the methylmercury concentrations (i.e., grams mercury per gram biomass) increase at successive trophic levels. Methylmercury concentrations have been shown to generally increase as mercury is passed from invertebrates and/or herbivorous fish to piscivorous fish, and ultimately humans [88]. As was illustrated for DDT residues above, methylmercury concentrations in marine apex predators can reach a level orders of magnitude higher than the concentration present in the water. The extent of concentration at the highest trophic level is influenced by the number of trophic levels in the system [88]. Methylmercury has a half-life of approximately 2 years in aquatic organisms. Such a prolonged residence time results in its bioaccumulation within aquatic food chains [79]. By contrast, methylmercury has a half-life of about 50 days in human blood [89].

Mercury Poisoning: Human Symptomologies and Historical Examples

Mercury is a neurotoxin. Many adverse health effects are associated with its accumulation in the human body. These effects vary depending on the amount of mercury one is exposed to, time of exposure, the mode of entry to the human body, the chemical form of the mercury, and the age of the subject.

Methylmercury is easily absorbed through the gastrointestinal tract when ingested. Mercury readily binds to proteins and therefore is not easily eliminated. Unlike essential metals such as iron, copper, or zinc, the human body does not require mercury for any purpose. The toxicity of mercury is closely related to its tendency to bind to sulfur and selenium, and in particular to amino acids that contain these elements: cysteine and methionine in the case of sulfur, and selenocysteine, selenohomocysteine, and selenomethionine in the case of selenium. Methylmercury, for example, can form a complex with cysteine that resembles methionine. In this methionine-resembling complex, methylmercury may be transported freely throughout the body, including across the placenta and blood–brain barrier, as well as through breast milk.

The uptake of mercury by humans generally occurs via two modes. One is inhalation. Metallic mercury (Hg^0) may be present in ambient air, particularly downwind from coal-burning power plants. Alternatively, mercury may be released from dental amalgams, and as methylmercury (CH_3Hg^+) from the consumption of foods such as fish. The human body is better adapted for mitigating the potential toxic effects of vaporous mercury, so health effects from this source are relatively rare and are generally limited to cases where the victim in question is in close and prolonged proximity to a source of elevated mercury in the atmosphere. Methylmercury, on the other hand, is the most biochemically active form of mercury. It affects the central nervous system and can cause irreversible damage to areas of the brain. Damage to the nervous system is the most prevalent of methylmercury effects on humans, but methylmercury exposure can also harm the lungs and kidneys [90] and has been linked to increased risk of cardiovascular disease [91–93]. The first symptom of methylmercury exposure in adults is generally paresthesia, a sensation of tingling on the skin. This usually manifests itself as numbness and tingling in the extremities, though paresthesia may also occur in other parts of the body as well. This is normally the first manifestation of damage to the central nervous system [94]. In cases where elevated, more extreme doses of methylmercury poisoning occur, the initial paresthesia symptoms may be followed by ataxia and generalized weakness. Ataxia refers to a wobbliness, unsteadiness, and general lack of coordination. It reflects dysfunctionality in the cerebellum and is manifested by the brain's failure to regulate body posture and/or direction of limb movements. Even higher doses may lead to dysarthria, i.e., the loss of vision, hearing, and finally severe tremors, coma, and/or death [95]. These severe symptoms have only been observed in people who repeatedly consumed fish that were contaminated directly by methylmercury from anthropogenic sources [94].

Because of the biochemical activity of methylmercury and thus its ability to cross the placenta, children exposed to methylmercury in utero display developmental impairments that reflect compromised brain function; these symptoms may include such things as lower IQ, decreased memory function, attention deficit, and/or impaired language skills. More serious physiological effects include motor difficulties, sensory problems, and mental retardation [96].

Physical and environmental parameters that influence methylmercury concentration in natural waters are water pH and dissolved organic carbon (DOC). Higher acidity and DOC increase methylation rates of mercury and enhance its mobility in the water, thus making it more likely to enter the food chain [97–99]. Additionally, algal blooms reduce the uptake of methylmercury in freshwater food webs [100].

Methylmercury released directly into aquatic systems can cause mass acute mercury poisoning in humans who consume contaminated fish and shellfish. Probably the most dramatic historical example of this occurred in Minimata, Japan. Minimata is considered to be one of the worst industrial disasters of all time. Between 1932 and 1968, a factory of the Chisso Corporation discharged roughly 27 tons of mercury into a waterway that fed directly into Minimata Bay. The factory was using mercury as a catalyst to produce acetaldehyde, a chemical employed in the manufacturing of plastics [101]. More than 3,000 people were "officially" recognized and reported by government officials as directly suffering from mercury intoxication. Later, however, estimates of the number of people from the contaminated area afflicted with health problems or left permanently disabled increased to two million [102]. The large difference between these two estimates of the number of Minamata Disease victims reflects the fact that a social stigma associated with the symptomologies caused denial and/or underreporting by the afflicted, the local populations, and the local/national governments.

The range and severity of the health effects from this incident of mercury poisoning were extreme. The afflictions included the various symptoms discussed above and extended to birth deformities and deaths. Public awareness of the incident became widespread at roughly the time of the birth of the environmental movement elsewhere in the world. The account of what happened was so highly publicized that mercury poisoning became known as "Minamata disease" [103, 104].

A similar event occurred in Niigata, Japan. Mercury waste from acetaldehyde production was released from the late 1950s to the early 1960s by the Shawa Denkó Corporation factory into the Agano River, 65 km upstream from Niigata. The quantity released remains unknown to this day [105]. The Minamata and Niigata examples illustrate instances where the use of inorganic mercury as a catalyst in an industrial process led to very serious impacts from methylmercury intoxication. Specifically, mercuric sulfate, $HgSO_4$, was used to convert acetylene into acetaldehyde, a precursor for, inter alia, acetic acid and acetone; and mercuric chloride, $HgCl_2$, was used to convert acetylene into vinyl chloride, a precursor of polyvinyl chloride (PVC). Unfortunately in both cases a side reaction of the catalytic cycle led to the production of methymercury, by far the most toxic form of mercury, and the methylmercury was discharged with the wastewater from the factories, with tragic consequence for segments of the local population. Such examples clarified and galvanized societies' awareness of mercury as a pollutant with potentially direct and dire human impacts. It is important also to understand that the Minamata and Niigata cases represent examples where large, direct inputs of mercury were being made to water basins of relatively small volume, and very close to large human populations. In these cases, the delivery:dilution ratio was heavily skewed toward the delivery side, and there was a high probability of frequent human interaction with the introduced mercury through both direct and indirect pathways.

A somewhat different example of mercury pollution occurred at Dryden, Ontario in Canada. In this case, the Dryden Chemical Company repeatedly discharged inorganic mercury into the Wabigoon River between 1962 and 1970. Total discharges have been estimated at ~9 tons. The mercury was discharged with the wastewater from a chlor-alkali plant, where once again inorganic mercury was being used, in this case to convert sodium chloride (NaCl) into chlorine and sodium hydroxide. The inorganic mercury subsequently became methylated in the natural environment (i.e., not within the plant itself), resulting in serious methylmercury pollution in the downstream freshwater system and its various aquatic components [106]. This case again involved substantial inputs of mercury directly into aquatic systems of relatively small volume and close to large human populations.

Fish and shellfish have not been the only contaminated food source associated with methylmercury poisoning in humans. Other foods such as grain [107, 108] and meat [109] have caused mercury-related health effects, the most notable of these examples being the Basra grain disaster. In the 1960s and 1970s, seed grain containing ~8 μg/g mercury was shipped to Basra, Iraq. In Basra, it was coated with methylmercury as a preservative/fungicide. The seed grain was then distributed throughout Iraq with the expectation that it would be planted. However, people fed the grain to livestock destined for human consumption. The mercury was biomagnified through the food chain from grain to livestock to humans. People also consumed the grain directly by using it to make flour for bread and related food products. The decision of people to consume the seed grain instead of planting it resulted in mass mercury poisoning. Ten thousand people died and 100,000 were severely and permanently brain damaged [107].

Humans are not the only victims of environmental methylmercury contamination. Due to a commonality of certain biochemical processes across all animal groups, methylmercury has the potential to adversely affect many wild and domesticated animals. Such effects have been most notable proximate to areas of contamination and have been most evidenced as diminished reproductive success in fish, fish-eating birds, and mammals [88, 110].

Etiology of Mercury Intoxication

As noted above, mercury has a strong tendency to bind to sulfur and especially to selenium, which are essential elements. Because of this tendency, mercury can compromise the functionality of essential sulfur-containing amino acids (cysteine and methionine) and selenium-containing amino acids (selenocysteine, selenohomocysteine, and selenomethionine). The term "essential" in this context refers to compounds that humans cannot synthesize. Such compounds must be obtained from the food people eat. The distinction between essential and non-essential amino acids is somewhat unclear, as some amino acids can be produced from others. Methionine, for example, can be converted into homocysteine, which in turn can be converted into cysteine. Thus, given an adequate supply of methionine from food, humans can produce all the cysteine they need. But humans cannot synthesize methionine and cysteine de novo, and for this reason the sulfur-containing amino acids are often considered to be a pool of nutritionally equivalent essential amino acids. Similar logic applies to the selenium-containing amino acids.

One of the important functions of these two groups of amino acids is their role in determining the so-called tertiary structure of proteins. Simplistically, proteins are polymers of amino acids, and the sequence of amino acids determines the primary structure of the protein. Hydrogen bonds between sequences of amino acids (peptides) determine the localized secondary structures of protein molecules. The tertiary structure is the overall three-dimensional structure of the protein and requires that certain parts of the molecule be locked into place by various mechanisms that include disulfide bonds, which is why cysteine and methionine are so important. Selenoproteins are analogs of ordinary proteins but contain one or more selenocysteines in place of cysteine. They are common in animals, and in humans $\sim$25 selenoproteins have been identified. Health effects associated with an inadequate supply of selenium in the diet are believed to reflect the body's inability to synthesize sufficient amounts of these selenoproteins.

Exposure Limits

The unfortunate, industrially related incidents discussed above focused the attention of both health officials and the public on the perils associated with mercury. These were cases were mercury was discharged directly into bodies of water from which the local populace removed and consumed seafood containing up to 40,000 μg Hg/kg. Such industrial events are extreme, and the resultant concentrations of mercury in fish and shellfish were far above background levels. But the fact is that mercury occurs naturally in the environment, and through perfectly natural processes it finds its way into food that people eat. Incidents such as Minamata and Niigata contributed to public awareness of health issues associated with mercury bioaccumulation and biomagnifications in food chains leading to humans and led to the question, how much is too much. Studies began showing developmental effects in children, in particular among island/coastal populations that primarily consume seafood [111–113]. Federal and global organizations (e.g., US Food and Drug Agency (FDA), US Environmental Protection Agency (EPA), United National Food and Agriculture Organization (FAO), and the World Health Organization

(WHO)) also took notice and began instituting guidelines for the consumption of fish and shellfish. Advisories, though intended for the general population, carried particular focus on pregnant woman, fetuses, and young children.

Currently the US EPA uses a Reference Dose of 0.1 μg methylmercury/kg body weight/day as an exposure limit without recognizably adverse effects. This limit carries a safety factor of 10 from the lowest observed adverse exposure level and corresponds to a blood mercury level of 5.8 μg/L or 5.8 parts per billion (ppb) mercury. Blood mercury levels below this value are considered to be without appreciable risk by the EPA [114]. The Joint FAO/WHO established a Provisional Tolerable Weekly Intake of Mercury (PTWI) of 1.6 μg of mercury/kg of body weight, somewhat higher than the EPA's recommendation. The WHO guideline of methylmercury intake is equivalent to a concentration in hair of 5 parts per million (ppm). Another study of mercury in hair conducted in the Seychelles Islands, where seafood consumption is significantly higher than average, found mercury levels in hair ranging from 7 to 26 ppm. The Seychelles study was supportive of the WHO guideline. Though the calculated FAO/WHO value for all adults was actually 3.3 μg Hg/kg of body weight, the published value of 1.6 μg of mercury/kg of body weight, which was weighted in favor of the most susceptible segment of the population (i.e., pregnant women, fetuses, young children) came to be commonly cited and used. Currently, the level of concern for mercury in fish is set by the FDA at 1.0 mg Hg/kg fish tissue.

Mercury in the Oceanic Ecosystem

Mercury from both natural and anthropogenic sources finds its way into the ocean, is taken up by plankton, and then transferred to small planktivorous fish. This begins a process of food chain magnification as the mercury is transferred to larger and larger predatory fish, ultimately finding its way into top-level carnivores like the tuna, swordfish, and sharks that people eat. Such large predatory fish tend to have the highest mercury concentrations (i.e., μg Hg/kg fish) and as such have attracted the most attention from the public, health officials, and scientists. The accumulation of methylmercury in high trophic level organisms results mainly from the ingestion of methylmercury-containing food (biomagnifications) rather than direct uptake of methylmercury from the water itself (bioaccumulation).

There are many factors that influence the concentrations of methylmercury in open ocean fish. These include size [115], age, trophic position [116, 117], physical and environmental parameters [100], and area of capture [118, 119]. There also remain large unexplained variations in the relationship of mercury concentration (i.e., μg Hg/kg fish tissue) versus fish size. In general, one would expect that, within a given species, larger fish are older fish and have consumed more prey to attain their size. A relationship between fish size and mercury content of the flesh is evident for some marine carnivorous species such as *Seriola rivoliana* (common name, bonita or jack), *Acanthocybium solandri* (common name, wahoo, ono), *Lutjanus syagris* (common name snapper), and *Epinephelis guttatus* (red hind, grouper) [120].

The age of the fish and preferred depth of feeding appear to explain more of the variance in mercury concentration than does size [117, 118, 121, 122]. Depth of habitat and foraging were found to be highly correlated with mercury concentrations [123]. This is because dissolved organic mercury concentrations increase with depth [123]. In addition, deeper waters (i.e., below the thermocline) tend to have lower oxygen, which leads to enhanced methylation rates [124–127]. This has led to the recent conclusion that the source of methylmercury in the open ocean is the deep water column [128, 129] and not coastal runoff [130] or the euphotic zone [131]. Many large oceanic predatory fish regularly dive to great depths to forage for prey from waters having lower oxygen concentration.

Mercury and Selenium

The confusion generated by the guidelines from the various agencies, along with repeated media attention to the presence of mercury in seafood has led to significant misunderstandings concerning the safety of seafood consumption and the role of seafood in a healthy diet. This has been to the detriment of both the commercial fishing industry and the consumers who have been confused and/or discouraged from regularly eating seafood.

What are some of the reasons that the perceived risk from eating seafood is overestimated? First, there has never been a confirmed case of mercury poisoning from the consumption of tuna or any other open ocean fish. The potential harmful effects of low levels of mercury from open ocean fish are undocumented and hypothetical. Mercury may be toxic at high levels, but not at the low levels typically found in oceanic fish. The joint advisory on seafood issued in 2004 by the FDA and EPA (www.cfsan.fda.gov/~dms/admehg3.html) was for pregnant women, nursing mothers, and young children. It directed pregnant women, nursing mothers, and young children to avoid eating swordfish, sharks, king mackerel, and tilefish, to limit consumption of other forms of fish to two meals per week, and to limit consumption of tuna to one meal per week. There has never been an advisory directed at the general consumer to limit seafood consumption.

Scientists are now realizing that earlier studies and resulting consumer guidelines were flawed. In fact, deliberately avoiding fish may increase the risk of adverse nutritionally related health effects. The benefits of fish oils (i.e., the omega-3 polyunsaturated fatty acids) on brain development are known and widely accepted. The basis of this realization began when adverse child brain development effects were not seen in multiple follow-up studies in populations consuming ocean fish [132–134]. Omega-3 polyunsaturated fatty acids (PUFAs) are critical to normal development of the human brain and nervous system.

Pregnant women and nursing mothers who avoid foodstuffs rich in omega-3 PUFAs without replacement from another source risk adverse effects on their child's brain development. The excessive caution associated with fish consumption reflected knowledge that the fetal brain is very sensitive to maternal exposure to methylmercury. It is true that methylmercury can impair selenoenzyme activities in the brain.

Further exploration, however, revealed that selenium, an essential trace element, is simultaneously present with mercury in most oceanic fish. The ratios of selenium to mercury in these fish probably explain why consumption of ocean fish is not associated with mercury intoxication [135]. In almost all species of oceanic fish tested the amount of selenium exceeded that of mercury on a molar basis [136, 137]. Selenium was first considered to be a toxin, which it is when the dose is sufficiently high. Its essential antioxidant function was discovered in 1957. It is now considered an essential trace element, with a recommended daily allowance for adults of 55 μg/day. It is found in virtually all commercial vitamin supplements. Selenium has been shown to protect against mercury toxicity in all animal species studied. Although historically the toxicity of mercury has been attributed to its great affinity to bond to sulfur-containing amino acids in protein, the binding affinity of mercury to selenium (to form inert mercury selenide) is a million times stronger than mercury's binding affinity to sulfur. In effect then, selenium can prevent mercury from causing health problems by sequestering mercury as mercury selenide. Alternatively, however, too much mercury in the diet can actually cause a selenium deficiency.

The conclusion then is that fish are unsafe to eat only when mercury concentrations in the fish are higher than the selenium concentrations [137]. In such cases the adverse health effects may reflect a combination of the directly toxic effects of the mercury and/or a selenium deficiency caused by the fact that selenium and mercury sequester one another. The only cases where mercury to selenium ratios exceed one is in some sharks and pilot whales. Selenium is necessary for cellular function in most animals. It is required for more than 25 enzymes with important functions [138, 139]. The ability of selenium to effectively detoxify metals such as mercury was first reported in 1967 [140]. The ability of selenium to decrease mercury toxicity has been established in all species examined to date [141, 142]. The high binding affinity between mercury and selenium [143] is very likely responsible for the ability of selenium to sequester and therefore detoxify mercury. As long as the ratio of selenium to mercury in the diet is sufficiently high, the selenium can sequester the mercury and effectively render the mercury harmless [136, 144]. Molar concentrations of selenium in fish are essential factors in evaluating risks associated with dietary mercury exposure. If the selenium level is higher in the fish than the mercury level, the fish is considered safe to eat [137]. The molar ratio of selenium to mercury is quite high in the muscle of most commercially important ocean fish [137, 145–148]. Pelagic oceanic fish are rich in selenium relative to mercury. Such fish appear more likely to prevent mercury toxicity than to cause it. The old thinking was that mercury was toxic by itself; the new thinking is that selenium and mercury sequester one another. Many of the symptomologies that were previously attributed to mercury toxicity are being reexamined as possible cases of selenium deficiency caused by mercury's role in sequestering selenium.

Future Directions

The prognoses for CFP and methylmercury are rather different. CFP is caused by a toxin that is naturally produced by certain species of marine dinoflagellates and passed up the food chain. Human activities appear to be largely unrelated to production of the gambiertoxins that are subsequently converted to various congeners of CTX. At the present time the best defense against CFP appears to be refraining from eating certain kinds of fish (e.g., barracuda) that have historically been associated with CFP. Based on extensive surveys in areas where CFP occurs, this policy would amount to erring on the side of safety but would apparently result in many harmless and nutritious fish being deleted from the menu. A preferable solution would be the development of a simple and inexpensive test to determine whether a fish contains CTX. Unfortunately, no such test currently exists, and because there are reported to be literally dozens of congeners of CTX, it is unclear that a single test could provide a reliable yes/no answer to the question. Ultimately an antibody-based assay may solve the problem, but at least at this point in time, development of such an assay is not imminent.

In the case of methylmercury, it is fair to say that very significant progress has been made since the discovery of Minamata Disease. Global production of mercury has declined by about 70% since 1970, a pattern that in no small part reflects recognition of the toxicity of mercury (especially methylmercury) and the efforts of government agencies to regulate mercury use and discharges and of industry to identify substitutes for mercury in many of its earlier applications and to reuse and recycle in cases where no substitutes exist. Probably the most important ongoing development is the controversy surrounding the health effects associated with consumption of fish and shellfish. With respect to this issue, public education is very important. The fact is that most fish and shellfish contain very little mercury. It is therefore straightforward to obtain the nutritional benefits of eating fish and shellfish without any risk from mercury intoxication. Remarkably, almost all fish reported to contain more than trace amounts of mercury also contain substantially more selenium than mercury (Fig. 3.5), the result being that consumption of those fish may actually safeguard the consumer against mercury intoxication from other sources of exposure as opposed to being a threat to human health. The exceptions to this last statement include pilot whales and mako sharks, which are not in the creel of most fishermen, not on the menu at many restaurants, and not on the dinner table of most people.

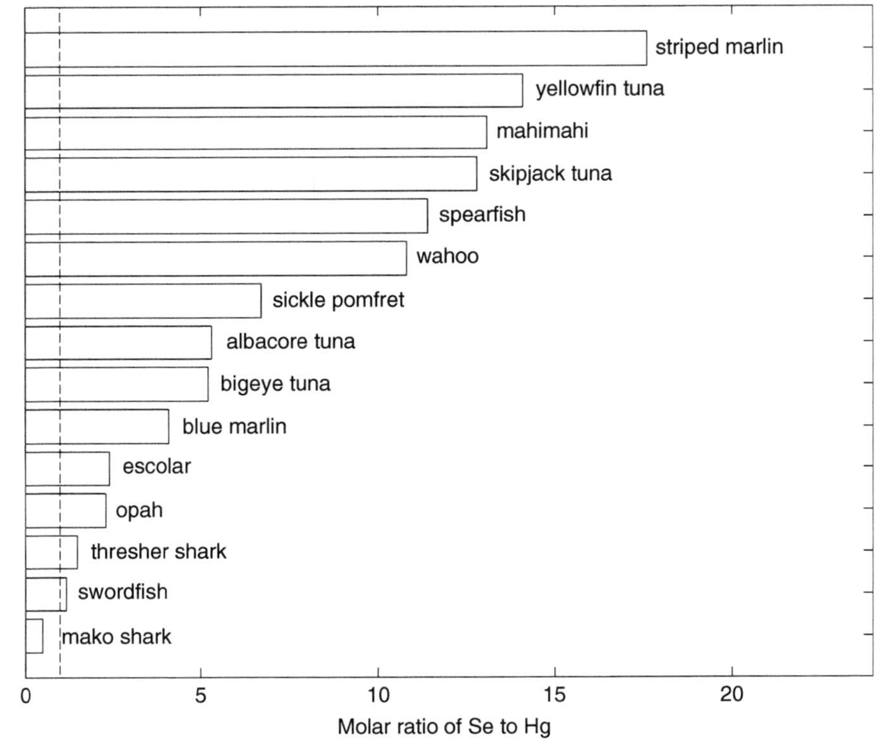

Fig. 3.5 Molar ratio of selenium to mercury in fish from the central North Pacific near Hawaii. Dashed line corresponds to a molar ratio of 1.0 (Data from Kaneko and Ralston [137])

Bibliography

Primary Literature

1. Gray JS (2002) Biomagnification in marine systems: the perspective of an ecologist. Mar Pollut Bull 45(1–12):46–52
2. Woodwell GM, Wurster CF, Isaacson PA (1967) DDT residues in an east coast estuary: a case of biological concentration of a persistent insecticide. Science 156:821–824
3. Bienfang PK et al (2011) Prominent human health impacts from several marine microbes: history, ecology, and public health implications. Int J Microbiol 2011:15, 152815
4. Hamelink JL, Waybrant RC, Ball RC (1971) Proposal – exchange equilibria control degree chlorinated hydrocarbons are biologically magnified in lentic environments. Trans Am Fish Soc 100(2):207–214
5. Harvey GR, Miklas HP, Bowen VT, Steinhauer WG (1974) Observations on distribution of chlorinated hydrocarbons in Atlantic Ocean organisms. J Mar Res 32(2):103–118
6. Giam CS, Hanks AR, Richards RL, Sackett WM, Wong MK (1972) DDT, DDE, and Polychlorinated Biphenyls in Biota from Gulf of Mexico and Caribbean Sea – 1971. Pestic Monit J 6(3):139–145

7. Isaacs JD (1973) Potential trophic biomasses and trace-substance concentrations in unstructured marine food webs. Mar Biol 22(2):97–104
8. Isaacs JD (1972) Unstructured marine food webs and pollutantanalogues. Fish Bull Natl Ocean Atmos Adm 70(3):1053–1059
9. Laws EA (1986) Unstructured food webs – a model for aquaculture. Aquacult Eng 5(2–4):161–170
10. Fraga S et al (2010) Ciguatera in macaronesia: an emerging problem. In Contribution to identification of gambierdiscus species. 14th conference of the international society for the study of harmful algae, Crete
11. Gudger EW (1930) Poisonous fishes and fish poisonings, with special reference to ciguatera in the West Indies. Am J Trop Med Hyg 10:43–55
12. Hessel DW, Halstead BW, Peckham NH (1960) Marine biotoxins. I. Ciguatera poison: some biological and chemical aspects. Ann NY Acad Sci 90:788–797
13. Beaglehole JC (1969) The journals of captain cook: the voyage of the resolution and adventure, vol 2. Cambridge University Press, Cambridge
14. Randall JE (1958) A review of ciguatera tropical fish poisoning with a tentative explanation of its cause. Bull Mar Sci Gulf Caribb 8:236–267
15. Banner AH, Scheuer P, Sasaki S, Helfrich P, Alender C (1960) Observations on ciguatera-type toxin in fish. Ann NY Acad Sci 90:770–787
16. Banner AH, Helfrich P, Piyakarn T (1966) Retention of ciguatera toxin by Red Snapper Lutjanus Bohar. Copeia 2:297–301
17. Banner AH, Helfrich P (1964) The distribution of ciguatera in the tropical Pacific. Hawaii Institute of Marine Biology, Honolulu, Hawaii
18. Pearn J (2001) Neurology of ciguatera. J Neurol Neurosurg Psychiatry 70(1):4–8
19. Arena P, Levin B, Fleming LE, Friedman MA, Blythe D (2004) A pilot study of the cognitive and psychological correlates of chronic ciguatera poisoning. Harmful Algae 3(1):51–60
20. Lewis RJ (1986) Disease and development: ciguatera fish poisoning. Soc Sci Med 23:983–993
21. Lange WR (1994) Ciguatera fish poisoning. Am Fam Physician 50(3):579–584
22. Fleming LE, Katz D, Bean JA, Hammond R (2001) Epidemiology of seafood poisoning. In: Hui YH, Kitts D, Standfield P (eds) Foodborne disease handbook. Marcel Dekker, New York, pp 287–310
23. Fleming LE, Baden DG, Bean JA, Weisman R, Blythe DG (1998) Seafood toxin diseases: issues in epidemiology and community outreach. In: Reguera B, Blanco J, Fernandez ML, Wyatt T (eds) Harmful algae. Zunta de Galicia and Intergovernmental Oceanographic Commission of UNESCO, Galicia, pp 245–248
24. Lehane L, Lewis RJ (2000) Ciguatera: recent advances but the risk remains. Int J Food Microbiol 61(2–3):91–125
25. Bienfang PK, Parsons ML, Bidigare RR, Laws EA, Moeller PDR (2008) Ciguatera fish poisoning: a synopsis from ecology to toxicity. In: Walsh PJ, Smith SL, Fleming LE, Solo-Gabriele HM, Gerwick WH (eds) Oceans and human health: risks and remedies from the sea. Elsevier, New York, pp 257–270
26. Dickey RW (2008) Ciguatera toxins: chemistry, toxicology and detection. In: Botana LM (ed) Seafood and freshwater toxins: pharmacology, physiology, and detection. CRC Press/Taylor & Francis, Boca Raton, pp 479–500
27. Dickey RW, Plakas SM (2010) Ciguatera: a public health perspective. Toxicon 56(2):123–136
28. Adachi R, Fukuyo Y (1979) Thecal structure of a marine toxic dinoflagellate Gambierdiscus-toxicus gen Et Sp-Nov collected in a ciguatera-endemic area. Bull Jpn Soc Sci Fish 45(1):67–71
29. Hokama Y, Yoshikawa-Ebesu JSM (2001) Ciguatera fish poisoning: a foodborne disease. J Toxicol Toxin Rev 20(2):85–139

30. Lewis RJ (2006) Ciguatera: Australian perspectives on a global problem. Toxicon 48(7):799–809
31. Dierking J (2007) Effects of the introduced predatory fish Cephalopholis argus on native reef fish populations in Hawaii. Dissertation Thesis, University of Hawaii, Honolulu, 125 pp
32. Steidinger KA, Baden DG (1985) Toxic marine dinoflagellates. In: Spector DL (ed) Dinoflagellates. Academic, Orlando, pp 210–261
33. Holmes MJ (1998) Gambierdiscus yasumotoi sp. nov. (Dinophyceae), a toxic benthic dinoflagellate from southeastern Asia. J Phycol 34:661–668
34. Chinain M, Feust MA, Pauillac S (1999) Morphology and molecular analysis of three toxic species of Gambierdiscus (dinophyceae): *G. pacificus* sp. Nov., *G. australes* sp. Nov. and *G. polynesiensis* sp. Nov. J Phycol 35:1282–1296
35. Richlen ML, Morton SL, Barber PH, Lobel PS (2008) Phylogeography, morphological variation and taxonomy of the toxic dinoflagellate *Gambierdiscus toxicus* (Dinophyceae). Harmful Algae 7:614–629
36. Litaker RW et al (2009) Taxonomy of Gambierdiscus including four new species, *Gambierdiscus caribaeus*, *Gambierdiscus carolineanus*, *Gambierdiscus carpenteri*, and *Gambierdiscus ruetzleri* (Gonyaulacales, Dinophyceae). Phycologia 48(5):344–390
37. Carlson RD (1984) Distribution, periodicity, and culture of benthic/epiphytic dinoflagellates in a ciguatera endemic region of the Caribbean. PhD thesis, Southern Illinois University, Carbondale
38. Bagnis R et al (1980) Origins of ciguatera fish poisoning: a new dinoflagellate, Gambierdiscus toxicus Adachi and Fukuyo, definitively involved as a causal agent. Toxicon 18:199–208
39. Holmes MJ, Lewis RJ, Gillespie NC (1991) Toxicity of Australian and French Polynesian strains of *Gambierdiscus toxicus* (Dinophyceae) grown in culture: characterization of a new type of maitotoxin. Toxicon 28:1159–1172
40. Faust MA (1995) Observation of sand-dwelling toxic dinoflagellates (Dinophyceae) from widely differing sites, including two new species. J Phycol 31:996–1003
41. Roeder K et al (2010) Characteristic profiles of Cigeatera toxins in different strains of *Gambierdiscus* spp. Toxicon 56(5):731–738
42. Murata M, Legrand AM, Ishibashi Y, Fukui M, Yasumoto T (1990) Structures and configurations of ciguatoxin from the moray eel Gymnothorax javanicus and its likely precursor from the dinoflagellate *Gambierdiscus toxicus*. J Am Chem Soc 112:4380–4386
43. Lewis RJ et al (1991) Purification and characterization of ciguatoxins from moray eel (Lycodontis javanicus, Muraenidae). Toxicon 29:1115–1127
44. Satake M, Fukui M, Legrand AM, Cruchet P, Yasumoto T (1998) Isolation and structures of new ciguatoxin analogs, 2,3-dihydroxyCTX3C and 51-hydroxyCTX3C, accumulated in tropical reef fish. Tetrahedron Lett 39(10):1197–1198
45. Satake M et al (1997) The absolute configuration of ciguatoxin. J Am Chem Soc 119(46):11325–11326
46. Satake M, Ishibashi Y, Legrand AM, Yasumoto T (1996) Isolation and structure of ciguatoxin-4A, a new ciguatoxin precursor, from cultures of dinoflagellate Gambierdiscus toxicus and parrotfish Scarus gibbus. Biosci Biotechnol Biochem 60(12):2103–2105
47. Satake M, Murata M, Yasumoto T (1993) The structure of CTX3C, a ciguatoxin congener isolated from cultured Gambierdiscus-toxicus. Tetrahedron Lett 34(12):1975–1978
48. Lewis RJ, Vernoux JP, Brereton IM (1998) Structure of Caribbean ciguatoxin isolated from Caranx latus. J Am Chem Soc 120:5914–5920
49. Scheuer PJ, Takahashi W, Tsutsumi J, Yoshida T (1967) Ciguatoxin isolation, and chemical nature. Science 155:1267–1268
50. Tachibana K, Nukina M, Joh YG, Scheuer PJ (1987) Recent developments in the molecular structure of ciguatoxin. Biol Bull 172:122–127
51. Yasumoto T, Nakajima I, Bagnis R, Adachi R (1977) Finding of a dinoflagellate as a likely culprit of ciguatera. Bull Jpn Soc Sci Fish 43:1021–1026

52. Murata M, Legrand AM, Ishibashi Y, Yasumoto T (1989) Structures of ciguatoxin and its congener. J Am Chem Soc 111:8929–8931
53. Legrand AM, Fukui M, Cruchet P, Ishibashi Y, Yasumoto T (1992) Characterization of ciguatoxins from different fish species and wild Gambierdiscus toxicus. In: Toteson TR (ed) Proceedings of the 3rd international conference on ciguatera fish poisoning, ciguatera fish poisoning. Polyscience Publications, Quebec, p 225
54. Hokama Y, Yoshikawa-Ebesu JSM (2001) Ciguatera fish poisoning: a foodborne disease. Toxin Rev 20(2):85–139
55. Lewis RJ, Jones A, Vernoux P-P (1999) HPLC/tandem electrospray mass spectrometry for the determination of sub-ppb levels of Pacific and Caribbean ciguatoxins in crude extracts of fish. Anal Chem 71:247–250
56. Lehane L, Lewis RJ (2000) Ciguatera: recent advances but the risk remains. Int J Food Microbiol 61:91–125
57. Dickey RW (2008) Ciguatera toxins: chemistry, toxicology and detection. In: Botana LM (ed) Seafood and freshwater toxins: pharmacology, physiology, and detection, 2nd edn. CRC Press/Taylor & Francis, Boca Raton, pp 479–500
58. Lewis RJ (2006) Ciguatera: Australian perspectives on a global problem. Toxicon 48:799–809
59. Dierking J (2007) Effects of the introduced predatory fish Cephalopholis argus on native reef fish populations in Hawaii. PhD thesis, University of Hawaii, 125 pp
60. Morton SL, Bomber JW, Tindall DR, Aikman KE (1993) Response of *Gambierdiscus toxicus* to light: cell physiology and toxicity. In: Smayda TJ, Shimizu Y (eds) Toxic phytoplankton blooms in the sea. Elsevier, Amsterdam, pp 541–546
61. Bomber JW, Tindall DR, Miller DM (1989) Genetic variability in toxin potencies among seventeen clones of *Gambiertoxicus toxicus* (Dinophyceae). J Phycol 25:617–625
62. Villareal TA, Morton SL (2002) Use of cell-specific PAM-fluorometry to characterize host shading in the epiphytic dinoflagellate *Gambierdiscus toxicus*. Mar Ecol 23:127–140
63. Parsons M, Settlemier C, Bienfang P (2010) A simple model capable of simulating the population dynamics of Gambierdiscus, the benthic dinoflagellate responsible for ciguatera fish poisoning. Harmful Algae 10(1):71–80
64. Cameron J, Capra MF (1993) The basis of the paradoxical disturbance of temperature perception in ciguatera poisoning. Clin Toxicol 31:571–579
65. Gordon CJ, Ramsdell JS (2005) Effects of marine algal toxins on thermoregulation in mice. Neurotoxicol Teratol 27:727–731
66. Anderson BS, Sims JK, Wiebenga NH, Sugi M (1983) The epidemiology of ciguatera in Hawaii, 1975–1981. Hawaii Med J 42(10):326–334
67. Tosteson TR, Ballantine DL, Winter A (1998) Sea surface temperature, benthic dinoflagellate toxicity and toxin transmission in the ciguatera food web in Reguera. In: Blanco BJ, Fernandez ML, Wyatt T (eds) Harmful algae. Xunta de Galicia and UNESCO, Paris, pp 48–49
68. Swift AEB, Swift TR (1993) Ciguatera. Clin Toxicol 31:1–29
69. Vernoux J-P, Lejeune J (1993) Ciguatera in the French West Indies. Mem Qld Mus 34:631–638
70. Friedman MA et al (2008) Ciguatera fish poisoning: treatment, prevention and management. Mar Drugs 6:456–479
71. Bienfang PK, Parson ML, Bidigare RR, Laws EA, Moeller PDR (2008) Ciguatera fish poisoning: a synopsis from ecology to toxicity. In: Walsh PJ, Smith SL, Fleming LE, Solo-Gabriele H, Gerwick WH (eds) Oceans and human health: risks and remedies from the sea. Elsevier Science, New York, pp 257–270
72. Steidinger KA, Baden DG (1985) Toxic marine dinoflagellates. In: Spector DL (ed) dinoflagellates. Academic, Orlando, pp 210–261

73. Campora CE, Hokama Y, Tamaru CS, Anderson B, Vincent D (2010) Evaluating the risk of ciguatera fish poisoning from reef fish grown at marine aquaculture facilities in Hawaii. J World Aquac Soc 41(1): 61–70
74. Helfrich P, Piyakarnchana T, Miles PW (1968) Occasional papers. Bernice P Bish Mus 23:305–369
75. Vernoux JP, Lahlou N, Elandaloussi SA, Riyeche N, Magras LP (1985) A study of the distribution of ciguatoxin in individual Caribbean Fish. Acta Trop 42(3):225–233
76. Helfrich P, Banner AH (1963) Experimental induction of ciguatera toxicity in fish through diet. Nature 197(487):1025–1026
77. Bagnis RA, Legrand A-M (1987) Clinical features on 12,980 cases of ciguatera (fish poisoning) in French polynesia. In: Gopalakrishnakone P, Tan CK (eds) Progress in venom and toxin research. Proceeding 1st Asia-Pacific congress on animal, plant and microbial toxins. National University of Singapore, Singapore, pp 372–384
78. Kodama AM, Hokama Y (1989) Variations in symptomatology of ciguatera poisoning. Toxicon 27(5):593–595
79. UNEP (2003) Global mercury assessment report: summary of the report. http://www.unep.org/gc/gc22/Document/UNEP-GC22-INF3.pdf85. Accessed 18 Apr 2011
80. Pacyna EG, Pacyna JM, Steenhuisen F, Wilson S (2006) Global anthropogenic mercury emission inventory for 2000. Atmos Environ 40(22):4048–4063
81. Solnit R (Sept/Oct 2006) Winged mercury and the golden calf. Orion Magazine
82. Tewalt SJ, Bragg LJ, Finkelman RB (2005) Mercury in U.S. coal – abundance, distribution, and modes of occurrence. U.S. Geological Survey Fact Sheet 095-01
83. Morel FMM, Kraepiel AML, Amyot M (1998) The chemical cycle and bioaccumulation of mercury. Annu Rev Ecol Syst 29:543–566
84. Keeler G, Glinsorn G, Pirrone N (1995) Particulate mercury in the atmosphere – its significance, transport, transformation and sources. Water Air Soil Pollut 80(1–4):159–168
85. Maprani AC et al (2005) Determination of mercury evasion in a contaminated headwater stream. Environ Sci Technol 39(6):1679–1687
86. March J, Smith MW (2001) March's advanced organic chemistry: reactions, mechanisms, and structure. Wiley, New York
87. Ullrich SM, Tanton TW, Abdrashitova SA (2001) Mercury in the aquatic environment: a review of factors affecting methylation. Crit Rev Environ Sci Technol 31(3):241–293
88. Wiener JG, Krabbenhoft DP, Heinz GH, Scheuhammer AM (2003) Ecotoxicology of mercury. In: Hoffman DJ, Rattner BA, Burton GA Jr, Cairns J Jr (eds) Handbook of ecotoxicology. CRC Press, Boca Raton, pp 409–463
89. Carrier G, Bouchard M, Brunet RC, Caza M (2001) A toxicokinetic model for predicting the tissue distribution and elimination of organic and inorganic mercury following exposure to methyl mercury in animals and humans. II. Application and validation of the model in humans. Toxicol Appl Pharmacol 171(1):50–60
90. Honda S, Hylander L, Sakamoto M (2006) Recent advances in evaluation of health effects on mercury with special reference to methylmercury: a minireview. Environ Health Prev Med 114(4):171–176
91. Salonen JT et al (1995) Intake of mercury from fish, lipid-peroxidation, and the risk of myocardial-infarction and coronary, cardiovascular, and any death in Eastern finnish men. Circulation 91(3):645–655
92. Guallar E et al (2002) Mercury, fish oils, and the risk of myocardial infarction. N Engl J Med 347(22):1747–1754
93. Choi AL et al (2009) Methylmercury exposure and adverse cardiovascular effects in Faroese whaling men. Environ Health Perspect 117(3):367–372
94. Clarkson TW, Magos L (2006) The toxicology of mercury and its chemical compounds. Crit Rev Toxicol 36(8):609–662
95. Shea KM, Perry KL, Shah M (2004) Health effects of methylmercury. Physicians for Social Responsibility, Washington, DC

96. Rice DC, Schoeny R, Mahaffey K (2003) Methods and rationale for derivation of a reference dose for methylmercury by the US EPA. Risk Anal 23(1):107–115
97. Miskimmin BM, Rudd JWM, Kelly CA (1992) Influence of dissolved organic-carbon, pH, and microbial respiration rates on mercury methylation and demethylation in lake water. Can J Fish Aquat Sci 49(1):17–22
98. Regnell O (1994) The effect of pH and dissolved-oxygen levels on methylation and partitioning of mercury in fresh-water model systems. Environ Pollut 84(1):7–13
99. Xun L, Campbell NER, Rudd JWM (1987) Measurements of specific rates of net methyl mercury production in the water column and surface sediments of acidified and circumneutral lakes. Can J Fish Aquat Sci 44(4):750–757
100. Pickhardt PC, Folt CL, Chen CY, Klaue B, Blum JD (2002) Algal blooms reduce the uptake of toxic methylmercury in freshwater food webs. Proc Natl Acad Sci USA 99(7):4419–4423
101. Nomura S (1968) Epidemiology of Minamata disease. Kumamoto Study Group of Minamata Disease, Kumamoto, Japan
102. McCurry J (2006) Japan remembers Minamata. Lancet 367(9505):99–100
103. Harada M (1995) Minamata disease – methylmercury poisoning in Japan caused by environmental-pollution. Crit Rev Toxicol 25(1):1–24
104. Anonymous (2002) Minamata disease the history and measures. http://www.env.go.jp/en/chemi/hs/minamata2002/. Accessed 18 Apr 2011
105. Myers GJ, Davidson PW, Weiss B (2004) Methyl mercury exposure and poisoning at Niigata, Japan. Seychelles Med Dent J 7:132–133
106. Fimreite N (1970) Mercury uses in Canada and their possible hazards as sources of mercury contamination. Environ Pollut 1:119–131
107. Bakir F et al (1973) Methylmercury poisoning in Iraq. Science 181(4096):230–241
108. Engleson G, Herner T (1952) Alkyl mercury poisoning. Acta Paediatr 41(MAY):289–294
109. Davis LE et al (1994) Methylmercury poisoning – long-term clinical, radiological, toxicological, and pathological-studies of an affected family. Ann Neurol 35(6):680–688
110. Scheuhammer AM, Meyer MW, Sandheinrich MB, Murray MW (2007) Effects of environmental methylmercury on the health of wild birds, mammals, and fish. Ambio 36(1):12–18
111. Grandjean P et al (1992) Impact of maternal seafood diet on fetal exposure to mercury, selenium, and lead. Arch Environ Health 47(3):185–195
112. Grandjean P et al (1997) Cognitive deficit in 7-year-old children with prenatal exposure to methylmercury. Neurotox Teratol 19(6):417–428
113. Grandjean P, Weihe P, White RF, Debes F (1998) Cognitive performance of children prenatally exposed to "safe" levels of methylmercury. Environ Res 77(2):165–172
114. WHO (1993) Evaluation of certain food additives and contaminants. Forty-first Report of the Joint FAO/WHO expert committee on food additives. Technical report series 837, World Health Organization, Geneva
115. Boush GM, Thieleke JR (1983) Total mercury content in Yellowfin and Bigeye Tuna. Bull Environ Contam Toxicol 30(3):291–297
116. Power M, Klein GM, Guiguer K, Kwan MKH (2002) Mercury accumulation in the fish community of a sub-Arctic lake in relation to trophic position and carbon sources. J Appl Ecol 39(5):819–830
117. Cai Y, Rooker JR, Gill GA, Turner JP (2007) Bioaccumulation of mercury in pelagic fishes from the northern Gulf of Mexico. Can J Fish Aquat Sci 64(3):458–469
118. Rivers JB, Pearson JE, Shultz CD (1972) Total and organic mercury in marine fish. Bull Environ Contam Toxicol 8(5):257–266
119. Colaco A, Bustamante P, Fouquet Y, Sarradin PM, Serrao-Santos R (2006) Bioaccumulation of Hg, Cu, and Zn in the Azores triple junction hydrotherinal vent fields food web. Chemosphere 65(11):2260–2267
120. Dewailly E, Rouja P (2008) Balancing the risks and the benefits of local fish consumption in Bermuda. Centre de Recherche Technical Report, Universite' Laval, Quebec, Canada

121. Yamashita Y, Omura Y, Okazaki E (2005) Total mercury and methylmercury levels in commercially important fishes in Japan. Fish Sci 71(5):1029–1035
122. Kojadinovic J, Potier M, Le Corre M, Cosson RP, Bustamante P (2006) Mercury content in commercial pelagic fish and its risk assessment in the Western Indian Ocean. Sci Total Environ 366(2–3):688–700
123. Choy CA, Popp BN, Kaneko JJ, Drazen JC (2009) The influence of depth on mercury levels in pelagic fishes and their prey. Proc Natl Acad Sci USA 106(33):13865–13869
124. Gill GA, Fitzgerald WF (1988) Vertical mercury distributions in the oceans. Geochim Et Cosmochim Acta 52(6):1719–1728
125. Mason RP, Fitzgerald WF (1990) Alkylmercury species in the equatorial pacific. Nature 347(6292):457–459
126. Mason RP, Fitzgerald WF (1993) The distribution and biogeochemical cycling of mercury in the equatorial Pacific-Ocean. Deep Sea Res I 40(9):1897–1924
127. Mason RP, Fitzgerald WF (1991) Mercury speciation in open ocean waters. Water Air Soil Pollut 56:779–789
128. Sunderland EM, Krabbenhoft DP, Moreau JW, Strode SA, Landing WM (2009) Mercury sources, distribution, and bioavailability in the North Pacific Ocean: insights from data and models. Glob Biogeochem Cycles 23:1–14
129. Cossa D, Averty B, Pirrone N (2009) The origin of methylmercury in open Mediterranean waters. Limnol Oceanogr 54(3):837–844
130. Fitzgerald WF, Lamborg CH, Hammerschmidt CR (2007) Marine biogeochemical cycling of mercury. Chem Rev 107(2):641–662
131. Monperrus M et al (2007) Mercury methylation, demethylation and reduction rates in coastal and marine surface waters of the Mediterranean Sea. Mar Chem 107(1):49–63
132. Davidson PW et al (1998) Effects of prenatal and postnatal methylmercury exposure from fish consumption on neurodevelopment – outcomes at 66 months of age in the Seychelles child development study. Jama J Am Med Assoc 280(8):701–707
133. Hibbeln JR et al (2007) Maternal seafood consumption in pregnancy and neurodevelopmental outcomes in childhood (ALSPAC study): an observational cohort study. Lancet 369(9561):578–585
134. Myers GJ, Davidson PW (1998) Prenatal methylmercury exposure and children: Neurologic, developmental, and behavioral research. Environ Health Perspect 106:841–847
135. Ganther HE et al (1972) Selenium – relation to decreased toxicity of methylmercury added too diets containing tuna. Science 175(4026):1122–1124
136. Ralston NVC, Blackwell JL, Raymond LJ (2007) Importance of molar ratios in selenium-dependent protection against methylmercury toxicity. Biol Trace Elem Res 119(3):255–268
137. Kaneko JJ, Ralston NVC (2007) Selenium and mercury in pelagic fish in the central north pacific near Hawaii. Biol Trace Elem Res 119(3):242–254
138. Behne D, Pfeifer H, Rothlein D, Kyriakopoulos A (2000) Cellular and subcellular distribution of selenium and selenium-containing proteins in the rat. In: Roussel AM, Favier AE, Anderson RA (eds) Trace elements in man and animals. Kluwer, New York, pp 29–34
139. Rayman MP (2000) The importance of selenium to human health. Lancet 356(9225):233–241
140. Parizek J, Ostadalo I (1967) Protective effect of small amounts of selenite in sublimate intoxication. Experientia 23(2):142–143
141. Cuvinaralar MLA, Furness RW (1991) Mercury and selenium interaction – a review. Ecotoxicol Env Saf 21(3):348–364
142. Ralston CR, Blackwell JL III, Ralston NVC (2006) Effects of dietary selenium and mercury on house crickets (*Acheta domesticus L.*): implications of environmental co-exposures: implications of environmental co-exposures. Environ Bioindic 1(1):98–109
143. Dyrssen D, Wedborg M (1991) The sulfur-mercury(Ii) system in natural-waters. Water Air Soil Pollut 56:507–519

144. Watanabe C, Yin K, Kasanuma Y, Satoh H (1999) In utero exposure to methylmercury and Se deficiency converge on the neurobehavioral outcome in mice. Neurotoxicol Teratol 21(1):83–88
145. Hall RA, Zook EG, Meaburn GM (1978) National marine fisheries service survey of trace elements in the fishery resource. US Dept of Commerce, National Oceanographic and Atmospheric Administration, National Marine Fisheries Service, Rockville, MD
146. Cabanero AI, Carvalho C, Madrid Y, Batoreu C, Camara C (2005) Quantification and speciation of mercury and selenium in fish samples of high consumption in Spain and Portugal. Biol Trace Elem Res 103(1):17–35
147. Robinson J, Shroff J (2004) Observations on the levels of total mercury (Hg) and selenium (Se) in species common to the artisanal fisheries of Seychelles. Seychelles Med Dent J 7(1):55–60
148. Plessi M, Bertelli D, Monzani A (2001) Mercury and selenium content in selected seafood. J Food Compos Anal 14(5):461–467

Books and Reviews

Atwell L, Hobson KA, Welch HE (1998) Biomagnification and bioaccumulation of mercury in an arctic marine food web: insights from stable nitrogen isotope analysis. Can J Fish Aquat Sci 55(5):1114–1121

Cox PA, Banack SA, Murch SJ (2003) Biomagnification of cyanobacterial neurotoxins and neurodegenerative disease among the Chamorro people of Guam. Proc Natl Acad Sci USA 100(23):13380–13383

Gobas F et al (1999) Mechanism of biomagnification in fish under laboratory and field conditions. Environ Sci Technol 33(1):133–141

Gobas F, Zhang X, Wells R (1993) Gastrointestinal magnification – the mechanism of biomagnification and food-chain accumulation of organic-chemicals. Environ Sci Technol 27(13):2855–2863

Sijm D, Seinen W, Opperhulzen A (1992) Life-cycle biomagnification study in fish. Environ Sci Technol 26(11):2162–2174

Chapter 4
Biomarkers and Metabolomics, Evidence of Stress

Young Soo Keum, Jeong-Han Kim, and Qing X. Li

Glossary

Stress	Any change(s) in physiology and biochemical process that causes deviation from a normal state of an organism and requires an adjustment to return to the normal state.
Stressor	An agent, condition, or other stimulus that causes stress.
Biomarker	A biological chemical or macromolecule used as an indicator of a biological state that is often reflected by changes in its concentration.
Toxicity	Degree of poisoning or damage caused by a substance to an exposed organism.
Metabolome	The whole set of metabolites, forming an extensive network of metabolic reactions, in a biological system (e.g., an organism, organ, or cell).

This chapter was originally published as part of the Encyclopedia of Sustainability Science and Technology edited by Robert A. Meyers. DOI:10.1007/978-1-4419-0851-3

Y.S. Keum (✉)
Division of Applied Life Sciences, Konkuk University,
College of Life and Environmental Science, 1 Hwayang-dong, Gwangjin-gu,
Seoul 143-701, South Korea
e-mail: rational@konkuk.ac.kr

J.-H. Kim
Department of Agricultural Biotechnology, Seoul National University, 599 Gwanak-ro,
Silim-dong, Gwanak-Gu, Seoul, South Korea
e-mail: kjh2404@snu.ac.kr

Q.X. Li
Department of Molecular Biosciences and Bioengineering, University of Hawaii, Honolulu,
HI 96822, USA
e-mail: qingl@hawaii.edu

E.A. Laws (ed.), *Environmental Toxicology: Selected Entries from the Encyclopedia of Sustainability Science and Technology*, DOI 10.1007/978-1-4614-5764-0_4,

Metabolomics — Comprehensive study of a metabolome or a set of metabolites in which one metabolite from a specific pathway affects one or more biochemical reactions, or a comprehensive and quantitative analysis of all metabolites.

Omics — Comprehensive study of a biological system. Omics fields include genomics, proteomics, metabolomics, and transcriptomics.

Definition of the Subject

To evaluate the biological effect(s) of a stress, it is necessary to identify and characterize the stressor(s). In general, stressors can be classified as toxicants, pathogens, and physical stimulants. Although the reactions and the degree of response may vary, stressors frequently induce extensive metabolic changes in a living organism. Comparative research on a large set of metabolites at different stress states can provide detailed insights into specific biochemical reactions and metabolic networks. Such information can be used for diagnosis of a disease or toxic effect, development of therapeutic remedies to relieve the stress or its detrimental effect, etc.

The rapid accumulation of genomic and proteomic information of various organisms and rapid advancement in analytical capability make new disciplines of omics possible. In comparison with conventional approaches that usually focus on specific biochemical pathways, a major interest of omics is to gain system-wide information about the target organism and biochemical events. The main goal of metabolomics is qualitative and quantitative differentiation of an entire set of metabolites and the biological networks in the target organism [1, 2]. Metabolite profiles, fingerprints, and related concepts have been applied in various biochemical and toxicological studies since the 1980s [3, 4]. Metabolomics is an emerging discipline and has been shown to be a powerful tool for biomarker discovery and stress research. A related term, metabonomics has also been widely used in the literature. In this chapter, both metabolomics and metabonomics are regarded as having the same meaning. This chapter discusses the metabolomics of animals, plants, and microorganisms relevant to environmental toxicology.

Introduction

Biomarkers provide direct evidence of stress or potential stress. There are different types of biomarkers. Some biomarkers indicate different stages of disease: early threat (i.e., risk indicator or predictive biomarkers), existence (i.e., diagnostic biomarker), and disease development (i.e., prognostic biomarker). Some biomarkers give an indication of chemical toxicity, while others indicate physical stresses. It is widely known that cholesterol values are a biomarker indicating coronary and vascular disease risks. Molecular biomarkers need to be naturally stable for chemical analysis, to be reliably quantified, and to be effective for diagnosis, prognosis, and risk assessment. In environmental toxicology, biomarkers play a major role in environmental risk assessment, exposure, and management. Biomarker-based biological monitoring aids in early risk diagnosis, risk prevention and minimization, action target identification, toxic response, bioremediation, etc. The analytical method for a biomarker must be accurate, simple, reproducible, and cost-effective. Many genomics, metabolomics, and proteomics techniques are available for biomarker discovery. It is noteworthy that discovery of a biomarker typically requires a large amount of research effort. Biomarkers can be difficult to validate and may require different levels of validation.

Metabolomics has been largely employed to analyze all metabolites (i.e., global analysis) in a biological sample. Metabolomics experiments typically involve sampling, sample preparation, extraction, fractionation, detection, measurement, and data mining and management. Although metabolomics concerns the entirety of metabolites as analytical targets, it is practically impossible to cover all with a single analytical method. A large portion of metabolites are water-soluble, whereas steroids and respiratory quinones, for example, are practically water insoluble. It is well recognized that metabolomic responses are more rapid than other omics targets upon a stimulus. Hence, cautious and rapid handling of samples is very important to preserve the integrity of the metabolome. Common sample preparation protocols include proper labeling and storage of biological specimens, homogenization, extraction, cleanup, and concentration (Fig. 4.1). Additional procedures are often used to enhance the data quality and instrument performance, including enzymatic treatment to remove proteins and excess metabolites. In a metabolomics study targeting a specific set of metabolites, specialized enrichment techniques are often utilized [5, 6].

Instrumental analyses are an integral step in metabolomics research. Metabolomics methodologies can be divided into chemical separations and measurements. Instrumentation commonly used in metabolomics includes nuclear magnetic resonance spectroscopy (NMR) and mass spectrometry (MS). MS is widely used in metabolomics because it can provide rapid and sensitive qualitative and quantitative analyses of metabolites [7, 8]. High selectivity and sensitivity can be achieved by using tandem mass spectrometer systems where a mass spectrometer is hyphenated with another mass spectrometer and/or coupled with a separation technique. Some mass spectrometers such as Fourier Transform-ion cyclotron resonance MS (FT-ICR MS) have enough resolution power, eliminating chromatographic or electrophoretic separation requirements [9–11]. NMR has been applied in metabolomics since the 1980s [4]. Its strengths include rapid data acquisition with minimal sample preparation time, high reproducibility, and rich structural information. These factors make NMR among the most popular tools for high throughput

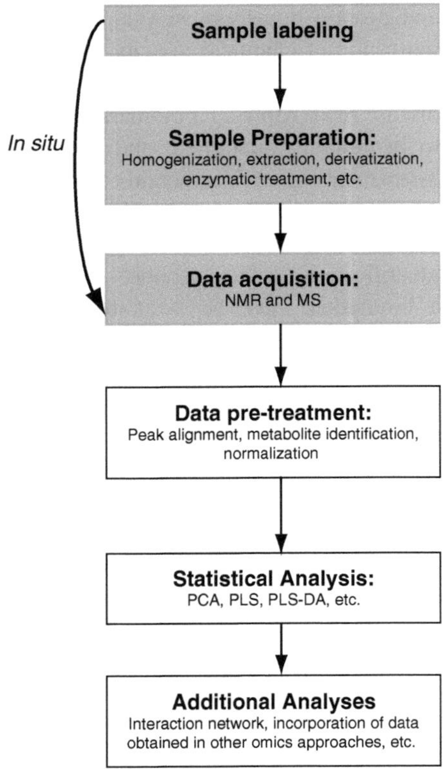

Fig. 4.1 A common metabolomics workflow. PCA, principal component analysis; PLS, partial least squares; DA, discriminant analysis

applications. Disadvantages in comparison with gas chromatography-MS (GC-MS) include low sensitivity and less comprehensive spectral libraries. Data from instrumental analysis require comprehensive interpretation, often accompanied by pretreatment of data (e.g., spectral deconvolution, peak alignment, and structural identification of specific metabolites).

A single metabolomics experiment can generate huge amounts of data leading to a comprehensive understanding of metabolic networks, whereas conventional approaches deal with a limited number of biochemical reactions. Complementary information from proteomics and genomics is often incorporated into a metabolomics data set to further interpretation of the biological events. Sophisticated statistical procedures are often necessary to uncover data of biological significance. Several data reduction methods, including principal component analysis (PCA), discriminant analysis (DA), partial least squares (PLS), and other newly introduced methods (e.g., Support Vector Machine and Neural Network) are used [12–15].

Integrated omics approaches are increasingly being used to verify and visualize metabolic differentiations. For example, Okuda et al. [16] reported the potential utility of the KEGG (Kyoto Encyclopedia of Genes and Genomes) atlas for global analysis of metabolic pathways and metabolomics. Many excellent reviews of metabolomics are available [17–22].

Applications of Metabolomics

Metabolomics in Animal Sciences

Metabolomics is considered to be the most promising omics tool for system-wide evaluation of stressors because primary metabolites are well conserved throughout the phylogenetic kingdoms [23, 24]. Chemical, physical, and biological stimuli have been studied. Much of the current research deals with mammals challenged by diseases, toxicants, and pathogens. Applications in this field are centered on the discovery of biomarkers for specific stressors, as exemplified in Table 4.1. Furthermore, metabolomics can provide insights into mechanisms of stress response, which can assist in the development of proper treatment methods [25, 26].

Mammalian metabolomes are more complex than those of plants and microorganisms. The concentrations of different metabolites in the same sample vary dramatically. In addition, both the concentrations and chemical identities of metabolites vary largely among different organs, sexes, ages, physiological stages, and numerous other factors. One of the most noticeable recent advancements was the construction of a metabolite database [27]. The human metabolome database (HMDB) is a comprehensive data resource of metabolites found in the human body (http://www.hmdb.ca). Many toxicological studies via metabolomics are approached with metabolic fingerprinting and pattern

Table 4.1 Examples of metabolomics studies of stressed animals

Type of study	Animal	Stressor	Description	Reference
Biomarker discovery	Rat	Drug (acetaminophen)	Depletion of antioxidants as marker	[37]
	Rat	Heavy metal (As, Hg)	Accumulation of creatine, lactate, and decrease of some amino acids	[46, 47]
	Human	Food (chocolate)	Decrease of stressor hormone (cortisol) by chocolate consumption	[55]
	Rat	Physical stress (shaking)	Accumulation of glucose, lipid and repression of acetate, amino acid	[56]
	Various	Environmental stress	Review of stress-metabolomics of non-model organism in ecosystem	[17]
Mode of action	Rat	Cancer and drug	Mode of action of anticancer drug	[33]
	Rat	Triazole fungicide	Unexpected toxicity of triazole fungicides	[43, 44]
Database	Human	Nontargeted	Identification of metabolites in the metabolome of healthy adults	[27]
	Human	Miscellaneous	Construction of metabolite database (http://www.hmdb.ca)	[115]

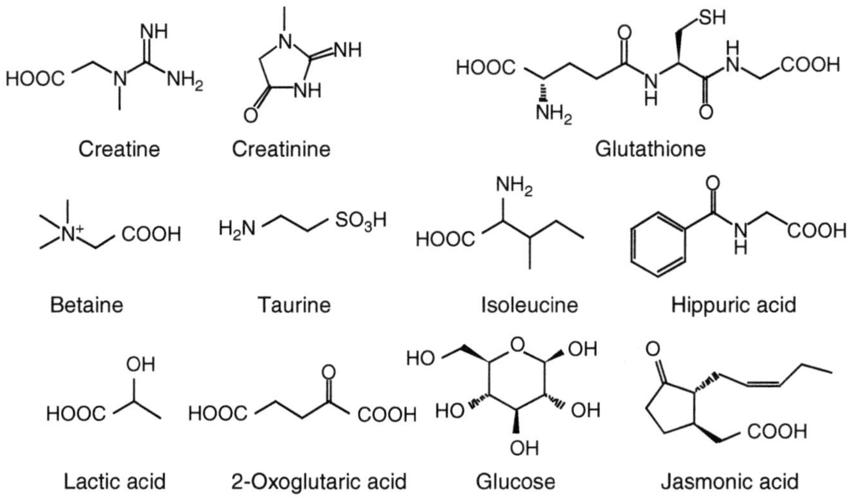

Fig. 4.2 Structures of selected biomarkers in stressed animals, plants, or microorganisms

recognition where knowing the exact identities of metabolites is not a prerequisite. Such applications of metabolomics have been used in clinical diagnostics. Biofluids such as blood and urine are commonly used samples. Figure 4.2 shows several common marker metabolites regardless of the stressor types (i.e., non-stress-specific biomarkers) [28–30]. In environmental toxicology, this type of biomarker can be used for the first tier of monitoring. The use of organic acids and creatine as biomarkers is often criticized due to their lack of stress specificity. It is noteworthy that the data from metabolomics should be interpreted on the basis of pattern changes rather than simple concentration changes of a few metabolites. The simple fingerprints along with structural identification can give comprehensive and detailed information of cellular status. Metabolomic profiles and structural analyses, for example, delineated a potential role for sarcosine in prostate cancer progression [31].

In drug discovery, many steps of evaluation are required, including screening for intended bioactivity (potency), adverse effects (toxicity), dose–response (efficacy), and metabolism (detoxification). Metabolic modulation of xenobiotics (drugs and pesticides) can trigger complicated problems because the extensive structural modification of xenobiotics through metabolism can alter the potency or toxicological profiles of the parent compounds. Holistic approaches, like metabolomics, may be a useful solution for drug discovery [32]. Metabolomic investigation of mode of action has been used to identify the exact targets and metabolic pathways of preclinical drugs. Watson et al. [33] identified the mode of action of an anticancer drug through global metabolic profiling. The mode of action of complex molecules (e.g., peptides and proteins) has been assessed with metabolomics. Angiotensin II (a well-known hormone) induces mitochondrial dysfunction. Mervaala et al. [34] used metabolomic methods to study the influence of angiotensin II on the metabolic profile and determine the mode of action of angiotensin II. They found that distinct patterns of cardiac substrate use in angiotensin II-induced cardiac hypertrophy are associated with mitochondrial dysfunction. In addition to natural primary metabolites, drug metabolites are profiled to obtain comprehensive data. The recent studies with naphthoquinone derivatives revealed that the toxicity of these chemicals is due to redox impairment, which causes strong oxidative stress [35]. It was found that oxidative stresses in response to acetaminophen in rats include depletion of antioxidants (e.g., ferulic acid), trigonelline, *S*-adenosyl-L-methionine, and energy-related metabolites [36, 37].

Pesticides are the cornerstones of pest management, food security, and public health. There are, however, public concerns about potential adverse effects of pesticides. Metabolomic evaluations of pesticide toxicity are much fewer than those of human drugs and diseases [38]. Similarly, a limited number of publications are available concerning metabolomics with organic pollutants. However, related research has gained some attention from scientific communities and regulatory organizations. Earlier studies of in vivo pesticide evaluation have indicated the complexity of the issue, particularly in higher animals. A specific pesticide can yield many metabolites with different toxicities [39–41]. In general, recently registered pesticides have much improved performance, including higher target selectivity, shorter half-lives, and fewer side effects (e.g., toxicity to nontarget organisms). In spite of these improvements, side effects are inevitable. Acetylcholine esterase is the target enzyme of organophosphorus insecticides (OPs). Some OPs are potent inhibitors of cytochrome P450 (CYP), a key enzyme of xenobiotic detoxification [42]. Azole fungicides target the fungal sterol 14[α]-demethylase used in ergosterol biosynthesis, but some of them can also inhibit mammalian sterol 14[α]-demethylase. Ekman et al. [43] analyzed metabolomic differences between normal and triazole fungicide-treated rats. The differential accumulation and depletion of some marker metabolites (creatine, choline, some osmolytes, and branched amino acids) in a dose-dependent manner were observed. Distinguishable metabolomic patterns indicated chemical-specific stresses of these fungicides [43, 44].

Mammalian metabolomics in response to heavy metals are frequently reported. Several heavy metals such as arsenic, cadmium, and mercury are commonly known environmental contaminants, especially in mining areas. Mally et al. [45] reported that 4-hydroxy-2(E)-nonenal is a potential biomarker of toxic mineral exposure, and its accumulation possibly indicates heavy metal-induced renal failure. Metabolomic analyses in the sera of cinnabar (mercury sulfide)-treated rats showed elevated concentrations of ketone bodies (3-D-hydroxybutyrate and acetoacetate), branched amino acids, choline, and creatine and decreased concentrations of glucose, lipids, and lipoproteins [46]. Similar metabolic profiles were also found in rats in response to realgar (arsenic sulfide) [47]. Accumulation of creatine, betaine, and some nitrogen-containing metabolites was the common, characteristic response to arsenic and mercury. The levels of glutathione and *N*-acyl amino acids, however, varied depending upon heavy metals and tissues [46, 47]. Toxicities of nano-materials are an emerging issue. Nano copper particles induced extensive changes of rat metabolomes [48]. An increase in triglycerides in the serum, liver, and kidney tissues could serve as a potentially sensitive biomarker for nano copper-induced lipidosis.

Target analytes in most metabolomics are naturally produced metabolites. However, recent studies of drug metabolism in conjunction with profiling natural metabolites suggested and confirmed novel toxicity biomarkers and toxicity mechanisms [49–52]. These studies produced extensive metabolic profiles of specific chemical stressors (e.g., drugs, pesticides, or environmental contaminants). Because the potency or toxicity of metabolites may differ from the parent chemicals, comprehensive profiling of metabolites can offer valuable information to evaluate chemical toxicity. Chen et al. [51] profiled both natural and drug metabolites and identified conjugates of acetaminophen metabolites as promising biomarkers of acetaminophen-induced toxicity. In addition, the oxidative stress elicited by CYP2E1-mediated acetaminophen metabolism might significantly contribute to its toxicity.

Recent studies have indicated several interesting aspects of nutritional metabolomics. Synergistic effects of co-application of anticancer drugs and methionine deprivation are well known for treatment of some cancers. However, the exact mechanism was not well understood. Guenin et al. [53] found that methionine deprivation affects several metabolite pools, including phospholipid and S-adenosylmethionine. Changes of these metabolites are accompanied with abnormal expression of some proteins and other key metabolites involved in tumor proliferation. Colon cancer is associated with diet, including higher intake of dietary sugar. Hansen et al. [54] applied metabolomics to investigate the relationships between the incidence of colon cancer and sucrose, glucose, and fructose. These sugars increased mutation rates in the colon and bulky adduct levels in the colon and liver to a similar extent. The metabolomics studies indicated disturbed amino acid metabolism and a decrease in plasma and urinary acetate to be a common feature for all dietary sugars and confirmed triglyceridemic effects of fructose. Accumulation of DNA-adducts was usually accompanied with a decrease of antioxidants, especially methionine. Increased colon cancer with associated changes of metabolic profiles indicated that a high sugar diet may affect the biochemical environment and may increase the probability of cancer development. In comparison with other stress-related metabolomic research, effects of food are often difficult to study, since food is a very complex mixture of chemicals. Routes of absorption, distribution, metabolism, and excretion (ADME) of specific constituents in the tested food can influence metabolic responses. A recent study on human consumption of chocolate has illustrated interesting metabolomic effects and a strong correlation with psychological activity [55]. Decreased excretion (or levels) of cortisol (stress hormone) was the most apparent effect of chocolate in reducing anxiety. This exemplifies the power of metabolomics to evaluate the effect of complex mixtures.

Animals are frequently exposed to physical stress (temperature, wound, etc.). Teague et al. [56] studied metabolomic responses of rats under physical stresses – namely, shaking the cage occasionally. They found several different marker metabolites, depending upon the shaking duration. As acute response, the accumulation of glucose and ketone bodies was accompanied with the repression of acetate and branched amino acids. Prolonged shaking induced additional metabolomic changes (e.g., increase of choline). A recent study with rats indicated that extensive metabolomic changes can occur during restraint stress [57]. Notable responses included differential levels of metabolites in glycolysis, the tricarboxylic acid (TCA) cycle, fatty acid oxidation, and urea cycles.

Environmental metabolomics, particularly targeting non-model animals, is more difficult because the needed information such as gene sequences and protein databases is limited, and the organisms are usually exposed to multiple stressors [17]. The recent social interests in environmental issues have accelerated such research needs. Metabolomes of various organisms were recently assessed, and fish were the most common model organisms. These species include Chinook salmon (*Onchorhynchuss tshawytscha*), flatfish (*Limanda limanda*), fathead minnow (*Pimephales promelas*), and Japanese medaka (*Oryzias latipes*) [43, 58–62]. Although multiple stressors are common in the environment, metabolomic studies with these animals often focus on a specific stimulus (disease or chemical stress). Ekman et al. [58] reported that metabolomic responses to the fungicide vinclozolin in fathead minnows included changes in the concentrations of several metabolites including creatine and betaine. Viant et al. [61] reported that several metabolites, most notably of energy-related intermediates (e.g., ATP), amino acids, and creatine, accumulated in Chinook salmon exposed to the pesticides dinoseb, diazinon, and esfenvalerate. Japanese medaka also gave a similar response to dinoseb [62]. In general, hydrophilic metabolites are more common metabolomics target molecules. Steroidal lipids were also proven to be good marker metabolites in rainbow trout (*Oncorhynchus mykiss*) in response to environmental estrogens [59].

Recent metabolomics studies revealed the effects of pesticides on several invertebrate species such as earthworms (*Eisenia* spp. and *Lumbricus rubellus*) [63–68]. In a multi-platform analysis of earthworm metabolomes, McKelvie et al. [68] found that DDT and endosulfan affected the levels of sugars and amino acids. The comparative analyses suggested the ratio of alanine to glycine as a potential biomarker of pesticide exposure. Rochfort et al. [69] observed a clear indication of different metabolomic patterns in earthworms being relevant to soil health.

Mollusks such as abalones and mussels are usually considered to be relevant marker species of chemical stress in the marine environment. Tuffnail et al. [70] investigated metabolomic responses of blue mussels (*Mytilus edulis*) under various physical and chemical stressors. In this case, alanine and some osmolytes were the important factors to discriminate different stress types. Research with red abalone (*Haliotis rufescens*) also indicated several osmolytes, amino acids, and storage carbohydrates (e.g., glycogen) to be the common indicators of environmental stresses [71]. Metabolomics studies with either freshwater or marine-dwelling crustaceans are quite limited. Recent examples with *Daphnia* and *Diporeia* concerned the metabolomic evaluation of atrazine and heavy metals [11, 72]. Several amino acids and amines were characterized as useful biomarkers through the analysis of metabolomes. However, detailed connections between this information and other biochemical observations are still missing.

Birds are important parts of ecosystems and food resources. However, metabolomic research has been very limited in this subject. Approximately 300 research articles have been published in other omics areas for chickens, but only one metabolomic publication was found as of April, 2010. Huber et al. [73] studied the effects of organically and conventionally produced feeds on a chicken model. Small but noticeable differences of metabolic profiles were observed, which may be correlated with animal productivity. Research is needed to fill in the knowledge gaps of metabolomic responses to stressors in birds.

Metabolomic research has also been very limited in insects, which are important parts of ecosystems. Recent studies with *Drosophila melanogaster* indicated that several metabolites (e.g., glutamate and choline) give differential responses against hypoxia and heat [74, 75]. Future metabolomics studies with insects are of great significance, particularly for ecosystems, agriculture, and food security.

Applications in Plant Sciences

Plants are fundamental to our life. Much research has been published on practically every aspect of plants. Most applications of metabolomics in plant sciences are phenotyping or simple mapping of metabolites. Many publications of plant metabolomics under various types of stresses have been published [76–79] although fewer than those concerning animals. Table 4.2 shows common physical, chemical, and biological stressors.

Temporal dynamics of metabolomes in heat and cold-stressed *Arabidopsis thaliana* have been monitored [80]. In general, cold shock influences metabolism far more profoundly than heat shock. However, a large portion of heat- and cold-shock responses overlaps. Those include accumulation of amino acids derived from pyruvate and oxaloacetate, polyamine precursors, and compatible solutes [80]. Rudell et al. [81] studied metabolomic changes of apples in response to cold and suggested that a wide array of metabolomes were changed under the specified conditions. Studies indicated that some antioxidants or plant growth regulators may alleviate the damage of cold stress. Although there were several unique responses, depending on the nature of stressors, physical stimuli usually induced accumulation of oxylipins [82]. Several jasmonate derivatives were identified as useful biomarkers of wound-induced response in *A. thaliana* [83] (Fig. 4.2).

Heavy metals are common stressors to plants. They are commonly used as toxicant models in metabolomics and other omics research. Plants that are highly tolerable to heavy metals usually produce a large amount of small peptides, called phytochelatins. Metabolic profiling of cadmium-treated *A. thaliana* has shown an accumulation of several phytochelatins [84]. Phytochelatins are oligomers of glutathione containing cysteine, glutamate, and glycine. Metabolomic differentiation in sulfur-containing amino acids is a commonly observed phenomenon in heavy metal-stressed plants, signifying an adaptive response to heavy metals.

Table 4.2 Examples of plant metabolomic studies under various stressors

Type of study	Plant	Stressor	Description	Reference
Biomarker discovery	*Arabidopsis thaliana*	Physical (heat, cold)	Accumulation of amino acid and osmolytes	[80]
	A. thaliana	Physical (wound)	Accumulation of oxylipins and jasmonates	[82, 83]
	A. thaliana	Heavy metal (Cd)	Accumulation of phytochelatins and precursors	[84]
	Hordeum vulgare (barley)	Salt stress	Accumulation of proline, sucrose, and glucose	[91]
	A. thaliana	Nutritional stress	Repression of many metabolites with reduced metabolic rates	[93]
Mode of action	*Zea mays* (corn)	Pesticides	Mode of action specific metabolomic response to 24 herbicides	[94]
	Lemna minor	Pesticides	Mode of action-dependent changes of amino acids and precursors	[96]
Toxicity	*Oryza sativa* (rice)	Ozone	Integrative approaches with metabolomics, transcriptomics, and proteomics	[103]
Biological stress	*Brassica rapa* (Chinese cabbage)	Insects	Differential accumulation of alanine, threonine, glucose, sucrose, feruloyl malate, sinapoyl malate, and gluconapin	[99]

Other common metabolites indicating adaptive changes include sugars and sugar alcohols. Plants can also modulate secondary metabolism in response to heavy metals such as copper and mercury. Jahangir et al. [85] observed an increase in the secondary metabolism and differentiation of amino acid and sugar levels in Chinese cabbage (*Brassica rapa*) when it was treated with copper, iron, and manganese. Rapid accumulation of glucosinolates (a secondary metabolite) was observed in *A. thaliana* under nutritional stresses [86]. These findings indicate that there may be a common feature of stress from exposure to heavy metal and nutrient deficiency.

High soil salinity significantly impacts plant growth and decreases crop yields. Salinity-related stresses are frequently accompanied with low water supplies in agricultural practices. The metabolic impact of salt stress was assessed on several crops and plant models (e.g., *A. thaliana*, barley, grape, rice, and tomato) [78, 87–91]. Accumulations of proline, sucrose/glucose, and inositol are the most frequently observed response in salt-stressed plants. Another common response includes up-regulation of proteins involved in polyamine and γ-aminobutyric acid metabolism and associated metabolites. The levels of organic acids in the TCA cycle usually decrease under salt stress, which may indicate reduced cellular activity. A general conclusion from metabolomics studies of salt-stressed plants includes changes in organic acid, amino acid and sugar metabolism.

Metabolomic profiles of plants under water stress share a common feature with those under salt stress, including increased sugar content [92]. Plants are primary sources of human foods. Therefore, plant metabolism under different nutritional conditions (i.e., fertilization) is of tremendous importance. However, only a limited number of papers have been published. *A. thaliana* is the most commonly used plant model. Hirai et al. [93] studied metabolite profiles and transcriptomes of *A. thaliana* under sulfur or nitrogen limitation. Several genes and metabolites were commonly expressed under both sulfur and nitrogen deficiencies. For example, nitrate reductase, a key enzyme of nitrogen assimilation, was down-regulated under both conditions. Reduction of this enzyme and the associated metabolites indicated that reduced metabolic rates may be a common response to nutrient deprivation. However, some metabolites such as glucosinolates were differently accumulated between nitrogen- and sulfur-deficient conditions. Metabolic consequences of sulfur-limitation in plants were further investigated by Nikiforova et al. [86]. Limited supplementation of sulfate leads to a decrease in sulfur metabolite pools (e.g., cysteine and methionine). Because these metabolites are an integral part of plant metabolism, sulfur deficiency induces a number of adaptive responses. Decreases in biomass, levels of proteins, chlorophylls, and total RNA suggest a general reduction of metabolic activity. These responses are compensated by a systemic adjustment of the major metabolic pathways. Representative responses include re-partitioning of sulfur, carbon, and nitrogen through the accumulation of metabolites.

Metabolomic investigations of pesticide effects on plants are limited. Ott et al. [94] studied metabolomic responses in corn (*Zea mays*) treated with 24 different herbicides. They successfully discriminated chemicals having different modes of action. Comparative metabolomic studies with pyrenophorol indicated that this fungal toxin may have different modes of action from the commercial herbicides used on the oat *Avena sterilis* [95]. Recently, the metabolic response of the aquatic weed *Lemna minor* was investigated in relation to herbicidal activities of glyphosate, atrazines, and other chemicals [96]. GC-MS based metabolomics of *A. thaliana* indicated that herbicides having different modes of action result in notably different metabolomes [97]. For example, sulfonylurea herbicides induced large differences in metabolite profiles, specifically those involved in glycolysis, the TCA cycle, and nitrogen assimilation. Applications of novel statistical tools in conjunction with LC-MS analysis indicated that the insecticide carbofuran can induce large changes of endogenous metabolomes in tomato fruits [98].

Interactions among plants, pest insects, and pathogens are another interesting subject of metabolomics. For example, *B. rapa* leaves, attacked by the diamondback moth (*Plutella xylostella*) or beet armyworm (*Spodoptera exigua*), differentially accumulated alanine, threonine, glucose, sucrose, feruloyl malate, sinapoyl malate, and gluconapin, depending upon the insect species and the developmental stages [99].

Metabolomics with aquatic plants or algae are scarce, although their environmental significance is as important as their terrestrial counterparts. Representative examples include metabolomic evaluations of the green algae *Chlamydomonas reinhardtii* under nutritional stress [100, 101]. This microalga is known to produce hydrogen under anaerobic conditions. Recent studies indicated that a large portion of metabolites are differentially expressed, depending upon the deprived nutrients. Time-resolved assays suggested that the kinetic responses also differ among nutrients. For example, the response to phosphorus deprivation was slower than to sulfur or nitrogen [100]. Matthew et al. [101] found that hydrogen overproduction in *C. reinhardtii* is induced by sulfur-deprivation through metabolic reprogramming or a build-up of the toxic fermentative products formate and ethanol. An excellent review has been published about metabolomic studies of marine macroalgae [102].

The most notable trend in recent plant metabolomics is integrated omics. By integrating transcriptomics, proteomics, and metabolomics, characteristic differentiation of metabolomes was identified as a defensive response in rice seedling (*Oryza sativa*) [103]. These findings provided insights into metabolic responses against stresses.

Another interesting advance in metabolomics is the development of novel data-mining technologies. Sato et al. [104] investigated rice foliage metabolomes through correlative approaches between metabolites and regulatory networks. In general, most metabolomic research gives numerical data with no correlative information among metabolites. Correlations among metabolites, genes, and proteins will give more comprehensive understanding of stress response.

Metabolomics in Microbiology

There have been a limited number of publications on metabolomics of microorganisms, particularly under stressed conditions. Microorganisms are more prone to metabolic changes than higher organisms because they are directly exposed to the stimuli. Tweeddale et al. [105] reported that *Escherichia coli* underwent interesting metabolic changes under oxidative stress. Paraquat, a well-known respiratory inhibitor, can induce the up-regulation of several oxidative stress-related proteins and an increase of branched amino acid and sulfur-containing metabolite pools (e.g., valine and glutathione) [105].

Elucidation of mode of action is a very important subject in pesticide and drug discoveries. However, this can be a laborious and time-consuming task because a biological system contains a wide array of metabolites, proteins, and genes. Metabolomics is a high throughput, comprehensive analytical option for this purpose. The first example can be found in metabolic phenotyping of the yeast *Saccharomyces cerevisiae* mutants [9]. Metabolic patterns of *S. cerevisiae* treated with 10 commercial fungicides were differentiated via metabolomics, where two large clusters were derived according to the inhibitory activities [106]. Such a study was extended to non-model diatoms under iron starvation [107].

Heavy metals are stressors to microorganisms [107–109]. Heavy metals can induce metabolic changes in glutathione and osmolytes (e.g., glycine and betaine) as well as other metabolites. Among these metabolites, glutathione and betaine are commonly found in stressed plants, animals, and microorganisms. Recent studies with the anaerobic bacteria *Shewanella oneidensis* indicated that the overall metabolisms are highly stable even under extremely stressed conditions. Only part of the metabolic pathways (e.g., sulfur amino acid metabolism) gave rapid responses to the external toxicants [110].

Bacterial metabolomics with environmental relevance were recently reported [111, 112]. For example, the bacterium *Sinorhizobium*, upon exposure to environmental contaminants, has shown a characteristic shift to a more hydrophobic fatty acid profile. In addition, the amount of storage metabolites such as polyhydroxyalkanoates decreased under stressed conditions. A large portion of the metabolites decreased, while trehalose, branched amino acids and sulfur-containing acids accumulated. The results indicated that oxidative stresses and metabolic confinement are a response to nutritional deficiency [112].

A recent trend in microbial metabolomics is integrative omics. Intracellular metabolism of benzoic acid in the white-rot basidiomycete *Phanerochaete chrysosporium* was investigated at the proteome and metabolome levels. Up-regulation of aryl-alcohol dehydrogenase, arylaldehyde dehydrogenase, and cytochrome P450s suggested that these enzymes play key roles in benzoic acid metabolism. Intracellular metabolic shifts to the TCA cycle indicated activation of heme biosynthesis and the production of NAD(P)H. Both metabolites are the most important components of oxidative enzymes. The analyses also indicated the role of trehalose as a storage disaccharide and the presence of an alternative energy-producing pathway [113]. The bacterium *Streptomyces tenjimariensis* can produce istamycins, displaying a high potency against many Gram-negative and Gram-positive bacteria. Denery et al. [114] used metabolic profiling and polymerase chain reaction (PCR) techniques as a means of studying the growth of *S. tenjimariensis*. The study showed metabolite profiles that are unique to growth phase and signify how they can be used for further applications in the understanding of microbial metabolism, antibiotic production, and physiology.

Future Directions

Metabolomics is an emerging field. It has started to play an important role in biological sciences, particularly in systems biology and biomarker studies. However, its applications are still limited by many challenges in methodologies and applications. A noteworthy example is "lipidomics," a branch of metabolomics focusing on lipids. A tremendous amount of research has been done with hydrophilic metabolites, whereas applications of lipidomics are very limited. It is no doubt that lipidomics will become an important tool in lipid research, particularly concerning the physiological significance and function of lipids. The metabolome is the down-stream product of genes and proteins. Metabolomics alone can give the final outcome of specific stresses, but does not provide direct information on the up-stream changes such in genes and proteins. This is both the strength and weakness of metabolomics and, therefore, integrated omics approaches seem to be inevitable. Many examples of integrated omics have recently been published, and this approach will become more popular in the future. Such a trend will improve both the quality and quantity of metabolomic data, especially in environmental toxicology. Technological challenges include data mining, analysis, and synthesis (i.e., data dimensionality and processing). Comprehensive networks of metabolites, proteins, and genes provide an overview and in-depth understanding of biological events and processes. Metabolomics tells what is happening at the metabolic level, provides direct evidences of stress, and will therefore contribute significantly to environmental toxicology.

Acknowledgments This work was supported in part by grants from the US Fish and Wildlife Service and Hawaii Energy and Environmental Technologies Initiative Award No. N00014-09-1-0709. We thank Adam Baker and Margaret R. Ruzicka for comprehensive review of this chapter.

Bibliography

Primary Literature

1. Fiehn O (2001) Combining genomics, metabolome analysis, and biochemical modeling to understand metabolic networks. Comp Funct Genom 2:155–168
2. Oliver SG, Winson MK, Kell DB, Baganz F (1998) Systematic functional analysis of the yeast genome. Trends Biotechnol 16(9):373–378
3. Blanchard J, Sawers SJ, Jonkman JH, Tang-Liu DD (1985) Comparison of the urinary metabolite profile of caffeine in young and elderly males. Br J Clin Pharmacol 19:225–232
4. Nicholson JK, Timbrell JA, Bales JR, Sadler PJ (1985) A high resolution proton nuclear magnetic resonance approach to the study of hepatocyte and drug metabolism. Application to acetaminophen. Mol Pharmacol 27:634–643
5. Carlson EE, Cravatt BF (2007) Chemoselective probes for metabolite enrichment and profiling. Nat Methods 4:429–435
6. Carlson EE, Cravatt BF (2007) Enrichment tags for enhanced-resolution profiling of the polar metabolome. J Am Chem Soc 129:15780–15782
7. Dunn WB, Ellis DI (2005) Metabolomics: Current analytical platforms and methodologies. Trend Anal Chem 24:285–294
8. Ramautar R, Somsen GW, de Jong GJ (2009) CE-MS in metabolomics. Electrophoresis 30:276–291
9. Allen J, Davey HM, Broadhurst D, Heald JK, Rowland JJ, Oliver SG, Kell DB (2003) High-throughput classification of yeast mutants for functional genomics using metabolic footprinting. Nat Biotechnol 21:692–696
10. Brown SC, Kruppa G, Dasseux J-L (2003) Metabolomics application of FR-ICR mass spectrometry. Mass Spectrom Rev 24:223–231
11. Taylor NS, Weber RJM, Southam AD, Payne TG, Hrydziuszko O, Arvanitis TN, Viant MR (2009) A new approach to toxicity testing in *Daphnia magna*: application of high throughput FT-ICR mass spectrometry metabolomics. Metabolomics 5:44–58

12. Beckonert O, Bollard ME, Ebbels TMD, Keun HC, Antti H, Holmes E, Lindon JC, Nicholson JK (2003) NMR-based metabonomic toxicity classification: hierarchical cluster analysis and k-nearest-neighbour approaches. Anal Chim Acta 490:3–15
13. Goodacre R, Vaidyanathan S, Dunn WB, Harrigan GG, Kell DB (2004) Metabolomics by numbers: acquiring and understanding global metabolite data. Trends Biotechnol 22:245–251
14. Qiu Y, Rajagopalan D, Connor SC, Damian D, Zhu L, Handzel A, Hu G, Amanullah A, Bao S, Woody N, MacLean D, Lee K, Vanderwall D, Ryan T (2008) Multivariate classification analysis of metabolomic data for candidate biomarker discovery in type 2 diabetes mellitus. Metabolomics 4:337–346
15. Wang Y, Bollard ME, Keun H, Antti H, Beckonert O, Ebbels TM, Lindon JC, Holmes E, Tang H, Nicholson JK (2003) Spectral editing and pattern recognition methods applied to high-resolution magic-angle spinning ^{1}H nuclear magnetic resonance spectroscopy of liver tissues. Anal Biochem 323:26–32
16. Okuda S, Yamada T, Hamajima M, Itoh M, Katayama T, Bork P, Goto S, Kanehisa M (2008) KEGG Atlas mapping for global analysis of metabolic pathways. Nucleic Acids Res 36(Web Server):W423–W426
17. Bundy JG, Davey MP, Viant MR (2009) Environmental metabolomics: a critical review and future perspectives. Metabolomics 5:3–21
18. Simpson MJ, McKelvie JR (2009) Environmental metabolomics: new insights into earthworm ecotoxicity and contaminant bioavailability in soil. Anal Bioanal Chem 394:137–149
19. Viant MR (2008) Recent developments in environmental metabolomics. Mol Biosyst 4:980–986
20. Viant MR (2009) Applications of metabolomics to the environmental sciences. Metabolomics 5(1):1–2
21. Weckwerth W (2003) Metabolomics in system biology. Annu Rev Plant Biol 54:669–689
22. Villas-Boas SG, Mas S, Akesson M, Smedsgaard J, Nielsen J (2005) Mass spectrometry in metabolome analysis. Mass Spectrom Rev 24:613–646
23. Griffin JL, Shockcor JP (2004) Metabolic profiles of cancer cells. Nat Rev Cancer 4:551–561
24. Miller MG (2007) Environmental metabolomics: a SWOT analysis (strengths, weaknesses, opportunities, and threats). J Proteome Res 6:540–545
25. Dix DJ, Gallagher K, Benson WH, Groskinsky BL, McClintock JT, Dearfield KL, Farland WH (2006) A framework for the use of genomics data at the EPA. Nat Biotechnol 24:1108–1111
26. Mench M, Schwitzguebel J-P, Schroeder P, Bert V, Gawronski S, Gupta S (2009) Assessment of successful experiments and limitations of phytotechnologies: contaminant uptake, detoxification and sequestration, and consequences for food safety. Environ Sci Pollut Res 16:876–900
27. Lawton KA, Berger A, Mitchell M, Milgram KE, Evans AM, Guo L, Hanson RW, Kalhan SC, Ryals JA, Milburn MV (2008) Analysis of the adult human plasma metabolome. Pharmacogenomics 9:383–397
28. Kleno TG, Kiehr B, Baunsgaard D, Sidelmann UG (2004) Combination of "omics" data to investigate the mechanism(s) of hydrazine-induced hepatotoxicity in rats and to identify potential biomarkers. Biomarkers 9:116–138
29. Masuo Y, Imai T, Shibato J, Hirano M, Jones OAH, Maguire ML, Satoh K, Kikuchi S, Rakwal R (2009) Omic analyses unravels global molecular changes in the brain and liver of a rat model for chronic Sake (Japanese alcoholic beverage) intake. Electrophoresis 30:1259–1275
30. Robertson DG (2005) Metabonomics in toxicology: a review. Toxicol Sci 85:809–822
31. Sreekumar A, Poisson LM, Rajendiran TM, Khan AP, Cao Q, Yu J, Laxman B, Mehra R, Lonigro RJ, Li Y, Nyati MK, Ahsan A, Kalyana-Sundaram S, Han B, Cao X, Byun J, Omenn GS, Ghosh D, Pennathur S, Alexander DC, Berger A, Shuster JR, Wei JT, Varambally S,

Beecher C, Chinnaiyan AM (2009) Metabolomic profiles delineate potential role for sarcosine in prostate cancer progression. Nature 457:910–914

32. Nicholson JK, Connelly J, Lindon JC, Holmes E (2002) Metabonomics: a platform for studying drug toxicity and gene function. Nat Rev Drug Discov 1:153–161
33. Watson M, Roulston A, Bélec L, Billot X, Marcellus R, Bédard D, Bernier C, Branchaud S, Chan H, Dairi K, Gilbert K, Goulet D, Gratton M-O, Isakau H, Jang A, Khadir A, Koch E, Lavoie M, Lawless M, Nguyen M, Paquette D, Turcotte E, Berger A, Mitchell M, Shore GC, Beauparlant P (2009) The small molecule GMX1778 is a potent inhibitor of NAD biosynthesis: strategy for enhanced therapy in nicotinic acid phosphoribosyltransferase 1-deficient tumors. Mol Cell Biol 29:5872–5888
34. Mervaala E, Biala A, Merasto S, Lempiainen J, Mattila I, Martonen E, Eriksson O, Louhelainen M, Finckenberg P, Kaheinen P, Muller DN, Luft FC, Lapatto R, Oresic M (2010) Metabolomics in angiotensin II-induced cardiac hypertrophy. Hypertension 55:508–515
35. Parry JD, Pointon AV, Lutz U, Teichert F, Charlwood JK, Chan P, Athersuch TJ, Taylor EL, Singh R, Luo JL, Phillips KM, Vetillard A, Lyon JJ, Keun HC, Lutz WK, Gant TW (2009) Pivotal role for two electron reduction in 2, 3-dimethoxy-1, 4-naphthoquinone and 2-methyl-1, 4-naphthoquinone metabolism and kinetics in vivo that prevents liver redox stress. Chem Res Toxicol 22:717–725
36. Coen M, Lenz EM, Nicholson JK, Wilson ID, Pognan F, Lindon JC (2003) An integrated metabonomic investigation of acetaminophen toxicity in the mouse using NMR spectroscopy. Chem Res Toxicol 16:295–303
37. Sun J, Schnackenberg LK, Holland RD, Schmitt TC, Cantor GH, Dragan YP, Beger RD (2008) Metabonomics evaluation of urine from rats given acute and chronic doses of acetaminophen using NMR and UPLC/MS. J Chromatogr B 871:328–340
38. Keum YS, Kim J-H, Li QX (2010) Metabolomics in pesticide toxicology. In: Krieger R (ed) Hayes' handbook of pesticide toxicology, 3rd edn. Academic, New York, pp 627–643
39. Coleman S, Linderman R, Hodgson E, Rose RL (2000) Comparative metabolism of chloroacetamide herbicides and selected metabolites in human and rat liver microsomes. Environ Health Persp 108:1151–1157
40. Hodgson E (2001) In vitro human phase I metabolism of xenobiotics I: pesticides and related chemicals used in agriculture and public health. J Biochem Mol Toxicol 15:296–299
41. Hodgson E, Levi PE (2010) Metabolism of pesticides. In: Krieger R (ed) Hayes' handbook of pesticide toxicology, 3rd edn. Academic, New York, pp 893–921
42. Abass K, Turpeinen M, Pelkonen O (2009) An evaluation of the cytochrome P450 inhibition potential of selected pesticides in human hepatic microsomes. J Environ Sci Health B 44:553–563
43. Ekman DR, Keun HC, Eads CD, Furnish CM, Murrell RN, Rockett JC, Dix DJ (2006) Metabolomic evaluation of rat liver and testis to characterize the toxicity of triazole fungicides. Metabolomics 2:63–73
44. Cherney DP, Ekman DR, Dix DJ, Collette TW (2007) Raman spectroscopy-based metabolomics for differentiating exposures to triazole fungicides using rat urine. Anal Chem 79:7324–7332
45. Mally A, Amberg A, Hard GC, Dekant W (2007) Are 4-hydroxy-2(E)-nonenal derived mercapturic acids and ^{1}H NMR metabonomics potential biomarkers of chemically induced oxidative stress in the kidney? Toxicology 230:244–255
46. Wei L, Liao P, Wu H, Li X, Pei F, Li W, Wu Y (2008) Toxicological effects of cinnabar in rats by NMR-based metabolic profiling of urine and serum. Toxicol Appl Pharmacol 227:417–429
47. Wei L, Liao P, Wu H, Li X, Pei F, Li W, Wu Y (2009) Metabolic profiling studies on the toxicological effects of realgar in rats by ^{1}H NMR spectroscopy. Toxicol Appl Pharmacol 234:314–325

48. Lei R, Wu C, Yang B, Ma H, Shi C, Wang Q, Wang Q, Yuan Y, Liao M (2008) Integrated metabolomic analysis of the nano-sized copper particle-induced hepatotoxicity and nephrotoxicity in rats: a rapid in vivo screening method for nanotoxicity. Toxicol Appl Pharmacol 232:292–301
49. Aubert S, Pallett KE (2000) Combined use of ^{13}C- and ^{19}F-NMR to analyze the mode of action and the metabolism of the herbicide isoxaflutole. Plant Physiol Biochem 38:517–523
50. Chen C, Gonzalez FJ, Idle JR (2007) LC-MS-based metabolomics in drug metabolism. Drug Metab Rev 39:581–597
51. Chen C, Krausz KW, Idle JR, Gonzalez FJ (2008) Identification of novel toxicity-associated metabolites by metabolomics and mass isotopomer analysis of acetaminophen metabolism in wild-type and Cyp2e1-null mice. J Biol Chem 283:4543–4559
52. Patterson AD, Gonzalez FJ, Idle JR (2010) Xenobiotic metabolism: a view through the metabolometer. Chem Res Toxicol 23:851–860
53. Guenin S, Morvan D, Thivat E, Stepien G, Demidem A (2009) Combined methionine deprivation and chloroethylnitrosourea have time-dependent therapeutic synergy on melanoma tumors that NMR spectroscopy-based metabolomics explains by methionine and phospholipid metabolism reprogramming. Nutr Cancer 61:518–529
54. Hansen M, Baunsgaard D, Autrup H, Vogel UB, Moller P, Lindecrona R, Wallin H, Poulsen HE, Loft S, Dragsted LO (2008) Sucrose, glucose and fructose have similar genotoxicity in the rat colon and affect the metabolome. Food Chem Toxicol 46:752–760
55. Martin F-PJ, Rezzi S, Pere-Trepat E, Kamlage B, Collino S, Leibold E, Kastler J, Rein D, Fay LB, Kochhar S (2009) Metabolic effects of dark chocolate consumption on energy, gut microbiota, and stress-related metabolism in free-living subjects. J Proteome Res 8:5568–5579
56. Teague CR, Dhabhar FS, Barton RH, Beckwith-Hall B, Powell J, Cobain M, Singer B, McEwen BS, Lindon JC, Nicholson JK, Holmes E (2007) Metabonomic studies on the physiological effects of acute and chronic psychological stress in Sprague-Dawley rats. J Proteome Res 6:2080–2093
57. Chen M, Wang Y, Zhao Y, Wang L, Gong J, Wu L, Gao X, Yang Z, Qian L (2009) Dynamic proteomic and metabonomic analysis reveal dysfunction and subclinical injury in rat liver during restraint stress. Biochim Biophys Acta 1794:1751–1765
58. Ekman DR, Teng Q, Jensen KM, Martinovic D, Villeneuve DL, Ankley GT, Collette TW (2007) NMR analysis of male fathead minnow urinary metabolites: a potential approach for studying impacts of chemical exposures. Aquat Toxicol 85:104–112
59. Samuelsson LM, Forlin L, Karlsson G, Adolfsson-Erici M, Larsson DGJ (2006) Using NMR metabolomics to identify responses of an environmental estrogen in blood plasma of fish. Aquat Toxicol 78:341–349
60. Stentiford GD, Viant MR, Ward DG, Johnson PJ, Martin A, Wei W, Cooper HJ, Lyons BP, Feist SW (2005) Liver tumours in wild flatfish: a histopathological, proteomic and metabolomic study. OMICS-J Integr Biol 9:281–299
61. Viant VR, Pincetich CA, Hinton DE, Tjeerdema RS (2006) Toxic actions of dinoseb in medaka (*Oryzias latipes*) embryos as determined by in vivo ^{31}P NMR, HPLC-UV and ^{1}H NMR metabolomics. Aquat Toxicol 76:329–342
62. Viant VR, Pincetich CA, Tjeerdema RS (2006) Metabolic effects of dinoseb, diazinon and esfenvalerate in eyed eggs and alevins of Chinook salmon (*Oncorhynchus tshawytscha*) determined by ^{1}H NMR metabolomics. Aquat Toxicol 77:359–371
63. Bundy JG, Osborn D, Weeks JM, Lindon JC, Nicholson JK (2001) An NMR-based metabonomic approach to the investigation of coelomic fluid biochemistry in earthworms under toxic stress. FEBS Lett 500:31–35
64. Bundy JG, Lenz EM, Bailey NJ, Gavaghan CL, Svendsen C, Spurgeon D, Hankard PK, Osborn D, Weeks JM, Trauger SA (2002) Metabonomic assessment of toxicity of 4-fluoroaniline, 3, 5-difluoroaniline and 2-fluoro-4-methylaniline to the earthworm Eisenia

veneta (Rosa): Identification of new endogenous biomarkers. Environ Toxicol Chem 21:1966–1972
65. Bundy JG, Lenz EM, Osborn D, Weeks JM, Lindon JC, Nicholson JK (2002) Metabolism of 4-fluoroaniline and 4-fluorobiphenyl in the earthworm *Eisenia veneta* characterized by high-resolution NMR spectroscopy with directly coupled HPLC-NMR and HPLC-MS. Xenobiotica 32:479–490
66. Bundy JG, Sidhu JK, Rana F, Spurgeon DJ, Svendsen C, Wren JF, Stürzenbaum SR, Morgan AJ, Kille P (2008) "Systems toxicology" approach identifies coordinated metabolic responses to copper in a terrestrial non-model invertebrate, the earthworm *Lumbricus rubellus*. BMC Biol 6:25
67. Guo Q, Sidhu JK, Ebbels TMD, Rana F, Spurgeon DJ, Svendsen C, Sturzenbaum SR, Kille P, Morgan AJ, Bundy JG (2009) Validation of metabolomics for toxic mechanism of action screening with the earthworm *Lumbricus rubellus*. Metabolomics 5:72–83
68. McKelvie JR, Yuk J, Xu Y, Simpson AJ, Simpson MJ (2009) ^{1}H NMR and GC/MS metabolomics of earthworm responses to sub-lethal DDT and endosulfan exposure. Metabolomics 5:84–94
69. Rochfort SJ, Ezernieks V, Yen AL (2009) NMR-based metabolomics using earthworms as potential indicators for soil health. Metabolomics 5:95–107
70. Tuffnail W, Mills GA, Cary P, Greenwood R (2009) An environmental ^{1}H NMR metabolomic study of the exposure of the marine mussel *Mytilus edulis* to atrazine, lindane, hypoxia and starvation. Metabolomics 5:33–43
71. Rosenblum ES, Tjeerdema RS, Viant MR (2006) Effects of temperature on host-pathogen-drug interactions in red abalone, *Haliotis rufescens*, determined by ^{1}H NMR metabolomics. Environ Sci Technol 40:7077–7084
72. Ralston-Hooper K, Hopf A, Oh C, Zhang X, Adamec J, Sepulveda MS (2008) Development of GCxGC/TOF-MS metabolomics for use in ecotoxicological studies with invertebrates. Aquat Toxicol 88:48–52
73. Huber M, van de Vijver LPL, Parmentier H, Savelkoul H, Coulier L, Wopereis S, Verheij E, van der Greef J, Nierop D, Hoogenboom RAP (2010) Effects of organically and conventionally produced feed on biomarkers of health in a chicken model. Br J Nutr 103:663–676
74. Feala JD, Coquin L, McCulloch AD, Paternostro G (2007) Flexibility in energy metabolism supports hypoxia tolerance in *Drosophila* flight muscle: metabolomic and computational systems analysis. Mol Syst Biol 3:99
75. Malmendal A, Overgaard J, Bundy JG, Sørensen JG, Nielsen NC, Loeschcke V, Holmstrup M (2006) Metabolomic profiling of heat stress: hardening and recovery of homeostasis in *Drosophila*. Am J Physiol Regul Integr Comp Physiol 291:205–212
76. Guy C, Kaplan F, Kopka J, Selbig J, Hinch DK (2008) Metabolomics of temperature stress. Physiol Plant 132:220–235
77. Hoefgen R, Nikiforova VJ (2008) Metabolomics integrated with transcriptomics: assessing systems response to sulfur-deficiency stress. Physiol Plant 132:190–198
78. Sanchez DH, Siahpoosh MR, Roessner U, Udvardi M, Kopka J (2008) Plant metabolomics reveals conserved and divergent metabolic responses to salinity. Physiol Plant 132:209–219
79. Shulaev V, Cortes D, Miller G, Mittler R (2008) Metabolomics for plant stress response. Physiol Plant 132:199–208
80. Kaplan F, Kopka J, Haskell DW, Zhao W, Schiller KC, Gatzke N, Sung DY, Guy CL (2004) Exploring the temperature-stress metabolome of *Arabidopsis*. Plant Physiol 136:4159–4168
81. Rudell DR, Mattheis JP, Hertog MLATM (2009) Metabolomic change precedes apple superficial scald symptoms. J Agric Food Chem 57:8459–8466
82. Thiocone A, Farmer EE, Wolfender J-L (2008) Screening for wound-induced oxylipins in *Arabidopsis thaliana* by differential HPLC-APCI/MS profiling of crude leaf extracts and subsequent characterisation by capillary-scale NMR. Phytochem Anal 19:198–205

83. Glauser G, Guillarme D, Grata E, Boccard J, Thiocone A, Carrupt P-A, Veuthey J-L, Rudaz S, Wolfender J-L (2008) Optimized liquid chromatography-mass spectrometry approach for the isolation of minor stress biomarkers in plant extracts and their identification by capillary nuclear magnetic resonance. J Chromatogr A 1180:90–98
84. Sarry J-E, Kuhn L, Ducruix C, Lafaye A, Junot C, Hugouvieux V, Jourdain A, Bastien O, Fievet JB, Vailhen D, Amekraz B, Moulin C, Ezan E, Garin J, Bourguignon J (2006) The early responses of *Arabidopsis thaliana* cells to cadmium exposure explored by protein and metabolite profiling analyses. Proteomics 6:2180–2198
85. Jahangir M, Abdel-Farid IB, Choi YH, Verpoorte R (2008) Metal ion-inducing metabolite accumulation in *Brassica rapa*. J Plant Physiol 165:1429–1437
86. Nikiforova VJ, Kopka J, Tolstikov V, Fiehn O, Hopkins L, Hawkesford MJ, Hesse H, Hoefgen R (2005) Systems rebalancing of metabolism in response to sulfur deprivation, as revealed by metabolome analysis of *Arabidopsis* plants. Plant Physiol 138:304–318
87. Cramer GR, Ergul A, Grimplet J, Tillett RL, Tattersall EAR, Bohlman MC, Vincent D, Sonderegger J, Evans J, Osborne C, Quilici D, Schlauch KA, Schooley DA, Cushman JC (2007) Water and salinity stress in grapevines: early and late changes in transcript and metabolite profiles. Funct Integr Genomics 7:111–134
88. Kim JK, Bamba T, Harada K, Fukusaki E, Kovayashi A (2007) Time-course metabolic profiling in *Arabidopsis thaliana* cell cultures after salt stress treatment. J Exp Bot 58:415–424
89. Sanchez DH, Lippold F, Redestig H, Hannah MA, Erban A, Kraemer U, Kopka J, Udvardi MK (2008) Integrative functional genomics of salt acclimatization in the model legume *Lotus japonicus*. Plant J 53:973–987
90. Smith AR, Johnson HE, Hall M (2003) Metabolic fingerprinting of salt-stressed tomatoes. Bulgarian J. Plant Physiol 29:153–163
91. Widodo PJH, Newbigin E, Tester M, Bacic A, Roessner U (2009) Metabolic responses to salt stress of barley (*Hordeum vulgare* L.) cultivars, Sahara and Clipper, which differ in salinity tolerance. J Exp Bot 60:4089–4103
92. Foito A, Byrne SL, Shepherd T, Stewart D, Barth S (2009) Transcriptional and metabolic profiles of *Lolium perenne* L. genotypes in response to a PEG-induced water stress. Plant Biotechnol J 7:719–732
93. Hirai MY, Yano M, Goodenowe DB, Kanaya S, Kimura T, Awazuhara M, Arita M, Fujiwara T, Saito K (2004) Integration of transcriptomics and metabolomics for understanding of global responses to nutritional stresses in *Arabidopsis thaliana*. Proc Natl Acad Sci USA 101:10205–10210
94. Ott KH, Aranı́bar N, Singh B, Stockton GW (2003) Metabonomics classifies pathways affected by bioactive compounds. Artificial neural network classification of NMR spectra of plant extracts. Phytochemistry 62:971–985
95. Aliferis KA, Chrysayi-Tokousbalides M (2005) Metabonomic strategy for the investigation of the mode of action of the phytotoxin (5 S, 8R, 13 S, 16R)-(-)-pyrenophorol using H nuclear magnetic resonance fingerprinting. J Agric Food Chem 54:1687–1692
96. Aliferis KA, Materzok S, Paziotou GN, Chrysayi-Tokousbalides M (2009) *Lemna minor* L. as a model organism for ecotoxicological studies performing 1 H NMR fingerprinting. Chemosphere 76:967–973
97. Trenkamp S, Eckes P, Busch M, Fernie AR (2009) Temporally resolved GC-MS-based metabolic profiling of herbicide treated plants treated reveals that changes in polar primary metabolites alone can distinguish herbicides of differing mode of action. Metabolomics 5:277–291
98. Perez IS, Culzoni MJ, Siano GG, Garcia MDG, Goicoechea HC, Galera MM (2009) Detection of unintended stress effects based on a metabonomic study in tomato fruits after treatment with carbofuran pesticide. Capabilities of MCR-ALS applied to LC-MS three-way data arrays. Anal Chem 81:8335–8346

99. Widarto HT, Meijden E, Lefeber AWM, Erkelens C, Kim HK, Choi YH, Verpoorte R (2006) Metabolomic differentiation of *Brassica rapa* following herbivory by different insect instars using two-dimensional nuclear magnetic resonance spectroscopy. J Chem Ecol 32:2417–2428
100. Bölling C, Fiehn O (2005) Metabolite profiling of *Chlamydomonas reinhardtii* under nutrient deprivation. Plant Physiol 139:1995–2005
101. Matthew T, Zhou W, Rupprecht J, Lim L, Thomas-Hall SR, Doebbe A, Kruse O, Hankamer B, Marx UC, Smith SM, Schenk PM (2009) The metabolome of *Chlamydomonas reinhardtii* following induction of anaerobic H_2 production by sulfur depletion. J Biol Chem 284:23415–23425
102. La Barre SL, Weinberger F, Kervarec N, Potin P (2004) Monitoring defensive responses in macroalgae – limitations and perspectives. Phytochem Rev 3:371–379
103. Cho K, Shibato J, Agrawal GK, Jung Y-H, Kubo A, Jwa N-S, Tamogami S, Satoh K, Kikuchi S, Higashi T, Kimura S, Saji H, Tanaka Y, Iwahashi H, Masuo Y, Rakwal R (2008) Integrated transcriptomics, proteomics, and metabolomics analyses to survey ozone responses in the leaves of rice seedling. J Proteome Res 7:2980–2998
104. Sato S, Arita M, Soga T, Nishioka T, Tomita M (2008) Time-resolved metabolomics reveals metabolic modulation in rice foliage. BMC Sys Biol 2:51
105. Tweeddale H, Notley-McRobb L, Ferenci T (1999) Assessing the effect of reactive oxygen species on *Escherichia coli* using a metabolome approach. Redox Rep 4:237–241
106. Allen J, Davey HM, Broadhurst D, Rowland JJ, Oliver SG, Kell DB (2004) Discrimination of modes of action of antifungal substances by use of metabolic footprinting. Appl Environ Microbiol 70:6157–6165
107. Allen AE, Laroche J, Maheswari U, Lommer M, Schauer N, Lopez PJ, Finazzi G, Fernie AR, Bowler C (2008) Whole-cell response of the pennate diatom *Phaeodactylum tricornutum* to iron starvation. Proc Natl Acad Sci USA 105:10438–10443
108. Tanaka Y, Higashi T, Rakwal R, Wakida S-I, Iwahashi H (2007) Quantitative analysis of sulfur-related metabolites during cadmium stress response in yeast by capillary electrophoresis-mass spectrometry. J Pharm Biomed Anal 44:608–613
109. Tremaroli V, Workentine ML, Weljie AM, Vogel HJ, Ceri H, Viti C, Tatti E, Zhang P, Hynes AP, Turner RJ, Zannoni D (2009) Metabolomic investigation of the bacterial response to a metal challenge. Appl Environ Microbiol 75:719–728
110. Tang YJ, Martin HG, Deutschbauer A, Feng X, Huang R, Llora X, Arkin A, Keasling JD (2009) Invariability of central metabolic flux distribution in *Shewanella oneidensis* MR-1 under environmental or genetic perturbations. Biotechnol Progr 25:1254–1259
111. Barsch A, Patschkowski T, Niehaus K (2004) Comprehensive metabolite profiling of *Sinorhizobium meliloti* using gas chromatography–mass spectrometry. Funct Integr Genomics 4:219–230
112. Keum YS, Seo JS, Li QX, Kim JH (2008) Comparative metabolomic analysis of *Sinorhizobium* sp. C4 during the degradation of phenanthrene. Appl Microbiol Biotechnol 80:863–872
113. Matsuzaki F, Shimizu M, Wariishi H (2008) Proteomic and metabolomic analyses of the white-rot fungus *Phanerochaete chrysosporium* exposed to exogenous benzoic acid. J Proteome Res 7:2342–2350
114. Denery J, Cooney M, Li QX (2010) Diauxic and antimicrobial growth phases of *Streptomyces tenjimariensis*: metabolite profiling and gene expression. J Biomed Biotechnol (submitted)
115. Wishart DS, Knox C, Guo AC, Eisner R, Young N, Gautam B, Hau DD, Psychogios N, Dong E, Bouatra S, Mandal R, Sinelnikov I, Xia J, Jia L, Cruz JA, Lim E, Sobsey CA, Shrivastava S, Huang P, Liu P, Fang L, Peng J, Fradette R, Cheng D, Tzur D, Clements M, Lewis A, De Souza A, Zuniga A, Dawe M, Xiong Y, Clive D, Greiner R, Nazyrova A, Shaykhutdinov R, Li L, Vogel HJ, Forsythe I (2009) HMDB: a knowledgebase for the human metabolome. Nucleic Acids Res 37(Database issue):D603–D610

Books and Reviews

Fiehn O (2001) Combining genomics, metabolome analysis, and biochemical modeling to understand metabolic networks. Comp Func Genom 2:155–168
Hodgson E, Levi PE (2010) Metabolism of pesticides. In: Krieger R (ed) Hayes' handbook of pesticide toxicology, 3rd edn. Academic, New York, pp 893–921

Chapter 5
Bioremediation and Mitigation

Ralph J. Portier

Glossary

Bioremediation	Bioremediation can be defined as a process using a microbial community and/or the related enzymes to return the natural environment altered by contamination to its original condition. Bioremediation can also be considered an industry that uses these techniques to solve real-world problems. Bioremediation is an applied field of science that combines advanced biotechnology and engineering approaches with basic microbiology to solve complicated challenges in soil water and groundwater contamination [1].
Biodegradation/ biomineralization	The process in the carbon cycle of using a microbe/microbial community (bacteria, fungi, yeasts, actinomycetes) or plants to convert complex carbon-based chemicals (organic matter) to simpler structures and biomass. The complete conversion of said chemical to its mineral form, namely, biomass, water, and CO_2 is called biomineralization. The process is described as aerobic if oxygen is required by the biological community and anaerobic if it is not required or present [1].
Mitigation	The effort to eliminate or reduce loss and/or impact to habitat, life, property, or natural resources. Hazard mitigation means any action taken to reduce or eliminate the

This chapter was originally published as part of the Encyclopedia of Sustainability Science and Technology edited by Robert A. Meyers. DOI:10.1007/978-1-4419-0851-3

R.J. Portier (✉)
Department of Environmental Sciences, School of the Coast & Environment,
Louisiana State University, Baton Rouge, LA 70803, USA
e-mail: rportie@lsu.edu

E.A. Laws (ed.), *Environmental Toxicology: Selected Entries from the Encyclopedia of Sustainability Science and Technology*, DOI 10.1007/978-1-4614-5764-0_5,

	long-term risk to human life and property from natural hazards (Stafford Act 44: CFR 206: 401) [1, 2].
Polycyclic aromatic hydrocarbons (PAHs)	Lipophilic, fused, aromatic rings typically found in compounds of oil, coal, tar deposits or residuals from burning any fossil fuel. Also known as polynuclear aromatic hydrocarbons, several are considered to be potential human carcinogens by USEPA. Benzo(a)pyrene is considered a benchmark regulatory toxicant for bioremediation of most hydrocarbon-contaminated sites [1].
Sustainable remediation	Practices or multitasked approaches for restoring a site to its native state with a frugal use of available resources so as to benefit human health and the environment [1].

Definition of the Subject

The high costs involved with conventional remedial techniques and frequently, the liability associated with the use of such techniques, has encouraged industry to search for innovative remediation technologies. The use of innovative technologies that provide a cost-effective and permanent treatment method is gaining the attention of regulatory agencies, the public, and, most importantly, the industries responsible for cleanup. In order to address these needs, research organizations and engineering firms are testing various innovative technologies in the laboratories and in the field. One such technology, bioremediation, is currently gaining more and more acceptance and has proven to be a cost-effective mitigation strategy [3].

The US Environmental Protection Agency has been at the forefront in promoting bioremediation as an innovative solution for addressing risk-based regulatory challenges for the cleanup of contaminated soils, sediments, groundwater, and industrial effluent. The establishment of specific criteria for acceptable levels of exposure in soils and sediments, groundwater, and/or discharge into rivers and streams has hastened the recognition of applying specific microbial cultures and/or communities to mitigate in a sustainable effort [4–6].

Introduction

The practice of using microorganisms to clean up contamination can be traced back as early as 600 BC. It was common practice to treat municipal wastewater by building intricate sewage systems that would transport wastewater to collection vats and lagoons where microbes would aid in the biodegradation of organic wastes. Today, this same principal can be used to clean up modern pollution problems like chemical spills and tank leaks [6]. This process is the utilization of a consortium of microbes to degrade organic pollutants. Through bioremediation, toxic pollutants can be broken down into nontoxic byproducts such as CO_2 and water. Aerobic heterotrophs have enzymatic systems capable of oxidizing suitable substrates by transfer of electrons to molecular oxygen. To put it in simple terms, this process is performed through the following reaction:

$$\text{Substrate} + O_2 \rightarrow \text{Biomass} + CO_2 + H_2O + \text{Other Organics}$$

Typically, in an aerobic or oxygen-requiring bioremediation, the contaminant degrading potential of these microorganisms is enhanced through the addition of essential reactants (inorganic and organic nutrients, water, and oxygen) [6, 7]. Biodegradation of complex organic pollutants may require a series of concerted metabolic events involving several species of bacteria or coupled interactions with bacteria and fungi [8]. Slader and Godwin [9] described the basic reactions in microbial metabolism as:

- Enzymatic catalyzed degradation of aliphatic hydrocarbons where the end products of beta-oxidation are organic alcohols and acids
- Enzymatic catalyzed hydrolysis, acidification, and ring cleavage of the benzene ring
- Production of homeostatic enzymes which regulate the well-being of the organisms by coordinated responses that automatically compensate for environmental changes
- Cometabolism, which involves use of readily degradable hydrocarbons in combination with fortuitous transformation of a desired pollutant

Microbial Parameters for Successful Mitigation

Bioremediation is clearly an interdisciplinary field involving expertise from engineering, geology, chemistry, and ecology [10]. Numerous approaches to soil and sediment remediation have been developed and implemented on an international basis. A common goal in all of the processes is to create the necessary environment for the growth of an optimal microbial community which can, in an effective manner, biodegrade contaminants of concern [11, 12].

The rate and extent of contaminant biodegradation using any natural process can be affected by numerous factors including, the actual structure of the contaminants of concern, its solubility and water bioavailability, co-oxidation/cold metabolism potential, relative toxicity, and its interaction with other properties in soil and sediment. Oxygen plays a dominant role in the selection of appropriate microbial communities for biodegradation to proceed. Water content in either bound water or groundwater is equally important in that water serves not only as a universal solvent but also has as a facilitator of movement of important nutrients and buffers for microbial biodegradation process initiation. Other limiting factors which affect the efficacious use of microorganisms to remediate contaminated soils or sediments include but are not limited to temperature, pH, organic and inorganic soil content, and the presence of metals such as arsenic, lead, and mercury [13–15].

Contaminated sites present the problem of attempting to biodegrade a diverse mixture of contaminants. A typical example is oil, the mixture of chemicals with different molecular weights, latent heat of vaporization, and relative water solubility. This necessitates the requirement of careful characterization of all constituents within the mixture. Therefore, remediation approaches are often dictated by the relative quantity of the most recalcitrant chemical in the mixture [14].

Bioremediation as a mitigation tool has the additional problem of project specificity, that is, requiring a case-by-case evaluation of the suitable bioremediation approach as dictated by the challenges of each specific site. Thus no one all-encompassing bioremediation tool or remedy will work for all contaminated sites. The remediation services market represents less than 5% of the $250 billion annual environmental industry market and is limited in its further expansion by the requirement of designing site-specific solutions using an adapted microbial community. However, the significant savings to both the public and private sector in using a microbiological approach to cleanup sites continues to be the major driver for looking at mature and evolving bioremediation technologies as a solution to mitigation challenges.

Candidate Contaminants for Bioremediation

Candidate contaminants for bioremediation processes have been successfully used to remediate petroleum hydrocarbons such as gasoline, diesel fuel, crude oil and creosote; pesticides and their derivatives such as phenoxyacetate herbicides, carbamates, and organophosphates; chlorinated solvents such as methylene chloride, trichloroethylene, and vinyl chloride; and halogenated aromatic hydrocarbons such as pentachlorophenol, chlorinated benzenes, and even some polychlorinated biphenyls (PCBs) [16–18].

Because of the nature of bioremediation, it is most applicable toward detoxification of organic wastes, most of which are at least theoretically biodegradable [19–21]. Among the most recalcitrant organic contaminants are the organohalogens, which, because of their rarity in nature, are persistent, as few biological systems have evolved to degrade them [19]. In contrast, petroleum hydrocarbons (PHCs) are relatively easily degraded [20, 21]. Historically PHCs entered the environment only via sporadic seepage and erosion, which allowed for the development of some natural microbial biodegradative pathways. In recent history, however, man's increasing reliance on fossil fuels has resulted in dramatic increases in petroleum hydrocarbon pollution [21]. Currently the combined chronic marine and terrestrial discharges of PHCs account for greater than 90% of anthropogenic environmental pollution [21].

Advantages of Bioremediation

Microbiological processes can be used as the keystone technology to develop cost-effective remediation systems for hazardous/nonhazardous waste sites. Bioremediation can be much cheaper than other technologies, approximately one third to one half the cost of transport and incineration of similar volumes of waste. In addition, byproducts that require landfilling such as incineration ash are not produced [23]. As the technology of bioremediation methods advances, the market for bioremediation products has also increased. Microbial pollution products to be used in remediation projects, including dried or liquid microbial inocula with or without nutrient additives, have annual sales of $7–10 million and the potential market may reach $200 million [13].

Additionally, in situ bioremediation reduces the risk of exposure during cleanups by avoiding the need for excavation [22]. Bioremediation shows promise for further reducing the low levels of contaminants left after excavation of high level contaminants [22]. Bioremediation is a natural process that has the potential of degrading toxics and other wastes to harmless products – carbon dioxide, water, and fatty acids – when the process is completed [17, 22]. An in situ bioremediation system can serve as a permanent in-place management system. The installation of an in situ bioremediation system can be accomplished with minimal site/business disruption and public exposure [23]. Bioremediation is currently applied to a wide range of site conditions and contaminants, from oil-contaminated beaches in Alaska to PCB-contaminated soils in Florida [24]. It is at the forefront of a larger group of innovative remediation technologies.

Bioremediation depends on the natural selection of organisms that have the capacity to metabolize xenobiotic chemicals. Many of these organisms occur naturally in contaminated areas, and the growth of these populations can be enhanced by the addition of nutrients and oxygen. Degradation of contaminants that resemble natural compounds is more rapid than with complex organic molecules such as dieldrin [6, 25].

Disadvantages of Bioremediation

Contaminated sites may differ greatly in terms of both microbiological and physiochemical parameters. Previous biodegradation at long-standing sites may result in increased proportions of components of greater recalcitrance. Greater assessment details are required for proposed biological treatment prior to cleanup to determine the impact of the chosen methodology upon the microbial population [26].

Actual experience with the bioremediation technology is limited when compared to conventional remediation technologies such as incineration, and competing technologies can usually be completed in less time. The present body of knowledge on bioremediation is widely dispersed and often inaccessible, since much of it is considered proprietary. Bioremediation is not effective in the destruction of metals [27–29].

Soils and sediments are extremely complex mixtures with many microenvironments and vast arrays of living organisms. This complexity and other factors enhance the efficient chemical or physical attack on a molecule, leading to its degradation. Consequently, it is often difficult to distinguish between the microbial, chemical, and physical factors that contribute to the removal or transformation of a molecule [17, 30, 31].

Environmental limitations to biodegradation include: (1) toxic levels of waste, (2) lack of oxygen, (3) unfavorable pH, (4) lack of nutrients, (5) lack of moisture, and (6) unfavorable temperatures [17, 32]. These limitations can be circumvented by environmental manipulation either ex situ as in a bioreactor or in situ using appropriate amendment strategies. Parameters such as waste loading, pH, nutrients, and size of organism populations can be controlled and lead to maximum degradation rates. This type of manipulation is most effective for liquid wastes. Groundwater containing the organochlorine, toxaphene, and the organophosphates dioxathion and methyl parathion have been biologically treated ex situ using two reactors containing adapted *Acinetobacter* spp. immobilized on a support material. The bioreactors were operated over a 90-day period at a flow rate of 80 gallons per day. The microorganisms degraded 60% and 35% of the chlorinated and phosphate wastes, respectively [33].

The physicochemical bioavailability of any polluting chemical mixture in soils/sediments can be affected by its sorption equilibrium, irreversible sorption, and degree of incorporation into humic material. Mass transfer limitations can include oxygen diffusion and solubility, the relative rate of diffusion of nutrients, and the solubility/miscibility with water. Oxidation/reduction potential and the availability of electronic receptors as dictated by the presence or absence of oxygen plays a role in microbial population speciation and density [4]. Finally, not only is the chemical structure of the contaminants of concern and its relative solubility important in the overall efficiency of the mitigation process, the net concentration of the chemical itself plays an important role [5]. The induction of suitable enzymes for effective bioremediation and subsequent growth and enrichment of a capable microbial community is often dictated by the limits of detection of each chemical in the mixture at the cellular level [17].

Single Species Versus Community Approaches

Many studies evaluating the microbial degradation of chemicals and soil, sediment and groundwater usually follow a reductive approach into a single microbial strain characterized on a chemical constituent basis [34]. Each organism is identified by either traditional selective culture approaches or amplified rDNA restriction (a simple method of identifying microbial community structure by characterizing

DNA sequences indicative of temporal changes in the indigenous microbial population).

Simple screening studies of site soils/water for important microbes at the species level, which may be effective, and degrading chemicals of concern are evaluated in shaker flash studies or microcosm studies (a variable specific simulation of the complex field remediation challenge). Data sets generated can give an indication as to the appropriate degradation pathway necessary for site remediation. The screening data sets at the single species level also sets the stage for optimizing site conditions to insure a multiple species or community approach will be viable [34].

The issue of releasing genetically engineered microorganisms (GEM) whose genetic material has been altered into an ecosystem has become a controversial one. Critics contend that an ecosystem imbalance caused by a genetically adjusted organism could lead to disastrous consequences. The genome of the microbial population would be altered and the results of the genetic manipulation, when known, would be too late and could not be changed [34]. There are questions of economics, production, quality control, application, host specificity, and safety. It may not be economically feasible to develop a stable population, ensure integrity by quality control and apply microbial pesticides on a large scale. These organisms should only affect a target population, but in many cases it is difficult to predict whether a microbial pesticide could affect nontarget organisms [35]. Engineered organisms can lose or alter the gene(s) that allow them to degrade a specific toxicant after several generations, rendering them useless in bioremediation [36]. The advantages of isolating an organism which has adapted naturally to the degradation of waste products are many, and eliminates the issues with using GEM.

Economics

There are several remedial technologies that may be considered when determining corrective action for any site. The technologies discussed are nonbiological approaches like soil venting (including vacuum extraction), soil flushing, hydraulic barriers, and excavation. The first three methods are considered in situ treatment methods, remediation of those soils in their original location. The fourth method, excavation, is normally combined with off-site disposal, incineration, soil washing, or enhanced volatilization. A comparison of costs associated with each of these technologies is shown in Table 5.1. When considering each of these technologies, costs as well as site-specific variables must be considered in determining the most effective method [37].

Table 5.1 An economic comparison of remediation approaches

Remedial methods	Treatment costs[a]
Soil venting (including vacuum extraction)	\$130–180/yd^3
Land farming	\$60–123/yd^3
Soil flushing	\$200/yd^3
Hydraulic barriers	\$67–88/ft^2
Off-site disposal (including excavation and backfill)	\$155–275/yd^3
Incineration (including excavation and backfill)	\$235–675/yd^3
Soil washing (including excavation and backfill)	\$185–235/yd^3
Enhanced volatilization (including excavation and backfill)	\$300/yd^3

[a]US Dollars, 2010

In situ and Ex situ Bioremediation Approaches

Bioremediation methods and/or technologies are broadly classified as either ex situ or in situ processes. Ex situ methods require the physical removal of contaminated material prior to the initiation of the treatment process. In situ approaches involve treatment of contaminated soil sediment and or groundwater "in place." Since bioremediation is a mitigation tool used for treating contamination in heterogeneous environments with complex phase separation of many materials by relative water solubility, the ability to successfully mitigate without physical removal and transportation to another location with its inherent risks of recontamination is desirable. The treatment of the Alaskan shoreline of Prince William Sound from the Exxon Valdez oil spill is an example of the successful use of an in situ approach to bioremediation [39].

Bioremediation Technologies Available for Sustainable Mitigation

The following technologies have been effectively used for sustainable remediation and are recognized as approved technologies by USEPA and other national environmental agencies [37, 40, 41]:

- *Bioaugmentation or Enhanced Bioremediation*: a process of amending contaminated media with microorganisms (bioaugmentation) and or nutrients (biostimulation) to degrade/immobilize/accumulate contaminants of concern.
- *Natural Attenuation (NA) or Monitored Natural Attenuation (MNA)*: the use of natural attenuation processes, that is, biodegradation/biomineralization, to achieve site-specific remediation objectives within a time frame that is reasonable compared to those offered by more active methods. As a natural biological and chemical process, NA is considered a natural process in achieving the goal

of reducing mass, toxicity, mobility, concentration, and net total volume of a contaminant or mixture of contaminants in soil or groundwater. Risk assessment guidelines are used to ascertain whether remediation objectives are realized.

- *Bioventing*: an in situ remediation technology that uses indigenous microorganisms to biodegrade organic constituents absorbed to soils in unsaturated zones. Soils in the capillary fringe and the saturated zone are not affected. In bioventing, the activity of the indigenous bacteria is enhanced by inducing air or oxygen flow into the unsaturated zone using extraction or injection wells and, if necessary, by adding nutrients.
- *Bioplugs and Bioconduits*: an in situ technology that uses preselected acclimated microorganisms permanently immobilized or attached to a porous matrix and inserted into soils/sediments as a seeding device for both saturated and unsaturated zones. Microbial activity is maintained by using contaminated site water or ground water as a carbon nutrient source.
- *Biopiles*: also known as biocells, biomounds, or bioheaps; are excavated soils/ sediments amended with minerals, nutrients, and water so as to reduce concentrations of contaminants in with indigenous microflora.
- *Composting or Compost Piles*: a variation of bioplies in which organic contaminants (e.g., PAHs) are mixed with lignocellulosic materials so as to biodegrade and/or cometabolize the recalcitrant constituents as the acclimated microbial population converts simpler carbon (under aerobic and anaerobic conditions) to innocuous, stabilized residuals.
- *Landfarming*: one of the oldest full-scale bioremediation technologies dating back to the treatment of process residuals from oil refinery operations. Also known as land treatment or land application, it is an above-ground remediation technology in which contaminated soils, sediments, or sludges are tilled or "turned" using standard farm cultivation equipment and allowed to slowly biodegrade as the acclimated soil microbial populations reduce the concentration of contaminants over time. It is one of the more widely used processes for handling large volumes of contaminated soils/sludges but requires significant acreage for repeated dispersal in layers or "lifts" over time.
- *Bioslurry or Liquid Solids Contact (LSC) Reactors*: an aqueous slurry in either a tank or lined impoundment that is created by combining soil, sediment, or sludge with water and other additives. The slurry is mixed to keep solids suspended (at a ratio of 30–40% volume:volume in water) and microorganisms in contact with the soil contaminants. Upon completion of the process, the slurry is dewatered and the treated soil is distributed in a land farm for further biodegradation and dewatering by evaporation.
- *Online Bioreactor Systems*: the treatment of contaminants in extracted ground water that are put into contact with microorganisms in attached biological contractors (e.g., rotating and trickling filters) or suspended growth biological reactors (e.g., activated sludge). The technology is sometimes referred to as "pump & treat" technology.

- *Phytoremediation*: the use of plants to remediate contaminated soil and groundwater. However, different plant systems have individual mechanisms by which they extract (or accumulate), stabilize, volatilize, and/or degrade the contaminants, which is discussed in detail below.
- *Constructed Wetlands or "Rock and Reed" Systems*: treatment systems that have been designed and constructed to utilize the natural processes involving rock, clays, and specific wetland vegetation with a large root rhizosphere so as to allow surface and subsurface attached, associated microbial population to assist in treating contaminated water.

Mitigation Case Studies with Petroleum and Related Hydrocarbons

Over the past several decades, humans have grown more and more reliant on fossil fuels. Many of the comforts that humans enjoy today are possible largely because of fossil fuels such as petroleum. This intensification in use has resulted not only in an increased dependence but an increase in petroleum-related pollution. It has been estimated that petroleum contamination into the environment is between 1.7 and 8.8 million metric tons annually, most from anthropogenic sources. The manufacturing, transportation, and distribution of petroleum-based products over the past century have led to an increase in the amount of petroleum hydrocarbon pollution. Leaks, spills, and other accidents have all contributed to the contamination of air, soils, and water bodies by petroleum. Pollution can come from oil tankers, pipelines, and above and below ground storage tanks [42].

The pollution poses risk to human health and the environment. Although considered a nonhazardous waste by law, petroleum is a regulated contaminant. Most of the danger associated with petroleum-contaminated sites comes from lighter weight hydrocarbons such as BTEX (benzene, toluene, ethylbenzene, and xylenes) and from fire hazards caused by vapors. Petroleum or any other type of contamination also makes initiating new construction projects on contaminated sites more challenging. It can impede the construction process as cleanup must occur and environmental standards must be met before construction begins. Petroleum hydrocarbons, as pollutants, fall into an intermediate category in their degradation between highly biodegradable, biogenic, and highly recalcitrant xenobiotics. Fortunately, these hydrocarbons can usually be readily degraded by microorganisms present in the environment. This fact makes treatment of petroleum hydrocarbon–contaminated areas more feasible [33].

Many treatment methods, however, are expensive and not complete in cleaning up petroleum waste. Partial contamination may still reside in soil or groundwater. Consequently, new techniques, such as bioremediation, are required to clean up petroleum wastes more completely and at a cheaper cost.

Categories of Petroleum Hydrocarbons

Petroleum hydrocarbons (PHCs) can be divided into four structural categories: saturated or aliphatic, aromatic, asphaltic or polar, and resins [18, 20, 24]. Aliphatic and saturated constituents can either be cyclic or acyclic compounds with or without carbon–carbon multiple bonds. Examples include pentene, pentane, and cyclohexane. Aromatic components are unsaturated cyclic compounds such as benzene and its derivatives [44]. Asphaltenes are highly condensed partially oxygenated compounds that occur in crude petroleum and are particularly resistant to biodegradation. Resins include oxygen, nitrogen, and sulfur-containing compounds that are for the most part recalcitrant [18, 23, 24]. Various characteristics of PHCs affect their inherent biodegradability; the chemical structure of individual components of petroleum products is most important, but physical state, toxicity, influence of other compounds, and presence of additives are also factors [18, 24]. Rates of biodegradation are the greatest for saturates, followed by light aromatics, high molecular weight aromatics and finally polar compounds. Increasing numbers of carbons in alkanes (homology), variations in chain length and ring condensation account for the wide variety of hydrocarbons that occur in crude petroleum [18, 43]. Because of these variations, the biodegradability of PHCs will differ from site to site. In general, PHCs are ranked in the following order of susceptibility to biodegradation: straight chain alkanes > branched alkanes > low molecular weight aromatics > cyclic alkanes [20].

Fate of Petroleum upon Entering Soils

Petroleum waste usually enters the ground via accidental spills, leaks, or through permitted or illegal dumping [20]. Once spilled onto the soils, infiltration into the vadose zone is likely. Infiltrated soils move vertically downward through unsaturated soils, where they may dissolve into the gaseous phase, or be otherwise dispersed, diluted, or sorbed onto particulate matter, then spread out laterally over the soil water table [18, 43]. Absorption of petroleum components onto particulate matter decreases the toxicity to microorganisms, but adds to PHC persistence [20, 43]. In cases where infiltration cannot occur, for example, freezing or saturated soils, evaporation and photooxidative losses contribute greatly to PHC removal [18]. Contaminants within the vadose zone may also undergo chemical and microbiological transformations that can influence partitioning [45].

Simple aliphatic compounds are biodegraded by enzymes catalyzing oxidation, reduction, and oxidative coupling reactions. Normal alkane biodegradation is initiated by complex monooxygenase enzyme systems, which oxidize the alkane to the corresponding primary alcohol [46]. The primary alcohol is oxidized via the aldehyde to the corresponding carboxylic acid by alcohol and aldehyde dehydrogenases. Beta-oxidation enzymes convert the acid to a simpler alcohol and acetic acid. The new alcohol is further degraded with the same sequence of enzymes to form a simpler aliphatic hydrocarbon with fewer carbons [47].

Table 5.2 Chemical and physical properties of seven key PAHs

Compound	Rings	MW	Solubility
Naphthalene	2	128	31.70
Phenanthrene	3	178	1.0
Anthracene	3	178	0.045
Pyrene	4	202	0.132
Benzo(a)anthracene	4	228	0.006
Chrysene	4	228	0.002
Benzo(a)pyrene	5	252	0.001

MW molecular weight
Solubility (mg/L) is aqueous solubility at room temperature

The Toxic Component of Petroleum

Simple polycyclic aromatic hydrocarbons or PAHs are considered by national environmental agencies as the toxic component of petroleum. PAHs are the basis for risk assessment-based mitigation approaches for environmental remediation. They constitute by volume only 0.2–7.0% of oil mixtures. PAHs, such as benzene, toluene, naphthalene, anthracene, and phenanthrene, and their alkyl-substituted derivatives are catalyzed by a class of enzymes known as dioxygenases. The products of initial metabolism of simple aromatic hydrocarbons are commonly catechols or substituted catechols [47]. Ring cleavage of these compounds occurs by one of two pathways depending on the species and the substrate. The ortho-pathway involves cleavage of the ring between the hydroxyl group while the meta-pathway involves cleavage of a bond between a carbon atom bearing a hydroxyl and an adjacent, non-hydroxylated carbon atom [20]. Both ortho- and meta-pathways are involved in aromatic hydrocarbon degradation, but substituted catechols are generally degraded via the meta-pathway [5].

Benzo(a)pyrene and similar high-molecular weight compounds are initially metabolized by the enzymatic catalyst monoxygenase, which converts the parent compound into a 7,8-epoxide [5]. An epoxide hydrase then catalyzes the epoxides to dihydrodiols; arylhydrocarbon hydroxylase in turn tranforms the dihydrodiols to phenols [35, 43]. The diastereomeric benzo(a)pyrene, 7,8-dihydrodiol and 9,10-epoxides, are presumed to be the ultimate carcinogens of benzo(a)pyrene, since they are (1) highly carcinogenic for newborn mice, (2) mutagenic and cytotoxic for both mammalian and bacterial cells, and (3) chemically reactive in binding to DNA [12].

Due to the large size and extreme insolubility of such PAHs as benzo(a)pyrene or benzo(a)anthracene, few soil/sediment microorganisms have the capability to utilize aromatic hydrocarbons containing more than three aromatic rings as the sole source of carbon and energy [48–50]. The chemical and physical characteristics of the seven key PAHs of interest are shown in Table 5.2.

Ex situ Approach for Soils/Sediments

Land Treatment Unit (LTU) Along an Active Railway

Rail operations have been and continue to be in the mainstream of industrial activity. Unfortunately, the use of fuels for rail operations is not without its environmental costs. At every stage of fossil fuel shipment and utilization, waste is generated and released. Ecosystems along transportation corridors have been damaged by exposure to the accumulation of high concentrations of petroleum hydrocarbon materials in soils and ballast material. One strategy for dealing with broadcast railroad lubricants is to apply adapted microbial consortia to the contaminated soil/ballast matrix. This procedure is called land treatment and is used to degrade waste hydrocarbons into the nontoxic, naturally occurring end products CO_2 and H_2O. Over a 12-month period, this soil remediation strategy was used to remediate diesel-contaminated soils at an active rail yard in McGehee, Arkansas. A LTU was designed and constructed to optimize petroleum hydrocarbon biodegradation rates and prevent waste mobilization into subsoils and groundwater [52, 53] (Fig. 5.1). The LTUs had the following features:

- Compacted clay subgrade and 40-mm HDPE (high density polyethylene) geomembrane liner leachate collection system comprised of a sand drainage layer, a gravel, and perforated pipe drain; a gravel sump and standpipe; and an 80,000-l leachate storage tank
- Leachate collection/irrigation pump, mobile sprinkler, and piping/hose
- 19-mm HDPE geomembrane cover and surface water outlet for the recycling of leachate collected from the LTU when the water content was above optimum conditions

Lifts of contaminated soil/ballast layered to 0.4-m depth were loaded into the LTU and treated until a target cleanup concentration of <200 mg/kg soil dry weight was reached. Approximately 7,500 m^3 of contaminated soil were successfully treated over a 6-month time frame. Soil total petroleum hydrocarbon (TPH) levels were reduced from an initial concentration of 2,500 $\pm$ 485 mg/kg soil dry weight to less than 140 $\pm$ 62 mg/kg soil dry weight in each of three lifts. Mean petroleum hydrocarbon reduction rates ranged from 17.3 mg/kg soil dry weight per day for cooler months to 42.5 mg/kg soil dry weight per day for warmer months. The data generated under varying climatic conditions demonstrate that the mitigation strategy of employing an adapted microbial population can be used to remediate rail TPH soils to acceptable regulatory levels for typical rail operations [55].

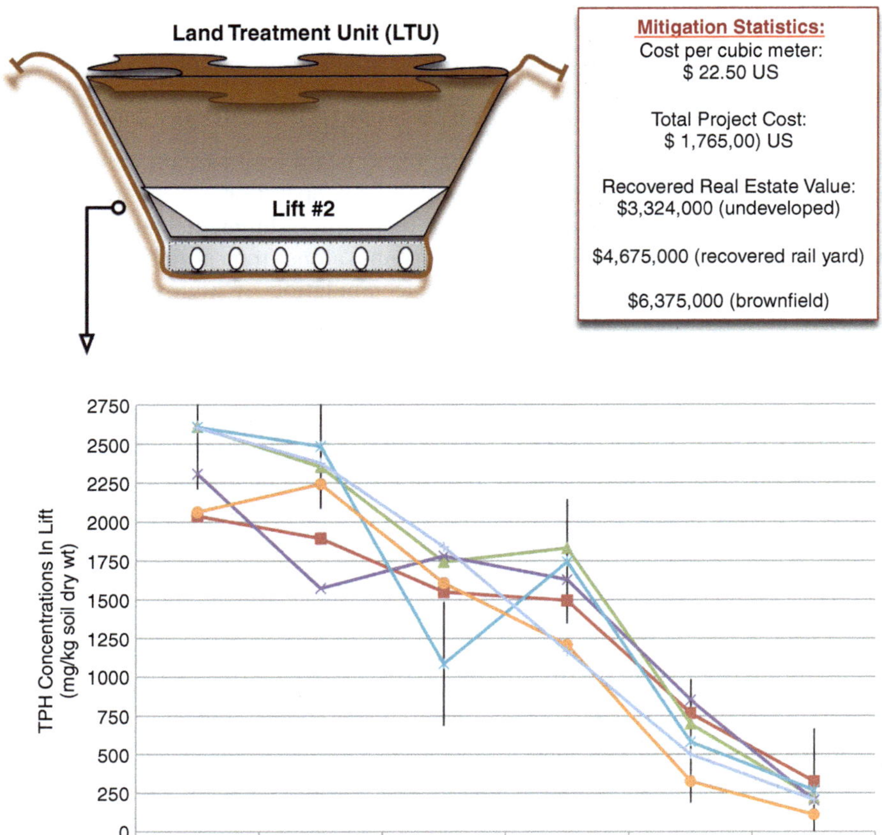

Treatment Zones / Time (days)	0	14	30	45	60	90
A	2038	1895	1550	1495	767	323
B	2610	2352	1744	1832	700	218
C	2308	1574	1782	1629	852	202
D	2608	2485	1085	1745	583	267
E	2064	2243	1608	1208	325	110
F	2603	2378	1841	1167	500	207

Fig. 5.1 Ex situ mitigation strategy for treatment of petroleum-based contamination along an active railway

Ex situ Approaches for Groundwater

Recovery and Biological Treatment of Wood Preservatives in Groundwater

Well-recovery networks coupled to immobilized microbe bioreactors (IMBRs) were installed at a 172-acre former wood preserving process facility in Baldwin,

Florida in 1995 for the bioremediation of toxic organic wood preservatives, both free phase and soluble, present in site groundwater. The well-recovery network was designed to recover contaminated components in two steps: (1) a pumping system to remove free phase creosote from the hardpan at 10–14 m below the surface and (2) a system for recovering subsurface-clarified groundwater to a holding tank, where trace creosote fractions and pentachlorophenol were further gravity separated. Immobilized microbial isolates evaluated in earlier laboratory and field pilot tests were established into two 40,000-l bioreactors for the biodegradation of all targeted constituents [56, 57]. Microbial growth, DO, pH, flow rate and temperature were monitored in this combined in situ/ex situ bioremediation system.

The immobilized packed bed reactor (IMBR) system is filled with bio-support media, and contains adapted microbes with enzymatic capability of mineralizing the organic compounds of interest [58]. The packed bed provides a large surface area for microbial colonization [59, 60]. The bio-support media is usually a chemically inert and physically stable quartz-based diatomaceous earth material, which has a good pore morphology and high surface area. Adsorption or covalent bonding to this media is the primary mechanism for immobilization. Bacterial immobilization involves the entrapment of cells onto the matrix. Once bound, the cells are then readily accessible to the surrounding substrate.

Ten-year operational data sets have indicated successful biodegradation of the clarified feed composed primarily of pentachlorophenol (PCP), targeted polycyclic aromatic hydrocarbons (PAHs), and minor amounts of other chlorinated phenols and nitrophenols. The recovery system retention time of approximately 3.2 days satisfied federal and state NPDES target levels (Fig. 5.2). Flow rates of 34.5 ± 3.8 L/min were maintained throughout the treatment time frame. Treated effluent was used during drought years to irrigate surface vegetative cover and allow for perculation down to the hardpan [59]. The site received a national award in 1996 for the innovative control and treatment of a groundwater plume from the American Forest and Paper Association. The site is currently zoned recreational and is close to meeting groundwater MCLs.

In situ Approaches for Soils, Sediments, and Groundwater

In situ remediation approaches are targeted at reducing groundwater concentrations without employing a “pump and treat” approach. Earlier technology approaches allowed for passive treatment of a contaminated plume as it moved through a biological barrier or trench. The concept has proven to be marginally successful for plume control, that is, preventing toxic groundwater from moving beyond the legal boundaries of the affected site, but does not address the real problem of the source of contamination in soils/sediments [55, 59].

Newer technologies attempt to focus treatment on the soil/sediment itself using direct addition of free microbial cells to the contaminated soil/sediment lens (bioaugmentation) or providing nutrients and allowing for the indigenous

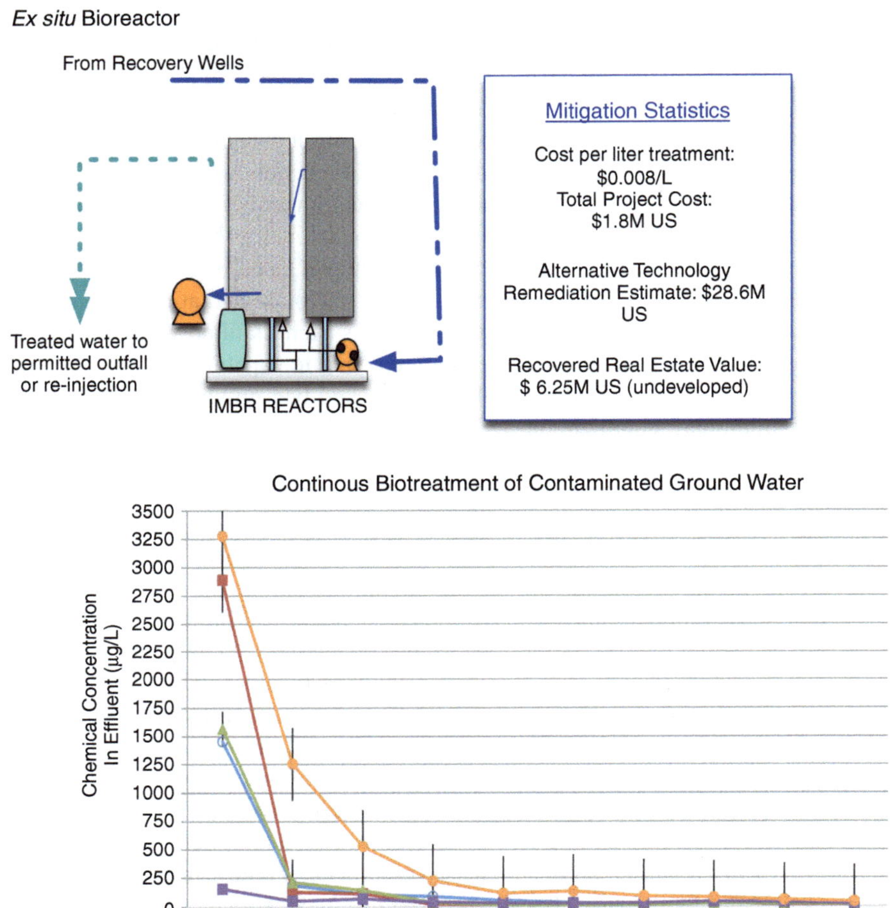

Time (Days)	0	3	7	14	30	60	90	120	240	360
Napthalene	1455	188	108.5	89.2	55.6	34.5	28.8	26.6	32.8	22.5
Phenanthrene	2890	129	112.8	22.5	18.5	28.8	12.5	32.5	16.8	15.9
Chrysene	1567	212	141.5	36.5	14.8	13.5	10.5	28.7	22	18.8
Benzo(a)pyrene	156	52.5	68.9	41.5	32.5	29.5	31.5	42.2	36.2	24.5
Pentachlorophenol	3278	1255	528.2	224.5	116.8	132.5	89.5	77.8	56.8	42.4

Fig. 5.2 Ex situ mitigation approach for continuous treatment and hydraulic control of a contaminated groundwater bloom in an urban area

microflora to reduce soil concentrations over time. This approach, called natural attenuation, models the rate of kinetic biodegradation/mineralization of soil/sediment contamination and provides an estimate for reaching MCLs [37, 39].

A variation of both approaches is to treat the subsurface groundwater and redirect a preferred acclimated biomass to the point sources of contamination in soils/sediments. The same immobilized microbe bioreactor concept mentioned for ex situ groundwater treatment has been used for treatment of petroleum contaminated soils and groundwater, in situ. Soil seeding bioreactors called bioplugs are installed vertically using conventional drilling equipment; bioconduits are horizontally placed bioreactors installed using directional drilling equipment [61].

Adapted indigenous microbial strains are used for immobilization in the bioreactors, having been identified as able to degrade the specific compounds/contaminants of interest. The use of indigenous microflora, a concept combining the positive aspects of bioaugmentation and natural attenuation, reduces the time of microbial acclimation to surrounding soil/sediment conditions and lowers the rejection rate of viable biomass in close proximity to areas of elevated contamination. Once placed on the subsurface, the reactors contain and sustain selected microbial populations required for effective degradation of highly contaminated soils. The in situ soil bioreactors are designed with ports for nutrient and air/oxygen amendment to maintain optimum conditions. The operational flow is maintained by initiating a pressure gradient in the reactors using compressed air/oxygen or nitrogen for anaerobic applications. The aeration flow rate is designed based on the site-specific conditions and the contaminants (volatile organic compounds or semi-volatile organic compounds) to be remediated [60, 61].

The use of contaminated groundwater or site water through the immobilized bed results in mineralization of organics in the water phase to biomass and generates whole cell bleed-off from the bed. The elevated biomass is allowed to escape from the bioplugs/bioconduits, thus introducing an enriched, adapted microflora moving layer into the surrounding soil/sediment strata in a radial pattern. Over time the biomass fronts, by physical movement and chemotaxis, infiltrate contaminated areas and reduce concentrations of chemical constituents in the soil/sediment matrix. The point source of the contaminated groundwater plume is therefore eliminated [60, 62].

Hydraulic control of the treatment area is quite important [62]. The bioplugs generate a hydraulic gradient at a targeted site, either as a result of the injection of water through the bioplugs or the injection of air, which can modify the pressure gradient. Hydraulic control generally consists of some method of collecting the water down gradient from the region of influence, that is, a recovery trench, and recycling it back to the site trough the bioplugs. There are two major reasons for recycling the water: (1) recycling will control the migration of contaminants from the affected area and (2) recycling will help maintain the nutrient concentration in the injection water and reintroduce adapted microorganisms back to the affected area while treating the groundwater to meet MCLs [62].

In situ Remediation of Contaminated Groundwater and Site Soils at a Natural Gas Pipeline Compressor Station

Over the last 30 years, leakage and accidental discharge of petroleum-based lubricating oil used to maintain the pipeline and related equipment of a gas transmission facility have resulted in petroleum hydrocarbon (PHC) contamination of the soil. The site evaluation conducted by a private engineering firm identified fifty areas at four different sites exhibiting PHC concentrations greater than regulatory acceptable levels. All fifty areas were successfully remediated using bioremediation techniques. Due to the depth of the contamination and soil conditions, fifteen areas were remediated by installing approximately 550 in situ immobilized bioreactors (bioplugs) (Fig. 5.3a). A monitoring protocol developed for the site was implemented during the remediation period to maintain optimal conditions and make operational adjustments, as required. Only one of the areas targeted for remediation is discussed in this case study [61].

The bioplug system was designed to facilitate petroleum hydrocarbon mineralization by placing PHC-degrading microorganisms in close contact with the contaminants [61]. Initial total petroleum hydrocarbon concentrations as analyzed by EPA Method 418.1 at the subject area were as high as 16,000 mg/kg with an average concentration of 4,400 mg/kg soil dry weight. Samples were collected at different radial distances (1, 2, and 3 m) and depths (1, 2, and 3 m below ground surface) from each bioplug location within the subject area. The samples were analyzed to determine the effective radius of influence for each plug. Analysis of covariance indicated that TPH degradation at 2-, 4-, and 5-m radial distances and depths occurred at an equivalent rate throughout the degradation period [64].

Changes in polycyclic aromatic hydrocarbon (PAH) composition were monitored by gas chromatography/mass spectrometry. Samples were analyzed for PAH content and concentration. Fourteen PAH compounds were identified in varying concentrations during the day-14 sampling. By day-90, PAHs with three or fewer rings were completely degraded. However, low concentrations of flouranthene, benzo(a)pyrene, benzo(b)flouranthene, and indeno(1,2,3cd)pyrene were identified after 90 days of remediation. The presence of 4+ ring PAHs was expected because of the fact that the ability of microbes to mineralize PAHs decreases with increasing ring number [60, 62, 63].

The assays were coupled with microbial measurements of total heterotrophic and total petroleum-degrading bacteria, and direct measurements using acridine orange florescent staining. Changes in the population of PHC-degrading microorganisms over time in the subject area were determined by averaging the values obtained from all samples collected within the area. The microbial growth and the degradation of the contamination occurred exponentially within the first 30 days. The microbial growth was observed decreasing after 30 days, when the majority of the easily degraded petroleum hydrocarbons, saturated alkanes and PAHs containing fewer than three rings had been depleted (approximately by 72%). Sampling and analysis during the remediation period indicated an increase in the heterotrophic and petroleum-degrading microbial population by 75% and 57%, respectively, from day 14 to 90. Within 180 days from the activation of the remediation system, targeted compliance TPH concentrations (100 mg/kg soil dry weight) were attained. The reduction of PAH concentrations for soils and groundwater over time is shown in Fig. 5.3b.

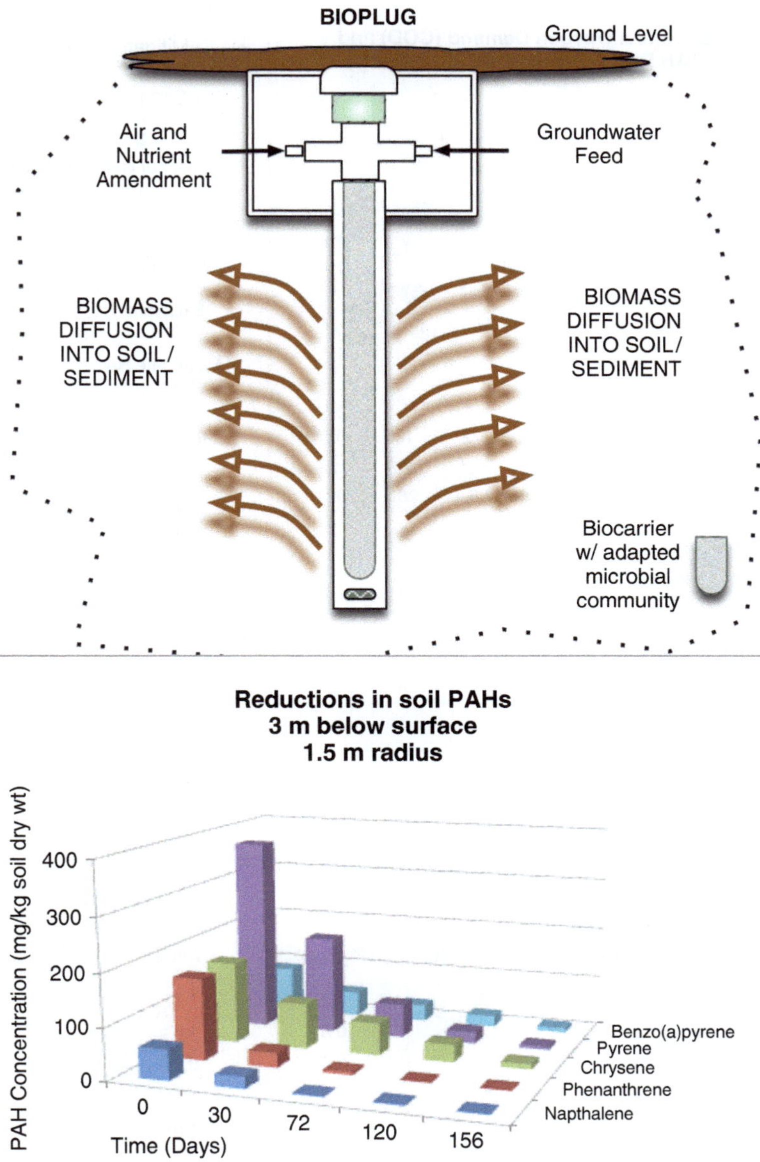

	0	30	72	120	156
Napthalene	58.8	22.5	1.8	2.1	1.2
Phenanthrene	158.2	28.5	5.5	0.5	0.5
Chrysene	158.5	88.4	62.5	33.5	9.6
Pyrene	370.5	188.4	65.8	22.5	6.2
Benzo(a)pyrene	88.5	48.5	32.2	21.5	8.2

a

Fig. 5.3 (continued)

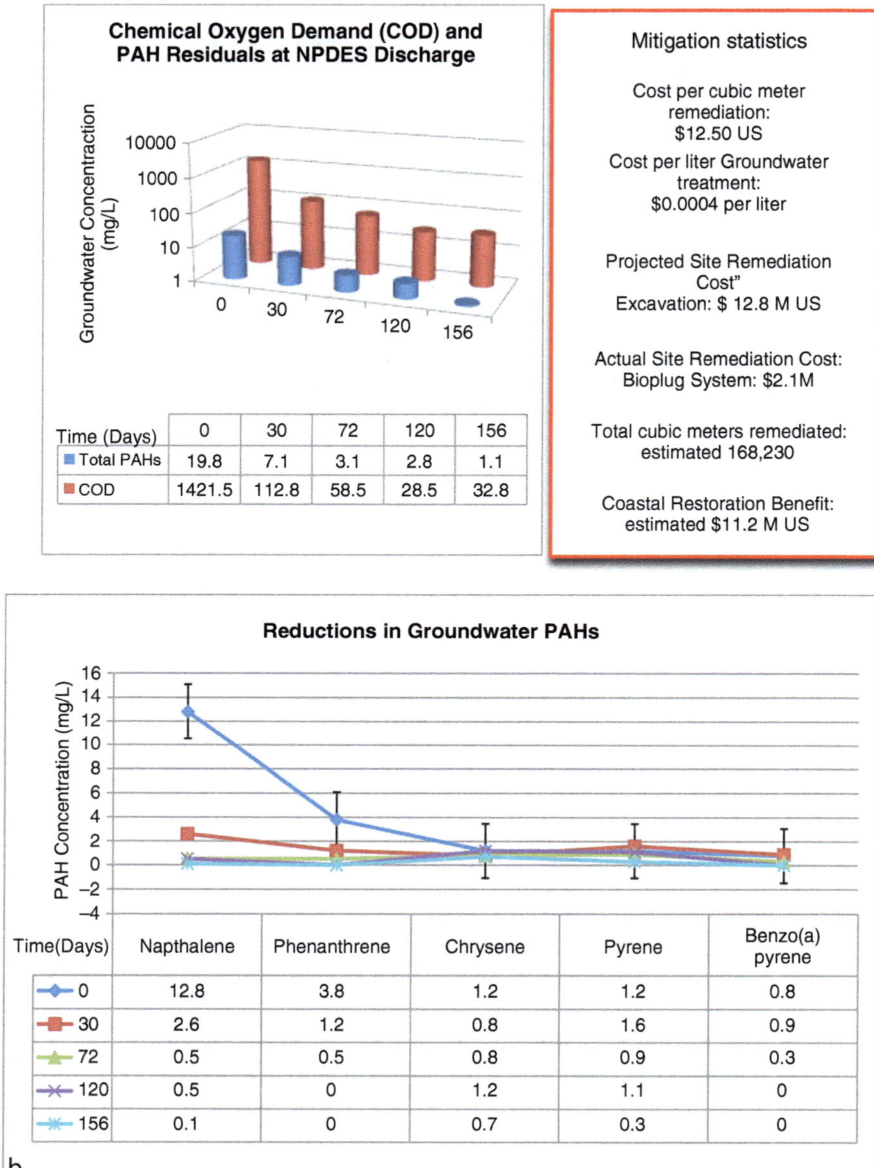

Fig. 5.3 (**a**) The application of an in situ mitigation strategy to reduce soil PAH concentrations with depth at a natural gas compressor station. (**b**) Reductions in groundwater PAH and total COD concentrations in a combined in situ mitigation approach to remediate a natural gas pipeline compressor station

Future Directions

Improved Mitigation Drivers

In the past few years, the goal of establishing US cleanup standards using bioremediation approaches has made significant progress as cities, municipalities, state agencies, and the federal government have focused on establishing bioremediation actions linked to accepted numerical metrics within a defined cost structure or economic price tag. The US Environmental Protection Agency has used maximum contamination level goals (MCLGs) and maximum contaminant levels (MCLs) as tools for establishing the highest allowable level of contamination in drinking water [40]. With the use of the MCL rules and MCLGs, US-EPA has set in place a framework for deciding "how clean is clean" for contaminated soils and sediments and affected groundwater under the Resource Conservation and Recovery Act (RCRA) program for contaminated industrial sites and Comprehensive Environmental Response, Compensation Liability Act (CERCLA), or "Superfund," for abandoned hazardous waste sites. Linking the MCL and MCLG criteria to "Best Available Remediation Technology," or BART, has become the driver for risk-based bioremediation mitigation [65]. Legal judgments and records of decision (RODs) now involve decisions affecting not only environmental protection, land use, and related economics, but human health and quality of life [66]. Risk-based cleanup standards have made significant progress in the past 10–15 years; most states now have in place a series of decision tiers (usually in the form of tables by specific contaminant) for assessing risk and reaching legally acceptable cleanup criteria for soil and groundwater in specific residential, light industry, or heavy commercial locations [67]. As mitigation criteria become more focused on the quality of life in determining "how clean is clean," Risk-Based Corrective Action Plans (RBCAs) will have additional social and economic mitigation criteria separate from technical engineering and economic indices in establishing sustainable remediation cleanup standards [68, 69].

Defining Stakeholders

Over the past decade, the selection of mitigation goals and remediation technologies has been driven by technical criteria and acceptance by regulatory agencies [70]. Only in recent years has a health and health protection criteria played the more dominant role in decision making. The development of the Internet and access to information has not only affected site owners, state and municipal agencies, and board industry, but has also led to an expansion in a broader inclusion of the general public as stakeholders in any bioremediation approach. Thus the concept of sustainable remediation, practices or multitask approaches for restoring a site to its native state with a frugal use of available resources, involves stakeholders at the regional, national, and international level [70–72].

Developing National and International Bioremediation Mitigation Programs

Executive Order 13123, "Greening of the Government Through Efficient Energy Management," by the US-EPA Office of Solid Waste and Emergency Response provided definable environmental goals for state and local municipalities in initiating sustainable mediation [65]. Smart energy resources management and green remediation can be incorporated into sustainable environmental practices. Remediation management practices of contaminated sites are outlined by Diamond et al., in which future initiatives and sustainable mitigation-based bioremediation are based [73]. State initiatives in Minnesota, California, Illinois, and Louisiana are early indications of how important federal tools are being assimilated into state and local programs [65, 74, 75]. The Canadian Environmental Protection Act implemented in 2,000 forms the basis for establishing risk-based criteria for contaminated soil and groundwater and air blue emissions in the various provinces of the country [76, 77]. The province of Québec has made significant progress in establishing written guidelines or soil protection and contaminated site restoration. Key components of this guideline include the concept of prevention, the necessity of reclamation and rehabilitation, and most importantly the idea of the polluter pays principle. However, Québec has achieved only modest success with many sites remaining under the category of off-site disposal [77].

The European Union adopted the Environmental Technology Action Plan in 2004 to demonstrate sustainable remediation practices at specific field locations in an attempt to follow US-EPA initiatives for standardization of risk-based criteria based on human health considerations [71, 78].

Asian countries have seen record economic growth over the past 2 decades which has resulted in significant progress in quality of life improvements at the expense of the deterioration of natural resources, increased pollution, and threat to human health [79]. Japan has made significant progress with the implementation of the soil contamination countermeasures law of 2003. China and India are attempting to emulate both the US-EPA initiatives and the European Union as they struggle to make difficult public health decisions within the world's fastest-growing economies [80].

Conclusions

One of the key axioms mentioned in Executive Order 13123 is that of "sustainability means change" [81]. Leadership at the local, state, national, and international level will be needed so as to realize the public perceptions as to what successful mitigation put into practice really means [70, 82]. Remediation selection and optimization assessments have become the norm in making good decisions for restoring impacted sites [83]. However, sustainability does not rank as the highest priority in making decisions on-site remediation. As market and government forces continue to seek economically efficient but socially responsible approaches to site cleanup, sustainability has been the driving criterion for mitigation-based bioremediation to become the norm [70].

Bioremediation mitigation strategies can be effectively designed by incorporating vadose zone and groundwater modeling [84]. Based on the availability of site hydrogeologic data and conditions, analytical modeling and/or numerical modeling can be used in designing the in situ bioremediation of the saturated zone.

An effectively designed, operated, and maintained bioremediation system offers an added advantage over conventional strategies like "pump and treat" by reducing the time of remediation. An advantage of using a technology such as the in situ immobilized bioreactors is to effectively increase the petroleum degrader population by placement of the adapted microorganisms in close contact with the organic residue. The in situ immobilized bioreactors can also be used to provide a cometabolite for degradation of hazardous by-products produced during the degradation process of some of the chlorinated solvents [57]. Also, the frequently encountered problem of ineffective indigenous microorganisms and/or low indigenous microbial population can be avoided [56].

The use of bioremediation mitigation approaches in conjunction with physical/chemical systems, that is, excavation, solidification, or vapor extraction, can provide a combination of mass reduction phenomenon and mass transfer phenomenon, thus reducing the time of remediation and the cost of cleanup. In many cases, due to the complex nature of the surface and subsurface conditions, a combination of in situ remediation techniques is recommended [83]. With rigorous engineering design, bioremediation technologies can be economical, viable, and safe solutions to real-world contamination problems [84].

Bibliography

Primary Literature

1. U.S. Environmental Protection Agency (2004) Terms of environment: glossary, abbreviations and acronyms. EPA 175-B-001 (Last Revised December 1997). http://www.epa.gov/OCEPAterms/
2. Robert T (2000) Stafford disaster relief and emergency assistance act, as amended by public law, pp 106–390, 30 October 2000. United States Code Title 42. The Public Health and Welfare
3. Bradford ML, Krishnamoorthy R (1991) Consider bioremediation for waste sitecleanup. Chem Eng Prog 12(1):80–85

4. Sayler GS, Fox R, Blackburn J (eds) (1991) Environmental biotechnology for waste treatment. Springer, New York
5. Tabak HH, Quave SA, Mashni CI, Barth EF (1981) Biodegradability studies with organic priority pollutant compounds. Water Qual 53(10):1503–1517
6. National Research Council (2005) Contaminants in the subsurface: source zone assessment and remediation. The National Academies Press, Washington, DC
7. Understanding Bioremediation: A Guidebook for Citizens (1991) United states environmental protection agency. Office of Research and Development, Office of Environmental Engineering and Technology
8. Worth M, Fernandes AL, Freshner R (1991) Bioremediation of contaminated soil: a case study in Canada, pp 326–343. Environmental Pollution, Canada.
9. Slater JH, Godwin D (1980) Contemporary microbial ecology. Academic, London
10. Carberry JB, Lee SW (1990) Care and transport of petroleum in the unsaturated soil zone under biotic and abiotic conditions. Water Sci Technol 22(6):45–52
11. Hughes JB, Neale CN, Ward CH (2000) Encyclopedia of microbiology, 2nd edn. Academic, New York
12. Prince RC (1998) Kirk-Othmer encyclopedia of chemical technology. Wiley, New York
13. Srinivasan U (2003) U.S. bioremediation markets. Frost and Sullivan research report, no. 7857. Frost and Sullivan, Palo Alto, CA
14. Portier RJ (1988) Enhanced biotransformation and biodegradation of polychlorinated biphenyls in the presence of aminopolysaccharides. Aquat Toxicol 10:503–516
15. Vance OB (1991) Onsite bioremediation of oil and grease contaminated soils. Nat Environ J 1(1):26–30
16. Roos D, Stroo HF, Bourquin AW, Sikes DJ (1988) Bioremediation of hazardous waste sites in the USA: case histories. The Association Dedicated to Air Pollution Control and Hazardous Waste Management, vol 38.6, pp 2–9
17. Walker WW, Cripe CR, Pritchard PH, Bourquin AW (1988) Biological and abiotic degradation of xenobiotic compounds in vivo estuarine water and sediment/water systems. Chemosphere 17(12):2255–2270
18. Bartha R (1986) Biotechnology of petroleum pollutant biodegradation. Microb Ecol 12:155–172
19. Frederickson JK, Bolton H Jr, Brockman FJ (1993) In Situ and on-site bioreclamation. Environ Sci Technol 27(9):1711–1716
20. Lyman WJ, Noonan DC, Reidy PJ (1990) Cleanup of petroleum contaminated soils at underground storage tanks. Pollut Technol Rev No.195: 38–65
21. Musts for USTs (1995) United States environmental protection agency. Office of Underground Storage Tanks. EPA 510-k-95-002
22. Skinner JH (1991) Bioremediation research issues. In: Sayler GS, Fox R, Blackburn JW (eds) Environmental biotechnology for waste treatment, vol 41. Plenum, New York, pp 61–70
23. Gabriel PF (1991) Innovative technologies for contaminated site remediation: Focus on bioremediation. Air Waste Manage Assoc 41(12):1657–1660
24. Riser-Roberts E (1992) Bioremediation of petroleum contaminated sites. CRC Press, Washington, DC
25. Iwamoto T, Nasu M (2001) Current bioremediation practice and perspective. J Biosci Bioeng 92(1):1–8
26. Grady L (1985) Biodegradation: its measurement and microbiological basis. Biotechnol Bioeng 27(5):660–674
27. States use of bioremediation: advantages, constraints, and strategies (1991) Natural resources policy studies center for policy research. National Governors' Association
28. Portier RJ, Friday DD (1991) Development of an immobilized microbe bioreactor for VOC applications. Environ Prog 10(1):30–39

29. Portier RJ (1982) Correlative field and laboratory microcosm approaches in ascertaining xenobiotic fate and effect in diverse aquatic environment. Ph.D. Dissertation, Louisiana State University, Baton Rouge, LA, 205 p
30. Grommen R, Verstraete W (2002) Environmental biotechnology: the ongoing quest. J Biotechnol 98(1):113–123
31. Ward OP, Singh A, VanHamme J (2003) Accelerated biodegradation of petroleum hydrocarbon waste. J Ind Microbiol Biotechnol 30(5):260–270
32. Halvorson HO, Pramer D, Rogul M (eds) (1985) Engineered organisms in the environment: scientific issues. American Society of Microbiology (ASM), Washington, DC
33. Atlas RM, Bartha R (1987) Microbial ecology: fundamentals and application, 2nd edn. The Benjamin/Cummings Publishing Company, Inc, Menlo Park, CA
34. Begley S, Waldrop T. (1989) Microbes to the rescue. Newsweek, June 19, pp 56–57.
35. McFarland MJ, Sims RC, Blackburn JW (1991) Use of treatability studies in developing remediation strategies for contaminated soils. In: Sayler G, Fox R, Blackburn JW (eds) Environmental biotechnology for waste treatment. Plenum, New York, pp 163–174
36. Johnson LM, McDowell CS, Krupha M (1985) Microbiology in pollution control from bugs to biotechnology. Dev Ind Microbiol 26:365–376
37. Van Hamme J, Singh A, Ward OP (2003) Recent advances in petroleum microbiology. Microbiol Mol Biol Rev 67(4):503–549
38. U.S. Environmental Protection Agency (2008) Green remediation: incorporating sustainable environmental practices into remediation of contaminated sites. Office of Solid Waste and Emergency Response, EPA 542-R-08-002. http://www.brownfieldstsc.org/pdfs/green-remediation-primer.pdf. Accessed 5 May 2009
39. Brune H, Schaat H (1991) Organic pollutants in soil and food chain. In: Nath B (ed) International conference on environmental pollution. Inderscience Enterprises, Ltd, Geneva, Switzerland, pp 335–343
40. Atlas RM (1984) Diversity of microbial communities. Adv Microb Ecol 7:1–47
41. Atlas RM (1984) Microbiology, fundamentals and applications. Macmillan, New York
42. Bond PL, Druschel GK, Banfield JF (2000) Comparison of acid mine drainage microbial communities in physically and geochemically distinct ecosystems. Appl Environ Microbiol 66(11):4962–4971
43. Morrison BA (1992) On-site bioreclamation of creosote-contaminated soil at the Scott Lumber site. Air Waste Manage Assoc 92(27):2–12
44. Diamond ML, Page CA, Campbell M, McKenna S, Lall R (1999) Life-cycle framework for assessment of site remediation options: method and generic survey. Environ Toxicol Chem 18:788–800
45. U.S. Environmental Protection Agency (1996) Semi-volatile organic compounds by gas chromatography/mass spectrometry (GC/MS). In: SW-846. Test methods for evaluating solid wastes physical/chemical methods (Method 8270C), 3rd edn, EPA Revision 3. Office of Solid Waste, Washington, DC
46. Venosa AD, Haines JR, Nisamaneepong W, Govind R, Pradhan S, Siddique B (1991) Protocol for testing bioremediation products against weathered Alaskan crude oil. Proceedings of 1991 international oil spill conference. American Petroleum Institute, Washington, DC, pp 563–570
47. U.S. Environmental Protection Agency (1999) Monitored natural attenuation of petroleum hydrocarbons, EPA 600-F-98-021. Office of Research and Development
48. Portier RJ, Martin AM (eds) (1991) Applications of adapted microorganisms for site remediation of contaminated soil and water biological degradation of wastes. Elsevier Science, New York
49. Ward OP, Singh A (2004) Applied bioremediation and phytoremediation, vol 1, Soil biology series. Springer, New York
50. Nelson CH (1993) A natural cleanup. Civil Engineering, American Society of Civil Engineers (ASCE), 63(3):57–59

51. Bianchini MC, Henry D, McMillin, Portier R (1986) Contribution to toxicant biotransformation by commercially available mutated bacterial cultures as compared to adapted indigenous microflora in the degradation of aromatic and aliphatic hydrocarbons. Abstract. American society for Microbiology/Canadian society for Microbiology symposium on environmental insult and recovery of stressed systems, Toronto, Canada
52. Portier RJ, Ahmed SI (1988) A marine biotechnical approach for coastal and estuarine site remediation and pollution control. Mar Technol Soc J 22(2):34–42
53. Portier RJ, Bianchini M (1988) Comparison of effective toxicant biotransformation by autochthonous microorganisms and commercially available cultures in the 'in situ' reclamation of contaminated sites. Biotechnologische In-Situ Sanierung. Gustav Fishscher Verlag, pp 273–292
54. Portier RJ (2003) Trends in soil science, technology and legislation in the USA. J Soils Sediments 3(4):257
55. Friday DD, Portier RJ (1989) Evaluation of a packed bed immobilized microbe bioreactor for the continuous biodegradation of halogenated-contaminated ground waters. AWMA/EPA international symposium on hazardous waste treatment: biosystems for pollution control, pp 97–112.
56. Sattler DL, Hoover DG, Davis TM, Portier RJ (1997) Recovery and biological treatment of organic wood preservatives in groundwater. In: In situ and on-site bioremediation, vol 5. Batelle Press, Columbus, OH
57. Portier RJ, Sattler DL, Hoover DG, Davis TM, Williams SE (1998) Recovery and microbial remediation of organic wood preservatives in groundwater at a former wood treating plant. Remediation 8(4):95–108
58. Portier RJ, Barton K, Koury J (1995) Low intervention soil bioremediation approaches. Remediation 5(2):57–75
59. Portier RJ, Metosh-Dickey C, Anas D (2004) Biological treatment of wood preservatives in site soils/sediments and associated groundwater: a 5-year study. Proceedings: session, WA1. Biorestoration strategies for contaminated sediments fourth SETAC world congress, Portland, OR, 3 p
60. Metosh-Dickey C, RJ Portier (2004) In situ bioremediation of contaminated soils and sludges. Presented at CPBR 2004 symposium, Washington, DC (poster and extended abstract)
61. Morgan P, Watkinson RJ (1990) Assessment of the potential for in situ biotreatment of hydrocarbon-contaminated soils. Water Sci Technol 22(6):63–68
62. Morris MS, Novak JT (1989) Mechanisms responsible for the biodegradation of organic compounds in the subsurface. J Hazard Mater 22(3):393–405
63. Einarson MD (2006) The practical handbook of environmental site characterization and ground-water monitoring, 2nd edn. CRC Press, Boca Raton, FL
64. Efroymson RA, Nicolette JP, Suter GW II (2004) A framework for net environmental benefit analysis for remediation or restoration of contaminated sites. Environ Manage 34:315–331
65. Interstate Technology and Regulatory Council (2008) Use of risk assessment in management of contaminated sites. ITRC Risk Assessment Resources Team. http://www.itrcweb.org/Documents/Risk Docs/RISK2 ExSum.pdf. Accessed 5 May 2009
66. Diamond ML, Page CA, Campbell M, McKenna S, Lall R (1999) Life-cycle framework for assessment of site remediation options: method and generic survey. Environ Toxicol Chem 18:788–800
67. Dresner S (2002) The principles of sustainability. London, Earthscan. Economist Intelligence Unit. http://a330.g.akamai.net/7/330/25828/20080208181823/graphics.eiu.com/upload/Sustainability allsponsors.pdf. Accessed 5 May 2009
68. Brundtland Commission (1987) Report of the world commission on environment and development: our common future. http://www.un-documents.net/wced-ocf.htm. Accessed 5 May 2009
69. Favara P, Nicolette JP, Rockel M (2008) Combining net environmental benefit analysis with economics to maximize remediation sustainability and value. Presented at the sixth

international conference on the remediation of chlorinated and recalcitrant compounds, Monterey, CA
70. Diamond ML, Page CA, Campbell M, McKenna S (1998) Life cycle framework for contaminated site remediation options: final report. Prepared for Ontario Ministry of Environment and Energy
71. Ellis DE, Ei TA, Butler PB (2008) Sustainability analysis for improving remedial action decisions. Presented at the sixth international conference on remediation of chlorinated and recalcitrant compounds, Monterey, CA
72. Elkington J (1994) Towards the sustainable corporation: win-win-win business strategies for sustainable development. Calif Manage Rev 36(2):90–100
73. Environment Canada (2008) Turning the corner: regulatory framework for industrial greenhouse gas emissions. http://www.ec.gc.ca/doc/virage-corner/2008-03/541eng.htm. Accessed 5 December 2008
74. Environment Canada (2008) Turning the corner: Canada's offset system for greenhouse gases. http://www.ec.gc.ca/doc/virage-corner/2008-03/526eng.htm. Accessed 5 December 2008
75. HOH Water Technology A/S, NIRAS Consulting Engineers and Planners A/S, Revisorsamvirket/Pannell Kerr-Forster, ScanRail Consult (2000) Environmental/economic evaluation and optimising of contaminated sites remediation. Prepared for Danish National Railway Agency and Danish State Railways
76. Nielsen DM, Nielsen GL, Preslo LM (2006) The practical handbook of environmental site characterization and ground-water monitoring, 2nd edn. CRC Press, Boca Raton, FL
77. Beinat E (1997) Value functions for environmental management. Kluwer, Dordrecht, Germany
78. Office of the Federal Environmental Executive (2007) Leading by example: Federal government environmental and energy efficiency accomplishments for 2004–2006. http://ofee.gov/leadingbyexample/LeadingbyExample2004–2006.pdf. Accessed 5 May 2009
79. Economist Intelligence Unit (2008) Doing good: business and the sustainability challenge. http://a330.g.akamai.net/7/330/25828/20080208181823/graphics. eiu.com/upload/Sustainability allsponsors.pdf. Accessed 9 May 2009
80. Volkwein S, Hurtig HW, Klopffer W (1999) Life cycle assessment of contaminated sites remediation. Int J Life Cycle Assess 4(5):263–274
81. Efroymson RA, Nicolette JP, Suter GW II (2004) A framework for net environmental benefit analysis for remediation or restoration of contaminated sites. Environ Manage 34(3):315–331

Books and Reviews

Atlas RM, Phlip J (eds) (2005) Bioremediation, applied microbial solutions for real-world environmental cleanup. ASM Press, Washington, DC

Baltz RH, Demain AL, Davies JE (eds) (2010) Manual of industrial microbiology and biotechnology, 3rd edn. ASM Press, Washington, DC

Sutter GW II (ed) (2007) Ecological risk assessment, 2nd edn. CRC Press/Taylor & Francis, Boca Raton, FL

U.S. Environmental Protection Agency (2005) Groundwater sampling and monitoring with direct push technologies. Office of Solid Waste and Emergency Response, EPA 540-R-04-005. http://www.clu-in.org/download/char/540r04005.pdf. Accessed 8 December 2008

Chapter 6
Biosensors and Bioassays for Ecological Risk Monitoring and Assessment

Scott D. Soelberg and Clement E. Furlong

Glossary

Analyte	The molecule or microorganism that is the detection target in an analytical procedure.
Recognition element	A protein or other biomolecule that can bind to an analyte with specificity and affinity.
ELISA	Enzyme Linked Immunosorbent Assay. An assay using an antibody immobilized on a solid phase, (usually microtiter plate) to capture analyte and a second antibody coupled to an amplifier to detect the specific analyte.
Lateral flow assay	An immunoassay in which a liquid sample is added to a dry porous carrier and wicked by capillary action to a recognition element immobilized on a specific area of the support material. A colored nanoparticle is used to detect the presence of analyte.
Nanoparticle	A particle with dimensions between 1 and 100 nm.
Hapten	A small molecular weight molecule (usually <1,000 Da) that cannot elicit an immune response by itself and must be attached to a larger carrier molecule prior to injection into the host animal.
Paramagnetic	Magnetism induced in the presence of an externally applied magnetic field.

This chapter was originally published as part of the Encyclopedia of Sustainability Science and Technology edited by Robert A. Meyers. DOI:10.1007/978-1-4419-0851-3

S.D. Soelberg (✉) • C.E. Furlong
Departments of Medicine-Division of Medical Genetics and Genome Sciences,
University of Washington, Seattle, WA 98195-5065, USA
e-mail: scottjs@u.washington.edu; clem@u.washington.edu

E.A. Laws (ed.), *Environmental Toxicology: Selected Entries from the Encyclopedia of Sustainability Science and Technology*, DOI 10.1007/978-1-4614-5764-0_6,

Definition of the Subject and Its Importance

This article describes the development and use of portable recognition element (RE)-based assays for environmental biosensing. It will focus on using portable optical biosensors; specifically surface plasmon resonance (SPR)-based biosensors for detecting a wide variety of analytes that may pose environmental risks. Portable SPR-based biosensor systems are suitable for real-time environmental monitoring as well as for many other applications including biodefense, medical diagnostic applications, food safety, and general laboratory research. Many of the detection strategies described here can also be used across other detection platforms that use recognition elements as the fundamental detection event, including lateral flow [1] and ELISA assay platforms [2].

Introduction

This article will provide an overview of portable SPR biosensor applications, including a description of the fundamental technology used in SPR biosensing, the assay development process including preparation of the sensor surface for specific analyte detection, environmental sample collection, and the many different protocols and approaches used for analyte detection and signal amplification. Examples of detection of environmental pathogens and pollutants as well as the limits of detection for these examples will also be illustrated.

Biosensor Introduction

A biosensor is generally made up of four component subsystems, (1) a means to deliver the sample (fluidics system) is a requirement for most aqueous-based biosensor systems; (2) a biological sensing component (recognition element), sometimes including a signal amplifier; (3) an electronic signal transducer to convert the detection event to a signal interpretable by the user; and (4) a user interface to display the result. Figure 6.1 shows a flow chart of a typical biosensor system. Portable SPR biosensors function by introducing a liquid sample via a fluidics delivery system to a sensor surface containing an immobilized bio-recognition element.

Most other RE-based assay systems require the addition of secondary REs attached to amplifiers to create detectable signals. Signal amplifiers include enzymes coupled to an RE that can convert a reporter substrate into a measurable colorimetric, luminescent, or fluorescent output. Fluorometric, luminometric, or spectrophotometric readers are then required to convert the signal to digital output. SPR-based biosensors convert molecular binding events directly to a detectable

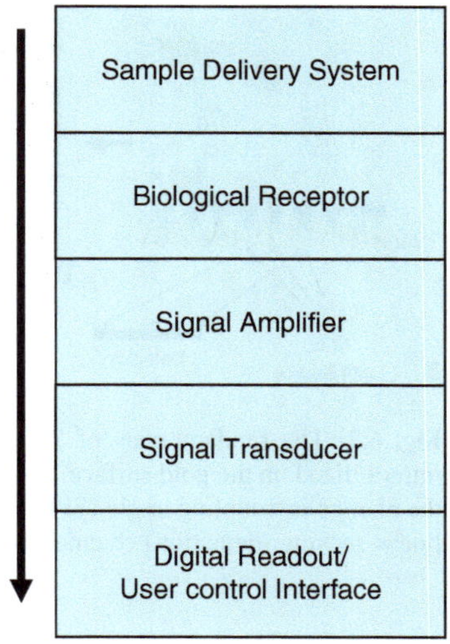

Fig. 6.1 Biosensor flow diagram. Arrow indicates the flow of information as the signal is acquired

signal in real time, thus a signal amplifier step is not always required for detecting the presence of analyte.

Commercial SPR systems first became available in the mid 1980s. Large bench-top instruments were developed for determining the binding rates of interaction between biomolecules in real time without the requirement for amplification. Currently most commercial SPR-based sensor systems are still used for this purpose, as well as for the study of interaction analyses in drug discovery. Detection strategies and protocols developed with bench-top SPR systems are readily implemented for use in portable devices. Bench-top SPR systems, however, are large and have delicate optical components that require sensitive calibrations and regular maintenance, precluding their use as portable devices. In addition, most commercially available SPR instruments are prohibitively expensive for use in environmental monitoring. In the late 1990s Texas Instruments developed a miniature, integrated SPR module (Spreeta) [3–5]. The robust nature and small size of this integrated chip made possible the development of portable SPR biosensor systems. Naimushin et al. [6] described a temperature-controlled SPR system based on an early Spreeta three-channel SPR module manufactured by Texas Instruments. The importance of temperature control and the use of reference channels were noted in this publication. Chinowsky et al. [7] described a portable 24-channel SPR-based system that could detect as many as 24 different analytes. This system is based on eight miniature 3-channel Spreeta sensor chips (further described in Soelberg et al. [8]).

Principles of SPR Sensors

SPR sensors are optical sensors that exploit the interaction of light with a gold surface for the signal transduction event. Figure 6.2 shows the Kretschmann [9] configuration of an SPR-based detector (Fig. 6.2a). Polarized light illuminating

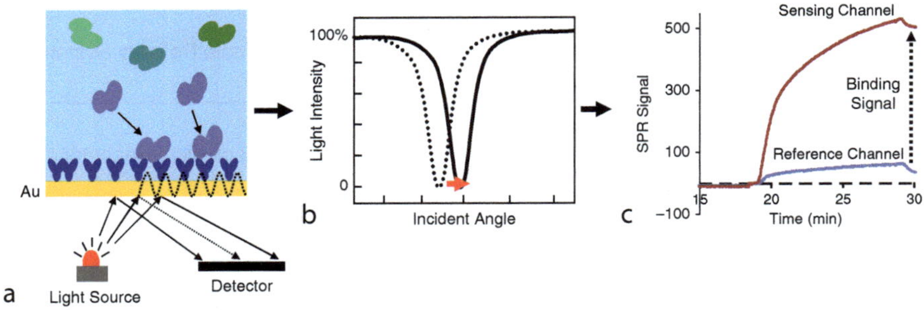

Fig. 6.2 The fundamentals of SPR detection. (**a**) Specific analyte molecules bind to the REs immobilized on the gold surface. (**b**) This binding causes a shift in the coupling angle. (**c**) Plotting the change in coupling angle (SPR signal) as a function of real time generates a sensorgram. Red line = sensing (detection) channel. Blue line = reference (nonspecific) channel

a thin (50 nM) gold surface at varying angles (or wavelengths) is reflected back to a detector element (Fig. 6.2b). Light of a specific angle (or wavelength) is not reflected, but couples to surface plasmon electrons to generate a surface plasmon wave across the metal surface of the sensor. This coupling angle or wavelength is dependent on the refractive index (RI) of the solution adjacent to the gold surface. SPR curves are shown in Fig. 6.1b. The coupling angle or coupling wavelength is dependent on the refractive index (RI) of the solution adjacent to the gold surface (Fig. 6.2b). The reflected light intensity reaches a minimum at a specific coupling angle or wavelength. Specific recognition elements (REs) (e.g., antibodies, receptors, nucleic acids, or small molecular weight analytes) are attached to the sensor surface. The blue "Y" shapes in Fig. 6.2a represent recognition elements. Binding of analyte to a surface-bound recognition element causes a change in RI at the sensor surface, and a resulting shift in the SPR angle or wavelength of minimum intensity (Fig. 6.2c). Analyte binding is followed in real time as the system software converts the shift in the coupling angle or wavelength to an RI value as a function of time.

For a general discussion of SPR principals, the reader is referred to the following excellent books on SPR technology [10–12].

Sensor Surface Preparation

Modifying the Gold Surface for Covalent Attachment of Recognition Elements

There are several methods for attaching recognition elements to sensor surfaces. The different approaches to surface attachment of the RE can result in different levels of interferent binding, and several attachment methods should be evaluated to determine the optimal conditions for specific assays and sample matrices.

The methods outlined below will be restricted to attaching recognition elements to gold surfaces, as most SPR sensors use gold as the surface metal. Several examples of surface functionalization are described below.

Self assembling monolayers (SAMs). Various SAMs have been described for use in coupling recognition elements to gold surfaces. Most exploit the strong affinity of sulfhydryl (thiol) or disulfide groups that can interact with the gold surface to form strong and stable bonds (also called chemisorption). The sulfur gold interaction is semi-covalent with a binding energy of 45 kcal/mol [13]. For a comprehensive review of these methods, the reader is referred to [14–16].

Examples of SAMs Used for SPR Sensors

Mixed alkane thiols. Alkane thiols are molecules with sulfhydryl head groups for binding to the gold surface, a carbon chain as the back bone, with a functional terminal group for attachment chemistries or passivation. Mixed alkane thiols are terminated at the tail end with either a hydroxyl (to reduce nonspecific binding (NSB) to the sensor surface) or with a functional group ($R\text{-}NH_2$ or $R\text{-}C00^-$) at a fixed percentage to allow even spacing of the functional groups to which an RE can be attached by standard coupling chemistries. For example, EDC/sulfo-NHS activate surface carboxyl groups, which in turn couple to amine groups of the RE [17]. Attachment of the alkane thiols to the gold surface is achieved by soaking the clean gold surface in mixtures of alkane thiols [18].

Hydrogel surfaces. One of the most widely used SPR surfaces is a carboxy-dextran hydrogel to which multiple receptors or ligands can be attached per strand of dextran [19, 20]. This protocol has the advantage of a fairly dense three-dimensional decoration of the sensor surface with receptors or ligands, but can have issues with bulkier analytes since the matrix can slow mass transfer to the receptors located throughout the hydrogel matrix. The carboxy-dextran layer can be built on a SAM to form a very stable surface for protein or ligand immobilization (Fig. 6.3).

Gold-binding peptide (GBP). GBP is an engineered protein domain with repeat sequences that binds tightly to gold surfaces [21]. This small peptide can be used as a foundation layer on the gold surface to which REs can be attached via standard EDC-NHS protein coupling [17]. A description of the protocols involved in using the gold-binding peptide for sensor surface preparation can be found in Woodbury et al. [22].

Physical adsorption of recognition elements. Antibodies and other proteins can also be directly adsorbed (also called physisorption) through charge interactions between the gold and protein surfaces without specific chemical modification. For this procedure, antibodies in solution are simply introduced in concentrated form (>1 mg/mL) to a gold surface cleaned with nitric acid and ethanol. The orientation of the antibody is not controlled in this procedure; therefore not all the antibodies

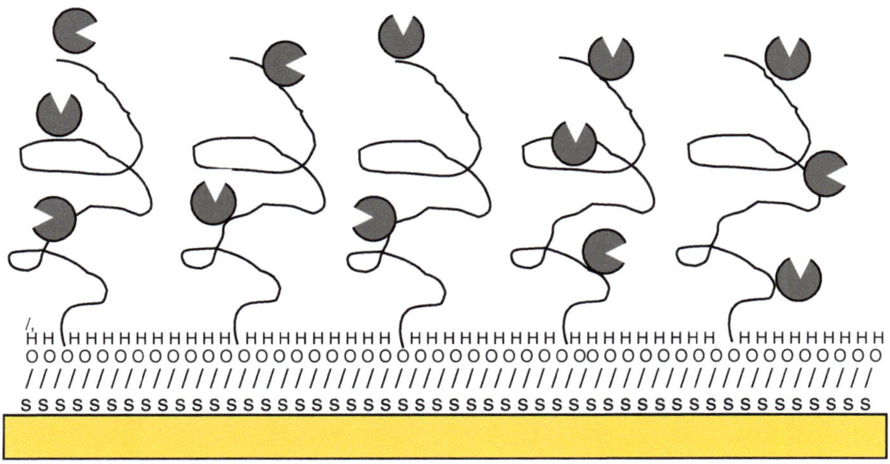

Fig. 6.3 Alkane thiol with terminal OH groups and functional groups with recognition elements attached

will have the functional arms oriented outward. However, this method has been shown to produce a surface that is robust and capable of sensitive detection and regeneration [7]. This simple approach to surface coating can be useful for many sensor applications.

Covalent attachment of recognition elements to modified sensor surfaces. Sensor surfaces modified by SAMs terminated with free amino or carboxyl groups can then be modified to covalently couple recognition elements (e.g., protein molecules) through standard protein coupling procedures. The most commonly used reaction is the amino to carboxyl protein coupling using the chemicals EDC or EDC/sulfo-NHS for carboxyl group activation and protein coupling [17].

Attachment of "adapter molecules." Generic surface foundation layers may be attached by adsorption or covalently coupled to SAM-modified sensor surfaces followed by binding of the RE to generic adapter molecules. Some examples of these generic receptor molecules include avidin or streptavidin to which can be attached biotin-labeled recognition elements such as antibodies, DNA, protein nucleic acids (PNAs) [23], locked nucleic acids (LNAs) [24], ligand receptors, or other biotin-labeled recognition elements [25] (Fig. 6.4, *Right*). Other types of receptors that can be attached to the sensor surface include proteins A or G. The bacterial proteins A and G that bind specifically to the Fc domain of immunoglobulins (Igs) (most notably many IgGs) can be adsorbed or covalently attached to a modified gold surface and used for construction of regenerable sensor surfaces for repeat assays. These surface-adsorbed molecules provide "sockets" for antibody binding to sensor surfaces. The Igs can be removed with regeneration buffer (e.g., low pH), and new antibodies, for example, with different specificity, can be added to the surface "sockets" (see Fig. 6.4, *Left*) [26, 27]. Covalent

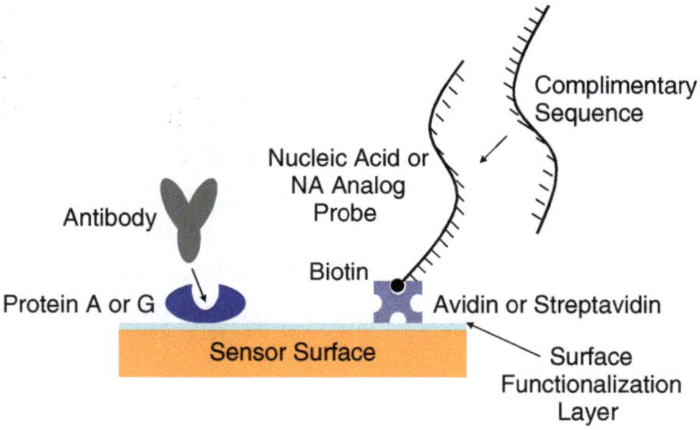

Fig. 6.4 Adaptor molecules allow for replaceable antibody surface (*left*) or binding of biotinylated probes (*right*)

attachment of antibody receptor molecules to robust surfaces such as SAMs should provide longer-lasting regenerable surfaces.

Surface functionalization for competition assays. For the competition-based assays described below in the section entitled "Small Molecule Detection," small molecular weight ligands are often conjugated directly to a SAM surface or to a carrier protein [28, 29] that can either be covalently attached to a SAM, adsorbed to the sensor surface, or chemically coupled to a surface foundation layer in a similar manner as with the antibody attachment protocols.

Surface passivation with blocking proteins. Another important aim in preparing a sensor surface for analyte detection is to generate a sensor surface that has a low degree of nonspecific, or background binding from the sample. Additional steps to passivate the sensor surface and minimize nonspecific binding may be required after attaching the RE [30]. For example, when analyzing a serum sample, proteins may nonspecifically bind to the sensor surface and interfere with or overwhelm the analyte-specific binding signals. By pretreating the sensor surface with serum proteins such as albumin or other blocking agents, nonspecific sample matrix proteins can be blocked from interaction with the sensor surface.

Referencing

Referencing is another way to compensate for the effects of nonspecific interactions in bioassays. Referencing involves monitoring an additional channel with an RE that is not specific for the analyte of interest. It is important to include a reference channel with the same antibody type, such as a rabbit polyclonal or mouse monoclonal, and use the signal from this channel to correct for nonspecific binding. In the example shown in Fig. 6.2c, the light blue line shows that there was little nonspecific binding signal during the sensing step. With proper reference channel treatment, subtraction of the reference channel value from the signal of the analyte-specific channel provides an accurate measure of the specific binding.

Sensor Storage and Preservation

It is important to make use of stable recognition elements and to develop protocols for maintaining activity of the sensor REs during storage. There is relatively little information in the literature regarding preservation of recognition elements on sensor surfaces. Drying a specific carbohydrate glass such as a trehalose and high molecular weight dextran layer over the RE has been shown to preserve antibody function for at least 1 year without the need for vacuum packing the sensor element [28].

Sample Acquisition from the Environment and Removal of Interferents

SPR-based biosensors for environmental monitoring require that the analytes are contained in an aqueous solution, often with some preprocessing to concentrate the sample and/or reduce the level of interferents and enhance the specific signal from the target analyte. The protocols for sample collection/processing are highly dependent on the target analyte and the sample matrix. An environmental sample will generally fall into one of three general classes: sample in liquid form, sample in solid form, such as in dirt or on a solid surface, and sample dispersed in air, such as aerosolized pesticides, microbes, or nerve agents. Once the sample is collected or transferred to an aqueous phase, it can be delivered to the biosensor for detection of the specific target analyte.

Airborne samples. For bioaerosols and airborne chemical samples, a number of collection systems have been described or are commercially available. Because the immunoassay SPR-based assays depends on the sample being in liquid form, the two most practical collectors for this medium are wetted-wall cyclonic collectors [31] and glass impingers [32, 33]. Samples can also be obtained by directing air flow through a filtration system; however the filter must be stripped with aqueous buffer to suspend the trapped particulates for analysis.

Bioaerosol wetted-wall cyclone collectors are those in which air samples are directed over a high-surface-area sample collecting liquid making use of cyclonic action. Environmental sampling using a cyclonic collector to analyze long distance airborne transport of both viral and bacterial porcine pathogens was demonstrated by Dee et al. [31]. In their example, aerosolized porcine reproductive and respiratory syndrome virus (PRRSV) and *Mycoplasma hyopneumoniae* were collected using a liquid cyclonic collector from as far as 4.7 km from the source of the swine pathogens.

Glass impingers are the simplest types of air-to-liquid sample collectors. Impingers are commercially available and operate by simply directing a stream of air into a liquid sample (Fig. 6.5). There are a number of examples in the literature of using impingers

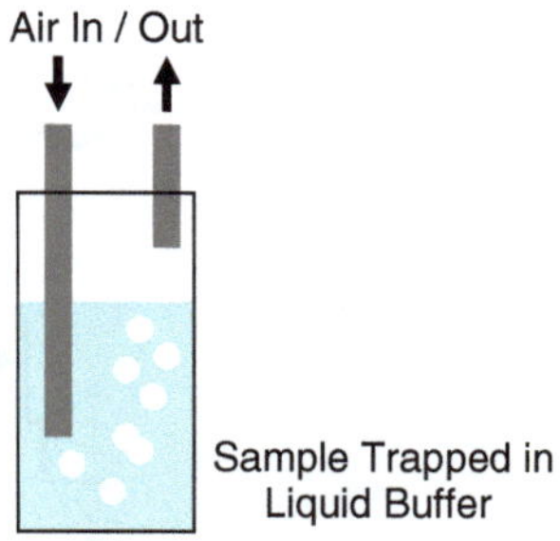

Fig. 6.5 Glass impinger for converting aerosolized samples to aqueous samples

for collecting airborne particles into a liquid for analysis, including for the detection of airborne nitrophenols (see, e.g., Bishop and Mitra, 2007) [32] and airborne bacterial spores dispersed in an aerosol chamber (see, e.g., Rosen, 2006) [33].

Filter sampling and resuspension in liquid medium is another approach to sample introduction for environmental biosensing. In these samplers, an air or water stream is driven through a filter device, and the filter will trap the small target analyte particles. Mixed cellulose ester (MCE), Teflon, poly-tetra fluoroethylene (PTFE) filters, and gelatin filters can trap particles such as bacteria and viruses from aerosolized or aqueous samples. The concentrated samples are then suspended in an aqueous solution for analysis. Fabian in 2009 [34] demonstrated a comparison of four aerosol collectors, including two types of filter collectors (gelatin and Teflon) for airborne influenza virus collection and viability analysis. Burton et al. in 2005 [35] compared the MCE, PTFE, and gelatin filter materials for detection of the *Bacillus anthracis* stimulant *B. subtilis* in bioaerosols. Filtration in which the filtrate is analyzed for specific molecules or organisms smaller than the filter pore size can also be employed. Figure 6.6 shows an example of a small molecule detection filter device that excludes larger protein molecules in a saliva sample, and smaller molecules that pass through the filter are then delivered to a sensor surface [29]. This approach has the advantage of eliminating components of the matrix that may bind nonspecifically to the sensor surface, such as mucins in saliva or biofilm forming molecules from seawater.

Liquid samples. Samples already in liquid such as freshwater or seawater samples are relatively easy to process and are amenable to processing with automated systems. Liquid collection systems can be as simple as a collection tube dipped by hand into the water for manual collection or a pump delivery system to the sensor. Filtration can also be an efficient way to collect a sample if the analyte of interest is large enough to be trapped with a filter. For example, *Giardia duodenalis* and *Cryptosporidium* species present in low levels in drinking water were filtered and detected via polymerase chain reaction (PCR) by Plutzer et al. [36]. In cases where the analyte may be present at low concentrations or where the matrix may contain many interferents, filtration can be a way to concentrate samples and select for specific sizes of particles through low or high molecular weight cutoff filters. The analytes can be washed from a filter with a buffer that is optimal for detection, or a hollow fiber flow-filter can be used to change out buffer conditions in real time (Fig. 6.6). More complex ocean sample collection

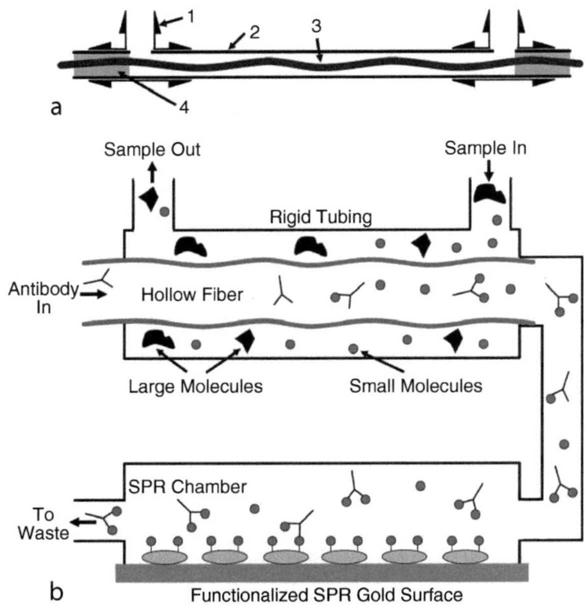

Fig. 6.6 An example of in-line filtration of samples to remove interferents before competition assay analysis with a portable Surface Plasmon Resonance biosensor. External, in-line filtering flow cell, which directs flow into the SPR system. (**a**) Construction of the in-line filtering flow cell required sliding a 1.6-mm Tee connector (1) (T210-1, Valve Plastics, Fort Collins, CO) over the ends of 0.75-mm inner diameter PEEK tube (2) (1533, Upchurch Scientific, Oak Harbor, WA), followed by drilling a hole into the PEEK tube to allow sample flow, and then sealing the PEEK tube to the tee with urethane adhesive (Lord Corp., Cary, NC). A hydrophilic hollow fiber (3) (Minntech, Minneapolis, MN) with a molecular weight cutoff of 20,000 was inserted into the PEEK tube, and both ends were sealed with urethane adhesive (4). Following sealing of the hollow fiber into the PEEK tube, the ends of the hollow fiber were cut flush with the tee preserving the opening to the hollow fiber. (**b**) Salivary samples were diluted 1:2 with buffer and flowed through tubing external to the hollow fiber. The antibody solution was flowed countercurrent through the hydrophilic hollow fiber and then through the SPR biosensor system. Diffusion may also be aided by bulk flow across the hollow fiber wall controlled by a pressure differential created by higher flow rates outside the hollow fiber (not drawn to scale). (From [29])

makes use of a suspended-particle multi-sampler (called a rosette) for obtaining discrete samples from given depths and locations [37].

Solid surface samples. Collecting environmental samples from surfaces for analysis can be as simple as swiping the surface with a moistened swab. The swab can then be added to a liquid buffer solution and vortexed to resuspend the sample in liquid. Particulates such as soil samples can be measured or weighed, then resuspended in liquid buffer. In "Surface Sampling of Spores in Dry-Deposition Aerosols" Edmonds et al. compared different swab materials used on four surfaces with different surface properties for sampling of bacterial spores [38]. Once the target analyte(s) has been collected in or transferred to an aqueous solution, a number of different strategies have been used for the specific detection of the target analyte. These strategies are outlined in the following section.

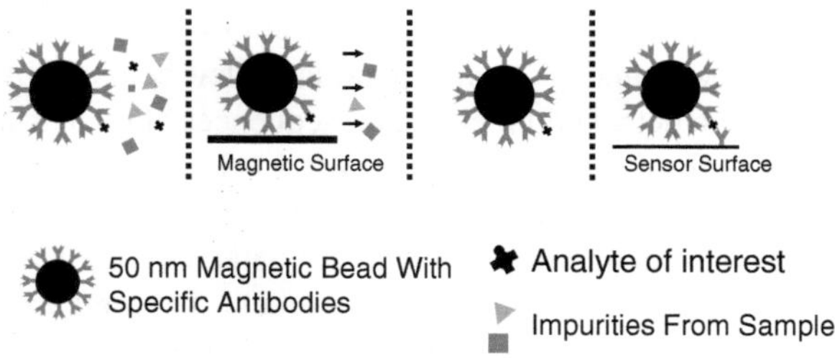

Fig. 6.7 Magnetic bead cleanup of sample matrix. Each frame (separated by a dotted line) is a separate step in the cleanup and detection process

Immunomagnetic Particle Purification of Specific Analyte

Paramagnetic antibody-conjugated particles can be used to rapidly concentrate and purify an analyte from a complex matrix of particles commonly found in environmental samples. Figure 6.7 shows an example of this type of cleanup procedure. Immunomagnetic beads (IMBs) decorated with analyte-specific antibodies are added to a sample and mixed. The IMBs bind the analyte of interest, then are immobilized in a magnetic field, and all impurities and potential interferents are washed away with analysis buffer. The IMBs with bound analyte in analysis buffer are then released by removal of the magnetic field and introduced to the sensor for detection [8]. Once the concentrated, purified target analyte(s) is collected in or transferred to analysis buffer, a number of different strategies have been used for the specific detection of the target analyte. These strategies are outlined in the following section.

Detection Strategies

Development of an SPR-based assay requires consideration of several factors that will affect the SPR signal: (1) the type and properties of the surface-immobilized recognition or specificity element; (2) size of the analyte of interest, reflecting its ability to generate a RI change large enough to be detected when bound by the RE immobilized on the sensor surface; (3) the specificity of the recognition element for target analyte and related analytes; (4) strategies for reducing nonspecific binding of other components of the matrix, and (5) any strategy for amplifying the signal from bound analyte.

Properties of the recognition element. A recognition element is any molecule that can bind with high specificity and affinity to the target analyte. The recognition element should be stable and ideally be capable of regeneration for many cycles of analyte binding and surface regeneration. Antibodies have been found to fulfill these requirements and have been used for many different sensing applications [39]. Receptors with high specificity and affinity for analyte such as metallothioneins binding to heavy metal ions (Cd^{2+} and Hg^{2+}) can also serve as suitable recognition elements [40]. The use of molecular imprinted polymers (MIPs) can also be used, for example, ochratoxin A, an *Aspergillus* that produced mycotoxin, was detected with a MIP surface with SPR at concentrations down to 0.05 ppm [41]. Nucleic acids or nucleic acid analogs such as peptide nucleic acids (PNAs) can be used on a biosensor surface to detect complementary sequences of nucleic acids such as ribosomal RNA (rRNA) from *Escherichia coli* [42]. Locked nucleic acids (LNAs) are other analogs with increased thermal stability and resistance to nuclease activity, which can also be used as recognition elements for detection of DNA or RNA [24].

There are several issues to consider when preparing an SPR sensor surface for the detection of a specific analyte. First, the recognition element that is attached to the sensor surface should be very specific for the target analyte. This specificity is often provided by the immobilization of an antibody raised with against all or part of the target analyte. Specificity can be determined by testing the antibody for cross reactivity with closely related molecules or microorganisms. It is important to test the specificity of the chosen recognition element against analytes with structures similar to those of the target analytes, particularly if related analytes may be present in the sample matrix. For example a bacterial antibody raised to *B. anthracis* spores can be tested against other Bacillus species for cross reactivity. This example is demonstrated in the multi-analyte detection shown in Fig. 6.13. Here *B. subtilis* (BG) spores are tested against *B. anthracis* spores, and against many other unrelated antigens. The specificity of the antibody for the bacterial analyte of interest, but not for a closely related bacterial species, is shown in this figure. The extent of cross reactivity with other antigens, especially related antigens, determines usefulness of a given antibody preparation for specific applications.

REs with high affinity are also desirable for achieving high sensitivity of analyte detection. Antibody target affinities can vary by as much as four orders of magnitude [43]. SPR and ELISA analysis can be used for determining the antibody association and dissociation rates (K_a and K_d) and thus determine the usefulness of the antibody for high sensitivity detection assays [44]. For example, SPR can be used to screen for high affinity monoclonal antibodies from a pool of different clones [45, 46]. The primary goal when developing a detection assay is to have a sensor surface that binds the target with very high affinity and specificity without binding other interferents present in the sample matrix.

Consideration of antibody properties and purity. The choice of an analyte-specific antibody may require evaluation of several different polyclonal or monoclonal antibody preparations. Significant differences in both specificity and affinity of analyte binding will greatly affect the ability of a given protocol to provide the levels of sensitivity and specificity required for a given application. Often, it may require the generation of monoclonal or polyclonal antibodies to the target analyte if the appropriate antibodies are not available from commercial sources or from collaborators/colleagues. An additional factor affecting the quality of signal is the extent of purity of the antibody being used for the assay. It is important to note that with a surface interaction method such as SPR, the greater the purity of the specific antibody on the SPR surface, the more robust will be the response to analyte binding. Affinity purified polyclonal or monoclonal antibodies are essential for optimal SPR-based protocols. When making use of amplification protocols, either

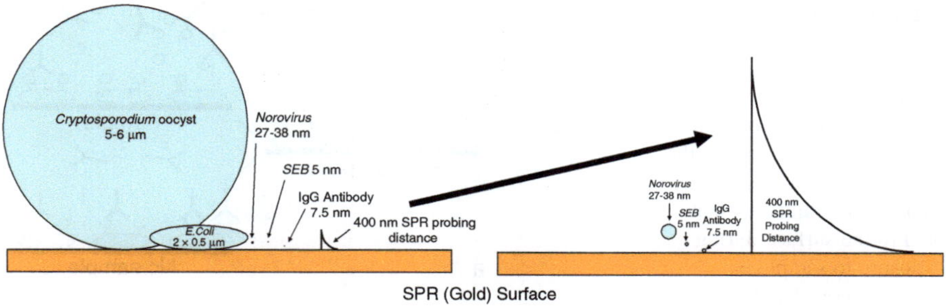

Fig. 6.8 Some examples of the size variation of different classes of analyte detectable by SPR biosensor assay compared to the effective detection range of an SPR sensor

monoclonal or polyclonal antibodies, or monoclonal antibodies with different epitope specificity may be useful. The amplification antibodies may be of lesser purity since any unrelated antibodies will wash through the system. Similar considerations will be obtained for other types of recognition elements.

Detection Limit by Analyte Type

Once a suitable recognition element has been acquired, the next step is determining the detection limit for the analyte of interest. Dividing analytes into four general classes is helpful for discussion. The categories of analyte detection with SPR are usually broken down into groups based on size. Small molecules do not usually affect generate a significant change in RI directly when bound to the sensor surface; very large analytes (e.g., whole cells) can be larger than the effective probing range of the plasmon wave, and thus only a fraction of the particle will affect the SPR signal. The categories for detection are usually classified as (1) small molecules such as organic pollutants, chemical toxins, and heavy metals (below MW 5,000) [14, p. 476] usually undetectable by direct SPR measurement (2) protein molecules >5,000 Da (e.g., protein toxins, protein biomarkers, or antibodies), which diffuse easily and are large enough to generate a significant surface change in RI, (3) viruses, and larger particles such as (4) microbes, and (5) single-celled eukaryotic organisms (algae and fungi). Whole cells, whether bacteria or eukaryotes, often diffuse slowly to the sensor surface, may not bind easily to surface recognition elements due to shear forces, and fall at least partly outside of the penetration depth of most SPR instruments (approximately 100–600 nm) [14], p 463. Figure 6.8 shows the sizes of several proteins and microorganisms relative to the antibody surface and SPR penetration depth. A *Cryptosporidium* oocyst [47], an *E. coli* cell [48], a norovirus particle [49], an SEB protein toxin molecule [50], and an IgG detection antibody [51] are displayed in relation to the probing distance from the SPR surface. Detection limits from direct detection protocols (no amplifiers) as well as with those using additional amplification steps are compared below.

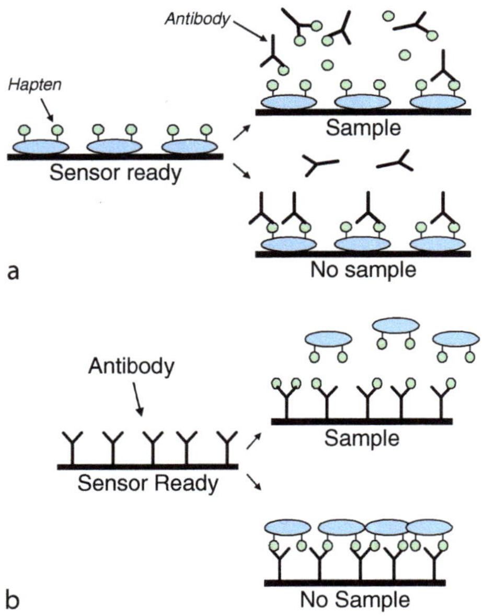

Fig. 6.9 (**a**) Competition detection of a small molecule using SPR. A schematic of the sensor surface shows the hapten-conjugated protein on the surface of a sensor in the "sensor ready" state. The quantity of antibodies that bind to the surface during exposure to sample is inversely correlated to the amount of analyte in the sample. (**b**) Alternatively, small molecule binds to the antibody on the sensor surface and inhibits binding of the larger carrier protein derivatized with the same small molecule. This protocol will conserve antibody if it is costly or scarce

Small Molecule Detection

Direct detection of small molecules with SPR. There are a few examples of SPR biosensors being used to directly detect the binding of metal ions. Wu et al., 2004 [52] used the heavy metal binding protein metallothionein immobilized on an SPR surface as a sensor for metal ions such as cadmium (Cd^{++}), zinc (Zn^{++}), and nickel (Ni^{++}). With this system, concentrations of Cd^{++} as low as 2 μM were detectable.

Competition (inhibition) for detection of small molecules. Since the binding of small analytes does not generally result in large changes in RI at the sensor surface, detection of small molecular weight analytes often makes use of an indirect assay in which the analyte itself is immobilized on the sensor surface, often through coupling to a carrier protein or tethered to the surface covalently. The most common method of antibody-based detection of small molecules is referred to as a competition or inhibition assay. The target analyte (hapten) molecule is attached to the sensor surface using one of the protocols described above in the sensor surface preparation section. Antibody in solution is then allowed to bind to the surface-immobilized target analyte, establishing a rate of antibody binding in the absence of analyte. Samples containing the target analyte are introduced, inhibiting antibody binding at levels proportional to the concentration of target analyte in the sample (Fig. 6.9a). As an alternative to rate determination, total binding signal can also be measured, and this signal is also inversely proportional to the concentration of analyte over a specific concentration range [53]. Another protocol (Fig. 6.9b) can be used in which anti-analyte antibody is immobilized on the sensor surface. With this protocol, sample containing the small molecule is first added and allowed to bind to the surface antibody. Analyte-protein conjugate is then introduced to the surface. The signal from this binding step is again inversely proportional to the amount of analyte bound to the surface from the original sample. This protocol is beneficial when antibody needs to be conserved due to low availability or high cost.

There are many examples in the literature of small molecule detection using the competition assay. For example, domoic acid (MW 311), a toxin produced by the algae *Pseudo-nitzschia* that can accumulate in marine filter feeders, was detected using a competition assay in both laboratory buffers and in spiked clam samples at levels as low as 4 ng/mL [28]. Yu et al. [53] described a domoic acid competition assay with a detection limit of 0.1 ng/mL in laboratory buffers using an SPR-based detection system. The organophosphate insecticide chlorpyrifos (MW 250) was detected in a variety of environmental water samples by SPR using a competition assay with a detection limit of 45–64 ng/mL [54]. TNT (Trinitrotoluene), a component of land mines and other explosive devices (MW 227) was detected by SPR analysis at concentrations ranging from 0.008 ng/mL (8 ppt) to 30 ng/mL (30 ppb) [55].

Direct Detection of Protein Analytes

One of the primary advantages of SPR-based systems is the ability to detect direct binding of analytes quickly and without amplifier reagents or other manipulation of the sample (Fig. 6.10). If the concentration of a larger analyte (>5,000 Da) such as a protein toxin is at an adequate level, and if there are not significant interferents in the sample, it is possible to not only directly detect the analyte, but also to quantify the analyte from either the initial rate of binding or the maximum shift in RI on

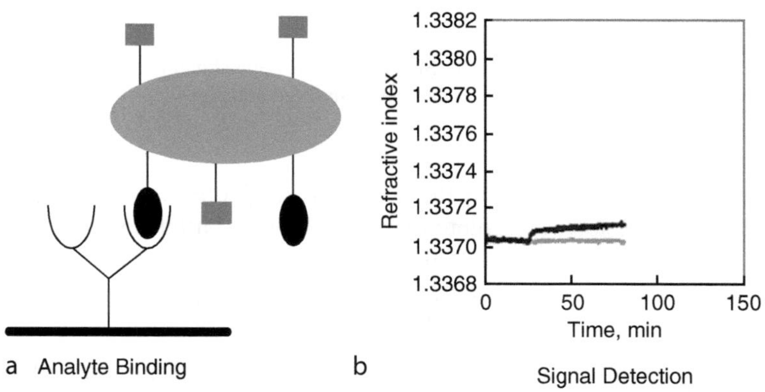

Fig. 6.10 (**a**) Binding of protein analyte to antibody immobilized on the gold sensor surface. (**b**) Real-time signal (refractive index) generated during the binding event

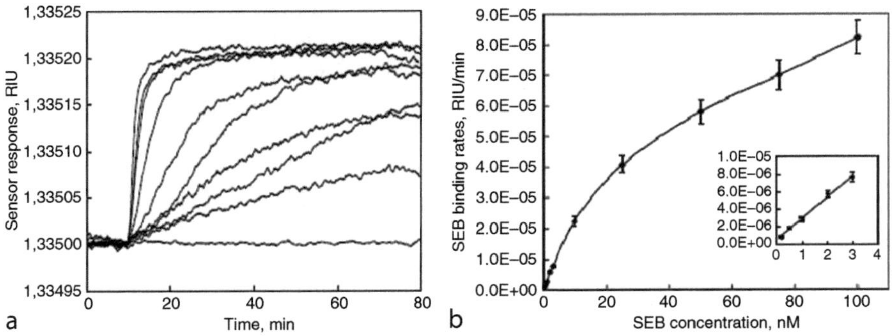

Fig. 6.11 (**a**) Binding of *Staphylococcal* enterotoxin B (SEB) to sensor surface at different concentrations of SEB (0.2, 0.5, 1, 2, 3, 10, 25, and 75 nM). (**b**) Plot of initial rates of binding of SEB as a function of SEB concentration (inset, expanded scale for the indicated low concentrations of SEB). (From [56])

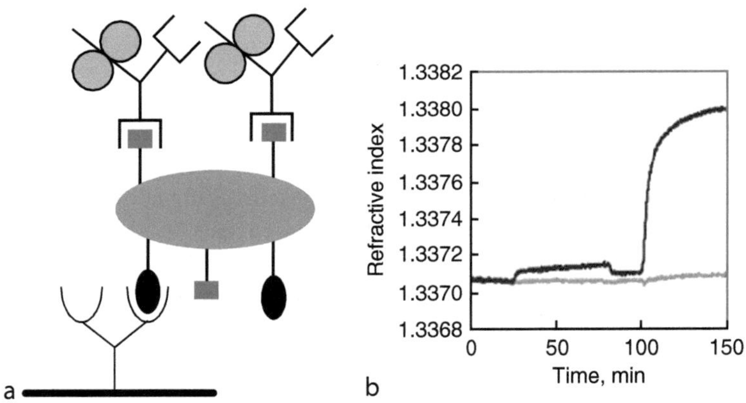

Fig. 6.12 (**a**) Binding of secondary antibodies with different epitope specificity coupled to amplifier particles to captured analyte. (**b**) Signals generated upon binding of analyte to the sensor surface followed by binding of the secondary antibody/bead complex

binding of the target analyte to the surface RE. Figure 6.11 shows the direct detection of *Staphylococcal* enterotoxin B (SEB) in buffer solution at several different analyte concentrations [56]. At low analyte concentrations (Fig. 6.11b, inset), the concentration of analyte is linearly dependent on the initial rate of binding of SEB to the surface-immobilized antibodies.

Amplification strategies. Amplification is generally carried out with addition of a second recognition element prior to or after the analyte is bound to the surface (Fig. 6.12). Also called a sandwich assay, this second element can be an antibody, or an antibody coupled to a larger particle to affect a greater increase in the SPR signal.

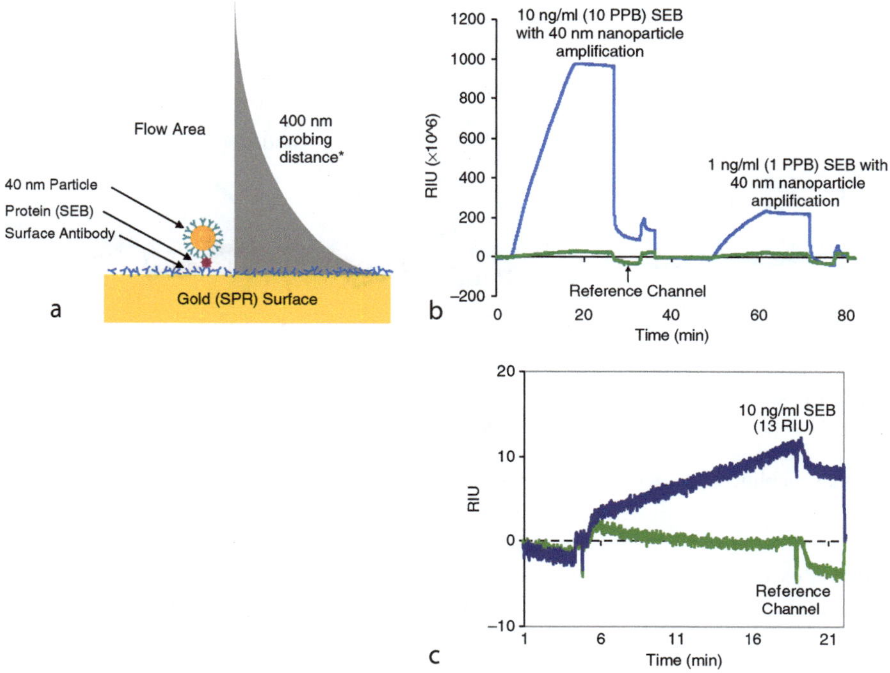

Fig. 6.13 (**a**) Cartoon of protein (SEB) detection with amplifier in relation to the SPR sensor probing distance. (**b**) Detection of SEB with amplifier at two different concentrations with a low pH wash to reset the sensor between detection events. Protein (SEB) was mixed with 40 nm particles prior to injection for amplification of the signal. (**c**) The 10 ng/mL signal without nanoparticle amplification is approximately 13 RIU. Nanoparticle amplification resulted in a 70-fold increase in the SPR signal

Sensitivity of analyte detection can be dramatically increased by using amplifiers. Proteins, including antibodies like the fish iridovirus antibody [57], microbial toxins (SEB) [56, 58], and many others are detected with amplification steps to increase the binding signal.

To increase the sensitivity of detection and at the same time validate the identity of analyte, a second, particle-bound antibody specific for an analyte epitope different than that of the surface-immobilized antibodies may be used. Nano-immunomagnetic particles or nano-gold particles have been used for this application. The advantage of the nano-immunomagnetic particles is that they may be used in sample processing steps to purify and concentrate the target analyte before introduction to the sensor [8]. Figure 6.13 shows an integrated amplification protocol, where the SEB is bound first to a magnetic particle and then allowed to bind to the SPR surface. The resulting signals for 10 ng and 1 ng/mL SEB are much larger than SEB alone. These amplifiers can be the same magnetic particle used for purification of the sample shown in Fig. 6.6. The concentration/purification step provides a significant increase in the sensitivity of the detection protocol and

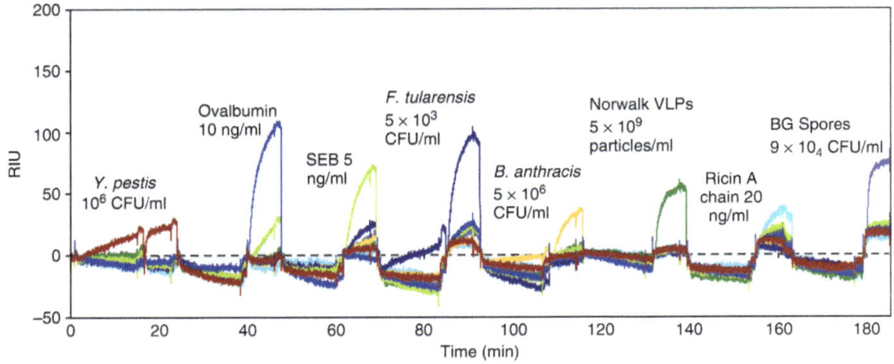

Fig. 6.14 Serial detection of *Yersinia pestis*, Hen egg ovalbumin, *Staphylococcal* enterotoxin B (SEB), *Francisella tularensis*, *Bacillus anthracis* spores, Norwalk virus-like particles (VLPs), ricin A chain, and *B. subtilis* (BG) spores at the indicated concentrations with single steps of antibody amplification after each direct binding step

eliminates possible interferents from the sample stream at the same time. Larger micron-sized particles conjugated with the secondary antibodies diffuse too slowly to the sensor surface to be useful in real-time detection. Nano-gold and nano-magnetic particles can also be coupled to nucleic acids for amplification of DNA or RNA binding to a sensor surface. A sulfur group-derivatized nucleic acid or a biotinylated nucleic acid can be attached to the gold or magnetite particles.

This sandwich protocol is useful for the detection of microbes, spores, viruses, and toxic proteins [7]. Figure 6.14 also shows sandwich type amplification of several different types of analytes.

Subtractive Inhibition Assays for Large Molecules (Whole Cells)

To detect large molecules (i.e., whole cells) that lie mostly outside the SPR evanescent wave (Fig. 6.8) a technique similar to the small molecule competition can be employed. This technique involves addition of antibody to a suspension of bacteria or other larger particles, and then removal of the antibody-complexed particles by centrifugation or filtration and quantification of the remaining antibodies by binding to anti-antibody RE or immobilized target epitope on the SPR surface. Depletion of antibody signal is proportional to the concentration of analyte in the sample. Simplifying the protocol by eliminating the filtration/centrifugation step and relying on the difference in diffusion rate between the large particles and the antibody can also be employed, although the sensitivity may be affected.

Multiplex Detection

For many environmental biosensing applications it is useful to be able to detect a number of different analytes from a single sample. These include testing for threats from deliberate contamination, from the many potential pathogens in water supplies, testing for different strains of a similar microbial organism, and testing for different chemical pesticides [59] in the field. Figure 6.14 provides an example of detecting eight different analytes with sequential injections into the same sensor system. The system incorporated eight 3-channel sensors with the same recognition elements on all three channels of each sensor element. This configuration provided binding data in triplicate in real time.

Future Directions

The future directions of portable environmental SPR biosensing have been partially defined with the development of the SPR microscope or imager. SPR imagers are a variation of the SPR sensor described here in which light incident at a fixed angle or wavelength illuminates a much larger surface area. The ability of the SPR imaging systems to interrogate multiple different spots or REs on a single chip surface in an array format can also be performed in real time, allowing for hundreds and even thousands of samples to be analyzed simultaneously. A 1,000-spot bench-top array system has been demonstrated [60], and a prototype portable SPR array imager has also been described [61]. The challenge will be to produce systems that are sufficiently small to be used as portable devices or small, inexpensive bench-top systems. For general laboratory use, the binding partners of interest will include a vast array of different biomolecules, drugs, and other small molecules.

Future enhancements of portable systems will depend on further development of microfluidic and nanofluidic systems, including microvalves and pumps, simplification of the user interface, smaller processors and control electronics, and perhaps the further reduction in size of the sensing element. The currently available SPREETA biosensor element is only about one-fourth the size of the original devices. Refractive index is highly sensitive to small temperature changes, and reduction in the size of the temperature control circuitry will allow for further miniaturization of portable sensor systems. Simply reducing the size of the existing portable systems to hand-held devices will provide instruments with a broad range of applications.

The continued development of recombinant antibody selection and expression systems such as recombinant *E. coli* single chain camelid antibodies [62] will be important for generating both wild-type and variant recombinant proteins that will be useful as inexpensive sensor recognition elements. This is an area that can benefit from increased efforts and funding. The reduction in antibody costs will be especially important for protocols that consume reasonable quantities of antibodies, such as those utilizing amplification approaches.

Bibliography

1. Ngom B et al (2010) Development and application of lateral flow test strip technology for detection of infectious agents and chemical contaminants: a review. Anal Bioanal Chem 397(3):1113–1135
2. Lequin RM (2005) Enzyme immunoassay (EIA)/enzyme-linked immunosorbent assay (ELISA). Clin Chem 51(12):2415–2418
3. Melendez J et al (1996) A commercial solution for surface plasmon sensing. Sens Actuators, B 35(1–3):212–216
4. Melendez J et al (1997) Development of a surface plasmon resonance sensor for commercial applications. Sens Actuators, B 39(1–3):375–379
5. Elkind JL et al (1999) Integrated analytical sensors: the use of the TISPR-1 as a biosensor. Sens Actuators, B 54(1–2):182–190
6. Naimushin AN et al (2003) A portable surface plasmon resonance (SPR) sensor system with temperature regulation. Sens Actuators, B 96(1–2):253–260
7. Chinowsky TM et al (2007) Portable 24-analyte surface plasmon resonance instruments for rapid, versatile biodetection. Biosens Bioelectron 22(9–10):2268–2275
8. Soelberg SD et al (2009) Surface plasmon resonance detection using antibody-linked magnetic nanoparticles for analyte capture, purification, concentration, and signal amplification. Anal Chem 81(6):2357–2363
9. Kretschm E (1971) Determination of optical constants of metals by excitation of surface plasmons. Z Phys 241(4):313
10. Davies J (1996) Surface analytical techniques for probing biomaterial processes, CRC series in chemistry and physics of surfaces and interfaces. CRC Press, Boca Raton, p 178
11. Homola J (2006) Surface plasmon resonance based sensors, Springer series on chemical sensors and biosensors. Springer, Berlin, p xii, 251
12. Schasfoort RBM, Tudos AJ (2008) Handbook of surface plasmon resonance. RSC Pub, Cambridge
13. Love JC et al (2005) Self-assembled monolayers of thiolates on metals as a form of nanotechnology. Chem Rev 105(4):1103–1169
14. Homola J (2008) Surface plasmon resonance sensors for detection of chemical and biological species. Chem Rev 108(2):462–493
15. Mrksich M, Whitesides GM (1996) Using self-assembled monolayers to understand the interactions of man-made surfaces with proteins and cells. Annu Rev Biophys Biomol Struct 25:55–78
16. Rusmini F, Zhong Z, Feijen J (2007) Protein immobilization strategies for protein biochips. Biomacromolecules 8(6):1775–1789
17. Staros JV, Wright RW, Swingle DM (1986) Enhancement by N-hydroxysulfosuccinimide of water-soluble carbodiimide-mediated coupling reactions. Anal Biochem 156(1):220–222
18. Bain CD, Evall J, Whitesides GM (1989) Formation of monolayers by the coadsorption of thiols on gold: variation in the head group, tail group, and solvent. J Am Chem Soc 111(18): 7155–7164
19. Lofas S, Johnsson B (1990) A novel hydrogel matrix on gold surfaces in surface-plasmon resonance sensors for fast and efficient covalent immobilization of ligands. J Chem Soc Chem Commun 21:1526–1528
20. Leidberg BaL (1993) Principles of biosensing with an extended coupling matrix and surface plasmon resonance. Sens Actuators, B 11:62–73
21. Brown S (1997) Metal-recognition by repeating polypeptides. Nat Biotechnol 15(3):269–272
22. Woodbury RG et al (1998) Construction of biosensors using a gold-binding polypeptide and a miniature integrated surface plasmon resonance sensor. Biosens Bioelectron 13(10): 1117–1126

23. Joung HA et al (2008) High sensitivity detection of 16s rRNA using peptide nucleic acid probes and a surface plasmon resonance biosensor. Anal Chim Acta 630(2): 168–173
24. Mohrle BP, Kumpf M, Gauglitz GN (2005) Determination of affinity constants of locked nucleic acid (LNA) and DNA duplex formation using label free sensor technology. Analyst 130(12):1634–1638
25. Yang N et al (2007) Evaluation of two- and three-dimensional streptavidin binding platforms for surface plasmon resonance spectroscopy studies of DNA hybridization and protein-DNA binding. Biosens Bioelectron 22(11):2700–2706
26. Jung Y et al (2007) Self-directed and self-oriented immobilization of antibody by protein G-DNA conjugate. Anal Chem 79(17):6534–6541
27. Bae YM et al (2005) Study on orientation of immunoglobulin G on protein G layer. Biosens Bioelectron 21(1):103–110
28. Stevens RC et al (2007) Detection of the toxin domoic acid from clam extracts using a portable surface plasmon resonance biosensor. Harmful Algae 6(2):166–174
29. Stevens RC et al (2008) Detection of cortisol in saliva with a flow-filtered, portable surface plasmon resonance biosensor system. Anal Chem 80(17):6747–6751
30. Huber D et al (2009) Effectiveness of natural and synthetic blocking reagents and their application for detecting food allergens in enzyme-linked immunosorbent assays. Anal Bioanal Chem 394(2):539–548
31. Dee S et al (2009) Evidence of long distance airborne transport of porcine reproductive and respiratory syndrome virus and *Mycoplasma hyopneumoniae*. Vet Res 40(4):39
32. Bishop EJ, Mitra S (2007) Measurement of nitrophenols in air samples by impinger sampling and supported liquid membrane micro-extraction. Anal Chim Acta 583(1):10–14
33. Rosen DL (2006) Airborne bacterial endospores detected by use of an impinger containing aqueous terbium chloride. Appl Opt 45(13):3152–3157
34. Fabian P et al (2009) Airborne influenza virus detection with four aerosol samplers using molecular and infectivity assays: considerations for a new infectious virus aerosol sampler. Indoor Air 19(5):433–441
35. Burton NC et al (2005) The effect of filter material on bioaerosol collection of *Bacillus subtilis* spores used as a *Bacillus anthracis* simulant. J Environ Monit 7(5):475–480
36. Plutzer J, Torokne A, Karanis P (2010) Combination of ARAD microfibre filtration and LAMP methodology for simple, rapid and cost-effective detection of human pathogenic *Giardia duodenalis* and *Cryptosporidium* spp. in drinking water. Lett Appl Microbiol 50(1):82–88
37. Breier JA et al (2009) A suspended-particle rosette multi-sampler for discrete biogeochemical sampling in low-particle-density waters. Deep Sea Res I 56(9):1579–1589
38. Edmonds JM et al (2009) Surface sampling of spores in dry-deposition aerosols. Appl Environ Microbiol 75(1):39–44
39. Conroy PJ et al (2009) Antibody production, design and use for biosensor-based applications. Semin Cell Dev Biol 20(1):10–26
40. Zhang Y et al (2007) Studies of metal ion binding by apo-metallothioneins attached onto preformed self-assembled monolayers using a highly sensitive surface plasmon resonance spectrometer. Sensor Actuator B Chem 123(2):784–792
41. Yu JCC, Lai EPC (2005) Interaction of ochratoxin A with molecularly imprinted polypyrrole film on surface plasmon resonance sensor. React Funct Polym 63(3):171–176
42. Joung HA et al (2008) High sensitivity detection of 16s rRNA using peptide nucleic acid probes and a surface plasmon resonance biosensor. Anal Chim Acta 630(2):168–173
43. Wark KL, Hudson PJ (2006) Latest technologies for the enhancement of antibody affinity. Adv Drug Deliv Rev 58(5–6):657–670
44. Nieba L, Krebber A, Pluckthun A (1996) Competition BIAcore for measuring true affinities: large differences from values determined from binding kinetics. Anal Biochem 234(2): 155–165

45. Friguet B et al (1985) Measurements of the true affinity constant in solution of antigen-antibody complexes by enzyme-linked immunosorbent-assay. J Immunol Meth 77(2):305–319
46. Pope ME et al (2009) Anti-peptide antibody screening: selection of high affinity monoclonal reagents by a refined surface plasmon resonance technique. J Immunol Meth 341(1–2):86–96
47. Achi R, Mata L (1997) Oocyst size of *Cryptosporidium* sp. (Apicomplexa: Cryptosporidildae) suggest the presence of *C. parvum* in Costa Rican children. Rev Biol Trop 45(1B):615–618
48. Kubitschek HE (1990) Cell-volume increase in *Escherichia coli* after shifts to richer media. J Bacteriol 172(1):94–101
49. Prasad BVV et al (2001) Structural studies on gastroenteritis viruses. In: Chadwick D, Goode JA (eds) Gastroenteritis viruses, pp 26–46
50. Papageorgiou AC, Tranter HS, Acharya KR (1998) Crystal structure of microbial superantigen staphylococcal enterotoxin B at 1.5 angstrom resolution: implications for superantigen recognition by MHC class II molecules and T-cell receptors. J Mol Biol 277(1):61–79
51. Ban N et al (1994) Crystal-structure of an idiotype antiidiotype Fab complex. Proc Natl Acad Sci USA 91(5):1604–1608
52. Wu CM, Lin LY (2004) Immobilization of metallothionein as a sensitive biosensor chip for the detection of metal ions by surface plasmon resonance. Biosens Bioelectron 20(4):864–871
53. Yu QM et al (2005) Detection of low-molecular-weight domoic acid using surface plasmon resonance sensor. Sens Actuators, B 107(1):193–201
54. Mauriz E et al (2006) Real-time detection of chlorpyrifos at part per trillion levels in ground, surface and drinking water samples by a portable surface plasmon resonance immunosensor. Anal Chim Acta 561(1–2):40–47
55. Kawaguchi T et al (2007) Fabrication of a novel immunosensor using functionalized self-assembled monolayer for trace level detection of TNT by surface plasmon resonance. Talanta 72(2):554–560
56. Naimushin AN et al (2002) Detection of *Staphylococcus aureus* enterotoxin B at femtomolar levels with a miniature integrated two-channel surface plasmon resonance (SPR) sensor. Biosens Bioelectron 17(6–7):573–584
57. Cho HS, Kim TJ (2007) Comparison of surface plasmon resonance imaging and enzyme-linked immunosorbent assay for the detection of antibodies against iridovirus in rock bream (*Oplegnathus fasciatus*). J Vet Diagn Investig 19(4):414–416
58. Hodnik V, Anderluh G (2009) Toxin detection by surface plasmon resonance. Sensors 9(3):1339–1354
59. Hennion MC, Barcelo D (1998) Strengths and limitations of immunoassays for effective and efficient use for pesticide analysis in water samples: a review. Anal Chim Acta 362(1):3–34
60. Boozer C et al (2006) Looking towards label-free biomolecular interaction analysis in a high-throughput format: a review of new surface plasmon resonance technologies. Curr Opin Biotechnol 17(4):400–405
61. Chinowsky TM et al (2007) Compact, high performance surface plasmon resonance imaging system. Biosens Bioelectron 22(9–10):2208–2215
62. Alvarez-Rueda N et al (2007) Generation of llama single-domain antibodies against methotrexate, a prototypical hapten. Mol Immunol 44(7):1680–1690

Chapter 7
CERCLA, Sustainability and Public and Environmental Health

Robert Davis Jewett and Michael W. Wascom

Glossary

One of the most useful ways to begin a study of the Comprehensive Environmental Response, Compensation, and Liability Act (CERCLA) is to engage the intriguing lexicon the act provides. As is demonstrated by the definitions below, terms are often defined under CERCLA in ways that are remarkably broad, granting a great deal of elbow room and flexibility for the US Environmental Protection Agency (EPA) to go about the work of protecting the public from hazardous materials, no matter when or how they have come to present a threat.

Environment — Under CERCLA, the term "environment" is defined broadly in a manner that is distinctly multimedia. An environment is anywhere or anything that can become contaminated with a hazardous substance, including all surface and ground waters, land surfaces or subsurface strata, or the ambient atmosphere. Any release of a hazardous material into any sort of environment is potentially covered under CERCLA.

This chapter was originally published as part of the Encyclopedia of Sustainability Science and Technology edited by Robert A. Meyers. DOI:10.1007/978-1-4419-0851-3

R.D. Jewett (✉) • M.W. Wascom
Department of Environmental Sciences, Louisiana State University, Baton Rouge, LA 70803, USA
e-mail: robertd42@gmail.com; coewas@lsu.edu

E.A. Laws (ed.), *Environmental Toxicology: Selected Entries from the Encyclopedia of Sustainability Science and Technology*, DOI 10.1007/978-1-4614-5764-0_7,

Facility	The term "facility" usually brings to mind an image of a building or series of buildings, but CERCLA defines the term more broadly. A "facility" under CERLCA includes any site or area where a hazardous material has been deposited, placed, stored, disposed of, or has somehow come to be located. It is just as likely to refer to a faulty storage tank or a leaking pipe as to a building or similar structure.
Hazardous substance	In defining hazardous substances, CERCLA makes use of all other definitions and lists of such substances in previous US environmental legislation. However, there is one important exception: petroleum and petroleum products, natural gas and synthetic gas used for fuel. This is the so-called Petroleum Exception. Interestingly, the definition of a hazardous substance does not refer to any particular or minimum amount of a material that must be present for it to become hazardous. In theory, there is no lower limit of materials that must be present or must have been released before a remedial action is authorized, or liability is imposed.
Contaminant or pollutant	CERCLA also authorizes cleanups of categories of substances that are broader than materials judged to be hazardous according to the standards of earlier federal regulations. Contaminants or pollutants, under CERCLA, are materials that may cause adverse health effects. While CERCLA allows the EPA to respond to sites polluted by these lesser threats, it authorizes recovery of cleanup costs from responsible parties only in cases of contamination by a documented hazardous substance.
Release	Again, a release of a hazardous substance is broadly defined. CERCLA authorizes response to any sort of a release, including all manner of spills, leaks, and discharges into the air, as well as leaching, dumping, injecting, emitting, or disposing. Certain exceptions are made, generally for releases that are covered under other federal laws. As an example, motor vehicle emissions are excluded from CERCLA but are covered by federal emission standards and other regulations.
Removal	Removal is also broadly defined, creating more potential opportunities to assess liability. Removal may refer to certain actions directly involved in

	cleanup of a site, such as removing the hazardous material from the area, or transporting waste to an authorized disposal facility. However, it may also refer to actions such as monitoring, assessing, and evaluating releases or potential releases. This allows the EPA to take preventive action to prevent a potential release.
Remedy or remedial action	As in the case of removal, the construction of the terms "remedy" or "remedial action" in CERCLA allows for response to both actual or potential or threatened releases. The terms are also broadly construed to allow for a wide range of response actions, including public health or environmental monitoring, transportation of hazardous materials, implementation of an alternative water supply, and even removal or relocation of communities to safe locations.
Potentially responsible party or PRP	Remedial actions that occur through CERCLA are often conducted not by the EPA directly, but by parties that are judged to be in some way accounle for the release. These parties, known as Potentially Responsible Parties, or PRPs, fall into four categories: owners, operators, arrangers or generators, and transporters. For more information on these categories and their responsibilities and liabilities under CERCLA, see section "Liability Scheme" below.
CERCLA	Is also a distinctly creative act, which has also introduced new entities and terms into the knowledge base of environmentally conscious people, those affected by CERCLA actions and even CERCLA's critics.
The Superfund	The Superfund has become a common name for all of CERCLA, but it also refers to the Hazardous Substance Superfund, established to provide financial resources to clean up hazardous sites. The Superfund was established by taxes imposed upon the chemical and petroleum industries and an environmental tax on corporations. It has since been supported by allocations from the federal general fund and through additional legislation, such as the American Recovery and Reinvestment Act.
The National Priority List or NPL	The NPL was added to an existing resource, the National Contingency Plan (NCP), which was brought into being under the Clean Water Act to

address the containment and removal of spilled oil and other hazardous substances. Under CERCLA, the NCP was revised to include direct provisions for responding to releases of hazardous substances, pollutants, and contaminants, including those released on land or in the air. CERCLA also revised the NCP to include a list of sites that were considered to be top priorities for response. This is the National Priority List, or NPL. Criteria for placement on the NPL are discussed in section "NCP, NPL, and Site Procedures" below.

Finally, CERLA brought into being a new federal agency

The Agency for Toxic Substances and Disease Registry or ATSDR. ATSDR is charged under CERCLA with assessing the presence and nature of health hazards at specific Superfund sites, as well as working to prevent or reduce further exposure and the illnesses that result from exposures. It compiles and maintains a prioritized list of hazardous materials at Superfund sites, the CERCLA Priority List of Hazardous Substances. ATSDR also has a range of other duties, including health surveillance and the development and distribution of information on toxic substances. For more information, see section "ATSDR and the CERCLA List" below.

Definition of the Subject

The Comprehensive Environmental Responses, Compensation, and Liability Act of 1980, or CERCLA, also known as the Superfund Act, is a landmark piece of environmental legislation [1]. Its notably bold design, discussed in detail below, fits the daunting work it was established to perform, the remediation of the worst and most dangerous releases of hazardous materials in the history of the USA. CERCLA also plays a role in the prevention of threatened releases. While it is subject to criticism and may not function perfectly, or even quite as intended, CERCLA remains an essential safeguard of environmental and public health in the USA.

Introduction to CERCLA

Sustainability may be defined as the capacity to endure. The achievement of sustainability suggests a relationship to time, in which the past is never so bad as to doom the present or erase the future. CERCLA is unique among the central pieces of environmental legislation in the USA in that it is largely oriented to the past, and ensuring that the legacy of environmental destruction that accompanied the growth of industrialism in the USA does not continue to jeopardize the future of the nation, its people, or the environment. As a practical measure, CERCLA casts environmental disasters and hazardous releases as problems that can be solved. These problems, however, are often of a scope and gravity that defy inexpensive or easily realized solutions. This is why CERCLA operates not only through the use of a public "Superfund" but also through partnerships, sometimes coerced, with other parties, who are held to be responsible for hazardous releases, even those that occurred many years ago.

Early in 2010, approximately 1,280 sites were listed on the National Priority List, either in the midst of a CERCLA remediation or awaiting one. An additional 340 sites had been delisted in the history of CERCLA since 1980, with more than 60 sites under consideration to be added to the list [2]. However, it was only a small group of well-publicized releases of hazardous substances that originally inspired CERCLA.

CERLCA was enacted by Congress in response to environmental and public health disasters, including, inter alia: (a) the environmental contamination at a site in Kentucky that came to be known as "The Valley of the Drums"; a site of dioxin contamination in Times Beach, Missouri, that eventually caused the abandonment of the town; and, most notably, (b) the Love Canal contamination near Niagara Falls, New York.

The Love Canal incident is one of the more famous environmental disasters to occur in the USA. In the summer of 1977, 200 families were evacuated from the area of Love Canal when waste chemicals began to seep into the basements of homes in the community. In 1980, the US government filed suit, under the Resource Conservation and Recovery Act and the Safe Drinking Water Act, against the company that had dumped tens of thousands of tons of hazardous chemicals at the Love Canal site and then abandoned it, years before.

CERCLA was designed to address cases such as Love Canal, in which the past intervenes into the present, clouding the future of an area and its people. As noted above, CERCLA is, in large measure, retroactive in nature, allowing the government to go back in time to find responsible parties for current environmental contaminations. In 1980, during the closing days of the 96th Congress, the original CERCLA legislation passed. It was signed by President Jimmy Carter as one of his last official acts in office.

This entry will review CERCLA, including its initial provisions and later revisions. It will focus on issues that include liability under CERCLA, defenses against liability, site selection including technical review of sites, and the CERCLA Priority List of Hazardous Materials under ATSDR.

Understanding CERCLA

Original CERCLA and Basic Elements

The original CERCLA legislation included four basic elements:

1. A system for gathering and analyzing information concerning potential or actual releases of hazardous materials, pollutants, or contaminants
2. Federal authority to respond to control and clean up releases of hazardous materials, pollutants, or contaminants
3. The Hazardous Substance Response Trust Fund, or Superfund, to underwrite response and cleanup activities
4. A strict liability scheme for those judged to be responsible for releases of hazardous substances

In the matter of notification of a release, CERCLA functions in concert with other regulations, including the Emergency Planning and Community Right to Know Act, in requiring any person in charge of a facility (as uniquely defined under CERCLA as noted above) who has knowledge of a release of a hazardous substance to notify the National Response Center, a federal clearinghouse for hazard and public and environmental health information, as well as other appropriate state, federal, or local agencies.

CERCLA permits the EPA to request from any individual or party information deemed relevant to a release of hazardous materials. As with many federal authorities under CERCLA, this power is broadly construed. As an example, the EPA can obtain financial information to discover a party's ability to pay for the costs of a cleanup, as well as information that more directly relates to the handling of hazardous materials or the cause or scope of the release [3].

The second basic authority under CERCLA is the authority to clean up a site at which a release has occurred or is occurring. CERCLA authorizes two different types of response actions to releases or threatened releases, removal actions, and remedial actions:

- Removal actions are short-term response actions for cases in which releases or threatened releases require prompt response. These actions may be emergency actions, time-critical actions, or non-time-critical actions. Removal responses are generally used to address localized risks; for instance, the removal of drums containing hazardous materials that have been found on a site may be necessary to reduce or eliminate a contamination and mitigate risks posed to public health. Removal may also prevent contamination from compromising other elements of the environment; for example, topsoil removal may prevent further contamination of the water supply.
- Remedial actions are generally longer-term actions, designed to permanently reduce the risks associated with a release or threatened release. While these actions may have some urgency, they may constitute a long-range project rather

than a time-intensive emergency response. For instance, a remedial action may involve taking actions to contain a hazardous substance or neutralize a toxin. Only sites listed on the National Priority List (NPL) are eligible for remedial action under CERCLA.

On some occasions, the type of response selected under CERCLA is subject to controversy: As one example, see the case of the Agriculture Street Landfill in New Orleans as discussed below, where a partial removal action may not have been sufficient to remediate a complex site.

CERCLA also created the Superfund, and is sometimes known as "The Superfund Act." This fund supports response, removal, and remediation activities. However, it is not the only source of funds for these activities. Despite its nickname, the Superfund lacks sufficient funds to clean up all the sites on the National Priority List, or even those that are being cleaned up at any particular time. Instead, the preferred federal strategy is to conserve Superfund resources and compel or persuade other parties to clean up a site themselves. These additional parties, known as "Potentially Responsible Parties" or PRPs, are held responsible under the CERCLA liability scheme. These parties may also sue or reach agreements with other PRPs concerning which parties are responsible for the cleanup and in what proportion. See section "Liability Scheme" below.

SARA Revisions of CERCLA

In 1986, CERCLA was revised by the passage of the Superfund Amendments and Reauthorization Act, or SARA [4]. SARA added standards for an acceptable cleanup, provisions concerning the legal settlement process, and rights for state and citizens including some of those listed under Citizen Roles above.

Before SARA, CERCLA did not include any particular standards regarding cleanup operations. SARA provides guidelines on matters such as the type of cleanup action, compliance with the National Contingency Plan, and cost-effectiveness. SARA also provides criteria for assessing proposed remedies. These are discussed in section "NCP, NPL, and Site Procedures" below.

SARA also engages the issue of how to promote voluntary settlements EPA claims against PRPs. SARA provides substantial guidance on issues such as contributions among PRPs, how to address shares of unknown PRPs, and what to do about PRPs who are responsible only for very small quantities of hazardous material. SARA also added the "Innocent Landowner" defense for potential PRPs, which is discussed under Liability Scheme below.

CERCLA was also revised by the Small Business Liability Relief and Brownfields Revitalization Act of 2002 [5]. Some of the changes made by this act are noted throughout the entry below.

Citizen Roles

SARA also established a series of important roles for citizens to play in legal actions under CERCLA. First, citizens or citizens groups affected by a Superfund site may apply for a Technical Assistance Grant. These grants are used by citizens to hire their own independent technical consultants to advise them about health and environmental issues and monitor the work of the EPA and others regarding the site.

Citizens may sue any party, except the EPA, that is alleged not to be in compliance with any order, standard, condition, or requirement that is in force under CERCLA. The EPA may choose to intervene in the suit and prosecute the alleged offender. In such a case, the citizen suit is barred.

Citizens may also petition for a preliminary assessment at a site if that site is not scheduled for response actions. This method may be useful in making a certain site a higher priority for a remediation [3]. Another central right reserved for citizens is the right to comment. For instance, citizens can comment on any settlement agreements or consent decrees prior to judicial approval.

In addition, citizens have the right to know about the chemical and substance hazards they may face. Specifically, this requirement includes emergency notification of chemical releases; notification of chemical use, storage, and production activities in a community; emission reporting requirements; state and local emergency planning, and the development of emergency response plans.

NCP, NPL, and Site Procedures

As mentioned in the Glossary above, CERCLA substantially revised the National Oil and Hazardous Substances Pollution Contingency Plan (NCP). The NCP provides guidelines and procedures governing the response to releases or threatened releases of hazardous substances, pollutants, or contaminants. Under CERCLA, NCP criteria for the remediation of a hazardous release include a preference for permanent solutions, as well as cost-effectiveness and a practical ability to implement the solution.

Under CERCLA, the NCP was also revised to include a National Priority List (NPL), which serves as an information and management tool to help the EPA prioritize sites for cleanup. Inclusion of a site on the NPL does not in itself assign liability or require potentially liable parties to initiate action to clean up the site. The NPL serves primarily informational purposes, identifying for the states and the public those sites, facilities, or releases that appear to warrant remedial actions.

Upon notification of a potentially hazardous release, or when a potentially hazardous site or release is otherwise discovered, the site is listed in the Comprehensive Environmental Response and Liability Information System (CERCLIS) for subsequent evaluation. EPA then assembles information on the site and conducts a Preliminary Assessment/Site Inspection (PA/SI) to determine the scope of any contamination or potential contamination. This process involves a review of records regarding the site and any hazardous substances associated with it, interviews of potentially knowledgeable parties, visual inspections by qualified personnel, and limited field sampling. The PA recommends further investigation, if warranted.

If the site is thought to present a threat to public or environmental health or safety, additional investigations are performed concerning the hazardous materials at the site, their potential for contacting humans, and possible migration pathways for spreading contamination. Information gathered in the investigative review process is then subject to a hazard analysis, using criteria that include the toxicity of hazardous substances, the location of possible receptors, and threats to the human food chain or the watershed or atmosphere. This analysis culminates in the development of a score on the Hazard Ranking System (HRS). A HRS score above 28.5 results in placement of the site on the NPL.

Sites that score at or above the threshold are then subject to a Remedial Investigation (RI) and Feasibility Study (FS). The RI includes an extensive sampling program and risk assessment in order to define the extent and degree of the contamination and the associated risks. It is designed to be a thorough analysis but can take years to complete. Typically, reports are compiled that describe the site's geology and hydrogeology, the sources of contamination, the type and degree of mobility of the contaminants, and all aspects of the threat to the public health or the environment. Any part of these data is potentially useful in developing and studying possible methods of remediation.

The next step is the FS, which is used to develop and evaluate various remediation alternatives. The FS details several remedial methods and analyzes each in light of the data from the RI. The work is often broken down into phased components, called Operable Units, to separate the remediation process, and often the physical or geographic area of the contamination, into more manageable units [6].

The preferred alternative identified by the FS is presented in a proposed plan for public review and comment. If the EPA stays with the proposed solution after this comment period, the selected alternative is approved in a Record of Decision (ROD). The site then enters into a Remedial Design phase and then the Remedial Action phase. Many sites require Long-Term Monitoring and 5-year reviews once the Remedial Action has been completed. If a site is not considered sufficiently hazardous for inclusion on the NPL, the EPA may still order a removal action to address an imminent threat.

EPA's nationwide strategic plan for addressing hazardous sites is found in another document, the Superfund Comprehensive Accomplishment Plan (SCAP). This document provides a list of those activities that are expected to occur at each site during a fiscal quarter. Information often appears in the SCAP before it is announced to the public [6].

Remediation Priorities and ARARs

As the EPA works toward a feasible solution for a site, various factors may become important. One is the level or degree of cleanup necessary. The central issue here is often what level of risk or remaining contamination is acceptable. As the EPA generally works with PRPs with a financial stake in the cleanup process, levels of remediation often become controversial. See the section on the "Agriculture Street Landfill site" for an example of a case where no real agreement was reached between affected citizens, the EPA and the PRPs, resulting in a lengthy court battle.

Section 121 of CERCLA is especially significant with respect to decisions about methods of remediation as it sets forth requirements for cleanup standards [7]. Remedial actions under CERCLA are consistent with the precepts of the NCP when possible. This section states that actions should be cost-effective. The guidelines also state a preference for permanent solutions that reduce the volume, toxicity, or mobility of the contamination. Less permanent solutions, such as the construction of barriers, are not favored.

For hazardous materials left on site, remedies under CERCLA are required to achieve all Applicable or Relevant and Appropriate Recommendations (ARARs). The level or degree of the hazard must remain less than any standard, requirement, criterion, or limit under any federal environmental law or any state or facility siting law that is more stringent than the federal guidelines. This means that the site must achieve the cleanup level (or safety level) set by the most stringent laws that are applicable. The NCP includes criteria for determining whether a law or standard applies. Laws or standards may not apply, for instance, if they target a certain medium, such as water, and the hazardous material at the site is found in a different medium, such as soil. Other differences include (a) the activities regulated by the regulation versus those undertaken at the site and (b) the type of facility considered in the regulation versus that at which the proposed remedial action is to take place.

Once determined to apply, an ARAR generally must be met unless it is waived. CERCLA does provide a few circumstances in which a remedy may be selected that does not meet an ARAR, for instance, if the remedy is only part of a larger solution that will meet the ARAR or if meeting the ARAR would lead to greater risks to human health or the environment than would alternative actions [6].

ATSDR and the CERCLA List

Another important aspect of site selection and development of remedial responses is the CERCLA Priority List of Hazardous Substances maintained by the Agency for Toxic Substances (ATSDR). This list established, in order of priority, the substances most commonly found at facilities on the NPL and which are determined to pose the most significant potential threat to human health due to their known or suspected toxicity and potential for human exposure. The list is revised and published every 2 years, with a yearly review and revision to reflect additional information on hazardous substances.

This priority list is not a list of "most toxic" substances, but rather a prioritization of substances based on their frequency, toxicity, and potential for human exposure at NPL sites. This priority list and the order of the materials on it are based on an algorithm that relies on three components: frequency of occurrence at NPL sites, toxicity, and potential for human exposure to the substances found at NPL sites. Some substances included on the list are low in toxicity but often appear at NPL sites. The items high on the list are subjects of toxicological profiles prepared by ATSDR [8].

Liability Scheme

The liability scheme in force under CERCLA is novel and contradicts even some conventional canons of common law. For instance, a party cannot avoid liability under CERCLA simply by entering into a contract with another party, such as the new owner of a site, that transfers liability or indemnity. CERCLA supersedes such agreements.

CERCLA delineates four classes of "Potentially Responsible Parties," or PRPs, any of which may be liable for the cleanup of a Superfund site. These four classes are:

- The current owners or operators of the facility
- Owner or operators of the facility at the time of disposal of any hazardous substance
- Any person who arranges for the disposal, treatment, or transport of a hazardous substance at the facility
- Any person who transported a hazardous substance, pollutant, or contaminant to a facility, or accepted the material for transport, if that person selected the facility

Within these categories, the liability is broad. As an example, both the current owner and the owner at the time of contamination may be held liable. Does this mean that an owner who knows nothing about the dumping of materials years before on a property the new owner purchased just last year may be held liable for the cleanup of the material? In some cases, the answer is "yes," although such an owner is not entirely without defenses. Owners have a better opportunity to avoid liability if the release did not occur during the time they owned the property, and if they performed due diligence in purchasing the property, handled any hazardous materials discovered with care, and cooperated fully with EPA inquiries. See section "Defenses" below.

The operator is generally defined as the party who controls the operations at a site or facility. Parties that hold a security stake in a facility or site but are not technically its owners may be considered operators if they make any control decisions about a business or property.

Transportation companies that move materials from place to place may or may not be held liable under CERCLA. Courts have ruled that transporters who select the site where the materials are brought (generally the Superfund site) are liable, while those simply moving materials to a site picked by another party, such as the generator, are not liable. However, in actual practice, generators have often hired transporters simply to remove the materials from the site, without knowing where the materials would go. In these cases, the transporter is liable for selecting the disposal site, and the generator is also liable under the category of "Arranger."

Arrangers, often referred to as "generators," are the largest category of parties held liable under CERCLA. Arrangers contract for the removal, disposal, or treatment of hazardous materials, which may or may not be owned or produced by the arranger. This category has somewhat fuzzy boundaries – for instance, what if the arranger sells the material to another party for a legitimate business reason, such as for use in an industrial process? Is this "arranger" liable when the material later is dumped at an unmarked site? However, what if this "sale" was simply a ruse to cover for an illegal dumping operation? Such questions are subject to careful investigation and often litigation.

Defenses

Defenses against liability under CERCLA are limited. CERCLA is considered a "strict liability" law, meaning that mitigating factors such as negligence or an absence of negligence are not always considered in reaching determinations of liability. Even the defenses provided for potential PRPs under the law may not be viable in actuality. As one example, the law makes a provision for an Act of God defense; in actual practice, however, this remedy has not been upheld. Courts have ruled that all the usual "acts of God," hurricanes, tornadoes, earthquakes, etc., are foreseeable, and the resulting release of hazardous materials is avoidable with the proper precautions. Similarly, an act of war provides a defense only when the act is demonstrated to be the sole cause of the release [9].

A more viable defense that is frequently raised is the "Omission of a Third Party" defense whereby a defendant attempts to escape liability by claiming that the release was solely the act of a third party, for which the PRP is not responsible. However, for this defense to be applicable, the PRP may not have an articulated relationship with the third party; for instance, the third party may not be a rogue employee of the PRP company, or a party working under a contract with the defendant.

Another related defense is referred to as the "Held Only for a Security Interest" defense, or the "Secured Creditor Exception," which may protect parties that hold a stake in a property only for a security interest, without participating in the management of the property. Lenders are among those who may be eligible for this defense.

The revisions of CERCLA under SARA codified an emerging defense related to the Omission of a Third Party defense, the "Innocent Landowner" defense. To qualify for this defense in most cases, the landowner must establish a justifiable ignorance of the contamination. For instance, it is not enough simply to have purchased the property and then discovered only afterward that the site was contaminated. Instead, owners are required to show that they did due diligence, including undertaking all appropriate inquiry into the conditions of the property before purchase. Landowners may also rely on an "innocent landowner" defense in cases in which a property is acquired through inheritance or bequest.

Under the Small Business Liability Relief and Brownfields Revitalization Act, another defense for another class of "innocent landowners" is designated. To qualify, a number of provisions have to be demonstrated, including:

- All disposal or release of hazardous materials occurred before the landowner purchased the property
- All appropriate inquires into the previous ownership and use of the facility were made
- Appropriate care was taken regarding the hazardous materials, including appropriate action to stop the release

The landowner is also required to cooperate fully with the EPA's investigation into the property, the release of hazardous materials, and any response actions.

The Small Business Liability Relief and Brownfields Revitalization Act also provided another category of defense for PRPs, called the "De Micromis Party Exception." This defense is for parties who have generated, transported, or arranged for transport of only small amounts of hazardous material. The exemption applies when amounts are below 110 gal of liquid materials or 200 lb of solid materials. However, the exemption only applies when these materials do not figure heavily in the cost of remediation or restoration of the site. This exemption also protects a "de micromis" party from other PRPs seeking contributions toward the expense of remedies: A PRP suing another potential PRP must show that the "de micromis" defense does not apply [3].

Enforcement Tools

The EPA has several powerful enforcement tools to direct toward PRPs. Two of these, civil judicial injunctive actions and unilateral administrative orders, fall under the "imminent hazard" section of CERCLA, which authorizes the EPA to act if it determines that a release or threatened release poses an "imminent hazard" to public health [10]. An administrative order, in particular, can be compelling, as courts have held that PRPs have no right to a hearing or review before the order is issued. Willful violation of a unilateral EPA order makes the PRP subject to a fine of up to $25,000/day [3].

Litigation is another potential remedy. Under CERCLA, monies that can be recovered through litigation include costs incurred by state or federal governments, costs incurred by another party, and damages to natural resources including the expense of damage assessment. Under these broad categories, the EPA has recovered expenses for investigation, monitoring, and testing of a site; planning a response; staff costs; attorney fees; as well as expenses related more directly to remedial actions at the site [3].

The EPA prefers not to take remedial action on its own and then attempt to recoup the costs of its actions from PRPs. The preferred response is for PRPs themselves to conduct the remediation, under the monitoring and supervision of the EPA or relevant state agencies. Further, the EPA prefers that these actions be taken "voluntarily," which generally means under threat of a potential administrative order or another action. The EPA has several advantages that make its consultations with PRPs likely to be convincing. If the PRP complies with the EPA, it may escape the fines imposed under administrative orders, the damages granted by courts, and the legal fees entailed in court battles. It may also reduce any negative publicity or quell the complaints of angry citizens. Rather than battle the EPA, a route that many PRPs find more fruitful is to engage in discussions with, or legal actions against, other PRPs involved with the same release or facility to attempt to spread fiscal responsibility for the cleanup.

States and Local Governments Under CERCLA

States are responsible for the future management of all removal and remedial actions, once those actions are judged to be complete. States must also provide at least one federally licensed hazardous waste disposal facility to receive wastes removed from the Superfund sites within the state. The facilities may also be privately owned or operated and contracted to the state.

Under SARA, states were given additional roles in federal cleanup actions. States have the opportunity to review and comment on many federal actions involved in a remediation, including the Remedial Investigation and Feasibility Study, the proposed remedial action, the engineering design of the action, technical data and reports related to the remedy, and any decisions to waive standards.

States and local governments may also be held liable under CERCLA if they have acted in the roles of the PRP categories. However, governmental agencies acting purely in a regulatory manner are generally exempt from liability.

Case Study: Agriculture Street Landfill Project

The Agriculture Street Landfill in the Ninth Ward of New Orleans was operated as a landfill for approximately 50 years, ending in 1958. In 1965, it was briefly reopened to receive a massive amount of debris from Hurricane Betsy. The landfill was locally referred to as Dante's Inferno because its contents often caught fire, and the area was frequently smoldering. It was a toxic environment to be sure.

In 1967, the city of New Orleans and the Housing Authority of New Orleans decided to develop the area as a community for low-income residents. Between 1969 and 1971, the Press Park townhomes were developed. Later, Gordon Plaza, a development of single-family homes was constructed. Several playgrounds were also built in the area. A tract of land was set aside for an elementary school, and, in 1986, Mouton Elementary School opened for its first group of students. Residents and employees working in the area often complained of unusual smells and illnesses.

The record of environmental testing at the site has been contested. According to the area citizens' group and the law firm that eventually filed a class action suit on behalf of area citizens, the results have often been kept from residents. The city of New Orleans tested the site in the late 1970s in anticipation of the development of Gordon Plaza. After the testing, the city required the builders to add topsoil to the building site, but the results of the tests themselves have never been made public. Before the construction of Mouton School, another round of testing revealed the presence of lead, mercury, and arsenic. The Orleans Parish School Board received a recommendation that topsoil be brought in and a clay barrier used to contain the contamination. However, no barrier was established when the school was built. Instead, during construction area residents were subjected to dusts that may have included hazardous materials [11].

In 1986, the EPA tested the area, but area residents said that they were never shown the results of the test. In 1993, the EPA tested again . This time, they reported to area residents that the former Agriculture Street Landfill was contaminated with approximately 150 different hazardous substances, including almost 50 that were known to cause cancer. In 1994, the EPA declared the area a potential Superfund site and placed it on the National Priority List.

In 1994, area residents formed Concerned Citizens of Agriculture Street Landfill. Utilizing an important provision of CERCLA, the group requested a technical assistant grant from the EPA, to fund an independent environmental consultant. The group lobbied for the closing of the Mouton school, and the school was soon closed.

In the mid-1990s, the city government broke ranks with the federal authorities; led by Mayor Marc Morial, the city joined residents in arguing for federal support for relocating area residents. The EPA believed that the site could be remediated and from 1998 to 2001 took action, including replacing topsoil from open areas, but without excavating structures or streets. Residents were given certificates stating that their properties had been partially remediated, and were also given written restrictions on what they could do with their properties [11].

A lawsuit filed on behalf of the residents was granted class action status in 1999. The case went to trail in 2004, with the residents prevailing. The landfill area was ruled to be unreasonably dangerous under state law. The residents were awarded damages and market value of their properties. Employees and students also received awards. The judge in the case used discretionary powers to increase the monies received by 100%. Upon appeal, judgments for emotional distress were halved, but the rest of the decision stood. As of 2010, residents were applying to receive their claims, some 40 years after the regrettable residential development of the former landfill began [11].

As for the landfill site, it is believed that Hurricane Katrina removed much of the new topsoil deposited in remedial actions and may have dispersed contaminants throughout the area. It may be that the EPA's efforts in this area, which many believe were not of sufficient boldness to respond effectively to the severity of the contamination, have come largely to naught [12].

Criticisms of CERCLA

CERCLA cuts a board swath through American society. It is an ambitious federal regulation that was designed to engage the enormous and complex problem of releases of hazardous materials. It also imposes new and in some ways novel legal requirements on those who may be deemed in any manner responsible for a release or threatened release, even in cases in which they would not be held responsible under conventional canons of common law. It is little wonder that CERLCA is controversial and comes under criticism from a range of viewpoints and for myriad reasons.

Business owners, industry and corporate interests have complained that the strict liabilities imposed by CERLCA inhibit economic growth, not only by tying up money in expensive remediations, but also by creating a chilling effect that prevents the purchase for redevelopment of any sites that could prove to be contaminated. One of the main charges laid against CERCLA from this perspective is that it simply is not "fair." It is unfair, for instance, for owners or operators who exercised all due caution in handling a material to be held liable for a release that occurred due to the unsafe actions of another party. It is also unfair to expect today's generation of business owners or shareholders to remediate a site that was contaminated years ago.

A particular sticking point has been the extension of liability to parties that may have sent only a small quantity of materials, comprising a hazard that is limited in scope and degree, to a disposal facility where a hazardous release eventually occurred. For instance, cases that involved pizzerias, donut shops, fitness facilities, and a local Elks Club have been publicized [13]. The revisions of defense provisions of CERCLA under the Small Business Liability Relief and Brownfields Revitalization Act were directed at ensuring that small business and other parties who are not large owners, operators, generators, handlers, or transporters of hazardous materials may escape liability.

Simultaneously, environmentalists and citizens' groups have criticized CERCLA on other grounds. These parties have argued that CERCLA remains tied to corporate interests, that the EPA is hesitant or unwilling to take on controversial projects, and that its preferred remedial actions often do not go far enough to solve the problem. The slow progress of cleanup projects is another common complaint [6].

As a large public project that uses taxpayer money, CERCLA has also drawn criticism from taxpayers and watchdog groups that have raised objections to what is perceived as inefficient use of public funds. However, other critics and public officials argue that the real problem is that there is simply not enough money in the system to perform the ambitious work CERCLA calls for, especially since the industry tax that originally funded the Superfund expired in 1995 and has not been renewed.

In the absence of public funding, private funding for remedies under CERCLA is all the more essential. However, businesses have also found a way to circumvent the law: declaring bankruptcy to avoid legal or financial responsibility [14]. The use of the legal system to settle questions of CERCLA liability, whether between PRPs and the EPA or among different PRPs, has also been criticized as a drain on resources [15].

Critics from a variety of perspectives have also argued over issues involving the intersection of science and politics, including methods for assessing and rating sites and placing certain projects of the National Priority List. Some have claimed that the risks of some sites are greatly exaggerated by the EPA, while the risks posed by other sites may be overlooked. Questioning the objective, scientific basis for placement on the NPL, some critics have said that there is no evidence that those projects included on the NPL are more hazardous than those that are not. Similarly, some critics have claimed that due to improper risk assessments, the EPA has performed unnecessary actions on sites, beyond those needed to control the hazards [15]. In the matter of the Agriculture Street Landfill, however, the opposite claim borne out by the courts is seen: that the EPA's actions did not go nearly far enough to ensure public health.

Perhaps one of the most resonant criticisms of CERCLA, offered over the years by people from different political persuasions and economic or social interests, is that its provisions make the revitalization or rehabilitation of older, often abandoned, industrial sites much less attractive. Not only have potential developers or owners been reluctant to take control of such properties for fear of liability, lenders are also reluctant to make loans for the purchase or redevelopment of the property. Provisions of the Small Business Liability Relief and Brownfields Revitalization Act address this issue.

Future Directions

While criticisms of CERCLA are many, they have not yet reached a pitch that would put the future of CERCLA into radical question. Americans are too concerned with issues of sustainability, public health, and the condition of the environment to allow hazardous sites to go without remediation. Alternative schemes have been proposed – for instance, market-based systems that would allow for the remediation and redevelopment of sites by private parties under market conditions [15]. However, such alternatives have not so far been seriously engaged in the political system, suggesting that at this point, Americans want governmental agencies to have a strong hand in remediation efforts, as well as the prevention of threatened hazardous releases.

The most pressing issues as CERCLA heads into the future appear to be funding cleanups and maximizing the effectiveness of funds spent. Historically, approximately 70% of cleanup activities under CERCLA have been paid for by PRPs. However, as parties find avoidance methods, whether through legal exceptions or by declaring bankruptcy, the federal share may increase. Originally, federal monies were provided through taxes on industry, but these have since expired. The Superfund is now increasingly dependent on general fund appropriations. The Superfund did receive a boost in funding under the American Recovery and Reinvestment Act of 2009, the federal stimulus meant to alleviate a recession. This act provided $600 million that was directed toward 50 Superfund sites, approximately doubling the federal investment in these cleanup operations. However, it is not clear how or if this level of funding can be sustained when the Recovery Act expires [16]. One hopeful sign, however, may be a decrease in litigation and resultant expenses [17]. President Obama has also proposed reinstating the corporate taxes that originally funded the Superfund [18].

Even as commitment to CERCLA wavers, at least some degree in the USA, the act is seen worldwide, in some instances, as an intriguing method for funding other environmental and sustainability projects. As one example, the proposed International Climate Change Fund, an alternative to the climate agreement set by the Kyoto Protocol, would use a trust fund similar to the Superfund to pay for the development and purchase of "clean" or low-carbon technologies for developing countries. All countries would be required to pay into the fund, with the payment based on their own carbon emissions, including greenhouse gases and "flow" emissions [19].

CERCLA is still a relatively young law, especially given the scale of the work it engages in the world. It is also novel, in purpose, orientation, and form. Even after 30 years, one may still be in the mode of adjusting to it and striving to understand if it is the right means for the job it set out to do. Thus, revisions of CERCLA to date, such as SARA and the Small Business Liability Relief and Brownsfield Revitalization Act, have performed formative functions, reshaping the law to be more effective and fair, rather than supplanting CERCLA altogether. CERCLA was boldly constructed for an enormous undertaking. Whether it can complete this work is unknown and under dispute. However, it seems fair to say that lesser or more conventional legislation would be even less likely to succeed.

Bibliography

Primary Literature

1. Comprehensive Environmental Response, Compensation, and Liability Act of 1980. United States Public Law, 96–510
2. National Priority List Site Totals by Status and Milestone. EPA. Retrieved 22 Mar 2010
3. Ferrey S (2007) Environmental law, 4th edn. Wolters Kluwer, Austin, pp 362–382
4. Superfund Amendments and Reauthorization Act. United States Public Law 99–499
5. Small Business Liability Relief and Brownfields Revitalization Act. United States Public Law 107–118
6. Sullivan TFP (ed) (2009) Environmental law handbook, 20th edn. Government Institutes, Lanham, pp 510–528
7. Cleanup standards. 42 United States Code § 9621
8. ADSTR http://www.atsdr.cdc.gov/cercla/. Retrieved 20 Mar 2010
9. Switzer CS, Gray P (2002) CERCLA, 2nd edn, Basic practice series. American Bar Association, Chicago, pp 42–43
10. 42 United States Code § 9606(a)
11. Bagneris Law Firm. http://www.bagnerislawfirm.com/landfill_litigation_history. Retrieved 19 Mar 2010
12. Knickerbocker B (2005) Katrina Lays Bare Superfund Woes. Christian Science Monitor, 15 Sept 2005
13. Steinzor R (1998) The legislation of unintended consequences. Duke Environ Law Policy Forum 95:95–113
14. Borgardus K (3 May 2007) Bankrupt companies avoid more than $700 million in cleanup costs. Center for Public Integrity. http://projects.publicintegrity.org/superfund/report.aspx?aid=867. Retrieved 22 Mar 2010
15. DeLong J (2010) Privatizing superfund: how to clean up hazardous wastes. Cato Institute. http://www.cato.org/pubs/pas/pa-247es.html. Retrieved 19 Mar 2010
16. EPA (2010) http://www.epa.gov/superfund/eparecovery/faq.html. Retrieved 22 Mar 2010
17. Should the U.S. resurrect superfund? (2 Mar 2009). Natl J. http://energy.nationaljournal.com/2009/03/should-the-us-resurrect-superf.php. Retrieved 27 Mar 2010
18. Superfund: Litigation has decreased and EPA needs better information on site cleanup and cost issues to estimate future program funding requirement (15 July 2009). United States Government Accounting Office
19. International Climate Change Fund. http://climatelab.org/International_Climate_Change_Fund. Retrieved 25 Mar 2010

Books and Reviews

Airst RL (1994). Fair Assessment for Contaminated Properties, Urban Land.

Applegate JS, Laitos JG (2005) Environmental law: RCRA, CERCLA and the management of hazardous waste. Foundation, New York

Austrian Z, Eichler H (1994) Urban brownfields site survey: preliminary analysis. Cleveland State University, Levin College of Urban Affairs, Cleveland

CERCLA/Superfund Orientation Manual (1992). EPA, Washington

Chemical Waste Litigation Reporter (Information Service)

Cooke SM (ed) (1992) Law of hazardous waste: management, clean-up, liability and litigation (updated annually). Michael Bender, Stony Brook, NY

Davis TS (ed) (2002) Brownfields: a comprehensive guide to redeveloping contaminated property, 2nd edn. American Bar Association, Chicago
Federal Environmental Laws 2009 edn. (2009). West, New York.
Fogelman V (1992) Hazardous waste cleanup, liability, and litigation: a comprehensive guide to superfund law. Quorum, Westport
Garrett TL (ed) (1992) The environmental law manual. American Bar Association, Chicago
Gochfeld M (1995) Hazardous waste. In: Rosenstock L, Cullen M (eds) Textbook of occupational and environmental medicine. W.B. Saunders, Philadelphia
Gochfeld M, Burger J (1995) Assessment and mediation of hazardous waste sites. In: Brooks S et al (eds) Environmental medicine. Mosby, St. Louis
Initial guidance for implementing the superfund remedial program provisions of the American reinvestment and recovery act of 2009 (2009). EPA, Washington
Introduction to the Hazard Ranking System EPA, Washington. http://www.epa.gov/superfund/programs/npl_hrs/hrsint.htm. Accessed 27 Mar 2010
LaGrega MD, Buckingham PL, Evans JC (2001) Hazardous waste management, 2nd edn. McGraw-Hill, New York
Preliminary Assessment/Site Inspection. EPA, Washington. http://www.epa.gov/superfund/cleanup/pasi.htm. Retrieved 27 Mar 2010
Robinson N, Reilly KA (1992) Environmental regulation of real property. Law Journals Seminar Press, New York
Rodgers WH (2001) Environmental law. West Group, New York
Spranking JG, Weber GS (1997) Law of hazardous wastes and toxic substances in a nutshell. West, New York
Stern C, McKinney JA, Graham DB (1996) CERCLA enforcement: a practitioner's compendium of essential EPA guidance and policy documents. American Bar Association, Chicago
Stever DW (2010) Law of chemical regulation and hazardous waste. West, New York
Sullivan TFP (ed) (1995) Environmental law handbook. Government Institutes, Rockville
Superfund 20th Anniversary Report (2000) EPA, Washington
Switzer CS, Gray PL, McCullough MS (2000) CERCLA enforcement: a practitioner's compendium of essential EPA guidance and policy documents, Vol 2. ABA, Chicago
Watts RJ (1997) Hazardous wastes: sources, pathways, receptors. Wiley, New York

Chapter 8
Ecological and Health Risks at Low Doses

Kristine L. Willett and Christy M. Foran

Glossary

Acclimation	A decline in response with previous exposure.
Adaptation	Physiological changes that increase the fitness of a population in response to chronic exposure.
Epigenetic	Mechanisms of altered gene expression that do not change primary DNA sequence.
Hormesis	A beneficial effect from a low dose of a "poison" or contaminant.
NOAEL	No-observed-adverse-effect level is the highest treatment level from which the response does not differ statistically from control groups.
Threshold dose	Dose required to produce a measurable response.

This chapter was originally published as part of the Encyclopedia of Sustainability Science and Technology edited by Robert A. Meyers. DOI:10.1007/978-1-4419-0851-3

K.L. Willett (✉)
Department of Pharmacology and Environmental Toxicology Research Program,
University of Mississippi, School of Pharmacy, University, MS 38677, USA
e-mail: kwillett@olemiss.edu

C.M. Foran
Engineer Research and Development Center-Environmental Lab,
Concord, MA 01742, USA
e-mail: Christy.M.Foran@usace.army.mil

E.A. Laws (ed.), *Environmental Toxicology: Selected Entries from the Encyclopedia of Sustainability Science and Technology*, DOI 10.1007/978-1-4614-5764-0_8,

Definition of Subject

The concept of "dose makes the poison" has been attributed to Paracelsus who lived in the 1500s. The goal of much of modern toxicology and pharmacology has since been to define the precise biological factors that dictate the relationship between dose and response following exposure to chemicals. One of the major challenges in risk assessment more recently appreciated is the difficulty in predicting with statistical certainty adverse effects at low doses when the probability of response in a population is low [1]. The nature of the risks at low doses may be qualitatively different from the nature of the risks at high doses. In other words, the risk at low doses is not just a smaller version of the risk at high doses, because the mechanism (s) responsible for the risk at low doses is often different.

Introduction

Paracelsus (1493–1541), or Theophrastus Phillippus Aureolus Bombastus von Hohenheim, is often credited as being the father of modern toxicology for development of the concept of a dose–response relationship. Paracelsus is quoted as saying "all things are poison and nothing is without poison; only the dose makes a thing not a poison" [2]. This principle tells us that even the most innocuous of substances, water for example, can be dangerous in excess. The public is generally aware that dehydration, or an excess of salt relative to water, can be life threatening. However, hyponatremia, an excess of water in the blood, is less widely accepted as a potentially fatal condition. As recently as 2007, a contestant in a radio contest died after consuming 6 l of water in 3 h [3]. The idea that the amount of the exposure or dose determines the frequency or magnitude of response is the fundamental basis for most toxicological research. In theory and practice, the principle that the dose determines the poison is the foundation for experimental toxicology even to this day.

Historically, the relationship between dose and response was considered to be in one direction (i.e., an increase in response with dose or a decrease in response with dose). This single direction, or monotonic, relationship is based on both an understanding of biological processes and mathematical descriptions of experimental outcomes [4]. The most common description of the relationship between dose and response is a sigmoidal curve (Fig. 8.1). This relationship describes the response expected from the log of the dose or concentration. Under this relationship, the lowest doses or exposures elicit no measureable response. In general, and in agreement with Paracelsus, organisms have some tolerance for exposure to poisons and toxicants. Some "threshold dose" is required to produce a measureable response. The response then increases exponentially until the midpoint of the curve, where it is half of the maximal response, a point referred to as the EC50 (effective concentration to produce 50% response). The rate of increase in the response decreases above the EC50 to zero where the response reaches its maximum. Increased doses or concentrations above the maximum (EC100) will not

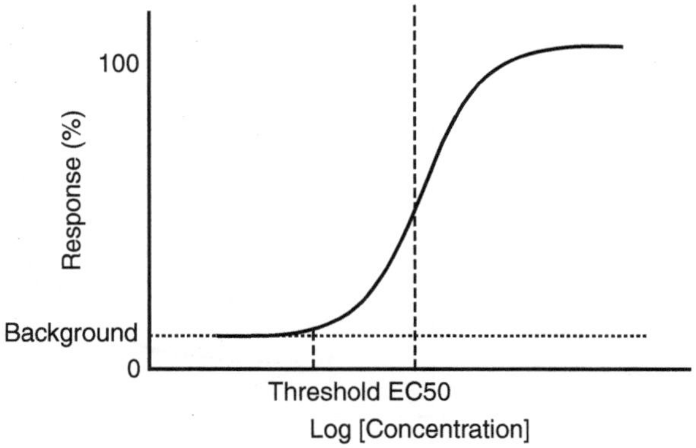

Fig. 8.1 The classical sigmoidal dose–response curve has doses or concentrations that elicit a 50% response (EC50). Theoretically, there is a threshold concentration or dose below which the response is the same as controls. This spontaneous response is the background rate in the absence of treatment

increase the response. The sigmoidal curve is the default assumption for the relationship between dose and response.

Despite this general convention, agreement on the shape of the dose–response curve at the low end has been controversial [4–6]. The classic sigmoidal dose–response relationship implies that there is a threshold for a biological response. Below that threshold value, exposure to a toxicant should have no measureable effect. Some responses to toxicants, such as the occurrence of tumors, occur spontaneously in control groups. In the absence of exposure, this rate of occurrence is considered the "background" level (Fig. 8.1). If an incremental dose increases this background rate of response, then the relationship between the incremental dose and response can be fitted linearly [4]. Furthermore, early studies of the biological basis of tumor formation led to the idea that a single molecule of a carcinogen could cause a mutation that led to tumor development. These "one hit" models have been disputed with both theoretical and physical lines of evidence [4]. For low-energy radiation, theoretical studies indicate that at least two hits of radiation are necessary for mutagenesis, potentially due to high DNA repair rates. Furthermore, cancer research indicates that tumor initiation requires a series of promotional changes even after mutations are incorporated into the genome.

For carcinogens, the precautionary assumption has been that every dose can elicit a response. In other words, there is not an observable no-effect threshold. The most common approach for estimating the low-dose effects of a carcinogen is to adequately describe the relationship between exposure and response over some range of treatments and then extrapolate to lower doses, generally using a linear function. If all poisons have a threshold dose, the linear extrapolation of low-dose response is a conservative assumption. There are reports of DNA adduct formation in proportion to dose even at extremely low concentrations. The US Environmental Protection Agency (EPA) has adopted a policy to utilize the formation of DNA and protein adducts as a means to extend the understanding of the shape of the dose–response relationship beyond the range that can be investigated with cancer bioassays [7]. The relationship between formation of DNA adducts and physiologically meaningful formation of tumors is not as clear. A review of the literature on biomarkers indicates that the dose–response relationship for biomarkers of exposure

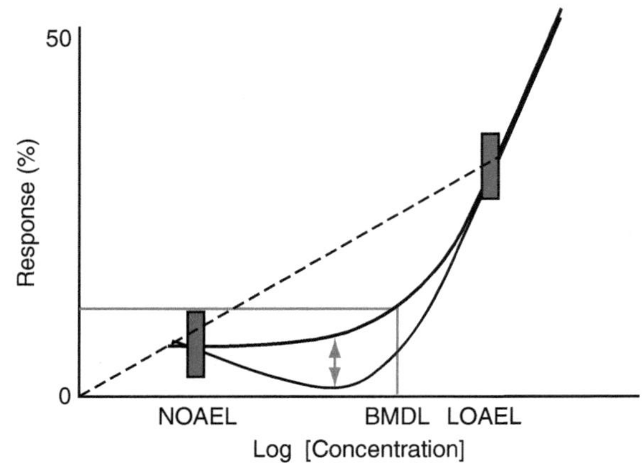

Fig. 8.2 The controversial low end of the dose–response curve is shown here. Experimental data determine the "lowest observed adverse effects level" (LOAEL) and the "no-observed-adverse-effects level" (NOAEL), indicated by the *vertical gray boxes*. Linear extrapolation from the LOAEL (*dashed line*) is a conservative estimation of the low-dose effects. Threshold dose can be estimated statistically using a lower bound benchmark dose (BMDL) through a proportion of the response (*gray bars*). In this hypothetical case, all extrapolations would overestimate toxicity if there were a hormetic response (*gray arrow*)

is not the same as the relationship for biomarkers of effect [7]. Recent descriptions of this debate include the historical progression of acceptance of linear extrapolation by the EPA [8].

In contrast to carcinogens, the assumption of a no-effect threshold has been used extensively for chemicals that are not considered carcinogenic. For these chemicals, the dose at which no adverse effects are detectable is a critical value incorporated into risk assessments. This dose is called the no-observed-adverse-effect level or NOAEL. The NOAEL is derived from experimental data; it is the highest treatment level which does not differ statistically from the control groups. NOAEL is the basis for regulatory guidance such as acceptable intake levels. Data limitations and experimental design can limit the utility of the NOAEL. For example, note the chosen test concentrations represented by the gray vertical boxes in Fig. 8.2 for the NOAEL and the lowest observed adverse effects level (LOAEL). The doses in this hypothetical experiment might have been chosen to produce the maximum spread of responses or cover the relevant dose range for this compound. However, a higher concentration may have been chosen which would have still resulted in no significant response – and therefore a different value for the NOAEL. An alternative approach is to use a benchmark dose (BMD) [9]. The BMD is a derived value based on the effective range of the dose–response curve. Calculated BMD have the benefit of using a maximum likelihood approach to determine a confidence interval for the value. The variance in the response to the LOAEL or the EC50 can be used to estimate the concentration resulting in 5% response (or another level).

Extensive evidence now indicates that often the relationship between dose and response is biphasic. In other words, low-dose treatments may cause a decrease in response while higher doses cause an increase in response. An extensive review of the toxicological literature [10] found thousands of cases of these types of nonmonotonic dose–response relationships. In many of these cases, low-dose treatments with a toxicant induced a beneficial response – fewer tumors than unexposed controls, for example. A beneficial effect from a low dose of a "poison" or contaminant is referred to as hormesis. These beneficial effects have been reported over a range of responses from growth rate to carcinogenesis. Modeling studies indicate that hormetic response can be driven by a summation of the dose–response relationships for different steps in the biological response, specifically receptor binding, transcriptional activation, rate of DNA repair, and rate of cell division [5]. An example of how multiple physiological responses work together to result in a hormetic response is described below.

Alternatively, low-dose exposure may stimulate an increase in response while higher dose produces a lesser response. This "inverted U" shaped dose–response curve has also been observed in many cases. This relationship between dose and response is characterized by a heightened response to low doses and a depressed response to higher doses. A hypothetical example might be an increase in reproduction observed in response to a low concentration of estrogen and suppressed reproductive rates observed with higher treatments. The more complicated relationships between dose and response demonstrated in cases of hormesis and "inverted U" curves are likely to be observed in systems that require multiple steps to produce the response [5]. Systems biology and genomic/proteomic studies provide an opportunity to test summation hypotheses in a rigorous manner [1, 11]. Modeling and circuitry analysis provide an opportunity to test, or to generate new predictions about the responses of cells, individuals, and populations.

Further complicating the relationship between dose and response is the issue of exposure to mixtures [12]. Some cocontaminants will have the same mechanisms of action and interact with the same receptors, while others will interact antagonistically through similar mechanisms of uptake, circulation, transformation, or elimination. EPA guidance recommends dose addition for chemicals which have similar mechanisms of action or exhibit response addition. However, predicting the toxicity of mixtures is difficult even when the mechanisms of action are understood.

The focus or target of a research study has a significant impact on predictions related to the dose–response relationship. Many cellular responses are either all or none switches (i.e., cell division). These cellular responses are described as being "quantal," and the sensitivity of the response in these cases will be described as a cumulative frequency of response at each dose. How the frequency of these incidences relates to specific biological response and adverse health outcomes is largely an unresolved issue [11]. By contrast, examination of a population of cells will produce a smooth dose–response curve as individual cells respond at different concentrations. An analogous dose–response relationship can be described in epidemiological studies [6]. Individuals participating in a study will either have a tumor or not; their response to their particular dose is quantal. However, different individuals will have different sensitivities based on their age, health, lifestyle, and previous exposure history [4]. Each individual is likely to have a different "threshold" for tumor incidence based on their situation. Summation of these different thresholds is expected to "smooth" the dose–response relationship in studies that include a diverse population. Indeed, most human epidemiological studies have a linear relationship between dose and response [6].

Physiological Response to Low Doses

As described above, the response of an organism to a toxicant may be nonmonotonic. One range of treatment may suppress a response while another may enhance the response [12]. Physiological mechanisms underlying these complicated hormetic responses are of much interest. The long history of these types of responses has been documented by Rozman and Doull [13]. Their review summarizes the notion that a beneficial response to a low dose of a poison is the result of homeostatic compensation in response to a stressor. Rozman and Doull make an analogy between low-dose treatment and physical exercise. Both are minor stressors and homeostatic response to those stimuli may ultimately enhance fitness. Adaptation to a low dose of a stressor, like recovery from physical exercise, may be the result of diverse physiological mechanisms. As a general phenomenon, adaptation or compensation induced by a low-dose treatment is believed to be the basis for hormesis [14].

In situations that result in a hormetic response, continuous exposure to a relatively constant dose should result in a continuous state of (over)compensation. This response has been described as steady-state hormesis induced by chronic exposure [15]. A critical cellular aspect involved in the maintenance of homeostasis is the suite of biotransformation and conjugation enzymes [16]. Three categories of enzymes, Phase I–III, are the basis for xenobiotic metabolism. Phase I enzymes include the cytochrome P450 family, which classically oxidize xenobiotics rendering them more water soluble. However, these Phase I enzymes also activate some procarcinogens and mutagens to reactive metabolites. Phase II enzymes conjugate solubilizing groups that further facilitate excretion. Phase III enzymes are membrane-bound transporters that export metabolized xenobiotics from the cell. A model of these xenobiotic metabolizing enzymes (XMEs) by Zhang and coworkers [17] has generated a mechanistic hypothesis for the hormetic response to many mutagens and carcinogens.

Zhang et al. [17] have described a multipart negative feedback and feedforward model of XMEs, including cross-induction of Phase I and II enzymes. Simulations of this model find a drop in the production of a reactive metabolite below baseline levels in response to 64 h of exposure to an exogenous procarcinogen. Activation of a homeostatic control mechanism, in this case transcriptional induction of XME in response to binding of a nuclear receptor (the aryl hydrocarbon receptor), will compensate for the poison and produce a new physiological state. Because of this hormetic compensation, the authors conclude that the response of biological systems will not be "linearly correlated with the dose of the stressing agent, especially in the low dose region" [17]. The regulation and induction of Phase I and II enzymes is a common cellular mechanism for coping with toxicants. In the cases where Phase I metabolism activates a toxicant, the feedforward induction of these enzymes is a potential mechanism for hormetic responses at the cellular level. This model provides a hypothesis for the cellular basis of suppression of background levels of tumor induction in response to low-dose exposure to carcinogens to the extent that reactive metabolites are correlated with mutagenesis, apoptosis, and cell proliferation.

Low-level or subacute doses have the potential to produce long-term changes in the magnitude of the response elicited from a toxicant. A decline in response with exposure is called acclimation. The process of acclimation is best studied in relation to exposure of fish to heavy metals (reviewed by Hamilton and Mehrle [18]). Pretreatment of animals with metals, specifically cadmium, copper, mercury, and zinc, is protective for subsequent exposures. Exposure induces synthesis of thionein, a protein that normally occurs only in trace amounts in certain tissues (blood, gills, liver, kidney, and intestine). Binding of metals to this protein, then called metallothionein, reflects the relative concentrations of the metals present and their binding affinity. Chronic toxicity of metals that are sequestered by metallothionein has been described as "spillover"; once metallothionein's metal binding sites are saturated, the excess will cause tissue damage. This example of competition for a target will influence the shape of the dose–response curve at the low end. Metallothionein is induced by metal, and to a lesser extent, by glucocorticoids. Therefore, some protection or acclimation to metals may be produced by a number of different stressors that produce a glucocorticoid response. In this case, exposure to some classes of metals or heightened stress response will alter the threshold dose for tissue damage from toxicants such as mercury, cadmium, and silver.

Different populations will vary in their response to a toxicant, especially at the low range of the dose–response curve. If a mathematical description of the response indicates that 5% of the population should be impacted, a change in sensitivity of the population (not to mention sample size and health-related factors) will result in more variability in the reported response rate. Populations may differ in age, or individuals may have encountered a previous exposure as described above. Additionally, populations may have a history of chronic exposure that results in transgenerational changes in the dose–response curve. Genetic or epigenetic adaptation to a chronic stressor will impact the response of a population to exposure. Resistant phenotypes may have enhanced fitness under chronic exposure conditions. Those phenotypes would be expected to increase in frequency in subsequent generations. Experimental studies of killifish from a creosote-contaminated Elizabeth River (Virginia) location have documented resistance to the teratogenic effects of creosote exposure which is maintained through two generations of animals reared in clean conditions [19]. Transgenerational resistance to the polycyclic aromatic hydrocarbons (PAHs), which make up the majority of creosote, is not associated with a general increase in resilience. In fact, offspring of Elizabeth River–collected killifish exhibited higher mortality in response to hypoxic conditions. This study reveals that the magnitude of resistance to PAHs decreases with subsequent laboratory-raised generations (F2 vs. F1); therefore the mechanism of adaptation might not be conferred genetic resistance, but instead be a heritable change in gene regulation.

Epigenetic mechanisms are molecular phenomena, such as DNA methylation of CpG nucleotides, which regulate gene expression without changing the DNA sequence itself [20]. Methylation patterns are normally established two times during development in somatic cells during gastrulation and in the germ cells after sex determination [21]. Many common diseases (e.g., diabetes, osteoporosis, atherosclerosis etc.) have a sex-related incidence bias that may be a result of sex-specific (e.g., X-chromosome inactivation) epigenetic control [22]. As reviewed by Gabory and coworkers, environmental factors including xenobiotics, nutrition, or social factors can influence epigenetic marks throughout development [22]. Furthermore, when germ cells are adversely epigenically reprogrammed, sex-specific transgenerational effects have been reported. For example, when pregnant rats were exposed to the antiandrogen, vinclozolin, all subsequent generation (F1–F4) males had sperm cell defects and subfertility [21, 23]. Epigenetic changes are also becoming well-recognized mechanisms contributing to cancer and have been associated with all stages of tumor formation and progression [24]. Losses of DNA methylation in oncogenes, hypermethylation of tumor suppressor genes, and altered histone modifications have all been noted during oncogenesis. The extent to which these epigenetic modifications of gene expression and/or their multigenerational consequences mediate xenobiotic low-dose effects are still largely unknown.

Specific Examples

While a list of xenobiotics that display different effects at high and low doses is likely to grow as "omic" tools and systems biology approaches are implemented in risk assessment, there are a number of examples already known. The bulk of the initial work on predicting low-dose effects was done with respect to carcinogen risk assessment [8]. Radiation and polycyclic aromatic hydrocarbons (PAHs) will be used to highlight issues in cancer low-dose predictions. Similar to cancer progression, neurotoxicological risk assessment has acknowledged that neurological effects are rarely unifactoral [25, 26]. Here, lead and organophosphate pesticides will be used as case studies. A focus of a National Toxicology Program Peer Review [27], certain endocrine disrupting chemicals including bisphenol A (BPA) also have been recognized for their potential to cause "biological change" at low doses. Accordingly, the specific examples described below will focus on carcinogens, neurotoxicants, and endocrine disruptors.

A recent review by Calabrese has provided the history of carcinogen risk assessment [8]. As he describes, the basis for the idea that carcinogen effects were linear at low doses was derived from the early work of Hermann Muller [28], wherein it was discovered that X-ray radiation induced mutations in *Drosophila*. The relationship between low-dose linearity and radiation-induced mutations in germ cells was further asserted in the 1956 Biological Effects of Atomic Radiation (BEAR) Report. The idea that mutations in somatic cells could lead to cancer also with a linear response at low doses was brought to public debate following publication of E.B Lewis' paper relating radiation risks and leukemia [29]. The criticisms of this manuscript are reviewed by Calabrese [8]; nonetheless, it was a key contribution to the adoption of a precautionary approach and low-dose linear risk assessment for both radiation and ultimately chemical carcinogens. The current risk assessment paradigm for radiation exposure includes a "single-cell origin, a radiation-induced mutation/aberration as the rate-limiting step, simple expectations of a linear non-threshold radiation response (LNT) at low doses and little or no other relevant contributions. . .to promotion or progression of the tumor" [30]. As pointed out by Goodhead, however, there are a number of radiobiological phenomena including bystander effects and genome instability that are not yet completely understood that may impact how risk of low doses should be assessed and ultimately how human radiation protection is performed. Bystander effects, effects in neighboring nonirradiated cells, are most apparent in the low-dose range where overt toxicity by direct radiation is not dominant [31]. The consequences of bystander effects would be supposed to be relevant to carcinogenicity in that they could be detrimental (toxic effects or mutations in adjacent cells) or beneficial (apoptosis of impacted cells, so no additional tumor progression possible); however, how to mechanistically account for their role in dose–response relationships and risk assessment is currently unknown [30]. Similarly, *de novo* genetic instability in cells generations after irradiation appears to display nonlinear dose responses, and the mechanisms are also unclear. Further research will certainly be necessary to establish the relevance of these more recently recognized potentially low-dose effects of radiation and whether a change in the radiation risk assessment paradigm is necessary [30].

With respect to chemical carcinogens, the Guidelines for Carcinogen Risk Assessment [32] establish a two-step process for dose–response assessment. First, animal or human effect data are used to model biological effects such as adduct, mutation, or tumor formation. The second step then is to extrapolate below the range of observation to the expected range of human exposure with considerations as appropriate or available to mode of action of the chemical, quality of data, and uncertainties therein. Because of the lack of human data, the high costs and low sensitivity of rodent cancer studies, and the goal to protect against one cancer per million, extrapolation of dose back five orders of magnitude is sometimes necessary [33]. Furthermore, it has recently become apparent that biomarkers of exposure (e.g., DNA adducts; linear response) do not have the same dose–response curve shape as biomarkers of effect (e.g., mutations; threshold response) [7]. As reviewed by Swenberg et al., acrylamide, methymethanesulphonate, and ethylene oxide all represent examples wherein different exposure and effect curves have been reported [7]. Similarly, in a recent 40,800 animal study investigating ultralow-dose effects of the PAH dibenzo[a,l] pyrene in rainbow trout, the initial carcinogen-DNA adduct and the hepatic tumor response curves were "distinctly different" leading the authors to conclude that DNA adducts did not "provide an accurate biomarker for ultralow dose cancer risk evaluation" [33]. Additionally, the study design necessitated significantly less extrapolation below the modeled data set (200-fold vs. typical 100,000-fold) and found that the virtual safe dose for liver tumor formation was 500–1,500 times higher than the standard linear extrapolation from the lower confidence limit on the dose causing 10% incidence typically used in carcinogen risk assessment [33]. Mechanistically, the difference in these curves is relatively easy to appreciate. A specific PAH-DNA adduct will reflect exposure to that parent PAH exclusively. Furthermore, there exist very sensitive analytical methods of DNA adduct measurement. However, it is known that many adducts are not promutagenic. When mutagenic adducts formed, the excretion and repair rates of a particular adduct may be variable because of multiple gene interactions; yet, a relationship between adduct formation and initial exposure will exist. In contrast, mutations are caused by a number of initiating events. There always exists some background of mutations caused by endogenous chemicals or mistakes in DNA replication. Therefore, while physiologically more significant, mutation measurement in exposure experiments will ultimately approach a background frequency at low doses [7]. As these recent reports indicate, the way to measure and predict risks due to exposure to carcinogens is still relevant and continually being reevaluated even though this was one of the first areas in toxicology research to appreciate the need to estimate effects at low doses [8].

Effects of low doses are also important for acute acting compounds. Humans have used and been exposed to lead for thousands of years, and at high doses, a range of adverse effects can occur [34]. In humans, effects depend on developmental stage with the central nervous system as a target in children and peripheral neuropathies, nephrotoxicity, and cardiovascular effects common in adults [35]. Recent emphasis has been on how childhood low-dose, environmentally relevant lead exposure affects cognitive function. As more data were collected on lead's low-dose toxicity, the regulatory limit of concern for childhood lead intoxication reduced from its initial value of <60 μg/dl set in the 1960s, to 40 μg/dl in 1971, to 30 μg/dl in 1978, and 25 μg/dl in 1985 [36]. The current limit recommended by the Center for Disease Control and World Health Organization is <10 μg/dl. Studies from around the world have found blood lead concentrations far exceeding this level in particular study populations (reviewed in [34, 37]). For example, children from a lead-contaminated area of Ecuador had blood lead concentrations of 52.6 μg/dl (range 9.9–110.0 μg/dl) [38]. Findings from the National Health and Nutrition Examination Surveys (NHANES) in the United States show that average children's blood lead levels have been declining since the 1970s, and the prevalence of elevated levels in children was 0.7% in the 1999–2002 survey [39]. However, recent evidence suggests that blood lead concentrations below 10 μg/dl are not without adverse effect.

The 10 μg/dl standard was based, in part, on a study wherein children were grouped based on umbilical-cord blood concentrations (either <3 μg/dl, between 6 and 7 μg/dl, or >10 μg/dl). At 2 years of age, Mental Development Index scores were significantly lower in children from the >10 μg/dl group [40]. However, using the NHANES III cohort, Lanphear and coworkers found that, after results were corrected for confounding factors, there was an inverse relationship between blood lead concentration and cognitive functioning. Specifically, they reported "for every 1 μg/dl increase in blood lead concentration, there was a 0.7-point decrement in mean arithmetic scores, an approximately 1-point decrement in mean reading scores, a 0.1-point decrement in mean scores on a measure of nonverbal reasoning, and a 0.5-point decrement in mean scores on a measure of short-term memory" [41]. Importantly, these results included children who had blood lead concentrations less than 5 μg/dl. A more recent study by this laboratory followed 172 children measuring their blood lead concentrations at 6, 12, 18, 24, 36, 48, and 60 months of age and tested their performance on the Stanford-Binet Intelligence Scale at 3 and 5 years of age [42]. They found that as lifetime average blood lead concentrations increased from 1 to 10 μg/dl, IQ declined by 7.4 points using a nonlinear model or 4.6 points with a linear model. Additionally, chelation therapy is largely ineffectual for children with blood lead concentrations between 20 and 44 μg/dl [43] and cognitive deficits due to lead exposure can last into adulthood [44]. These studies suggest that there is not a clear threshold for lead effects in children.

Unfortunately, the molecular targets and mechanistic reasons to explain low-dose effects of lead exposure are currently lacking. Clearly, critical neurodevelopmental processes occur in the early years of life that have profound effects on learning and memory. A current area of emphasis with respect to mechanisms of lead toxicity is focused on N-methyl-D-aspartate receptor (NMDAR)-mediated synaptic plasticity, specifically hippocampal long-term potentiation [37]. As reviewed by Toscano and Guilarte [37], in rodent models, environmentally relevant lead exposure alters NMDAR subunit expression and function including decreased calcium and nitric oxide signaling, both critical signaling pathways involved in learning and memory. Better understanding of the molecular targets of lead toxicity may in turn offer more effective therapeutic approaches to lead exposure in the future.

Organophosphate pesticides (OP), such as chlorpyrifos, have a well-recognized biomarker of exposure and effect in the irreversible inhibition of acetylcholinesterase (AChE). Significant inhibition of AChE causes a recognizable "cholinergic syndrome" causing accumulation of acetylcholine at cholinergic synapses and overstimulation of cholinergic receptors of the muscarinic and nicotinic type [45]. In addition to the acute cholinergic crisis, high-dose OP exposure can lead to OP-induced delayed polyneuropathy. Percent decreases in plasma pseudocholinesterase or red blood cell cholinesterase (e.g., 30% from baseline) are used in workplace surveillance even though the physiological significance of circulating cholinesterase activity is not known. Furthermore, signs of intoxication typically do not occur until cholinesterase inhibition is >70% [25]. Accordingly, organophosphate pesticides are another class of neurotoxicant that can exhibit strikingly different low- versus high-dose toxic effects.

Chlorpyrifos is one of the most widely applied OPs. In the United States, registrations for its use in residential settings were canceled in June 2000 [46]. Prior to this ban, chlorpyrifos was detected in 99.7% of personal air samples and 100% of indoor air in an urban pregnant woman cohort [47]. When pregnant rats are given chlorpyrifos at doses below those affecting acetylcholinesterase, their offspring have permanent deficits in learning and memory [25]. Work in the Slotkin laboratory has established persistent changes in serotonergic and dopaminergic systems in adult rats exposed either prenatally or neonatally to chlorpyrifos at doses that are devoid of systemic toxicity [48–50]. Furthermore, in rats exposed neonatally on gestational days 17–20 (during peak neurogenesis), there were sex-dependent behavioral impairments persistent to adulthood [51]. Intriguingly, in addition to the neurotoxicological effects of chlorpyrifos, adverse effects on adenyl cyclase cellular signaling cascades were noted in heart and liver of developmentally exposed rats [52]. As concluded by Slotkin, "The mechanisms for neurodevelopmental effects from toxins such as chlorpyrifos cannot be presumed to be based on known mechanisms for systemic toxicity in the adult" [25]. More recently, an epidemiologic study of inner-city children compared umbilical cord chlorpyrifos plasma concentrations and cognitive and motor development (at 12, 24, and 36 months of age) and behavior (at 36 months). A significantly higher proportion of children from the "highly exposed" group showed signs of mental and motor delays and were more likely to score in the clinical range for attention problems compared to the lower exposure group at 3 years of age [53]. Further insight into the mechanisms of chlorpyrifos and other organophosphates at low doses may further tighten regulatory control of this class of chemicals and/or offer therapeutic or psychological interventions to remediate low-dose and developmental exposures.

Endocrine disrupting chemicals (EDCs) can change endocrine function and cause adverse effects at the level of the organism, its progeny and/or (sub) populations of organisms. There have been a number of research conferences since the initial Wingspread Conference in 1991, spearheaded by Dr. Theo Colburn, to address the effects of EDCs. Germane to this chapter, in 2000, the EPA and the National Institute of Environmental Health Sciences (NIEHS) held a peer review "aimed at evaluating the scientific evidence on reported low-dose effects and dose-response relationships for endocrine disrupting chemicals in mammalian species that pertain to assessments of effects on human health" [27]. Nine EDCs, not including dioxins or phthalates esters, were considered. More recently, as reviewed by Vandenberg and coworkers [54], bisphenol A (BPA) individually has been a focus of expert analysis and regulatory review.

Bisphenol A is a high-volume chemical used in polycarbonate plastic. Use of this chemical has recently received intense media scrutiny because of its occurrence in consumer products such as food container liners and baby bottles and ubiquitous occurrence in human samples [55]. Based on relatively low in vitro binding affinity to the estrogen receptor compared to estradiol, BPA was initially considered a weak environmental estrogen. However, more recently lower-dose effects are being noted at concentrations below those needed for nuclear estrogen receptor binding [56]. A number of in vitro studies measuring BPA-mediated effects in cell types as diverse as adipose explants, and pituitary, prostate cancer, and pancreatic cells have reported U-shaped nonmonotonic dose–response curves (reviewed in [54]). For example, cell proliferation was lower at both low and high doses of BPA compared to intermediate doses. In vivo, BPA-mediated effects on CD-1 mouse uterine endpoints also indicated nonmonotonic dose–response curves for some endpoints (e.g., vaginal opening and uterine wet weight) [57]. Fetal exposure to low doses of BPA causes changes in tissue organization and histoarchitecture of developing mammary gland in mice [58, 59]. In males, BPA causes adverse effects on the male reproductive system and sex-related behaviors [60, 61]. Important observations revealed from BPA studies, with particular relevance to low-dose effects, are that experimental design and study controls are critical. For example, compounds present in animal feeds can mask treatment effects, and different animal strains can have different sensitivities [62]. The exact mechanisms underlying BPA low-dose effects are still under investigation. It is important to remember that estrogen is a critical physiological mediator of many diverse tissue-dependent cellular processes. Furthermore, appropriate timing of sex steroid exposure is key in establishing the hypothalamus-pituitary-gonad feedback necessary for successful development and reproduction. Issues of timing of exposure would be expected to have profound impacts on effects of BPA.

Conclusions from the 2000 peer review included that relevant low-dose effects of BPA included increased prostate weight in male mice and advanced puberty in female mice after *in utero* exposure. Effects in rats (e.g., uterine growth and serum prolactin) were strain specific. They concluded "Data are insufficient to establish the shape of the dose-response curve for BPA in the low dose region, and the mechanism and biological relevance of reported low dose effects are unclear" [27]. In 2006, the NIEHS organized an additional meeting specifically addressing BPA. A product of this meeting was the Chapel Hill Consensus Statement [63], which concluded "…that human exposure to BPA is within the range that is predicted to be biologically active in over 95% of people sampled. The wide range of adverse effects of low doses of BPA in laboratory animals exposed both during development and in adulthood is a great cause for concern with regard to the potential for similar adverse effects in humans." Meanwhile, the National Toxicology Program's Center for the Evaluation of Risks to Human Reproduction (CERHR) evaluated the scientific data and concluded that some of the previously attributed effects of BPA (e.g., changes in prostate weight, reproductive tract malformations etc.) were not conclusive [64]. This report was highly criticized (see review [54]), and NTP released its own report finally concluding "there is *some concern* for effects on the brain, behavior, and prostate gland in fetuses, infants, and children at current human exposures" but "*minimal concern* for effects on the mammary gland and an earlier age for puberty for females" [65]. Importantly in August 2008, and in an interim update issued January 2010, the US Food and Drug Administration, which regulates food packaging etc., released a draft report finding that BPA remains safe in food contact materials [66]. Some of the questions plaguing the debates on BPA risk assessment include different routes of exposure and metabolism of BPA in animals and humans, lack of consistency in results between laboratory studies, and the relevancy of the animal models to the human condition. Clearly, this is a timely ongoing debate that absolutely hinges on elucidating the consequences of low-dose environmentally relevant exposure to BPA.

Future Directions

The mechanisms of toxicity and the few selected xenobiotics used as examples here emphasize how timely and relevant the issue of effects at low doses is. Furthermore, answers to the issues raised are key for more informed and appropriate risk assessment. Toxicology studies historically have been limited to testing concentrations representing the high end of environmental relevance. The need to test higher doses was necessitated both by inability to measure/recognize subtle effects, and the need to achieve statistically significant results with the minimal number of test organisms and cost. Now with the realization that, in general, mechanisms of toxicity are not consistent from high to low doses, new robust experiments are needed to explain the more relevant low-dose effects. However, care must be taken so that the cost of regulatory measures that conservatively estimate the harm of low-dose effects is balanced with the cost of potential mechanisms that alter the response of current or future populations. Furthermore, when a biomarker of effect is recognized to not represent true risk of adverse effect (e.g., DNA adduct formation), then more appropriate endpoints (e.g., mutation) should be incorporated into risk assessment. As described above, many low-dose effects are critically linked to the developmental age at the time of exposure. Populations which differ in age or previous exposure are likely to have different responses to low-dose treatment of any contaminant. The World Health Organization has developed a multiple exposure/multiple effects children's health model incorporating the influence of social, economic, and demographic contexts on health outcomes and exposure in addition to the relationship between exposure and effect [67]. As the underlying system biology of test organisms becomes better understood, better evaluation of effects of xenobiotic exposure will be possible [1]. Finally, the field of risk assessment and society in general may ultimately have to deal with the ethical questions that arise associated with potential hormetic effects of chemicals that are toxic at high doses [31].

Bibliography

Primary Literature

1. Andersen ME, Thomas RS, Gaido KW, Conolly RB (2005) Dose-response modeling in reproductive toxicology in the systems biology era. Reprod Toxicol 19:327–337
2. Krieger WC (2001) Paracelsus dose response. In: Krieger WC (ed) Handbook of pesticide toxicology. Academic, New York
3. Ballantyne C (2007) Strange but true: drinking too much water can kill. Scientific American June 21, 2007. http:// www.scientificamerican.com/article.cfm?id=strange-but-true-drinking-too-much-water-can-kill&print=true. Accessed 24 January 2010
4. Zeise L, Wilson R, Crouch EA (1987) Dose-response relationships for carcinogens: a review. Environ Health Perspect 73:259–306
5. Conolly RB, Lutz WK (2004) Nonmonotonic dose-response relationships: mechanistic basis, kinetic modeling, and implications for risk assessment. Toxicol Sci 77:151–157
6. Lutz WK (1990) Dose-response relationship and low dose extrapolation in chemical carcinogenesis. Carcinogenesis 11:1243–1247
7. Swenberg JA, Fryar-Tita E, Jeong YC, Boysen G, Starr T, Walker VE, Albertini RJ (2008) Biomarkers in toxicology and risk assessment: informing critical dose-response relationships. Chem Res Toxicol 21:253–265
8. Calabrese EJ (2009) The road to linearity: why linearity at low doses became the basis for carcinogen risk assessment. Arch Toxicol 83:203–225
9. Sand S, Victorin K, Filipsson AF (2008) The current state of knowledge on the use of the benchmark dose concept in risk assessment. J Appl Toxicol 28:405–421
10. Calabrese EJ, Baldwin LA (2001) Hormesis: a generalizable and unifying hypothesis. Crit Rev Toxicol 31:353–424
11. Andersen ME, Dennison JE, Thomas RS, Conolly RB (2005) New directions in incidence-dose modeling. Trends Biotechnol 23:122–127
12. Borgert CJ, Quill TF, McCarty LS, Mason AM (2004) Can mode of action predict mixture toxicity for risk assessment? Toxicol Appl Pharmacol 201:85–96
13. Rozman KK, Doull J (2003) Scientific foundations of hormesis. Part 2. Maturation, strengths, limitations, and possible applications in toxicology, pharmacology, and epidemiology. Crit Rev Toxicol 33:451–462
14. Stebbing AR (2003) Adaptive responses account for the beta-curve-hormesis is linked to acquired tolerance. Nonlinearity Biol Toxicol Med 1:493–511
15. Calabrese EJ, Bachmann KA, Bailer AJ, Bolger PM, Borak J, Cai L, Cedergreen N, Cherian MG, Chiueh CC, Clarkson TW, Cook RR, Diamond DM, Doolittle DJ, Dorato MA, Duke SO, Feinendegen L, Gardner DE, Hart RW, Hastings KL, Hayes AW, Hoffmann GR, Ives JA, Jaworowski Z, Johnson TE, Jonas WB, Kaminski NE, Keller JG, Klaunig JE, Knudsen TB, Kozumbo WJ, Lettieri T, Liu SZ, Maisseu A, Maynard KI, Masoro EJ, McClellan RO, Mehendale HM, Mothersill C, Newlin DB, Nigg HN, Oehme FW, Phalen RF, Philbert MA, Rattan SI, Riviere JE, Rodricks J, Sapolsky RM, Scott BR, Seymour C, Sinclair DA, Smith-Sonneborn J, Snow ET, Spear L, Stevenson DE, Thomas Y, Tubiana M, Williams GM, Mattson MP (2007) Biological stress response terminology: integrating the concepts of adaptive response and preconditioning stress within a hormetic dose-response framework. Toxicol Appl Pharmacol 222:122–128
16. Parkinson A, Ogilvie BW (2008) Biotransformation of xenobiotics. In: Klaassen CD (ed) Casarett & Doull's toxicology: the basic science of poisons. McGraw Hill, New York, pp 161–304
17. Zhang Q, Pi J, Woods CG, Andersen ME (2009) Phase I to II cross-induction of xenobiotic metabolizing enzymes: a feedforward control mechanism for potential hormetic responses. Toxicol Appl Pharmacol 237:345–356

18. Hamilton SJ, Mehrle PM (1986) Metallothionein in fish: review of its importance in assessing stress from metal contaminants. Trans Am Fish Soc 115:596–609
19. Meyer JN, DiGiulio RT (2003) Heritable adaptation and fitness costs in killifish (*Fundulus heteroclitus*) inhabiting a polluted estuary. Ecol Adapt 13:490–503
20. Jones PA, Takai D (2001) The role of DNA methylation in mammalian epigenetics. Science 293:1068–1070
21. Anway MD, Skinner MK (2006) Epigenetic transgenerational actions of endocrine disruptors. Endocrinology 147:S43–S49
22. Gabory A, Attig L, Junien C (2009) Sexual dimorphism in environmental epigenetic programming. Mol Cell Endocrinol 304:8–18
23. Anway MD, Cupp AS, Uzumcu M, Skinner MK (2005) Epigenetic transgenerational actions of endocrine disruptors and male fertility. Science 308:1466–1469
24. Baylin SB, Ohm JE (2006) Epigenetic gene silencing in cancer – a mechanism for early oncogenic pathway addiction? Nat Rev Cancer 6:107–116
25. Connors SL, Levitt P, Matthews SG, Slotkin TA, Johnston MV, Kinney HC, Johnson WG, Dailey RM, Zimmerman AW (2008) Fetal mechanisms in neurodevelopmental disorders. Pediatr Neurol 38:163–176
26. Weiss B, Cory-Slechta D, Gilbert SG, Mergler D, Miller E, Miller C, Newland MC, Rice D, Schettler T (2008) The new tapestry of risk assessment. Neurotoxicology 29:883–890
27. National Toxicology Program (2001) National toxicology program's report of the endocrine disruptors low-dose peer review. NTP Office of Liaison and Scientific Review, Research Triangle Park
28. Muller HJ (1927) Artificial transmutation of the gene. Science 66:84–87
29. Lewis EB (1957) Leukemia and ionizing radiation. Science 125:965–972
30. Goodhead DT (2010) New radiobiological, radiation risk and radiation protection paradigms. Mutat Res 687(1–2):13–16
31. Hoffmann G, Stempsey W (2008) The hormesis concept and risk assessment: are there unique ethical and policy considerations? Hum Exp Toxicol 27:613–620
32. EPA US (2005) Guidelines for carcinogen risk assessment. US EPA, Washington DC
33. Bailey GS, Reddy AP, Pereira CB, Harttig U, Baird W, Spitsbergen JM, Hendricks JD, Orner GA, Williams DE, Swenberg JA (2009) Nonlinear cancer response at ultralow dose: a 40800-animal ED (001) tumor and biomarker study. Chem Res Toxicol 22:1264–1276
34. Tong S, von Schirnding YE, Prapamontol T (2000) Environmental lead exposure: a public health problem of global dimensions. Bull World Health Organ 78:1068–1077
35. Liu J, Goyer RA, Waalkes MP (2008) Toxic effects of metals. In: Klaassen CD (ed) Casarett and Doull's toxicology: the basic science of poisons. McGraw Hill, New York, pp 931–979
36. Landrigan PJ (2000) Pediatric lead poisoning: is there a threshold? Public Health Rep 115:530–531
37. Toscano CD, Guilarte TR (2005) Lead neurotoxicity: from exposure to molecular effects. Brain Res Brain Res Rev 49:529–554
38. Counter SA, Vahter M, Laurell G, Buchanan LH, Ortega F, Skerfving S (1997) High lead exposure and auditory sensory-neural function in Andean children. Environ Health Perspect 105:522–526
39. CDC (2005) Blood lead levels – United States, 1999–2002. MMWR 54:513–516
40. Bellinger D, Leviton A, Waternaux C, Needleman H, Rabinowitz M (1987) Longitudinal analyses of prenatal and postnatal lead exposure and early cognitive development. N Engl J Med 316:1037–1043
41. Lanphear BP, Dietrich K, Auinger P, Cox C (2000) Cognitive deficits associated with blood lead concentrations <10 microg/dL in US children and adolescents. Public Health Rep 115:521–529
42. Canfield RL, Henderson CR Jr, Cory-Slechta DA, Cox C, Jusko TA, Lanphear BP (2003) Intellectual impairment in children with blood lead concentrations below 10 microg per deciliter. N Engl J Med 348:1517–1526

43. Rogan WJ, Dietrich KN, Ware JH, Dockery DW, Salganik M, Radcliffe J, Jones RL, Ragan NB, Chisolm JJ Jr, Rhoads GG (2001) The effect of chelation therapy with succimer on neuropsychological development in children exposed to lead. N Engl J Med 344:1421–1426
44. White RF, Diamond R, Proctor S, Morey C, Hu H (1993) Residual cognitive deficits 50 years after lead poisoning during childhood. Br J Ind Med 50:613–622
45. Costa LG (2008) Toxic effects of pesticides. In: Klaassen CD (ed) Casarett & Doull's toxicology: the basic science of poisons. McGraw Hill, New York, pp 883–930
46. EPA US (2002) Interim reregistration eligibility decision for chlorpyrifos. US EPA, Washington DC, pp 1–235
47. Whyatt RM, Barr DB, Camann DE, Kinney PL, Barr JR, Andrews HF, Hoepner LA, Garfinkel R, Hazi Y, Reyes A, Ramirez J, Cosme Y, Perera FP (2003) Contemporary-use pesticides in personal air samples during pregnancy and blood samples at delivery among urban minority mothers and newborns. Environ Health Perspect 111:749–756
48. Aldridge JE, Meyer A, Seidler FJ, Slotkin TA (2005) Alterations in central nervous system serotonergic and dopaminergic synaptic activity in adulthood after prenatal or neonatal chlorpyrifos exposure. Environ Health Perspect 113:1027–1031
49. Roy TS, Sharma V, Seidler FJ, Slotkin TA (2005) Quantitative morphological assessment reveals neuronal and glial deficits in hippocampus after a brief subtoxic exposure to chlorpyrifos in neonatal rats. Brain Res Dev Brain Res 155:71–80
50. Slotkin TA, Seidler FJ (2005) The alterations in CNS serotonergic mechanisms caused by neonatal chlorpyrifos exposure are permanent. Brain Res Dev Brain Res 158:115–119
51. Levin ED, Addy N, Baruah A, Elias A, Christopher NC, Seidler FJ, Slotkin TA (2002) Prenatal chlorpyrifos exposure in rats causes persistent behavioral alterations. Neurotoxicol Teratol 24:733–741
52. Meyer A, Seidler FJ, Slotkin TA (2004) Developmental effects of chlorpyrifos extend beyond neurotoxicity: critical periods for immediate and delayed-onset effects on cardiac and hepatic cell signaling. Environ Health Perspect 112:170–178
53. Rauh VA, Garfinkel R, Perera FP, Andrews HF, Hoepner L, Barr DB, Whitehead R, Tang D, Whyatt RW (2006) Impact of prenatal chlorpyrifos exposure on neurodevelopment in the first 3 years of life among inner-city children. Pediatrics 118:e1845–e1859
54. Vandenberg LN, Maffini MV, Sonnenschein C, Rubin BS, Soto AM (2009) Bisphenol-A and the great divide: a review of controversies in the field of endocrine disruption. Endocr Rev 30:75–95
55. Calafat AM, Ye X, Wong LY, Reidy JA, Needham LL (2008) Exposure of the US population to bisphenol A and 4-tertiary-octylphenol: 2003–2004. Environ Health Perspect 116:39–44
56. Welshons WV, Nagel SC, vom Saal FS (2006) Large effects from small exposures III. Endocrine mechanisms mediating effects of bisphenol A at levels of human exposure. Endocrinology 147:S56–S69
57. Markey CM, Michaelson CL, Veson EC, Sonnenschein C, Soto AM (2001) The mouse uterotrophic assay: a reevaluation of its validity in assessing the estrogenicity of bisphenol A. Environ Health Perspect 109:55–60
58. Markey CM, Luque EH, De Munoz TM, Sonnenschein C, Soto AM (2001) In utero exposure to bisphenol A alters the development and tissue organization of the mouse mammary gland. Biol Reprod 65:1215–1223
59. Vandenberg LN, Maffini MV, Wadia PR, Sonnenschein C, Rubin BS, Soto AM (2007) Exposure to environmentally relevant doses of the xenoestrogen bisphenol-A alters development of the fetal mouse mammary gland. Endocrinology 148:116–127
60. Farabollini F, Porrini S, Della SD, Bianchi F, Dessi-Fulgheri F (2002) Effects of perinatal exposure to bisphenol A on sociosexual behavior of female and male rats. Environ Health Perspect 110(Suppl 3):409–414
61. Richter CA, Birnbaum LS, Farabollini F, Newbold RR, Rubin BS, Talsness CE, Vandenbergh JG, Walser-Kuntz DR, vom Saal FS (2007) In vivo effects of bisphenol A in laboratory rodent studies. Reprod Toxicol 24:199–224

62. Vom Saal FS, Richter CA, Ruhlen RR, Nagel SC, Timms BG, Welshons WV (2005) The importance of appropriate controls, animal feed, and animal models in interpreting results from low-dose studies of bisphenol A. Birth Defects Res A Clin Mol Teratol 73:140–145
63. vom Saal FS, Akingbemi BT, Belcher SM, Birnbaum LS, Crain DA, Eriksen M, Farabollini F, Guillette LJ Jr, Hauser R, Heindel JJ, Ho SM, Hunt PA, Iguchi T, Jobling S, Kanno J, Keri RA, Knudsen KE, Laufer H, LeBlanc GA, Marcus M, McLachlan JA, Myers JP, Nadal A, Newbold RR, Olea N, Prins GS, Richter CA, Rubin BS, Sonnenschein C, Soto AM, Talsness CE, Vandenbergh JG, Vandenberg LN, Walser-Kuntz DR, Watson CS, Welshons WV, Wetherill Y, Zoeller RT (2007) Chapel Hill bisphenol A expert panel consensus statement: integration of mechanisms, effects in animals and potential to impact human health at current levels of exposure. Reprod Toxicol 24:131–138
64. National Toxicology Program (2007) NTP-CERHR expert panel report on the reproductive and developmental toxicity of bisphenol A. In: Center for the Evaluation of Risks to Human Reproduction (ed) National toxicology program, pp 1–396
65. National Toxicology Program (2008) NTP-CERHR monograph on the potential human reproductive and developmental effects of bisphenol A. In: Center for the Evaluation of Risks to Human Reproduction (ed) National toxicology program, pp 1–321
66. US FDA (2010) Update on Bisphenol A for use in food contact applications. In: FDA (ed) US food and drug administration, pp 1–7
67. Briggs D (2003) Making a difference: indicators to improve children's environmental health. World Health Organization, Geneva, pp 1–17

Chapter 9
Ecological Risk Assessment and Animal Models

Lisa J. Bain

Glossary

Bioaccumulation factor (BAF)	The concentration of contaminant in the organism divided by the concentration of the contaminant from all exposure sources, including water, air, soil, sediment, or food.
Bioconcentration factor (BCF)	The concentration of contaminant in the organism divided by the concentration of the contaminant in the water.
Biomarkers	A quantifiable change in a biochemical, physiological, or histological parameter that can be used as a measure of response to a toxicant.
Biological survey	A systemic gathering of information on the presence, abundance, and condition of species at a given site to assess the health of those organisms.
EPT index	The number of organisms in the orders *Ephemeroptera* (mayflies), *Plecoptera* (stonefiles), and *Trichoptera* (caddisflies). These orders are typically considered to be sensitive to contaminants.
LC_{50}	The median lethal concentration, or the concentration of the medium that was lethal to 50% of the test organisms.

This chapter was originally published as part of the Encyclopedia of Sustainability Science and Technology edited by Robert A. Meyers. DOI:10.1007/978-1-4419-0851-3

L.J. Bain (✉)
Department of Biological Sciences, Clemson University, 132 Long Hall,
Clemson, SC 29634, USA
e-mail: LBAIN@clemson.edu; LBAIN@exchange.clemson.edu

E.A. Laws (ed.), *Environmental Toxicology: Selected Entries from the Encyclopedia of Sustainability Science and Technology*, DOI 10.1007/978-1-4614-5764-0_9,

LOEC	The lowest observed effect concentration, or the lowest concentration of test medium in which statistically significant toxic effects are seen.
NOEC	No observed effect concentration, or the highest concentration of test medium in which no significant toxic effects are seen.
Toxicity test	An assay in which organisms are exposed to soil, water, or air, and the effects of contaminants on processes such as survival, growth, and reproduction are determined.
Toxicogenomics	Examining changes in gene expression, protein expression, and/or metabolite profiles in a response to an exposure by a toxicant.

Definition of the Subject

An ecological risk assessment attempts to determine the effects of contaminants or other environmental stressors on individuals, populations, and communities of species. Although ecological risk assessments can be used for regulating chemicals and managing watersheds, the most typical use is for the study and remediation of contaminated sites. The organismal effects that can be examined vary considerably, depending on the ultimate use of the assessment, but most often focus on the organisms' ability to survive, grow and develop appropriately, and reproduce. The choice of which organisms to use should be based on easily evaluated criteria, such as the ease of sampling a particular organism, ensuring that the organism can both respond to the toxicant or mixture of toxicants, and that the organismal response is graded with respect to increasing exposure concentrations and/or times. The specific endpoints measured in the risk assessment with respect to organisms can therefore include lethality, weight gain, population numbers, birth and death rates, alterations in community structure, among others. Some endpoints are easily quantified, such as lethality. Other endpoints may be much more difficult to measure, such as reproduction, development, productivity, and immune function. For these more subtle outcomes, one often uses biological markers to indirectly assess these endpoints, or surrogate species in the laboratory to find out more about the health of the population or community under study. Other methods to better refine the risk assessment include the use of models and the use of genomics. The underlying goal to conduct an effective assessment relies on the choice of the organism and the endpoint examined so that management decision is appropriate and protective of organismal health.

Introduction

The selection of which organisms to monitor or use is an important part of the risk assessment process. This entry first discusses the role of organismal selection and response in the basic framework of risk assessment. The criteria used when selecting an appropriate organism to use and the common organisms routinely used in assessments will be discussed. Laboratory toxicity tests, using either standard organisms as surrogate species or nontraditional organisms that are more representative of the site can be used. Field studies are also routinely conducted, which can monitor the presence and the abundance of a variety of species. Beyond typical field assessment, biotic indices are used as measures of community structure, while biomarker data are used to assess sublethal impacts that often cannot be determined during standard laboratory toxicity testing. Finally, the entry will cover future directions in risk assessment, which may include the use of genomics, proteomics, other multi-end point analysis, and physiologically-based pharmacokinetic (PBPK) modeling.

Animal Models in the Basic Framework of Risk Assessment

Overview

The basic framework for risk assessment consists of four phases: problem formulation, analysis, risk characterization, and risk management [1, 2]. Much of this entry will focus on the first two phases, where the selection and analysis of organismal responses is most crucial. There is an additional step prior to the risk assessment itself, which is termed planning [3]. At the planning stage, the risk manager, risk assessors, and stakeholders should work together to identify the goals of the risk assessment. These goals should include developing clear objectives for what organisms the manager is most interested in protecting, determining what contaminants are most important to consider for the remediation effort, and indicating the acceptable levels of contaminants or the desired condition of the environment and organismal health after the remediation is completed [4].

During the problem formulation stage, a conceptual model should be developed to describe the relationships between the contaminants at the site and the organisms that may be affected by them. Although many species can be found at a site, the risk assessment is only concerned with those that could be adversely affected. Additionally, since there are always factors that mitigate or exacerbate the concentration of chemical that an organism encounters, these should be taken into account during model development. The risk assessor should identify the key ecological characteristics that allow or prohibit certain species to thrive. Potential confounding factors should be identified, such as physical manipulations, building and construction, other human disturbances, or competition by invasive species [4]. There is also the issue of indirect effects, in which the impact on a species affects another. Examples include a loss of a prey or predator species, or loss of vegetation that provides food and shelter. Once these factors are taken into account, then appropriate selection of the animal models to use in the risk assessment can occur.

During the analysis stage, there is both a characterization of the exposure and a characterization of the effects. Organisms may be sampled from the exposed site and changes due to the exposure determined. If suitable native species cannot be found or cannot be found in great enough numbers, outside organisms may be brought in and caged at the site to determine the amount of toxicant being taken into the organism and its resultant effects. More frequently, if a suitable field study cannot be performed on site, laboratory tests are conducted using surrogate species in standardized tests to determine the effects of contaminant exposure.

After data collection, one must determine whether there is a relationship between the contaminants and effects on organisms. One useful way to categorize the strength of the evidence is based upon the Bradford-Hill criteria [5], which provide a basic framework for assessing the strength of causal relationships. Although Hill originally published these criteria for human epidemiological work to determine the causes of diseases, they have been modified for ecotoxicology and risk assessments [6, 7]. The criteria or types of evidence a risk assessor is looking for to support a cause-effect relationship include:

1. *Spatial co-occurrence* – the effect should be observed only where the contaminant is present.
2. *Temporality* – there should be a cause-effect time relationship, in that the effect has to occur after the exposure to the contaminant.
3. *Biological Gradient* – this criterion describes the classical exposure-response relationship, in which higher exposure concentrations should lead to a greater incidence or more severe effects if the effect is due to the contaminant.
4. *Complete Exposure Pathway* – the contaminant has to be able to reach the organism in order to cause an effect. If the toxicant is in media that are inaccessible to the organism of interest, than exposure will not occur.
5. *Consistency of Association* – if similar observations are documented by multiple investigators in multiple locations, this increases the likelihood that the effects seen are due to the contaminant.
6. *Experiment* – if the exposure is manipulated or modified, this should cause a decrease in effects. The manipulation can be conducted by using site media in the laboratory and altering the contaminant's concentration by dilution, or by conducting field experiments in which the exposure can be modified, contained, or eliminated.
7. *Plausibility* – this criterion asks whether the effects seen in the organisms can be explained because of a known or plausible biological mechanism.

Additionally, the risk assessor should understand whether the effects would be expected at the concentration of contaminant seen at the site.

8. *Analogy* – this criterion examines uses case studies from contaminants that are similar in structure to the suspected contaminant, and asks whether the expected effects be the same.
9. *Specificity of Cause* – is the observed effect seen consistently associated with the contaminant or likely to only be caused by the contaminant. If so, the cause-effect relationship is quite strong.
10. *Predictive Performance* – this criterion asks whether side effects occurred as a result of the exposure, and if so, could they have been predicted based upon the contaminant's mode of action.

The first four provide the strongest evidence, since they are directly based upon the site being assessed. The next six are often considered less strong, since the data for these criteria typically come from laboratory studies or other field-based studies, which may or may not be completely applicable for the site in question. However, they provide corroborating evidence that, when coupled with the first four, can support the risk characterization process.

Monitoring Trends in the Environment

Knowing the baseline levels of contaminants and their impact on the health of populations and communities can support or act as a precursor to a risk assessment. Baseline levels are found by monitoring trends in the environment. Although these are not risk assessments, they can provide valuable information to the risk assessor. The US EPA's Environmental Monitoring and Assessment Program (EMAP) is one such large-scale endeavor that seeks to monitor sites in a particular region for many years, thus providing data on multiple spatial and temporal scales [8, 9]. EMAP data can include such information as contaminant levels, bacterial loads, and species distributions. Thus, the data can confirm trends in a particular area or provide baseline levels to serve as a reference site to aid risk assessors.

Biological surveys may also provide valuable information to support a risk assessment. Biological surveys most often are conducted in aquatic systems and often have data bases that span multiple sites and years. They are typically used to determine if an aquatic system is meeting its designated aquatic life uses, but have also been used to evaluate the effectiveness of controls and restoration activities, and characterize regional biotic attributes of reference sites. These surveys determine the presence or abundance of species at a given point in time. Information may also be collected on the numbers of sensitive or tolerant species, and may provide information on the overall community composition. These data is often converted and used in as a biotic index, which assigns numbers to analyzed sites to indicate the level of impact that external factors have on the community [10]. The survey can provide baseline data on the health of a particular system or region, which can be used as a reference during the risk assessment to determine whether the impacted site is similar or not.

Ecological epidemiology is another method to monitor trends in the environment. This is often used prior to a risk assessment, when the sources of contamination are either unknown or due to nonpoint sources [11]. Most often, this endeavor is initiated after reports of a noticeable environmental effect, such as the mortality of a large number of individuals, an increase in the number of deformities, or the decline of a species. The goal of the epidemiological study is to try to form a causal chain from source to exposure to effects. Once the cause is identified, those parties involved can then formulate the problem and continue on with the traditional risk assessment.

Considerations When Determining Which Organisms to Use for an Ecological Risk Assessment

The basic purpose of the risk assessment is either to estimate the exposure to the consumers or to the endpoint organisms, such as a carnivore higher up the food chain. Therefore, the relevance of the assessment heavily depends on choosing the appropriate organisms to monitor or study, or the appropriate species assemblages. There are certain criteria that are desirable when considering which organisms to use in the risk assessment [4]. First, the organism should be easily sampled. This suggests that the organism should be abundant in the area under study and amenable to capture and holding. Second, the organism should be sensitive to the contaminants in the system. If a particular species is known to be insensitive to PAHs, for example, than selecting it to determine effects of creosote contamination will not be protective of the environment. Third, the organism should respond to the toxicant under question in a predictable manner. Ideally, one wants a gradual change in response when the contaminant levels are increased. Ideally, a threshold should exist in which contaminant concentrations below these levels should be of little concern because they will not cause adverse effects [4, 12].

Levels of Biological Organization

There are different levels of biological organization that can be used in an ecological risk assessment: (1) monitoring or testing individual organisms; (2) testing or monitoring populations of a particular species; or (3) community level assessment (Table 9.1). The choice of level will ultimately depend upon a variety of factors, such as the types of organisms present, their abundance, and their types of interactions with one another. Using the assessment to investigate impacts on a rare species might necessitate only examining a few individual organisms. However, if during the problem formulation step, there is concern about indirect effects,

Table 9.1 Levels of biological organization and typical endpoints assessed during a risk assessment

Organism	Population	Community
Survival	Survival	Survival
Growth	Population density	Population density
Condition factor	Reproductive success	Community structure changes
Reproductive success	Productivity	Trophic imbalances
Tumors	Changes in genetic structure	Biomass
Behavioral alterations	Development of resistance	
Acetylcholinesterase inhibition	Biomass	
Stress proteins		
Changes in metabolism		
Immune system alterations		

in which the impact on one species affects another, than population or community levels may be most appropriate. For example, if the contaminant is thought to reduce levels of a prey species, then using a community level approach to determine predator–prey interactions might be in order.

The spatial characteristics of the site and the specific agent or agents being assessed will clearly influence the choice of biological organization, as will any temporal variations in the exposure. For example, if one is doing an assessment involving an effluent stream released once per month, this will necessitate a different strategy for testing and monitoring than an assessment for remediation of a contaminated site in which the organisms are continuously exposed. Along with temporal variations, the concentration of the agents may make a difference in the choice of scale as well. An exposure that is extremely high in concentration and pulsatile in nature would warrant different tests than a low level, chronic exposure [3].

The first level of organization is to monitor or test individual organisms. The endpoints garnered from these tests are typically survival, growth, and/or reproduction (Table 9.1). Depending on the organism used, these tests can take anywhere from hours to days. Population level tests typically are conducted via microcosms or mesocosms, using multiple organisms. The endpoints assessed are usually survival and growth/production, which takes days to months to conduct. The effects on populations can also be determined in the field by either a designed test or, more often, by simply monitoring the environment. Examples of these types of investigations might include examining organisms in a lake or pond, or in gridded and staked terrestrial areas. Again, the endpoints often assessed are survival, growth, reproduction, and production, which may take days to years to conduct. Community assessments are conducted either using a mesocosm (a small community), field tests, or environmental monitoring. Beyond the typical endpoints of survival and growth, one can now examine changes in community structure and trophic balance.

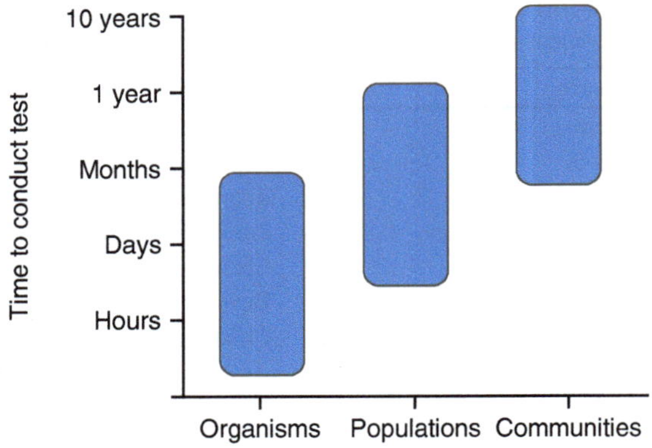

Fig. 9.1 Time scale for collecting field or laboratory data to support the risk assessment

Exposure Time, Organismal Characteristics, and Exposure Routes

Once the level of biological organization is decided upon, the next step is to determine the types organism(s) to use and how long to expose or monitor them. Influencing factors for the exposure time and type of organism include the type of media that are contaminated (soil, sediment, water), the concentrations of the contaminants, and what types of organisms are present at the site, and how many of them are at the site.

The length of exposure time for each of these tests depends on the agent and how it is released or occurs in the area. If assessing the effects of a pesticide application, one may only need to conduct the test or monitor the environment for a few days. If assessing the effects of an oil spill, monitoring for months to years might be more appropriate (Fig. 9.1). The timing of when the test is conducted also needs to take into account issues like seasonal variability. Water and food may be limited at certain times during the year, and thus the organisms' exposure could be reduced. Additionally, breeding and migration times may pose issues for accurate monitoring and should be taken into account before beginning the study.

Most risk assessments focus on a limited number of organisms – typically no more than just a few species or a small community, if looking at invertebrates or algae. The choice of organisms to examine therefore becomes critical. Among the specific factors and organismal characteristics that should be taken into consideration is the mobility of the organism. Ideally, one would want to monitor an organism with limited mobility, such as earthworms or invertebrates, or to monitor those organisms with low mobility at some stage in their life cycle [12]. Another concern is the sensitivity of species toward the contaminant of interest. Ideally, one would want to use the most sensitive species, which would be protective of all other species. However, given that most sites have a suite of contaminants, finding the species most sensitive to the mixture may be difficult. A third consideration is the ecological function or role of the organism. Often, it is good to choose at least one primary producers to study because of their large

Table 9.2 Routes of uptake in different organisms

Organism	Common uptake routes
Terrestrial invertebrates	Oral, dermal
Aquatic invertebrates	Oral, respiratory surfaces
Fish	Oral, gills, dermal
Mammals	Oral >> dermal, inhalation
Birds	Oral, inhalation
Amphibians	Dermal, oral
Reptiles	Oral

Routes listed are in the order of importance for each class of organism

contribution to the food chain. Other factors to take into account are choosing an organism that is easy to collect and choosing an organism of high relative abundance. Finally, one also needs to be concerned with a species that has a certain recognized value, either economic or symbolic, as the one that you want to protect. It may be necessary to monitor these types of species, even if they do not fit the other criteria for choice of an organism to use [12].

The last issue is to determine the most likely route of uptake, be it oral, dermal, or inhalation (for a recent review, see [13]). The oral route is the most likely exposure route (Table 9.2). However, organisms can also swim in water, absorb contaminants through dermal contact, and take up compounds via inhalation, so that depending upon the contaminants at the site, these minor routes of uptake may be more important. For terrestrial invertebrates, the most common type of organism to monitor is the earthworm. For these animals, exposure occurs either dermally or orally. Other arthropods can be examined, but these typically have an exoskeleton, so a food-borne exposure is more likely than dermal exposure. Aquatic invertebrates are most likely exposed through the oral route. Rather than conducting a test or monitoring these organisms, it is sometimes more desirable to use water or sediment bioaccumulation factors (BAFs) to estimate exposure. The uptake route of fish is typically oral but can also be via the gills or as a dermal exposure. As with aquatic invertebrates, one sometimes estimates the exposure to fish using BAFs or bioconcentration factors (BCFs). BAFs tend to be more consistent across different water chemistries and different organisms. They are also considered more realistic of an exposure since diet and food chain multipliers are included within the BAF. BCFs are determined in laboratory-based studies and are considered more precise. Therefore, a BCF is often what is used to estimate exposure, if available [3]. Exposure factors for many organisms can be found in the *Wildlife Exposure Factors Handbook* [14].

For most terrestrial organisms, the oral uptake route tends to be most important (for a recent review, see [13]). In mammals, the uptake of contaminants tends to be through the diet, while the dermal and inhalation routes tend to be negligible. Mammals have extremely varied feeding strategies, both in terms of prey preference and in feeding frequency. This can result in high exposures to some mammals while no exposure to others, so care must be taken in the risk assessment to choose a species that will come into contact with the contaminants of concern. Periods of fasting, hibernation, and migration can mobilize stored contaminants, resulting in toxicity without changing the body burden. Additionally, lacatational and placental transfer may occur from mothers to the developing offspring. In birds, the oral route is the most important uptake pathway, most commonly through the ingestion of contaminated prey items. During pesticide exposure, uptake via inhalation may contribute significantly to their body burdens. The transfer of contaminants from gravid females into their eggs may also represent a significant exposure source to embryos. In reptiles, the oral route is most important via ingestion of contaminated food items. The skin tends to inhibit some compounds from passing through dermally. As with avian species, contaminant transfer into eggs also represents an important exposure route. In contrast, the dermal route may be extremely important to amphibians due to their life cycle, which is spent in contact with both water and soil for parts of their life span.

When quantifying how much contaminant the organism had received during the exposure, one can determine its concentration in the whole body of the organism, its burden in an organ in which the toxicant is likely to exert a toxic effect, such as the liver, kidney, or brain, or in organs of accumulation, such as fat, hair, or feathers. As stated above, for aquatic organisms, BAFs or BCFs are often used to estimate exposure.

Field Studies and Toxicity Testing

Toxicity tests involve collecting contaminated soil, sediment, or water from the site, and then exposing organisms to that medium or those media in a laboratory setting. There are several different advantages of toxicity tests over field studies. First, a toxicity test can evaluate whether the contaminant was bioavailable to the organisms. Second, a toxicity test can enable a more informed assessment of the impact of mixtures of compounds than the use of data in the literature on single compounds. They can determine whether characteristics of the medium itself, such as hardness or pH, are influencing toxicity. Third, they can examine substances whose effects on wildlife are not well understood. Fourth, they are typically used for monitoring to determine whether the risk mitigation strategy is working [2, 15].

Although a variety of endpoints can be measured from toxicity tests, the most common ones are survival, growth, and reproduction. Survival is most often assessed using a short term, acute toxicity test. An acute test typically lasts 24–96 h, during which the organisms are exposed to a medium from the site, plus a reference medium, and the number of deaths is recorded. Outcomes of this test include the medium lethal concentration or LC_{50}, which is the concentration of the medium that killed 50% of the test organisms. Other values that are typically obtained are the lowest observed effect concentration (LOEC), which is the lowest concentration of test medium in which statistically significant toxic effects are seen, and the no observed effect concentration (NOEC), which is the highest concentration of test medium in which no significant toxic effects are seen.

Sublethal endpoints, such as growth and reproduction, are best measured using a chronic toxicity test. These tests take a longer period of time to conduct, because they encompass either a critical developmental phase or the organism's entire life cycle, but give better resolution as to whether a toxic effect is occurring on site. Similar to an acute test, the values obtained are typically a median effective concentration (EC_{50}), or the concentration of medium that caused effects, such as reduced growth or reproduction, on 50% of the organisms. A LOEC and NOEC are typically calculated for chronic tests.

Generally, values such as the LC_{50}, NOEC, and LOEC for a single contaminant using standardized test organisms can be easily be obtained from the literature or using databases such as the EPA's ECOTOX database (http://cfpub.epa.gov/ecotox/) or the USGS's Contaminant Exposure and Effects-Terrestrial Vertebrates database (CEE-TV) (http://www.pwrc.usgs.gov/contaminants-online/pages/ceetv/ceetvintro.htm). This means that oftentimes, a *de novo* test may not be necessary. However, one should consider conducting a toxicity test if the data for the compound or interest is absent in the literature, if there is reason to suspect that conditions at the site, such as pH or organic matter, may be influencing bioavailability or toxicity to organisms, or if the interactions between mixtures of compounds are considered significant.

Standard Toxicity Tests

When the risk assessor decides that a *de novo* toxicity test is needed, there are several elements that need to be decided upon. First, the assessor should decide upon the level of effort needed to achieve the goals of the study. If he or she only wants to know whether the medium is toxic or not, using two test organisms and collecting medium from just a few sites may be sufficient to obtain the desired information. However, to fully characterize the site's toxicity, the study might use test organisms at different trophic levels, a number of sampling locations, and several dilutions of each test medium [15]. Second, a reference site should be included that is similar in characteristics to the site under study and is located as close to the study site as possible. Third, the test medium must be decided upon. Although it is most typical to use the water, sediment, or soil directly from the site, one can use soil or sediment elutriate to expose the test organisms. Finally, the test organisms themselves need to be selected. Whenever possible, a standardized organism should be used that is representative of the onsite organisms and is sensitive to the contaminant of concern [15, 16].

Several types of acute aquatic tests involving fish, macroinvertebrates, mysids, Daphnia, and algae have been developed (for a good overview, see [17]). Most of these are 48–96 h LC_{50} tests that measure survival as the endpoint (Table 9.3).

Table 9.3 Standard toxicity tests

Test	Organism	Test duration	Endpoints	Reference
Freshwater acute	*Daphnia magna* or *Daphnia pulex*	48–96 h	Survival	[18, 19]
	Fathead minnow	96 h	Survival	[18, 19]
	Rainbow trout	96 h	Survival	[18, 19]
	Brook trout	96 h	Survival	[18, 19]
Freshwater chronic	*Cerodaphnia dubia*	7 days	Survival, reproduction	[20]
	Daphnia magna	21 days	Survival, reproduction	[21]
	Fathead minnow larvae	7 days	Survival, growth	[22]
	Algae (*Selenastrum capricornutum*)	96 h	Cell density	[23]
Saltwater acute	Bivalve mollusk larvae	48 h	Abnormal shell development	[24]
	Silverside (*Menidia* sp)	48 h	Survival	[18, 19]
	Sheepshead minnow (*Cyprinodon variegates*)	48 h	Survival	[18, 19]
Saltwater chronic	Silverside larvae	7 days	Survival, growth	[18, 19]
	Sheepshead minnow	7 days	Survival, growth	[18, 19]
	Mysid shrimp	7 days	Survival, growth, reproduction	[25]
	Sea urchin (*Arbacia punctulata*)	20 min	Egg fertilization	[25]
	Algae (*Champia parvula*)	5–7 days	Fertilization	[26]
Freshwater sediment chronic	*Hyalella azteca*	10–30 days	Survival, growth, reproduction	[26]
	Midge (*Chironomid tetans* or *C. riparius*)	10–14 days	Survival, growth	[26]
Marine sediment chronic	Amphipod	10 days	Burrowing	[25]
Terrestrial	Earthworm (*Eisenia foetida*)	14 days	Survival	[27]
	Bobwhite quail	8 days	Survival, growth	[28]
	Mallard duck	8 days	Survival, growth	[28]

Chronic tests generally involve either early life stage tests with juveniles, but there are some partial or full life cycle tests. Most of these measure endpoints such as survival, growth, and reproduction (Table 9.3).

Nonstandard Tests and Field Tests

If a laboratory test is needed to assess site medium, the risk assessor should try to choose a standardized test and a standardized organism (Table 9.3). This is not always possible, as the biota in a particular site may not be similar enough to a standardized organism, or the standardized organisms are not sensitive to the contaminant at the site. In these cases, using an alternative species is completely appropriate [2, 15]. Additionally, field studies are routinely conducted at the actual site. These studies are more concerned with effects on populations and communities, rather than on individual organisms [12]. Reasons for doing a field study might include the fact that a current or future effect is likely, but there is not enough information to support a management decision, or it can provide additional evidence to link a site's contaminant and adverse effects. The key to conducting a good nonstandardized laboratory test or a field test is to find an appropriate reference site.

Studies of sites in which the contaminants of concern are found in the water should include the examination of the effects on periphyton, plankton, benthic macroinvertebrates, or fish [12]. Periphyton and plankton, such as algae, protozoa, and some crustaceans, are primary producers. Periphyton is typically collected by scraping or suction, or an artificial substrate may be deployed for a number of days to allow attachment. Plankton is collected by pumping, netting, or trapping. Endpoints examined usually involve measures of community composition, such as species numbers, diversity, and richness.

Benthic macroinvertebrates are indicative of the effects of sediment contamination and are typically a good choice to examine if there are lipophilic contaminants at the site. They are collected by digging, netting, dredging, or grab sampling. Endpoints measured include the number and types of species, and other measures of species diversity. Macroinvertebrates can also easily be used in laboratory sediment toxicity tests, using either spiked artificial sediments or sediments taken from the site. Fish are good to examine because they can occupy different trophic levels, and they are relatively easy to collect by seining, passive netting, trawling, or electrofishing. Endpoints examined are often numbers and diversity, or sublethal effects such as behavior or tumor burden. Like benthic macroinvertebrates, fish are often used in laboratory toxicity tests.

Studies of sites in which the contaminants of concern are found in the soil often involve examination of effects on earthworms, nematodes, amphibians, mammals, or reptiles [29]. Earthworms and other soil nematodes are good for both metals and organics since they are in constant contact with the soil. Nematodes are collected by coring, sieving, or they can be driven from the soil by heat. Worms can be tested using spiked soil as the medium, or are examined directly on site, with typical endpoints being survival and reproduction [27]. Organisms such as amphibians, reptiles, and mammals are not used or collected as frequently. When they are, the risk assessor generally chooses a small species, since its home range is likely to be smaller and it tends to be more abundant than larger species. These types of organisms are generally collected by trapping, or their presence can simply be determined by examining surrogate measurements such as footprints, droppings, or carcasses [12]. Birds are not typically used or examined in a risk assessment because their home ranges can be quite large and their collection tends to be difficult and time intensive. When they are examined, they can be trapped to determine their numbers, or surrogate measures, such as auditory signals, used to determine their presence. Alternatively, the number of nests can be counted and the success of each nest can be documented.

Other Assays Providing Support to Risk Assessments

Beyond using standard toxicity tests to assess survival, growth, and reproduction, or looking for the presence of an organism during a field survey, there are several different ways to determine other types of adverse impacts, such as altered community composition, to estimate body burdens of toxicants, and to examine markers or exposure and/or effects.

Biological Surveys

Two such methods to examine altered community composition are biotic indices and rapid bioassessment protocols. These two types of surveys compare population and community compositions of macroinvertebrates, fish, or periphyton between a reference site and the site of interest. For both of these protocols, an initial survey is conducted to quantify the presence and abundance of different species at both sites. These quantities are then converted into numerical values that relate to the overall health of the community.

More than 50 different indices have been developed to assess community health [10]. Parameters examined include species richness, or the number of species in a community, density, dominance, diversity (the abundance of individuals in one taxon versus all other individuals), as well as the number of sensitive species, such as the number of species belonging to the Ephemeroptera (mayflies), Plecoptera (stonefiles), and Trichoptera (caddisflies) or the EPT index [30]. Other metrics include percent dominant taxon, percent tolerant taxa, and percent Chironomidae, all of which are predicted to increase in response to increased perturbation [10]. The Family Biotic Index is also a good measure of perturbation, as it uses tolerance values to weight abundance in an estimate of overall pollution [10]. Other indices such as the Simpson's Diversity Index and the Shannon-Wiener Diversity Index are also used to account for the richness and the evenness of the different species that make up the total community [31]. Finally, determining the percentage of individuals in each of the five functional feeding groups can be used to determine whether the trophic structure is balanced [32].

Since many of this indices require a lot of time, expertise, and money, rapid bioassessment protocols (RBPs) have been developed since the 1980s to obtain these same measurements in a more cost effective and efficient way. These protocols were developed from previously existing bioassessment methods that were used by many state agencies. The RBPs examine other parameters besides numbers of organisms. In the case of streams, these include the available cover, embeddedness, velocity and depth of the river or stream, sediment deposition, channel flow status, channel alteration, frequency of riffles, bank stability, vegetative protection, and riparian vegetative zone width [10]. RBPs are appropriate for distinguishing the existence and severity of impairment of the water body, as well as identifying the causes of impairment. They are also used to determine if a stream is supporting a designated aquatic life use specified in state Water Quality Standards, evaluating the effectiveness of controls and various restoration activities, and characterizing regional reference conditions. Additional information on both types of bioassessments and their uses is provided in another entry.

Biomarkers and Toxicokinetic Models

A second method to estimate exposure of an organism to a contaminant is to use a biomarker. Biomarkers are defined as a quantifiable change in a biochemical, physiological, or histological parameter that can be used as a measure of response to a xenobiotic. The idea behind biomarkers is that they can detect more subtle adverse effects than lethality at lower chemical concentrations, which would better protect the population or community. Biomarkers can be indicators of either exposure or effects, although exposure biomarkers are much more common. Examples of biomarkers routinely used in environmental toxicology studies include the induction of cytochrome P4501A (CYP1A) to indicate PAH exposure, acetylcholinesterase activity to indicate organophosphate or carbamate pesticide exposure, metallothionein induction to indicate exposure to several metals, vitellogenin to indicate exposure to estrogenic compounds, and eggshell thinning to indicate organochlorine exposure. Additional information on biomarkers is provided in another entry.

A final strategy is to incorporate physiologically-based pharmacokinetic (PBPK) models into the risk assessment. These model the time course of disposition of toxicants in the whole organism, which gives an estimate of body burdens.

In a PBPK model, the organism is represented by several different compartments in which toxicants might be expected to be absorbed in, distributed to, biotransformed in, and excreted from, such as the alimentary canal, blood, fat, liver, and kidney. Thus, one can use the model to predict concentrations of contaminants in an organism, rather than having to measure actual body burdens of contaminants, which can be time consuming and extremely expensive. Beyond cost, the models have other advantages, such as being able to provide a time course of distribution of toxicants to any organ or tissue, being able to estimate effects by changing physiological parameters such as food consumption, mobility, or reproductive status, and being able to use the same model to predict toxic action in different species. Although PBPK models can improve estimates in determining whether the contaminant will accumulate to high enough levels in the organism of interest to cause an adverse effect, there are a number of distinct disadvantages to their use compared to simply measuring body burdens. First and foremost, one needs information to build the model, such as blood volume, heart rate, breathing rate, and organ volumes. The fact that these parameters are often ill-defined for many species detracts from the power and utility of the model. Second, the mathematics, computational tools, and training needed to build and run the models may be out of reach for many risk assessors.

In spite of these obstacles, there are several PBPK models available for wildlife species, such as fish and polar bears, which can describe the body and tissue burdens of contaminants. For example, fish PBPK models have been built for rainbow trout and have been used to estimate tissue concentrations of paraoxon and cadmium [33–35]. Likewise, there is a PBPK model that estimates arsenic concentrations in the different organs of tilapia (*Oreochromis mossambicus*) [36]. Additionally, there are also models to assess ionizing radiation doses to a variety of non-human species [37]. The models for polar bears are a bit more sophisticated than those built for fish, in that they can estimate tissue concentrations of organohalogenated compounds, such as dieldrin and PCBs, but can also be used to estimate risk of effects, such as adverse impacts on reproduction [38]. It is likely that they can be extended to other at-risk populations, such as orcas and the arctic fox [39].

There are a few models in reptiles and birds that were developed to be more applicable to veterinary uses, but could be used to estimate contaminant concentrations in wildlife. Examples include a model developed in red-eared sliders (*Trachemys scripta elegans*) to determine the pharmacokinetics and tissue distribution of buprenorphine, a potential analgesic drug for reptiles [40]; a model for midazolam developed in chickens, but able to accurately predict tissue residues in turkey, pheasant, and quail [41]; and a model used to determine tissue levels of lipophilic pesticides in chickens, ducks, geese, and turkeys [42]. Alternatively, there are many rodent models for environmental contaminants that could potentially be adapted for use in other species. One such recent example includes a PBPK model in rats for developmental exposure to polybrominated diphenyl ethers (PBDEs) [43].

Future Directions Using Animal Models in Risk Assessments

Some future directions for increasing the quality of risk assessment is to refine our understanding of sublethal effects rather than focusing on survival as an endpoint, and to better determine the mode of action of contaminants. One potential way to accomplish this goal is through the use of genomics – analyzing changes in gene, protein, and metabolite expression. Genomics has the advantage of examining thousands of genes or proteins simultaneously as opposed to traditional biomarkers, which often measure a single gene or protein at a time. Like biomarkers, genomic changes can be linked to both exposure and effect outcomes. For example, rather than looking for reproductive failure in a toxicity test, one could examine changes in genes involved in steroidogenesis. These genes should be differentially expressed at much lower toxicant concentrations, yet can still provide a predictive measure that adverse reproductive effects are likely to occur [44]. Thus, genomic techniques are more sensitive than traditional toxicity testing or biomonitoring and can be used to detect early indicators of environmental perturbation.

One of the issues encountered when using genomic data in a risk assessment is that baselines or control patterns of expression need to be established for a variety of species. This is true for those organisms routinely used in risk assessments, such as fathead minnows and Daphnia, but even more so for nontraditional species and those organisms collected in field studies, since the variation in natural populations can be quite significant [45]. Second, the data generated from a genomic study cannot be easily translated into terms familiar to many risk assessors, such as a NOEL or EC_{50} [46, 47]. Some recent studies have developed no observed transcriptional effect levels (NOTEL) for *Daphnia magna* exposed to copper, and these NOTELs were validated after exposing Daphnia to water samples collected upstream and downstream from two copper mines in California [48, 49]. However, there are almost no published studies using a NOTEL and relating that value to an adverse event. Third, there is often a lack of phenotype associated with the changes in mRNA levels, making it hard to interpret just what types of adverse effects would be occurring to organisms at the site [46, 47, 50]. Studies linking adverse effects, such as altered growth and reproduction, with transcriptomic changes have also been conducted (reviewed in [51]), but there are too few of these published to support the current use of genomics in risk assessments. Finally, genomic studies are expensive and the computational needs are quite high.

The current policy at EPA is that toxicogenomics data may be used in a weight-of-evidence approach for ecological risk assessment, but must be provided in conjunction with other information. At this point, genomics data alone are insufficient for a risk assessment and any subsequent management decision [52, 53]. However, some potential uses for genomics in risk assessment for the foreseeable future are to guide the experimental design of monitoring programs so as to reduce the number of animals used, as an initial screening to rank or set priorities for further toxicity testing, and to help extrapolate the effects of contaminants across species [54].

Bibliography

Primary Literature

1. EPA (1992) Framework for ecological risk assessment. EPA/630/R-92/001, Risk Assessment Forum, Washington
2. EPA (1998) Guidelines for Ecological Risk Assessment. EPA/630/R-95/002F. Risk Assessment Forum, Washington
3. Suter GW II (2007) Ecological Risk Assessment, 2nd edn. CRC, Boca Raton, p 643
4. Dale VH, Mulholland PJ, Olsen LM, Feminella JW, Maloney KO, White DC, Peacock A, Foster T (2004) Selecting a suite of ecological indicators for resource management. In: Kapustka L et al (eds) Landscape ecology and wildlife habitat evaluation: critical information for ecological risk assessment, land-use mangement activities, and biodiversity enhancement practices. ASTM International, West Conshohocken
5. Hill AB (1965) The environment and disease: association or causation? Proc R Soc Med 58:295–300
6. Fox GA (1991) Practical causal inference for ecoepidemiologists. J Toxicol Environ Health 33:359–373
7. EPA (2000) Stressor identification guidance document. EPA/822/B-00/025. Office of Water, Washington
8. EPA (1997) Environmental monitoring and assessment program (EMAP): research strategy. EPA/620/R-98/001. Office of Research and Development, Washington
9. EPA (2002) Research strategy: environmental monitoring and assessment program. EPA 620/R-02/002. Office of Research and Development, Research Triangle Park
10. Barbour MT, Gerristen J, Snyder BD, and Stribling JB (1999) Rapid bioassessment protocols for use in streams and wadeable rivers: periphyton, benthic macroinvertebrates, and fish, 2nd edn. EPA 841-B-99-002, ed. U.S.E.P.A.O.o. Water, Washington
11. Suter GW II (2006) Ecological risk assessment and ecological epidemiology for contaminated sites. Hum Ecol Risk Assess 12:31–38
12. EPA (1994) ECO update: field studies for ecological risk assessment. Publication 9345.0-05I. Office of Emergency and Remedial Response, Washington
13. Smith PN, Cobb GP, Godard-Codding C, Hoff D, McMurry ST, Rainwater TR, Reynolds KD (2007) Contaminant exposure in terrestrial vertebrates. Environ Pollut 150:41–64
14. EPA (1993) Wildlife exposure factors handbook. EPA/600/R-93/187. Office of Research and Development, Washington
15. EPA (1994) ECO update: using toxicity tests in ecological risk assessments. Publication 9345.0-05I. Office of Solid Waste and Emergency Response, Washington
16. 04 AE (2004) Standard Guide for Selection of Resident Species as Test Organisms for Aquatic and Sediment Toxicity Tests. ASTM International, West Conshohocken, DOI: 10.1520/E1850-04
17. Landis WG, Yu M-H (2003) Introduction to environmental toxicology: impacts of chemicals upon ecological systems, 3rd edn. CRC, Boca Raton
18. 96 AE (2007) Standard guide for conducting acute toxicity tests on test materials with fishes, macroinvertebrates, and amphibians. ASTM International, West Conshohocken, DOI: 10.1520/E0729-96R07
19. 97 AE (2008) Standard guide for conducting acute toxicity tests on aqueous ambient samples and effluents with fishes, macroinvertebrates, and amphibians. ASTM International, West Conshohocken, DOI: 10.1520/E1192-97R08
20. 01 AE (2006) Standard guide for conducting three-brood, renewal toxicity tests with *Ceriodaphnia dubia*. ASTM International, West Conshohocken, DOI: 10.1520/E1295-01R06
21. E1193-97 AS (2004) Standard guide for conducting daphnia magna life-cycle toxicity tests. ASTM International, West Conshohocken, DOI: 10.1520/C0033-03

22. E1241-05 A (2005) Standard guide for conducting early life-stage toxicity tests with fishes. ASTM International, West Conshohocken, DOI: 10.1520/E1241-05
23. D3978-80 (1987) e1 (1997) Standard practice for algal growth potential testing with *Selenastrum capricornutum*. ASTM International, West Conshohocken
24. 98 AE (2004) Standard guide for conducting static acute toxicity tests starting with embryos of four species of saltwater bivalve molluscs. ASTM International, West Conshohocken, DOI: 10.1520/E0724-98R04
25. 03 AE (2008) Standard test method for measuring the toxicity of sediment-associated contaminants with estuarine and marine invertebrates. ASTM International, West Conshohocken, DOI: 10.1520/E1367-03R08
26. 05e1 AE (2005) Standard test method for measuring the toxicity of sediment-associated contaminants with freshwater invertebrates. ASTM International, West Conshohocken, DOI: 10.1520/E1706-05E01
27. 04 AE (2004) Standard guide for conducting laboratory soil toxicity or bioaccumulation tests with the lumbricid earthworm *Eisenia Fetida* and the Enchytraeid Potworm *Enchytraeus albidus*. ASTM International, West Conshohocken, DOI: 10.1520/E1676-04
28. 05e1 AE (2005) Standard practice for conducting subacute dietary toxicity tests with avian species. ASTM International, West Conshohocken, DOI: 10.1520/E0857-05E01
29. Golden NH, Rattner BA (2003) Ranking terrestrial vertebrate species for utility in biomonitoring and vulnerability to environmental contaminants. Rev Environ Contam Toxicol 176:67–136
30. Wallace JB, Grubaugh JW, Whiles MR (1996) Biotic indices and stream ecosystem processes: results from an experimental study. Ecol Appl 6:140–151
31. Lampert W, Sommer U (1997) Limnoecology: the ecology of lakes and streams, 2nd edn. Oxford University Press, New York
32. Merritt RW, Cummins KW (1995) An introduction to the aquatic insects of North America, 3rd edn. Kendall/Hunt, Iowa
33. Abbas R, Hayton WL (1997) A physiologically based pharmacokinetic and pharmacodynamic model for paraoxon in rainbow trout. Toxicol Appl Pharmacol 145:192–201
34. Thomann RV, Shkreli F, Harrison S (1997) A pharmacokinetic model of cadmium in rainbow trout. Environ Tox Chem 16:2268–2274
35. Franco-Uria A, Otero-Muras I, Balsa-Canto E, Alonso AA, Roca E (2010) Generic parameterization for a pharmacokinetic model to predict Cd concentrations in several tissues of different fish species. Chemosphere 79:377–386
36. Liao CM, Liang HM, Chen BC, Singh S, Tsai JW, Chou YH, Lin WT (2005) Dynamical coupling of PBPK/PD and AUC-based toxicity models for arsenic in tilapia *Oreochromis mossambicus* from blackfoot disease area in Taiwan. Environ Pollut 135:221–233
37. Beresford NA, Barnett CL, Brown JE, Cheng JJ, Copplestone D, Filistovic V, Hosseini A, Howard BJ, Jones SR, Kamboj S, Kryshev A, Nedveckaite T, Olyslaegers G, Saxén R, Sazykina T, Vives I Batlle J, Vives-Lynch S, Yankovich T, Yu C (2008) Inter-comparison of models to estimate radionuclide activity concentrations in non-human biota. Radiat Environ Biophys 47:491–514
38. Sonne C, Gustavson K, Rigét FF, Dietz R, Birkved M, Letcher RJ, Bossi R, Vorkamp K, Born EW, Petersen G (2009) Reproductive performance in East Greenland polar bears (*Ursus maritimus*) may be affected by organohalogen contaminants as shown by physiologically-based pharmacokinetic (PBPK) modelling. Chemosphere 77:1558–1568
39. Sonne C (2010) Health effects from long-range transported contaminants in Arctic top predators: An integrated review based on studies of polar bears and relevant model species. Environ Int 36:461–491
40. Kummrow MS, Tseng F, Hesse L, Court M (2008) Pharmacokinetics of buprenorphine after single-dose subcutaneous administration in red-eared sliders (*Trachemys scripta elegans*). J Zoo Wildl Med 39:590–595

41. Cortright KA, Wetzlich SE, Craigmill AL (2009) A PBPK model for midazolam in four avian species. J Vet Pharmacol Ther 32:552–565
42. MacLachlan DJ (2010) Physiologically based pharmacokinetic (PBPK) model for residues of lipophilic pesticides in poultry. Food Addit Contam A Chem Anal Control Expo Risk Assess 27:302–314
43. Emond C, Raymer JH, Studabaker WB, Garner CE, Birnbaum LS (2010) A physiologically based pharmacokinetic model for developmental exposure to BDE-47 in rats. Toxicol Appl Pharmacol 242:290–298
44. Kennedy SW, Euling S, Huggett DB, De Coen LW, Snape JR, Zacharewski TR, Kanno J (2008) Application of genomics to screening. In: Ankley GT et al (eds) Genomics in regulatory ecotoxicology. CRC, Pensacola
45. Oleksiak MF, Churchill GA, Crawford DL (2002) Variation in gene expression within and among natural populations. Nature Genet 31:261–266
46. Ankley GT, Daston GP, Degitz SJ, Denslow ND, Hoke RA, Kennedy SW, Miracle AL, Perkins EJ, Snape J, Tillitt DE, Tyler CR, Versteeg D (2006) Toxicogenomics in regulatory ecotoxicology. Environ Sci Technol 40:4055–4065
47. Boverhof DR, Zacharewski TR (2006) Toxicogenomics in risk assessment: applications and needs. Toxicol Sci 89:352–360
48. Poynton HC, Loguinov AV, Varshavsky JR, Chan S, Perkins EJ, Vulpe CD (2008) Gene expression profiling in Daphnia magna part I: concentration-dependent profiles provide support for the No Observed Transcriptional Effect Level. Environ Sci Technol 42:6250–6256
49. Poynton HC, Zuzow R, Loguinov AV, Perkins EJ, Vulpe CD (2008) Gene expression profiling in Daphnia magna, Part II: validation of a copper specific gene expression signature with effluent from two copper mines in California. Environ Sci Technol 42:6257–6263
50. Degitz SJ, Hoke RA, Bradbury S, Brennan R, Ferguson L, Klaper R, Orban L, Spurgeon D, Tilton S (2008) Application of genomics to regulatory ecological risk assessments for pesticides. In: Ankley GT et al (eds) Genomics in regulatory ecotoxicology. CRC, Pensacola
51. Van Aggelen G, Ankley GT, Baldwin WS, Bearden DW, Benson WH, Chipman JK, Collette TW, Craft JA, Denslow ND, Embry MR, Falciani F, George SG, Helbing CC, Hoekstra PF, Iguchi T, Kagami Y, Katsiadaki I, Kille P, Liu L, Lord PG, McIntyre T, O'Neill A, Osachoff H, Perkins EJ, Santos EM, Skirrow RC, Snape JR, Tyler CR, Versteeg D, Viant MR, Volz DC, Williams TD, Yu L (2010) Integrating omic technologies into aquatic ecological risk assessment and environmental monitoring: hurdles, achievements, and future outlook. Environ Health Perspect 118:1–5
52. Dix DJ, Gallagher K, Benson WH, Groskinsky BL, McClintock JT, Dearfield KL, Farland WH (2006) A framework for the use of genomics data at the EPA. Nat Biotechnol 24:1108–1111
53. EPA (2007) Interim guidance for microarray-based assays: data submission, quality, analysis, management, and training considerations. Draft, Science Policy Council, Washington
54. Perkins EJ, Denslow N, Chipman JK, Guiney PD, Oris JT, Poynton H, Robidoux PY, Scroggins R, Van Der Kraak GJ (2008) Application of genomics to assessment of the ecological risk of complex mixtures. In: Ankley GT et al (eds) Genomics in regulatory ecotoxicology. CRC, Boca Raton

Books and Reviews

Barnthouse LW, Munns JWR, Sorensen MT (eds) (2008) Population-level ecological risk assessment. CRC, Boca Raton
Kapustka L, Biddinger G, Luxon M, Galbraith H (eds) (2004) Landscape ecology and wildlife habitat evaluation: critical information for ecological risk assessment, land-use management activities, and biodiversity enhancement practices. ASTM International, West Conshohocken

Solomon KR, Brock TCM, de Zwart D, Dyer SD, Posthuma L, Richards SM, Sanderson H, Sibley PK, van den Brink PJ (eds) (2008) Extrapolation practice for ecotoxicological effect characterization of chemicals. CRC, Boca Raton

Ware GW (2006) Reviews of environmental contamination and toxicology, vol 187. Springer, New York

Chapter 10
Environmental Toxicology: Carcinogenesis

Vincent L. Wilson

Glossary

Complex dose	The biologically effective (e.g., carcinogenic) dose of an agent that takes into account additive, subtractive, and synergistic interactions in a complex mixture
DNA adduct	Any covalent addition or modification of a nucleotide or the phosphate backbone in DNA
Driver mutation	Oncogenic mutations that are primary to the further advancement and development of a cancer
Genotoxic	Causes DNA damage and/or mutations
Immortal stranding	The retention of a specific strand of the DNA double helix by a stem cell regardless of how many times the cell divides
Initiation	The formation of the first oncogenic mutation in a cell
Oncogenic process	Any process that is involved in the development of a tumor
Proto-oncogene	A normal gene that upon mutation produces an active oncogenic component in the development of cancer
Promotion	The effect of encouraging the development of cancer by increasing cellular proliferation and turnover
Tumor suppressor gene	A normal gene whose expression product appears to block the development of cancer

This chapter was originally published as part of the Encyclopedia of Sustainability Science and Technology edited by Robert A. Meyers. DOI:10.1007/978-1-4419-0851-3

V.L. Wilson (✉)
Department of Environmental Sciences, Louisiana State University,
Baton Rouge, LA 70803, USA
e-mail: monster77@aol.com

E.A. Laws (ed.), *Environmental Toxicology: Selected Entries from the Encyclopedia of Sustainability Science and Technology*, DOI 10.1007/978-1-4614-5764-0_10,

Definition of the Subject

Carcinogenesis is the induction of cancer by exposure to exogenous agents, chemical or physical carcinogens. Cancer is a large family of life-threatening environmental diseases. This definition is supported by the plethora of published evidence that cancer results from the accumulation of DNA mutations, which in turn depends upon progressive cell divisions. This process takes time. The accumulation of a sufficient allotment of oncogenic genetic errors for cancer to develop occurs over a large number of cell divisions (10s to 1,000s), which may take 10–50 years or more. Even without the intentional introduction of carcinogens into the human body by occupation or habits such as tobacco smoking, alcoholic use, and other life-style choices, our environment provides numerous opportunities and sources of DNA-damaging and potentially mutagenic lesions in our cells. Given enough time, the risk of cancer is inherent with life, while exposure(s) to carcinogenic agents not only shortens the time required to achieve tumor development, but also increases the probability of cancer occurrence.

In the present document, the term environmental carcinogenesis encompasses the induction of cancer by exposure to carcinogenic agents whether natural or man-made, present in our environment. The term environmental carcinogens will refer to all compounds foreign to the human or mammalian body, that is, xenobiotic and physical carcinogens. However, the development of cancer is a long and involved process that requires numerous cellular perturbations generally produced by both exogenous and endogenous reactive agents. As will be clarified in this chapter, for many cancers it is exceedingly difficult to separate out the contributions of the various etiological factors, including natural environmental carcinogens from man-made chemicals (see Table 10.1). For instance, a number of carcinogenic xenobiotics have been identified that are inherent in our food supply, for which the best-known examples are aflatoxins. Aflatoxins are present in fresh corn, peanuts (and peanut butter), and other grains, while other carcinogens such as PhIP (2-Amino-1-methyl-6-phenylimidazo[4,5-*b*]pyridine), are produced from cooking meat [1, 2]. The potent nephrotoxic aristolochic acid component of certain herbs sometimes used in herbal remedies produces urinary tract cancers [3–5]. To add to this disparaging viewpoint, the lack of proper nutrition, such as folate deficiency, also leads to genomic instabilities that are involved in carcinogenesis [6–8].

The contribution of a few environmental carcinogens is more clearly evident in selective cancer cases. Skin cancer induced by the ultraviolet (UV) radiation from sunlight is one of the more prominent environmental cancers in our world [9, 10]. Mesothelioma may result from inhalation of asbestos fibers [11]. Exposures due to personal life-style choices and habits, such as tobacco smoking and alcohol use have been well studied and also provide clear examples of environmental carcinogenesis [12–16]. The carcinogenic impact of occupational exposures is another area that offers substantial clarification of mechanisms and risks involved in cancer development. Similarly, medical exposures, especially those involving chemotherapeutic agents and radiation therapy provide further risks of carcinogenic impacts, although these agents comprise risks of a spectrum of cancer types.

Table 10.1 The causes of cancer

Agent//Source	Examples	Associated cancer(s)
Chemical carcinogens[a]	Aflatoxins	Hepatocellular carcinoma
	Alcohol	Hepatic, breast, stomach, colorectal, oral, ovarian, pancreatic
	Aristolochic Acid	Urinary tract transitional cell carcinoma (renal, ureter, bladder)
	Asbestos	Mesothelioma and lung cancers
	Benzo(a)pyrene	Lung cancers (tobacco smoke associated)
	Benzene	AML and non-Hodgkin's lymphoma
	Formaldehyde	Nasopharyngeal
	Metals	
	Arsenic	Skin, liver, lung, kidney, bladder
	Beryllium	Lung cancer
	Cadmium	Lung
	Chromium (VI)	Lung, stomach
	Nickel compounds	Lung, nasopharyngeal
	Asbestos	Mesothelioma and lung cancers
	2-Naphthylamine	Bladder
	Tobacco Smoke	Lung cancers (also nasopharyngeal, esophageal, bladder, pancreatic, kidney)
	Tobacco Dip/Chew	Oral cancers
	Vinyl chloride	Hepatic angiosarcoma
Physical carcinogens[a]	Ionizing radiation (α, β, γ, X-rays, etc.)	multiple types of tumors
	Sunlight (Solar Radiation)	Skin (BCC, SCC, and melanoma)
Biological carcinogens	Hepatitis B, chronic infection	Hepatocellular carcinoma
Viruses	Hepatitis C, chronic infection	Hepatocellular carcinoma
	Epstein-barr virus	Burkits lymphoma, nasopharyngeal carcinoma
	HPV (high risk types)	Cervical cancer
Parasites (chronic inflammation)	Schistosoma hematobium	Bladder cancer
Chronic inflammation	Inflammatory bowel disease	Colon or intestinal cancers
	Crohn's disease	
	Ulcerative colitis	
	Chronic pancreatitis	Pancreatic cancer
	Barrett's esophagus	Esophageal carcinoma
	Helicobacter pylori, chronic gastritis	Stomach cancer
Fidelity limits of enzymatic processes	DNA Replication errors (1 to 3 errors per genome replication)[b]	Possibly all or most cancers
	Imperfect fidelity of DNA repair systems	Possibly all or most cancers

(continued)

Table 10.1 (continued)

Agent//Source	Examples	Associated cancer(s)
Genetics (inherited predisposition)		
Cancer family syndromes	Breast cancer family (BRCA1 or BRCA2 gene mutation)	Breast cancer
	FAP (Familial adenomatous polyposis) (APC gene mutation)	Colorectal carcinoma
	Hereditary melanoma syndromes Dysplastic nevus syndrome Familial atypical mole-malignant melanoma syndrome Melanoma-astrocytoma syndrome	Melanoma
	HNPCC (Hereditary nonpolyposis colorectal carcinoma) (MSH2 or MLH1 gene mutation)	Colorectal carcinoma
	Li-Fraumeni cancer family syndrome (TP53 gene mutation in 70% cases)	Osteosarcoma, adrenocortical carcinoma, brain, breast, colon, leukemia, lung, pancreatic cancers, various sarcomas
	MEN2A and 2B (Multiple endocrine neoplasia, types 2A and 2B) (RET gene mutation)	Pheochromocytoma, thyroid carcinoma
	Neurofibromatosis types 1 and 2 (NF1 and NF2 gene mutations)	Neurofibromas, schwannomas, meningiomas, ependymomas, acoustic neuromas
	Retinoblastoma (RB gene mutation)	Retinoblastoma
	Von Hippel–Lindau Syndrome (VHL gene mutation)	Renal cell carcinoma, pheochromocytoma, hemangioblastomas
Other genetic diseases	Ataxia telangiectasia (AT)	Leukemias, lymphomas
	Bloom's syndrome	Leukemias, lymphomas, carcinomas
	Xeroderma pigmentosum (XP)	Skin (BCC, SCC, and melanoma)

[a]All chemical and physical carcinogens listed are classified as Group I: Carcinogenic to Humans, by the International Agency for Cancer Research (IARC), update January 16, 2009; http://monographs.iarc.fr/ENG/Classification/index.php
[b]DNA synthesis and the replication machinery has an astonishing fidelity of one error per 10^9 base pairs, which results in one to three errors every time the human genome is replicated [31, 32].

Since the same principles and mechanisms involved in the development of human cancer apply to various animal species, and the primary focus of environmental carcinogenic risk calculations is designed to protect humans, the human will be the primary focus of this chapter on environmental carcinogenesis.

Introduction

The biophysical aspects of this planet support and sustain life, while also being conducive to the development of cancer in most multicellular species. Fortunately, the risk of cancer is low without the assistance of acute exposures to carcinogenic agents. However, the risk and incidence of cancer increases with age in the mammalian organism [17, 18]. The adage that "cancer is a disease of old age" has its foundings in empirical observation, while also demonstrating the requirement of extensive cell divisions (10s to 1,000s) for the development of tumors [19, 20]. The mixtures of gases that make up Earth's atmosphere sustain the carbon-based metazoan life forms on this planet due to the cellular oxidative metabolic processes. Low carbon dioxide (less than 0.04%) and high oxygen (20.9%) is crucial to the life-sustaining respiration of mammals. Unfortunately, the same life-sustaining oxidative metabolism functioning in the human cell also produces reactive oxygen species (ROS) that damage our cellular components including membranes, lipids, microfilaments, enzymes, and DNA. Additional ROS are produced by exposures to exogenous toxicants, including ionizing radiation, UV light, numerous toxic chemicals and heavy metals. The level of oxidative damage to mammalian DNA has been reported to be as high as 10,000 insults per cell per day [21, 22]. DNA replication (cell division) in the presence of oxidative damage leads to fixed mutations (permanent mutations). Endogenous DNA damage occurs in the form of modified nucleotides (e.g., 8-hydroxy-2′-deoxyguanosine, thymidine glycol, 5-hydroxymethyluracil, etc.), depurination, depyrimidination, deamination, and single and double strand breaks on a daily basis from oxidative processes [23–27]. Tallying up all these insults suggests that human DNA receives greater than 10,000 insults per cell per day. Other mammals receive similar insults at the level proportional to their basal metabolic rates, that is, mice and rats receive approximately tenfold more DNA damage per cell per day as humans. The DNA repair systems of the human cell have a phenomenally high fidelity, but even an error rate of only 0.0001% (the equivalent of one in a million) may lead to the accumulation of genetic errors on the order of 1 every 100 days for each and every cell [28, 29]. To add further insult to injury, the fidelities of DNA polymerases are also less than perfect, although remarkably precise in the human, such that errors are incorporated during DNA synthesis [30]. The reported occurrence of one to three mutations every time the human genome is duplicated further supports the limitations of cellular functions and endurances [31, 32]. Thus the processes involved in sustaining life also lead to DNA mutations. These endogenous life processes coupled with the environment that is so crucial in supporting and nurturing life leads to the accumulation of DNA mutations and, over many years, a significantly increased risk of cancer.

Nature maintains a balance between productive life and cancer risk. As with any genetic disease that terminates life before reaching a reproductive age or during the fertile years, the early occurrence of terminal cancer would be self-limiting. Barriers to cancer development are necessary for the survival and proliferation of the species. As advanced life forms evolved, so did a number of protective mechanisms that are quite effective in derailing tumor development. In addition to detoxifying enzymes for reactive species, glutathione to quench reactive electrophiles, and numerous DNA repair mechanisms, mammalian and human cells become terminally senescent or undergo self-suicide (apoptosis) upon self-recognition of excess DNA damage or perturbation of cell growth controls [33]. The processes involved in cellular aging, that is, loss of telomere length of chromosomes, shifts in gene expression, and responsiveness to external signals that finally lead to replicative senescence after a finite number of cell divisions, may actually be a means to block the development of cancer. Terminal differentiation programs, such as the short 4–5 day lifespan of the colonic mucosal cells, and the approximately 14 day lifespan of the squamous differentiating skin keratinocytes ensures that these cells will not be around if they pick up too many oncogenic DNA mutations during the rapid cell divisions that take place immediately prior to the final terminal differentiation processes [34]. Should these intracellular anticancer mechanisms fail, the local tissue microenvironment plays a role in containing or dampening the expansion of abnormal cells, and the body's immune system may take over and remove aberrant cells, thereby removing the threat of cancer [35–37]. Considering the low frequency of occurrence of cancer until old age, these endogenous protective anticancer mechanisms are advantageous for the human race, albeit not necessarily for the less fortunate individual statistical outlier.

Accumulation of Mutations

Avenues are available for the determined cell to find ways to circumvent these protective barriers against cancer development. With each protective mechanism, one or more DNA mutations enable cells to bypass this blockade and progress toward a neoplastic state. These "driver" mutations are the primary oncogenic errors involved in tumor development. Due to the number and variety of checks and balances opposing oncogenic development, a cell must accumulate a sufficient number of "driver" mutations to circumvent all the blockades for a tumor to develop, without meeting its demise by one of these protective mechanisms. This process is a statistical probability that weighs heavily on the healthy life side until old age, unless perturbed by carcinogenic exposures whereby the probability of tumor development increases (see Table 10.1). The rate of tumor development is increased sometimes many orders of magnitude above the spontaneous background rate by exposure to exogenous carcinogens. The timing of the onset of cancer is due to the dose and duration of carcinogen exposure and not to either chronological age or the age that exposure began [38, 39].

Properties of Cancer Cells

The product of carcinogenesis is a thriving cancer cell. The successful cancer cell has collected driver mutations that enable it to: (1) ignore antigrowth signals in the tissue microenvironment; (2) be at least partially independent of the requirements for exogenous mitogenic factors for cellular growth and proliferation; (3) block or inhibit differentiation programs; (4) evade apoptotic and senescence programs; (5) establish an extended replicative potential or immortality; (6) secrete factors that enhance and sustain angiogenesis; (7) acquire the ability to evade the immune system; and (8) acquire genomic instability [33, 35–37]. These are the minimal requirements for the development of a benign tumor. A malignancy has acquired one or more additional mutations that confer a ninth property, that of invasive and metastatic capabilities. For each of these nine properties of cancer, one or more mutations are needed to alter the intracellular signaling pathways and/or response to the tissue microenvironment. In some cases, only one driver mutation may be required to confer a specific property, but many of these nine properties are conferred only after multiple oncogenic mutations have been established in the developing cancer cell genome.

Oncogenic Mutations

The establishment of each driver mutation requires the successful negotiation of several protective molecular mechanisms. First of all, the formation of a DNA adduct, single or double strand break, or other form of DNA damage must occur due to exposure to a physical carcinogen (i.e., UV light or ionizing radiation), a chemical carcinogen, or the presence of an endogenous reactive species that has successfully negotiated the gamut of detoxifying enzymes and glutathione. This DNA damage must also have occurred in an oncogenically important base, gene, and/or sequence. Next this DNA damage must be converted into a permanent genetic alteration, a fixed mutation. The formation of a fixed mutation requires the combined failure of the cell's genome surveillance system to recognize and/or repair the DNA lesion, incorporation of an incorrect nucleotide (miss-pairing), and the survival of this mistake up to and through a second round of DNA synthesis, or through faulty repair [40–42] (Fig. 10.1). The frequency by which this occurs is exceedingly rare, as noted above by the high fidelity of the DNA repair system alone. The bombardment of cells in the human body by acute and chronic exposures to exogenous carcinogens increases the frequency of DNA damaging events and cellular turnover, both of which in turn increase the occurrence of fixed mutations and tips the balance of these processes toward an increased risk of cancer development (Fig. 10.1). The higher the dose received and the longer the exposure, the greater the risk of cancer.

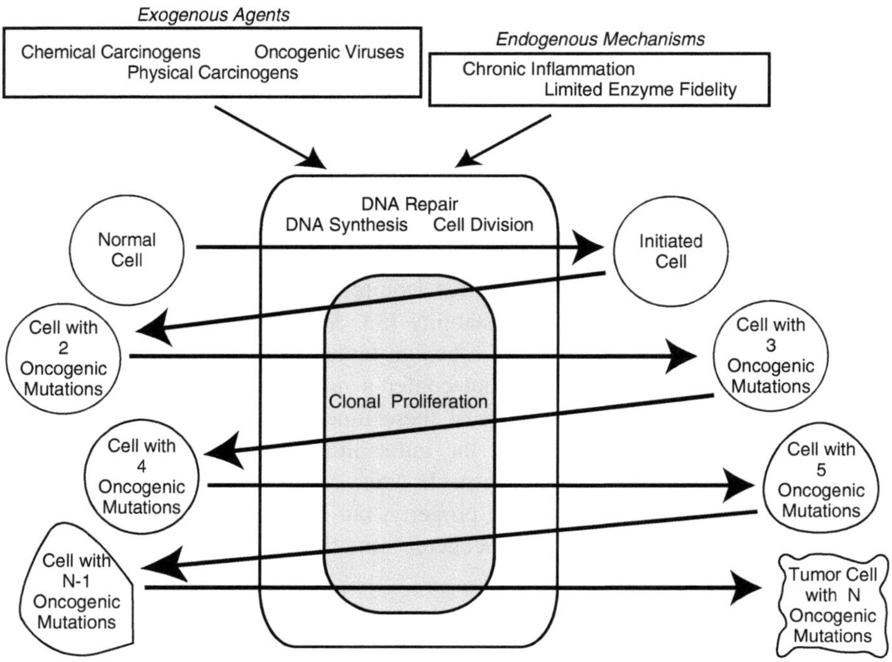

Fig. 10.1 A schematic representation of the accumulation of oncogenic mutations and the influences of exogenous agents and endogenous mechanisms on this process. Each time a cell completes the cell cycle and divides, there is a good chance that one or more mutations will occur. Fortunately, most of these DNA mutations are not oncogenic. However, the probability of an initiated cell or a cell with two or more oncogenic mutations, picking up another oncogenic mutation increases as clonal expansion occurs. A tumor develops once a sufficient number of oncogenic mutations have accumulated to bypass all of the protective barriers of the normal human cell

There are several mechanisms by which carcinogens can influence the formation of DNA mutations, and the type of mutation(s) formed is dependent upon the mode of action of the carcinogen (see Table 10.2). Exogenous and endogenous agents can produce DNA mutations by directly damaging DNA, perturbing DNA replication machinery, polymerases, etc., inhibiting DNA repair systems, critically shortening telomere lengths, altering the cytoskeletal proteins, mitotic spindle fibers, enzymes involved in mitotic processes, and by stimulating cellular turnover and proliferation. These different mechanistic impacts on cells encourage the formation of a variety of mutations, generally classified as single base mutations, chromosomal aberrations, insertions and deletions (indels), and epigenetic mutations.

Table 10.2 Genotoxic and carcinogenic mechanisms

Mode of action	Effects	Mutations	Instability
1. Direct DNA damage (covalent alterations)	Base pair mismatch	Single base mutation (SBS and Indels)	NSI
	Depurination and depyrimidination	Indels (larger than one base) Epigenetic mutations	EIS
	Single strand break Double strand break	Chromosomal aberrations (deletions, insertions, translocations)	CIN
2. Interference with DNA polymerases and replication machinery	Base pair mismatch	Single base mutation (SBS and Indels)	NSI
	Microsatellite slippage	Microsatellite length changes	MSI
	Perturbation of epigenetic patterns	Epigenetic mutations	EIS
3. Interference with cytoskeleton and mitotic spindle Mitotic proteins and enzymes	Nondisjunction	Aneuploidy	CIN
4. Interference with DNA repair systems	Faulty mismatch repair	Microsatellite length changes	MSI
	Improperly or unrepaired	Single base mutation Indels	NSI
	Single and double strand breaks	Chromosomal aberrations (translocation, inversions, deletions, etc.)	CIN
5. Interference with maintenance of telomeres	Critically shortened telomeres	Dicentric chromosomes aneuploidy	CIN
6. Cytotoxicity	Increased cell turnover Increased opportunity for accumulating mutations	(Most types of mutations)	Any possible (NIS, MSI, CIN, EIS)
7. Stimulate cellular proliferation (promotion)	Clonal expansion Increased probability for accumulating mutations	(Most types of mutations)	Any possible (NIS, MSI, CIN EIS)

CIN chromosomal instability, *EIS* epigenetic instability, *MSI* microsatellite instability, *NIS* nucleotide instability (also known as single base instability)

Single Base Mutations

Single base mutations encompassing single base substitutions and single base indels represent a significant proportion of cancer-associated genetic errors. It is generally accepted that the more transformed a cell becomes, the more likely a single base substitution will occur in an activating site of a proto-oncogene or in deactivating site in a tumor suppressor gene. Most of the alkylating and arylating carcinogens listed in Table 10.1 are known to produce single base mutations, especially single base substitution mutations. Oxidation and the production of ROS and reactive nitrogen species (RNS), common with inflammatory processes, are also well known to produce single base mutations [25, 43]. The vast majority of errors produced by DNA polymerases and the DNA replication machinery are single base mutations [31, 32].

Single base substitution (SBS) mutations represent 85% or more of the activating and deactivating mutations in proto-oncogenes and tumor suppressor genes, respectively, in human tumors [44–48]. Although single base deletions and insertions are sometimes involved, SBSs represent the vast majority of these oncogenic errors [47, 49, 50]. This has been clearly demonstrated in human tissues for a number of proto-oncogenes and tumor suppressor genes, for example, Braf, H-, K-, and N-Ras, RB and TP53 [45–47, 49–51]. Missense SBS mutations alone comprise more than 75% of all identified disease-associated TP53 mutations [49].

Chromosomal Aberrations

Cancer-associated chromosomal errors have been known since the middle of the last century. Cytogenetic studies identified numerous leukemia-specific (and lymphoma-specific) chrosomal translocations [52]. The best-known example is the chronic myeloblastic leukemia (CML)-specific chromosomal translocation between chromosomes 9 and 22 [t(9;22)(q34:q11)],which forms the chimeric bcr-abl gene. Chromosomal errors are prominent in many types of human tumors and most likely crucial to the development of at least some leukemias and lymphomas [52]. Aneuploidy is the most frequently identified mutation in human cancers, albeit cytogenetic analysis is routine medical care for cancers.

Disruption of cellular cytoskeleton, mitotic spindle apparatus, and/or enzymes involved in the processes of mitosis commonly lead to chromosomal errors. For example, asbestos fibers spear cells and bind proteins that are involved in mitosis, leading to non-disjunction and other chromosomal aberrations [53]. Polycyclic aromatic hydrocarbons (PAHs) also produce chromosomal errors by causing the rapid shortening of the telomere length [54]. Tobacco smoke constituents play a prominent role in the chromosomal aberrations [55–57].

Indel Mutations

Chemical and physical carcinogens are well known to cause single and double strand breaks in DNA. Some DNA adducts, such as the alkyl-N7-deoxyguanosine formed by alkylation (or aryl-N7-deoxyguanosine by arylation) may lead to a depurinic site, which weakens the strand and easily leads to a DNA strand break. The formation of the covalent adduction on the N7-position of purine residues in DNA destabilizes the guanidine-N9-1′-deoxyribose glycosidic linkage [58–60]. Single and double strand breaks also result from ionizing radiation. Two DNA strand breaks some distance apart can lead to deletions, the size of which is dependent upon the separation distance of the breaks. These carcinogen-initiated deletions may disrupt oncogenes such as RB or TP53, produce loss of heterozygosity (LOH), or result in chromosomal translocations [39, 45, 49]. Large deletions (and insertions) of one million bases or more are detectable by cytogenetics and will fall into the chromosomal aberration category.

The explanation for the formation of insertions is less clear, but may entail the same basic mechanisms as noted above for deletions, or by retrotransposition of viral or repeat sequences. The highly repeated Alu sequence, of approximately 300 nucleotides in length, is present in over a million copies in the human genome [61, 62]. Every so often, an Alu repeat sequence inserts itself in a new section of the DNA. This has been observed as a major oncogenic mutation in retinoblastoma and numerous genetic diseases [63]. However, these events are rare and retrotransposition is probably not a prominent mechanism in cancer development.

Epigenetic Mutations

Another form of DNA mutation that has been demonstrated in recent years to be crucially involved in cancer development is the alteration of patterns of DNA 5-methyldeoxycytidine (5mdC) and/or changes in DNA chromatin modification. This type of mutation is referred to as an "epigenetic mutation," since these changes do not alter the basic DNA coding sequence. To clarify firstly DNA 5mdC patterns, selective deoxycytidine residues in DNA are enzymatically modified by endogenous DNA methyltransferases (DNMTs) that catalyze the transfer of a methyl group from S-adenosylmethionine to the 5-position on the cytosine base, forming 5mdC [64]. In adult somatic cells, this normal DNA modification occurs almost exclusively at 5′-CpG-3′ sequences and duplex DNA is maintained in a symmetrically methylated state, so that the complementary 5′-CpG-3′ sequence is also methylated on the opposing strand [65, 66]. It should be noted that in stem cells, especially embryonic stem cells, a significant portion of the 5mdC also occurs in non-CpG sequences [65, 66]. The meaning of these non-CpG methylated sequences is presently unknown. Maintenance of symmetrically methylated duplex DNA ensures the mitotic heritability of these DNA 5mdC patterns in somatic cells by providing the 5mdC pattern template for each newly synthesized strand following semiconservative DNA synthesis. The level of 5mdC on a specific DNA locus or gene promoter region determines whether the sequences are expressed or not. Highly methylated promoter sequences 5′ to the coding sequence of a given gene indicate that the gene is not available for transcription [67]. Today this is commonly referred to as gene hypermethylation, which is identified frequently as an epigenetic mutation in one or more tumor suppressor genes in human cancers [67, 68]. Hypomethylation of these same promoter sequences opens the gene sequences up for transcription, but does not indicate that the gene is being transcribed since additional DNA binding proteins and transcription factors are required for expression of the gene. Not surprisingly, DNA 5mdC patterns change during tumor development [69, 70], and chemical carcinogens are known to alter DNA 5mdC patterns [71–75].

Chromatin modifications are more complex than DNA 5mdC patterns and include a number of different molecular moieties. The basic repeating unit of chromatin is the nucleosome, which is composed of 147 base pairs of DNA wrapped around a core containing two copies each of four core histone proteins H2A, H2B, H3, and H4 [76, 77]. Various molecular modifications of the N-terminal tails of these core histones occur in mammalian cells. In 1964, Allfrey et al. [78] first reported histone N-terminal tail modification, acetylation, but it was many years later before the complexity and importance of this posttranslational modification became known. As it turns out, multiple histone modifications occur, consisting of methylation [79–81], acetylation [82], phosphorylation [83], ubiquitinylation [84], and sumoylation [85]. The patterns and combinations of these modifications are referred to as the histone code [86], which influences the chromatin structure and remodeling for either active expression or silencing of genes [87]. Euchromatin (active chromatin) along with active gene expression was associated with methylation of the lysine residue in the fourth position of the N-terminal tail of histone H3 (denoted as H3K4me), while H3K9me and K3K27me formation were linked to heterochromatin (silent chromatin) and repression of gene expression [79–81]. These patterns become even more complex as each of these histone lysine residues may be mono-, di-,or tri-methylated (i.e., $H3K4me^1$, $H3K4me^2$, and $H3K4me^3$) [88–91]. The importance of multiple methyl groups on lysine residues is presently only poorly understood. Histone methylation also takes place on other lysine residues on histone H3 K36 and K79, and histone H4 K20, and on arginine residues histone H3 R2, R17, and R26, and histone H4 R3.

Acetylation of the positively charged lysine residues decreases the effective ionic charge, reducing the affinity to the negatively charged phosphate backbone of DNA and creates a binding side for regulatory DNA proteins that activate transcription [92]. Essentially, histone acetylation alters the chromatin structure to a more transcriptionally supportive architecture. Acetylation of the amino group of lysine residues occurs on histone H2A K5 and K9, histone H2B K5, K11, K12, K15 and K20, histone H3 K9, K14, K18, K23 and K56, and histone H4 K5, K8, K12 and K16 [86]. Acetylation patterns are established by highly regulated histone acetyltransferases (HATs) and histone de-acetylases (HDACs) [93]. HAT enzymes are grouped into different family classes, GNAT family, MYST family, and CBP/p300 family [94]. Some transcription factors such as yeast protein Gcn5, a putative transcriptional adaptor, are HAT enzymes [95]. HDACs catalyze the removal of acetyl groups from histones by activation of a water molecule through a zinc ion in the active site [96, 97], and are involved in formation of heterochromatin and the silencing of genes. In human cells, 18 nonredundant HDACs have been identified [98], which have been divided into classes I, II, and III [99]. Class I HDACs are localized within the nucleus while Class II shuttle between the nucleus and cytoplasm [98]. Regulation of HDACs consists of protein–protein interactions, posttranslational modifications, cellular localization, and metabolic cofactors [100].

In 1967, Gutierrez and Hnilica [101] first reported histone phosphorylation of serine residues on H3S10 and H3S28. Histone phosphorylation appears to be involved in cell cycle chromatin condensation, active transcription, apoptosis, and DNA repair mechanisms [102, 103]. Histone H3T11 and H3S28 have been identified as mitosis-specific sites for phosphorylation [104, 105]. Histone phosphorylation is also involved with activation of early response genes such as c-fos, c-myc, and c-jun. This is a stimulant-dependent H3 phosphorylation, resulting in quick transcriptional activation of a subset of immediate-early genes [106]. Although, acetylation and phosphorylation are both linked to transcription activation, it is unknown whether the two modifications work independently, in parallel, or synergistically [107, 108]. Phosphorylation also occurs on serine and threonine residues on histone H2A S1, histone H2B S14, and histone H4 S1.

Histone ubiquitinylation and sumoylation are the least understood of the chromatin modifications [84, 85]. However, these modifications may be important in DNA damage response and repair, V(D)J recombination, meiotic and spermatogenesis remodeling, and chromosomal stability [109–114].

It is becoming clear that histones are major carriers of epigenetic information. Histone modifications provide mechanisms to generate and stabilize chromatin structures. The study of chromatin modification patterns is a very young field, which will undoubtedly become very important to the delineation of carcinogenesis. In fact, evidence is already beginning to accumulate regarding alterations in chromatin histone modifications due to exogenous chemical carcinogen exposures [75, 115].

Cytotoxicity and Cellular Turnover

Most exogenous carcinogens are cytotoxic at recognized carcinogenic doses, and probably at much lower doses to impacted cells. This concept is easy to envision when considering agents that directly damage cellular macromolecules (i.e., ionizing radiation, alkylating and arylating carcinogens). Interference with cellular functions, enzymatic processes, cytoskeletal structure, mitochondrial electron transport and oxidative phosphorylation, etc. may lead to cell senescence or death. If the DNA is damaged to the extent that successful repair is doubtful, the cell undergoes apoptosis as mediated by TP53 [39, 116]. Both senescence and apoptotic pathways are protective means by which cells may avoid accumulating DNA mutations [33].

Concentrations of chemical carcinogens and doses of physical carcinogens that produce mutagenicity in experimental studies generally also produce a significant percentage of cell lethality [117]. Whether cell death is the result of cellular damage inflicted directly by the carcinogen or from triggering apoptosis, the loss of the cell leaves a gap to be filled by cellular turnover. In tissue, this leads to increased cell division to replace these lost cells. This damage-induced cell proliferation offers increased opportunities for the accumulation of mutations during DNA synthesis and mitotic division. Cells adjacent to the vacated space will have the chance to divide to replace the missing cell. One or more of these adjacent cells might harbor a mutation that confers a growth advantage enabling it to dominate over its neighbors and progress through cell division. Thus, the potential for clonal expansion and the continued accumulation of oncogenic mutations is encouraged by the impact of carcinogens on tissues.

The opportunity or encouragement of carcinogen exposed cells to undergo cell division provides yet another enhancement for mutation occurrence. Many of the cells that have the opportunity to divide have also received a carcinogen-induced damaging event, which if not repaired prior to DNA synthesis may lead to a fixed mutation [117]. For example, the presence of an alkyl-O^6dG adduct during DNA synthesis will result in the formation of a G to T mismatch approximately 50% of the time. Alkyl-O^6dG base pairs equally efficiently with either a thymidine or a cytidine residue [60, 118, 119]. Different DNA adducts and forms of damage will result in different mutational events following DNA synthesis. But the cytotoxic action of these carcinogens couples with the DNA damaging activity to induce mutations.

When carcinogen induced cell death involves a tissue stem cell, the dead cell is replaced by another stem cell. Most of cancer arises from tissue stem cells. The accumulation of mutations in a transit-amplifying progenitor cell will usually result in no consequence as the developing cancer cell will be terminally lost either by terminal differentiation and cell death or by being sloughed off the epithelial layer [34]. Stem cells have the capability of surviving long enough to accumulate the necessary mutations to become cancerous. Even in rapidly turning over tissues, stem cells appear to be long lived, with estimated lifetime division numbers totaling in the hundreds and higher [120]. But these stem cells also enjoy a slow or delayed cell cycle, dividing intermittently (only once every 7 or 8 days), and often sitting quiescent and in a nonproliferative or reserve stem cell state [121–123]. Stem cell division is generally considered to be asymmetric, giving rise to both a replacement stem cell and a progenitor cell that will undergo numerous symmetrical cell divisions and then terminally differentiate [120, 124–126]. This asymmetric cell division appears to also include the retention of the original DNA strand by the daughter stem cell replacement. Or stated differently, the original template DNA strand of the mother stem cell selectively segregates to the daughter stem cell replacement, while the newer DNA strand segregates to the transit-amplifying daughter cell. This is the Cairns' "immortal stranding" hypothesis [127], which appears to have experimental validity [124, 126]. DNA synthesis in human cells accumulates one to three mutations every time the genome is copied [31, 32]. "Immortal stranding" enables the stem cell to minimize the accumulation of DNA mutations, and thus limit the number of mutations that are passed on to the transit-amplifying differentiating epithelial cells.

Adult stem cells are very sensitive to DNA damage, and rather than perform DNA repair, stem cells appear to undergo apoptosis in response to external exposure to DNA damaging agents [128]. Discarding the cell rather than risk faulty repair ensures that the accumulation of DNA mutations in stem cells does not occur. However, the loss of one or more stem cells due to external toxic exposure(s) (UV light, ionizing radiation, arylating or alkylating chemicals, or inflammation or infection produced ROS), requires the replacement of these stem cells. Stem cell replacement requires a symmetrical cell division to give rise to two daughter stem cells, which will produce one "aged" DNA template strand regardless of "immortal stranding." This process increases the number of mutations arising in the stem cells and increases the risk of disease. Alternatively, a transit-amplifying cell may undergo dedifferentiation, characteristic of mesenchymal-epithelial transition [129]. Unfortunately, this dedifferentiation process again produces a stem cell without the benefit of "immortal stranding" and increases the risk of mutations and disease developing in the tissue stem cell population.

Genomic Instability

Genomic instability is the loss of genomic stability and the increased probability for accruing DNA mutations. Restated, genomic instability is the inability of cells to maintain the integrity of their own genomic inheritance.

It is generally accepted that most cancers express one or more forms of genomic instability. However, among the scientific community there is still controversy as to whether the loss of genomic stability is required for the development of cancer. One argument states that genomic instability is not necessary for the accrual of a sufficient number of oncogenic mutations to reach a malignancy [130, 131]. The other argument states that human cells would not have the opportunity to accumulate a sufficient allotment of driver mutations to achieve a cancerous cell without an additional navigating mechanism(s) [132]. The statistical probability during a normal human life span has been proposed to be too low based on theoretical cell division numbers alone. However, many oncogenic mutations, such as in "caretaker" genes, result in the enhanced probability of initiated cells achieving more mutations and progressing along the road of tumor development [31, 40, 42, 133]. Caretakers are genes that code for cellular proteins and factors that protect the genome from mutation. Disruption of the finely balanced cellular network of monitoring and repair systems will lead to a loss of fidelity of these genome protection systems, increases in macromolecular damage, increased longevity of DNA damage, faulty repair, and an increased accumulation of DNA mutations. For example, a mutation that disrupts the normal function of a mismatch DNA repair gene leads to upward of a 100-fold increased rate of cellular DNA mutation [40].

Genomic instability is a hallmark of cancer and one of the trademarks of a number of inherited genetic diseases that predispose individuals to cancer [131] (see Table 10.1). By today's laboratory techniques, genomic instability manifests as chromatin instability (CIN), microsatellite instability (MSI) (sometimes referred to as MIN), nucleotide instability (NIS) (also called single base instability), and epigenetic instability (EIS) (see Table 10.2).

Chromosomal Instability (CIN)

The inability of the cell to properly repair single or double strand DNA breaks, perform appropriate recombination activities in somatic cells, or correctly perform the mitotic spindle functions during mitosis will lead to the development and accumulation of chromosomal aberrations, which is the manifestation of CIN [33, 134]. Cells expressing CIN miss-segregate chromosomes 10–100 times more frequently than chromosomally stable diploid cancer cells [135]. CIN may promote tumor evolution by enabling the clonal expansion of cells with proliferative advantages and/or metastatic potential [136–138].

Chemical and physical carcinogens can interfere with the cytoskeleton, mitotic spindle apparatus, or the various proteins and enzymes involved in mitosis. The impact of such carcinogenic damage is the increased probability of chromosomal mutations. As noted above, asbestos produces aneuploidy in skewered cells by interfering with the cytoskeleton and mitotic proteins [53]. Tobacco smoke PAHs, BPDE, and 4-aminobiphenyl enhance telomere shortening and produce CIN in vivo and in vitro studies [54–57].

Microsatellite Instability (MSI)

Microsatellites are relatively short sequences composed of tandemly repeated units of one to six bases in length [139]. The repeated unit of sequence may be only a mono-nucleotide repeat (e.g., AAAAAAA…), di-nucleotide repeat (e.g., CACACACACACA…), tri-nucleotide repeat, etc., although microsatellites of imperfect and compound sequences also exist in the human genome (e.g., Bat25 and Bat26) [140]. The most common class of microsatellites in the human genome are composed of various tandemly repeated lengths of $(CA)_n$. Microsatellites are widely prevalent throughout the human genome, occurring more frequently in noncoding regions on all chromosomes, and act as gene promoters, as target sites for recombination, and as binding sites for DNA topoisomerases [139]. The length of a specific microsatellite is an inherited trait and therefore the same in all tissues. Insertion or deletion of one or more repeat units within a microsatellite occurs on rare occasion by slippage during DNA replication, resulting in a longer or shorter microsatellite sequence, respectively. These polymerase slippage errors are normally repaired by a post-replication mismatch repair system, and the presence of a microsatellite variant generally results from the lack of fidelity in the mismatch repair mechanism [141–143]. MSI is denoted by the presence of microsatellite variants of different lengths in tumors compared to normal tissue, and is a hallmark of mismatch repair deficient familial and sporadic cancers [144, 145].

In theory, a carcinogenic agent that interferes with DNA polymerases and/or the mismatch repair system would lead to MSI. For many types of cancer, MSI is the one genomic instability that is consistently acquired late in the tumorigenic process [139]. The major exception to this rule is the hereditary nonpolyposis colorectal cancer (HNPCC) syndrome, for which individuals inherit an error in the mismatch repair system and are at risk of developing colon cancer [146, 147]. However, airborne particulates ($PM_{2.5}$) have been associated with the development of MSI in human lung epithelial cells [148, 149]. Cancers induced by betel-quid chewing and chronic occupational exposure to chromium have been found to express MSI significantly more frequently than noncarcinogen-associated tumors [150, 151].

Nucleotide Instability (NIS)

Nucleotide instability may be one of the more easily induced instabilities by either exogenous or endogenous influences. Increases in the frequencies of various forms of DNA nucleotide damage and/or loss of cellular DNA damage surveillance and repair capacity, or the perturbation of the DNA polymerases or synthesis complex machinery result in increases in single base mutations [19, 46–48, 132]. With the possible exception of chromosomal translocations in leukemias and lymphomas, NIS is probably the most common genomic instability expressed in human cancer [47, 49, 50]. NIS may also be predominant in the earliest mechanisms of carcinogenesis as roughly 85% of the activating and deactivating mutations in proto-oncogenes and tumor suppressor genes, respectively, are single base substitution mutations [44–48]. Much of the NIS data in the literature has been accumulated from clonal expansion of mutations in tumor tissue, that is, TP53 mutation database and retinoblastoma gene mutations [49, 50]. The analysis of NIS has been hampered by the requirement of exceedingly sensitive techniques, and the limited viable ultrasensitive approaches are labor intensive [47]. However, with the present availability of the ultra-sensitive methods, the presence of NIS in non-tumor tissues and the induction of NIS by exogenous environmental carcinogens are beginning to unfold [48, 152].

Epigenetic Instability (EIS)

Although any alteration from normal of DNA 5mdC and/or chromatin modification patterns constitutes epigenetic instability, the majority of the scientific literature has focused on hypermethylation, that is, the transcriptional suppression of important anticancer genes by high levels of 5mdC present in the gene promoter sequences

[67, 68]. In human cancer, a hypermethylation phenotype, referred to as CIMP (CpG island methylator phenotype), has been proposed [153]. The CIMP phenotype may have clinical significance, but the cause or mechanism underlying the CIMP tumor phenotype has yet to be identified [154–156]. Regardless of the gene specific hypermethylation ascertainment biased approach, cancer cells are globally hypomethylated and harbor 20–60% less genomic 5mdC than their normal counterpart. The hypomethylated sequences in cancer cells may be mainly due to the loss of 5mdC levels in repetitive sequences, which are normally heavily methylated and comprise 20–30% of the human genome [157]. For at least some cancers, the level of global hypomethylation is associated with progression, becoming progressively more hypomethylated from normal to benign, to tumor and metastasis [158–160]. Indeed, transient global hypomethylation is inducible by either chemical carcinogens or hypoxia [71, 72, 161]. Interestingly, increased NO production that may occur in inflammatory disease, or in response to hypoxia, reduces S-adenosylmethionine levels by inactivating methionine adenosyltransferase, and produces hypomethylation [162]. Dietary deficiencies in either folic acid or methyl donor compounds, that is, methionine, also cause hypomethylation [6–8]. Hypomethylation of gene-specific sequences in cancer cells leads to ectopic gene expression, such as neuron-specific γ-synuclein expression in gastric cancers [163]. The degree of hypomethylation of the XIST gene correlates with prostate cancer aggressiveness [164].

There may be several mechanisms at work in the alterations of epigenetic signals during carcinogenesis as these patterns are intimately involved in all biological processes, that is, genomic imprinting, gametogenesis, embryogenesis, differentiation, aging, and carcinogenesis. As noted above, dietary deficits are known to alter DNA 5mdC levels and patterns [6–8]. Alkylation, arylation and other forms of DNA damage induced by carcinogens have been clearly shown to disrupt DNA 5mdC patterns [71, 72]. DNA epigenetic patterns may also be impacted by many other environmental factors, including tobacco use, drug use, physical exercise, and health status as DNA 5mdC patterns diverge with age in monozygotic twin pairs [165].

Chromatin modification patterns are well known to be involved in carcinogenesis and the development of cancer [166, 167]. However, this field is still developing, and the data regarding exogenous agent-induced changes in histone modifications are presently limited. Several in vitro tissue culture studies have been performed that suggest toxicants induced changes in selective histone residue methylation or acetylation (see review [75]), but the interpretation of these specific histone modification changes amid the complex arrays of epigenetic patterns and extrapolation to in vivo will require further analyses.

To summarize genomic instability, there are a large number of genes involved in the maintenance of the genome. Many of these genes, such as the DNA repair genes, have been referred to as caretaker genes for their role in genomic stability, while others such as the RB and TP53 genes have been denoted as gatekeeper genes due to their involvement in cell cycle control and the pathway to cancer development [168]. Each of these instabilities enhances the probability of accumulating oncogenic mutations and progression along the road of tumor development, the acquisition of invasive or metastatic capability, and/or resistance to chemotherapeutic regimens. This last effect has also been observed in non-chemotherapeutic carcinogens, whereby cells exhibiting selective genomic instabilities are resistant to the cytotoxic effects of some carcinogens. Cells expressing CIN were found to be resistant to the lethal effects of PhIP but not to the methylating carcinogen, *N*-methyl-*N′*-nitro-*N*-nitrosoquanidine (MNNG), while cells expressing MSI were resistant to MNNG but not to PhIP [169]. This provides yet another important mechanism by which chemical carcinogens may select for mutant cells and promote their clonal expansion by exerting cytotoxic effects on neighboring cells in the tissue.

Susceptible Individuals

Based on the above discussion of genomic instabilities, it is not surprising that some individuals in the human population have inherited genetic errors that render them more prone to developing cancer [47, 131] (see Table 10.1). For a portion of these susceptible individuals, the risk of cancer is quite high. For example, individuals who have inherited a mutation in one of the mismatch repair genes (i.e., MSH2, MLH1, MSH3, MSH6, PMS1, and PMS2) are carriers of hereditary nonpolyposis colorectal cancer (HNPCC) syndrome and are at high risk of colorectal carcinoma [139, 146, 147]. Inheritance of any one of several DNA repair gene mutations confers a genetic disease (i.e., xeroderma pigmentosum, ataxia telangiectasia, Fanconi anemia, Cockayne syndrome, and others), which is associated with adverse clinical symptoms including a substantially elevated risk of cancer. Individuals harboring a germ line mutation in the TP53 gene, referred to as the guardian of the genome gene, generally present as a Li-Fraumeni Cancer Family Syndrome and are at a substantially elevated risk of multiple types of cancers during their lifetime [170]. More than 100 Mendelian diseases that confer predisposing cancer syndromes have been recorded in "McKusick's Online Mendelian Inheritance in Man" (OMIM), each with varying levels of cancer risk [171].

There are however heritable mutations that convey more subtle increases in cancer risk, especially when environmental exposures to exogenous carcinogens are involved [47, 172]. Polymorphisms in genes involved in the deactivation and elimination of toxic metabolites and carcinogens may confer to the individual a decreased ability to remove offending toxicants and/or reactive electrophiles before they can cause DNA damage, which provides a slightly increased probability of developing cancer. Examples of these susceptibility traits are found in polymorphisms among the phase I and phase II biotransformation enzymes, such as cytochrome P450 and glutathione-S-transferase genes [172–174]. These individuals are at an increased risk of cancer from exposures to polycyclic aromatic hydrocarbon (PAH) carcinogens commonly found in tobacco smoke and other combustion products [173, 175]. Depending upon the metabolic pathway used to activate, deactivate, and/or eliminate carcinogens, gene polymorphisms may render individuals more susceptible or less susceptible to cancer from exposure to selective carcinogenic agents.

Dose and Complex Dose

Humans are exposed to a plethora of toxic and carcinogenic materials in the environment. Exposure to complex toxic mixtures poses a higher risk of detrimental health effects than that represented by the toxicity of each separate individual constituent. When an individual or animal is exposed to a complex mixture of carcinogenic agents, the dose required to produce significant health problems and tumor development will be lower than that for each separate individual constituent. In toxicology, it is well recognized that exposures to significant levels of complex mixtures of toxic compounds results in increased detrimental health effects. This is also true for carcinogenic agents. The combined toxic effect of separate toxicants is generally predicted to be at a minimum additive, unless specific information exists to demonstrate a synergistic (above additive) or inhibitory (below additive) action. Many of these carcinogens target the same organs or tissues, and induce similar toxic effects in humans and animals. For example, alcohol use, foodborne aflatoxins and hepatitis B (or C) all adversely impact the liver and promote the development of hepatocellular carcinoma. Similarily, asbestos, arsenic, and tobacco smoke drive the development of lung cancer. For each of these examples, some differences in agent-specific direct DNA damage may exist, but similar mechanisms of cellular damage, enhanced cellular turnover, inflammatory responses, and ROS and RNS species involvement contribute to the development of cancer.

Although people are not generally exposed to an acute carcinogenic dose for each one of these environmental carcinogens, they may receive an accumulated dose sufficient to induce cancer due to long-term chronic low dose exposure. Damage to the skin from lifetime exposure to UV radiation from sunlight is one of the more obvious examples. However, the same may be true for a number of environmental chemical carcinogens, especially if one has been chronically exposed to tobacco smoke, whether as a smoker or nonsmoker. What is known regarding what constitutes a carcinogenic dose is based mostly on laboratory studies involving a single agent, not a complex mixture. It should note, however, that tobacco smoke is a complex mixture, and epidemiological studies have provided estimates of carcinogenic risk based on the number of cigarettes smoked per day and the number of pack years of smoking [176, 177]. Estimates of carcinogenic risk are generally derived from the carcinogenic dose of individual agents. The combined effects of exposures to multiple environmental carcinogens must certainly lower the carcinogenic dose threshold for individual carcinogens. Thus the chronic long-term low-dose exposure to a multitude of carcinogenic agents in our food, air, water, from sunlight, x-rays, and biological pathogens, enhances the risk of cancer as our bodies reach old age. Whether cancer is inherent with life or not, moderating one's exposure by avoiding habitual exposures, that is, tobacco smoke and alcohol abuse, and maintaining a healthy lifestyle significantly lowers one's lifetime risk of cancer.

Environmental Carcinogens

To reiterate, the rate of tumor development is increased sometimes many orders of magnitude above the spontaneous background rate by exposure to exogenous carcinogens. The timing of the onset of cancer is due to the dose and duration of carcinogen exposure and not to either the chronological age or the age that exposure began [38, 39]. Regardless of whether a cancer develops from carcinogen exposure (s) or endogenous processes, or the combination of exogenous and endogenous attacks (the most likely explanation), the accumulation of oncogenic driver mutations and a large number of cell divisions are required.

Separating the effects of exogenous versus endogenous genotoxic damage and cancer induction may not be an easy undertaking. More than likely, a tumor originating only from exogenous carcinogen exposure is rare if nonexistent. However, the impact of exogenous exposures to genotoxic agents is easily demonstrated by the significant increases in lung cancer of smokers [176], and by the patterns of oncogenic single base substitution mutations in the TP53 gene in lung cancers and breast carcinomas from smokers. Lung and breast tumors arising in smokers are twice and ten times, respectively, as likely to harbor a TP53 gene driver mutation of the G to T transversion type than similar tumors arising in nonsmokers [178, 179]. Benzo(a)pyrene and other PAHs are known to target deoxyguanosine in the most commonly mutated sites (codons 157, 248, and 273) observed in lung tumors, and to produce G to T transversion mutations in these same base sites [180–182]. Similar findings have reported for aflatoxin B1 adduction and G to T mutation of codon 249 of TP53 [183–185]. Sunlight (UV light) produces signature CC to TT mutations in TP53 and other DNA sequences that have been identified in sun exposed skin, which is well correlated with the occurrence of skin cancers [186–188]. Unfortunately, few carcinogens presently provide such clearly apparent tumor-associated agent-induced genotoxic damage. Signature diseases that aid in identifying the exogenous impact of toxicant exposure are also limited to only a few carcinogens, for which asbestos and mesothelioma are the best known examples. Many other cancers have multiple etiological causes, and delineation of specific carcinogen exposures in the causation is not easily identified (see Table 10.1).

The following are brief descriptions of the genotoxic processes and carcinogenic mechanisms for a few examples of common environmental carcinogens, which are provided here to clarify the discussions of the impact of these exposures on the development of human cancer. The discussions are not intended to be fully comprehensive. Extensive literature is available of each of these agents and numerous references are provided for the reader.

Alcohol

Chronic alcohol consumption is associated with oral cavity, pharynx, larynx, esophageal, liver, and colorectal cancer, and breast cancer in women [12, 13, 16]. Alcohol use is also suspected to be involved in the etiology of pancreatic and lung cancers. The development of alcohol-induced cirrhosis of the liver is a well-known etiology for hepatocellular carcinoma development [16], and the risk of hepatocellular carcinoma increases with the synergistic impact of tobacco smoke and alcohol, or of hepatitis (B or C) and alcohol [12]. However, pure ethanol itself is not genotoxic, but its primary metabolite acetaldehyde is mutagenic [189, 190].

The present theories regarding the carcinogenic mechanisms of alcohol are based on DNA damaging events, tissue inflammation and perturbation of nutrition and physiological functions [13, 14]. Acetaldehyde forms DNA adducts [189, 190], and damages hepatocytes and tissue architecture, which leads to increased cellular proliferation [191]. Alcohol induces CYP2E1, whose gene product (P450 2E1) adds to the metabolism of alcohol to acetaldehyde as well as activating tobacco carcinogens (i.e., nitrosamines) [12]. Alcohol also induces ROS & RNS [192], which leads to lipid peroxidation and further DNA damage from reactive electrophilic products of the lipid peroxidation reactions [193]. Alcohol intake interferes with folate metabolism [194] and folate deficiency increases the risk of colorectal carcinoma by two to five times [195]. Alcohol may also alter the absorption and metabolism of vitamins, such as vitamin B_{12}, B_6, and retinoic acid, and often results in malnourishment in the alcoholic [13, 14, 194]. Thus chronic alcohol use utilizes at least three different avenues of carcinogenesis; DNA damage and mutation, inflammatory processes that promote proliferation, and vitamin deficiencies that are known to perturb DNA epigenetic patterns. It is not surprising that alcohol works synergistically with other carcinogens such as tobacco smoke to increase the occurrence of cancers [12].

Sunlight

UV light from the sun is probably the most important naturally occurring environmental carcinogen to which humans are exposed. Chronic exposure to sunlight and repeated sunburns are strongly correlated with nonmelanoma skin cancers (mainly squamous cell carcinoma and basal cell carcinoma), especially on the head, neck, forearms, and hands, which clothing generally leaves uncovered [196]. Skin cancer is one of the more frequent cancers in aging populations, and is the most common cancer in the USA [9, 10]. The lifetime risk of developing skin cancer in the USA is one in five. Fortunately, most skin cancers are amenable to early detection and a high cure rate. Melanoma is the exception, leading to a lethal disease course in greater than 10% of cases.

UV light is comprised of electromagnetic radiation wavelengths below 400 nm, and has been broken down to vacuum UV (100–200 nm), UVC (200–290 nm), UVB (290–320 nm), and UVA (320–400 nm). [Note that some authors refer to the wavelength ranges for UVB and UVA as 280–315 nm and 315–400 nm, respectively (e.g., Ridley et al. 2009 [197]).] The upper layers of the Earth's atmosphere filter out much of the harmful UV shorter wavelength, although this will become less effective and a potentially major health issue as the ozone layer thins over parts of the planet. The remaining UVA and UVB radiation reaches the ground. Both UVA and UVB damage cells and macromolecules in cells. UVB provides the major DNA mutational impact producing signature cyclobutane pyrimidine dimers and pyrimidine 6–4 pyrimidone photoproducts, which if not repaired, result in signature CC to TT and C to T mutations, respectively [186, 187, 198]. Cyclobutane pyrimidine dimers are responsible for approximately 80% of the UV light signature mutations, since the 6–4 photoproducts are repaired much more efficiently than the cyclobutane pyrimidine dimers [188, 197]. Cyclobutane pyrimidine dimers are only slowly repaired by a transcription-coupled repair system. These UV signature mutations are known to occur in at least one important tumor suppressor gene for human cancer, the TP53 gene [186]. UV-induced TP53 mutations probably represent an early step in the skin cancer carcinogenesis pathway for basal cell carcinoma and squamous cell carcinoma [186, 199].

UVA radiation also contributes to the carcinogenic impact of sunlight exposure. UVA radiation leads to DNA damage and mutations, albeit at more than a 1,000-fold lower extent than UVB. However, UVA penetrates more deeply into the skin than UVB and produces UV signature mutations in the basal layer, while the majority of UVB produces signature mutations in suprabasal keratinocytes [200]. UVA also initiates substantial cellular macromolecular damage, ROS, and lipid peroxidation [197].

UV light exposure may enhance the development of skin tumors both by inducing mutations and by interfering with the turnover of keratinocytes [201–203]. UV irradiation induces apoptosis in severely damaged keratinocytes. However, insulin-like growth factor-1 (IGF-1) produced by dermal fibroblasts opposes UV-induced apoptosis in keratinocytes and induces replicative senescence by NFκβ and TP53 signalling [201–204]. UV damaged keratinocytes that survive apoptosis still retain the capacity to divide unless held in check by IGF-1. Aging skin fibroblasts produce significantly less IGF-1, leading to the increased potential of damaged and mutant keratinocytes replicating. This may explain the increased incidence of nonmelanoma skin cancer in geriatric patients [205].

The combined effects of UVB and UVA to multiple cells in the epidermal layer result in localized tissue inflammation, which contributes to the melee of oxidative reactions and includes reactive nitrogen species [25, 27, 43]. The setting of cellular damage and inflammation results in a tissue microenvironment that encourages cellular turnover, promotes clonal expansion of mutated cells that leads to the accumulation of oncogenic mutations and ultimately to skin cancer development.

Tobacco Smoke

Tobacco smoke is the most frequent cause of lung cancer and probably the most important anthropogenic carcinogen of the last century. Whether by lifestyle choice of smoking or by exposure to side-stream smoke for the nonsmoker, tobacco smoke has been a primary etiological factor in lung cancer in the USA [176]. On average, lung cancer develops in cigarette smokers after 50 pack-years [177]. Tobacco smoke contains a large complex mixture of carcinogens, including many that may both initiate and promote the development of cancer. Cigarette smoke contains between 3,000 and 4,000 toxic compounds, more than 300 of which are known or suspected carcinogens, including the majority of the IARC Group I chemical carcinogens discussed in this chapter [206] (see Table 10.1). Many of these toxic and carcinogenic components of tobacco smoke are readily absorbed systemically, and smoking is well associated with diseases such as cardiovascular disease and cancers in addition to lung [176, 177, 207]. Tobacco smoking is believed to be an important etiological factor in many cases of oral cancers, esophageal, bladder, pancreatic, kidney, and possibly breast and cervical cancers [208].

Many of the carcinogens in tobacco smoke alkylate or arylate DNA, producing promutagenic lesions. For example, benzo(a)pyrene is activated by the host cells' cytochrome P450 1A1 enzyme activity to the ultimate carcinogen, benzo (a)pyrene-7,8-dihydrodiol-9,10-epoxide (BPDE) form, which reacts with nucleophiles including the exocyclic amino group on deoxyguanine of DNA [209]. The BPDE-N^2dG adduct is a promutagenic lesion that leads most frequently to a G to T transversion mutation. Numerous other tobacco smoke carcinogenic PAHs, heterocyclic amines, and nitrosamines, produce DNA adducts that can lead to primarily G to T transversions. As noted above, G to T transversions are the most common single base mutations in lung tumors [178, 179]. A few of these carcinogenic nitrosamines are tobacco smoke specific, that is, 4-(methylnitrosamino)-1-(3-pridyl)-1-butanone (NNK) and N'-nitrosonornicotine (NNN) [207, 210]. The list of carcinogens in tobacco smoke is extensive and ensures multiple types of DNA damage and adverse impacts on other cellular macromolecules, enzymes, and structural components is complex [208]. It has been estimated that a developing preneoplastic clone of lung cells accumulates one DNA mutation for every 15 cigarettes smoked [211]. In addition to damaging DNA, most of these tobacco smoke toxicants also promote cellular turnover and inflammatory processes in lung tissue, providing another avenue of ROS and RNS species and altered microenvironments for clonal expansion of mutant cells [25, 43]. Particulates from tobacco smoke interfere with pulmonary function independent of the plethora of toxic chemicals, recovery and repair following exposure [148, 149]. In spite of all these genome destabilizing and tumor-promoting activities, many years of smoking history are generally required before a tumor develops, which is a testimony to the resilience of the human body to the bombardment of such a powerful combination of the carcinogens and toxic agents present in tobacco smoke.

Future Directions

It would be very difficult if not impossible for life on this planet to completely avoid exposure to environmental carcinogens. One can purposely avoid alcohol, tobacco smoke, and many occupational exposures, but the presence of carcinogenic substances in our food, air, water, and sunlight coupled with health, nutritional issues, and biological carcinogenic agents make avoiding all exposure to carcinogens very difficult. The extensive years of educating the public regarding the hazards of smoking, along with the banning of smoking in public buildings and other establishments has had a dramatic effect on the reduction in lung cancer and probably other cancers in recent years [176, 177]. Increased safety precautions and worker awareness to reduce occupational exposures remain a very important component in reducing cancer risk. Vaccinations against biological carcinogens, such as hepatitis B and C, may be important as well, and perhaps more important depending upon where one travels in the world. As people live longer and healthier lives, reductions in carcinogenic contaminants in our food, air, and water appears to be the only other viable avenue that society as a whole can take to diminish cancer risk. Individuals remain responsible for their own nutritional health, exercise, and personal habits in avoiding carcinogenic exposures.

Fortunately, our food also contains anticarcinogens [212–214]. For example, the isothiocyanates in broccoli are suggested to aid in diminishing the carcinogenic impacts of a number of chemical carcinogens [215]. This is a relatively young, but growing field of research that has the potential to provide viable strategies for long-term reductions in cancer risks. Individuals may only have to monitor their dietary intake of selected vegetables and herbs to enjoy significantly reduced cancer risk. However, this may work for reducing some types of cancer and not others.

The relative contribution of endogenous versus exogenous causes of human cancer is still a matter of some debate. Since the carcinogenic impacts of endogenous and exogenous are completely intertwined, separating out individual contributions is very difficult. Only in selective cases where signature diseases and/or signature biomarkers such as specific DNA adducts and/or mutations are identifiable can the contributions be estimated in rough manner [116]. This is a problem that is likely to remain for many years to come. Learning and understanding many of the xenobiotics in our food and environment to which humans are presently exposed continues to be a prominent field of safety research. Add in the large influx of synthetic chemical compounds joining our environmental melee each year, many of which may contribute to the carcinogenic impacts, and identifying all carcinogens in our environment becomes a rather daunting undertaking. For example, the explosion in designed nanoparticles for industrial and medical purposes, introduces potentially genotoxic agents into the realm of human exposure. At least some synthetic nanomaterials have thus far been found to cause chromosomal aberrations, DNA breaks, point mutations, and oxidative DNA damage [216]. Control of occupational exposures and environmental releases of these new genotoxic synthetic compounds will be important in opposing additions to the causes of environmental carcinogenesis.

The lessons learned from the classical examples of environmental carcinogenesis (sunlight and skin cancer, tobacco smoke and lung cancer, and alflatoxin, hepatitis and liver cancer) have guided research studies and advances in understanding cancer development. Continued advances in understanding the tissue, cellular and molecular processes of carcinogenesis will enable the design of strategies to further diminish cancer incidence and reduce risks throughout life.

Bibliography

1. Ferguson LR, Philpott M (2008) Nutrition and mutagenesis. Annu Rev Nutr 28:313–329
2. Dybing E, O'Brien J, Renwick AG, Sanner T (2008) Risk assessment of dietary exposures to compounds that are genotoxic and carcinogenic–an overview. Toxicol Lett 180(2):110–117
3. Hashimoto K, Higuchi M, Makino B, Sakakibara I, Kubo M, Komatsu Y, Maruno M, Okada M (1999) Quantitative analysis of aristolochic acids, toxic compounds, contained in some medicinal plants. J Ethnopharmacol 64(2):185–189
4. Nortier JL, Martinez MC, Schmeiser HH, Arlt VM, Bieler CA, Petein M, Depierreux MF, De Pauw L, Abramowicz D, Vereerstraeten P, Vanherweghem JL (2000) Urothelial carcinoma associated with the use of a Chinese herb (Aristolochia fangchi). N Engl J Med 342(23):1686–1692
5. Lai MN, Wang SM, Chen PC, Chen YY, Wang JD (2010) Population-based case-control study of Chinese herbal products containing aristolochic acid and urinary tract cancer risk. J Natl Cancer Inst 102(3):179–186
6. Christman JK, Sheikhnejad G, Dizik M, Abileah S, Wainfan E (1993) Reversibility of changes in nucleic acid methylation and gene expression induced in rat liver by severe dietary methyl deficiency. Carcinogenesis 14(4):551–557
7. Pufulete M, Al-Ghnaniem R, Rennie JA, Appleby P, Harris N, Gout S, Emery PW, Sanders TA (2005) Influence of folate status on genomic DNA methylation in colonic mucosa of subjects without colorectal adenoma or cancer. Br J Cancer 92(5):838–842
8. Pogribny IP, Ross SA, Wise C, Pogribna M, Jones EA, Tryndyak VP, James SJ, Dragan YP, Poirier LA (2006) Irreversible global DNA hypomethylation as a key step in hepatocarcinogenesis induced by dietary methyl deficiency. Mutat Res 593(1–2):80–87
9. Robinson JK (2005) Sun exposure, sun protection, and vitamin D. JAMA 294(12):1541–1543
10. Rogers HW, Weinstock MA, Harris AR, Hinckley MR, Feldman SR, Fleischer AB, Coldiron BM (2010) Incidence estimate of nonmelanoma skin cancer in the United States, 2006. Arch Dermatol 146(3):283–287
11. Kao SC, Reid G, Lee K, Vardy J, Clarke S, van Zandwijk N (2010) Malignant mesothelioma. Intern Med J 40(11):742–750
12. Boffetta P, Hashibe M (2006) Alcohol and cancer. Lancet Oncol 7(2):149–156
13. Seitz HK, Stickel F (2007) Molecular mechanisms of alcohol-mediated carcinogenesis. Nat Rev Cancer 7(8):599–612
14. Pelucchi C, Gallus S, Garavello W, Bosetti C, La Vecchia C (2008) Alcohol and tobacco use, and cancer risk for upper aerodigestive tract and liver. Eur J Cancer Prev 17(4):340–344
15. Hecht SS, Kassie F, Hatsukami DK (2009) Chemoprevention of lung carcinogenesis in addicted smokers and ex-smokers. Nat Rev Cancer 9(7):476–488
16. McKillop IH, Schrum LW (2009) Role of alcohol in liver carcinogenesis. Semin Liver Dis 29(2):222–232

17. He S, Nakada D, Morrison SJ (2009) Mechanisms of stem cell self-renewal. Annu Rev Cell Dev Biol 25:377–406
18. United States Cancer Statistics (USCS) (1999–2006) Cancer Incidence and Mortality Data, National Program of Cancer Registries (NPCR), Center for Disease Control and Prevention, Department of Health and Human Services, USA. http://apps.nccd.cdc.gov/uscs/index.aspx. Accessed March 2010
19. Bielas JH, Loeb LA (2005) Mutator phenotype in cancer: timing and perspectives. Environ Mol Mutagen 45(2–3):206–213
20. Stratton MR, Campbell PJ, Futreal PA (2009) The cancer genome. Nature 458(7239):719–724
21. Ames BN (1989) Endogenous oxidative DNA damage, aging, and cancer. Free Radic Res Commun 7(3–6):121–128
22. Loft S, Poulsen HE (1998) Estimation of oxidative DNA damage in man from urinary excretion of repair products. Acta Biochim Pol 45(1):133–144
23. Shigenaga MK, Gimeno CJ, Ames BN (1989) Urinary 8-hydroxy-2'-deoxyguanosine as a biological marker of in vivo oxidative DNA damage. Proc Natl Acad Sci USA 86(24):9697–9701
24. Cooke MS, Evans MD, Dizdaroglu M, Lunec J (2003) Oxidative DNA damage: mechanisms, mutation, and disease. FASEB J 17(10):1195–1214
25. Hussain SP, Harris CC (2007) Inflammation and cancer: an ancient link with novel potentials. Int J Cancer 121(11):2373–2380
26. Dahlmann HA, Vaidyanathan VG, Sturla SJ (2009) Investigating the biochemical impact of DNA damage with structure-based probes: abasic sites, photodimers, alkylation adducts, and oxidative lesions. Biochemistry 48(40):9347–9359
27. Sedelnikova OA, Redon CE, Dickey JS, Nakamura AJ, Georgakilas AG, Bonner WM (2010) Role of oxidatively induced DNA lesions in human pathogenesis. Mutat Res.
28. Yuan J, Narayanan L, Rockwell S, Glazer PM (2000) Diminished DNA repair and elevated mutagenesis in mammalian cells exposed to hypoxia and low pH. Cancer Res 60(16):4372–4376
29. Camenisch U, Naegeli H (2009) Role of DNA repair in the protection against genotoxic stress. EXS 99:111–150
30. Goldsby RE, Lawrence NA, Hays LE, Olmsted EA, Chen X, Singh M, Preston BD (2001) Defective DNA polymerase-delta proofreading causes cancer susceptibility in mice. Nat Med 7(6):638–639
31. Loeb LA, Bielas JH, Beckman RA (2008) Cancers exhibit a mutator phenotype: clinical implications. Cancer Res 68(10):3551–3557, discussion 3557
32. Zegerman P, Diffley JF (2009) DNA replication as a target of the DNA damage checkpoint. DNA Repair (Amst) 8(9):1077–1088
33. Hanahan D, Weinberg RA (2000) The hallmarks of cancer. Cell 100(1):57–70
34. van der Flier LG, Clevers H (2009) Stem cells, self-renewal, and differentiation in the intestinal epithelium. Annu Rev Physiol 71:241–260
35. Weinberg RA (2008) Mechanisms of malignant progression. Carcinogenesis 29(6):1092–1095
36. Weinberg RA (2008) Coevolution in the tumor microenvironment. Nat Genet 40(5):494–495
37. Chatten C, Bathe OF (2010) Immunoregulatory cells of the tumor microenvironment. Front Biosci 15:291–308
38. Peto J (2001) Cancer epidemiology in the last century and the next decade. Nature 411(6835):390–395
39. Loeb LA, Harris CC (2008) Advances in chemical carcinogenesis: a historical review and prospective. Cancer Res 68(17):6863–6872
40. Jiricny J (2006) The multifaceted mismatch-repair system. Nat Rev Mol Cell Biol 7(5):335–346

41. McCulloch SD, Kunkel TA (2008) The fidelity of DNA synthesis by eukaryotic replicative and translesion synthesis polymerases. Cell Res 18(1):148–161
42. Kunkel TA (2009) Evolving views of DNA replication (In)fidelity. Cold Spring Harbour Symp Quant Biol 74:91–101
43. Hussain SP, He P, Subleski J, Hofseth LJ, Trivers GE, Mechanic L, Hofseth AB, Bernard M, Schwank J, Nguyen G, Mathe E, Djurickovic D, Haines D, Weiss J, Back T, Gruys E, Laubach VE, Wiltrout RH, Harris CC (2008) Nitric oxide is a key component in inflammation-accelerated tumorigenesis. Cancer Res 68(17):7130–7136
44. Fearon ER, Vogelstein B (1990) A genetic model for colorectal tumorigenesis. Cell 61(5):759–767
45. Blanquet V, Turleau C, Gross-Morand MS, Senamaud-Beaufort C, Doz F, Besmond C (1995) Spectrum of germline mutations in the RB1 gene: a study of 232 patients with hereditary and non hereditary retinoblastoma. Hum Mol Genet 4(3):383–388
46. Hussain SP, Harris CC (1998) Molecular epidemiology of human cancer: contribution of mutation spectra studies of tumor suppressor genes. Cancer Res 58(18):4023–4037
47. Wilson VL (2001) Detecting rare mutations associated with cancer risk. Am J Pharmacogenomics 1(4):283–293
48. Tapp RA, Feng J, Jones JW, Carlson JA, Wilson VL (2007) Single base instability is promoted in vulvar lichen sclerosus. J Invest Dermatol 127(11):2563–2576
49. Petitjean A, Mathe E, Kato S, Ishioka C, Tavtigian SV, Hainaut P, Olivier M (2007) Impact of mutant p53 functional properties on TP53 mutation patterns and tumor phenotype: lessons from recent developments in the IARC TP53 database. Hum Mutat 28(6):622–629
50. Stenson PD, Mort M, Ball EV, Howells K, Phillips AD, Thomas NS, Cooper DN (2009) The Human Gene Mutation Database: 2008 update. Genome Med 1(1):13
51. Kumar R, Angelini S, Hemminki K (2003) Activating BRAF and N-Ras mutations in sporadic primary melanomas: an inverse association with allelic loss on chromosome 9. Oncogene 22(58):9217–9224
52. Aplan PD (2006) Causes of oncogenic chromosomal translocation. Trends Genet 22(1):46–55
53. MacCorkle RA, Slattery SD, Nash DR, Brinkley BR (2006) Intracellular protein binding to asbestos induces aneuploidy in human lung fibroblasts. Cell Motil Cytoskeleton 63(10):646–657
54. Pavanello S, Pesatori AC, Dioni L, Hoxha M, Bollati V, Siwinska E, Mielzynska D, Bolognesi C, Bertazzi PA, Baccarelli A (2010) Shorter telomere length in peripheral blood lymphocytes of workers exposed to polycyclic aromatic hydrocarbons. Carcinogenesis 31(2):216–221
55. Hirao T, Nelson HH, Ashok TD, Wain JC, Mark EJ, Christiani DC, Wiencke JK, Kelsey KT (2001) Tobacco smoke-induced DNA damage and an early age of smoking initiation induce chromosome loss at 3p21 in lung cancer. Cancer Res 61(2):612–615
56. Saletta F, Matullo G, Manuguerra M, Arena S, Bardelli A, Vineis P (2007) Exposure to the tobacco smoke constituent 4-aminobiphenyl induces chromosomal instability in human cancer cells. Cancer Res 67(15):7088–7094
57. Shinmura K, Iwaizumi M, Igarashi H, Nagura K, Yamada H, Suzuki M, Fukasawa K, Sugimura H (2008) Induction of centrosome amplification and chromosome instability in p53-deficient lung cancer cells exposed to benzo[a]pyrene diol epoxide (B[a]PDE). J Pathol 216(3):365–374
58. Groopman JD, Donahue PR, Zhu JQ, Chen JS, Wogan GN (1985) Aflatoxin metabolism in humans: detection of metabolites and nucleic acid adducts in urine by affinity chromatography. Proc Natl Acad Sci USA 82(19):6492–6496
59. Groopman JD, Wild CP, Hasler J, Junshi C, Wogan GN, Kensler TW (1993) Molecular epidemiology of aflatoxin exposures: validation of aflatoxin-N7-guanine levels in urine as a biomarker in experimental rat models and humans. Environ Health Perspect 99:107–113

60. Essigmann JM, Wood ML (1993) The relationship between the chemical structures and mutagenic specificities of the DNA lesions formed by chemical and physical mutagens. Toxicol Lett 67(1–3):29–39
61. Batzer MA, Deininger PL (2002) Alu repeats and human genomic diversity. Nat Rev Genet 3(5):370–379
62. Konkel MK, Batzer MA (2010) A mobile threat to genome stability: The impact of non-LTR retrotransposons upon the human genome. Semin Cancer Biol 20(4):211–221
63. Callinan PA, Batzer MA (2006) Retrotransposable elements and human disease. Genome Dyn 1:104–115
64. Jones PA, Liang G (2009) Rethinking how DNA methylation patterns are maintained. Nat Rev Genet 10(11):805–811
65. Lister R, Pelizzola M, Dowen RH, Hawkins RD, Hon G, Tonti-Filippini J, Nery JR, Lee L, Ye Z, Ngo QM, Edsall L, Antosiewicz-Bourget J, Stewart R, Ruotti V, Millar AH, Thomson JA, Ren B, Ecker JR (2009) Human DNA methylomes at base resolution show widespread epigenomic differences. Nature 462(7271):315–322
66. Schubeler D (2009) Epigenomics: Methylation matters. Nature 462(7271):296–297
67. Guil S, Esteller M (2009) DNA methylomes, histone codes and miRNAs: tying it all together. Int J Biochem Cell Biol 41(1):87–95
68. Lopez-Serra L, Esteller M (2008) Proteins that bind methylated DNA and human cancer: reading the wrong words. Br J Cancer 98(12):1881–1885
69. Esteller M (2008) Epigenetics in cancer. N Engl J Med 358(11):1148–1159
70. Sharma S, Kelly TK, Jones PA (2010) Epigenetics in cancer. Carcinogenesis 31(1):27–36
71. Wilson VL, Jones PA (1983) Inhibition of DNA methylation by chemical carcinogens in vitro. Cell 32(1):239–246
72. Wilson VL, Smith RA, Longoria J, Liotta MA, Harper CM, Harris CC (1987) Chemical carcinogen-induced decreases in genomic 5-methyldeoxycytidine content of normal human bronchial epithelial cells. Proc Natl Acad Sci USA 84(10):3298–3301
73. Chanda S, Dasgupta UB, Guhamazumder D, Gupta M, Chaudhuri U, Lahiri S, Das S, Ghosh N, Chatterjee D (2006) DNA hypermethylation of promoter of gene p53 and p16 in arsenic-exposed people with and without malignancy. Toxicol Sci 89(2):431–437
74. Bollati V, Baccarelli A, Hou L, Bonzini M, Fustinoni S, Cavallo D, Byun HM, Jiang J, Marinelli B, Pesatori AC, Bertazzi PA, Yang AS (2007) Changes in DNA methylation patterns in subjects exposed to low-dose benzene. Cancer Res 67(3):876–880
75. Baccarelli A, Bollati V (2009) Epigenetics and environmental chemicals. Curr Opin Pediatr 21(2):243–251
76. Luger K, Mader AW, Richmond RK, Sargent DF, Richmond TJ (1997) Crystal structure of the nucleosome core particle at 2.8 A resolution. Nature 389(6648):251–260
77. Malik HS, Henikoff S (2003) Phylogenomics of the nucleosome. Nat Struct Biol 10(11):882–891
78. Allfrey VG, Faulkner R, Mirsky AE (1964) Acetylation and Methylation of Histones and Their Possible Role in the Regulation of Rna Synthesis. Proc Natl Acad Sci USA 51:786–794
79. Kondo Y, Shen L, Yan PS, Huang TH, Issa JP (2004) Chromatin immunoprecipitation microarrays for identification of genes silenced by histone H3 lysine 9 methylation. Proc Natl Acad Sci USA 101(19):7398–7403
80. McKittrick E, Gafken PR, Ahmad K, Henikoff S (2004) Histone H3.3 is enriched in covalent modifications associated with active chromatin. Proc Natl Acad Sci USA 101(6):1525–1530
81. Kondo Y, Shen L, Ahmed S, Boumber Y, Sekido Y, Haddad BR, Issa JP (2008) Downregulation of histone H3 lysine 9 methyltransferase G9a induces centrosome disruption and chromosome instability in cancer cells. PLoS ONE 3(4):e2037
82. Yamada T, Mizuno K, Hirota K, Kon N, Wahls WP, Hartsuiker E, Murofushi H, Shibata T, Ohta K (2004) Roles of histone acetylation and chromatin remodeling factor in a meiotic recombination hotspot. EMBO J 23(8):1792–1803

83. Toh GW, O'Shaughnessy AM, Jimeno S, Dobbie IM, Grenon M, Maffini S, O'Rorke A, Lowndes NF (2006) Histone H2A phosphorylation and H3 methylation are required for a novel Rad9 DSB repair function following checkpoint activation. DNA Repair (Amst) 5(6):693–703
84. Shahbazian MD, Zhang K, Grunstein M (2005) Histone H2B ubiquitylation controls processive methylation but not monomethylation by Dot1 and Set1. Mol Cell 19(2):271–277
85. Iniguez-Lluhi JA (2006) For a healthy histone code, a little SUMO in the tail keeps the acetyl away. ACS Chem Biol 1(4):204–206
86. Biel M, Wascholowski V, Giannis A (2005) Epigenetics–an epicenter of gene regulation: histones and histone-modifying enzymes. Angew Chem Int Ed Engl 44(21):3186–3216
87. Henikoff S (2005) Histone modifications: combinatorial complexity or cumulative simplicity? Proc Natl Acad Sci USA 102(15):5308–5309
88. Talasz H, Lindner HH, Sarg B, Helliger W (2005) Histone H4-lysine 20 monomethylation is increased in promoter and coding regions of active genes and correlates with hyperacetylation. J Biol Chem 280(46):38814–38822
89. Schneider J, Wood A, Lee JS, Schuster R, Dueker J, Maguire C, Swanson SK, Florens L, Washburn MP, Shilatifard A (2005) Molecular regulation of histone H3 trimethylation by COMPASS and the regulation of gene expression. Mol Cell 19(6):849–856
90. Khalil AM, Driscoll DJ (2006) Histone H3 lysine 4 dimethylation is enriched on the inactive sex chromosomes in male meiosis but absent on the inactive X in female somatic cells. Cytogenet Genome Res 112(1–2):11–15
91. Kondo Y, Shen L, Cheng AS, Ahmed S, Boumber Y, Charo C, Yamochi T, Urano T, Furukawa K, Kwabi-Addo B, Gold DL, Sekido Y, Huang TH, Issa JP (2008) Gene silencing in cancer by histone H3 lysine 27 trimethylation independent of promoter DNA methylation. Nat Genet 40(6):741–750
92. Taverna SD, Li H, Ruthenburg AJ, Allis CD, Patel DJ (2007) How chromatin-binding modules interpret histone modifications: lessons from professional pocket pickers. Nat Struct Mol Biol 14(11):1025–1040
93. Kurdistani SK, Grunstein M (2003) Histone acetylation and deacetylation in yeast. Nat Rev Mol Cell Biol 4(4):276–284
94. Roth SY, Denu JM, Allis CD (2001) Histone acetyltransferases. Annu Rev Biochem 70:81–120
95. Brownell JE, Zhou J, Ranalli T, Kobayashi R, Edmondson DG, Roth SY, Allis CD (1996) Tetrahymena histone acetyltransferase A: a homolog to yeast Gcn5p linking histone acetylation to gene activation. Cell 84(6):843–851
96. Finnin MS, Donigian JR, Cohen A, Richon VM, Rifkind RA, Marks PA, Breslow R, Pavletich NP (1999) Structures of a histone deacetylase homologue bound to the TSA and SAHA inhibitors. Nature 401(6749):188–193
97. Gray SG, Ekstrom TJ (2001) The human histone deacetylase family. Exp Cell Res 262(2):75–83
98. Dokmanovic M, Clarke C, Marks PA (2007) Histone deacetylase inhibitors: overview and perspectives. Mol Cancer Res 5(10):981–989
99. Verdin E, Dequiedt F, Kasler HG (2003) Class II histone deacetylases: versatile regulators. Trends Genet 19(5):286–293
100. Sengupta N, Seto E (2004) Regulation of histone deacetylase activities. J Cell Biochem 93(1):57–67
101. Gutierrez RM, Hnilica LS (1967) Tissue specificity of histone phosphorylation. Science 157(794):1324–1325
102. Wei Y, Mizzen CA, Cook RG, Gorovsky MA, Allis CD (1998) Phosphorylation of histone H3 at serine 10 is correlated with chromosome condensation during mitosis and meiosis in Tetrahymena. Proc Natl Acad Sci USA 95(13):7480–7484
103. Hirota T, Lipp JJ, Toh BH, Peters JM (2005) Histone H3 serine 10 phosphorylation by Aurora B causes HP1 dissociation from heterochromatin. Nature 438(7071):1176–1180

104. Goto H, Tomono Y, Ajiro K, Kosako H, Fujita M, Sakurai M, Okawa K, Iwamatsu A, Okigaki T, Takahashi T, Inagaki M (1999) Identification of a novel phosphorylation site on histone H3 coupled with mitotic chromosome condensation. J Biol Chem 274(36):25543–25549
105. Preuss U, Landsberg G, Scheidtmann KH (2003) Novel mitosis-specific phosphorylation of histone H3 at Thr11 mediated by Dlk/ZIP kinase. Nucleic Acids Res 31(3):878–885
106. Barratt MJ, Hazzalin CA, Cano E, Mahadevan LC (1994) Mitogen-stimulated phosphorylation of histone H3 is targeted to a small hyperacetylation-sensitive fraction. Proc Natl Acad Sci USA 91(11):4781–4785
107. Cheung P, Tanner KG, Cheung WL, Sassone-Corsi P, Denu JM, Allis CD (2000) Synergistic coupling of histone H3 phosphorylation and acetylation in response to epidermal growth factor stimulation. Mol Cell 5(6):905–915
108. Clayton AL, Rose S, Barratt MJ, Mahadevan LC (2000) Phosphoacetylation of histone H3 on c-fos- and c-jun-associated nucleosomes upon gene activation. EMBO J 19(14):3714–3726
109. Panier S, Durocher D (2009) Regulatory ubiquitylation in response to DNA double-strand breaks. DNA Repair (Amst) 8(4):436–443
110. Galanty Y, Belotserkovskaya R, Coates J, Polo S, Miller KM, Jackson SP (2009) Mammalian SUMO E3-ligases PIAS1 and PIAS4 promote responses to DNA double-strand breaks. Nature 462(7275):935–939
111. Morris JR, Boutell C, Keppler M, Densham R, Weekes D, Alamshah A, Butler L, Galanty Y, Pangon L, Kiuchi T, Ng T, Solomon E (2009) The SUMO modification pathway is involved in the BRCA1 response to genotoxic stress. Nature 462(7275):886–890
112. An JY, Kim EA, Jiang Y, Zakrzewska A, Kim DE, Lee MJ, Mook-Jung I, Zhang Y, Kwon YT (2010) UBR2 mediates transcriptional silencing during spermatogenesis via histone ubiquitination. Proc Natl Acad Sci USA 107(5):1912–1917
113. Grazini U, Zanardi F, Citterio E, Casola S, Goding CR, McBlane F (2010) The RING domain of RAG1 ubiquitylates histone H3: a novel activity in chromatin-mediated regulation of V(D) J joining. Mol Cell 37(2):282–293
114. Lu LY, Wu J, Ye L, Gavrilina GB, Saunders TL, Yu X (2010) RNF8-dependent histone modifications regulate nucleosome removal during spermatogenesis. Dev Cell 18(3):371–384
115. Mabruk MJ, Toh LK, Murphy M, Leader M, Kay E, Murphy GM (2009) Investigation of the effect of UV irradiation on DNA damage: comparison between skin cancer patients and normal volunteers. J Cutan Pathol 36(7):760–765
116. Wogan GN, Hecht SS, Felton JS, Conney AH, Loeb LA (2004) Environmental and chemical carcinogenesis. Semin Cancer Biol 14(6):473–486
117. Lee WR, Perantie DC, Clark KB, Guillot DA, Wilson VL (2001) Effect of mutagen-induced cell lethality on the dose response of germline mutations. Environ Mol Mutagen 37(4):340–344
118. Ellison KS, Dogliotti E, Connors TD, Basu AK, Essigmann JM (1989) Site-specific mutagenesis by O6-alkylguanines located in the chromosomes of mammalian cells: influence of the mammalian O6-alkylguanine-DNA alkyltransferase. Proc Natl Acad Sci USA 86(22):8620–8624
119. Dosanjh MK, Singer B, Essigmann JM (1991) Comparative mutagenesis of O6-methylguanine and O4-methylthymine in Escherichia coli. Biochemistry 30(28):7027–7033
120. Potten CS, Booth C (2002) Keratinocyte stem cells: a commentary. J Invest Dermatol 119(4):888–899
121. Bapat SA (2007) Evolution of cancer stem cells. Semin Cancer Biol 17(3):204–213
122. Orford KW, Scadden DT (2008) Deconstructing stem cell self-renewal: genetic insights into cell-cycle regulation. Nat Rev Genet 9(2):115–128
123. Yahata T, Muguruma Y, Yumino S, Sheng Y, Uno T, Matsuzawa H, Ito M, Kato S, Hotta T, Ando K (2008) Quiescent human hematopoietic stem cells in the bone marrow niches organize the hierarchical structure of hematopoiesis. Stem Cells 26(12):3228–3236

124. Karpowicz P, Morshead C, Kam A, Jervis E, Ramunas J, Cheng V, van der Kooy D (2005) Support for the immortal strand hypothesis: neural stem cells partition DNA asymmetrically in vitro. J Cell Biol 170(5):721–732
125. Lechler T, Fuchs E (2005) Asymmetric cell divisions promote stratification and differentiation of mammalian skin. Nature 437(7056):275–280
126. Smith GH (2005) Label-retaining epithelial cells in mouse mammary gland divide asymmetrically and retain their template DNA strands. Development 132(4):681–687
127. Cairns J (1975) Mutation selection and the natural history of cancer. Nature 255(5505):197–200
128. Potten CS, Grant HK (1998) The relationship between ionizing radiation-induced apoptosis and stem cells in the small and large intestine. Br J Cancer 78(8):993–1003
129. Lee JM, Dedhar S, Kalluri R, Thompson EW (2006) The epithelial-mesenchymal transition: new insights in signaling, development, and disease. J Cell Biol 172(7):973–981
130. Rubin H (2001) The role of selection in progressive neoplastic transformation. Adv Cancer Res 83:159–207
131. Negrini S, Gorgoulis VG, Halazonetis TD (2010) Genomic instability–an evolving hallmark of cancer. Nat Rev Mol Cell Biol 11(3):220–228
132. Beckman RA, Loeb LA (2006) Efficiency of carcinogenesis with and without a mutator mutation. Proc Natl Acad Sci USA 103(38):14140–14145
133. Chu WK, Hickson ID (2009) RecQ helicases: multifunctional genome caretakers. Nat Rev Cancer 9(9):644–654
134. Lengauer C, Kinzler KW, Vogelstein B (1998) Genetic instabilities in human cancers. Nature 396(6712):643–649
135. Ganem NJ, Godinho SA, Pellman D (2009) A mechanism linking extra centrosomes to chromosomal instability. Nature 460(7252):278–282
136. Weaver BA, Silk AD, Montagna C, Verdier-Pinard P, Cleveland DW (2007) Aneuploidy acts both oncogenically and as a tumor suppressor. Cancer Cell 11(1):25–36
137. Thompson SL, Compton DA (2008) Examining the link between chromosomal instability and aneuploidy in human cells. J Cell Biol 180(4):665–672
138. Torres EM, Williams BR, Amon A (2008) Aneuploidy: cells losing their balance. Genetics 179(2):737–746
139. Hussein MR, Wood GS (2002) Microsatellite instability and its relevance to cutaneous tumorigenesis. J Cutan Pathol 29(5):257–267
140. Dietmaier W, Wallinger S, Bocker T, Kullmann F, Fishel R, Ruschoff J (1997) Diagnostic microsatellite instability: definition and correlation with mismatch repair protein expression. Cancer Res 57(21):4749–4756
141. Boland CR, Thibodeau SN, Hamilton SR, Sidransky D, Eshleman JR, Burt RW, Meltzer SJ, Rodriguez-Bigas MA, Fodde R, Ranzani GN, Srivastava S (1998) A National Cancer Institute Workshop on Microsatellite Instability for cancer detection and familial predisposition: development of international criteria for the determination of microsatellite instability in colorectal cancer. Cancer Res 58(22):5248–5257
142. Hoeijmakers JH (2001) Genome maintenance mechanisms for preventing cancer. Nature 411(6835):366–374
143. Jascur T, Boland CR (2006) Structure and function of the components of the human DNA mismatch repair system. Int J Cancer 119(9):2030–2035
144. Thibodeau SN, French AJ, Roche PC, Cunningham JM, Tester DJ, Lindor NM, Moslein G, Baker SM, Liskay RM, Burgart LJ, Honchel R, Halling KC (1996) Altered expression of hMSH2 and hMLH1 in tumors with microsatellite instability and genetic alterations in mismatch repair genes. Cancer Res 56(21):4836–4840
145. Hussein MR, Wood GS (2002) Microsatellite instability in human melanocytic skin tumors: an incidental finding or a pathogenetic mechanism? J Cutan Pathol 29(1):1–4

146. Fishel R, Lescoe MK, Rao MR, Copeland NG, Jenkins NA, Garber J, Kane M, Kolodner R (1993) The human mutator gene homolog MSH2 and its association with hereditary nonpolyposis colon cancer. Cell 75(5):1027–1038
147. Leach FS, Nicolaides NC, Papadopoulos N, Liu B, Jen J, Parsons R, Peltomaki P, Sistonen P, Aaltonen LA, Nystrom-Lahti M et al (1993) Mutations of a mutS homolog in hereditary nonpolyposis colorectal cancer. Cell 75(6):1215–1225
148. Garcon G, Dagher Z, Zerimech F, Ledoux F, Courcot D, Aboukais A, Puskaric E, Shirali P (2006) Dunkerque City air pollution particulate matter-induced cytotoxicity, oxidative stress and inflammation in human epithelial lung cells (L132) in culture. Toxicol In Vitro 20(4):519–528
149. Saint-Georges F, Garcon G, Escande F, Abbas I, Verdin A, Gosset P, Mulliez P, Shirali P (2009) Role of air pollution Particulate Matter (PM(2.5)) in the occurrence of loss of heterozygosity in multiple critical regions of 3p chromosome in human epithelial lung cells (L132). Toxicol Lett 187(3):172–179
150. Hirose T, Kondo K, Takahashi Y, Ishikura H, Fujino H, Tsuyuguchi M, Hashimoto M, Yokose T, Mukai K, Kodama T, Monden Y (2002) Frequent microsatellite instability in lung cancer from chromate-exposed workers. Mol Carcinog 33(3):172–180
151. Zienolddiny S, Aguelon AM, Mironov N, Mathew B, Thomas G, Sankaranarayanan R, Yamasaki H (2004) Genomic instability in oral squamous cell carcinoma: relationship to betel-quid chewing. Oral Oncol 40(3):298–303
152. Parsons BL, Marchant-Miros KE, Delongchamp RR, Verkler TL, Patterson TA, McKinzie PB, Kim LT (2010) ACB-PCR quantification of K-RAS codon 12 GAT and GTT mutant fraction in colon tumor and non-tumor tissue. Cancer Invest 28(4):364–375
153. Toyota M, Ahuja N, Ohe-Toyota M, Herman JG, Baylin SB, Issa JP (1999) CpG island methylator phenotype in colorectal cancer. Proc Natl Acad Sci USA 96(15):8681–8686
154. Teodoridis JM, Hardie C, Brown R (2008) CpG island methylator phenotype (CIMP) in cancer: causes and implications. Cancer Lett 268(2):177–186
155. Tanemura A, Terando AM, Sim MS, van Hoesel AQ, de Maat MF, Morton DL, Hoon DS (2009) CpG island methylator phenotype predicts progression of malignant melanoma. Clin Cancer Res 15(5):1801–1807
156. Yagi K, Akagi K, Hayashi H, Nagae G, Tsuji S, Isagawa T, Midorikawa Y, Nishimura Y, Sakamoto H, Seto Y, Aburatani H, Kaneda A (2010) Three DNA methylation epigenotypes in human colorectal cancer. Clin Cancer Res 16(1):21–33
157. Gaudet F, Hodgson JG, Eden A, Jackson-Grusby L, Dausman J, Gray JW, Leonhardt H, Jaenisch R (2003) Induction of tumors in mice by genomic hypomethylation. Science 300(5618):489–492
158. Gama-Sosa MA, Slagel VA, Trewyn RW, Oxenhandler R, Kuo KC, Gehrke CW, Ehrlich M (1983) The 5-methylcytosine content of DNA from human tumors. Nucleic Acids Res 11(19):6883–6894
159. Goelz SE, Vogelstein B, Hamilton SR, Feinberg AP (1985) Hypomethylation of DNA from benign and malignant human colon neoplasms. Science 228(4696):187–190
160. Kim YI, Giuliano A, Hatch KD, Schneider A, Nour MA, Dallal GE, Selhub J, Mason JB (1994) Global DNA hypomethylation increases progressively in cervical dysplasia and carcinoma. Cancer 74(3):893–899
161. Shahrzad S, Bertrand K, Minhas K, Coomber BL (2007) Induction of DNA hypomethylation by tumor hypoxia. Epigenetics 2(2):119–125
162. Avila MA, Carretero MV, Rodriguez EN, Mato JM (1998) Regulation by hypoxia of methionine adenosyltransferase activity and gene expression in rat hepatocytes. Gastroenterology 114(2):364–371
163. Yanagawa N, Tamura G, Honda T, Endoh M, Nishizuka S, Motoyama T (2004) Demethylation of the synuclein gamma gene CpG island in primary gastric cancers and gastric cancer cell lines. Clin Cancer Res 10(7):2447–2451

164. Laner T, Schulz WA, Engers R, Muller M, Florl AR (2005) Hypomethylation of the XIST gene promoter in prostate cancer. Oncol Res 15(5):257–264
165. Fraga MF, Ballestar E, Paz MF, Ropero S, Setien F, Ballestar ML, Heine-Suner D, Cigudosa JC, Urioste M, Benitez J, Boix-Chornet M, Sanchez-Aguilera A, Ling C, Carlsson E, Poulsen P, Vaag A, Stephan Z, Spector TD, Wu YZ, Plass C, Esteller M (2005) Epigenetic differences arise during the lifetime of monozygotic twins. Proc Natl Acad Sci USA 102(30): 10604–10609
166. Kampranis SC, Tsichlis PN (2009) Histone demethylases and cancer. Adv Cancer Res 102:103–169
167. Kristensen LS, Nielsen HM, Hansen LL (2009) Epigenetics and cancer treatment. Eur J Pharmacol 625(1–3):131–142
168. Kinzler KW, Vogelstein B (1997) Cancer-susceptibility genes. Gatekeepers and caretakers. Nature 386(6627):761–763
169. Bardelli A, Cahill DP, Lederer G, Speicher MR, Kinzler KW, Vogelstein B, Lengauer C (2001) Carcinogen-specific induction of genetic instability. Proc Natl Acad Sci USA 98(10):5770–5775
170. Palmero EI, Achatz MI, Ashton-Prolla P, Olivier M, Hainaut P (2010) Tumor protein 53 mutations and inherited cancer: beyond Li-Fraumeni syndrome. Curr Opin Oncol 22(1):64–69
171. Amberger J, Bocchini CA, Scott AF, Hamosh A (2009) McKusick's Online Mendelian Inheritance in Man (OMIM). Nucleic Acids Res. 37(Database issue): D793–D796
172. Cazier JB, Tomlinson I (2010) General lessons from large-scale studies to identify human cancer predisposition genes. J Pathol 220(2):255–262
173. Gresner P, Gromadzinska J, Wasowicz W (2007) Polymorphism of selected enzymes involved in detoxification and biotransformation in relation to lung cancer. Lung Cancer 57(1):1–25
174. Cote ML, Wenzlaff AS, Bock CH, Land SJ, Santer SK, Schwartz DR, Schwartz AG (2007) Combinations of cytochrome P-450 genotypes and risk of early-onset lung cancer in Caucasians and African Americans: a population-based study. Lung Cancer 55(3):255–262
175. Schwartz AG, Prysak GM, Bock CH, Cote ML (2007) The molecular epidemiology of lung cancer. Carcinogenesis 28(3):507–518
176. Cokkinides V, Bandi P, McMahon C, Jemal A, Glynn T, Ward E (2009) Tobacco control in the United States–recent progress and opportunities. CA Cancer J Clin 59(6):352–365
177. Lubin JH, Alavanja MC, Caporaso N, Brown LM, Brownson RC, Field RW, Garcia-Closas M, Hartge P, Hauptmann M, Hayes RB, Kleinerman R, Kogevinas M, Krewski D, Langholz B, Letourneau EG, Lynch CF, Malats N, Sandler DP, Schaffrath-Rosario A, Schoenberg JB, Silverman DT, Wang Z, Wichmann HE, Wilcox HB, Zielinski JM (2007) Cigarette smoking and cancer risk: modeling total exposure and intensity. Am J Epidemiol 166(4):479–489
178. Robles AI, Linke SP, Harris CC (2002) The p53 network in lung carcinogenesis. Oncogene 21(45):6898–6907
179. Olivier M, Hollstein M, Hainaut P (2010) TP53 Mutations in Human Cancers: Origins, Consequences, and Clinical Use. Cold Spring Harb Perspect Biol 2(1):a001008
180. Denissenko MF, Pao A, Tang M, Pfeifer GP (1996) Preferential formation of benzo[a]pyrene adducts at lung cancer mutational hotspots in P53. Science 274(5286):430–432
181. Greenblatt MS, Bennett WP, Hollstein M, Harris CC (1994) Mutations in the p53 tumor suppressor gene: clues to cancer etiology and molecular pathogenesis. Cancer Res 54(18):4855–4878
182. Bennett WP, Hussain SP, Vahakangas KH, Khan MA, Shields PG, Harris CC (1999) Molecular epidemiology of human cancer risk: gene-environment interactions and p53 mutation spectrum in human lung cancer. J Pathol 187(1):8–18
183. Hsu IC, Metcalf RA, Sun T, Welsh JA, Wang NJ, Harris CC (1991) Mutational hotspot in the p53 gene in human hepatocellular carcinomas. Nature 350(6317):427–428
184. Soini Y, Chia SC, Bennett WP, Groopman JD, Wang JS, DeBenedetti VM, Cawley H, Welsh JA, Hansen C, Bergasa NV, Jones EA, DiBisceglie AM, Trivers GE, Sandoval CA, Calderon

IE, Munoz Espinosa LE, Harris CC (1996) An aflatoxin-associated mutational hotspot at codon 249 in the p53 tumor suppressor gene occurs in hepatocellular carcinomas from Mexico. Carcinogenesis 17(5):1007–1012

185. Gouas D, Shi H, Hainaut P (2009) The aflatoxin-induced TP53 mutation at codon 249 (R249S): biomarker of exposure, early detection and target for therapy. Cancer Lett 286(1): 29–37
186. Brash DE, Rudolph JA, Simon JA, Lin A, McKenna GJ, Baden HP, Halperin AJ, Ponten J (1991) A role for sunlight in skin cancer: UV-induced p53 mutations in squamous cell carcinoma. Proc Natl Acad Sci USA 88(22):10124–10128
187. Armstrong BK, Kricker A (2001) The epidemiology of UV induced skin cancer. J Photochem Photobiol B 63(1–3):8–18
188. Runger TM (2007) How different wavelengths of the ultraviolet spectrum contribute to skin carcinogenesis: the role of cellular damage responses. J Invest Dermatol 127(9):2103–2105
189. Fang JL, Vaca CE (1997) Detection of DNA adducts of acetaldehyde in peripheral white blood cells of alcohol abusers. Carcinogenesis 18(4):627–632
190. Vaca CE, Nilsson JA, Fang JL, Grafstrom RC (1998) Formation of DNA adducts in human buccal epithelial cells exposed to acetaldehyde and methylglyoxal in vitro. Chem Biol Interact 108(3):197–208
191. Clemens DL, Forman A, Jerrells TR, Sorrell MF, Tuma DJ (2002) Relationship between acetaldehyde levels and cell survival in ethanol-metabolizing hepatoma cells. Hepatology 35(5):1196–1204
192. Molina PE, Hoek JB, Nelson S, Guidot DM, Lang CH, Wands JR, Crawford JM (2003) Mechanisms of alcohol-induced tissue injury. Alcohol Clin Exp Res 27(3):563–575
193. el Ghissassi F, Barbin A, Nair J, Bartsch H (1995) Formation of 1,N6-ethenoadenine and 3, N4-ethenocytosine by lipid peroxidation products and nucleic acid bases. Chem Res Toxicol 8(2):278–283
194. Giovannucci E (2004) Alcohol, one-carbon metabolism, and colorectal cancer: recent insights from molecular studies. J Nutr 134(9):2475S–2481S
195. Giovannucci E (2002) Epidemiologic studies of folate and colorectal neoplasia: a review. J Nutr 132(8 Suppl):2350S–2355S
196. Leiter U, Garbe C (2008) Epidemiology of melanoma and nonmelanoma skin cancer–the role of sunlight. Adv Exp Med Biol 624:89–103
197. Ridley AJ, Whiteside JR, McMillan TJ, Allinson SL (2009) Cellular and sub-cellular responses to UVA in relation to carcinogenesis. Int J Radiat Biol 85(3):177–195
198. Matsunaga T, Hieda K, Nikaido O (1991) Wavelength dependent formation of thymine dimers and (6–4) photoproducts in DNA by monochromatic ultraviolet light ranging from 150 to 365 nm. Photochem Photobiol 54(3):403–410
199. Lacour JP (2002) Carcinogenesis of basal cell carcinomas: genetics and molecular mechanisms. Br J Dermatol 146(Suppl 61):17–19
200. Agar NS, Halliday GM, Barnetson RS, Ananthaswamy HN, Wheeler M, Jones AM (2004) The basal layer in human squamous tumors harbors more UVA than UVB fingerprint mutations: a role for UVA in human skin carcinogenesis. Proc Natl Acad Sci USA 101(14):4954–4959
201. Kuhn C, Hurwitz SA, Kumar MG, Cotton J, Spandau DF (1999) Activation of the insulin-like growth factor-1 receptor promotes the survival of human keratinocytes following ultraviolet B irradiation. Int J Cancer 80(3):431–438
202. Lewis DA, Spandau DF (2007) UVB activation of NF-kappaB in normal human keratinocytes occurs via a unique mechanism. Arch Dermatol Res 299(2):93–101
203. Lewis DA, Yi Q, Travers JB, Spandau DF (2008) UVB-induced senescence in human keratinocytes requires a functional insulin-like growth factor-1 receptor and p53. Mol Biol Cell 19(4):1346–1353
204. Tavakkol A, Elder JT, Griffiths CE, Cooper KD, Talwar H, Fisher GJ, Keane KM, Foltin SK, Voorhees JJ (1992) Expression of growth hormone receptor, insulin-like growth factor 1

(IGF-1) and IGF-1 receptor mRNA and proteins in human skin. J Invest Dermatol 99(3):343–349
205. Ferber A, Chang C, Sell C, Ptasznik A, Cristofalo VJ, Hubbard K, Ozer HL, Adamo M, Roberts CT Jr, LeRoith D et al (1993) Failure of senescent human fibroblasts to express the insulin-like growth factor-1 gene. J Biol Chem 268:17883–17888
206. Agents Classified by the IARC Monographs 1–100, Updated 30 August 2010; IARC Monographs on the Evaluation of Carcinogenic Risks to Humans. http://monographs.iarc.fr/ENG/Classification/index.php. Accessed November 2010
207. Hecht SS, Carmella SG, Murphy SE, Akerkar S, Brunnemann KD, Hoffmann D (1993) A tobacco-specific lung carcinogen in the urine of men exposed to cigarette smoke. N Engl J Med 329(21):1543–1546
208. Smoking and Tobacco Control Monograph 10, Health Effects of Exposure to Environmental Tobacco Smoke (August 1999), National Cancer Institute, National Institutes of Health, Bethesda, MD
209. Alexandrov K, Rojas M, Kadlubar FF, Lang NP, Bartsch H (1996) Evidence of anti-benzo[a] pyrene diolepoxide-DNA adduct formation in human colon mucosa. Carcinogenesis 17(9):2081–2083
210. Hecht SS, Hoffmann D (1988) Tobacco-specific nitrosamines, an important group of carcinogens in tobacco and tobacco smoke. Carcinogenesis 9(6):875–884
211. Pleasance ED, Stephens PJ, O'Meara S, McBride DJ, Meynert A, Jones D, Lin ML, Beare D, Lau KW, Greenman C, Varela I, Nik-Zainal S, Davies HR, Ordonez GR, Mudie L J, Latimer C, Edkins S, Stebbings L, Chen L, Jia M, Leroy C, Marshall J, Menzies A, Butler A, Teague JW, Mangion J, Sun YA, McLaughlin SF, Peckham HE, Tsung EF, Costa GL, Lee CC, Minna JD, Gazdar A, Birney E, Rhodes MD, McKernan KJ, Stratton MR, Futreal PA, Campbell PJ (2010) A small-cell lung cancer genome with complex signatures of tobacco exposure. Nature 463(7278):184–190
212. McCullough ML, Giovannucci EL (2004) Diet and cancer prevention. Oncogene 23(38):6349–6364
213. Kensler TW, Groopman JD (2009) Is it time to advance the chemoprevention of environmental carcinogenesis with microdosing trials? Cancer Prev Res (Phila Pa) 2(12):1003–1007
214. Ju J, Picinich SC, Yang Z, Zhao Y, Suh N, Kong AN, Yang CS (2010) Cancer-preventive activities of tocopherols and tocotrienols. Carcinogenesis 31(4):533–542
215. Fowke JH (2007) Head and neck cancer: a case for inhibition by isothiocyanates and indoles from cruciferous vegetables. Eur J Cancer Prev 16(4):348–356
216. Singh N, Manshian B, Jenkins GJ, Griffiths SM, Williams PM, Maffeis TG, Wright CJ, Doak SH (2009) NanoGenotoxicology: the DNA damaging potential of engineered nanomaterials. Biomaterials 30(23–24):3891–3914

Chapter 11
Environmental Toxicology: Children at Risk

Maria J. Carroquino, M. Posada, and P.J. Landrigan

Glossary

Adult	The time of life usually starting at 18 years (some systems such as skeleton and brain may continue to develop).
Adverse effect	A treatment-related alteration from baseline that diminishes an organism's ability to survive, reproduce, or adapt to the environment.
Carcinogen	Any substance that can cause cancer.
Critical period	A specific phase during which a developing system is particularly susceptible.
Developmental disorders/effects	Adverse effects such as altered growth, structural abnormality, functional deficiency, or death observed in a developing organism.
Dose (exposure)–response relationship	Characterization of the relationship between administered dose or exposure and the biological change in organisms.
Embryonic period	The period from fertilization to the end of major organogenesis.

This chapter was originally published as part of the Encyclopedia of Sustainability Science and Technology edited by Robert A. Meyers. DOI:10.1007/978-1-4419-0851-3

M.J. Carroquino (✉) • M. Posada
Instituto de Salud Carlos III, Madrid, Spain
e-mail: mjcarroquino@isciii.es

P. Landrigan
Department of Preventive Medicine, Mount Sinai School of Medicine,
1 Gustave L. Levy Place, P.O. Box 1057, New York, NY 10029, USA

E.A. Laws (ed.), *Environmental Toxicology: Selected Entries from the Encyclopedia of Sustainability Science and Technology*, DOI 10.1007/978-1-4614-5764-0_11,

Environmental exposures and environmental hazards	For this entry, these terms refer to specific environmental chemicals and environmental pollutants.
Exposure	Contact with a chemical by swallowing, by breathing, or by direct contact (such as through the skin or eyes). Exposure may be short term (acute) or long term (chronic).
Fetus	The period from 8 weeks of pregnancy to birth.
Gestation	Length of time between conception and birth.
Infant	The period from 28 days of age to 1 year.
Lowest observed adverse effect level (LOAEL)	The lowest concentration of a chemical in a study, or group of studies, that produces statistically or biologically significant increase in the frequency or severity of adverse effects between the exposed population and an appropriate control.
Mechanism of action	The detailed molecular knowledge of the key events leading to an adverse effect in an organism.
Neonate	The period from birth to 28 days of age.
Perinatal stage	The period of 29 weeks of pregnancy to 7 days after birth.
Pregnancy	The condition of having an implanted embryo or fetus in the body, after fusion of an ovum and spermatozoon.
Preterm birth	A birth occurring at 24–37 weeks of pregnancy.
Risk assessment	An empirically based paradigm that estimates the risk of adverse effect(s) from exposure of an individual or population to a chemical, physical, or biological agent. It includes the components of hazard identification, assessment of dose–response relationships, exposure assessment, and risk characterization.
Route of exposure	Exposure route refers to the different ways a substance may enter the body. The route may be dermal, ingestion, or inhalation.
Sexual maturation	Achievement of full development of the reproductive system and sexual function.
Susceptibility	An individual's intrinsic or acquired traits that modify the risk of illness (e.g., high susceptibility to cancer).
Toxicokinetics	The process of the uptake of potentially toxic substances by the body, the biotransformation they undergo, the distribution of the substances and their metabolites in the tissues, and the elimination of the substances and their metabolites from the body. Both the amounts and the concentrations of the substances and their metabolites are studied. (Pharmacokinetics is the term used to study pharmaceutical substances.)

Vulnerability — A matrix of physical, chemical, biological, social, and cultural factors that result in certain communities and subpopulations being more susceptible to environmental factors because of greater exposure to such factors or a compromised ability to cope with and/or recover from such exposure.

Definition of the Subject

Children today live in a world that is vastly different from a few generations ago. While industrialization has maximized (for many) children's opportunities to survive, develop and enjoy high levels of health, education, recreation, and fulfillment, it has also added significant challenges to their development.

In the countries in the Organization for Economic Co-operation and Development (OECD), an infant born today has about 20 years longer life expectancy than one born at the beginning of the twentieth century [1]. Infant mortality has gone down by over 90%. Despite AIDS, SARS, West Nile virus, and the constant threat of other emerging infections, the ancient epidemics of smallpox, yellow fever, cholera, bubonic plague, polio, and measles are no longer the dominant causes of disease and death. However, an increase in the incidence (new cases) of many chronic diseases in children has been observed. These include asthma, cancer (which is the second leading cause of death in children after injuries), birth defects, developmental disabilities, and autism. For some of these conditions, an environmental origin has been established, and for others it is hypothesized.

Environmental threats to children's health range from asthma-inducing air pollution and lead-based paint in older homes, to treatment-resistant microbes in drinking water and persistent industrial chemicals that may cause cancer or result in reproductive or developmental changes.

Patterns of illness among children in the industrially developed nations have changed substantially. Infant mortality has declined. Life expectancy has increased. However, an increase in the incidence (new cases) of some diseases, such as asthma, birth defects, neurodevelopmental disorders and certain types of childhood and adolescence cancers, and obesity has been observed. Evidence is accumulating that toxic chemicals are responsible for at least some of these changing patterns of disease.

Children are uniquely susceptible to chemicals. This great vulnerability reflects the juxtaposition of two phenomena early in life: first, that infants and children have disproportionately greater exposures than adults to environmental chemicals, and second, that children are exquisitely sensitive to these exposures, because they are poorly equipped to metabolize many toxic compounds and because they are progressing through the complex, delicate, and easily disrupted stages of early development. The protection of children from environmental health hazards requires the consideration of their exposure patterns and susceptibility factors when conducting risk assessments, development of child protective legislation, and wider application of the Precautionary Principle in the face of early warning of danger.

Introduction: Brief History of the Issue of Children's Vulnerability to Environmental Hazards

Children's risks from environmental health threats have received considerable political attention during the last two decades. In 1989, the United Nations Convention on the Rights of the Child laid down basic standards for the protection of children and proclaimed that they are entitled to special care and assistance. A year later, the World Summit for Children (WSC) adopted a Declaration on the Survival, Protection and Development for Children, in which the signatories agreed to join efforts on taking measures to protect the environment, so that all children can enjoy a safer and healthier future.

Agenda 21, adopted in 1992 at the United Nations Conference on Environment and Development ("the Earth Summit") gives attention to the protection of children from the effects of a deteriorating environment in several chapters. Chapter 6 of Agenda 21 "Protecting and Promoting Human Health" emphasizes the need to pay special attention to protecting vulnerable groups, particularly infants, young people, women, indigenous people, and the poor. Agenda 21 urges governments to develop programs to protect children from the effects of environmental and occupational toxic compounds.

In the USA, the National Research Council report [2] "Pesticides in the Diet of Infants and Children" was critical in raising awareness about the importance in risk assessment of children's environmental health. This report elevated concern on a broad national level about children's special vulnerabilities to environmental agents. It made clear that protection of the health of vulnerable populations would require a new approach to risk assessment. The NRC report recommended an approach to risk assessment that moved beyond consideration of average exposures based primarily on adult characteristics to one that accounted for the heterogeneity of exposures and for potential differential sensitivities of various life stages, particularly during prenatal development, infancy, and childhood.

Responding to recommendations in the NRC report [2], the US Government took decisive steps to attend to the growing concern on children susceptibility to environmental toxicants. In 1995, the EPA issued a National Policy to consistently and explicitly take into account health risks to children and infants from environmental hazards when conducting assessments of environmental risks. In 1997, the Clinton Administration issued an Executive Order, Executive Order 13045: Protection of Children from Environmental Health Risks and Safety Risks. The Executive Order requires all federal agencies to address health and safety risks to children, to coordinate research priorities on children's health, and to ensure that their standards take into account special risks to children. To implement the order, the US EPA established the Office of Children's Health Protection (OCHP) (renamed the Office of Children's Health Protection and Environmental Education – OCHPEE – in 2005), whose job is to work with Program and Regional Offices within the US EPA to promote a safe and healthy environment for children by ensuring that all regulations, standards, policies, and risk assessments take into account risks to children. Legislation, such as the Food Quality Protection Act and the Safe Drinking Water Act amendments, has made coverage of children's health issues more explicit, and research on children's health issues is continually expanding. As a result of the emphasis on children's risk, the US EPA Office of Research and Development (ORD) developed a strategy for research on environmental risks to children. The goal of this research agenda is to discover the environmental causes of disease in children and then to convert these research findings into blueprints for disease prevention and health promotion.

The 1997 Declaration of the Environmental Leaders of the Eight on Children's Environmental Health intensified their commitment to protecting children's health from environmental hazards. The Environment Ministers of the G8 countries acknowledged the special vulnerabilities of children and committed their countries to taking action on several specific environmental health issues such as lead, microbiologically safe drinking water, endocrine disrupting chemicals, environmental tobacco smoke (ETS), and air quality. They called on financial institutions, the World Health Organization (WHO), the United Nations Environment Program (UNEP), and other international bodies to continue ongoing activities and to pay further attention to children's environmental health, in particular the economic and social dimensions of children's health. In addition, they committed their countries to fulfilling and to promoting the Organization for Economic Co-operation and Development (OECD) Declaration on Risk Reduction for Lead. The underlying rationale for each of these actions is that disease of environmental origin in children can be prevented by controlling harmful exposures in the environment.

International organizations such as the World Health Organization (WHO), the United Nations Environment Program (UNEP), and the European Union (EU) responded to this call and also developed initiatives on Children, Environment and Health. Children's Health and Environment has been the central focus of Europe wide Ministerial Conferences organized by WHO, and other EU initiatives, such as SCALE (Science, Children, Awareness, Legal Instrument, Evaluation). Thus, the protection of children from environmental hazards has been a subject of intense political and scientific attention and effort during the last two decades. Since the initial reports on this issue, research has intensified, and scientific literature abounds with new findings on children's differential vulnerability to environmental health threats that have provided the basis to maintaining the political attention on children's health.

Toxicological Basis of Children's Vulnerability to Environmental Hazards: Susceptibility and Exposure Factors Affecting Children's Vulnerability

Over the past 2 decades, a number of scientific reports have integrated the knowledge available on the susceptibility of children to environmental toxicants, have highlighted knowledge gaps, and have pointed to future directions of research to fill these gaps. This has expanded and deepened the knowledge of the differential susceptibility of children to environmental toxicants, and at the same time has allowed the firm establishment of certain fundamentals concerning children's susceptibility to environmental toxicants.

The International Life Science Institute (ILSI) and the EPA, recognizing the need for examining the scientific evidence on the broad question of the differential susceptibility of children to environmental hazards, organized in 1990 a conference titled: "Similarities and Differences Between Children and Adults: Implications for Risk Assessment." The results from this conference were summarized in the publication "Similarities and Differences Between Children and Adults" [3] and was incorporated into ongoing work of the National Research Committee on Pesticides in the Diets of Infants and Children. The National Academy of Science (NAS) published the NRC report "Pesticides in the Diets of Infants and Children" [2], which was critical in raising awareness of the importance of considering the vulnerable life stages of children when conducting risk assessment of exposures to children. In 2001, the International Life Sciences Institute convened a number of scientific experts to develop a conceptual framework for conducting health risk assessments of children exposures, which takes into consideration their unique characteristics and special vulnerabilities [4].

The ILSI report highlighted the fact that "children are not little adults," but a unique subpopulation that needs to be considered in risk assessment due to differential exposure patterns, immaturity in physiological development, or differential toxicant metabolism.

In 2006, the WHO published the Principles for Evaluating the Health Risks in Children Associated with Exposure to Chemicals in the Environmental Health Criteria 237 [5]. This publication constitutes the most recent monographic work of the subject of children susceptibility to environmental toxicants, and provides comprehensive information on children's developmental stages and the critical windows of susceptibility that appear through the course of their development.

In summary, it is clear from these three major scientific reports – NAS, ILSI, and WHO – that there is broad scientific consensus that risks to children from environmental health threats differ qualitatively or quantitatively from those of adults for a variety of reasons:

- Differential exposure patterns: Compared to adults, children have heavier exposures in relation to body weight because they drink more water, eat more food, and have higher breathing rates per unit of body weight than adults do. As a consequence, children have substantially heavier exposures than adults to any toxicants that are present in water, food, or air.

- Children's ability to metabolize, detoxify, and excrete chemicals is different from that of adults. During the first months after birth, their metabolic pathways are immature. In some cases, children may actually have a higher metabolic capacity for some toxicants than adults. Commonly, however, they are less able to deal with toxic chemicals and thus are more vulnerable to them.
- Children undergo rapid growth and development, and their development phases are perfectly scheduled to achieve complete functional development. If a development phase is disturbed at a given time, its time to take place may be lost definitively. Thus, interferences with certain phases of development may have irreversible effects. If cells in a child's developing brain are destroyed by chemicals such as lead, mercury, or solvents, or if false signals are sent to the developing reproductive organs by endocrine disruptors, there is a high risk that the resulting dysfunction will be permanent and irreversible.
- Children's exposures are affected qualitatively and quantitatively by their behavior and the unique microenvironment in which they spend their time. Their hand-to-mouth behavior brings contaminated items or soil to their mouth, and living and playing closer to the ground also exposes them to pollutants in a different pattern than adults, both in quantitative and qualitative terms.
- Because children generally have more years of life ahead than adults, they have more time to develop chronic diseases triggered by early exposures. Many diseases, such as cancer and neurodegenerative diseases, are thought to arise through a series of stages that require years or even decades from initiation to actual manifestation of disease. Carcinogenic and toxic exposures, sustained early in life, including prenatal exposures, would then be more likely to lead to disease than similar exposures encountered later.

The purpose of this chapter has been to select, integrate, and summarize the most relevant and illustrative information that explains and highlights the potential susceptibility of children to environmental exposures. The goal of the chapter is to promote understanding of the issue of differential susceptibility and exposures of children to environmental threats among scientists from adjacent fields, and to encourage scientists to consider the issue of children's differential susceptibility and exposures in their research. Recognition of children's exquisite vulnerability to toxic exposures in the environment is critical to child-protective risk assessment and to disease prevention.

For a more in-depth study of any of the particular health issues or particular toxicants, extensive and rigorous reviews that are cited through this document can be consulted. The following sections will elaborate on the general toxicological aspects of children's susceptibility.

Exposure

There are several factors that influence the differential exposures of children to environmental toxicants. These are: (1) their higher ingestion, drinking, and breathing rates per unit of body weight, (2) their unique behaviors, (3) where they spend their time, and (4) their unique microenvironments.

Ingestion, Breathing, and Drinking Rates

In toxicology, exposure is defined as the contact that occurs between a receptor (a living organism) and an environmental agent. The most commonly considered exposure pathways are inhalation, ingestion, and dermal absorption. For infants in the womb, transplacental transfer of toxic chemicals is another unique route of exposure, and for nursing infants, the proportion of the toxicant that enters the body through each of these pathways will depend on both external, environmental factors, and on biological factors that will determine the rate of uptake of the chemical by the organism. Exposure can also be referred to as the concentration that a target organ receives, once the chemical has been absorbed.

The environmental concentration and duration of exposure will determine the extent of exposure. In addition, other environmental factors such as for example, temperature, humidity, and pH, may alter pollutant concentrations. For example, the concentration of some volatile chemicals may vary under different temperature conditions. Sometimes, chemical reactions between pollutants in the environment may change qualitatively the nature of exposure, such as in the case of ozone formation in the air from nitrogen oxides (NOx), carbon monoxide (CO), and volatile organic compounds (VOCs) in the presence of sunlight.

The patterns of exposure to ingested and inhaled toxicants are different between children and adults. Young children drink more water on a body weight basis than adults do (seven times as much water per kilogram of body weight). For the first 6 months, for example, children consume on average 88 mL/kg/day of tap water directly or indirectly (water added in the preparation of formula or fruit juice) compared to adults who drink 17 mL/kg/day [6]. As a result of this greater consumption of water, children may be disproportionately exposed to chemicals found in drinking water, including the water used to make infant formula [2].

Infants and young children have a higher resting metabolic rate and rate of oxygen consumption per unit body weight than adults. The oxygen consumption of a resting infant aged between 1 week and 1 year is 7 mL/kg body weight per minute, compared to that of an adult in the same condition, which is 3–5 L/kg/min [5]. Thus, on a body weight basis, the volume of air passing through the lungs of a resting infant is twice that of an adult, and therefore twice as much of any chemical in the atmosphere would pass through the lungs of an infant. An additional consideration is the smaller lung surface area per kilogram in the early stages of development. Thus, the higher amount of inspired air will affect a relatively smaller area of lung tissue. In addition, children have narrower airways than those of adults. Thus, irritation caused by air pollution that would produce only a slight response in an adult, can result in potentially significant obstruction in the airways of a young child [6]. Furthermore, the fact that children spend more time engaged in vigorous activities than adults exacerbates the differential effects. Children also ingest more food per unit of body weight than do adults. A 1–5 year old child eats three to four times more food per kilogram than the average adult, resulting in larger amounts of chemicals and infectious agents per unit of body mass [2, 7]. A child's diet is very different from an adult's. The diet of children contains more milk products and more fruits and vegetables per unit of body weight than adults. During the first year of life, human milk or cow milk–based products constitute most of their energy and nutrient source. The NRC report [2] reported that over the first year, cow milk products comprise 36% and 58% of the diets of nursing and non-nursing infants, respectively, compared to adults, where milk and milk products constitute only about 29% of their diet. Environmental pollutants, such as PCBs and dioxins, accumulate in fat and have been found in breast milk, although it is broadly accepted that the benefits of breast-feeding still outweigh the risks of exposure [5].

After the first few months of life, fruits and fruit juices constitute a large proportion of infants' diet. For example, the average consumption of apples for children between birth and 5 months of age is almost ten times that of an adult older than 20 years old [6]. Fruits can be contaminated with pesticides and other toxicants, causing children to be exposed to these chemicals.

Behavioral Patterns Influencing Exposure in Children

Two characteristic behaviors of children have been considered and studied in relation to children's exposure to chemicals: mouthing behavior and pica behavior. Infants and toddlers pass through a developmental phase, characterized by intense oral exploratory behavior, when they indiscriminately bring their hands and objects to their mouths. Mouthing behavior can result in oral exposures to chemicals that may be part of the surfaces of objects, or adhered to them in dust particles.

Pica is a behavior of some children that has been defined as the craving or ingestion of nonfood items. The cravings found in patients with pica have been associated with a nutritional deficiency state, such as iron-deficient anemia; with pregnancy or with mental illness. Pica tends to disappear as children grow older, except for mentally retarded children [8]. Pica behavior can give rise to the ingestion of soil, paint, and other possibly contaminated substances. About 95% of children ingest 0.2 g of soil per day or less, but studies have shown that children with pica behavior ingest up to 60 g of soil per day [8]. Children living in neighborhoods built on soil contaminated with heavy metals (such as lead), aromatic hydrocarbons, and other compounds were found to be at increased risk for elevated blood levels [8].

Children may also have higher exposures because of their higher levels of physical activity, at least during certain stages of their development. When they are outdoors, they are more prone to run, jump, or play vigorously than adults do. Their breathing rates may reach exercise levels more frequently during a day than in adults. When near water, for example, children tend to spend more time in the water than adults. This can lead to higher exposures of air pollutants, such as ozone, or to toxicants in water due to the inhalation of volatilized chemicals, ingestion of pool water, and dermal exposures.

Other behaviors characteristic of children such as overexposure to sunlight for sun tanning among adolescents, or drug and alcohol ingestion, can also lead to disproportionately higher risks. However, in this chapter, the central focus is on those risks that are completely involuntary, and that would be difficult to change without altering the normal behavior of a child.

The Unique Microenvironment of Children

The children's microenvironment is different from that of adults from the very beginning of conception. The womb, where the fetus is exposed to environmental chemicals and other agents from previous exposures of the mother is a unique environment to this developmental stage. In fact, humans are among the living mammals that have the longest gestational period, resulting in a long in-utero exposure. Many pollutants are known to cross the placenta and enter the fetal circulation, resulting in fetal exposure to toxicants present in the mother. Once the baby is born, breast-feeding can also result in exposures to chemicals accumulated in breast milk.

Children spend a large proportion of their time indoors, and because of their smaller stature and because they crawl and play on the floor, they are more highly exposed to ground-level contaminants. In indoor environments, for example, formaldehyde exposure from carpeting would be higher at ground level than about 4 or 6 ft above, the breathing zone of adults. House dust or particles to which pollutants adhere accumulate near the floor both in the indoor and outdoor air. Further, outdoors where there is traffic, their breathing zone is close to the height at which car exhaust systems expel their fumes.

Indoor air pollutants can originate indoors, such as formaldehyde from carpeting, fumes from home cooking, and solvents from freshly painted rooms, or can enter the inside of homes through windows or open airways. Sometimes, pollutants may be carried inside by persons entering the indoor environment. For example, children whose parents work on farms where pesticides are applied can be exposed to pesticides carried inside in the clothes and skin of adults who walk into the house. Studies have found higher levels of pesticides in farmworker households in agricultural areas compared with those in non-farmworker households in the same area [9]. In addition, some studies have demonstrated that degradation of pesticides is slower in the indoor environment, so pesticide exposure persists longer [9].

Toxicokinetics

Once exposed, the amount of toxicant being absorbed, its metabolism, and excretion from children's bodies may be very different from that of adults. Rates of absorption through the oral, dermal, and respiratory pathways vary between children and adults. For example, children have a larger lung surface area per kilogram of body weight than adults and, under normal breathing, breathe 50% more air per kilogram of body weight than adults [6]. Therefore infants potentially receive a greater internal exposure to airborne compounds on a body weight basis.

Children's absorption of chemicals through the gastrointestinal route is also different from that of adults. For example, the absorption of lead from the intestine was found to be 40 times higher for children than for adults [3]. Gastric pH is higher in newborns than in adults, thus causing differences in ionization and absorption of certain chemicals. The alkaline gastric pH in newborns and infants may enhance the absorption of certain compounds.

The skin surface area of children relative to body weight is greater for children than for adults, resulting in a higher potential dose absorbed through the skin of about three times greater for infants than for adults [10]. In addition, hydration of the skin in neonates is greater than in older children, which potentially could result in them having a greater absorption of some hydrophilic chemicals. The permeability of the epidermal is incomplete in the preterm infant, resulting in greater percutaneous absorption of chemical agents [5].

Once absorbed, the different body composition of children will result in a different distribution and accumulation of chemicals compared to adults. For example, the relatively larger extracellular fluid volume of the infant means somewhat greater dilution of water-soluble chemicals. However, the lipid-soluble substances would be distributed in a smaller volume of fat in infants relative to adults. The body composition in terms of water, fat, and protein changes from birth to adulthood resulting in corresponding changes in the bioaccumulation, metabolism, and excretion of contaminants from the body throughout development. Water content, for example, decreases rapidly during the first 6 months of life and then remains fairly constant. Body fat, on the contrary, increases rapidly up to 6 months and then decreases, accounting for similar percentages of body weight at ages 4 and 12 months [5]. This can result in changes in the concentration of chemicals in different tissues and organs that can override the child's ability to metabolize and excrete the chemical.

Metabolism and elimination rates are generally lower in neonates than in adults. Many metabolic pathways are not fully developed in the infant. Renal clearance is lower in neonates than in older children and adults for all classes of chemicals. These factors will lead to differences in the bioaccumulation and persistence of environmental agents in children's and adult's bodies and the concentration of the agents in specific organs. The maturation of these metabolic and elimination rates can result in the mobilization of chemicals, and higher exposures of the target organs or systems to the products of metabolism. In terms of excretion, certain life stages such as pregnancy, lactation, and also menopause, will result in the mobilization of chemicals from fat or bone stores, and the exposure of other organs or the developing fetus and lactating child to the toxicants that were accumulated in the mother.

Developmental Aspects of Children's Susceptibility: Critical Periods of Development

Effects During Gestation

From conception to birth, the human organism advances rapidly through a complex set of developmental processes that culminate in the newborn. These development processes include cell division, organ formation, and growth as well and functional development. During development, some biological processes occur only during certain stages of development and not in others, or they occur at a different rate in different developmental stages. For example, cell division in most organs takes place rapidly during early development and much more slowly at later stages. Other processes such as apoptosis, or programmed cell death, occur more widely during development and are less prominent during adulthood.

These biological processes need to be effectively coordinated and require the cellular and intercellular signaling systems to work correctly. Because of the complexity and speed at which these processes take place and the intricate relation between them, interference at the sequence of any of these processes can have damaging and irreversible effects. Exposure to environmental toxicants can have a completely different effect depending on whether it occurs at one developmental stage or another. In addition, damage due to environmental exposures may occur and manifest itself immediately, or may not appear until subsequent stages of development, after development is complete, including late adulthood.

Identifying and understanding these "critical" periods of unique susceptibility is essential to developing strategies to protect children from adverse health effects associated with environmental exposures. The following sections describe the current state of knowledge about the windows of susceptibility during childhood development.

Effects on Germ Cells

Germ cells (sperm and egg cells) carry the genetic information from each parent that will provide the unique genetic blueprint for each child. In the male fetus, primordial germ cells develop in utero. From puberty to adulthood, these cells undergo cell division, mitosis, and meiosis, to produce mature sperm and continue to be produced from stem cells through adulthood. In females, primordial germ cells undergo mitosis and the first phase of meiosis during fetal life, and in women, mature oocytes are produced every month from follicular cells.

Environmental toxicants that harm germ cells can affect an adult's own fertility as well as the health of the offspring. In animal models, preconceptional carcinogenesis has been demonstrated for a variety of types of radiation and chemicals, with demonstrated sensitivity for all stages from fetal gonocytes to postmeiotic germ cells [11]. Although this link in humans is not demonstrated, it is theoretically possible that environmental toxicants that harm germ cells can affect an adult's own fertility as well as the health of the offspring.

Results of environmental damage to germ cells may include reduced fertility later in life or offspring with congenital health problems [12, 13]. For example, men exposed to diethylstilbestrol (DES) in utero had lowered sperm count and increased frequency of abnormal sperm [14]. Women exposed to cigarette smoke during their mother's pregnancy had reduced fertility [13]. There is a substantial body of evidence demonstrating that exposures to environmental agents and medical radiation can injure germ cells in such a way as to cause increased incidence of cancer, particularly leukemia, among offspring of the exposed individuals. For example, paternal exposures to benzene have been linked to leukemia and lymphoma in children [15]. Animal studies support these findings [11].

Embryonic and Fetal Development During Pregnancy

Several stages of embryonic and fetal development are susceptible to environmental harm. During development, gene expression is very active because a large number of genes are being "switched on" or "switched off" to control cellular activities. This high level of metabolic activity provides for a wide range of opportunities for environmental agents to interfere with cell development and growth.

Environmental toxicants may interact directly with DNA (e.g., alkylating agents) and disturb gene expression or may interact with the products of gene expression, such as enzymes and control molecules. A toxicant that interferes with gene expression may prevent the synthesis of enzymes necessary for toxicant metabolism, resulting in the accumulation of the toxicant in the body. DNA activation by a chemical may result in excessive synthesis of enzymes that catalyze the bioactivation of a toxicant (production of a toxicant metabolite more toxic than the compound originally present).

Chemicals can interfere with the activation or inactivation of genes that occur during early fetal life and that may be essential for the protection of the organism to external or internal harmful processes. For example, interference with genes involved in DNA repair, such as p53, a tumor suppressor gene important for DNA repair, may result in enhanced vulnerability to specific toxicants during development. Studies with transgenic mice that are missing this gene have shown an increased sensitivity of mice fetuses to benzo[a]pyrene exposure and an increased death rate when exposed to the chemical during gestation [16]. Damage to p53 in humans could likewise increase sensitivity to agents that damage genetic material.

If a toxicant interferes with cell differentiation, cells may not reach their specific form and function necessary for their final role in the body, and organ function may be compromised. Also, undifferentiated cells may be more vulnerable than differentiated cells to toxic effects. Some chemicals such as ethanol [17] and TCDD [18] have been demonstrated to affect specific types of undifferentiated cells.

Apoptosis or programmed cell death is a critical biological process for healthy development. Apoptosis involves the removal of certain cell types when they are no longer necessary. In some instances, one type of cell is succeeded by another during a specific developmental period. Apoptosis is involved, for example, in the elimination of cells in the immune system that, if they survived, could cause autoimmune disease [19]. Apoptosis is also critical in the development of the nervous system, where phases of cell proliferation alternate with phases of apoptosis on the basis of the progression of neuronal development [20] and remains active through the postnatal period because of ongoing nervous system development.

Normal patterns of apoptosis may be altered through altered gene expression or failure of signaling mechanisms resulting from environmental exposures. Certain autoimmune lympho-proliferative diseases and certain cancers have been related to the disruption of normal patterns of apoptosis. For example, Wilms' tumor, a relatively common childhood cancer, may arise from the transformation by postnatal exposures of renal stem cells that fail to disappear 4–6 weeks prior to birth [21].

Neuronal migration is an important process in nervous system development and its alteration may result in irreversible damage. For example, schizophrenia is thought to result, in part, from abnormal neuronal migration [22], but the role of prenatal exposures to environmental agents in causing this disease is not clear. Exposures to ionizing radiation and methylmercury, for example, have been shown to affect the migration of neurons during development [20, 23].

During the period of organ development, which occurs (varying according to organ system) between the 3rd and 16th week, disruption of development can disrupt the large-scale structure of organs, often resulting in physical malformations (congenital anomalies). The best known example of such gestational damage is exposure to diethylstilbestrol (DES). DES caused genital anomalies among male children born of women who took the medication before the 11th week of gestation twice as often as among those who were exposed later in gestation [14].

Other effects such as low birth weight, pregnancy complications, or late fetal death have been shown to be a result of environmental exposures during later stages of prenatal development [24]. Disinfection by-products have been linked to the risk of spontaneous abortion for some time. There is now fairly consistent evidence for associations between early and late fetal deaths and indices of transplacental exposure to disinfection by-products [25–27]. Maternal smoking during pregnancy increases the risk of pregnancy loss, stillbirth, and infant mortality [28].

Development During Childhood

Several organs and systems continue to grow and develop during childhood and in some cases almost until adulthood. For example, neuron migration, cell proliferation, and synapse formation are very active until 3 years of age, and myelination continues until adolescence [29] and possibly well into adulthood [20].

The immune response is also immature at birth and develops during infancy and childhood until about 1 year of age, while establishment of immunologic memory is not fully established until 18 years of age [30]. Exposure to environmental agents during early childhood may affect immune system development and may contribute to the development of certain diseases such as asthma and cancer later in life.

Physical growth and maturation of organ systems continues through adolescence. The process of sexual maturation is accompanied by complex interactions between the central nervous system and hormone-secreting organs, which can be affected by environmental exposures. For example, the risk of breast cancer has been found to be greater among women who were exposed to radiation before 20 years of age [31].

Cellular Metabolism and Biotransformation

Many important metabolic and biotransformation processes are poorly developed in the fetus, and full metabolic activity is not fully developed until after childbirth. Metabolism can increase or decrease the toxicity of a chemical, depending on the metabolic products of the chemical and pathway involved. Metabolism may also make elimination from the body easier or harder, although the most common metabolic pathways usually render chemicals more hydrophilic and thus, more easily excreted. In some cases, the adult biotransformation of a certain chemical may consist of a bioactivation pathway that makes the compound more hazardous than the one originally present. The absence of a metabolic pathway may result in the bioaccumulation of the chemical in the body and a later bioavailability and disposition to exert its toxic effects. Immaturity could be an advantage if the activation pathway is not present in the fetus or child and there is an alternate pathway for the toxicant to be metabolized. However, according to [32], given the primary evolutionary function of detoxifying and eliminating potentially toxic chemicals, immature or underdeveloped metabolic pathways are likely to render infants and children more sensitive to common environmental contaminants.

Major Groups of Pollutants to Which Children are Exposed

Heavy Metals

Heavy metals are natural elements that have been extracted from the earth and used in human industry and products for centuries. As a consequence of human activity, concentrations of heavy metals in air, water, and surface soil today are hundreds of times higher than in the preindustrial era. Some metals are naturally found in the body and are essential to the functioning of critical enzyme systems. Iron, for example, prevents anemia, and zinc is a cofactor in over 100 enzyme reactions. Magnesium and copper are other familiar metals that, in minute amounts, are necessary for proper metabolism to occur. The body has need for approximately 70 trace elements, but there are others, such as lead, mercury, aluminum, arsenic, cadmium, and nickel, that have no roles in human physiology and can be toxic at even trace levels of exposure. Nutritionally, heavy metals can compete with nutrient elements, such as the case of lead, which is stored in the bones in the place of calcium.

Metals are notable for their wide environmental dispersion, their tendency to accumulate in select tissues, and their overall potential to be toxic at even relatively minor levels of exposure. In general, heavy metals are systemic toxins with specific *neurotoxic*, *nephrotoxic*, *fetotoxic*, and *teratogenic* effects. Heavy metals can directly influence behavior by impairing mental and neurological function, influencing neurotransmitter production and utilization, and altering numerous metabolic body processes.

Exposure to heavy metals can occur through drinking water, air, or ingestion of heavy metal–contaminated soil. The amount that is actually absorbed from the digestive tract can vary widely, depending on the chemical form of the metal and the age and nutritional status of the individual. Once a metal is absorbed, it distributes in tissues and organs. Excretion of metals typically occurs through the kidneys and digestive tract, but they tend to persist in some storage sites, like the liver, bones, and kidneys, for years or decades.

Lead is one of the best known heavy metals in terms of its toxicity. Exposure to lead can occur in the prenatal stage through the placenta, and in infants through the mother's milk and the water used in milk formula [3]. During pregnancy, body stores of lead may be mobilized and transferred from the mother to the fetus [33]. Behavioral characteristics of children later on, such as the hand-to-mouth behavioral pattern of 1–3 year olds, can result in high exposure and internal levels of lead. Lead paint is a major source of environmental exposure for children who ingest flaking paint, paint chips, and weathered powdered paint (mostly from deteriorated housing units in urban areas). Lead can leach into drinking water from lead-based solder used in water pipes. Lead also leaches into foods or liquids stored in ceramic containers made with lead glazing, which is still used in some countries.

The absorption of lead from ingestion of lead-contaminated water is higher for children than for adults, so that for a given level of exposure, the resultant internal dose is higher in children than in adults [3]. Children are also more sensitive than adults to the toxicological effects of lead at a given internal exposure level. The lowest observed adverse effect levels (LOAELs) for several health endpoints occur at lower blood lead levels in children than in adults. The most sensitive targets for lead toxicity are the developing nervous system, the hematological and cardiovascular systems, and the kidney. There appear to be no safe exposure "thresholds" for lead or for other metals in early development.

Mercury is a ubiquitous heavy metal of both natural and anthropogenic sources. Mercury occurs in both inorganic and organic forms, and it is most hazardous in its organic form of methylmercury. The nervous system is very sensitive to all forms of mercury. Methylmercury and metallic mercury vapors are more harmful than other forms, because methylmercury can cross the blood brain barrier. Methylmercury in the marine and freshwater environment is absorbed by fish and shellfish and bioaccumulates in the food chain. Increased risk is of particular concern in children and in populations that have an increased dietary exposure to fish.

Arsenic occurs naturally in the environment and in some areas of the world is a natural contaminant of underground water that is used as drinking water. It is also an anthropogenic contaminant. Once absorbed into the body, arsenic undergoes some accumulation in soft tissue organs such as the liver, spleen, kidneys, and lungs, but the major long-term storage site for arsenic is keratin-rich tissues, such as skin, hair, and nails.

Cadmium is another chemical that is toxic to adults, although it has not been extensively studied in children. The US Department of Health and Human Services has determined that cadmium and cadmium compounds are known human carcinogens. Cadmium has been linked to diminished kidney function lung disease, chronic bronchitis, and lung, kidney, and prostate cancers. In the USA, smoking is the primary source of cadmium exposure, although high levels of cadmium can also be found in organ meats, shellfish, and vegetables.

Pesticides

Pesticides are substances that are used to prevent, repel, or destroy pests – organisms that compete for food supply, adversely affect comfort, or endanger human health (FIFRA 1996). More than 20,000 pesticide products with nearly 900 active ingredients are registered for use as insecticides, miticides, fumigants, wood preservatives, and plant growth regulators. It cannot be denied that pesticides have beneficial economic and also public health impacts. Pesticide usage helps improve human nutrition through greater availability, longer storage life, and lower costs of food. It also reduces human labor requirements and attendant risks of injury. Pesticides also assist in the control of food-borne and vector-borne diseases, such as malaria, which kill millions of persons in the world. Pesticides also pose human health concerns because they are toxic substances and widely spread in the environment. Although the toxic mechanisms on targeted pest species are well characterized, the potential for adverse health effects in humans is not fully known.

Pesticides are composed of several classes of chemicals with different mechanisms of action. Most insecticides work by interfering with nervous system function. Organophosphates, which account for approximately one-half of the insecticides used in the USA, and carbamates, which are widely used in homes and gardens, inhibit the activity of acetyl cholinesterase at nerve endings, resulting in an excess of acetylcholine in the synapsis and a depolarizing blockage of neural transmission. The effects of carbamates are readily reversible and of shorter duration. Organochlorines, such as dichlorodiphenyltrichloroethane (DDT) and lindane, interfere with nerve cell membrane cation transport, resulting in neural irritability and excitation of the central nervous system. Herbicides, including the chlorophenoxy compounds 2,4 D and 2,4,5 T are primarily irritative to the skin and respiratory tract during acute exposures and work by different mechanisms. Some substances, such as paraquat, are highly corrosive and can cause multisystem injury and progressive pulmonary failure [34].

Arsenical pesticides, such as copper chromium arsenate, have been used, until recently, as wood preservatives to prolong the useful life of exterior wooden structures. These compounds cause central nervous system depression at sufficient doses.

Pesticides are ubiquitous in the environment. They are found in food, water, homes, schools, workplaces, lawns, and gardens. They are present in soils that have been spread with pesticides or where pesticides from adjacent agricultural areas have drifted, and may reach water supplies from agricultural runoff. In the USA alone, more than 0.45 billion kilograms of pesticides are applied each year, in agriculture, in homes and gardens, and in schools and hospitals. In developing countries, many highly toxic and biologically persistent pesticides such as parathion, DDT, and paraquat, which are no longer permitted in many developed countries, are still in wide use and result in chronic exposures and acute, too often fatal poisonings of thousands of young children each year.

Most children in the world are exposed to some degree to pesticides. Children in rural and agricultural areas and especially children whose parents are farmworkers or pesticide applicators are at highest risk of having increased exposures to pesticides. Pesticides may reach their homes due to the drifting of pesticides that are applied to the ground through aerial spraying. Children may work or play near their parents in the fields where pesticides have been used. Parents who work with pesticides may bring pesticides to their homes, impregnated in their clothes and bodies. In countries where residential housing with gardens and lawn predominate, homes and garden pesticide use may result in significant levels of exposure [34].

Exposure of children to pesticides may occur through inhalation, ingestion, and dermal absorption. Ingestion of pesticides occurs either through accidental exposure due to pesticides stored in food containers (i.e., soft drink bottles), or through ingestion of pesticide-treated foods, particularly fruits and vegetables. Foods grown in pesticide-contaminated soils and fish from pesticide-contaminated water can also carry significant amounts of pesticides. Children may also ingest pesticides adhered to the surface of toys or other objects through their hand-to-mouth behavior. The potential of dermal exposure of children to pesticides is higher than that of adults because of their relatively large body surface area and extensive contact with lawns, gardens, and floors by crawling and playing on the ground.

Prenatal and early childhood exposures are of special concern because of the susceptibility of the developing organ systems to pesticides (the central nervous system in particular) as well as behavioral, physiological, and dietary characteristics of children. Breast-feeding infants may ingest pesticides or pesticide metabolites present in the breast milk. The quantity of pesticide that is passed to the infant via breast milk is influenced by many variables such as maternal age and parity, maternal body burden of the chemical, and breast-feeding patterns. As infants are weaned and progress to solid foods, they consume, per unit of body weight, proportionally more fruit and more fruit juice than adults. The NAS in 1993 reported that children's dietary exposures to pesticides differed from adults both quantitatively and qualitatively and questioned the protection provided to infants and children from pesticide tolerances in effect at the time. The report estimated that 50% of lifetime pesticide exposure occurs during the first 5 years of life [2].

Environmental Tobacco Smoke

Environmental tobacco smoke (ETS), also known as second-hand smoke, is exhaled smoke and sidestream smoke emitted from the burning of the tip of the cigarette. The inhalation of ETS is known as "involuntary smoking" or "passive smoking." ETS contains more than 4,000 different chemical compounds, many of which are toxic. In 1992, the US Environmental Protection Agency (EPA) declared ETS as a Group A carcinogen.

The effects of passive smoking begin in utero, where constituents of tobacco smoke, such as PAHs, nicotine, and carbon monoxide, cross the placenta and are concentrated in the fetal circulation [35]. Children are also exposed during childhood if any of the parents smoke.

Persistent Organic Compounds: PCBs, Dioxins, and Related Organohalogens

Polychlorinated biphenyls are synthetic compounds with two linked phenyl rings and variable degrees of chlorination. They have been used for many years because of their thermal and chemical stability. They are nonvolatile, hydrophobic oils that are not easily biotransformed in the environment or metabolized by the human organism, so they are very persistent in the environment, and bioaccumulate in the food chain and in the fat compartment of the human body. Although they have been banned in the USA for more than 30 years, they are still widely present in the environment. They have been found in wildlife, human tissue, and human milk. Polychlorinated dibenzofurans (PCDFs) are partially oxidized PCBs that appear as contaminants of PCBs. Polychlorinated dibenzodioxins (PCDDs), commonly referred as dioxins, are formed during paper bleaching and waste incineration. One dioxin congener, 2,3,7,8-tetrachlorodibenzo-p-dioxin (TCDD), is considered to be the most toxic synthetic chemical known.

The most significant source of exposure is contaminated food, particularly fish at the top of their trophic chain. Of greatest concern are the populations that consume high amounts of fish. Because PCBs and PCDFs are not metabolized or excreted, they can accumulate in the fat tissue of the body, as well as in human milk, resulting in high exposures of the developing fetus and the newborn.

Disinfection By-products (DBPs)

Disinfection by-products (DBPs) form when disinfectants are added to drinking water and react with naturally occurring organic matter. Chlorine, the most widely used primary disinfectant, reacts with naturally occurring organic matter to form a range of unwanted by-products such as the trihalomethanes (THMs) which include chloroform, bromodichloromethane (BDCM), chlorodibromomethane (DBCM), and the haloacetic acids (HAAs), such as monochloroacetate, dichloroacetate, and trichloroacetate.

Exposure to DBPs occurs through ingestion of water or through inhalation and absorption during showering, bathing, and swimming. While there is some concern that these chemicals may pose a health risk, the potential risks arising from not treating drinking water are considerably greater, and the disinfection of water should never be compromised as a result.

Environmental Threats to Children on Specific Organ Systems

Nervous System

The developing nervous system is more susceptible than the adult brain to the disrupting effect of toxic chemicals [5]. The lengthy period of brain development and the extensive number of processes needed to take place contribute to the susceptibility of the developing nervous system. In the 9 months of pregnancy, the human brain and spinal cord must develop from a thin strip of cells along the back of the embryo into a complex organ comprised of billions of precisely located, highly interconnected, and specialized cells. Brain development requires that neurons move along precise pathways from their points of origin to their assigned location, that they establish connections with other cells near and distant, and that they learn to intercommunicate. Each connection between and among neurons must be precisely established at a particular point in development, and redundant connections need to be pruned away through programmed cell death, apoptosis. All these processes must take place within a tightly controlled time frame, in which each developmental state must be reached on schedule and in the correct sequence.

Critical windows of vulnerability, which exist only in the 9 months of pregnancy and to a lesser extent in early childhood, are unique to early brain development. They have no counterpart in adult life. Exposure to toxic chemicals during these windows of vulnerability can cause devastating damage to the brain and nervous system.

Any toxic or other environmental exposure that interferes with the tightly orchestrated sequence of events involved in brain formation is likely to have profound effects on intellect, behavior, and other functions. If a developmental process in the brain is halted or inhibited, if cells fail to migrate the proper sequence to their assigned locations, if synapses fail to form, or if pathways are not established, there is only limited potential for late repair, and the consequences can be permanent [36].

Environmental toxicants can affect both the structural and functional development of the nervous system. Depending on the developmental stage at which exposure occurs, sensory development, intelligence, or behavior will be affected differentially. While early-developing neural systems have been considered the most vulnerable to chemical insult, scientists have called attention to the importance of chemical exposure that occurs late in childhood, as it has been recently suggested that behavioral and physiological foundations of cognition continue to develop during childhood and adolescence [5, 37].

Exposure to environmental toxicants such as lead, methylmercury, and certain pesticides and PCBs even at very low levels have been shown to produce neurobehavioral (functional) deficits, and increased susceptibility to neurodegenerative diseases much later in life [38]. Of critical concern is the possibility that developmental exposure to neurotoxicants may result in an acceleration of age-related decline in function that could lead to Parkinson Disease, Alzheimer Disease, and other forms of brain degeneration. This concern is compounded by the fact that developmental neurotoxicity that results in small effects on the individual at a particular stage can have a profound societal impact when considered in the whole population or across the life span of the individual [36]. For example, a five-point decline in average population intelligence (IQ score) can result in reduction by more than 50% in the number of children with superior IQ (>130) and concomitant doubling in the number of children with IQ scores in the retarded range (<70). Behavior is also disrupted, sometimes permanently, by such exposures.

The following sections will give a short overview of some neurotoxicological effects of the best known chemicals for their effect on the developing nervous system. Some toxicants such as ethanol have been excluded, as the focus of this chapter is on exposures to toxicants to which the majority of children are exposed.

Lead

Lead is one of the best studied toxicants and one of the few pollutants for which susceptibility of children has been clearly established. The best known health effects on children are its neuropsychological effects although other effects have also been studied and are partially documented.

The neurotoxic effects of lead in children have been extensively studied. One distinctive characteristic of the research findings on lead neurotoxicity over the past 2–3 decades is that it has led to a progressive decline in the LOAELs (lowest observed adverse effect levels). The early finding in the 1970s by Landrigan et al. [39] and Needleman [40] that low-dose exposure to lead is associated with a significant decrease in intelligence quotient (IQ) has been confirmed in many studies. The recommended action level of blood lead levels established by the Centers for Disease Control is 10 μg/dL of blood. However, epidemiological studies during the last decade have found strong evidence for cognitive deficits among school-aged children at blood lead levels below the current CDC action level [41, 42]. Further, in a pooled analysis of seven prospective longitudinal studies, the average IQ deficit associated with an increase in concurrent blood lead concentration from <1 to 10 μg/dL was about threefold that for an increase from 10 to 20 μg/dL [42]. Birth cohort studies have shown inverse dose-response relationships between transplacental lead exposure indices and central auditory processing indices among infants and children at maternal blood lead levels below 10 μg/dL [43].

Research has also demonstrated a link between developmental lead exposure and behavioral outcome. In a prospective study, the behavior of lead-exposed children at 8 years of age was significantly related to tooth dentine levels [44], suggesting that social and emotional difficulties correlate with lead exposure. In another prospective, longitudinal study, both prenatal and postnatal lead exposure was related to antisocial and delinquent behavior in adolescents [45]. It is generally accepted in the scientific and medical community that the adverse neurobehavioral consequences of lead are not reversible and remain in place across the life span [41]. Further, the possibility that a threshold level for the effects of lead does not exist has been suggested [42].

Mercury

Methylmercury is a well-established neurotoxicant that can cause serious adverse effects on the development and functioning of the human central nervous system, especially when exposure occurs prenatally. The well-known episodes of community-wide poisoning in Japan and Iraq revealed the particular sensitivity of the fetus to the toxic effects from mercury exposure. In these communities, pregnant women exposed to methylmercury and who themselves had no or minimal symptoms, had babies with devastating neurological handicaps, including delayed attainment of developmental milestones, blindness, deafness, and cerebral palsy.

At levels of exposure lower than those encountered in the Minamata Bay, there is limited epidemiological evidence of neuropsychological effects. A birth cohort of 1,000 children was established at the Faroe Islands in 1986–1987, and the methylmercury exposure was determined from the mercury concentration in the cord blood [46]. More than 90% of these children were then examined at age 7 years.

Neuropsychological effects in the areas of language, attention, and memory and to a lesser extent in visuospatial and motor functions were observed [46]. In Brazil, cross-sectional studies of Amazonian children aged 7–12 years show mercury-associated effects in agreement with the Faroe findings [47]. However, a large cohort study conducted in the Seychelles did not reveal consistent associations between perinatal methylmercury exposure indices and developmental milestones or neuropsychologic test scores up to age 5 years old [48].

Some studies have found an association between central auditory processing deficits and current childhood hair mercury levels as low as 5 μg/g [49]. These findings are supported by animal studies, showing that transplacental or postnatal methylmercury exposure affects auditory systems at the cortical level [50].

The important question is to which degree these findings relate to fish-consuming populations in general. Although this question cannot be answered with any confidence at this time, it looks like the recommended one to two fish meals per week during pregnancy would be very unlikely to cause any risk to the fetus, unless the seafood is severely contaminated. The National Academy of Sciences recommended that a limit of about 0.1 μg/kg of body weight per day should not be exceeded by pregnant women [51].

PCBs, Dioxins, and Related Organohalogens

Polychlorinated biphenyls are known to interfere with thyroid hormones, some of which are critical for normal brain development, and this effect of PCBs on the thyroid function of the newborn has been postulated as the mechanism of action of neuropsychological effects [52].

The earliest evidence of PCB-related neurotoxicity comes from the poisoning episodes in Japan (Yusho) and Taiwan (Yucheng) where people became ill after ingesting rice oil that was highly contaminated with PCBs. Infants born to mothers who consumed PCB-contaminated rice oil during pregnancy were at increased risk for low birth weight, abnormal brown pigmentation of the skin, and clinical abnormalities of the gingival, skin, nails, teeth, and lungs [53]. In addition, children of both cohorts had various neurobehavioral deficits such as delayed attainment of developmental milestones, lower scores on intelligence tests, and higher activity levels [54]. Children were followed and examined 6 years later and showed persistent behavioral abnormalities and ectodermal defects [55].

Many studies after the Taiwan event provide evidence of neuropsychological effects of PCBs [56]. The evidence indicates that PCBs cause neurobehavioral deficits in children who are exposed prenatally, while the evidence of effects of exposure to background levels commonly found in the general population is not conclusive. Studies support the association between low-level transplacental PCB exposure and childhood deficits in cognitive, psychomotor, memory, language, and attention functions [56, 57]. None of the studies up to date has been able to document an adverse effect of PCBs exposure from breast-feeding. It has been postulated that the larger fat compartment of the nursing infant than the fetus may make it possible to dilute somewhat the lipid-soluble contaminants absorbed through human milk. Thus, despite the occurrence of these contaminants in human milk, the advantages of breast-feeding for 4–6 months apparently override any limited neurotoxin damage due to the contaminants. Still, of particular concern would be the populations who frequently eat contaminated fish or who reside in contaminated areas and may still be exposed at levels that have been associated with adverse effect.

Pesticides

Current knowledge about the neurotoxicological effects of pesticides is limited, despite the extensive knowledge available of the mechanisms of pesticide toxicity in animals. One observational study in children from a region in Mexico with intensive pesticide use found a variety of developmental delays compared with otherwise similar children living in a region where the population had not adopted a pesticide-based agriculture. The children were similar in growth and physical development, but significant delays were noted among the exposed children in physical stamina, gross and fine hand-eye coordination, short-term memory, and ability to draw a human figure [58]. However, these conclusions have been questioned, as pesticide levels were not reported for the individual children who received neuropsychological testing.

Fetal and neonatal animals are often more sensitive than adults to the neurotoxic effects of some organophosphates [2], and levels presumed to be nontoxic in adults may not be adequately protective of the developing organisms. For example, the young rat is deficient in the actions of two enzymes that detoxify chlorpyrifos, and young rats have increased sensitivity to chlorpyrifos toxicity [59]. In rats, chlorpyrifos appear to affect cholinergic function [60]. Infants may be particularly vulnerable to reductions in brain cholinesterase given that acetyl choline plays an important role in normal brain development and plasma and erythrocyte (and therefore probably brain) cholinesterase do not reach adult values until 6–12 months of age [61].

Animal studies suggest that exposure to pesticides such as DDT and its metabolites DDE and dichlorodiphenyldichloroethane at the levels found in the environment affect the developing brain. Ten-day old mice treated once with DDT showed behavioral changes compared with controls when tested at 4 months of age. These effects occur only after DDT dosing at 10 days of age and not after similar dosing at 3 or 19 days of age [62].

In spite of the absence of direct conclusive evidence of the neuropsychological effects of pesticides, the known subclinical effects observed in adults through neuropsychological testing, and what is known on the mechanism of action of pesticides and susceptibility of the developing brain, has led many investigators to infer that chronic low-dose exposure to certain pesticides might pose a potential hazard to the health and development of infants and children. Other investigations have concluded such inferences can be neither supported nor refuted at the present time.

Endocrine System

The endocrine system is one of the body's main communication networks and is responsible for controlling and coordinating numerous body functions. Hormones are first produced by the endocrine tissues, such as the ovaries, testes, pituitary, thyroid, and pancreas, and then secreted into the blood to act as the body's chemical messengers. They direct communication and coordination among other tissues throughout the body, and they exert a powerful influence over growth and development of every organ system in the body.

The endocrine system regulates metabolic, nutritional, reproductive, and behavioral processes, as well as growth, responses to stress, and the function of the digestive, cardiovascular, renal, and immune system. Programming of endocrine set points is a unique aspect of endocrine system development that takes place during fetal/neonatal development, and exposure to toxicants during this critical period of programming can result in permanent abnormalities in endocrine function [5]. Disruption of endocrine function can have severe health consequences in adults, and exposures that interfere with the development of the endocrine system during early life stages can have even more far-ranging consequences [63].

Endocrine disruptors are natural compounds or man-made chemicals that may alter the production or activity of hormones of the endocrine system leading to adverse health effects. Although there is limited scientific information on the potential adverse human health effects, concern arises because endocrine disrupting chemicals, while present in the environment at very low levels, have been shown to have adverse effects in wildlife species, as well as in laboratory animals at these low levels. Many of these chemicals have been linked with developmental, reproductive, neural, immune, and other problems in wildlife and laboratory animals. The potential adverse effects of EDCs include neurodevelopmental and neurobehavioral abnormalities, reproductive disorders such as declined fertility, immune impairment, and certain hormone-related cancers [1].

An example of a phthalate is di(2-ethylhexyl) phthalate (DEHP). DEHP is a high-production-volume chemical used in the manufacture of a wide variety of consumer food packaging, some children's products, and some polyvinyl chloride medical devices. Recently, an independent panel of experts assembled by the National Toxicology Program (NTP) found that DEHP may pose a risk to human development, especially critically ill male infants.

In 2000, an independent panel of experts convened by the NIEHS and the National Toxicology Program (NTP) found that there was "credible evidence" that hormone-like chemicals can affect test animals' bodily functions at very low levels – well below the "no effect" levels determined by traditional testing. Although there is little evidence to prove that low-dose exposures are causing adverse human health effects, there is a large body of evidence in experimental animals and wildlife suggesting that endocrine disruptors may cause [1]:

- Reductions in male fertility and declines in the numbers of males born
- Abnormalities in male reproductive organs

- Female reproductive diseases including fertility problems, early puberty, and early reproductive senescence
- Increases in mammary, ovarian, and prostate cancers

Endocrine disruptors may interfere with the endocrine system through several mechanisms. Some mimic or partially mimic occurring hormones in the body like estrogens and androgens and thyroid hormones, potentially producing overstimulation. Another group of natural and synthetic substances interferes with the hormones at receptors. Substances that compete with a hormone at the receptor, and imitate its effect are called agonists, those that block the receptor are antagonists. Other chemicals interfere, or block the way natural hormones or their receptors are made or controlled, for example, by blocking their metabolism in the liver. Environmental chemicals with estrogenic activity are probably the most well studied; however chemicals with antiestrogen, androgen, antiandrogen, progesterone, or thyroid-like activity have also been identified.

A wide range of substances are thought to cause endocrine disruption. Chemicals that are known endocrine disruptors include diethylstilbestrol (DES), dioxin and dioxin-like compounds, PCBs, DDT, and some other pesticides. Some chemicals, particularly pesticides and plasticizers, such as Bisphenol A are suspected endocrine disruptors based on animal studies.

Phthalates, a class of chemicals that soften and increase the flexibility of polyvinyl chloride plastics, are considered to be endocrine disruptors on the basis of extensive animal research plus emerging studies in humans. An example of a phthalate is di(2-ethylhexyl) phthalate (DEHP). DEHP is a high-production-volume chemical used in the manufacture of a wide variety of consumer food packaging, some children's products, and some polyvinyl chloride medical devices. Recently, an independent panel of experts assembled by the National Toxicology Program (NTP) found that DEHP may pose a risk to human development, especially critically ill male infants.

Research shows that endocrine disruptors may pose the greatest risk during prenatal and early postnatal development when organ and neural systems are developing. In animals, adverse consequences, such as subfertility, premature reproductive senescence, and cancer, are linked to early exposure, but they may not be apparent until much later in life.

There is some evidence that endocrine disruptors may not only impact the individual directly exposed but also future generations. It has been found that animals exposed to low doses of the natural human estrogen estradiol, or the environmental estrogen bisphenol A (BPA) (a chemical used in great quantities in the production of polycarbonates and epoxy resins), during fetal developmental and estradiol as adults were more likely to develop a precursor of prostate cancer than those who were not exposed [64]. This suggests that exposure to environmental and natural estrogens during fetal development could affect the way prostate genes behave, and may lead to higher rates of prostate disease during aging. It has also been shown that the adverse effects of diethylstilbestrol in mice can be passed to subsequent generations even though they were not directly exposed [64].

The increased susceptibility for tumors was seen in both granddaughters and grandsons of mice who were developmentally exposed to DES [65]. One study found that endocrine disruptors caused fertility defects in male rats that were passed down to nearly every male in subsequent generations [66]. These intergenerational effects may be the consequence of epigenetic changes caused by endocrine disruptors.

There is concern that alterations in thyroid hormone signaling by endocrine disruptors during fetal and neonatal development could disrupt central nervous system development. There have been several epidemiological studies of the neurobehavioral effects of in utero exposure to PCBs that have been mentioned in the section on nervous system effects. Thus far, few studies have directly linked neurobehavioral effects of exposure to polyhalogenated hydrocarbons to disruption of thyroid hormone signaling. One example is a study that showed that low-frequency hearing loss caused by developmental exposure in rats to PCBs could be partially reversed by replacement of T4 [67]. In animals, many polyhalogenated aromatic hydrocarbons such as dioxins and PCBs alter thyroxin levels via an increased metabolism and excretion of these hormones [52].

Perchlorates are another class of environmental chemicals that affect thyroid function via inhibition of iodine uptake by the thyroid gland, reducing T4 and T3 synthesis. There has been a concern that contamination of drinking water with perchlorates from industrial sites would suppress fetal thyroid hormone synthesis, disrupting central nervous system development. Increased rates of congenital hypothyroidism have been found in communities with detectable perchlorate levels in the drinking water [68], although others have not detected such elevated rates [69].

Evidence from epidemiological studies demonstrates that exposures during early life can result in greater susceptibility to diabetes and obesity later in life [70]. In the course of development of the pancreas and pancreatic function, for example, several windows of susceptibility have been identified [71]. Studies about the Dutch Famine Winter have shown that poor maternal nutrition, especially during the last trimester of pregnancy had an impact on glucose tolerance and insulin resistance, and that in terms of obesity, those individuals born to mothers who were exposed to the famine around the first half of pregnancy were more obese than those of mothers exposed during the last trimester [72]. A number of subsequent studies have documented the role of the intrauterine environment, and low birth weight in the development of diabetes [71, 73] and other metabolic disorders.

Reproductive System

The development of the reproductive system is a long process that takes place from the beginning of gestation, when organ development starts to take place, to maturation of the reproductive system during puberty. This long developmental period provides for several windows of susceptibility. Adverse effects of reproductive toxicants can become manifest at birth (e.g., hypospadias and cryptorchidism in humans), in puberty (as delay or precocity), or in adulthood (e.g., infertility, alterations in accessory sex organs, disturbances in pregnancy maintenance, endometriosis, or premature reproductive senescence) [74].

The developmental susceptibility of the reproductive system was clearly demonstrated through the diethylstilbestrol (DES) exposure. In utero exposure of men to DES has been linked to increased incidence of meatal stenosis, epididymal cysts, testicular hypoplasia, cryptorchidism, microphallus, and sperm abnormalities. In females, DES exposure resulted in adenosis, clear cell adenocarcinoma, and structural defects of the cervix, vagina, uterus, and fallopian tubes [75].

Environmental tobacco smoke has been associated with decreased fecundity and earlier menopause in women smokers [76]. Women whose mothers smoked while pregnant also had reduced fecundity compared with women whose mothers did not smoke [13]. In men whose mothers smoked tobacco during pregnancy, reduced semen quality, smaller testis size, and reduced fecundability odds ratios have been observed [77].

Phthalates such as diethylhexyl phthalate (DEHP) are developmental toxicants in experimental animals. The organ system most sensitive to phthalates is the reproductive tract of immature males [78]. In utero exposure of male rats to some phthalate esters results in changes in the reproductive tract, such as decreased anogenital distance, hypospadias, cryptorchidism, disturbed development of prostate, epididymis, vas deferens, and seminal vesicles, retained nipples and decreased sperm production [79–82]. In humans, similar dysgenetic changes in the histology of the testis have been found in patients with testicular cancer, subfertility, or cryptorchidism [83, 84]. It has been hypothesized that all these human disorders (testicular germ cell cancer, cryptorchidism, hypospadias, and low sperm counts) have common origins in fetal life and that they all represent different symptoms testicular dysgenesis syndrome [87].

Neonatal Mortality, Growth Restriction, and Birth Defects

Birth defects are normally defined as structural problems in the newborn, attributable to faulty development or deformation, but defects in function, metabolism, or body chemistry that lead to physical or mental problems or to death may also be considered birth defects. The broader term "developmental disorders" is generally used when considering all effects observed on the conceptus from fertilization to sexual maturity. Developmental disorders include both structural birth defects and functional defects, such as blindness, deafness, or neurobehavioral disabilities.

The majority of birth defects are considered the result of multiple environmental and/or genetic causes acting together. Environmental toxicants studied for their relation with birth defects include maternal smoking and alcohol use, pesticides, disinfection by-products, plastics and plastic components, solvents, metals and numerous air pollutants [88]. Other environmental causes such as maternal diseases (e.g., rubella), and use of pharmaceuticals (e.g., valproic acid, an anticonvulsant, and mood stabilizers), have served to document the potential environmental role in developmental birth defects.

Many epidemiological studies have attempted to evaluate whether some chemical exposures are linked to increased rates of spontaneous abortions. For example, occupational exposure of the mother to organic solvents has been associated with spontaneous abortion in several studies [89, 90]. A review of six occupational studies found suggestive evidence for an association between toluene exposure and spontaneous abortion, although most workers were simultaneously exposed to multiple chemicals, which may have confounded the interpretation of results. According to this review, spontaneous abortion has generally not been observed as a major problem among highly exposed women who abuse toluene during pregnancy.

Many studies have investigated the effects of *disinfection by-products* (DBPs) on the developing fetus and children. Review of epidemiologic studies reveal that there is fairly consistent evidence for associations between early and late fetal deaths and indices of transplacental DBP exposure [25, 26, 89]. Associations between late fetal deaths and drinking water chloroform, bromodichloromethane, and total THM concentration at levels below the Canadian drinking water guideline of 100 μg/L have been found in Canada [90]. In England, increased risks of late fetal deaths among women living in regions with THM concentrations above 30 μg/dL [91]. These findings are supported by animal studies, where high prenatal maternal exposure to chloroform, bromodichloromethane (BDCM), haloacetonitriles, or haloacetic acids produced fetal resorption and reduced fetal survival [26].

Evidence on the effects of maternal *DBP* exposure and SGA (small-for-gestational-age) effects is mixed. Recent reviews have found limited evidence for association with this effect [25, 26], but a cohort study in Massachusetts demonstrated associations between SGA and drinking water trihalomethanes and mutagenic activity levels during the third trimester in the town of maternal residence [92]. In experimental animals, high prenatal maternal exposure to chloroform, BDCM, haloacetonitriles, or haloacetic acids reduced fetal weight [26, 27].

The evidence of the effects of chlorination disinfection by-products in drinking water and birth defects is inconclusive. Some studies have shown an association between neural tube defects and prenatal maternal DBP [25, 26]. The evidence of effects of water disinfection from five previous studies with a cross-sectional study of all Taiwanese births in years 2001–2003 was reviewed [93] and evidence of an effect of exposure to chlorination by-products on the risk of neural tube defects, urinary system defects, and ventricular septal defects was concluded. However, a recent review of a small number of recent studies reported inconsistent results for an association between drinking water chlorination by-products and risk of al congenital anomalies combined and of specific groups of anomalies, and there was little evidence of an exposure–response relationship [94].

Maternal *smoking* during pregnancy increases the risk of pregnancy loss, stillbirth, and infant mortality [28]. There is some strong evidence that exposure to ETS increases the probability of preterm birth [95]. A large birth cohort study in California found borderline statistical association between maternal serum cotinine concentrations during early pregnancy and late fetal deaths (borderline statistical significance), preterm birth, and a significant association with reduced birth weight (significant) [24].

The association between sudden infant death syndrome (SIDS) and maternal smoking has been firmly established. As reported by Wigle et al. [98], a meta-analysis of 39 epidemiological studies and an expert panel review concluded that there is sufficient evidence of a causal association between sudden infant death syndrome and postnatal ETS exposure, independent of prenatal maternal active smoking [95]. Sudden infant death syndrome was also associated with paternal smoking, even when the mothers did not smoke.

Active *smoking* by mothers has been shown to significantly reduce the rate of fetal growth, and the effect was shown to be dose-dependent. DiFranza and Lew [97] using data from 23 studies, calculated an odds ratio of 1.82 for the association between maternal smoking and low birth weight (<2,500 g). The risk of intrauterine growth retardation caused by maternal smoking appears to increase with maternal age, from twofold for mothers aged 17 years to fivefold for mothers aged 35 years [98]. Although smoking throughout pregnancy is known to affect birth weight, there is some evidence that the final trimester may be particularly important [99]. In some studies, babies of smoking mothers have been reported to be shorter and to have smaller head circumferences at birth than babies of non-smoking mothers [98].

Studies *on pesticide* exposures and birth defects have found evidence of associations between paternal pesticide exposure and cryptorchidism [99], although other studies have failed to find a similar association. Cryptorchidism was associated also with maternal serum DDT/DDE and hexachlorobenzene levels [100] and with adipose tissue or maternal serum DDE concentrations [101]. In animal studies, trans-placental exposure to the pesticides DDT/DDE vinclozolin, procymidone, or linuron, produce hypospadias, cryptorchidism, and other abnormalities [102]. The epidemiological evidence of the link between pesticide exposure and hypospadias is not conclusive enough to affirm a link between pesticide exposure and this type of birth defect.

Respiratory System

Lung development is a continuous process from embryonic life to adolescence. At birth, about 85% of the human alveoli are present. Alveolar number and lung surface area increase through childhood and begin to level off between 2 and 4 years of age, whereas lung expansion continues up to 8 years of age [5]. Immature (neonatal) differentiating cells of the respiratory tract are more sensitive to injury following exposure to respiratory toxicants than mature cells, and at dose levels that cause no effects in adult cells. Children are usually physically active, and have greater exposure to air pollutants. Because of their higher metabolic rates, they breathe more rapidly and inhale more pollutants per kilogram of body weight than adults. Their narrower airway passages compared to adults can be more easily obstructed in a greater proportion than in adults. Thus, because of the immaturity of the lungs during childhood and the greater exposures relative to body weight in children than in adults, it is inferred that children may be more susceptible to the effects of respiratory toxicants than adults, whose lung growth is complete.

Air pollution has been extensively studied in relation to respiratory health and in general it has been easier to demonstrate the effects of air pollution on exacerbation of certain conditions than on the causing of disease in previously normal individuals. Both outdoor and indoor air pollution have been identified as potential risk factors for both the initiation/induction and the exacerbation of respiratory diseases, especially asthma. Evidence of the effect of air contaminants such as ozone and ETS on lung function has been demonstrated in both animal and human studies. Air pollution is now linked to SIDS [103].

Indoor Air Pollution

Indoor air pollution is the most important source of respiratory exposure to toxicants in children, given the fact that children spend up to 90% of their time indoors and the large range of pulmonary irritants that can be found in the home. Exposure to pollutants in the home environment in developed countries has increased with improved insulation and reduced ventilation and the use of chemical detergents and building or furnishing constituents that contain noxious pulmonary irritants [5]. Infants and young children in particular have little control over their exposure to pollutants in the home environment and are vulnerable to the activities of the adults (particularly ETS). Common indoor air pollutants include nitrogen dioxide, formaldehyde and other volatile organic compounds (VOCs), and ETS.

Environmental Tobacco Smoke

Environmental tobacco smoke has been extensively studied in relation to children's respiratory health. The developing fetus can be involuntarily exposed to tobacco during pregnancy through a smoking mother or by the pregnant mother's exposure to environmental tobacco smoke. Exposure may continue throughout childhood if any of the parents smoke. The early exposures may have persistent adverse effects throughout life.

There is evidence of ETS effects on the increased risk of lower respiratory illness rates, especially in the first year of life, increased rates of chronic middle ear effusion in children, impairment of lung function and exacerbation of certain conditions such as asthma. Exposure to environmental tobacco smoke is a risk factor for sudden infant death syndrome. Children exposed to ETS are more likely to suffer from respiratory illness (bronchitis, pneumonia) and to be hospitalized because of the illness than unexposed children [103]. EPA estimates that between 150,000 and 300,000 of bronchitis and pneumonia cases annually in infants and young children up to 18 months of age are attributable to exposure to ETS. Of these, between 7,500 and 15,000 will result in hospitalization. Furthermore, there is evidence that exposure to ETS leads to increased infant mortality from respiratory illness as well as increased morbidity [104].

Exposure to pollutants during critical periods of lung development may have effects that would not be seen if exposure occurred during adulthood [105]. Maternal smoking during pregnancy was related to impaired lung function in newborn infants [106]. Permanent effects of parental smoking on lung function were found in adults of 30–59 years of age [107], indicating that the impairment of lung function persists through life.

The exacerbating effect of ETS on asthma has been known for some time, and many studies evidence this effect. ETS exposure increases the frequency of episodes and severity of symptoms and the rates of hospitalization in asthmatic children. In addition, ETS exposure is a risk factor for new cases of asthma in children who have not previously displayed symptoms (EPA http://www.epa.gov/smokefree/healtheffects.html) [69].

Another group of environmental agents that have been studied in relation to respiratory health are the bioaerosols. Bioaerosols include inhaled allergens (house dust mites and other insects, molds, pets, pollens) and bacterial (lipopolysaccharide) and fungal (glucans) products. Sensitization to one or more common inhalant allergens is consistently associated with childhood asthma especially in developed countries [107]. An expert panel concluded that house-dust mite allergens produce incident (new onset) asthma and that cat, cockroach, and house-dust mite allergens induce episodes in sensitized asthmatics [96]. However, the relationship between exposure to inhaled allergens in early life and the development of asthma or wheeze in childhood is controversial [107]. Contrarily, it has been hypothesized that early exposure to some allergens may be protective of later development of asthma [109], although this hypothesis was discarded in later studies [110].

Outdoor Air Pollution

Air pollution, from both vehicular and stationary sources, is associated with an increase in respiratory symptoms, a decrease in lung function, and the exacerbation of asthma symptoms among asthmatic children. Recent studies have evidenced the relationship between traffic air pollution and incident development of asthma. Most of these effects occur at levels within the ambient air quality standards of most countries that have the potential to affect a large population of children. Air pollution effects contribute significantly to children's respiratory morbidity and have a significant economic impact in terms of health expenses. Most studies attribute respiratory symptoms to particulate matter, although the close correlation of particulate matter levels with nitrogen dioxide and sulfur dioxide levels makes the contribution of individual pollutants difficult to determine [5].

There is a consistent body of evidence that outdoor air pollution is associated with increased respiratory symptoms, such as cough, bronchitis, respiratory infections, and upper respiratory tract illness in children [111]. Several studies have demonstrated an increase in respiratory symptoms such as cough and bronchitis associated with increases in PM10. A recent study in Eastern Germany evaluated the association between respiratory symptoms and air pollution levels before and after political reunification. A reduction in air pollution since reunification is associated with reductions in the rates of chronic cough and bronchitis symptoms in a new cohort of children, suggesting a potentially reversible effect of air pollution [112]. A similar dramatic effect was observed with children who moved out of an area to other areas of higher and lower PM10 concentrations. Those children who moved to areas of lower PM10 concentrations showed higher rates of growth in lung function, whereas the opposite effect was observed in children who moved to areas of higher PM10 concentrations than before [113].

The role of air pollution in the exacerbation of asthma is well established, and there is mounting evidence that air pollution, particularly from traffic, is implicated in the pathogenesis of asthma [114]. Because children with asthma have increased airway reactivity, the effects of air pollution on the respiratory system can be more serious for them. Children with asthma have been shown to experience more respiratory symptoms, use extra medication, produce chronic phlegm, and have more bronchitis following exposure to high levels of particulate pollution [115, 116]. Air pollution is also associated with increased school absenteeism due to respiratory illness [117], and increased admissions to hospital emergency department [118].

The role of the different components of air pollution in the development and exacerbation of asthma is difficult to elucidate. Oxidant gases such as nitrogen oxide and ozone have been associated with asthma prevalence [119]. Other studies suggest that traffic pollution, but probably not NO_2 from traffic, is associated with atopy and wheezing [120]. The authors suggested that diesel particles or some component of those particles, such as polycyclic aromatic hydrocarbons, may be the most important etiologic component. A recent review of the short-term effects of PM10 and NO_2 on respiratory health among children with asthma or *asthma*-like symptoms concluded there were clear effects of PM10 on the occurrence of asthma symptom episodes, and to a lesser extent on cough and PEF (pulmonary expiratory function). Results with respect to NO_2 were inconclusive [121].

Of great relevance have been recent studies that have related proximity to heavily trafficked roads with asthma and reduced lung functions as well as other respiratory symptoms such as wheeze and dry cough. Lin et al. [122] found significant odds ratios for living within 200 m of a street with the highest tertile of traffic density and asthma prevalence, and the children with asthma were more likely to have truck traffic on their street. Although some studies showed no increased risk [123], the weight of evidence suggests that traffic pollution is associated with the risk of developing asthma [114].

Lung function is also affected by air pollution. Acute exposure to ozone, nitrogen dioxide, sulfur dioxide, and particulate matter is known to cause transient reversible decreases in lung function [124, 125]. Some recent studies indicate that long-term exposure to ozone and related co-pollutants (individually and synergistically) is associated with impairment of lung function capacity in children and adolescents [126], although other studies are not consistent. Further, it has been hypothesized that a fraction of the increases in the prevalence of chronic obstructive lung disease in adults who live in more polluted areas could be the result of exposures that occurred during childhood [127].

Cancer

Childhood cancers are relatively rare diseases during the first 2 decades of life, but are the leading cause of death of children in many countries. The most common types of childhood cancer are lymphoid neoplasms (leukemia, lymphoma) and cancers of the central nervous system. Other kinds of childhood tumors include embryonal tumors of the retina, sympathetic nervous system, kidney, and liver; tumors of bone and soft connective tissues; and certain gonadal neoplasms. Carcinomas in epithelial tissues, the most frequent type of cancer in adults, are rare in children. Neoplasms in adults resulting from known or iatrogenic exposure typically have latency periods of 20 years or more. Thus, cancers in children are qualitatively distinct from those in adults.

Although childhood cancer is a rare disease, the occurrence of new cancer cases among children has continued to rise during the last 2 decades. While medical advances have led to a sharp reduction in mortality from childhood cancer, it is still a group of diseases with potentially devastating outcomes, which can be at least partially prevented; so the study of the environmental causes of cancer is essential. A recent report from the European Automated Childhood Cancer Information System (EACCIS) provides evidence of a 1% increase per year in childhood cancers and 1.5% increase per year in adolescent cancer in Europe for the period 1970–1999. All of the common types of neoplasms showed significant increases, including leukemias, lymphomas, central nervous system tumors, neuroblastomas, soft tissue sarcomas, retinoblastoma, and germ cell, renal, hepatic, and bone tumors in children and carcinomas, lymphomas, soft-tissue sarcomas, and germ cell and CNS tumors in adolescents. There were in addition significant differences between Eastern and Western Europe with regard to leukemias, lymphomas, carcinomas, and central nervous system tumors in children, and carcinomas, lymphomas, leukemias, and soft-tissue tumors in adolescents. Earlier reports of European populations also noted increases in several of these types of childhood cancers. Similarly, in the USA, there was an overall increase of about 25% in childhood cancers between 1975 and 2000. While the reported increases in the 1980s may have reflected improvements in diagnosis and reporting, it is unlikely that the most recently reported trends reflect the same bias. Evidence from both epidemiological and mutagenicity/carcinogenicity studies suggest that environmental toxicants may be involved at least in the causation of some forms of cancers such as acute lymphoblastic leukemia (ALL), which is most common in industrialized countries.

Most cancers result from the interaction between genetic factors and the environment [128]. In this context, environment is defined broadly and includes diet, alcohol, drugs, tobacco, and all other nongenetic factors as well as classic environmental toxins. It has been estimated that about 80–90% of all cancers are attributable to environmental factors acting in conjunction with both genetic and acquired susceptibility. For cancer in adults, different opportunities for environmental exposure have commonly been considered a major (albeit not exclusive) reason for the geographical distribution of cancer. Geographical differences in incidence of most cancers (except lymphomas) are less marked for childhood than for adult cancer, thus suggesting that the fraction of childhood cancers due to environmental factors is probably lower than for adults [128].

The potentially higher susceptibility of the developing fetus and children to the effect of environmental carcinogens is based on several aspects. Chemical carcinogenesis is a multistep process involving genetic and epigenetic changes in susceptible cells that gain a selective growth advantage and undergo clonal expansion as the result of activation of protooncogenes and/or inactivation of tumor-suppressor genes. The occurrence of these events is modulated by several factors that themselves change in the course of development. Thus, DNA damage may be repaired by the action of DNA repair enzymes, whose presence and activity may change with age. Similarly, changes in metabolism will determine whether carcinogenic metabolites are formed, thus allowing for the initiation of the cancer process.

DNA repair mechanisms play an important role in cancer protection. DNA repair enzymes have been found to be well expressed in embryos and fetuses. In fact, there are a number of DNA repair genes/activities that have higher expression in fetuses or embryos than in adults. Interference with the synthesis of DNA repair enzymes during development may result in increased susceptibility to cancer during childhood or later in life. An example of the important role of DNA repair mechanisms is the p53 gene, which encodes a protein that modulates DNA repair and cell division. Mutations of the p53 genes are involved in at least 50% of all cancers [129]. p53 mutations have been linked with tobacco smoking [130], and it is possible that p53 mutations occur in the offspring of smoking mothers.

The higher rates of cell proliferation during development can contribute to increased likelihood of carcinogenesis. For example, PAHs and aflatoxin B1 (AFB1) produce liver tumors when administered to newborn rodents but not when administered to older animals, presumably because the liver proliferates rapidly in the developing system but more slowly in older animals [131]. Women who were in their teens at the time of atomic bombings had the greatest risk of radiation-induced breast cancer [132]. Some organs, such as the brain, are fully developed in early childhood, whereas others such as the skeletal system do not achieve maturity until adolescence

[131]. This may be the reason why osteosarcoma, the most common bone cancer, peaks in late adolescence, a period of rapid bone growth.

The immune system of the newborn is not fully developed until about 6 months of age. Chemicals that affect the immune system, such as halogenated aromatic hydrocarbons that bind to the Ah receptor [133], may modify the host defense mechanism against infection and cancer [134].

The susceptibility of children to cancer may be influenced by the presence/absence of metabolizing enzymes. The role of these enzymes is potentially complex, especially for those that carry out the phase I (usually detoxifying) reactions. Usually, chemical metabolism protects the adult and fetus from carcinogenicity, but the activity of enzymes can also result in bioactivation, such as in the case of bioactivation of benzo[a]pyrene to its carcinogenic form, which has been demonstrated by the observation of the formation of DNA adducts in mice and monkey fetuses, which probably originated in maternal liver [135].

The first step in metabolism is usually oxidation, carried out by cytochrome P450 enzymes. This group of enzymes are named as CYP enzymes and grouped in families according to their genotype and activity. Several groups of these enzymes are involved in the metabolism of exogenous carcinogens, such as the CYP families 1–3, which are thought to be the most relevant to effects of exogenous carcinogens. Other examples are the CYP4B1, which activates aromatic amines, and CYP2E1, which metabolizes nitrosamines and organic solvents that have been implicated in causation of childhood cancers.

Phase II detoxification enzymes have received less attention than CYPs in the perinatal context, though they are also likely to be important. For example, among the glutathione S-transferases (GSTs), the I (P) form is apparently expressed at the highest levels and in many tissues in human fetuses from early in gestation, and levels of total GST activity in all fetal tissues studied was comparable to that in the corresponding adult organ [135]. An increasing body of evidence implicates GST polymorphisms in risk of childhood leukemias. Glucuronidation, a type of Phase II reaction, catalyzed by uridine diphosphate-glucuronosyltransferase (UDG), is a key detoxification step for solubilization of high molecular weight carcinogens such as polycyclic aromatic hydrocarbons, heterocyclic amines, and tobacco-specific nitrosamines. Low expression of UDG in the fetus and neonate, and as a result of polymorphisms in children, could well increase sensitivity to carcinogenesis by some chemicals. Sulfotransferases (SULT) have been proposed to constitute a major enzymatic detoxification system in the fetus [136, 137]. However, these enzymes also catalyze activation of several types of chemical carcinogens to DNA-damaging forms, including aromatic amines, polycyclic aromatic hydrocarbons, heterocyclic amines, 3-nitrobenzanthrone, and benzylic alcohols. This illustrates how alteration of enzymatic metabolism can greatly influence susceptibility to perinatal carcinogenesis.

While mutagenicity/carcinogenicity studies, as well as the study of enzymatic systems and DNA repair enzymes, provide mounting evidence that the interplay between chemical agents and these biological factors are significant in the development of cancer, epidemiological studies have only been of limited usefulness to confirm this hypothesis. The increasing trends over the last 20–30 years and the geographical distribution of childhood cancers, clearly point to environmental causes. However, few studies have been able to firmly establish the environmental origin of childhood cancers. Cumulative epidemiological evidence pertaining to childhood cancers, including international variation, time trends, and risk factor studies was reviewed and analyzed by Bunin [138]. This review concluded that ionizing radiation and a variety of genetic conditions are thought to explain 5–10% of childhood cancers. There are clear associations between Epstein-Barr virus infection and Burkitts lymphoma in Africa, and between human immunodeficiency virus and Kaposi's sarcoma [138].

Other risk factors have not been conclusively identified. Among the nongenetic causes, the pattern of international variation and associations with surrogates of infection suggested an infectious etiology for acute lymphoblastic leukemia, although no agent has been identified. For brain tumors, cured meats, polyomaviruses, and farm exposures were pointed out as potential causes. Changes in the incidence and characteristics of children with hepatoblastoma as well as risk factor studies suggest a role for an exposure of very low birth weight babies. High birth weight, tea or coffee consumption, and certain paternal occupations have shown some consistency in their association with Wilms' tumor. For most of the other cancers, very few epidemiologic studies have been conducted, so it is not surprising that nongenetic risk factors have not been detected. The most important difference between the cancers for which there are good etiologic clues and those for which there are not may be the number of relevant studies.

This section summarizes the evidence found in the recent literature of the associations between environmental factors and childhood cancer. Attention is given only for those agents for which there is at least limited evidence of an association between environmental agent and disease.

Ionizing Radiation

Evidence of the potential of ionizing radiation to induce childhood cancer comes from events such as Hiroshima and Nagasaki, populations affected by accidents at nuclear plants, as well as investigations on the longer effects of indoor radon and on the consequences of exposure to x-ray for diagnostic or therapeutic reasons.

Japanese children exposed to the atomic bombings of Hiroshima and Nagasaki had increased risks of adult cancers, including leukemia and solid tumors. The highest risk was that of children who were exposed in utero. Radiation-related leukemia started to occur 2–3 years after the bombing, reached its peak within 6–8 years, and has declined steadily since then. For people exposed as adults, the excess risk was lower than that of people exposed as children, but the excess risk appears to have persisted throughout the follow-up period [5].

Thyroid adult carcinomas were also increased in survivors who were children at the time of the bombings at Hiroshima and Nagasaki. The greatest risk for thyroid cancer occurred in individuals who received a radiation dose to the thyroid greater than 1 Sv before age of 10 years. Similarly, breast cancer risk among women survivors to the atomic bombings was highest for the women who were less than 10 years of age or between 10 and 20 years of age, lower for women who were between 20 and 40, and even lower for women exposed after 40 years of age [132].

Prenatal diagnostic X-irradiation has also been linked to increased risk of leukemia in offspring, as has therapeutic, high-dose, ionizing radiation in childhood for other cancers and for various non-neoplastic conditions [139].

The Chernobyl nuclear reactor accident in 1996 caused a significant increase in the incidence of thyroid cancer in children since 1990. Most of the tumors have been observed among individuals who were very young at the time of the accident [31]. In Belarus, over half of the tumors occurred in people who were less than 6 years old at the time of the accident. In a series of 472 children with thyroid cancer diagnosed up to 1995 in Belarus, only 2% had been conceived after the accident, 9% had been exposed in utero, and 88% were under 15 years of age at the time of diagnosis [140]. The very early age at which the thyroid cancers have begun to be diagnosed is one of the most striking examples of the special sensitivity to a carcinogen other than an infectious agent occurring exclusively during preadult life.

UV Light

Sun exposure is known to be a risk factor for the development of skin cancer later in life. The occurrence of sunburn is an indicator of risk [141]. Numerous studies have assessed the carcinogenic effect of sunburn at different ages and concluded exposure to solar radiation before 10 years of age was a primary contributor to risk of melanoma. The IARC concluded that childhood is an especially vulnerable life stage [142] to the carcinogenic effect of UV radiation. While the use of sunscreen reduces the risk of sunburn, an expert working group convened by the International Agency for Research warned against the risk of relying solely on sunscreens for protection from ultraviolet radiation [143], as the use of sunscreens may lead to an extension of the duration of intentional sun exposure, which can increase the risk of melanoma.

Environmental Tobacco Smoke

Environmental tobacco smoke is an established human carcinogen by the International Agency for Research on Cancer [144]. Over 50 epidemiological studies have reported risk ratios of lung cancer for secondhand smoking in adults. A recent meta-analysis of epidemiological studies of lung cancer and adult exposure to environmental tobacco smoke resulted in risk ratios of 1.22 in women and 1.36 in men from workplace exposure. Other meta-analyses have produced similar results. The evidence of a causal relationship between ETS exposure and cancers in organs other than the lung is inconclusive.

Tobacco smoke contains many carcinogens, such as PAHs and 4-aminobiphenyl, that can cross the placenta and be transferred to the fetus. The genotoxicity of tobacco smoke to the fetal liver has been tested in an animal study. Sister chromatid exchange in the liver cells of fetal mice was analyzed at the 16th day of gestation after short-term exposure (twice, on the 15th and 16th days of gestation), long-term exposure (starting 4 weeks before mating and stopping on the 16th day of gestation), and prepregnancy exposure (4 weeks before mating). The number of sister chromatid exchanges was significantly increased in all exposure groups, and long-term exposure caused a significantly higher increase than did short-term exposure [145].

Most epidemiological studies on the link between exposure to ETS and childhood cancer have focused mainly on pregnant women. While most of studies of smoking by mothers reveal no effects on childhood cancer, many studies of smoking by fathers have shown a significant association with risk of cancer in their children, and when both mothers and fathers are included in the study, the effect appears to be greater for exposure of fathers than of mothers [135]. This effect has been attributed to germ-cell mutations during spermatogenesis caused by tobacco products [146]. Three reports from the Oxford Survey of Childhood Cancers have suggested that paternal but not maternal cigarette smoking is associated with increased risks for the generality of childhood cancers. Some, however, have produced conflicting findings. A large case control study conducted in the UK (the United Kingdom Childhood Cancer Study) concluded that there was no evidence that paternal smoking is a risk factor for childhood cancer in general [147]. A significant association was reported in this study, however, between hepatoblastoma risks and smoking by both parents relative to neither parent smoking. Sorahan and Lancashire speculated that the importance of both parents smoking in the etiology of hepatoblastoma might arise from the combination of oxidative damage to sperm DNA and damage to the fetal liver from carcinogenic metabolites in the blood of the pregnant mother [148]. Further, in a study where both preconception and postnatal smoking by the fathers was quantified, and few of the mothers smoked, the association with childhood cancer related significantly only to preconceptional paternal smoking levels [149]. Thus, second-hand exposure of the infants to smoke was less likely. Transplacental effects of sidestream smoke were one possibility. In rats, sidestream smoke constituents received transplacentally caused oxidative DNA damage in fetal tissues [136]. In one epidemiological study, paternal smoking had a larger possible effect when mothers were nonsmokers [137], suggesting protective effects of detoxification enzymes induced in the placenta and in maternal tissues by maternal smoking.

Some studies have examined the relationship between exposure to environmental tobacco smoke during childhood and cancer risk. Sandler and colleagues [150] found that the overall cancer risk was greater for individuals with exposures to environmental tobacco smoke during both childhood and adulthood than for individuals with exposure during only one period. When specific cancer sites or types were considered, leukemia and lymphoma among adults were significantly related to exposure to maternal passive smoke before 10 years of age [150].

Pesticides and Cancer

Pesticides are biologically active molecules that are commonly used to destroy unwanted organisms in agricultural and residential environments. The widespread use of these chemicals has raised concerns over the potential of pesticides to cause childhood cancer.

Although the biochemical mechanisms relating pesticide exposures to childhood cancer have not been fully described, some evidence suggests that pesticides may promote the formation of chromosomal aberrations known to be associated with an increased cancer risk [149]. Studies in adult populations suggest that pesticide exposures may have a direct effect on chromosome structure, and therefore a causal relationship between pesticide exposures and childhood cancer is plausible [152].

Epidemiological studies have reported associations between childhood cancer and either parental or child exposures to pesticides. Research reviews have suggested an increase in the risk of brain cancer, leukemia, non-Hodgkins's lymphoma (NHO), Wilms' tumor, Ewing's sarcoma, and germ cell tumors associated with parental occupational and nonoccupational exposure to pesticides [153, 154]. The exposures observed occurred prior to and during pregnancy, as well as after the childbirth, thus involving different potential modes of action. Zahm and Ward [154] concluded that at least some childhood cancer could potentially be prevented by reducing or eliminating pesticide exposure, although methodological limitations common to many studies limit the possibility of making conclusions regarding the role of pesticides in the etiology of childhood cancers [153, 154].

Two extensive reviews were conducted after the Zahm and Ward review [154] that attempted to provide conclusions on the basis of cumulative evidence. In a review that evaluated 18 new studies conducted between 1998 and 2004, it was concluded that while collectively all studies seem to suggest an increase in the risk of different cancer types associated with exposure to pesticides, no conclusions could be drawn with respect to cancer types as well as to specific causative factors across studies [153]. Infante-Rivard and Weichenthal [152] reviewed studies conducted between 1999 and 2004 and critically evaluated the evidence on the associations between pesticide exposures and leukemia (12 studies), brain cancer (10 studies), neuroblastoma (4 studies), non-Hodgkin's lymphoma (3 studies), Wilm's tumor (2 studies), and Ewing's sarcoma (1 study), as in the Zahm and Ward review. The authors found recent studies to be consistent with the suggestion of Zahm and Ward of an association between pesticide exposure and childhood leukemia, brain tumors, neuroblastoma, and also non-Hodgkin's lymphoma and Wilm's tumor. Specifically childhood exposure to household insecticides and prenatal exposure to pesticides seemed to pose the greatest risks of leukemia and brain tumors. Exposure-response gradients were observed in some studies on pesticide exposure and risk of leukemia. For neuroblastoma, the authors found that all four recent studies reviewed showed an association between exposure of pesticides and risk of neuroblastoma that supported earlier findings by Zahm and Ward. A recent study also indicates that residential use of pesticides, and herbicides specifically, may increase the risk of neuroblastoma in children [153]. The risk of non-Hodgkin's lymphoma was found to be associated with residential exposure to pesticides, and two of the studies reviewed provided evidence of exposure response gradients in two of the three reviewed studies. With respect to Wilm's tumor, recent studies did not provide additional evidence of an association between insecticide exposures and parental pesticide exposure before birth and Wilms' tumor that had been indicated in Zahm and ward's review. The evidence on Ewing's sarcoma was inconclusive.

A systematic review and meta-analysis of studies on childhood leukemia and parental occupational exposure to pesticides concluded that there was not sufficient evidence to affirm an overall association between childhood leukemia and any paternal occupational pesticide exposure among all studies combined or subgroups of studies [155]. An elevated childhood leukemia risk was found in relation to paternal occupational exposure to the broad pesticide classes of insecticides and herbicides, but the authors considered that the small number of studies showing this effect and the lack of exposure–risk relationships did not allow making firm conclusions. However, an association was found between childhood leukemia and prenatal maternal occupational pesticide exposures. The studies also showed associations between childhood leukemia and maternal occupational exposure to insecticides and herbicides. It was concluded that the overall evidence, though limited, warranted exposure prevention measures on the basis of the precautionary principle [152].

Many authors have pointed out the major shortcomings of epidemiological studies on pesticides exposures and childhood cancer [156]. One of the major shortcomings is exposure assessment, as many of the studies use very unspecific measures of exposure, such as "farming" as a measure of "exposures to pesticides." In most of the studies, the specific pesticide/s, timing and other concomitant exposures to which the population was exposed is not known. The authors recommended directing research efforts to the characterization of pesticide exposures in future research. Further they have remarked that "another 40 epidemiological studies similar to the majority of those conducted thus far will not provide clarity."

In spite of the limitations of epidemiological research on pesticide exposures and childhood cancers, many researchers find that there is sufficient evidence to be concerned about the potential role of pesticides in childhood cancers, and to recommend a precautionary approach. Infante-Rivard [152] contrasted the overall evidence against the Bradford Hill's causality criteria (strength, consistency, specificity, temporality, biological gradient, plausibility, coherence, experiment and analogy [157]). The strongest evidence in support of a causal relationship between pesticide exposure and childhood cancer was considered to be the repeated detection of statistically significant increased risks between childhood pesticide exposure and cancer. The authors concluded that there is sufficient evidence to conclude that there is at least some association between pesticide exposure and childhood cancer. In addition, the biological gradients observed in recent studies also suggest that there may be a causal relationship between childhood insecticide exposures and the development of ALL and NHL [158, 159]. Infante-Rivard [152] indicated that the development of childhood cancer probably depends on the presence of many factors, including genetic predisposition, and recommended the use of improved exposure assessments that include separate parental interview, specific pesticide exposure questions, and semiquantitative exposure measures that can be used to confirm exposure information obtained through questionnaires.

Disinfection By-products and Cancer

Disinfection by-products (DBPs), such as trihalomethanes (THMs), are regulated carcinogens in drinking water and have been detected in the blood and breath of swimmers and of nonswimmers at indoor pools. There are a few data on the effects of low doses on humans, particularly infants and children.

Villanueva et al. [160] conducted a pooled analysis of six epidemiological studies and calculated a summary relative risk of bladder cancer equal to 1.18 (95% CI 1.06, 1.32) for exposure above 1 μg/L of trihalomethanes. Boffetta [161] estimated the attributable fraction of bladder cancer on the basis of these figures to be about 10.3%.

Arsenic

Inorganic arsenic is a human carcinogen [162] which causes bladder, skin, and lung cancers in humans. At least in mice, inorganic arsenic is a much more potent carcinogen to the fetus than to adults. The modes of carcinogenic action of inorganic arsenic in rodents and in humans are not yet fully understood, but the possibility exists that inorganic arsenic in drinking water poses a special carcinogenic concern for pregnant women and their unborn infants.

Air Pollution and Childhood Cancer

There have been several studies during the last decade that have linked children's exposure to air pollutants to childhood cancer, leukemia in particular. Infante-Rivard [163] evaluated the results of epidemiological studies conducted between

1998 and 2008 on diverse chemical exposures and childhood leukemia. The review included nine case-control studies and four ecological studies. The latter showed an association between incidence rates of childhood leukemia and levels of air pollution in the area of residence of the population at the time of diagnosis. The lack of information about the place of residence of subjects prior to diagnosis (including during the preconception and gestation period) made it difficult to conclude an association between the exposure levels and risk of childhood leukemia. The review concluded that the weight of evidence from both case-control and ecological studies indicated no increased risk for childhood leukemia associated with exposure to traffic-related residential air pollution. The same conclusion was reached in another review conducted by Raaschou-Nielsen [164].

Endocrine Disruptors and Cancer

Increases in the incidence of cancer in certain parts of the world are often cited as evidence that widespread exposure to endocrine disruptors has adverse effects on human health. Of particular concern are the observed increased incidences of cancer at hormonally sensitive sites, such as breast, uterus, prostate, and testis in Europe and North America. These increases cannot be explained only on the basis of improved diagnostic techniques, and it has been argued that these trends coincide roughly with the increasing use and release of industrial chemicals into the environment. Concerns are also based on plausible mechanisms of action because both human and experimental animal studies show that these cancers are modulated hormonally.

Many epidemiological and experimental animal studies have attempted to evaluate the link between exposure to endocrine disruptors and increased risk of breast cancer. A direct association between these chemicals and increased risk of breast cancer has not been established. However, some researchers claim that the time of life when exposure takes place (e.g., prenatal, neonatal, childhood, adolescence) may have a major influence on the appearance of this and other health effects. The development of the mammary gland occurs in multiple stages. Fetal development of the mammary gland rudiment is governed by tissue interactions in both males and females. In females, the pubertal period drives ductal morphogenesis, and pregnancy results in massive differentiation of the mammary gland. Thus, the perinatal period and the period between age at menarche and age at first full-term pregnancy may be particularly important for breast tumor development and latency [165]. Young girls exposed to carcinogenic agents during puberty may be at high risk of future breast cancer due to susceptibility of rapidly growing breast tissue mediated by hormonal changes during this time. This claim is supported by data from atomic bomb survivors, where an increased risk of breast cancer was found in women exposed before 20 years of age [132]. Similarly, an elevated risk was found for women irradiated during childhood for medical reasons [166].

DES was extensively prescribed in developed countries from the late 1940s through the 1970s to women with high-risk pregnancies to prevent miscarriages and other complications of pregnancy. In the early 1970s, a rare form of female reproductive tract cancer, clear cell adenocarcinoma began to be detected among women whose mothers had taken DES during pregnancy [167]. Although clear cell adenocarcinoma occurs in only 0.1% of women who were exposed to DES in utero, this represents a 40-fold increased risk in comparison with the nonexposed population. In contrast, men who were exposed to DES in utero do not have a clear increased risk of any cancer, although a statistically nonsignificant threefold increased risk of testicular cancer has been reported [168]. While there is no epidemiological evidence for a link between exposure to estrogenic or anti-androgenic compounds and testicular cancer, some authors have hypothesized that the similar increases in the incidence of testicular cancer and in the incidence of cryptorchidism and hypospadias in similar geographical areas, suggest a common cause of similar environmental origin in both health effects [85].

Some studies have shown that women taking DES during pregnancy to prevent miscarriage have been shown to have a slightly increased risk of developing breast cancer 30 years after taking the drug [169]. Data on the risks of breast cancer in daughters of DES-exposed women are not yet available.

Future Directions

During the last 2 decades, there has been an exponential increase of scientific literature about the susceptibility of children to the effects of environmental exposures. Evidence is accumulating that toxic chemicals are responsible for at least some of the changing patterns of disease. Well-studied examples include adenocarcinoma of the vagina in girls exposed prenatally to DES; asthma and pneumonia caused by smoke and particulate air pollutants; neurodevelopmental toxicity in infants exposed to lead, PCBs, methylmercury; and small head circumference at birth in infants exposed in utero to organophosphate pesticides.

Past discoveries of etiologic associations between toxic environmental exposures and diseases in children have led to successful programs of exposure control and disease prevention. Examples include reductions in the use of alcohol and tobacco during pregnancy, minimization during pregnancy of diagnostic x-rays, and removal of lead from gasoline. Sadly, though, the interval between initial recognition or suspicion of effects and their eventual control has been typically been far too long. Early warnings have frequently been ignored. The price of delayed action has been widespread increases in the incidence of certain diseases such as asthma and cancer.

When evaluating the health and social impact of environmental threats on children, it is necessary to take into account, not only their effects on childhood health, but also the long-term potential health effects throughout the lifelong span of the individual. It has been hypothesized that early exposure to environmental toxicants could affect the brain later in life. Consensus among scientists is based on experimental studies on associations between early life exposures to pesticides and Parkinson's disease, as well as on epidemiologic studies of the toxic and apparently irreversible effects on the developing brain of in utero exposure to lead, methylmercury, and polychlorinated biphenyls. A mechanistic hypothesis proposed that early exposure to neurotoxic chemicals reduce the number of neurons in critical areas of the brain such as the substantia nigra to levels below those needed to sustain function in the face of neuronal attrition associated with advancing age. In addition, as some researchers have pointed out, the effects at population level would add a substantial burden to society, in terms of health, economic, and human costs.

The difficulties in conducting environmental health research have been pointed out by many authors during the last 3 decades. Human populations are exposed to hundreds of chemicals through air, water, and food, under many different exposure situations. Environmental exposures are usually low, and often occur concomitantly with occupational exposures, smoking or naturally occurring agents. And when an association between with an exposure and a disease in found, such as for example, air pollution and asthma, a long path lies ahead to determine the specific causative agent of the disease, and the population and individual susceptibility factors that makes some individuals and not others become ill. The potential for interaction of different pollutant exposures makes the question with respect to the role of environmental exposures impact on health even more difficult to answer. The need for taking into account multiple and cumulative exposures has been clearly affirmed by the EPA and other government and research bodies.

Slowly, evidence is found to confirm or discard certain effects, but it seems that progress is too slow to catch up to the increases in incidence of certain diseases, such as asthma and cancer. It is possible that the statement of Olshan and Daniels [170] with respect to pesticide research, that "another 40 epidemiological studies similar to the majority of those conducted thus far will not provide clarity," should be applied to other groups of chemicals, such as disinfection by-products, or endocrine disruptors, and that we need to think of a more effective way of examining evidence. Many authors have pointed out the need of better exposure assessment methods. In addition to continuing toxicological research, which is essential to elucidate the mechanisms of action, epidemiological research should probably incorporate exposure assessment methods that could enable detection of at least the specific exposures and exposure levels. Exposure assessment can be improved in many ways. In the extreme, if the body burden of a list of suspected carcinogens in the blood or adipose tissue of each child diagnosed with cancer were determined, more specific associations between disease and toxic agent may be determined. The path to confirm and rule out some proposed environmental causes would probably be at least somewhat shorter. While this approach may be costly, it may be cost-efficient in the long run. There have been some efforts to integrate an environmental health perspective in the medical practice. These efforts should be further developed and supported.

Bibliography

Primary Literature

1. Tamburlini G, von Ehrenstein OS, Bertollini R (eds) (2002) Children's health and environment: A review of evidence. A joint report from the European Environment Agency and the WHO Regional Office for Europe. Environmental issue report No. 29. Office for Official Publications of the European Communities, Luxembourg
2. NRC (1993) Pesticides in the diets of infants and children. US National Research Council, National Academy Press, Washington, DC
3. ILSI (1992) Similarities and differences between children and adults: implications for risk assessment. International Life Sciences Institute Press, Washington, DC
4. ILSI (2003) Final report: workshop to develop a framework for assessing risks to children from exposure to environmental agents. International Life Sciences Institute Press, Washington, DC
5. WHO (2006) Principles for evaluating health risks in children. World Health Organization, Geneva
6. Moya J, Bearer CF, Etzel RA (2004) Children's behavior and physiology and how it affects exposure to environmental contaminants. Pediatrics 113(4 Suppl):996–1006
7. Landrigan PJ (2004) Children as a vulnerable population. Int J Occup Med Environ Health 17(1):175–177
8. Calabrese EJ, Stanek EJ, James RC, Roberts SM (1997) Soil ingestion: a concern for acute toxicity in children. Environ Health Perspect 105(12):1354–1358
9. Simcox NJ, Fenske RA, Wolz SA, Lee IC, Kalman DA (1995) Pesticides in household dust and soil: exposure pathways for children of agricultural families. Environ Health Perspect 103(12):1126–1134
10. Clewell HJ, Teeguarden J, McDonald T, Sarangapani R, Lawrence G, Covington T et al (2002) Review and evaluation of the potential impact of age- and gender-specific pharmacokinetic differences on tissue dosimetry. Crit Rev Toxicol 32(5):329–389
11. Anderson LM, Diwan BA, Fear NT, Roman E (2000) Critical windows of exposure for children's health: cancer in human epidemiological studies and neoplasms in experimental animal models. Environ Health Perspect 108(Suppl 3):573–594
12. Loeffler IK, Peterson RE (1999) Interactive effects of TCDD and p, p′-DDE on male reproductive tract development in in utero and lactationally exposed rats. Toxicol Appl Pharmacol 154(1):28–39
13. Weinberg CR, Wilcox AJ, Baird DD (1989) Reduced fecundability in women with prenatal exposure to cigarette smoking. Am J Epidemiol 129(5):1072–1078
14. Wilcox AJ, Baird DD, Weinberg CR, Hornsby PP, Herbst AL (1995) Fertility in men exposed prenatally to diethylstilbestrol. N Engl J Med 332(21):1411–1416
15. Buckley JD, Robison LL, Swotinsky R, Garabrant DH, LeBeau M, Manchester P et al (1989) Occupational exposures of parents of children with acute nonlymphocytic leukemia: a report from the Children's Cancer Study Group. Cancer Res 49(14):4030–4037
16. Nicol CJ, Harrison ML, Laposa RR, Gimelshtein IL, Wells PG (1995) A teratologic suppressor role for p53 in benzo[a]pyrene-treated transgenic p53-deficient mice. Nat Genet 10(2):181–187
17. Mullikin-Kilpatrick D, Mehta ND, Hildebrandt JD, Treistman SN (1995) Gi is involved in ethanol inhibition of L-type calcium channels in undifferentiated but not differentiated PC-12 cells. Mol Pharmacol 47(5):997–1005
18. Murante FG, Gasiewicz TA (2000) Hemopoietic progenitor cells are sensitive targets of 2, 3, 7, 8-tetrachlorodibenzo-p-dioxin in C57BL/6J mice. Toxicol Sci 54(2):374–383
19. Amsen D, Kruisbeek AM (1998) Thymocyte selection: not by TCR alone. Immunol Rev 165:209–229

20. Rodier PM (2004) Environmental causes of central nervous system maldevelopment. Pediatrics 113(4 Suppl):1076–1083
21. Sharpe CR, Franco EL (1995) Etiology of Wilms' tumor. Epidemiol Rev 17(2):415–432
22. Bunney BG, Potkin SG, Bunney WE Jr (1995) New morphological and neuropathological findings in schizophrenia: a neurodevelopmental perspective. Clin Neurosci 3(2):81–88
23. Rodier PM (1995) Developing brain as a target of toxicity. Environ Health Perspect 103(Suppl 6):73–76
24. Kharrazi M, DeLorenze GN, Kaufman FL, Eskenazi B, Bernert JT Jr, Graham S et al (2004) Environmental tobacco smoke and pregnancy outcome. Epidemiology 15(6):660–670
25. Bove F, Shim Y, Zeitz P (2002) Drinking water contaminants and adverse pregnancy outcomes: a review. Environ Health Perspect 110(Suppl 1):61–74
26. Graves CG, Matanoski GM, Tardiff RG (2001) Weight of evidence for an association between adverse reproductive and developmental effects and exposure to disinfection by-products: a critical review. Regul Toxicol Pharmacol 34(2):103–124
27. Nieuwenhuijsen MJ, Toledano MB, Eaton NE, Fawell J, Elliott P (2000) Chlorination disinfection byproducts in water and their association with adverse reproductive outcomes: a review. Occup Environ Med 57(2):73–85
28. Platt RW, Joseph KS, Ananth CV, Grondines J, Abrahamowicz M, Kramer MS (2004) A proportional hazards model with time-dependent covariates and time-varying effects for analysis of fetal and infant death. Am J Epidemiol 160(3):199–206
29. Paus T, Zijdenbos A, Worsley K, Collins DL, Blumenthal J, Giedd JN et al (1999) Structural maturation of neural pathways in children and adolescents: in vivo study. Science 283(5409): 1908–1911
30. Dietert RR, Lee JE, Bunn TL (2002) Developmental immunotoxicology: emerging issues. Hum Exp Toxicol 21(9–10):479–485
31. International Agency for Research on Cancer (2001) IARC monographs on the evaluation of carcinogenic risks to humans: ionizing radiation, Part 2: some internally deposited radionuclides, vol 78. International Agency for Research on Cancer, Lyon, France
32. Altshuler K (2003) Critical periods in development. EPA . Paper Series on Children's Health and the Environment
33. Mahaffey KR (1991) Biokinetics of lead during pregnancy. Fundam Appl Toxicol 16(1):15–16
34. American Academy of Pediatrics Committee on Environmental Health (2003) Pediatric environmental health, 2nd edn. American Academy of Pediatrics, Elk Grove Village
35. Perera FP, Jedrychowski W, Rauh V, Whyatt RM (1999) Molecular epidemiologic research on the effects of environmental pollutants on the fetus. Environ Health Perspect 107(Suppl 3): 451–460
36. Rice D, Barone S Jr (2000) Critical periods of vulnerability for the developing nervous system: evidence from humans and animal models. Environ Health Perspect 108(Suppl 3):511–533
37. Casey BJ, Tottenham N, Liston C, Durston S (2005) Imaging the developing brain: what have we learned about cognitive development? Trends Cogn Sci 9(3):104–110
38. Cory-Slechta DA, Thiruchelvam M, Barlow BK, Richfield EK (2005) Developmental pesticide models of the Parkinson disease phenotype. Environ Health Perspect 113(9):1263–1270
39. Landrigan PJ, Whitworth RH, Baloh RW, Staehling NW, Barthel WF, Rosenblum BF (1975) Neuropsychological dysfunction in children with chronic low-level lead absorption. Lancet 1(7909):708–712
40. Needleman HL (1979) Lead levels and children's psychologic performance. N Engl J Med 301(3):163
41. Bellinger DC (2004) Lead. Pediatrics 113(4 Suppl):1016–1022
42. Lanphear BP, Hornung R, Khoury J, Yolton K, Baghurst P, Bellinger DC et al (2005) Low-level environmental lead exposure and children's intellectual function: an international pooled analysis. Environ Health Perspect 113(7):894–899

43. Rothenberg SJ, Poblano A, Schnaas L (2000) Brainstem auditory evoked response at five years and prenatal and postnatal blood lead. Neurotoxicol Teratol 22(4):503–510
44. Bellinger D, Hu H, Titlebaum L, Needleman HL (1994) Attentional correlates of dentin and bone lead levels in adolescents. Arch Environ Health 49(2):98–105
45. Dietrich KN, Ris MD, Succop PA, Berger OG, Bornschein RL (2001) Early exposure to lead and juvenile delinquency. Neurotoxicol Teratol 23(6):511–518
46. Grandjean P, Weihe P, White RF, Debes F, Araki S, Yokoyama K et al (1997) Cognitive deficit in 7-year-old children with prenatal exposure to methylmercury. Neurotoxicol Teratol 19(6):417–428
47. Grandjean P, White RF, Nielsen A, Cleary D, Oliveira Santos EC (1999) Methylmercury neurotoxicity in Amazonian children downstream from gold mining. Environ Health Perspect 107(7):587–591
48. Davidson PW, Myers GJ, Weiss B (2004) Mercury exposure and child development outcomes. Pediatrics 113(4 Suppl):1023–1029
49. Murata K, Weihe P, Budtz-Jorgensen E, Jorgensen PJ, Grandjean P (2004) Delayed brainstem auditory evoked potential latencies in 14-year-old children exposed to methylmercury. J Pediatr 144(2):177–183
50. Newland MC, Paletz EM (2000) Animal studies of methylmercury and PCBs: what do they tell us about expected effects in humans? Neurotoxicology 21(6):1003–1027
51. National Academy of Sciences CotTEM (2000) Toxicological effects of methylmercury. National Academy Press, Washington, DC
52. Brouwer A, Longnecker MP, Birnbaum LS, Cogliano J, Kostyniak P, Moore J et al (1999) Characterization of potential endocrine-related health effects at low-dose levels of exposure to PCBs. Environ Health Perspect 107(Suppl 4):639–649
53. Rogan WJ, Gladen BC, Hung KL, Koong SL, Shih LY, Taylor JS et al (1988) Congenital poisoning by polychlorinated biphenyls and their contaminants in Taiwan. Science 241(4863):334–336
54. Guo YL, Lambert GH, Hsu CC, Hsu MM (2004) Yucheng: health effects of prenatal exposure to polychlorinated biphenyls and dibenzofurans. Int Arch Occup Environ Health 77(3):153–158
55. Chen YC, Yu ML, Rogan WJ, Gladen BC, Hsu CC (1994) A 6-year follow-up of behavior and activity disorders in the Taiwan Yu-Cheng children. Am J Public Health 84(3):415–421
56. Ribas-Fito N, Sala M, Kogevinas M, Sunyer J (2001) Polychlorinated biphenyls (PCBs) and neurological development in children: a systematic review. J Epidemiol Community Health 55(8):537–546
57. Grandjean P, Weihe P, Burse VW, Needham LL, Storr-Hansen E, Heinzow B et al (2001) Neurobehavioral deficits associated with PCB in 7-year-old children prenatally exposed to seafood neurotoxicants. Neurotoxicol Teratol 23(4):305–317
58. Guillette EA, Meza MM, Aquilar MG, Soto AD, Garcia IE (1998) An anthropological approach to the evaluation of preschool children exposed to pesticides in Mexico. Environ Health Perspect 106(6):347–353
59. Moser VC, Chanda SM, Mortensen SR, Padilla S (1998) Age- and gender-related differences in sensitivity to chlorpyrifos in the rat reflect developmental profiles of esterase activities. Toxicol Sci 46(2):211–222
60. Slotkin TA, Tate CA, Cousins MM, Seidler FJ (2002) Functional alterations in CNS catecholamine systems in adolescence and adulthood after neonatal chlorpyrifos exposure. Brain Res Dev Brain Res 133(2):163–173
61. EPA (1997) Background document on cholinesterates. Attachment 4E presented to Environmental Protection Agency Scientific Advisory Panel Meeting of June 3–4, 1997. US EPA, Arlington
62. Weiss B, Amler S, Amler WR (2004) Pesticides. Pediatrics 113: 1030–1036
63. Damstra T (2002) Potential effects of certain persistent organic pollutants and endocrine disrupting chemicals on the health of children. J Toxicol Clin Toxicol 40(4):457–465

64. Ho SM, Tang WY, Belmonte DF, Prins GS (2006) Developmental exposure to estradiol and bisphenol A increases susceptibility to prostate carcinogenesis and epigenetically regulates phosphodiesterase type 4 variant 4. Cancer Res 66(11):5624–5632
65. Newbold RR, Hanson RB, Jefferson WN, Bullock BC, Haseman J, McLachlan JA (2000) Proliferative lesions and reproductive tract tumors in male descendants of mice exposed developmentally to diethylstilbestrol. Carcinogenesis 21(7):1355–1363
66. Anway MD, Cupp AS, Uzumcu M, Skinner MK (2005) Epigenetic transgenerational actions of endocrine disruptors and male fertility. Science 308(5727):1466–1469
67. Goldey ES, Crofton KM (1998) Thyroxine replacement attenuates hypothyroxinemia, hearing loss, and motor deficits following developmental exposure to Aroclor 1254 in rats. Toxicol Sci 45(1):94–105
68. Brechner RJ, Parkhurst GD, Humble WO, Brown MB, Herman WH (2000) Ammonium perchlorate contamination of Colorado River drinking water is associated with abnormal thyroid function in newborns in Arizona. J Occup Environ Med 42(8):777–782
69. Kelsh MA, Buffler PA, Daaboul JJ, Rutherford GW, Lau EC, Barnard JC et al (2003) Primary congenital hypothyroidism, newborn thyroid function, and environmental perchlorate exposure among residents of a Southern California community. J Occup Environ Med 45(10):1116–1127
70. Lau C, Rogers JM (2004) Embryonic and fetal programming of physiological disorders in adulthood. Birth Defects Res C Embryo Today 72(4):300–312
71. Sadler TW (2000) Susceptible periods during embryogenesis of the heart and endocrine glands. Environ Health Perspect 108(Suppl 3):555–561
72. Ravelli AC, van der Meulen JH, Michels RP, Osmond C, Barker DJ, Hales CN et al (1998) Glucose tolerance in adults after prenatal exposure to famine. Lancet 351(9097): 173–177
73. Bo S, Cavallo-Perin P, Scaglione L, Ciccone G, Pagano G (2000) Low birthweight and metabolic abnormalities in twins with increased susceptibility to Type 2 diabetes mellitus. Diabet Med 17(5):365–370
74. Pryor JL, Hughes C, Foster W, Hales BF, Robaire B (2000) Critical windows of exposure for children's health: the reproductive system in animals and humans. Environ Health Perspect 108(Suppl 3):491–503
75. Stillman RJ (1982) In utero exposure to diethylstilbestrol: adverse effects on the reproductive tract and reproductive performance and male and female offspring. Am J Obstet Gynecol 142(7):905–921
76. Parente RC, Faerstein E, Celeste RK, Werneck GL (2008) The relationship between smoking and age at the menopause: a systematic review. Maturitas 61(4):287–298
77. Jensen TK, Joffe M, Scheike T, Skytthe A, Gaist D, Christensen K (2005) Time trends in waiting time to pregnancy among Danish twins. Hum Reprod 20(4):955–964
78. Shea KM (2003) Pediatric exposure and potential toxicity of phthalate plasticizers. Pediatrics 111(6 Pt 1):1467–1474
79. Weisbach V, Koch HM, Angerer J, Eckstein R (2006) Di(2-ethylhexyl)phthalate exposure of apheresis donors is procedure-related. Transfusion 46(8):1457–1458
80. Skakkebaek NE, Jorgensen N, Main KM, Rajpert-De Meyts E, Leffers H, Andersson AM et al (2006) Is human fecundity declining? Int J Androl 29(1):2–11
81. Lottrup G, Andersson AM, Leffers H, Mortensen GK, Toppari J, Skakkebaek NE et al (2006) Possible impact of phthalates on infant reproductive health. Int J Androl 29(1): 172–180
82. Mylchreest E, Wallace DG, Cattley RC, Foster PM (2000) Dose-dependent alterations in androgen-regulated male reproductive development in rats exposed to Di(n-butyl) phthalate during late gestation. Toxicol Sci 55(1):143–151
83. Hoei-Hansen CE, Holm M, Rajpert-De Meyts E, Skakkebaek NE (2003) Histological evidence of testicular dysgenesis in contralateral biopsies from 218 patients with testicular germ cell cancer. J Pathol 200(3):370–374

84. Skakkebaek NE, Holm M, Hoei-Hansen C, Jorgensen N, Rajpert-De Meyts E (2003) Association between testicular dysgenesis syndrome (TDS) and testicular neoplasia: evidence from 20 adult patients with signs of maldevelopment of the testis. APMIS 111(1):1–9
85. Fisher JS, Macpherson S, Marchetti N, Sharpe RM (2003) Human 'testicular dysgenesis syndrome': a possible model using in-utero exposure of the rat to dibutyl phthalate. Hum Reprod 18(7):1383–1394
86. Weinhold B (2009) Environmental factors in birth defects: what we need to know. Environ Health Perspect 117:A440–A447
87. Lindbohm ML, Taskinen H, Sallmen M, Hemminki K (1990) Spontaneous abortions among women exposed to organic solvents. Am J Ind Med 17(4):449–463
88. Lipscomb JA, Fenster L, Wrensch M, Shusterman D, Swan S (1991) Pregnancy outcomes in women potentially exposed to occupational solvents and women working in the electronics industry. J Occup Med 33(5):597–604
89. Nieuwenhuijsen MJ, Toledano MB, Elliott P (2000) Uptake of chlorination disinfection by-products: a review and a discussion of its implications for exposure assessment in epidemiological studies. J Expo Anal Environ Epidemiol 10(6 Pt 1):586–599
90. Dodds L, King W, Allen AC, Armson BA, Fell DB, Nimrod C (2004) Trihalomethanes in public water supplies and risk of stillbirth. Epidemiology 15(2):179–186
91. Toledano MB, Nieuwenhuijsen MJ, Best N, Whitaker H, Hambly P, de Hoogh C et al (2005) Relation of trihalomethane concentrations in public water supplies to stillbirth and birth weight in three water regions in England. Environ Health Perspect 113(2):225–232
92. Wright JM, Schwartz J, Dockery DW (2004) The effect of disinfection by-products and mutagenic activity on birth weight and gestational duration. Environ Health Perspect 112(8):920–925
93. Hwang BF, Jaakkola JJ (2003) Water chlorination and birth defects: a systematic review and meta-analysis. Arch Environ Health 58(2):83–91
94. Nieuwenhuijsen MJ, Martinez D, Grellier J, Bennett J, Best N, Iszatt N et al (2009) Chlorination disinfection by-products in drinking water and congenital anomalies: review and meta-analyses. Environ Health Perspect 117(10):1486–1493
95. California Environmental Protection Agency (2010) Proposed identification of environmental tobacco smoke as a toxic air contaminant. California Environmental Protection Agency, Sacramento
96. Wigle DT, Arbuckle TE, Walker M, Wade MG, Liu S, Krewski D (2007) Environmental hazards: evidence for effects on child health. J Toxicol Environ Health B Crit Rev 10(1–2):3–39
97. DiFranza JR, Lew RA (1995) Effect of maternal cigarette smoking on pregnancy complications and sudden infant death syndrome. J Fam Pract 40(4):385–394
98. Charlton A (1994) Children and passive smoking: a review. J Fam Pract 38(3):267–277
99. Wang J, Wang B (2002) Study on risk factors of cryptorchidism. Zhonghua Liu Xing Bing Xue Za Zhi 23(3):190–193
100. Waliszewski SM, Infanzon RM, Arroyo SG, Pietrini RV, Carvajal O, Trujillo P et al (2005) Persistent organochlorine pesticides levels in blood serum lipids in women bearing babies with undescended testis. Bull Environ Contam Toxicol 75(5):952–959
101. Longnecker MP, Klebanoff MA, Dunson DB, Guo X, Chen Z, Zhou H et al (2005) Maternal serum level of the DDT metabolite DDE in relation to fetal loss in previous pregnancies. Environ Res 97(2):127–133
102. Gray LE, Ostby J, Furr J, Wolf CJ, Lambright C, Parks L et al (2001) Effects of environmental antiandrogens on reproductive development in experimental animals. Hum Reprod Update 7(3):248–264
103. Woodruff TJ, Zeise L, Axelrad DA, Guyton KZ, Janssen S, Miller M et al (2008) Meeting report: moving upstream-evaluating adverse upstream end points for improved risk assessment and decision-making. Environ Health Perspect 116(11):1568–1575
104. DiFranza JR, Aligne CA, Weitzman M (2004) Prenatal and postnatal environmental tobacco smoke exposure and children's health. Pediatrics 113(4 Suppl):1007–1015
105. Dietert RR, Etzel RA, Chen D, Halonen M, Holladay SD, Jarabek AM et al (2000) Workshop to identify critical windows of exposure for children's health: immune and respiratory systems work group summary. Environ Health Perspect 108(Suppl 3):483–490

106. Stick SM, Burton PR, Gurrin L, Sly PD, LeSouef PN (1996) Effects of maternal smoking during pregnancy and a family history of asthma on respiratory function in newborn infants. Lancet 348(9034):1060–1064
107. Upton MN, Watt GC, Davey SG, McConnachie A, Hart CL (1998) Permanent effects of maternal smoking on offsprings' lung function. Lancet 352(9126):453
108. Sporik R, Platts-Mills TA (2001) Allergen exposure and the development of asthma. Thorax 56(Suppl 2):ii58–ii63
109. Hesselmar B, Aberg N, Aberg B, Eriksson B, Bjorksten B (1999) Does early exposure to cat or dog protect against later allergy development? Clin Exp Allergy 29(5):611–617
110. Merchant JA, Naleway AL, Svendsen ER, Kelly KM, Burmeister LF, Stromquist AM et al (2005) Asthma and farm exposures in a cohort of rural Iowa children. Environ Health Perspect 113(3):350–356
111. Schwela D (2000) Air pollution and health in urban areas. Rev Environ Health 15(1–2):13–42
112. Heinrich J (2003) Nonallergic respiratory morbidity improved along with a decline of traditional air pollution levels: a review. Eur Respir J Suppl 40:64s–69s
113. Avol EL, Gauderman WJ, Tan SM, London SJ, Peters JM (2001) Respiratory effects of relocating to areas of differing air pollution level. Am J Respir Crit Care Med 164(11): 2067–2072
114. Schwartz J (2004) Air pollution and children's health. Pediatrics 113(4 Suppl):1037–1043
115. Ostro B, Lipsett M, Mann J, Braxton-Owens H, White M (2001) Air pollution and exacerbation of asthma in African-American children in Los Angeles. Epidemiology 12(2):200–208
116. McConnell R, Berhane K, Gilliland F, Molitor J, Thomas D, Lurmann F et al (2003) Prospective study of air pollution and bronchitic symptoms in children with asthma. Am J Respir Crit Care Med 168(7):790–797
117. Gilliland FD, Berhane K, Rappaport EB, Thomas DC, Avol E, Gauderman WJ et al (2001) The effects of ambient air pollution on school absenteeism due to respiratory illnesses. Epidemiology 12(1):43–54
118. Tolbert PE, Mulholland JA, MacIntosh DL, Xu F, Daniels D, Devine OJ et al (2000) Air quality and pediatric emergency room visits for asthma in Atlanta, Georgia, USA. Am J Epidemiol 151(8):798–810
119. Studnicka M, Hackl E, Pischinger J, Fangmeyer C, Haschke N, Kuhr J et al (1997) Traffic-related NO_2 and the prevalence of asthma and respiratory symptoms in seven year olds. Eur Respir J 10(10):2275–2278
120. Kramer U, Koch T, Ranft U, Ring J, Behrendt H (2000) Traffic-related air pollution is associated with atopy in children living in urban areas. Epidemiology 11(1):64–70
121. Weinmayr G, Genuneit J, Nagel G, Bjorksten B, van Hage M, Priftanji A et al (2010) International variations in associations of allergic markers and diseases in children: ISAAC phase two. Allergy 65(6):766–775
122. Lin S, Munsie JP, Hwang SA, Fitzgerald E, Cayo MR (2002) Childhood asthma hospitalization and residential exposure to state route traffic. Environ Res 88(2):73–81
123. English P, Neutra R, Scalf R, Sullivan M, Waller L, Zhu L (1999) Examining associations between childhood asthma and traffic flow using a geographic information system. Environ Health Perspect 107(9):761–767
124. Vedal S, Petkau J, White R, Blair J (1998) Acute effects of ambient inhalable particles in asthmatic and nonasthmatic children. Am J Respir Crit Care Med 157(4 Pt 1):1034–1043
125. Pekkanen J, Pearce N (1999) Defining asthma in epidemiological studies. Eur Respir J 14(4):951–957
126. Gauderman WJ, Gilliland GF, Vora H, Avol E, Stram D, McConnell R et al (2002) Association between air pollution and lung function growth in southern California children: results from a second cohort. Am J Respir Crit Care Med 166(1):76–84
127. American Academy of Pediatrics Committee on Environmental Health (2003) Air pollutants, outdoor. In: Etzel R (ed) Pediatric environmental health. American Academy of Pediatrics, Elk Grove Village, pp 69–86

128. Perera FP (1997) Environment and cancer: who are susceptible? Science 278(5340):1068–1073
129. Bartsch H, Hietanen E (1996) The role of individual susceptibility in cancer burden related to environmental exposure. Environ Health Perspect 104(Suppl 3):569–577
130. Husgafvel-Pursiainen K (2004) Genotoxicity of environmental tobacco smoke: a review. Mutat Res 567(2–3):427–445
131. National Research Council (1982) Diet, nutrition and cancer. National Academy Press, Washington, DC
132. Tokunaga M, Land CE, Tokuoka S, Nishimori I, Soda M, Akiba S (1994) Incidence of female breast cancer among atomic bomb survivors, 1950–1985. Radiat Res 138(2):209–223
133. Schiestl RH, Aubrecht J, Yap WY, Kandikonda S, Sidhom S (1997) Polychlorinated biphenyls and 2, 3, 7, 8-tetrachlorodibenzo-p-dioxin induce intrachromosomal recombination in vitro and in vivo. Cancer Res 57(19):4378–4383
134. Trizio D, Basketter DA, Botham PA, Graepel PH, Lambre C, Magda SJ et al (1988) Identification of immunotoxic effects of chemicals and assessment of their relevance to man. Food Chem Toxicol 26(6):527–539
135. Anderson LM (2006) Environmental genotoxicants/carcinogens and childhood cancer: bridgeable gaps in scientific knowledge. Mutat Res 608(2):136–156
136. Maciag A, Bialkowska A, Espiritu I, Powell D, Alvord WG, Kasprzak KS et al (2003) Gestation stage-specific oxidative deoxyribonucleic acid damage from sidestream smoke in pregnant rats and their fetuses. Arch Environ Health 58(4):238–244
137. Schuz J, Kaletsch U, Kaatsch P, Meinert R, Michaelis J (2001) Risk factors for pediatric tumors of the central nervous system: results from a German population-based case-control study. Med Pediatr Oncol 36(2):274–282
138. Bunin GR (2004) Nongenetic causes of childhood cancers: evidence from international variation, time trends, and risk factor studies. Toxicol Appl Pharmacol 199(2):91–103
139. Ron E, Modan B, Boice JD Jr, Alfandary E, Stovall M, Chetrit A et al (1988) Tumors of the brain and nervous system after radiotherapy in childhood. N Engl J Med 319(16):1033–1039
140. Pacini F, Vorontsova T, Demidchik EP, Molinaro E, Agate L, Romei C et al (1997) Post-chernobyl thyroid carcinoma in Belarus children and adolescents: comparison with naturally occurring thyroid carcinoma in Italy and France. J Clin Endocrinol Metab 82(11):3563–3569
141. Whiteman DC, Whiteman CA, Green AC (2001) Childhood sun exposure as a risk factor for melanoma: a systematic review of epidemiologic studies. Cancer Causes Control 12(1): 69–82
142. International Agency for Research on Cancer (1992) IARC monographs on the evaluation of carcinogenic risks to humans: solar and ultraviolet radiation, vol 55. International Agency for Research on Cancer, Lyon, France
143. Vainio H, Bianchini F (2000) Cancer-preventive effects of sunscreens are uncertain. Scand J Work Environ Health 26(6):529–531
144. International Agency for Research on Cancer (2004) IARC monographs on the evaluation of carcinogenic risks to humans: involuntary smoke, vol 83. International Agency for Research on Cancer, Lyon, France
145. Karube T, Odagiri Y, Takemoto K, Watanabe S (1989) Analyses of transplacentally induced sister chromatid exchanges and micronuclei in mouse fetal liver cells following maternal exposure to cigarette smoke. Cancer Res 49(13):3550–3552
146. Wyrobek AJ, Adler ID (1996) Detection of aneuploidy in human and rodent sperm using FISH and applications of sperm assays of genetic damage in heritable risk evaluation. Mutat Res 352(1–2):173–179
147. Pang D, McNally R, Birch JM (2003) Parental smoking and childhood cancer: results from the United Kingdom Childhood Cancer Study. Br J Cancer 88(3):373–381
148. Sorahan T, Lancashire RJ (2004) Parental cigarette smoking and childhood risks of hepatoblastoma: OSCC data. Br J Cancer 90(5):1016–1018

149. Ji BT, Shu XO, Linet MS, Zheng W, Wacholder S, Gao YT et al (1997) Paternal cigarette smoking and the risk of childhood cancer among offspring of nonsmoking mothers. J Natl Cancer Inst 89(3):238–244
150. Sandler DP, Everson RB, Wilcox AJ, Browder JP (1985) Cancer risk in adulthood from early life exposure to parents' smoking. Am J Public Health 75(5):487–492
151. Abdollahi M, Ranjbar A, Shadnia S, Nikfar S, Rezaie A, Abdollahi M, Ranjbar A, Shadnia S, Nikfar S, Rezaie A (2004) Pesticides and oxidative stress: a review. Med Sci Monit 10(6): RA141–RA147
152. Infante-Rivard C, Weichenthal S (2007) Pesticides and childhood cancer: an update of Zahm and Ward's 1998 review. J Toxicol Environ Health B Crit Rev 10(1–2):81–99
153. Daniels JL, Olshan AF, Teschke K, Hertz-Picciotto I, Savitz DA, Blatt J et al (2001) Residential pesticide exposure and neuroblastoma. Epidemiology 12(1):20–27
154. Zahm SH, Ward MH (1998) Pesticides and childhood cancer. Environ Health Perspect 106(Suppl 3):893–908
155. Wigle DT, Turner MC, Krewski D, Shannon M, Graef JW (2009) A systematic review and meta-analysis of childhood leukemia and parental occupational pesticide exposure. Lead intoxication from lead-contaminated water used to reconstitute infant formula. Environ Health Perspect 117(10):1505–1513
156. Nasterlack M (2006) Do pesticides cause childhood cancer? Int Arch Occup Environ Health 79(7):536–544
157. Hill AB (1965) The environment and disease: association or causation? Proceedings of the Royal Society of Medicine. 58:295–300
158. Buckley JD, Meadows AT, Kadin ME, Le Beau MM, Siegel S, Robison LL (2000) Pesticide exposures in children with non-Hodgkin lymphoma. Cancer 89(11):2315–2321
159. Infante-Rivard C, Sinnett D (1999) Preconceptional paternal exposure to pesticides and increased risk of childhood leukaemia. Lancet 354(9192):1819
160. Villanueva CM, Cantor KP, Cordier S, Jaakkola JJ, King WD, Lynch CF et al (2004) Disinfection byproducts and bladder cancer: a pooled analysis. Epidemiology 15(3):357–367
161. Boffetta P (2006) Human cancer from environmental pollutants: the epidemiological evidence. Mutat Res 608(2):157–162
162. International Agency for Research on Cancer (2004) Some drinking-water disinfectants and contaminants, including arsenic, vol 84. International Agency for Research on Cancer, Lyon, France
163. Infante-Rivard C (2008) Chemical risk factors and childhood leukaemia: a review of recent studies. Radiat Prot Dosim 132(2):220–227
164. Raaschou-Nielsen O, Reynolds P (2006) Air pollution and childhood cancer: a review of the epidemiological literature. Int J Cancer 118(12):2920–2929
165. Snedeker SM, Diaugustine RP (1996) Hormonal and environmental factors affecting cell proliferation and neoplasia in the mammary gland. Prog Clin Biol Res 394:211–253
166. Hildreth NG, Shore RE, Dvoretsky PM (1989) The risk of breast cancer after irradiation of the thymus in infancy. N Engl J Med 321(19):1281–1284
167. Herbst AL (1999) Diethylstilbestrol and adenocarcinoma of the vagina. Am J Obstet Gynecol 181(6):1576–1578
168. Strohsnitter WC, Noller KL, Hoover RN, Robboy SJ, Palmer JR, Titus-Ernstoff L et al (2001) Cancer risk in men exposed in utero to diethylstilbestrol. J Natl Cancer Inst 93(7):545–551
169. Colton T, Greenberg ER, Noller K, Resseguie L, Van Bennekom C, Heeren T et al (1993) Breast cancer in mothers prescribed diethylstilbestrol in pregnancy. Further follow-up. JAMA 269(16):2096–2100
170. Olshan AF, Daniels JL (2000) Invited Response: pesticides and childhood cancer. Am J Epidemiol 151:647–649

Chapter 12
Environmental Toxicology: Oxidative Stress

Dean P. Jones

Glossary

Antioxidant A term loosely defined as something that stops oxidation when present at a low amount. The term is used most specifically to mean a free radical-scavenging chemical that accepts or donates an electron to a radical to terminate radical chain reactions. The term is also used for agents that inactivate radical initiation and catalysis, such as metal ion chelators, agents that block radiation-induced oxidation, and agents that counter oxidation of a substance with reduction.

Disulfide An intermediate oxidation state of sulfur that is readily interconverted with the reduced (thiol) state and commonly found in biomolecules. Disulfides are important in protein structure, function, regulation, and translocation. Disulfide formation in proteins and low molecular mass sulfur-containing chemicals (glutathione, cysteine, Coenzyme A) is a common and sensitive indicator of oxidative stress.

Electrophile (or electrophilic chemical) A chemical with a functional group that is deficient in electron density, such as a quinone or conjugated aldehyde, and therefore reactive with a nucleophilic chemical containing electron-rich functional groups, such as a thiol.

This chapter was originally published as part of the Encyclopedia of Sustainability Science and Technology edited by Robert A. Meyers. DOI:10.1007/978-1-4419-0851-3

D.P. Jones (✉)
Division of Pulmonary, Allergy and Critical Care Medicine, Department of Medicine,
Emory University School of Medicine, Atlanta, GA 30322, USA
e-mail: dpjones@emory.edu

E.A. Laws (ed.), *Environmental Toxicology: Selected Entries from the Encyclopedia of Sustainability Science and Technology*, DOI 10.1007/978-1-4614-5764-0_12,

	Electrophiles are important in toxicology because they react with nucleophilic centers in DNA, causing mutations and cancer, and with proteins, disrupting enzyme, receptor, transporter, gene expression, and other critical functions.
Lipid peroxidation	A radical chain reaction involving polyunsaturated fatty acids (PUFA) and O_2. The process is initiated by agents that abstract a hydrogen atom from the carbon skeleton to form a carbon-centered radical, propagated by addition of O_2 and abstraction of a hydrogen atom from another carbon skeleton with creation of a lipid hydroperoxide and another radical, and ultimately terminated by reaction of radicals with each other to create non-radical species. The process causes fats and oils to become rancid and is common in decaying organic matter.
Oxidative stress	A process or outcome in which an imbalance in prooxidant and antioxidant reactions causes macromolecular damage and/or disruption of biologic redox signaling and control. All classes of macromolecules and any biologic function can be affected. Effects on DNA are important in mutagenesis and cancer, effects on proteins contribute to many acute and chronic toxicities, effects on lipids contribute to dysfunctional energy metabolism and immunity, and effects on carbohydrates affect biological lubricants in joints and cellular identity.
Prooxidant	An agent that stimulates aberrant electron transfer in biologic systems and causes oxidative stress. This includes agents that initiate radical reactions, oxidize biologic components, or interfere with normal reductive and antioxidant functions.
Radical (or free radical)	An organic molecule with an unpaired electron. Radicals are important in biologic systems because they are reactive and can disrupt biologic functions. Radicals participate in a unique type of chemical reaction sequence termed a chain reaction. A radical chain reaction, such as that illustrated by lipid peroxidation, is initiated by an agent that removes (or adds) an electron from a non-radical chemical to form a radical. The chemical structure is modified, and the radical product becomes a non-radical by accepting or donating an electron to another non-radical species, thereby propagating the reaction sequence. The process is ultimately terminated by reaction of radicals with each other to create non-radical species.

Reactive oxygen species (ROS), sometimes reactive oxygen intermediates (ROI)	One or more oxygen-containing chemicals that are reactive with organic molecules. This term is often used to refer to superoxide anion radical, hydrogen peroxide, lipid hydroperoxide, and related oxygen-centered radicals, but may also include reactive nitrogen species (RNS), such as nitric oxide, peroxynitrite, and other oxides of nitrogen (NOx), which also contain oxygen. The term is most often used because the chemistry is too complex and/or the analytic methods are insufficient to identify specific reactive species.
Redox cycling	A process in which a chemical accepts an electron from a biological reductant and transfers that electron to O_2 or other acceptor to create aberrant radical generation. Many quinones and other aromatic chemicals do this by interacting with flavoproteins. The process is considered "cycling" because the chemical functions only as a catalyst. After donating the electron to an acceptor, the original form is regenerated so that it can accept another electron. If electrons are transferred between redox-signaling pathways, this creates a "short circuit," disrupting cell regulation.
Redox signaling	A mechanism for cellular communication involving an oxidation–reduction reaction. This process commonly involves generation of a small diffusible redox-active chemical that transfers a signal to a biologic receptor. Redox signaling is highly integrated with kinase signaling and other central cell communication mechanisms, and difficult to discriminate from a more general process of redox sensing, which coordinates cellular functions via widely distributed redox-sensitive cysteine residues in proteins.
Thiol	An organic form of sulfur containing a sulfur-hydrogen covalent bond. This is the reduced form of sulfur in the amino acid cysteine, in the antioxidant glutathione, and in most cysteine residues in cellular proteins. In proteins, thiols are also termed "sulfhydryl groups." The thiol functional group supports diverse reactivity and catalytic functions of proteins. Interconversion with disulfides and other forms allows thiols to serve as structural transducers as well as chemical signal transducers.
Xenobiotic	A chemical that is foreign to an organism. Although often used to refer to man-made chemicals, the term is also used to discriminate chemicals derived exogenously from the environment from those that are generated endogenously as a consequence of intermediary metabolism.

Definition of the Subject

Oxidative stress is a process or outcome in which an imbalance in prooxidant and antioxidant reactions causes macromolecular damage and/or disruption of biologic redox signaling and control (Fig. 12.1) [1]. Oxidation–reduction reactions, or "redox" reactions, are reactions involving electron transfer. These reactions are central to energy metabolism and maintenance of metabolic and physical organization of living organisms. Environmental agents that interfere with redox reactions or promote abnormal redox reactions are highly destructive to biologic macromolecules and metabolic organization. Electron transfer reactions in the atmosphere, water, and soil are common and important subjects of environmental sciences. In many cases, physical processes and inorganic chemicals are relevant causes of oxidative stress to living organisms. Oxidative stress can also occur endogenously within an organism due to organic chemicals derived from the environment. This entry focuses on oxidative stress within living organisms,

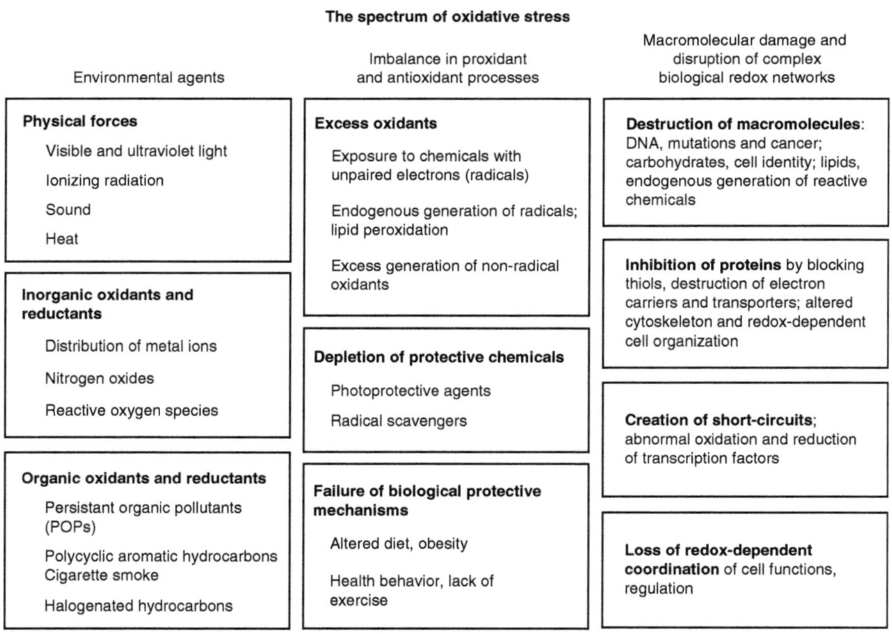

Fig. 12.1 The spectrum of oxidative stress in environmental toxicology. Oxidative stress refers to adverse effects (*right*) as well as the processes (*center*) that cause adverse effects in biologic systems. Environmental factors that contribute to oxidative stress (*left*) include physical, inorganic, and organic agents. These agents create imbalances in specific prooxidant and antioxidant reactions or reaction pathways (*center*) due to excess oxidants, depletion of protective chemicals, or failure of endogenous protective mechanisms. The imbalances result in the generation of reactive radicals and non-radical oxidants that cause macromolecular damage, inhibit thiol-dependent proteins, short-circuit redox pathways, and disrupt regulatory mechanisms (*right*)

recognizing that the environmental exposures causing oxidative stress in an organism are highly variable (Fig. 12.1, left). Many of these are addressed in detail in other sections of this Encyclopedia.

Living organisms depend upon the capability to maintain chemical and physical homeostasis despite variations in environmental exposures. Exposures that cause oxidative stress include chemical oxidants such as ozone and nitrogen oxides of the atmosphere [2, 3], transition metals such as iron [4] and cadmium [5–7], visible and ultraviolet radiation from the sun [8, 9], ionizing radiation from radioactive decay [10, 11], and a large number of pharmaceuticals [12] and chemicals from commercial products and industrial wastes [13, 14]. In some cases there is no real change in the amount of environmental toxicant, only a change in distribution that affects exposure of living organisms [15]. In other cases, there is an accumulation of environmental toxicant due to human activity. Some of these are recognized as persistent organic pollutants (POPs) [16–18] and other times they are persistent products of organic pollutants (P-POPs), some of which are not currently known to accumulate or represent a hazard. For instance, the recent investigation of PFOA, a product of fluorocarbons that is of unknown origin [19, 20], suggests that other unidentified chemicals derived from human activity may exist. Biologic defense mechanisms are diverse and include physical barriers, chemical detoxification mechanisms, repair mechanisms, and avoidance behaviors. Considerable scientific data are available in each of these areas, and efforts are currently being initiated to assemble this knowledge in a systematic description of the "exposome," i.e., the summation of all environmental exposures from conception throughout the life of an individual [21, 22].

The concepts of oxidative stress are currently undergoing revision from the earlier definition as an imbalance of prooxidants and antioxidants that cause macromolecular damage [23] to one that also includes oxidative stress as a disruption of vital signaling and control processes within a biologic system [1, 24]. In human health, the accumulating data suggest that the latter is more relevant. However, in environmental toxicology, both macromolecular damage and disruption of redox signaling and control mechanisms are relevant. Importantly, new developments in systems biology [25, 26], including powerful new molecular approaches and analytic tools to study genomics [27] and epigenomics [28], gene expression (transcriptomics) [29], protein modifications (proteomics) [30, 31] and chemical profiles (metabolomics) [32, 33] provide capabilities to improve surveillance, mechanistic understanding, and risk reduction.

Introduction

Two-electron and one-electron transfers in biologic systems. Living organisms obtain the energy necessary to sustain organization, delineation from their external environment, and reproduction, through electron transfer reactions. Loss of electrons is termed "oxidation," while gain of electrons is termed "reduction" (Fig. 12.2a–c). Because of the conservation of matter in chemical reactions, the processes are always coupled, so the terms "oxidation–reduction" reactions and "redox" (short for "reduction–oxidation") reactions are usually more correct. However, the common process of oxidation of chemicals in the presence of atmospheric O_2 (with water as the reduction product) explains the bias in terminology. Excessive reduction can also occur, termed "reductive stress" [34, 35], and is especially relevant to hypoxic and anoxic environmental conditions.

The major elements of life, carbon, hydrogen, oxygen, nitrogen, phosphorus, and sulfur form multiple stable biochemical structures in which pairs of electrons are shared. The major macromolecules (protein, nucleic acids, lipids, carbohydrates) are formed from smaller biochemicals (building blocks: e.g., amino acids, nucleotides, fatty acids, sugars) that include these elements. The biochemical structures are interconverted through chemical reactions that involve conservation of these pairs of electrons. The interconversions are often slow without a catalyst. Biologically, this involves two-electron transfers by NADH or NADPH, with the electrons being transferred as a hydride (H^-) ion (Fig. 12.2c–e). Enzymes serve as catalysts to enhance rates of reactions so that the presence of enzymes largely determines the biochemical pathways for interconversion of the carbon-containing biochemicals. Many of these interconversions involve reaction of "electrophiles," chemicals with orbitals deficient in electrons, and "nucleophiles," chemicals with orbitals with excess electron density.

The two-electron (2-e^-) transfer reactions are contrasted by one-electron (1-e^-) transfers, which are predominant reactions of many biologically important metals, such as iron and copper, environmentally important metals, such as cadmium, chromium, and tin, and atmospheric reactions driven by ionizing and non-ionizing radiation. Living organisms are dependent upon 1-e^- chemistry because the central energy capture and transfer reactions of photosynthesis and mitochondrial respiration involve 1-e^- transfers. However, these 1-e^- transfers are physically contained within subcellular structures (chloroplasts and mitochondria), so that both 1-e^- and 2-e^- chemistries can occur simultaneously and be used to provide critical reaction pathways in biological systems. The incompatibility is illustrated by the combination of reactions A and B in Fig. 12.2, where superoxide donates an electron to Fe^{3+} to form Fe^{2+}, and Fe^{2+} reacts with hydrogen peroxide to form the highly reactive and destructive hydroxyl radical. The 1-e^- systems of mitochondria and chloroplasts are important in environmental toxicology because they represent a loaded gun; these systems have very high electron transfer rates, are sensitive to visible and ultraviolet light, trace metals, oxidants and electrophiles, and cause extensive destruction to other biologic components when triggered by environmental toxicants.

The 2-e^- and 1-e^- pathways are connected through flavoproteins and hydroquinone/quinone structures that exist in interconvertible fully oxidized, 1-e^--reduced and 2-e^--reduced forms. Flavoproteins are widely used in biology to support 2-e^- transfer reactions involving NADH and NADPH. In addition, they are critical to connect 2-e^- pathways to 1-e^- pathways, accepting 2 e^- from NADH or NADPH and transferring 1 e^- to electron acceptors. An illustration is given in Fig. 12.2c,

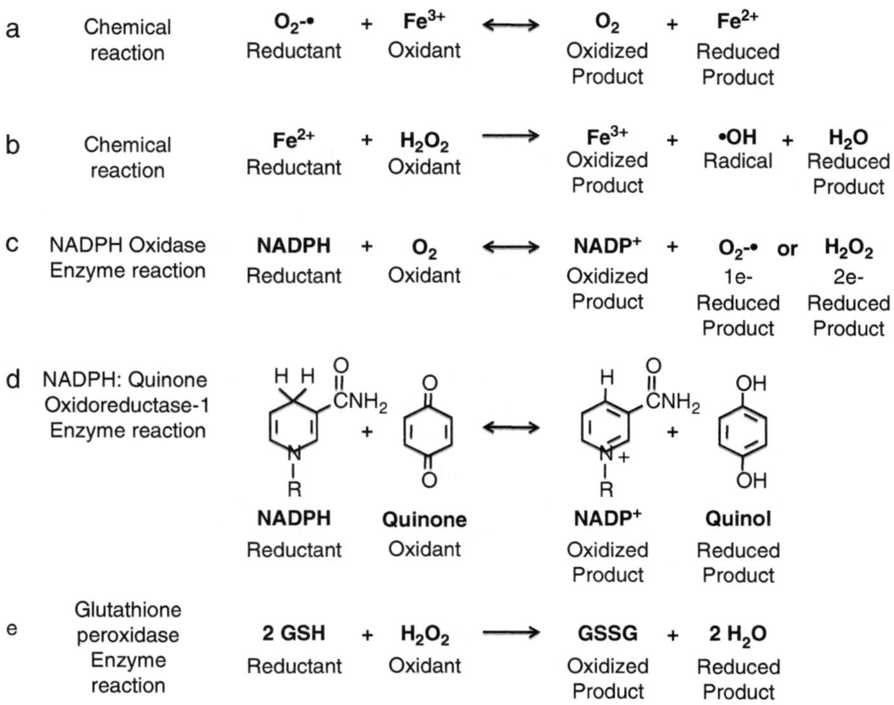

Fig. 12.2 Oxidation–reduction reactions. (**a**) One-electron transfer reaction from superoxide anion radical ($O_2^{-\bullet}$) to Fe^{3+}, as occurs in transfer from superoxide to oxidized cytochrome *c*. In this reaction, superoxide anion radical is oxidized to molecular oxygen (O_2), while the iron is reduced from Fe^{3+} to Fe^{2+}. Although the reaction can occur in the opposite direction, such as the generation of superoxide by electron transfer in oxy-hemoglobin, energetics in biologic systems usually favor the forward direction as written. (**b**) One-electron transfer from Fe^{2+} to hydrogen peroxide (H_2O_2). This energetically favorable reaction generates three products, Fe^{3+}, hydroxyl radical ($^{\bullet}OH$), and water. The hydroxyl radical is extremely reactive and destructive to biochemicals. Combination of reaction A and B constitutes the iron-catalyzed Haber–Weiss reaction, also known as the Fenton reaction, which has a central role in radical-mediated macromolecular damage. This process is very destructive when appropriate concentrations of superoxide, H_2O_2, and iron are present. (**c**) Flavin-containing enzymes, such as NADPH oxidases, catalyze both 1-e^- and 2-e^- transfers to O_2, thereby forming both superoxide and H_2O_2. Aerobic cells contain NADPH, O_2, and flavoproteins so that conditions are poised to support Fenton chemistry (A + B) when these enzymes are active. The most potent NADPH oxidase, Nox2, is present in phagocytic cells and activated to kill the invading microorganisms. (**d**) Two-electron transfer, shown for the detoxification enzyme NADPH:quinone reductase-1 (NQO1), is used to support chemical interconversions in intermediary metabolism. The niacinamide group shown for NADPH is identical to that in NADH, and serves to transfer a pair of electrons as a hydride ion (H^-) without radical formation. The reaction shown is an important detoxification reaction by which common environmental quinones are reduced to quinols for phase 2 conjugation and excretion. Quinones can function either as oxidants or electrophiles, and can also generate radicals through redox cycling (see Fig. 12.3). These adverse reactions are prevented by the NQO1 reaction. (**e**) Thiols are used as reductants to eliminate H_2O_2. In the glutathione (GSH) peroxidase reaction shown, two molecules of GSH are used to reduce H_2O_2 to water without forming radicals as intermediates. Reduction of the oxidized product, GSSG, by an NADPH-dependent reductase (not shown) completes a detoxification cycle for peroxides. This reaction cycle is complemented by a thioredoxin/peroxiredoxin system to maintain peroxide concentrations very low in cells despite continuous generation. In some cases thiol radicals are formed, but these rapidly react to form disulfides, thereby minimizing toxic effects

where different NADPH oxidases support both 1- and 2-e^- reduction of O_2. Flavoproteins catalyze 1-e^- transfer to the mitochondrial respiratory chain, the major enzyme systems for oxidation of xenobiotics, and to O_2 to form ROS for signaling and defense against invading microorganisms. They also are a major source of toxic oxidant generation by initiating lipid peroxidation (Fig. 12.3). The planar aromatic structure interacts with polycyclic aromatic structures, facilitating the generation of reactive chemicals with an unpaired electron, termed "radicals," in a process known as redox cycling [36]. Flavins and flavoproteins are also photoreactive because they absorb visible light, and this activates electrons for transfer to other chemicals.

Antioxidants include chemicals that are relatively stable as radicals and donate or accept an electron to reactive radicals to decrease their destructive chemistry. This is highly beneficial by preventing radical-mediated destruction of macromolecules. Vitamin C (ascorbic acid) and vitamin E (α-tocopherol) are radical-scavenging antioxidants and function to terminate free radical chain reactions (See Fig. 12.3). Vitamin C is water soluble and synthesized by most animals but not by humans. Vitamin E is fat soluble and one of several lipophilic chemicals that blocks radical reactions in biologic systems. Plants are rich in radical-scavenging antioxidants, and also in carotenoids, light-absorbing chemicals that protect against light-induced oxidative damage.

Evolution of "Free Radical" and "Radical-Free" Concepts of Oxidative Stress

Joseph Priestley, an eighteenth century theologian and chemist credited with the first publication demonstrating production of O_2 [37], intimated that "dephlogisticated air" was toxic even before Lavoisier demonstrated that "dephlogisticated air" was oxygen [38]. Priestley said of breathing dephlogisticated air "...I fancied that my breast felt peculiarly light and easy for some time..." and "a candle burns out much faster in dephlogisticated air than in common air, so we might, as may be said, live out too fast, and the animal powers be too soon exhausted in the pure kind of air. A moralist, at least, may say, that the air which nature has provided for us is as good as we deserve" [39].

This conclusion of the late eighteenth century presaged a cliché of late twentieth century science, that "oxygen is a double-edged sword." While science is usually better than cliché, the cliché was highly effective in promoting science, and in this case, an almost religious fervor to introduce "antioxidants" in dietary, cosmetic, and other commercial product development. Failure of large-scale, double-blind interventional trials with antioxidants in humans [40–47] has led to a backlash [48]. Some scientists now have a dismissive attitude concerning the importance of oxidative stress. However, the scientific knowledge of the mechanisms of oxidative stress is very strong and clearly establishes oxidative stress as relevant to

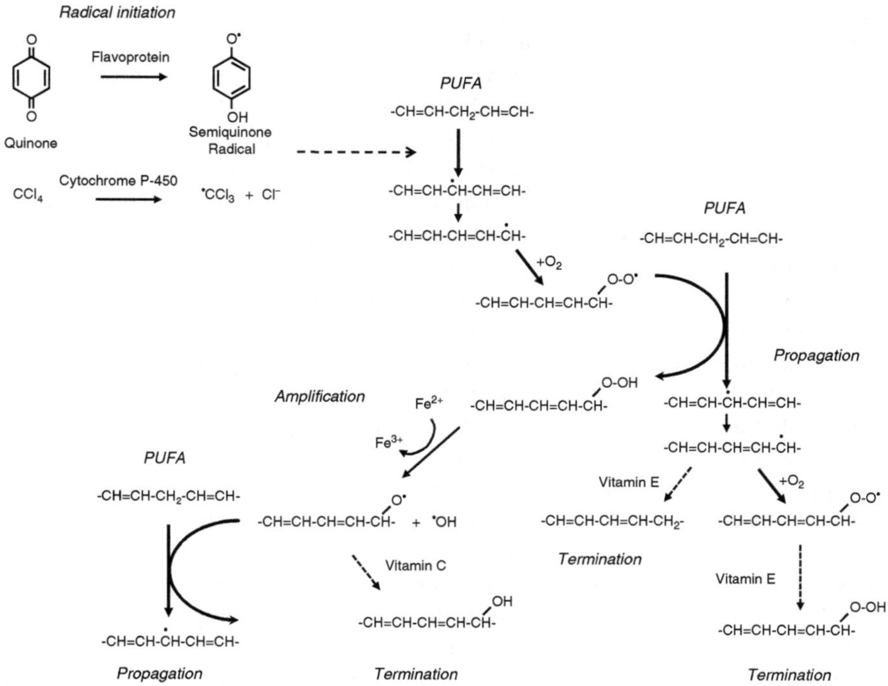

Fig. 12.3 Lipid peroxidation is a common radical chain reaction that occurs in biologic systems. Radical chain reactions include initiation, propagation, and termination events. In biologic systems, initiation (*top left*) often occurs by redox cycling of quinone compounds and reductive activation of halogenated hydrocarbons (e.g., CCl_4). In redox cycling, quinones accept an electron from a reduced flavoprotein, forming a radical. The radical can donate or accept an electron through a hydrogen abstraction from a polyunsaturated fatty acid (PUFA) (*top middle*). Normally this is most energetically favorable at a carbon adjacent to carbons with double bonds. The resulting radical rearranges to form a conjugated double bond system with an adjacent carbon-centered radical. Under aerobic conditions, O_2 rapidly adds to form a peroxyl radical (*center*). The peroxyl radical reacts with another PUFA (*right middle*) to propagate the free radical chain reaction. In this reaction, the peroxyl radical is reduced to a lipid hydroperoxide, which is normally reduced by GSH peroxidases or peroxiredoxin-6 (not shown). When activities of the latter systems are impaired and free iron is present, the process can amplify lipid peroxidation by Fe^{2+}-dependent initiation of another radical chain reaction (*bottom left*). In the presence of efficient radical scavengers, such as vitamin C and vitamin E, the chain reactions are terminated. Note that vitamins C and E are most active with different types of radicals and thus complement each other in function. Free radical chain reactions are limited in normal living tissues due to the presence of antioxidants and very high protein concentration. Lipid peroxidation in oils, fats, and decaying matter is very active when antioxidants are depleted

environmental toxicology. A brief look at the evolution of concepts facilitates resolution of apparent conflicts between the extensive biomedical literature on oxidative stress and implications for environmental health.

Catalase was one of the first enzymes to be studied [49], and its association with aerobic life was recognized by early biochemists [50]. Hydrogen peroxide (H_2O_2)

was recognized as a toxic reactive oxygen species (ROS) associated with O_2 exposure [51]. Early studies also included 1-e^--transfer reactions of the mitochondrial electron transfer pathway, begun by Keilin in the 1920s [52]. During this time, the environmental agent paraquat (methyl viologen) was introduced as an "electron mediator" to transfer electrons into the mitochondrial pathway [53]. The concepts of oxidative stress, though, are perhaps better traced to the concept of "free radicals," organic molecules with an unpaired electron, and their damaging effects on biological systems. The chemical terminology for a free radical has been changed to "radical," but the former is still more commonly used.

The study of radicals in biologic systems was popularized by the discovery of lipid peroxidation (Fig. 12.3) and antioxidant actions of vitamin E [54] and selenium-dependent glutathione peroxidase [55, 56]. Additional toxicologic studies of carbon tetrachloride (Fig. 12.3) showed that radical reactions of lipid peroxidation were responsible for hepatotoxicity [57, 58]. Finally, Denham Harmon attracted attention through his hypothesis that free radicals contributed to aging [59].

Lipid peroxidation is a free radical chain reaction contributing to rancidity of oils and fats and invariably present in dead and decaying matter [60]. Radical chain reactions involve an initiation event in which a radical is formed, subsequent propagation events, and finally termination reactions that eliminate the radicals (Fig. 12.3). Chemicals, such as quinones, that interact with flavoproteins and redox cycle [36] are a common source of initiation (Fig. 12.3, top left). Another process involves activation of halogenated hydrocarbons and other environmental chemicals to free radicals by the hemoproteins of the cytochrome P450 (CYP) family (Fig. 12.3, top left). These radicals abstract a hydrogen atom from polyunsaturated fatty acids (PUFA; Fig. 12.3, top center), which result in the rearrangement of the newly formed fatty acid radical to a more stable form with conjugated double bonds (Fig. 12.3, center). O_2 readily adds to this intermediate to form a peroxyl radical, which then propagates the chain reaction by the abstraction of a hydrogen atom from another PUFA (Fig. 12.3, right center). In the presence of trace Fe^{2+}, the process can be amplified by the creation of a new radical from the lipid hydroperoxide (Fig. 12.3, lower left). Vitamins C and E terminate the process by reducing the reactive intermediates and breaking the chain reaction.

While this constellation of findings and conceptual developments advanced the biochemical study of free radicals, the unanticipated finding of an enzyme that detoxifies superoxide attracted the attention of the broad spectrum of respiratory biologists and invigorated a new field of free radical biology and medicine. Study of the O_2-dependent reduction of cytochrome *c* by xanthine oxidase led to the discovery that the abundant blood protein erythrocuprein was an enzyme that converted the 1-e^- reduction product of O_2, i.e., the superoxide anion radical ($O_2^{-\cdot}$) to O_2 plus H_2O_2 [61]. The discovery of this enzyme, superoxide dismutase [62], was seminal because it provided evidence that biological systems had evolved an enzyme to eliminate radicals. This provided clear evidence that radicals must be commonly produced and be a threat to biological systems.

Chance et al. [63] provided an authoritative review on hydroperoxides that served as a platform for the integration of concepts of biologic damage by 1-e^-

and 2-e^- oxidative processes by Sies [23] in the first monograph entitled "Oxidative Stress." In this volume, Sies [23] defined oxidative stress as "an imbalance of prooxidants and antioxidants in which the former predominates and causes macromolecular damage." This clear and simple definition proved useful to stimulate research on oxidative stress as a mechanism of disease and aging, and antioxidants as means to protect against these deleterious processes.

The contemporary definition of oxidative stress has been refined based upon research findings of the turn of the century [1, 24, 64, 65]. Discovery of a family of enzymes, NADPH oxidases, that generates $O_2^{-\cdot}$ and H_2O_2 as signaling molecules [66, 67] showed that one cannot simply equate ROS with harmful chemistry, i.e., oxidative stress.

At the same time, results of large-scale, double-blind interventional studies with radical-scavenging antioxidants began to appear in the literature [40–47]. These studies showed that supplementation with antioxidants, i.e., shifting the "balance" toward protection, had little or no health benefit in humans. Consequently, the original definition of oxidative stress needs to be qualified in the sense that an "imbalance" is only considered in terms of specific reactions or pathways and not in terms of a global imbalance. In other words, an imbalance in a specific reaction or pathway can cause damage without an overall imbalance in the system, and a change in overall balance of prooxidants and antioxidants can occur without causing damage.

The contemporary view of oxidative stress has also been impacted by the development of new –omic technologies and systems biology. Instead of being limited to overall measures such as cell death [68], lipid peroxidation [69], or protein carbonylation [70], one can now measure effects on expression of specific genes [29], modification of specific lipids [71], and oxidation of specific proteins [72]. This research now shows that oxidative stress as it applies to environmental toxicology involves two processes: one related to earlier concepts of radical-mediated macromolecular damage and the other related to disruption of redox signaling and control mechanisms, which are largely non-radical in nature.

Radical Mechanisms of Environmental Toxicity

The lipid peroxidation mechanisms described above (Fig. 12.3) are more complex in living organisms because of the high concentrations of protein, the abundance of free radical scavenging antioxidants, and the processes of biologic turnover, which eliminate and replace damaged macromolecules and cells. In addition to redox cycling agents and halogenated hydrocarbons, a number of other initiators occur, including transition metals, ionizing and non-ionizing radiation, and sound. Because of the high concentration of protein in cells, the number of propagation reactions tends to be small, emphasizing the importance of initiation events in toxicity.

In pure lipid systems, free radical chain reactions can occur in which an initiation event causes modification of 200–400 molecules of fatty acid prior to being terminated by radicals reacting with each other. However, in biologic systems,

Ozone	O_3
Singlet oxygen	$^1O_2$
Hydrogen peroxide	H_2O_2
Hypochlorous acid	HOCl
Hydroperoxide	ROOH
Nitrous acid	HNO_2
Dinitrogen tetraoxide	N_2O_4
Peroxynitrite	$ONOO^-$
Alkylperoxynitrite	ROONO
Disulfide	RSSR′
Sulfenate	RSO^-
Sulfinate	RSO_2^-

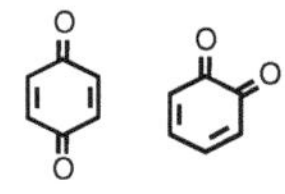

Quinones

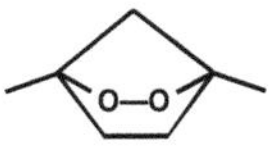

Endoperoxide

Epoxide

Fig. 12.4 Non-radical oxidants in biologic systems. Difficulties in quantifying oxidants and their rates of generation, as well as complexities in adaptive and toxic responses to long-term oxidant exposure and a generally conservative approach to terminology, have resulted in widespread use of the term "reactive oxygen species" (ROS). Although convenient, this term does not discriminate between radical and non-radical species and carries a bias that relevant biologic oxidants are closely related to O_2. A large number of non-radical oxidants, some of which are listed in this figure, exist in the environment and in biologic systems. Ozone and singlet oxygen (triplet O_2 is ground state) are common in the atmosphere, while H_2O_2, hypochlorous acid and lipid hydroperoxides are generated enzymatically and non-enzymatically in biologic systems. A number of other reactive carbon oxidation species occur, such as endoperoxides, epoxides and quinones (*right*). Reactive nitrogen oxidation products are also relevant biological oxidants. Nitrous acid and dinitrogen tetroxide are common in the atmosphere and also in biologic systems. Peroxynitrite is a reactive oxidant produced by reaction of two radicals, nitric oxide and superoxide. Nitric oxide (NO·) is a protective signaling molecule in the vascular system and is converted to peroxynitrite in the presence of superoxide. Alkylperoxynitrites can similarly form from nitric oxide and lipid peroxyl radicals. Other oxidants can also occur, such as disulfides of cysteine and other low molecular weight thiols. Other sulfur oxidation products, such as sulfenates and sulfinates, can also function as oxidants

such a chain reaction does not occur for several reasons. One is that initiation events are prevented by keeping free metal ion concentrations very low, and blocking other common initiation events. Another is that the concentration of protein is so high that H· is often abstracted from proteins rather than other polyunsaturated lipids, thereby decreasing propagation. Biologic systems also contain high concentrations of radical scavengers, such as vitamin C and vitamin E, which function as chain-terminating antioxidants. Amplification is also effectively prevented by the removal of lipid hydroperoxides by the peroxidases.

The propagation reactions are also not simple because multiple rearrangement products can occur. In addition to the rearrangement shown in Fig. 12.3, PUFA radicals can rearrange to eliminate a conjugated aldehyde, such as 4-hydroxynonenal (HNE). HNE is a reactive conjugated aldehyde that can react with proteins or DNA. The reactions generating HNE also generate carbon-centered radicals that propagate chain reactions. A number of other products can be formed, including epoxides (Fig. 12.4), which can be important as more stable forms with greater specificity in reaction with macromolecules, and isoprostanes [73], rearrangement products that are useful as biomarkers of this chemical reaction sequence. Conjugated aldehydes react with thiols and amines of proteins, creating modified proteins and degradation products that are universally detectable in biologic systems. Thus, lipid peroxidation appears to be an ongoing process. These are measured as protein carbonyls, another useful way to monitor oxidative stress in biologic systems [70].

Lipid peroxides are also formed enzymatically and are metabolized by several peroxidase systems. The most extensively studied enzymes are the glutathione (GSH) peroxidases [74], but more recent data indicate that a thioredoxin-dependent peroxidase, peroxiredoxin-6, is more active in the elimination of lipid hydroperoxides [75, 76]. The lipid peroxides can be very important in lipid peroxidation by providing a mechanism for amplification of lipid peroxidation [60]. Trace metal ions, such as Fe^{2+} or Cu^{1+}, can donate an electron to the peroxide to form another radical initiation event.

Because of these processes, radical chain reactions are almost completely prevented in living cells. In contrast, oxidative processes are a component of cellular destruction mechanisms used to kill microorganisms, to establish physical barriers to prevent infection, and to degrade and remove dead and dying cells. Phagocytes have a respiratory burst in which NADPH oxidase-2 generates ROS to kill microorganisms [67]. During apoptosis, a mitochondrial respiratory burst is created by the release of mitochondrial cytochrome *c* and switching the electron transfer chain to one that generates O_2^- [77]. Thus, unlike the interpretation that continuous generation of biomarkers of oxidative stress indicates that O_2 is a double-edged sword, the evidence shows that oxidative activities are part of normal biologic processes and that there is reserve capacity to protect against catastrophic collapse. Chain reactions of lipid peroxidation do not provide a major, central toxicologic process, but rather, when lipid peroxidation occurs, it mostly occurs by radical initiation events directly from environmental cause rather than as a consequence of chain reactions.

In summary, radical mechanisms are important for environmental toxicology because environmental agents can directly cause initiation of radical reactions that damage all types of macromolecules. Methods are available to evaluate radical-dependent mechanisms of oxidative stress in biological materials, especially through measurement of isoprostanes, protein carbonyls, and other well-documented oxidation products. Low levels of these products are found in normal cells, and healthy organisms are well protected against these reactions. Based upon extensive double-blind interventional studies with free radical scavengers in humans, these reactions do not contribute significantly to long-term health even though many studies show short-term reductions in biomarkers of oxidative stress. Thus, in the absence of high concentrations of acute environmental exposure, the radical mechanisms appear to be secondary to non-radical oxidative mechanisms of toxicity [78].

Non-radical Mechanisms of Environmental Toxicity

The principles of non-radical mechanisms of oxidative stress are outlined in detail in "Radical-free biology of oxidative stress" [78]. Research has provided extensive knowledge of radical chemistry and antioxidants studied in purified systems and subcellular fractions of biologic systems, often with little regard to biologic relevance [79]. Most critically, biologic systems generate more non-radical oxidants (Fig. 12.4) than radicals, and the effective radical scavenging processes largely convert radicals to non-radicals or otherwise limit their destructive potential [78]. Important non-radical oxidants (Fig. 12.4) include H_2O_2, lipid hydroperoxides, quinones, disulfides, and reactive nitrogen species (RNS), especially peroxynitrite [80]. Peroxynitrite is a major oxidant formed by reaction of superoxide with nitric oxide, a protective signaling molecule controlling microcirculation and immune function [80, 81].

In the xanthine oxidase system, which provided the discovery of superoxide dismutase, the predominant product is H_2O_2, not the radical superoxide anion [82]. Under optimal conditions for univalent flux to form superoxide, this represents only 30% of the total flux, i.e., under all conditions divalent flux to form H_2O_2 predominated [82]. The univalent flux was considerably less under most conditions, consistent with partitioning of radical and non-radical oxidant production as depicted in Fig. 12.5. Many redox proteins that generate superoxide show rapid, sequential univalent transfer so that H_2O_2 is a common product of these reactions. Furthermore, dismutation of superoxide anion to H_2O_2 and O_2 occurs at a rapid rate (46) and, as indicated above, reaction of radicals with radical scavengers results in efficient conversion to non-radical oxidants. Consequently, the 2-e^-, non-radical oxidant, H_2O_2, is produced at a higher rate and appears to represent the major oxidant burden. This rate has been estimated to be 1–4% of the rate of O_2 consumption in mammalian systems [63], but rates can be higher in some compartments like the peroxisomes [83], or lower, such as in the cytoplasm or nuclei [84].

With non-radical oxidants being quantitatively most important and radical scavenging antioxidants converting some of the radicals to non-radical oxidants, it is not surprising that radical-scavenging antioxidants do not provide protection against disease [40–47]. Perhaps a better way to view this in terms of environmental toxicity is that radical mechanisms are likely to contribute to acute toxicities but not as likely to contribute to chronic toxicities. In these reactions, macromolecular damage is of major concern, and all classes of macromolecules are potentially vulnerable. On the other hand, non-radical mechanisms can contribute to both; as described below, these are most relevant to chronic toxicity and can have effects through disruption of redox signaling and control, whether or not macromolecular damage occurs.

The major mechanisms of non-radical oxidative stress involve effects on thiols in proteins [78], although mutagenic damage to DNA also occurs [11, 85, 86]. Non-radical oxidants can interfere with reversible oxidation–reduction reactions of thiols by causing abnormal oxidation or irreversible modification. These changes alter physiological function in receptor signaling, transcriptional regulation, cell

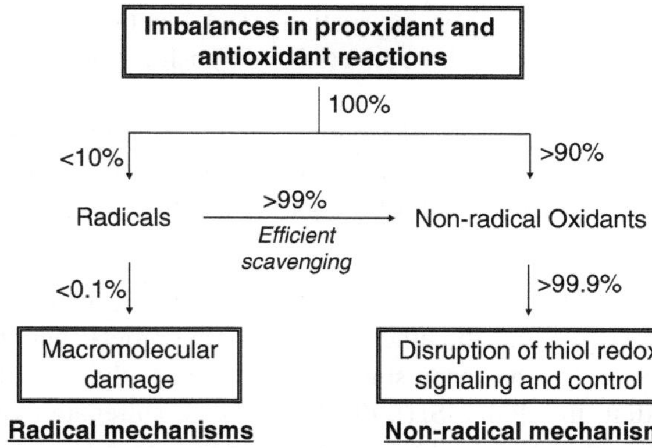

Fig. 12.5 Non-radical mechanisms of oxidative stress are quantitatively important in chronic and delayed toxicities. Although toxicity due to radical-induced macromolecular damage occurs under some conditions of acute toxicity, accumulating evidence indicates that non-radical mechanisms are important under other conditions. In this figure, the total rate of O_2 reduction to generate reactive oxygen species is expressed as 100%. Two-electron transfers predominate at biologic active sites that reduce O_2 to superoxide and H_2O_2, so that radicals are always produced at lower rates (e.g., <10% of total) than 2-electron products, such as H_2O_2 (>90% of total). Radical scavengers, including superoxide dismutase, are highly efficient (e.g., >99%) so that the radicals are largely converted to non-radical oxidants. This results in a low level of radical-mediated macromolecular damage and oxidative stress predominantly caused by non-radical oxidants. Non-radical oxidants preferentially target reactive thiols of redox signaling and control pathways, resulting in a disruption of signaling and control. Disruption of signaling and control mechanisms result in long-term dysfunction and reproductive and developmental toxicities. These can be insidious and pose a risk for environmental sustainability

proliferation, angiogenesis, and apoptosis. Whether or not macromolecular damage occurs, the disruption of these functions alters cell fate and impacts physiologic health. Such changes can be adaptive and protective, in the process known as hormesis [87], but also can cause maladaptation and disease. These processes can occur due to a number of specific environmental agents, can occur in combination, and can be insidious in their long-term effects.

These concepts have been formalized in the redox hypothesis of oxidative stress [78], which provides four postulates to guide investigation into detailed mechanisms, ways to detect oxidative stress, and strategies to intervene to avoid or minimize oxidative stress. These postulates point to a need to incorporate systems biology, modern –omic technologies and bioinformatics into environmental health research, specifically to address environmental toxicities. These four postulates are:

1. All biologic systems contain redox elements (e.g., redox-sensitive cysteine, Cys, residues) that function in cell signaling, macromolecular trafficking, and physiologic regulation.
2. Organization and coordination of the redox activity of these elements occurs through redox circuits dependent upon common control nodes (e.g., thioredoxin, GSH).

3. The redox-sensitive elements are spatially and kinetically insulated so that "gated" redox circuits can be activated by translocation/aggregation and/or catalytic mechanisms.
4. Oxidative stress is a disruption of the function of these redox circuits caused by specific reaction with the redox-sensitive thiol elements, altered pathways of electron transfer, or interruption of the gating mechanisms controlling the flux through these pathways.

Proteins contain three functional groups that undergo reversible oxidation, the thiol in cysteine (Cys), the thioether in methionine (Met), and the selenol in selenocysteine (Sec). Cys has been studied most extensively, with relevant oxidation states including the thiol (-SH), disulfide (-SS-), sulfenate ($-SO^-$), sulfinate ($-SO_2^-$), and sulfonate ($-SO_3^-$). Thiyl radicals (-RS·) generated from thiols in the presence of oxygen-centered radicals [88], as well as other reactive sulfur species [89], can also be considered as toxic species in oxidative stress. Thiyl radicals rapidly react to form disulfides [90], and this may serve as a convergence point between $1\text{-}e^-$ and $2\text{-}e^-$ events. Relatively unstable sulfenates are converted to disulfides in the presence of thiols, and in some protein structures are stabilized as sulfenamides [91, 92]. Higher oxidation states, sulfinates and sulfonates, are typically not reversible in mammalian systems. A yeast enzyme (sulfiredoxin) reduces sulfinate in peroxiredoxin [93] and could be important in redox signaling mechanisms [94, 95].

Reversible oxidations of Met residues [96, 97] and the less common amino acid, selenocysteine (Sec) [98, 99], can also be important in toxicologic mechanisms. Met oxidation to methionine sulfoxide occurs in oxidative stress and aging [97, 100]. A good example, studied extensively in association with cigarette smoking, is the increased lung rigidity associated with loss of α1-antitrypsin inhibitor activity upon Met oxidation [101]. The loss of inhibition of elastase results in damage to lung structures and contributes to obstructive lung disease. Additional research shows that two types of methionine sulfoxide reductases (MSR) are important in protecting against distinct S- and R-sulfoxides in methionine residues [96, 102, 103]. These MSR activities are dependent upon thioredoxins [103] and are associated with longevity [104–107]. This less frequently studied type of oxidation involves both Met and Sec, and probably deserves much greater attention in environmental oxidative stress due to the variable environmental availability of Met and Sec. The selenol of Sec undergoes reversible oxidation–reduction during catalytic functions of thioredoxin reductases and Se-dependent glutathione peroxidases [98, 99], enzymes present at key positions in both thioredoxin and GSH pathways and critically important to protect against non-radical oxidative stress.

Key points of oxidative stress according to the redox hypothesis include recognition of mechanisms of oxidative stress other than an imbalance of prooxidants and antioxidants. One of these involves altered function of redox circuits caused by specific reaction with redox-sensitive thiol elements (Fig. 12.6). This differs from the earlier concepts of macromolecular damage in that it could affect adaptability or tolerance of an organism without overt toxicity. For instance, redox-sensing thiols

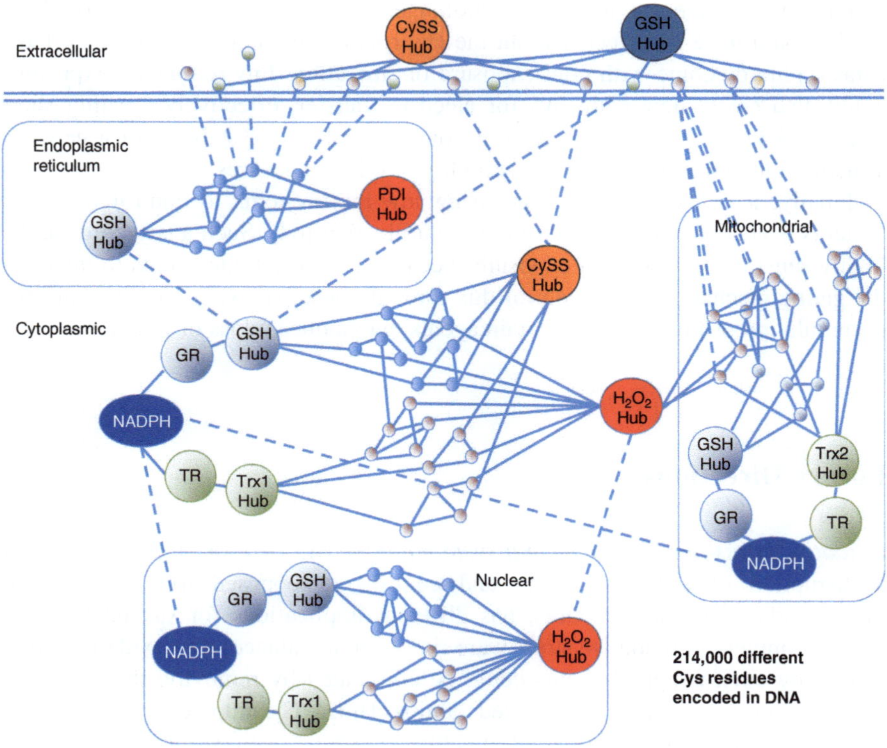

Fig. 12.6 Targets of oxidative stress include redox-sensitive thiols functioning in subcellular compartments. The human genome encodes 214,000 different cysteine residues. The resulting cysteine proteome is distributed among cellular compartments, with extracellular protein thiols (indicated as nodes) interacting with free cysteine and cystine, designated as a CySS redox-control hub, in the plasma and interstitial space. Solid connectors (edges) indicate direct interactions in a network structure. On the alveolar surface of the lung and lining of the gastrointestinal tract, GSH and GSSG appear to serve as the predominant redox-control hub. The extracellular thiol/disulfide systems are connected to intracellular systems (designated by broken lines) principally through transport systems or other indirect mechanisms. Within the cytoplasm, a CySS hub and a H_2O_2 hub function in the oxidation of cytoplasmic proteins. These oxidative reactions are ongoing and are countered by two major NADPH-dependent thiol reduction systems, represented by thioredoxin-1 (Trx1) and GSH/GSSG hubs. The relative rates for different proteins vary, with the majority of Cys residues being largely reduced under steady-state conditions. Trx1 is translocated into nuclei during oxidative stress; in the nuclei, Trx1 is maintained at a more reduced state than in the cytoplasm. Both nuclear Trx1 and nuclear GSSG hubs function are at least partially independent of respective cytoplasmic hubs. These hubs are important for redox control of transcription and are, therefore, critical nuclear targets of oxidative stress. Mitochondria contain Trx2 and GSH/GSSG hubs that control redox states of mitochondrial proteins. Lipophilic cationic oxidants can be especially toxic because they accumulate in mitochondria and disrupt bioenergetic control. In the secretory pathway, protein disulfide isomerase (PDI) is oxidized by an endoplasmic reticulum oxidase system (EROS) and functions as a redox hub controlling Cys oxidation in proteins destined for the plasma membrane or extracellular compartments. The range of organelle function in specialized tissues results in organ system–specific oxidative stress.

could govern tolerance to cold or prolonged starvation. Modification of these thiols would have no effect without the environmental challenge but would have catastrophic outcome with cold exposure or starvation. Environmental exposures could also create new pathways for electron transfer, effectively creating short circuits. Such short circuits may be accommodated by the stability of a network structure but create an overall vulnerability to other exposures.

Finally, a redox system design using low-flux systems to control high-flux systems is economical in terms of energy utilization [64]. Such a structure can be highly sensitive to low level exposures because it allows toxic agents to interrupt the gating mechanisms that control flux through these pathways. Because of this, accumulation of low levels of certain agents may induce a loss of function.

Future Directions

Oxidative stress and environmental sustainability. The view of oxidative stress as a disruption of redox signaling and control, as opposed to an imbalance of prooxidants and antioxidants, has different implications for sustainability in the environment. If oxidative stress were simply an imbalance, then oxidative stress caused by environmental agents could be eliminated by removing the source or directly compensated by increased antioxidants. Cleanup efforts generally approach the problem as though eliminating the source were sufficient. However, this simplistic approach, like the antioxidant supplementation trials, does not effectively address the problem with regard to environmental sustainability.

A critical issue is that complex systems respond to challenges with adaptation to maintain function. Because of this, subtoxic exposures can accumulate without overt signs of adverse effects. In the long term, this is not sustainable. Consequently, studies to evaluate impact can provide reassuring evidence that specific environmental challenges are not a problem, when the data may really mean that the biologic systems are effectively engineered to conceal hazards. While this may seem overly paranoid, millions of new chemicals have been synthesized in research efforts, and tens of thousands are in commercial use. Consequently, to consider sustainability in terms of oxidative stress, one must evaluate sustainability in terms of the impacts of cumulative exposures and acknowledge the reality that one cannot correct oxidative stress simply by shifting the balance with antioxidants.

Unfortunately, the scientific method is not well designed to deal with complex systems. The range of responses to multiple exposures within complex systems over long periods of exposure cannot be evaluated or predicted accurately with contemporary methods. Sustainability in the environment necessitates new ways to evaluate exposures and biologic responses.

From the standpoint of oxidative stress, a redox proteomics approach provides means for systems analysis. With a currently available liquid chromatography-mass spectrometry approach, one can measure the percent oxidation of specific cysteine residues in hundreds of proteins [108]. These Cys residues represent only a fraction of the total Cys encoded by the genome, which is about 200,000 in the 18,000 proteins in humans, but none-the-less provide an indication of the steady-state oxidation of the individual. The Cys residues change in oxidation state in response to exposures so that one could, in principle, establish sentinel species to examine environmental oxidative stress.

Two types of data can be obtained, one giving maps of electron transfer circuits and the other giving functional pathway maps linked by Cys oxidation [108]. Such a biomonitoring approach could be used to establish cumulative databases, providing a strategy to quantify deterioration of environmental conditions in terms of disruption of protein redox systems in the sentinel species. Of course, humans could be one of the sentinel species, and evaluation of protein oxidation in human samples could be useful to document oxidative stress. Similar information could be obtained from gene expression profiling; complete genomic information is available now for several species, so that multiple options are available to test the feasibility of such approaches.

Unfortunately, changes in redox proteomics or gene expression only reveal whether deterioration has occurred, and do not provide insight into the underlying causes. Exposure information will be needed to link environmental deterioration to specific causes. Current environmental surveillance methods are poorly equipped to address this problem because analyses are highly targeted to a relatively small number of hazardous agents. Newer methods are available to profile chemicals in biologic extracts, which show that thousands of chemicals can be detected in a single analysis [33]. Less than half of the chemicals detected in human plasma have been identified, meaning that humans are exposed to a large number of unidentified chemicals. Some of these may be environmental degradation products of known hazardous and nonhazardous chemicals. Whether these chemicals are helpful or harmful is not known, but multivariate analysis and modern bioinformatic methods can support studies linking alteration in metabolic characteristics with abundance of specific chemicals in biologic samples. Clearly, investment in new methods as well as cumulative environmental health data registries will be needed to address critical questions of environmental sustainability with regard to oxidative stress.

During the past decade, methods have been developed to allow study of oxidative stress within specific subcellular compartments [84, 109]. Mammalian cells contain about 214,000 Cys residues encoded in the genome, and redox biochemical approaches are beginning to provide an understanding of the organization of the redox network structure [108, 110, 111]. This research has revealed that oxidative stress is not uniform within cells but rather selectively affects specific pathways in compartments. The extracellular compartments of plasma and interstitial space are generally more oxidized than cellular compartments (Fig. 12.6) [84], have fewer antioxidant systems, and are vulnerable to oxidants. The lung alveoli, oral mucosa, and intestinal lumen are supplied with the thiol antioxidant GSH, and have protective enzymes associated with the mucous-lining layer. Thus, the maintenance of these surface barriers is critical for the protection of an organism because the lung lining is first exposed to airborne toxicants, while the intestine is first exposed to ingested toxicants.

Within cells, different redox systems function in the organelles. The endoplasmic reticulum (ER) and secretory pathway use protein disulfide isomerase (PDI) and an oxidase system (EROS) to oxidize proteins during processing for secretion ([112]; Fig. 12.6). Disruption of this oxidative pathway activates cell death through an ER stress mechanism [112, 113]. Nuclei and mitochondria are more reduced [84], and each has specific redox-sensitive proteins. Mitochondria have a unique thioredoxin-2, Trx2, and glutaredoxin-2, Grx2, while thioredoxin-1 and glutaredoxin-1 are found in the cytoplasm. Trx1 is translocated into nuclei during oxidative stress [114], and nuclei also contain Grx2, at least when over-expressed in cells [110]. Accumulating evidence indicates that most agents that cause oxidative stress do so by disrupting specific redox control pathways associated with specific organelles (Fig. 12.6).

Finally, there is a possibility that artificial intelligence may provide a more cost-effective means to monitor environmental sustainability. Irregularity of biologic data often reflects adaptability and health of the system. In essence, the accommodation of oxidative challenges can be expected to decrease the flexibility of the system. Thus, if appropriate methods were developed, it may be possible to evaluate health (or sustainability of a system) in terms of irregularity of metabolic or protein data [115]. Such an approach could potentially be accomplished using monofractality or multifractality, such as expressed by Hurst or Holder exponents [116]. While these are not traditional approaches used by experimental or environmental biologists, they could be very useful to approach the rather difficult problem of environmental sustainability in the face of such a large number of newly introduced and accumulating abundance of environmental agents.

In summary, oxidative stress is an important concern in environmental toxicology, both as a cause of acute toxicity and as a contributor to chronic toxicity. In terms of sustainability, agents causing acute toxicity due to oxidative stress are more readily identified and managed by containment and cleanup, in some cases complemented by use of metal chelators and free radical scavenging antioxidants. Agents that cause chronic toxicity or delayed reproductive or developmental toxicity are of greater concern because of the difficulty of identification and association with the toxicity. At present, there is no way to evaluate whether this is a minor or major problem, but halogenated hydrocarbons are commonly used. Some of these could accumulate in the biosphere, and most of these are likely to contribute to oxidative stress. Safety testing mostly is performed only on commercial materials and not on products formed by the modification of these materials after release into the environment. Mass spectrometry analyses of human plasma reveal thousands of unidentified ions with unknown biologic effects. Currently there is no environmental exposure census to reveal whether or not such chemicals are accumulating due to human activity. Thus, there is considerable opportunity and need to enhance understanding of the environmental load of chemicals that could accumulate and cause adverse effects due to oxidative stress [117–122].

Acknowledgment Research support provided by grants from the National Institutes of Environmental Health Sciences ES009047 and ES011195.

Bibliography

1. Sies H, Jones DP (2007) Oxidative stress. In: Fink G (ed) Encyclopedia of stress. Elsevier, San Diego, pp 45–48
2. Ruidavets JB, Cournot M, Cassadou S, Giroux M, Meybeck M, Ferrieres J (2005) Ozone air pollution is associated with acute myocardial infarction. Circulation 111:563–569
3. Seinfeld JH, Pandis SN (2006) Atmospheric chemistry and physics – from air pollution to climate change, 2nd edn. Wiley, New York, 1225 pp
4. Becana M, Moran JF, Iturbe-Ormaetxe I (1998) Iron-dependent oxygen free radical generation in plants subjected to environmental stress: toxicity and antioxidant protection. Plant Soil 201:137–147
5. Das P, Samantaray S, Rout GR (1997) Studies on cadmium toxicity in plants: a review. Environ Pollut 98:29–36
6. Klaassen CD, Liu J, Choudhuri S (1999) Metallothionein: an intracellular protein to protect against cadmium toxicity. Annu Rev Pharmacol Toxicol 39:267–294
7. Ragunathan N, Dairou J, Sanfins E, Busi F, Noll C, Janel N, Dupret JM, Rodrigues-Lima F (2010) Cadmium alters the biotransformation of carcinogenic aromatic amines by arylamine n-acetyltransferase xenobiotic-metabolizing enzymes: molecular, cellular, and in vivo studies. Environ Health Perspect 118:1685–1691
8. Hideg E, Vass I (1996) Uv-b induced free radical production in plant leaves and isolated thylakoid membranes. Plant Sci 115:251–260
9. Jurkiewicz BA, Buettner GR (1996) Epr detection of free radicals in uv-irradiated skin: mouse versus human. Photochem Photobiol 64:918–922
10. Bernhard WA (2009) Radical reaction pathways initiated by direct energy deposition in DNA by ionizing radiation. In: Greenberg MM (ed) Radical and radical ion reactivity in nucleic acid chemistry. Wiley, Hoboken, pp 41–68
11. Jackson SP, Bartek J (2009) The DNA-damage response in human biology and disease. Nature 461:1071–1078
12. Tafazoli S, Spehar DD, O'Brien PJ (2005) Oxidative stress mediated idiosyncratic drug toxicity. Drug Metab Rev 37:311–325
13. Araujo JA, Barajas B, Kleinman M, Wang X, Bennett BJ, Gong KW, Navab M, Harkema J, Sioutas C, Lusis AJ, Nel AE (2008) Ambient particulate pollutants in the ultrafine range promote early atherosclerosis and systemic oxidative stress. Circ Res 102:589–596
14. Siddique HR, Gupta SC, Mitra K, Bajpai VK, Mathur N, Murthy RC, Saxena DK, Chowdhuri DK (2008) Adverse effect of tannery waste leachates in transgenic drosophila melanogaster: role of Ros in modulation of Hsp70, oxidative stress and apoptosis. J Appl Toxicol 28:734–748
15. Matschullat J, Birmann K, Borba RP, Ciminelli V, Deschamps EM, Figueiredo BR, Gabrio T, Haßler S, Hilscher A, Junghänel I, de Oliveira N, Raßbach K, Schmidt H, Schwenk M, de Oliveira Vilhena MJ, Weidner U, Abmjbrz BP, Richard HL (2007) Long-term environmental impact of arsenic-dispersion in Minas Gerais, Brazil. In: Bundschuh J, Zevenhoven R, Loeppert RH (eds) Trace metals and other contaminants in the environment. Elsevier, Great Lakes, pp 365–382
16. Jones KC, de Voogt P (1999) Persistent organic pollutants (pops): state of the science. Environ Pollut 100:209–221
17. Ruzzin J, Petersen R, Meugnier E, Madsen L, Lock EJ, Lillefosse H, Ma T, Pesenti S, Sonne SB, Marstrand TT, Malde MK, Du ZY, Chavey C, Fajas L, Lundebye AK, Brand CL, Vidal H, Kristiansen K, Froyland L (2010) Persistent organic pollutant exposure leads to insulin resistance syndrome. Environ Health Perspect 118:465–471
18. Wania F, Mackay D (1996) Tracking the distribution of persistent organic pollutants. Environ Sci Technol 30:A390–A396
19. Apelberg BJ, Witter FR, Herbstman JB, Calafat AM, Halden RU, Needham LL, Goldman LR (2007) Cord serum concentrations of perfluorooctane sulfonate (PFOS) and

perfluorooctanoate (PFOA) in relation to weight and size at birth. Environ Health Perspect 115:1670–1676
20. Wallington TJ, Hurley MD, Xia J, Wuebbles DJ, Sillman S, Ito A, Penner JE, Ellis DA, Martin J, Mabury SA, Nielsen OJ, Sulbaek Andersen MP (2006) Formation of C7F15COOH (PFOA) and other perfluorocarboxylic acids during the atmospheric oxidation of 8:2 fluorotelomer alcohol. Environ Sci Technol 40:924–930
21. Rappaport SM (2010) Implications of the exposome for exposure science. J Expo Sci Environ Epidemiol 21:5–9
22. Wild CP (2005) Complementing the genome with an "Exposome": the outstanding challenge of environmental exposure measurement in molecular epidemiology. Cancer Epidemiol Biomark Prev 14:1847–1850
23. Sies H (1985) Oxidative stress: introductory remarks. In: Sies H (ed) Oxidative stress. Academic, London, pp 1–8
24. Jones DP (2006) Redefining oxidative stress. Antioxid Redox Signal 8:1865–1879
25. Voit E, Neves AR, Santos H (2006) The intricate side of systems biology. Proc Natl Acad Sci USA 103:9452–9457
26. Voit EO, Qi Z, Miller GW (2008) Steps of modeling complex biological systems. Pharmacopsychiatry 41(Suppl 1):S78–S84
27. Schuster SC (2008) Next-generation sequencing transforms today's biology. Nat Meth 5:16–18
28. Callinan PA, Feinberg AP (2006) The emerging science of epigenomics. Hum Mol Genet 15(Spec No 1):R95–R101
29. Bammler T, Beyer RP, Bhattacharya S, Boorman GA, Boyles A, Bradford BU, Bumgarner RE, Bushel PR, Chaturvedi K, Choi D, Cunningham ML, Deng S, Dressman HK, Fannin RD, Farin FM, Freedman JH, Fry RC, Harper A, Humble MC, Hurban P, Kavanagh TJ, Kaufmann WK, Kerr KF, Jing L, Lapidus JA, Lasarev MR, Li J, Li YJ, Lobenhofer EK, Lu X, Malek RL, Milton S, Nagalla SR, O'Malley JP, Palmer VS, Pattee P, Paules RS, Perou CM, Phillips K, Qin LX, Qiu Y, Quigley SD, Rodland M, Rusyn I, Samson LD, Schwartz DA, Shi Y, Shin JL, Sieber SO, Slifer S, Speer MC, Spencer PS, Sproles DI, Swenberg JA, Suk WA, Sullivan RC, Tian R, Tennant RW, Todd SA, Tucker CJ, Van Houten B, Weis BK, Xuan S, Zarbl H (2005) Standardizing global gene expression analysis between laboratories and across platforms. Nat Meth 2:351–356
30. Li H, Grigoryan H, Funk WE, Lu SS, Rose S, Williams ER, Rappaport SM (2010) Profiling Cys34 adducts of human serum albumin by fixed-step selected reaction monitoring. Mol Cell Proteomics (Ahead of print)
31. Funk WE, Li H, Iavarone AT, Williams ER, Riby J, Rappaport SM (2010) Enrichment of cysteinyl adducts of human serum albumin. Anal Biochem 400:61–68
32. Evans AM, DeHaven CD, Barrett T, Mitchell M, Milgram E (2009) Integrated, nontargeted ultrahigh performance liquid chromatography/electrospray ionization tandem mass spectrometry platform for the identification and relative quantification of the small-molecule complement of biological systems. Anal Chem 81:6656–6667
33. Johnson JM, Yu T, Strobel FH, Jones DP (2010) A practical approach to detect unique metabolic patterns for personalized medicine. Analyst 135:2864–2870
34. Namakkal Soorappan R, Varadharaj S, Khanderao GD, Davidson CJ, Kannan S, Firpo MA, Zweier J, Benjamin I (2010) Sustained activation of nrf2/are signaling promotes reductive stress in the human mutant protein aggregation cardiomyopathy in mice. Antioxid Redox Signal (Ahead of print)
35. Zhang X, Min X, Li C, Benjamin IJ, Qian B, Ding Z, Gao X, Yao Y, Ma Y, Cheng Y, Liu L (2010) Involvement of reductive stress in the cardiomyopathy in transgenic mice with cardiac-specific overexpression of heat shock protein 27. Hypertension 55:1412–1417
36. Kappus H, Sies H (1981) Toxic drug effects associated with oxygen metabolism: redox cycling and lipid peroxidation. Experientia 37:1233–1241
37. Priestley J (1775) Letter to sir john pringle. Philos Trans 65:384–390

38. Bitterman N, Bitterman H (2006) Oxygen toxicity. In: Mathiew D (ed) Handbook on hyperbaric medicine. Dordrecht, Springer
39. Silverman WA (2004) A cautionary tale about supplemental oxygen: the albatross of neonatal medicine. Pediatrics 113:394–396
40. The Alpha-Tocopherol, Beta Carotene Cancer Prevention Study Group (1994) The effect of vitamin E and beta carotene on the incidence of lung cancer and other cancers in male smokers. N Engl J Med 330:1029–1035
41. Gruppo Italiano Per Lo Studio Della Sopravvivenza Nell'infarto Miocardico (1999) Dietary supplementation with n-3 polyunsaturated fatty acids and vitamin e after myocardial infarction: results of the gissi-prevenzione trial. Lancet 354:447–455
42. Age-Related Eye Disease Study Research Group (2001) A randomized, placebo-controlled, clinical trial of high-dose supplementation with vitamins C and E, beta carotene, and zinc for age-related macular degeneration and vision loss: AREDS Report no. 8. Arch Ophthalmol 119:1417–1436
43. de Gaetano G (2001) Low-dose aspirin and vitamin e in people at cardiovascular risk: a randomised trial in general practice. Collaborative group of the primary prevention project. Lancet 357:89–95
44. Hennekens CH, Buring JE, Manson JE, Stampfer M, Rosner B, Cook NR, Belanger C, LaMotte F, Gaziano JM, Ridker PM, Willett W, Peto R (1996) Lack of effect of long-term supplementation with beta carotene on the incidence of malignant neoplasms and cardiovascular disease. N Engl J Med 334:1145–1149
45. Lonn E, Bosch J, Yusuf S, Sheridan P, Pogue J, Arnold JM, Ross C, Arnold A, Sleight P, Probstfield J, Dagenais GR (2005) Effects of long-term vitamin E supplementation on cardiovascular events and cancer: a randomized controlled trial. J Am Med Assoc 293:1338–1347
46. Yusuf S, Dagenais G, Pogue J, Bosch J, Sleight P (2000) Vitamin e supplementation and cardiovascular events in high-risk patients. The heart outcomes prevention evaluation study investigators. N Engl J Med 342:154–160
47. Zureik M, Galan P, Bertrais S, Mennen L, Czernichow S, Blacher J, Ducimetiere P, Hercberg S (2004) Effects of long-term daily low-dose supplementation with antioxidant vitamins and minerals on structure and function of large arteries. Arterioscler Thromb Vasc Biol 24:1485–1491
48. Williams KJ, Fisher EA (2005) Oxidation, lipoproteins, and atherosclerosis: which is wrong, the antioxidants or the theory? Curr Opin Clin Nutr Metab Care 8:139–146
49. Loew O (1901) Catalase, a new enzyme of general occurrence. U.S. Department of Agriculture Report no. 68, pp 47–55
50. Chelikani P, Fita I, Loewen PC (2004) Diversity of structures and properties among catalases. Cell Mol Life Sci 61:192–208
51. Maehly AC, Chance B (1954) The assay of catalases and peroxidases. Meth Biochem Anal 1:357–424
52. Keilin D (1925) On cytochrome, a respiratory pigment, common to animals, yeast, and higher plants. Proc R Soc Lond B – Containing Pap Biol Character 98:312–339
53. Dutton PL, Wilson DF (1974) Redox potentiometry in mitochondrial and photosynthetic bioenergetics. Biochim Biophys Acta 346:165–212
54. Wolf G (2005) The discovery of the antioxidant function of vitamin E: the contribution of Henry A. Mattill J Nutr 135:363–366
55. Mills GC (1957) Hemoglobin catabolism. I. Glutathione peroxidase, an erythrocyte enzyme which protects hemoglobin from oxidative breakdown. J Biol Chem 229:189–197
56. Rotruck JT, Pope AL, Ganther HE, Swanson AB, Hafeman DG, Hoekstra WG (1973) Selenium: biochemical role as a component of glutathione peroxidase. Science 179:588–590
57. Recknagel RO, Ghoshal AK (1966) Lipoperoxidation of rat liver microsomal lipids induced by carbon tetrachloride. Nature 210:1162–1163

58. Recknagel RO, Ghoshal AK (1966) Lipoperoxidation as a vector in carbon tetrachloride hepatotoxicity. Lab Invest 15:132–148
59. Harman D (1956) Aging: a theory based on free radical and radiation chemistry. J Gerontol 11:298–300
60. Tribble DL, Aw TY, Jones DP (1987) The pathophysiological significance of lipid peroxidation in oxidative cell injury. Hepatology 7:377–386
61. Fridovich I (1997) Superoxide anion radical ($O_2^{-\bullet}$), superoxide dismutases, and related matters. J Biol Chem 272:18515–18517
62. McCord JM, Fridovich I (1969) Superoxide dismutase. An enzymic function for erythrocuprein (hemocuprein). J Biol Chem 244:6049–6055
63. Chance B, Sies H, Boveris A (1979) Hydroperoxide metabolism in mammalian organs. Physiol Rev 59:527–605
64. Jones DP (2006) Disruption of mitochondrial redox circuitry in oxidative stress. Chem Biol Interact 163:38–53
65. Jones DP (2006) Extracellular redox state: refining the definition of oxidative stress in aging. Rejuvenation Res 9:169–181
66. Griendling KK (2006) Nadph oxidases: new regulators of old functions. Antioxid Redox Signal 8:1443–1445
67. Lambeth JD (2004) Nox enzymes and the biology of reactive oxygen. Nat Rev Immunol 4:181–189
68. Orrenius S (2004) Mitochondrial regulation of apoptotic cell death. Toxicol Lett 149:19–23
69. Morrow JD, Roberts LJ 2nd (1999) Mass spectrometric quantification of F2-isoprostanes in biological fluids and tissues as measure of oxidant stress. Meth Enzymol 300:3–12
70. Levine RL, Garland D, Oliver CN, Amici A, Climent I, Lenz AG, Ahn BW, Shaltiel S, Stadtman ER (1990) Determination of carbonyl content in oxidatively modified proteins. Meth Enzymol 186:464–478
71. Kagan VE, Borisenko GG, Tyurina YY, Tyurin VA, Jiang J, Potapovich AI, Kini V, Amoscato AA, Fujii Y (2004) Oxidative lipidomics of apoptosis: redox catalytic interactions of cytochrome c with cardiolipin and phosphatidylserine. Free Radic Biol Med 37:1963–1985
72. Go YM, Park H, Koval M, Orr M, Reed M, Liang Y, Smith D, Pohl J, Jones DP (2010) A key role for mitochondria in endothelial signaling by plasma cysteine/cystine redox potential. Free Radic Biol Med 48:275–283
73. Morrow JD, Roberts LJ 2nd (2002) Mass spectrometric quantification of F2-isoprostanes as indicators of oxidant stress. Meth Mol Biol 186:57–66
74. Brigelius-Flohe R (1999) Tissue-specific functions of individual glutathione peroxidases. Free Radic Biol Med 27:951–965
75. Fisher AB, Dodia C, Manevich Y, Chen JW, Feinstein SI (1999) Phospholipid hydroperoxides are substrates for non-selenium glutathione peroxidase. J Biol Chem 274:21326–21334
76. Wang Y, Phelan SA, Manevich Y, Feinstein SI, Fisher AB (2006) Transgenic mice overexpressing peroxiredoxin 6 show increased resistance to lung injury in hyperoxia. Am J Respir Cell Mol Biol 34:481–486
77. Cai J, Jones DP (1998) Superoxide in apoptosis. Mitochondrial generation triggered by cytochrome C loss. J Biol Chem 273:11401–11404
78. Jones DP (2008) Radical-free biology of oxidative stress. Am J Physiol Cell Physiol 295: C849–C868
79. Halliwell B (2007) Oxidative stress and cancer: have we moved forward? Biochem J 401:1–11
80. Thomas DD, Liu X, Kantrow SP, Lancaster JR Jr (2001) The biological lifetime of nitric oxide: implications for the perivascular dynamics of NO and O_2. Proc Natl Acad Sci USA 98:355–360
81. Thomas DD, Ridnour LA, Espey MG, Donzelli S, Ambs S, Hussain SP, Harris CC, DeGraff W, Roberts DD, Mitchell JB, Wink DA (2006) Superoxide fluxes limit nitric oxide-induced signaling. J Biol Chem 281:25984–25993

82. Fridovich I (1970) Quantitative aspects of the production of superoxide anion radical by milk xanthine oxidase. J Biol Chem 245:4053–4057
83. Jones DP, Eklow L, Thor H, Orrenius S (1981) Metabolism of hydrogen peroxide in isolated hepatocytes: relative contributions of catalase and glutathione peroxidase in decomposition of endogenously generated H_2O_2. Arch Biochem Biophys 210:505–516
84. Go YM, Jones DP (2008) Redox compartmentalization in eukaryotic cells. Biochim Biophys Acta 1780(11):1273–1290
85. Barja G (2002) Endogenous oxidative stress: relationship to aging, longevity and caloric restriction. Ageing Res Rev 1:397–411
86. Doudican NA, Song B, Shadel GS, Doetsch PW (2005) Oxidative DNA damage causes mitochondrial genomic instability in saccharomyces cerevisiae. Mol Cell Biol 25:5196–5204
87. Calabrese EJ, Baldwin LA (2003) The hormetic dose-response model is more common than the threshold model in toxicology. Toxicol Sci 71:246–250
88. Winterbourn CC, Metodiewa D (1994) The reaction of superoxide with reduced glutathione. Arch Biochem Biophys 314:284–290
89. Giles GI, Tasker KM, Jacob C (2001) Hypothesis: the role of reactive sulfur species in oxidative stress. Free Radic Biol Med 31:1279–1283
90. Starke DW, Chock PB, Mieyal JJ (2003) Glutathione-thiyl radical scavenging and transferase properties of human glutaredoxin (thioltransferase). Potential role in redox signal transduction. J Biol Chem 278:14607–14613
91. Salmeen A, Andersen JN, Myers MP, Meng TC, Hinks JA, Tonks NK, Barford D (2003) Redox regulation of protein tyrosine phosphatase 1B involves a sulphenyl-amide intermediate. Nature 423:769–773
92. van Montfort RL, Congreve M, Tisi D, Carr R, Jhoti H (2003) Oxidation state of the active-site cysteine in protein tyrosine phosphatase 1B. Nature 423:773–777
93. Biteau B, Labarre J, Toledano MB (2003) Atp-dependent reduction of cysteine-sulphinic acid by S. cerevisiae sulphiredoxin. Nature 425:980–984
94. Woo HA, Chae HZ, Hwang SC, Yang KS, Kang SW, Kim K, Rhee SG (2003) Reversing the inactivation of peroxiredoxins caused by cysteine sulfinic acid formation. Science 300:653–656
95. Woo HA, Kang SW, Kim HK, Yang KS, Chae HZ, Rhee SG (2003) Reversible oxidation of the active site cysteine of peroxiredoxins to cysteine sulfinic acid. Immunoblot detection with antibodies specific for the hyperoxidized cysteine-containing sequence. J Biol Chem 278:47361–47364
96. Kim HY, Gladyshev VN (2004) Methionine sulfoxide reduction in mammals: characterization of methionine-r-sulfoxide reductases. Mol Biol Cell 15:1055–1064
97. Stadtman ER, Van Remmen H, Richardson A, Wehr NB, Levine RL (2005) Methionine oxidation and aging. Biochim Biophys Acta 1703:135–140
98. Gladyshev VN, Hatfield DL (1999) Selenocysteine-containing proteins in mammals. J Biomed Sci 6:151–160
99. Kryukov GV, Castellano S, Novoselov SV, Lobanov AV, Zehtab O, Guigo R, Gladyshev VN (2003) Characterization of mammalian selenoproteomes. Science 300:1439–1443
100. Stadtman ER, Levine RL (2003) Free radical-mediated oxidation of free amino acids and amino acid residues in proteins. Amino Acids 25:207–218
101. Carp H, Miller F, Hoidal JR, Janoff A (1982) Potential mechanism of emphysema: alpha 1-proteinase inhibitor recovered from lungs of cigarette smokers contains oxidized methionine and has decreased elastase inhibitory capacity. Proc Natl Acad Sci USA 79:2041–2045
102. Hansel A, Kuschel L, Hehl S, Lemke C, Agricola HJ, Hoshi T, Heinemann SH (2002) Mitochondrial targeting of the human peptide methionine sulfoxide reductase (MSRA), an enzyme involved in the repair of oxidized proteins. FASEB J 16:911–913
103. Weissbach H, Etienne F, Hoshi T, Heinemann SH, Lowther WT, Matthews B, St John G, Nathan C, Brot N (2002) Peptide methionine sulfoxide reductase: structure, mechanism of action, and biological function. Arch Biochem Biophys 397:172–178

104. Miller RA, Buehner G, Chang Y, Harper JM, Sigler R, Smith-Wheelock M (2005) Methionine-deficient diet extends mouse lifespan, slows immune and lens aging, alters glucose, T4, IGF-I and insulin levels, and increases hepatocyte MIF levels and stress resistance. Aging Cell 4:119–125
105. Moskovitz J, Bar-Noy S, Williams WM, Requena J, Berlett BS, Stadtman ER (2001) Methionine sulfoxide reductase (MSRA) is a regulator of antioxidant defense and lifespan in mammals. Proc Natl Acad Sci USA 98:12920–12925
106. Richie JP Jr, Leutzinger Y, Parthasarathy S, Malloy V, Orentreich N, Zimmerman JA (1994) Methionine restriction increases blood glutathione and longevity in F344 rats. FASEB J 8:1302–1307
107. Sanz A, Caro P, Ayala V, Portero-Otin M, Pamplona R, Barja G (2006) Methionine restriction decreases mitochondrial oxygen radical generation and leak as well as oxidative damage to mitochondrial DNA and proteins. FASEB J 20:1064–1073
108. Jones DP, Go YM (2010) Mapping the cysteine proteome: analysis of redox-sensing thiols. Curr Opin Chem Biol 15:1–10
109. Hansen JM, Go YM, Jones DP (2006) Nuclear and mitochondrial compartmentation of oxidative stress and redox signaling. Annu Rev Pharmacol Toxicol 46:215–234
110. Gallogly MM, Starke DW, Mieyal JJ (2009) Mechanistic and kinetic details of catalysis of thiol-disulfide exchange by glutaredoxins and potential mechanisms of regulation. Antioxid Redox Signal 11:1059–1081
111. Kemp M, Go YM, Jones DP (2008) Nonequilibrium thermodynamics of thiol/disulfide redox systems: a perspective on redox systems biology. Free Radic Biol Med 44:921–937
112. Cenci S, Sitia R (2007) Managing and exploiting stress in the antibody factory. FEBS Lett 581:3652–3657
113. Todd DJ, Lee AH, Glimcher LH (2008) The endoplasmic reticulum stress response in immunity and autoimmunity. Nat Rev Immunol 8:663–674
114. Go YM, Jones DP (2010) Redox control systems in the nucleus: mechanisms and functions. Antioxid Redox Signal 13:489–509
115. Goldberger AL, Amaral LA, Hausdorff JM, Ivanov P, Peng CK, Stanley HE (2002) Fractal dynamics in physiology: alterations with disease and aging. Proc Natl Acad Sci USA 99(Suppl 1):2466–2472
116. Ivanov PC, Amaral LA, Goldberger AL, Havlin S, Rosenblum MG, Struzik ZR, Stanley HE (1999) Multifractality in human heartbeat dynamics. Nature 399:461–465
117. Greenberg M (2009) Radical and radical ion reactivity in nucleic acid chemistry. Wiley, New York, 466 pp
118. Sarkar D, Datta R, Hannigan R (2007) Concepts and applications in environmental geochemistry. Elsevier, Amsterdam, 761 pp
119. Keilin D (1966) The history of cell respiration and cytochrome. In: Keilin D (ed) The history of cell respiration and cytochrome. Cambridge University Press, Cambridge
120. Finkel T (2003) Oxidant signals and oxidative stress. Curr Opin Cell Biol 15:247–254
121. Wu JJ, Quijano C, Chen E, Liu H, Cao L, Fergusson MM, Rovira II, Gutkind S, Daniels MP, Komatsu M, Finkel T (2009) Mitochondrial dysfunction and oxidative stress mediate the physiological impairment induced by the disruption of autophagy. Aging 1:425–437
122. Swanson RA, Siesjo B (2002) Oxidative stress in cerebrovascular disease. In: Swanson RA, Siesjo B (eds) Oxidative stress in cerebrovascular disease. Cambridge University Press, Cambridge

Chapter 13
Harmful Algal Blooms

Timothy I. McLean and Geoffrey A. Sinclair

Glossary

Bioaccumulation	The retention of substances, various phycotoxins in this context, in the living tissues of organisms at concentrations that are higher than are found in the environment.
Cyanobacteria	A phylum of bacteria, once called blue-green algae, that appears in the fossil record as far back as 3.5 billion years all species are photosynthetic, but different species have adapted to aquatic (fresh or marine) or terrestrial (soil, bare rock, or symbiotic associations with plants) environments some species can form colonies, and some colony-forming cells can specialize or differentiate, e.g., to form nitrogen-fixing heterocysts.
Depuration	The cleansing or purification of impurities (toxins) from the tissues of the body.
Diatoms	A large group of microalgae distinguished by having siliceous cell walls called frustules single cells range in size from 2 μm to over 2 mm and are unicellular, often forming colonies of many cells they contain the pigments chlorophylls a and c, betacarotene, fucoxanthin, diatoxanthin, and diadinoxanthin they

This chapter was originally published as part of the Encyclopedia of Sustainability Science and Technology edited by Robert A. Meyers. DOI:10.1007/978-1-4419-0851-3

T.I. McLean (✉)
Department of Biological Sciences, University of Southern Mississippi,
118 College Dr., Hattiesburg, MS 39406, USA
e-mail: timothy.mclean@usm.edu

G.A. Sinclair
DeFelice Marine Center, Louisiana Universities Marine Consortium,
8124 Highway 56, Chauvin, LA 70344, USA
e-mail: gsinclair@lumcon.edu

E.A. Laws (ed.), *Environmental Toxicology: Selected Entries from the Encyclopedia of Sustainability Science and Technology*, DOI 10.1007/978-1-4614-5764-0_13,

constitute the predominant fraction of the phytoplankton community in fresh water and marine environments, found in the water column and in the benthos, being both epiphytic and edaphic.

Dinoflagellates	A large group of motile microalgae differentiated from other algal groups by their thecal morphology (cell covering) having either an armored form or a smooth form armored forms contain distinctly ornamented polysaccharide plates they contain two dissimilar flagellae, and species may be exclusively heterotrophic, mixotrophic, or exclusively phototrophic – those that are capable of photosynthesis contain the pigments chlorophylls a and c, dinoxanthin, diadinoxanthin, and either fucoxanthin or peridinin as their major carotenoid, depending upon the species.
Eutrophication	The introduction of excessive levels of nutrients such as nitrates and phosphorous into an aquatic ecosystem.
Harmful algae	A small fraction of algal species that can have disruptive or devastating effects to an ecosystem, public health, and/or various economies the harmful nature of different species of algae can be manifested in a variety of ways but primarily through the production of one or more phycotoxins.
Phycotoxins	Toxic molecules produced by harmful algae collectively, these molecules are chemically diverse and can produce an array of effects in intoxicated organisms ranging from gastrointestinal, neurotoxic, hepatotoxic, carcinogenic, and more.
Trophic transfer	The transfer of energy and/or substances, e.g., toxins, from one (trophic) level of a food chain to another, typically higher, (trophic) level within the food chain.

Definition of the Subject and Its Importance

Harmful algal blooms (HABs) pose threats to the environment, public health, and a variety of commercial interests and industries. A single bloom can lead to devastating outcomes, including large mortalities of marine organisms (e.g., fish kills); toxic contamination of filter-feeding organisms such as bivalve shellfish that subsequently enter the market for distribution to consumers; economic hardships for fisheries, aquaculture, and recreational- and tourism-related industries; and a compromised quality of life for people living or working along affected shorelines. Depending upon the size of the bloom, its duration, and the number and types of impacts produced, a single bloom can generate multimillion-dollar losses spanning from local to international economies (see Oceans and Human Health, Social and Economic Impacts for additional information on HAB-related economic impacts).

As well as a current concern, the issue of HABs will continue to grow in the future, as HABs are being reported at higher frequencies, lasting for longer periods of time, covering larger areas of shoreline or acreage of water, and spreading to regions that historically have not experienced blooms [1–8]. A number of factors have been cited as contributing to this global increase, but each algal species responds to a different set of variables. For many species, degraded water quality is one of the major reasons for the increase in HABs and HAB effects [9, 10]. As eutrophication, urbanization, and climate change continue to escalate, the increasing impact of HABs is a threat to the economic sustainability of many coastal and marine industries, public health, and the health of natural resources.

Introduction

"Algae" is not a true taxonomic designation. Algae are polyphyletic, i.e., species under this umbrella term can be found on numerous branches within the eukaryotic lineage. The diversity of organisms within this assemblage make it difficult to provide a concise definition of algae (reviewed in Ref. [11]), but in general terms algae can be described as simple (i.e., lacking the structures and vasculature characteristic of land plants), mostly aquatic, photosynthetic, and ranging in form from unicellular (microalgae) to multicellular (macroalgae). For historical and physiological reasons, this artificial grouping also includes the prokaryotic cyanobacteria, formerly known as blue-green algae. The inclusion of cyanobacteria is further strengthened in the context of discussions about harmful algae (see below).

Algal blooms denote an increase in the population of a single species of algae within a given area. The increase may be a result of an increased growth rate or the accumulation/aggregation of cells due to, respectively, an infusion of nutrients or physical features such as fronts, a stratified water column, or current entrainment [9, 12, 13]. Algal blooms can form for a multitude of reasons, and the reasons for and the characteristics of a bloom vary by the species. Algal blooms occur worldwide in fresh, brackish, and marine environments. If the cell concentration becomes sufficiently dense, the bloom may discolor the water to produce a "red tide," although, depending on the organism, the water may actually appear green, brown, or gold, instead of red. Although the term red tide has taken on a negative connotation, the majority of red tides occur as natural, seasonal phenomenon that produce beneficial rather than harmful consequences. For example, in the Southern Ocean, successions of blooms produced by different algal species during the spring and early summer months anchor food chains that feed a multitude of organisms from birds to whales [14]and sustains an important fishery that nets >100,000 t of krill per year [15]. Less than a tenth of the 3,400–4,100 identified algal species are known bloom formers, and less than a quarter of those (2% of the total) have the capacity to form a harmful algal bloom [16, 17].

Algal blooms are designated as harmful when they have detrimental effects to other organisms or ecosystem functions either directly or indirectly. In some cases, the mere presence of a high number of cells in the water column increases cell–cell collisions that may damage or kill other planktonic organisms. For example, some microalgae, particularly diatoms with extensive silicate setae, can cause physical trauma when they pierce or tear the cellular structure of other microbes or the gills of fish and crustaceans [1, 18, 19]. Indirectly, harmful effects include changes in resource competition or decreasing ecosystem health by producing biomass whose decomposition fuels the creation of hypoxic or anoxic conditions [20, 21]. Other harmful species may produce allelopathic molecules [22], allelochemicals [23], high amounts of ammonium [24], or extracellular polymeric secretions that create viscous or mucilaginous barriers in the water [25, 26]. The majority of our attention, however, has been focused on those algae that produce one or more toxic molecules referred to as phycotoxins. Direct contact with, inhalation of, or ingestion of the phycotoxins by humans (or other animals) can cause a range of effects from mild irritation to severe gastrointestinal distress to death. Some of the more hazardous phycotoxins can have hemolytic, hemoagglutinating, neurotoxic, or hepatotoxic effects, or they can promote the formation of tumors, e.g., in the colon (see below for specific examples and references). As described below, these toxins can vector through food webs as they pass from one trophic level to another. During these trophic transfers, toxins may bioaccumulate in tissues or undergo changes in potency. The effects of phycotoxins in the natural environment can lead to massive mortalities of fish, birds, marine mammals, and other less well-characterized marine organisms creating miles of coastline and hectares of water littered with dead animals.

A growing concern about HABs is the real or perceived threat that they are increasing year to year as evidenced by: (1) areas that have historically experienced blooms sporadically, or even rarely, are seeing an increase in frequency; (2) when blooms do occur, they are lasting longer and/or covering larger expanses of coastline; (3) new species of toxic algae are being discovered; (4) species that were previously described as nontoxic are now showing the capacity to be toxic; and (5) the biogeography of certain species is expanding [1–8]. Certainly, our ability and impetus to detect harmful species and their respective toxins has increased over the last 15 years, but the increased awareness and concern about HABs cannot fully explain the explosion in the number and extent of blooms around the world. The most likely culprits, although not for each species, are the introduction of species to new environments through mechanisms such as transport in ballast water and the alteration of the natural environment through mechanisms such as urbanization of our coastlines, diversion of natural water flows, draining of wetlands, introduction of new or increased sources of nutrients from aquaculture practices and runoff of urban and agricultural regions, and climate change. Any of these factors, or combinations of these factors, may produce an environment conducive for HAB species to bloom.

The following sections will highlight four of the better-studied HAB taxa (two genera: *Alexandrium* and *Pseudo-nitzschia*, one species: *Karenia brevis*, and one phylum: cyanobacteria) in terms of why they are harmful, the economic costs and types of impacts associated with each respective bloom, the methods used to detect the cells and/or associated toxin(s), and what is known in regard to what initiates and/or maintains the blooms. For comprehensive overviews of other HAB species or more information about the taxa described herein, refer to recommended readings at the end or the embedded reviews in each section.

Alexandrium

Description

The dinoflagellate genus *Alexandrium* currently contains greater than 30 species [27]. Members of the genus are armored, have a defined cellular and nuclear morphology, occur as either single cells or as chains of two or more cells, and are widely distributed in temperate regions of the Mediterranean and northern and southern Atlantic and Pacific Oceans [27]. Molecular data support the monophyletic grouping of *Alexandrium* spp. [28]. The same data challenge the division of species within the genus, however, as DNA-based assays show that cells tend to cluster based on geography rather than morphospecies designations [29–33]. Twelve *Alexandrium* spp. are confirmed to produce paralytic shellfish toxins ([34] and references therein). Members of the *Alexandrium tamarense* morphospecies complex (*A. tamarense, A. catenella*, and *A. fundyense*) are responsible for the majority of blooms along northeastern and northwestern North American coastlines.

Toxicity

Paralytic shellfish toxins (PSTs) consist of saxitoxins and at least 24 saxitoxin-like congeners (reviewed in Ref. [35]). (It is important to note that, while not discussed here, other dinoflagellates, principally *Gymnodinium catenatum* and *Pyrodinium bahamense* var. *compressum*, and some cyanobacterial species can also produce PSTs (reviewed in Ref. [35]).) The potency of different PSTs ranges over a scale of one to three orders of magnitude [36], and those with a sulfocarbamoyl group are the most toxic (reviewed in Ref. [35, 37]). The name of this grouping of toxins derives from the syndrome, paralytic shellfish poisoning (PSP), which is produced as a result of consuming seafood, primarily bivalves, containing PSTs. PSTs, such as saxitoxin, bind voltage-gated sodium channels on the membranes of electrically excitable cells such as neurons and block the flow of sodium ions through the channel [38, 39]. The blockage prevents depolarization of the cell [40], which causes a range of gastrointestinal and neurological symptoms that vary with the exact composition of various PSTs and the susceptibility of the individual. Symptoms include tingling around the mouth, numbness of the extremities, weakness, nausea, and vomiting, among others. If a sufficient dose is ingested, neurons will be unable to stimulate muscle contractions. Complete muscle relaxation leads to paralysis, and if the paralysis extends to the muscles controlling breathing, individuals can die from asphyxiation [41].

PSTs can vector through marine food webs, having been found in zooplankton and zooplanktivorous fish (reviewed in Ref. [42]). The bioaccumulation of PSTs in these lower trophic level organisms has been implicated in mortalities of large numbers of fish, oceanic and coastal birds, and marine mammals, including humpback and right whales, which consume whole fish [4, 43–48]. Human intoxication from eating fish is rare since the PSTs accumulate in the viscera, which is typically not eaten by people. The most common route for human intoxication is through consumption of contaminated bivalves (e.g., mussels and clams), although, in rare cases, other organisms (e.g., gastropods, crustaceans, and fish) have served as vectors [49, 50]. Shellfish are indiscriminate filter feeders; they can consume large quantities of microalgae, are relatively immune to the PSTs, and bioaccumulate the toxins in their tissues. Different bivalve species have different tolerances to PSTs, and having a high tolerance allows an organism to accumulate more toxins. An example of an organism with a high tolerance is the blue mussel (*Mytilus edulis*). In Alaskan waters, *M. edulis* can bioaccumulate greater than 20 mg of saxitoxin/100 g mussel tissue – more than enough saxitoxin to kill a person [51]. Typically, 1–2 months after moderate blooms have terminated, the majority of bivalves will have depurated the toxins and will be considered safe for harvest and consumption. By contrast, butter clams bind to saxitoxin with high affinity and can remain toxic for up to 3 years after a bloom has occurred [51]. In some bivalves, PSTs can undergo biotransformation, possibly becoming even more toxic than the original source toxins [52].

Economics

PSTs are heat and acid stable, so cooking not only does not detoxify seafood but also may convert some congeners to more toxic forms. The only way to avoid PSP incidents is to prevent the harvesting of contaminated marine organisms or process them in such a way to reduce or minimize toxicity. For example, adductor muscles of sea scallops, which typically contains very low concentrations of PSTs, are considered safe if separated from the rest of the scallop while at sea before being frozen [53]. If the whole organism is frozen, then PSTs may leech from other tissues into the adductor muscle. For a review of other methods to reduce toxicity in seafood, see [53].

Alternatively, the only mechanism for determining whether to harvest/not harvest or to allow shellfish beds to be open or closed is to monitor for the presence of PSTs in the relevant shellfish tissues. Unlike other HAB species, detection and quantifying of PSP-producing microalgal cells are not sufficient for such management decisions – the collection of tissues and analysis for PSTs is required. These steps add costs to monitoring efforts performed by the relevant agencies. Costs for monitoring can become expensive especially for states with extensive coastlines. Alaska, e.g., only samples random sites and, therefore, discourages recreational harvesting, which leads to a severe underutilization of the endemic shellfish resources [8]. Of course, in a grand economic sense, the lost income and sales are, in part, balanced by reduced monitoring and public health costs.

In terms of economic consequences of PST-producing blooms, only a few case studies have been performed to quantify in dollars how extensive these impacts are. In the Pacific northwest, an *Alexandrium* bloom in 1997 created toxic oysters along the coast of Washington state that resulted in approximately $9.45 million of lost revenue during a 2- to 3-week industry closure. An extended closure (2 months) of shellfish beds in South Puget Sound resulted in about $27 million in losses (Nosho in Ref. [54]). (Note: All dollar figures throughout this article are represented in 2011 values.) A similar bloom that closed beds along Washington, Oregon, and California state coastlines for a month in 1980 cost oyster growers $2.5 million [55]. During the same year in the northeastern United States, a bloom of *A. tamarense* closed the entire coastline of Maine to shellfish harvesting. The value of the lost harvest was placed at $6.6 million, and the total economic impact was placed at $20 million [56]. In 2005, a massive bloom extended over the majority of the New England coastline, resulting in $3.1 million/week in lost revenue for the affected seafood fisheries [57]. Later analysis determined that the bloom impacted the state of Massachusetts by greater than $20.7 million and at least $2.7 million to the state of Maine [58].

Detection

A number of molecular techniques have been developed to rapidly and correctly identify cells as a particular *Alexandrium* species. The majority of the efforts to differentiate species have focused on molecular sequence data, such as the sequence of the ribosomal RNA gene for either the small or large ribosomal subunit or the internal transcribed spacer [29, 30, 32, 59–61]. Depending upon the marker or technique used, the number of defining ribotypes for *Alexandrium* varies from 5 to 11 [30, 32, 60]. Despite the discrepancy in the number of ribotypes, the results of these assays show that *Alexandrium* cells from the same geographic region are more closely related to each other than to cells from different regions, even if the cells from the different areas are of the same species. The conclusion drawn from the data is that *Alexandrium* is monophyletic, but geographically isolated populations evolved independently subsequent to global dispersion [31].

More effort has gone to developing methods for detecting and quantifying PSTs. The standard assay is the mouse bioassay in which acid extracts of tissue samples from intoxicated organisms are injected into three mice and the average time to death is recorded [62, 63]. The results are converted into a concentration of saxitoxin, or saxitoxin equivalents, which then serves as the basis for determining whether a particular shellfish site stays open or is closed. The current action level in the United States set by the Food and Drug Administration (FDA) is 0.8 ppm (80 μg saxitoxin/100 g tissue) [64]. The action level in other countries may vary slightly and can be found in FAO Marine Biotoxins Paper 80 [65].

Despite monitoring and informing the public of the dangers of PSP, there are still sporadic poisoning incidents from recreational or subsistence harvesting in closed or unmonitored harvest sites [35]. The use of the mouse bioassay has always been fraught with concerns and problems (e.g., poor specificity for different congeners, low sensitivity, high variability, and the sacrifice of a large number of animals), and a number of alternative techniques have been and continue to be developed. Some of these techniques include neuroblastoma tissue culture bioassays [66, 67], a hemolysis assay [68], receptor-binding assays [69–71], enzyme-linked immunosorbent assays (ELISAs) [72, 73], and electrochemical biosensors [74–77]. Most of these techniques have offered improvement in some areas, but they have not been sufficiently developed or tested to replace the mouse bioassay. Of those listed, the ELISAs have received the most attention in terms of lowering the limit of detection and increasing broad-spectrum detection or developing congener-specific antibodies, e.g., [78–82]. An offshoot of the ELISA technique has the development of lateral flow immunochromatography assays [83, 84]. One such assay, the quick (less than 20 min), one-step MIST Alert™ (Jellet Rapid Testing Ltd.), uses a combination of highly specific antibodies to overcome cross-reactivity issues that have plagued immunoassay development. MIST Alert™ has been field tested, and the biggest concern with its use is the high rate of false positives [35, 85, 86]. The US FDA has approved the test kits for use in the field to prescreen samples – negative samples do not need further analysis whereas positives must be confirmed via mouse bioassays [35]. The prescreening will save time, money, and animal lives. A lot of effort has also been expended in regard to developing liquid chromatography-based techniques. Originally developed by the US FDA [87], a number of derivations and modifications have been implemented including tandem linkage to mass spectrometry and pre- and post-column oxidation of the toxins [75, 76, 88–95]. A particularly useful development was the creation and commercialization of PST standards for use in analyses such as liquid chromatography [96, 97]. Recently, the use of liquid chromatography with post-column oxidation and fluorescence detection has been approved as an official method of analysis by the AOAC [98]. The establishment and validation of this methodology is often a prerequisite for adoption by regulatory agencies and may pave the way for the replacement of the mouse bioassay as the standard assay for detection and quantitation of PSTs.

Origins and Nutrient Interactions

PSP events were reported in Canada over 100 years ago (Ganog in Ref. [99]), but the first documented case in the United States was not until 1958 [100]. In the New England area of the United States, specifically the western Gulf of Maine, PSP events were relatively rare, but they are increasing in frequency and becoming more widespread [101–103]. Similarly, *Alexandrium* and PSP events are now common in historically pristine Pacific waters such as those around Alaska and northern Japan [101]. From these observations, in conjunction with the results from laboratory and field experiments, the spreading of *Alexandrium* blooms and toxifying events can best be explained by examination of cyst deposition and cell transport via storms and currents. A combination of extensive spatial and temporal sampling of cyst seed beds, environmental monitoring and observations, and laboratory experimentation has produced a biophysical model of *Alexandrium* in the Gulf of

Maine [57, 104]. The model successfully recreated ("hindcasted") the formation and movement of the 2005 *Alexandrium fundyense* bloom that started in the northern Gulf of Maine and spread southward as far as Long Island, New York [57, 104]. The model has highlighted the role of three factors of particular import for determining bloom impacts to the coastline. First, heavy rainfall and snow melt created substantial freshwater inputs. The inflow provided some nutrients to the coastal ecosystem, but it also helped stratify the water and provided buoyant plumes that helped transport cells in coastal currents. Second, upwelling-favorable winds shifted to become downwelling-favorable, which promoted transport of offshore blooms onshore. Lastly, bloom intensity is directly correlated to the abundance of cysts present in one or both of two seed beds – in the Bay of Fundy or offshore of Penobscot and Casco Bays [57, 104–106]. Cysts are deposited at the end of a bloom where they overwinter (i.e., lie dormant in the sediment) until the spring, at which time the warmer water temperatures and increased light, nutrients, and oxygen will cause the cysts to germinate [107]. Upon germination the resulting vegetative cells rise to the surface, and they can be pushed onshore under downwelling-favorable conditions and transported along the coastline by prevailing winds and currents [57]. As blooms spread to new areas, the seeding of those areas with cysts likely expands the range of blooms [55, 57].

Understanding the role and importance of these variables and the creation of a model provides significant value as a year to year predictor of bloom likelihood and bloom severity. The model has been applied to the bloom seasons since 2006 and has provided guidance for shellfish harvesters, stakeholders, and managers to make more-intelligent decisions about when and where to harvest shellfish, when and where to monitor for toxin, and how best to preserve public health [8]. In light of the importance of cyst seed beds, similar beds are being mapped and monitored in Puget Sound, Washington, as potential initiating sites [8]. The information also provides a warning against indiscriminant dredging activities that may unearth buried, dormant cysts. Almost all cysts (90%) are buried below the surface in anoxic conditions [108] – conditions that sustain their encystment unless resuspended. Dredging may "reawaken" these cysts to form a bloom or enrich a bloom with additional cells the following spring. The transporting and depositing of marine sediment that may contain cysts to new locations must be monitored lest these activities serve to seed new areas and spread the occurrence of *Alexandrium* blooms [102].

Karenia brevis

Description

The harmful dinoflagellate *Karenia brevis* is an athecate dinoflagellate, i.e., it does not possess any armored plates, and it is about 20 μm in diameter. The size of *Karenia brevis* varies over a diel cycle increasing in size as the day progresses [109]. *Karenia brevis* swims about 250 μm per second or about 1 m an hour.

This swimming ability allows *Karenia brevis* to undergo a diel vertical migration, ascending during the day and descending at night [110]. The traditional explanation for this pattern is that cells are maximizing exposure to light during the day to support photosynthesis and maximizing exposure to nutrients at night. *K. brevis* does not, however, undergo a banded migration characterized by synchronous ascent and descent of the entire population, and therefore there appears to be other factors involved in its behavior other than external stimuli.

K. brevis has been intensively studied due to the socioeconomic impacts that blooms have on coastlines throughout the Gulf of Mexico but particularly on the west coast of Florida, where they now occur almost annually [111, 112]. From January to October 2006, Florida red tides killed thousands of fish and numerous manatees and caused incidents of human respiratory distress, which ultimately reduced local tourism. While cell densities of greater than 1×10^5 cells per liter near shore are responsible for these problems, the *K. brevis* populations appear to originate farther offshore in oligotrophic water columns [112, 113].

Toxicity

Karenia brevis produces a suite of cyclic polyether compounds referred to as brevetoxins (PbTxs). Brevetoxins are a type of neurotoxin that disrupts normal neurological processes by binding to voltage-gated sodium channels in nerve cells. Consumption of this toxin results in an illness described as neurotoxic shellfish poisoning [114]. For decades, brevetoxins produced by *K. brevis* have been implicated in animal mortalities throughout the food chain [115, 116]. Blooms not only result in massive fish kills, but brevetoxins also bioaccummulate in fish and result in mortality of bottlenose dolphins that regularly feed on those fish [116, 117]. Other mammals, like the protected manatee, experience mortality either by inhalation of aerosols or by consumption of contaminated sea grasses [115].

Unlike toxins produced by other species, brevetoxins can be aerosolized, affecting the human population along the coast [118, 119]. During a Florida red tide, aerosolized PbTxs produced by *Karenia brevis* cause both immediate and prolonged airway symptoms in humans, especially in those with preexisting airway disease (e.g., asthma). Environmental monitoring indicates that toxins remain airborne for up to four consecutive days. Repeated exposure to PbTx-3 can result in prolonged airway hyperresponsiveness and lung inflammation, and these pathophysiologic responses may contribute to the prolonged respiratory symptoms observed in humans after red tides [120].

Economics

Karenia brevis is a toxic dinoflagellate that is responsible for extensive ecological and socioeconomic impacts from Florida to Texas [121]. Brevetoxins, produced by *K. brevis*, have been implicated in mortalities of marine life including fish, dolphins, and manatees [122, 123], as well as causing respiratory illness in people and affecting local tourism [124–126]. The economic impacts in Florida have been estimated to range from $20 to $34 million per year [127] with losses incurred from restaurant closures to emergency room visits to shellfishery closures [128–131].

Detection

Despite the regularity of impacts of *K. brevis* blooms on the West Florida Shelf (WFS), predicting where and when *K. brevis* blooms occur remains problematic. The difficulty in predicting the occurrence of these populations resides in the phenomenon that dense populations of *K. brevis* (greater than 1×10^5 cells per liter) often appear more quickly than can be explained by growth rates alone [112, 132]. Increases in chlorophyll *a* biomass observed at frontal boundaries through satellite imagery can be tenfold greater over a 1- to 2-day period than can be explained by in situ growth [132]. Such unexplained increases have led to hypotheses of stealthy bloom initiation, defined as persistence of a population near the bottom of the water column [133] and below the first optical depth, therefore undetected by satellites over broad areas of the WFS [134]. Stumpf et al. [132] hypothesized that nutrients derived from the Mississippi River support similar near-bottom populations between the 25- and 50-m isobaths. As winds shift in the fall and promote upwelling conditions, these populations are subsequently advected onshore to areas of accumulation such as frontal zones [132].

Numerical models of *K. brevis* populations that are transported in the bottom Ekman layers replicate observations of in situ aggregation patterns in frontal zones [135, 136]. These models were based upon an extended Eulerian model of biologically and physically driven *K. brevis* population dynamics [137] and demonstrate the importance of incorporating the behavior and physiology of *K. brevis* into transport models. *K. brevis* undergoes a diel vertical migration (DVM) ascending just before sunrise and descending at night [138]. Recent observations of the diel vertical migration of *K. brevis* in mesocosms suggest that this diel vertical migration is to some degree under metabolic control [133, 139, 140]. Numerous scenarios of a biophysical model suggest that the trapping mechanism at frontal boundaries results from a combination of fluid advection and swimming behavior in response to internal metabolic state and ambient environmental stimuli (such as light and nutrients [135]).

Origins and Nutrient Interactions

Impacts are particularly troublesome in the region between Tampa Bay and Charlotte Harbor, a region that was labeled the red tide epicenter since *K. brevis* blooms were observed in 22 of the last 23 years compared to 5–10 outbreaks north and south of that region [127]. Blooms have been hypothesized to initiate 18–74 km offshore and in oligotrophic water columns [112, 113]. These oligotrophic water columns are characterized by nitrogen and phosphorus concentrations that are commonly about 0.1–0.2 μM within 2–4 km of shore in the Gulf of Mexico [141, 142] and less than 0.01 μM for NO_3^- and NH_4^+ [143]. Nitrate concentrations in both the surface and bottom waters at the 25-m and 50-m isobaths are often at or below 0.1 μM and are not able to sustain large populations of *K. brevis* associated with bloom events [113]. The importance of nutrients to the development and maintenance of bloom events has resulted in numerous hypotheses about potential sources of nutrients that may support populations of *K. brevis* [113, 144]. Potential nutrient sources for *K. brevis* in oligotrophic water columns include aerial deposition [113], N-fixation by *Trichodesmium* sp. blooms [145, 146], water column regeneration from grazers [147], near-bottom regeneration by diatoms responding to upwelling events [113], and regenerated nutrients derived from Mississippi River water [132]. Despite inputs of new nitrogen into the system, nitrate concentrations in both the surface and bottom waters at the 25-m and 50-m isobath are often at or below 0.1 μM [113]. Nutrients in the lower part of the water column, derived either from the sediment or from upwelling plumes below the pycnocline, may also be particularly important in supporting vertically migrating populations [148, 149].

Pseudo-nitzschia

Description

Diatoms are very proficient at utilizing available nutrients and, hence, at forming blooms [12]; however, relatively few diatoms are known to produce harmful blooms, and some species of the genus *Pseudo-nitzschia* are among them. *Pseudo-nitzschia* spp. are cosmopolitan, pennate diatoms found in nearly every major marine and estuarine environment on earth (polar to tropical, coastal to oceanic) [150, 151]. The growth of different *Pseudo-nitzschia* species varies based on each species' optimal ranges to a variety of environmental conditions. So, while *Pseudo-nitzschia* is detectable at the genus level year-round in many locales, the detectability is often a consequence of successions of multiple species in which any one species predominates based on the time of year or the prevailing conditions [152–154]. High levels of inter- and intraspecific variability have been detected at the genetic level [155–159]; this variability is often mirrored in regard to toxicity as well [160]. Twelve of the roughly 30 species of *Pseudo-nitzschia* are capable of producing the neurotoxin domoic acid [161–163]. Domoic acid-producing *Pseudo-nitzschia* species have produced significant blooms along the Pacific Northwestern coastlines of the United States and Canada, and they have recently been responsible for blooms in the Gulf of Mexico.

Toxicity

Domoic acid is structurally similar to glutamic acid, an excitatory neurotransmitter, and mimics its ability to bind to a subclass of the cognate neurotransmitter receptors at neuronal synapses [163]. Domoic acid not only binds with greater than 20 times the affinity to these receptors as the native glutamic acid [164], but it remains bound to the receptors rather than being recycled. These two characteristics allow domoic acid to create a sustained depolarization of the stimulated neuron. As a consequence, intracellular calcium levels in the stimulated neuron continually increase causing the cell to swell and burst [163]. These effects are particularly pronounced in the hippocampus, the region of brain associated with learning and memory that contains neurons that express high levels of glutamate receptor [165]. For this reason, people who consume sublethal amounts of domoic acid exhibit symptoms of amnesiac shellfish poisoning (ASP), a condition which produces short-term memory loss along with gastrointestinal sickness, vomiting, confusion/disorientation, and convulsions [166, 167]. In some patients, the memory loss lasted as long as 5 years [168].

The first, and so far only, identified ASP event that affected humans was in late 1987 around Prince Edward Island, Canada [162, 169]. Of the many people who got sick, 143 had symptoms befitting ASP and 4 of them died. It was determined that the causative agent was mussels contaminated with domoic acid [162, 167]. Subsequent analysis identified the source of the domoic acid found in the mussels as *Pseudo-nitzschia pungens* f. *multiseris* [169].

Pseudo-nitzschia produces domoic acid, and the toxin can be transferred to higher trophic levels, such as humans, through the consumption of filter-feeding bivalves that bioaccumulate domoic acid in their tissues [169–173]. Alternatively, zooplankton, krill, or planktivorous fish, e.g., anchovy, may consume toxic *Pseudo-nitzschia* spp. directly from the water column. Domoic acid may be biomagnified as secondary consumers, e.g., larger fish, birds, and marine mammals, eat the primary consumers that have digestive tracts full of toxic *Pseudo-nitzschia*. Domoic acid poisoning events have been documented in Monterey Bay that affected pelicans, cormorants, and sea lions [166, 172, 174, 175], in Mexico that affected pelicans [176], in the Gulf of California that affected loons and multiple species of marine mammals [177], and in the central California coast that affected California sea lions [178]. Because humans rarely consume whole fish, the primary and continuing concern is intoxication through consumption of contaminated shellfish or crabs.

Based on the Prince Edward Island incident and domoic acid dosage-response studies with a variety of animal models, it has been determined that acute gastrointestinal symptoms in humans develop within 24 h of ingestion followed by neurological symptoms within another 24 h. Monitoring and sampling have so far prevented additional ASP incidents in humans, so more attention has been devoted to outstanding questions regarding chronic or long-term domoic acid intoxication. Sea lions with chronic exposure to domoic acid exhibit symptoms that are distinct from those of ASP [179], which suggests similar long-term exposure could produce an additional array of detrimental effects in humans. Furthermore, domoic acid can be passed to pups in the breast milk of rodents [180]. If similar passage occurs in humans, what are the possible risks to the mental development of infants who are breast-fed from mothers who eat contaminated shellfish? Epidemiological studies are being conducted on native populations in the Pacific Northwest of the United States and Canada [181]. The diet of these native peoples relies heavily on seafood products, primarily razor clams, which are traditionally vulnerable to being toxified with varying levels of domoic acid. The results of such studies will elucidate what, if any, effects may result from consuming low levels of domoic acid over time [181].

Economics

Like other HAB events, domoic acid poisoning incidents will cause economic impacts if people and animals experience severe illness or even death, or if recreational areas and harvesting sites are closed. Beaches and shellfish-harvesting areas have faced closings in Washington state since 1991 (references within [182]). The first coastal closure in 1991 is estimated to have resulted in $23–28 million and $9 million in lost revenue to razor clam harvesters and Dungeness crab harvesters, respectively [173] (Nosho in Ref. [183]). From 1991 to 2008, about 23% of the potential razor clam fishing days along the Washington coastline have been lost due to closures from high domoic acid detected in shellfish samples (Ayers unpublished data in [184]). During an extensive domoic acid incident in 1998, the coastal tourism industry is estimated to have lost $15 million (Nosho in Ref. [183]). The first beach closure within Puget Sound occurred in 2003 [185], followed by a second closure in 2007 [182]. Ayers and Reed [186] estimate that the first closure resulted in $11.8 million in lost revenue to the local economy. The potential threat of domoic acid spreading to intracoastal regions and affecting commercial operations could amplify future losses [182]. Dyson and Huppert [184] provided a more detailed estimation of the economic effects stemming from closures of razor clam harvest sites. Based on responses of visitors and razor clam harvesters to a questionnaire, closing four beaches stretching over Grays Harbor and Pacific counties in Washington state during the 2007–2008 season would have resulted in over $25.3 million in lost expenditures. Apart from the recreational aspects of razor clam harvesting, the harvests are important to Native American cultures that rely on the razor clams and other products from the sea, not only for food but also for cultural identity.

In addition to the lost expenditures that result from closures, national, state, and local governments invest in monitoring and detection. Monitoring phytoplankton community composition, correctly identifying and enumerating *Pseudo-nitzschia* species, and measuring levels of domoic acid are all important measures for providing early warning to stakeholders located near possible blooms. Equally important is ongoing sampling of tissues from commercially and recreationally important species to monitor accumulated levels of domoic acid. The majority of methods in use for monitoring and detecting both species and toxin requires expensive equipment and can be costly to perform.

Detection

Species Identification

A major hurdle for the study of *Pseudo-nitzschia* is rapid and accurate methods for species determination. Early systematics of *Pseudo-nitzschia* relied on light microscopy to separate the genus into two groups based on the width of the cells [187]. This simple categorization belied significant diversity particularly in terms of toxicity. Later, tunneling and scanning electron microscopies were employed to better resolve species complexes by analyzing such subcellular features as the striae and pore density. These methods are labor intensive, time consuming, require access to and working knowledge of electron microscopes, and still do not always adequately resolve the question of whether a sample contains toxic or nontoxic cells [188–190].

To improve detection and species identification, efforts have been made to develop molecular reagents for use in lectin-binding assays, immunoassays, sandwich hybridization, and PCR. Lectins are glycoproteins that bind sugar molecules, and lectin-binding probes have been developed that recognize species-specific glycosylated proteins on the surface of *Pseudo-nitzschia* cells [191–193]. These probes could discriminate six of seven *Pseudo-nitzschia* species endemic to New Zealand waters [192]. They showed some cross application in testing of samples from Spain and Korea [194, 195], but it was noted that the location of sample isolation and physiological state of cells can confound identification or differences between strains of the same species [196]. DNA-based probes that recognize genetic determinants such as the ribosomal RNA genes and the internal transcribed spacer (ITS) have received the most attention and have been employed in assays such as whole-cell hybridizations, FISH, sandwich hybridization, and PCR [30, 158, 159, 197–207]. Chris Scholin and his group at Monterey Bay Aquarium Research Institute have developed DNA-based probes specific for the large subunit of ribosomal RNA of *Pseudo-nitzschia* species isolated from the west coast of the United States. Similar to the above reagents, the use of these probes has met with mixed results when applied to samples at other geographic locations [203, 208–211]. For example, probes that work for *Pseudo-nitzschia pseudodelicatissima* found on the west coast of the United States did not recognize the same "species" found in the Gulf of Mexico [210], presumably due to high levels of inter- and intraspecific variability. Supporting that conclusion are results based on analyses of microsatellite markers of some *Pseudo-nitzschia* lineages or species that show high level of genetic diversity [155–159]. Additionally, one might be at fault in not identifying or recognizing the relevant markers or characteristics that determine true species differences. The Scholin probes have been further developed for use in sandwich hybridization that has been automated and incorporated into an environmental sample processor (ESP) for undersea deployment. The ESP samples seawater and quantitates both the concentration of specific harmful algal cells and their corresponding toxin concentration via the sandwich hybridization assay and automated competitive enzyme-linked immunosorbent assay, respectively [212–214]. Autonomous and continual sampling should not only provide an early warning to the presence of toxic cells but also provide important environmental data that may be used to create or enhance models for predicting the presence of such cells and their toxins.

Domoic Acid Detection and Quantitation

Immediately following the Prince Edward Island incident, Canada took the lead in developing and implementing food safety standards for domoic acid. In 1989, Health and Welfare Canada instituted an action level at 20 μg domoic acid/g shellfish tissue [64]. That same level has now been adopted by the European Union, New Zealand, the United States, and Australia. One exception to this limit is the United States allows up to 30 μg/g of Dungeness crabs. If tested tissue samples exceed the limit, then the relevant beach/harvest area is closed until the domoic acid can be depurated by the organisms and repeated sampling shows that domoic acid is below the limit.

After the Prince Edward Island incident, Health and Welfare Canada relied on the same mouse bioassays as used for detecting PSTs [215]. For the reasons listed in the previous section, particularly in this case where the detection limit (~40 μg domoic acid/g of tissue) is higher than the action level, other more sensitive and reliable methods have been developed. Some of these alternatives offer one or more advantages to the mouse bioassay but typically at the expense of time, cost, or ease of performance. Different assays also show better efficacy in analyzing domoic acid from particular sources, e.g., a receptor binding assay developed by Van Dolah et al. [216] can analyze domoic acid from shellfish and from seawater extracts of algae. A variety of antibodies have been raised and incorporated into ELISAs and competitive immunoassays for the detection of domoic acid in human body fluids or shellfish extracts [217–220]. The AOAC has a validated ELISA method (AOAC Method 2006.02) for detecting domoic acid in shellfish, and the ELISA method is used in the EU for screening purposes. A variety of modifications to a liquid chromatography (LC) methodology allow for the sensitive detection of domoic acid, but routine detection of domoic acid is often carried out using liquid chromatography coupled with diode-array/UV absorbance or fluorometic detection [221]. An LC-UV method was developed for the routine determination of domoic acid in shellfish [222]. The method was validated in a multi-lab AOAC collaborative study [223] and had an AOAC method established (AOAC Method 991.26), but a number of improvements to the extraction/clean-up procedures, alternative tandem detection systems, or other modifications have been developed and continue to be validated, e.g., [224–226]. The EU Reference Laboratory for Marine Biotoxins has created a standard operating procedure for determining or confirming domoic acid in mollusks using ultraperformance LC-MS [227]. The National Research Council of Canada was also the first to provide certified domoic acid standards for use in calibrating liquid chromatographs [97]. Similar for the detection of PSTs, a promising development is the Jellett Rapid Test or Maritime In Vitro Shellfish Test (MIST™) (Jellett Biotek Ltd.). The immunochromatographic kits can be used quickly and cheaply at sample sites in the field for the detection of domoic acid [228, 229] that must be later confirmed in the laboratory.

It is necessary to individually sample each species of interest since different species of shellfish take up and depurate domoic acid at different rates (reviewed in Ref. [163]). For example, mussels and oysters consume *Pseudo-nitzschia* at higher rates than clams, especially razor clams, but mussels and oysters also depurate much more quickly. Once domoic acid accumulates in razor clams, it can take up to a year for domoic acid levels to fall below the action level [173]. Domoic acid levels in anchovy and sand crabs rise and fall almost contemporaneously (a slight lag) with the initiation and termination of a *Pseudo-nitzschia* bloom. Scallops also depurate slowly, but the only scallop tissue that humans eat is the adductor muscle which has very low levels of domoic acid. Most domoic acid in fish, crabs, and cephalopods is in the viscera, which is typically not consumed by humans. This type of discriminatory feeding behavior is not true for animals in the wild, however, which are more likely to be affected by the presence of domoic acid moving within the food web. It is also a concern that ground whole fish is used in animal and fish feeds, and those feeds could sicken domesticated animals or fish in aquaculture far from the coasts where blooms have formed.

Origins and Nutrient Interactions

As mentioned above, *Pseudo-nitzschia* is cosmopolitan, and it is capable of taking advantage of nutrients when they are present to grow and form blooms. *Pseudo-nitzschia* has been modeled to bloom under two different scenarios of nutrient introduction. One is upwelling events during early spring and summer that bring water from the bottom up to the surface. The nutrient-rich water coupled with high turbulence from storms are generally conditions that favor diatom growth [12]. The second scenario is rivers carry natural and introduced nutrients from a variety of point and nonpoint sources to the ocean. Corresponding to the second scenario is the detection of *Pseudo-nitzschia* spp. in the Gulf of Mexico. *Pseudo-nitzschia* was considered rare in the Gulf of Mexico prior to 1950, but the numbers of several *Pseudo-nitzschia* species have increased concurrently with elevated nutrient loading into the Gulf from the Mississippi River [230]. Similar connections have been documented in the Yellow and Yangtze rivers and in the South China Sea [153, 154]. Mesocosm experiments corroborate these correlations – *Pseudo-nitzschia* show enhanced growth after pulses of nutrients are added [231].

Off the west coast of the United States, upwelling brings nutrient-rich water in contact with enormous plumes of water exiting the Puget Sound via the Strait of Juan de Fuca. A large gyre, the Juan de Fuca eddy, is formed or enhanced each spring at the junction of these two phenomena. Throughout the summer and into the fall, when it dies back, the gyre is a site of persistent upwelling that acts as a giant bioreactor for many phytoplankton species. The use of drifters to track hydrological movement has enlightened our understanding of the role of the Juan de Fuca eddy, coastal currents, and upwelling versus downwelling conditions in the growth and transport of *Pseudo-nitzschia* to various beaches and coastlines along the Pacific northwest of the United States. Modeled on the movement of drifters, *Pseudo-nitzschia* cells within the Juan de Fuca eddy are believed periodically to spin out into offshore currents. Under upwelling favorable conditions, cells will migrate southward along the Washington and Oregon state coastlines in these offshore currents. A shift to downwelling favorable conditions, such as from a storm or other events leading to a wind reversal, in regions where there is low to no river outflows will cause cells/blooms to move onshore and cause shellfish to become contaminated with domoic acid [183, 232].

The requirements for cell growth are clearly not the same as those for producing domoic acid. While the availability of nutrients is important for growth and increases in cell number, those same factors do not always coincide with domoic acid production. Different toxic species (or even the same toxic species from different geographic regions) will produce or not produce domoic acid in response to different environmental conditions (including the availability of one or more limiting nutrients such as phosphorus or silicate), stages of culture growth (e.g., exponential vs stationary growth phases), and the composition of the associated bacterial community [161, 233–237]. It has also been suggested that iron limitation or copper stress can induce domoic acid to act as either a chelator of iron (to enhance its uptake into the cell) or copper (to alleviate its toxicity) [238]. Iron stimulation experiments in the Southern Ocean show increased *Pseudo-nitzschia* cell abundance, but not a concomitant increase in domoic acid [239, 240]. Axenic cultures of *Pseudo-nitzschia* produce less domoic acid than nonaxenic cultures [241–243] indicating a role for bacteria, although no bacterium has been shown capable of producing domoic acid on its own.

Cyanobacteria

Description

Cyanobacteria, originally classified as "blue-green algae," are prokaryotes capable of photosynthesis (reviewed in Ref. [244]). They underwent a worldwide adaptive radiation that has allowed them to fill a wide range of ecological niches including terrestrial and aquatic systems (both freshwater and marine) as well as forming numerous endo- and ectosymbiotic associations [245]. Their distribution and diversity are likely a function of their ancient origins – it is estimated that they have existed for at least 2 billion years and possibly as long as 3.5 billion years or more. They may have been one of the first forms of life on Earth (reviewed in Ref. [246]). For this reason, the diversity of cyanobacterial species is also reflected in the wide range of metabolites and secondary molecules that are produced within this phylum. Some of these unique molecules act as toxins, carcinogens, and/or allergens when ingested, inhaled, or contacted by humans and animals. Coupled with the ability of some fresh and marine water species to form blooms, cyanotoxins pose a wide range of threats to human health.

Toxicity

The first recorded deaths from cyanotoxins occurred in Australia in 1878 when livestock drank from a brackish, estuarine lake covered with a *Nodularia spumigena* bloom [247]. Since then, there have been many other reported instances of sickness or death of livestock, pets, and humans attributed to recreational exposure to or ingestion of cyanobacteria-contaminated water (reviewed in Ref. [248–253]). A comprehensive description, or even listing, of all identified or suspected toxic molecules from cyanobacteria is beyond the scope of this article. Listed in Table 13.1 are a few representative toxins from cyanobacteria.

Hepatotoxins

Microcystins are cyclic heptapeptides that include nonprotein amino acids and unique covalent linkages. They are the most frequently found cyanotoxin, and over 70 variants have been described [254]. They are highly stable and water-soluble. Ingestion of microcystins can lead to a rapid symptomatology (within 30 min to 24 h) that may ultimately lead to hemorrhaging in the liver and death [255, 256]. It is also possible that inhalation of microcystins (or other cyanotoxins) could be a route of exposure since microcystins have been detected, albeit at low levels, in aerosol droplets [257]. A major research focus for microcystins, indeed for all the cyanotoxins, is determining what effects low level or chronic exposure may produce. Mounting evidence suggests that microcystins promote tumor growth and certain cancers [258–262]. The World Health Organization has set a provisional consumption limit (for drinking water) of 1 μg/L for the most toxic form of microcystin [263]. A real concern is the fact that "finished" water from some water treatment facilities has tested above that limit (reviewed in Ref. [264–266]). Microcystin levels can be greater than 24 mg/L in waterbodies with very high cell counts [267], which can pose risks to recreational swimmers based on an

Table 13.1 Representative toxins listed with their genera of origin and toxicity

Toxin	Representative genera[a]	Ecosystem	Toxicity
Microcystins	*Microcystis, Anabaena, Nostoc, Planktothrix*	Fresh water	Hepatotoxin
Nodularins	*Nodularia*	Brackish water Marine water	Hepatotoxin
Cylindrospermopsin	*Cylindrospermopsis, Anabaena*	Fresh water	Hepatotoxin
Anatoxin-a	*Anabaena, Planktothrix*	Fresh water	Neurotoxin
Anatoxin-a(s)	*Anabaena*	Fresh water	Neurotoxin
Saxitoxins	*Anabaena*	Fresh water	Neurotoxin
	Lyngbya	Marine water	
	Cylindrospermopsins	Fresh water	
	Trichodesmium	Marine water	
LPS	All cyanobacteria	All	Endotoxin, Dermatoxin
Aplysiatoxin	*Lyngbya*	Marine water	Dermatoxin, Tumor promoter
	Planktothrix	Fresh water	
Lyngbyatoxin-a	*Lyngbya*	Marine water	Dermatoxin

[a]Not all species within the listed genera necessarily produce the associated toxin, and species in other genera not listed may also produce these toxins

average accidental ingestion of 100–200 mL of water. Freshwater-derived microcystins have also been found in marine shellfish located at sites where rivers discharge into marine waters, and these toxins are the implicated source for causing hepatotoxic shellfish poisoning (HSP) that led to the death of 21 southern sea otters in Monterey Bay, California [268]. Microcystins can be bioaccumulated to high levels in the tissue of shellfish that, when coupled with a slow depuration rate [268], poses additional health risks to shellfish consumers, including humans.

To date, nodularins are known to be produced by the single species, *Nodularia spumigena*, a species of cyanobacteria that can form blooms in brackish and marine waters. Nodularins are structurally similar to microcystins (cyclic pentapeptides), and their toxicity and mode of action are similar as well. One distinction between the two classes of hepatotoxins, however, is that nodularins are carcinogenic, i.e., they can initiate tumor formation, as well as act as potent tumor promoters [254].

Cylindrospermopsins primarily produce effects in the liver, but they can also damage other organs. These toxins are guanidine alkaloids that take days, rather than hours, to produce symptoms of intoxication. Additional toxic effects include the inhibition of protein synthesis and genotoxicity [269].

Neurotoxins

Saxitoxins have been discussed in a previous section of this article (see "Toxicity" under "*Alexandrium*"). Anatoxin-a and anatoxin-a(s) are structurally unrelated, but both can act swiftly (within minutes to hours) to disrupt the function at nerve synapse leading to muscle paralysis and respiratory arrest [270]. Anatoxin-a can be inhaled, ingested, or absorbed through the skin. The estimated toxic dose for a human adult male is less than 5 mg. Anatoxin-a(s) has only been found in two species of *Anabaena, A. flos-aque* and *A. lemmermannii* [271, 272]. It is the only known natural organophosphate. As such, it acts by inhibiting acetylcholinesterase, and it is ten times more potent than anatoxin-a [272, 273]. Fortunately, the occurrence and concentration of these two toxic molecules are much less than are typically detected for microcystins. While humans will not likely ingest enough cells/neurotoxin(s) in a mouthful of water, some animals that may drink liters at a time may be at risk.

Other Cyanobacterial Toxins

Lipopolysaccharides (LPS) are components of the outer membrane of gram-negative bacteria and cyanobacteria. They are capable of forming complexes with several biomolecules to elicit irritant and allergic responses [274]. Cyanobacterial LPS can cause skin irritation, gastroenteritis, and fever, or act as an endotoxin that can kill mice in mouse bioassays or exacerbate the effects of other cyanotoxins [275].

Aplysiatoxin and lyngbyatoxin-a are only produced in marine species of cyanobacteria. Generally, the concern surrounding toxins from marine cyanobacteria is not ingestion from drinking water but either trophic transfer and biomagnifications in fish, shellfish, and other consumables or contact exposure/inhalation from recreating around blooms. Aplysiatoxin and lyngbyatoxin-a can cause severe dermatitis through acute inflammatory reactions [276, 277]. They can also activate protein kinase C, a serine/threonine kinase that is central to diverse signal transduction pathways, to potently promote tumor formation [278, 279]. Sea turtles that consume seaweed overgrown or associated with *Lyngbya*, a genus of marine cyanobacteria, are suspected to bioaccumulate lyngbyatoxins in their tissues, and there has been one potential human death, and other suspected cases, linked to the consumption of sea turtle meat contaminated with lyngbyatoxins [280, 281].

β-*N*-Methylamino-L-Alanine (BMAA)

In addition to any of the acute effects listed above, long-term chronic effects, e.g., cancer development, are linked or suspected to several cyanotoxins [258–262]. Additional cyanobacterial metabolites that are receiving attention include unusual amino acids such β-*N*-methylamino-L-alanine (BMAA). BMAA has been proposed as a causative agent of neurotangle diseases, e.g., Parkinson's disease, amyotrophic lateral sclerosis, or Alzheimer's disease. Such a role is controversial, but it is supported by a number of correlative studies including direct detection of BMAA in brain tissue samples from native individuals from the island of Guam who died from lytico-bodig, a type of neurotangle disease, and a small sampling of Caucasian patients in Canada who died from Alzheimer's disease (reviewed in Ref. [282]). In vitro and in vivo laboratory studies using BMAA support the purported role as a neurotoxin in that BMAA leads to selective motor neuron death, produces behavioral and/or neurological differences, and interferes with learning and memory [283–288].

A second, similar biomagnification mechanism may be responsible for the occurrence of avian vacuolar myelinopathy (AVM) in aquatic and predatory birds. In the southeastern United States, *Hydrilla verticillata*, is an aquatic plant that often contains a BMAA-producing epiphytic cyanobacteria [289, 290]. Herbivorous birds, e.g., ducks and coots, which consume *H. verticillata* (and bald eagles that consume the herbivorous birds), contract high rates of AVM and die [290]. Although the majority of the evidence for vectoring of BMAA to humans and birds is correlative, some direct tests of the hypotheses support the proposed mechanisms (e.g., [291–293]). Also of note, and concern, is the fact that BMAA has been detected in almost all cyanobacteria in all environments worldwide (e.g., [294–298]). Intriguingly, a recent epidemiological study implicates a lake, noted for frequent cyanoblooms, as being responsible for or linked to a 10–25$\times$ higher incidence of sporadic ALS, compared to the general US population, in lake-adjacent areas [299].

Economics

In addition to the public health consequences, additional economic and sociocultural impacts caused by cyanoblooms include negative effects to: (1) recreation and tourism due to closure of waterbodies (often at times when recreational use is most desired or when tourism would be at peak, i.e., summer months); (2) aquaculture due to loss of product (death) or production of unwanted, tainted product (some cyanobacterial metabolites can alter the color or taste of the fish) [275]; and (3) the quality of life of people living and working around affected waterbodies due to the reduced aesthetic appeal and/or the unwanted health risks. Few studies have quantified the losses or impacts of cyanobacterial blooms. Tucker et al. [300] estimated that greater than $80 million was lost per year in catfish aquaculture facilities due to cyanobacterial metabolites that produce off-color or off-flavor in the catfish. In addition to toxic effects, cyanobacteria can form nuisance blooms that generate costs associated with cleaning biomass that blocks water intakes used for draining and industrial purposes such as hydroelectric power generation. A conservative estimate of economic costs of all freshwater HABs in the Unites States is $2.2–4.6 billion a year [301].

Where species of known or potential bloom formers are present, monitoring is essential as both a tool for aiding in possible prevention of blooms and to alert the public of existing or impending cyanoblooms. Public guidance should include precautions about avoiding areas of visible algal biomass, about avoiding direct contact with blooms, and to avoid swallowing water containing algae as those interactions are associated the highest risks to human/animal health. A bloom may reach densities of >100,000 cells/mL, but even at lower levels, the concentration of toxic metabolites may be sufficient to cause sickness or death. The WHO had published recommendations for recreational exposure, based in part upon results from [249, 255]: Up to 20,000 cyanobacterial cells/mL requires a warning to visitors but has a low/mild risk to adverse health effects, up to 100,000 cells/mL carries a moderate risk in the short-term as well as potential long-term risks if swallowed, and when a visible bloom forms, then a high risk of inhalation, ingestion, or body contact is likely at that site [302]. Monitoring and public alerts about the presence and dangers of cyanoblooms are recurring costs that state and local agencies will be forced to carry. Atech (2000) (in Ref. [303]) estimated costs related to monitoring for cyanobacteria in Australia at greater than $12 million dollars per year (in 2011 US dollars).

Another significant expense is the detection and removal of cyanotoxins from drinking water sources. It is axiomatic that preventing blooms and/or toxin production should be cheaper, easier, and safer than eliminating or removing cells and/or toxins from the water after a bloom. The primary and essential goal of cyanoHAB management, if not elimination, is identifying the type and source of nutrient loading in order to reduce cyanobacterial proliferation. To remove cells once a bloom has formed, however, has typically relied on common algaecides such as copper sulfate. Unfortunately, the resulting cell lysis can exacerbate the problems as intracellular cyanotoxins are released into the water, increasing their concentration. Additionally, the indiscriminate or repetitive use of copper-based reagents can cause inadvertent ecological damage and induce copper resistance in surviving algae [304–306]. Other methods for bloom control include destratification, which only works for small waterbodies, and increased flushing, which only works if upstream sources of water are available. Conventional water treatments such as flocculation, coagulation, chlorination, and filtration can remove cells; but not only do they not remove toxins, they also cause cell lysis leading to increased toxin levels in the water [307–312]. While some methods of oxidation, such as chlorination, do not appear effective [307–312], others such as UV disinfection, ozonation, and potassium permanganate treatment are more promising [265, 313]. Pipelines of multiple methodologies, e.g., use of activated charcoal (particularly wood-based, powdered activated carbon), oxidation via ozonation, and nanofiltration (with nominal molecular weight cutoff of 200) appear effective at removing microcystins and anatoxin-a [265, 313]. Titanium dioxide photocatalysis is being developed for both purification and disinfection. This method can destroy microcystins found in concentrations up to 5,000 μg/L [8, 314]. Not all of the methods listed above are universally employed and are absolutely lacking in sources of water such as private wells. Also, these treatments do not protect wildlife from consuming water at natural waterbodies where cyanoblooms may be occurring, all of which argue for preventive management of cyanoblooms rather than reactive management.

Detection

Microscopic inspection of water samples is still primarily used for identification and quantitation of cyanobacterial cells. Assaying for chlorophyll-*a* is a quick alternative to microscopy but may overestimate the amount of cyanobacteria in a sample because chlorophyll-*a* is also present in many other algae [255]. It is also a presumptive indicator of cyanotoxins rather that a direct detection method. Immunoassays [315] following standard practices outlined in Chorus and Bartram [255] have also been developed for these purposes. More recently, genetic identification using polymerase chain reaction and modifications thereof, e.g., quantitative polymerase chain reaction, are being used to detect (and quantify) specific species or toxin-synthesizing genes, e.g., a nodularin synthetase gene in *Nodularia* [316] or a microcystin synthetase gene in *Microcystis* [317], within an environmental sample.

For the detection of cyanotoxins, the mouse bioassay [256] is the official approved method, but a multitude of more-advanced, sensitive, and humane methodologies have been and continue to be developed. Some methods have been adapted to detect more than one type of cyanotoxin, while other methods are specific for one class or variant, e.g., in vitro biological assays such as acetylcholine esterase inhibition (to detect anatoxin-a(s)), protein phosphatase inhibition (to detect microcystins or nodularins), and protein synthesis inhibition (to detect cylindrospermopsins) [318]. Modifications to the standard high-pressure liquid chromatography methodology, e.g., UV detection at the respective absorption maximum of each cyanotoxin, following separation allows for the detection of microcystins, cylindrospermopsins, and anatoxin-a [254]. Liquid chromatography (LC) with post-column oxidation and fluorescence detection is now an official method of analysis by the AOAC for saxitoxins [98]. Similar coupling of LC with tandem mass spectrometry is also useful for detecting anatoxin-a(s) [319]. Antibodies against microcystin, nodularin, saxitoxin, and cylindrospermopsin have been useful for developing ELISAs [72, 320, 321] that show good correlation with HPLC methods. The detection methods have greatly benefited from the creation and commercialization of kits and cyanotoxin standards. Standard hemoagglutination assays, e.g., the Limulus amoebocyte lysate assay, can be used to quantify LPS in aqueous samples [322].

Origins and Nutrient Interactions

Cyanobacteria are ubiquitous [323, 324], and the threat of cyanoblooms is pervasive. Similar to other HAB species, the frequency and geographic distribution of cyanoHABs is increasing worldwide [8, 325]. A 1999–2000 survey in Florida sampled 75 bodies of water and 88/167 samples had cyanotoxins [326]. Eighty percent of the positive samples had levels sufficient to kill a mouse in a mouse bioassay [326]. In a survey of 45 water sources in the United States and Canada between 1996 and 1998, 80% had detectable levels of microcystin – a few above the WHO advisory limit [266]. Other surveys of national or international scope indicate that about 60% of bodies of fresh water contain toxic species of cyanobacteria [266, 297, 327]. The North American Great Lakes and associated water systems are a source of drinking water for ~22 million people, and during late summer months, microcystin sampling routinely exceeds the WHO threshold [328]. Over the long term, e.g., a 75-year lifetime, it has been estimated using the data from Carmichael [266] that cyanotoxin exposure from drinking water is an order of magnitude below the WHO guideline of 1 μg/L [329]. Such a study is somewhat reassuring, but the effects of low-level chronic exposure are still not understood; nor can it be expected that cyanotoxin levels in finished drinking water will not increase unless measures are taken to decrease nutrient loading to decrease bloom formation or methods to more effectively remove cyanotoxins from the water are developed.

Some cyanobacterial blooms occur naturally under specific conditions, but the frequencies are much higher in eutrophic bodies of water [330]. Other significant factors that support bloom formation are warmer temperatures, increased sunlight, and decreased flow/flushing of a waterbody [305]. The induction of bloom formation in stagnant waterbodies or slow-moving, low-flushing systems is clearly linked to nutrient loading, particularly phosphorous [331]. In such systems, gas-filled vesicles allow cyanobacteria to be buoyant and remain suspended near the surface to be exposed to light [332]. Even under conditions of limiting nitrogen or phosphorous, cyanobacteria can outcompete other planktonic organisms due to higher affinities for these nutrients and the capacity to store high amounts of phosphorous inside their cells [333]. In waterbodies that are mixed and/or flushed, factors such as warmer temperatures, high amounts of sunlight, and elevated nutrients will all contribute to different degrees in leading to cyanobloom formation (reviewed in Ref. [255, 333]). The convergence of the optimal conditions for a particular cyanobacterial species will allow it to form a "scum" or surface bloom. Where applicable or possible, efforts by watershed managers to reduce excess nutrient loading, to increase or maintain water flow/flushing rates, and to prevent the removal of grazers have all been effective strategies to reduce cyanoblooms (reviewed in Ref. [305]). The introduction of solar-powered circulation (SPC) devices into nutrient-enriched, cyanobacteria-containing reservoirs has been effective at suppressing the number of cyanobacterial cells and decreasing the need for supplemental copper sulfate (algaecide) applications [334]. Broad-based deployment of such devices would not only decrease the occurrence of cyanoHABs but also do so in an ecologically and environmentally sound manner.

Future Directions

Significant effort has generated a lot of biological and ecological knowledge of selected harmful algae. This knowledge has produced few methods to prevent bloom formation or mitigate the effects of blooms that have formed. Few attempts to control HABs have proven to be practical or free of adverse side effects that would further disrupt ecosystems. The difficulty in preventing HABs is that each species reacts to a different suite of factors; some of these factors can include a complex mixture of biochemical, ecological, geophysical, hydrological, and meteorological conditions. In the majority of cases, any of these conditions may be beyond our control except for one. The one almost universal requirement for any algal blooms is a supply of nutrients. Without nutrients, the likelihood of any species forming a bloom is low, so identifying the source(s) of relevant nutrients and then minimizing those nutrients may be the best strategy for preventing and mitigating harmful algal blooms.

The most advances in HAB research have been in the ability to detect/monitor HAB species and toxins. Despite these advances, continued development of appropriate standards and techniques for regulatory and management practices is imperative, especially because of the continuing discovery of new harmful species. An emerging line of research to aid in detection and modeling cell/bloom behavior is at the molecular level. A lot of new research is beginning to elucidate the genetic composition of these different species and the mechanisms that regulate the genetic responses, and subsequent physiological responses, to a variety of environmental parameters. These data sets will prove invaluable to refine or develop current or future models, respectfully, that will be useful for predicting bloom formation and behavior.

Bibliography

Primary Literature

1. Hallegraeff GM (1993) A review of harmful algal blooms and their apparent global increase. Phycologia 32:79–99
2. Sandifer P et al (2007) Interagency oceans and human health research implementation plan: a prescription for the future. Joint Subcommittee on Ocean Science and Technology. Interagency Working Group on Harmful Algal Blooms, Hypoxia and Human Health, Washington, DC
3. Smayda TJ (1990) Novel and nuisance phytoplankton blooms in the sea: evidence for a global epidemic. In: Graneli E (ed) Toxic marine phytoplankton. Elsevier, New York, pp 29–40
4. Smayda TJ (1992) Global epidemic of noxious phytoplankton blooms in the sea: evidence for a global epidemic. In: Sherman K, Alexander LM, Gold BD (eds) Food chains: models and management of large marine ecosystems. Westview Press, San Francisco, pp 275–307
5. Smayda TJ, White AW (1990) Has there been a global expansion of algal blooms? If so is there a connection with human activities? In: Graneli E (ed) Toxic marine phytoplankton. Elsevier, New York, pp 516–517
6. Hallegraeff GM (2010) Ocean climate change, phytoplankton community responses, and harmful algal blooms: a formidable predictive challenge. J Phycol 46(2):220–235
7. Ramsdell JS, Anderson DM, Glibert PM (eds) (2005) HARRNESS, Harmful algal research and response: a national environmental science strategy 2005–2015. Ecological Society of America, Washington, DC

8. Lopez CB et al (2008) Scientific assessment of marine harmful algal blooms. Joint Subcommittee on Ocean Science and Technology. Interagency Working Group on Harmful Algal Blooms, Hypoxia and Human Health, Washington, DC
9. Heisler J et al (2008) Eutrophication and harmful algal blooms: a scientific consensus. Harmful Algae 8(1):3–13
10. Anderson DM et al (2008) Harmful algal blooms and eutrophication: examining linkages from selected coastal regions of the United States. Harmful Algae 8(1):39–53
11. Huisman JM, Saunders GW (2007) Phylogeny and classification of the algae. In: McCarthy PM, Orchard AE (eds) Algae of Australia: introduction. Australian Biological Resources Study/CSIRO, Melbourne, pp 66–103
12. Smayda TJ, Reynolds CS (2001) Community assembly in marine phytoplankton: application of recent models to harmful dinoflagellate blooms. J Plankton Res 23(5):447–461
13. Vargo GA et al (2001) The hydrographic regime, nutrient requirements, and transport of a *Gymnodinium breve* Davis red tide on the west Florida shelf. In: Hallegraeff GM et al (eds) Harmful algal blooms 2000. Intergovernmental Oceanographic Commission of UNESCO, Paris, pp 157–159
14. Laws RM (1985) The ecology of the southern ocean. Am Sci 73(1):26–40
15. Croxall JP, Nicol S (2004) Management of southern ocean fisheries: global forces and future sustainability. Antarct Sci 16(4):559–584
16. Smayda TJ, Reynolds CS (2003) Strategies of marine dinoflagellate survival and some rules of assembly. J Sea Res 49(2):95–106
17. Sournia A (1974) Circadian periodicities in natural populations of marine phytoplankton. Mar Biol 12:325–389
18. Kent ML, Whyte JNC, Latrace C (1995) Gill lesions and mortality in seawater pen-reared Atlantic salmon Salmo-Salar associated with dense bloom of *Skeletonema costatum* and *Thalassiora* species. Dis Aquat Organ 22(1):77–81
19. Bell GR (1961) Penetration of spines from a marine diatom into gill tissue of Lingcod (*Ophiodon elongatus*). Nature 192(479):279–280
20. Dagg MJ et al (2007) A review of water column processes influencing hypoxia in the northern Gulf of Mexico. Estuaries Coast 30(5):735–752
21. Rabalais NN et al (2002) Nutrient-enhanced productivity in the northern Gulf of Mexico: past, present and future. Hydrobiologia 475(1):39–63
22. Legrand C et al (2003) Allelopathy in phytoplankton – biochemical, ecological, and evolutionary aspects. Phycologia 42:406–419
23. Kubanek J et al (2005) Does the red tide dinoflagellate *Karenia brevis* use allelopathy to outcompete other phytoplankton? Limnol Oceanogr 50(3):883–895
24. Okaichi T, Nishio S (1976) Identification of ammonia as the toxic principle of red tide of *Noctiluca miliaris*. Bull Plankton Soc Jpn 23:75–80
25. Lancelot C et al (2011) Cost assessment and ecological effectiveness of nutrient reduction options for mitigating Phaeocystis colony blooms in the southern North Sea: an integrated modeling approach. Sci Total Environ 409(11):2179–2191
26. Ciglenecki I et al (2003) Mucopolysaccharide transformation by sulfide in diatom cultures and natural mucilage. Mar Ecol Prog Ser 263:17–27
27. Guiry MD (2011) AlgaeBase. http://www.algaebase.org [cited 10 May 2011]
28. Daugbjerg N et al (2000) Phylogeny of some of the major genera of dinoflagellates based on ultrastructure and partial LSU rDNA sequence data, including the erection of three new genera of unarmoured dinoflagellates. Phycologia 39:302–317
29. Scholin CA, Anderson DM (1994) Identification of group-specific and strain-specific genetic-markers for globally distributed *Alexandrium* (Dinophyceae). 1. RFLP Analysis of SSU ribosomal-RNA genes. J Phycol 30(4):744–754
30. Scholin CA, Anderson DM (1996) LSU rDNA-based RFLP assays for discriminating species and strains of *Alexandrium* (*Dinophyceae*). J Phycol 32(6):1022–1035

31. Scholin CA, Hallegraeff GM, Anderson DM (1995) Molecular evolution of the *Alexandrium tamarense* species complex (Dinophyceae)-dispersal in the North American and West Pacific regions. Phycologia 34:472–485
32. Scholin CA et al (1994) Identification of group-specific and strain-specific genetic-markers for globally distributed *Alexandrium* (Dinophyceae) 2 sequence-analysis of a fragment of the Lsu ribosomal-RNA gene. J Phycol 30(6):999–1011
33. Lilly EL, Halanych KM, Anderson DM (2007) Species boundaries and global biogeography of the *Alexandrium tamarense* complex (Dinophyceae). J Phycol 43:1329–1338
34. Kalaitzis JA et al (2010) Biosynthesis of toxic naturally-occurring seafood contaminants. Toxicon 56:244–258
35. Etheridge SM (2010) Paralytic shellfish poisoning: seafood safety and human health perspectives. Toxicon 56:108–122
36. Oshima Y, Blackburn SI, Hallegraeff GM (1993) Comparative study on paralytic shellfish toxin profiles of the dinoflagellate *Gymnodinium catenatum* from three different countries. Mar Biol 116:471–476
37. Hall S et al (1990) The saxitoxins: sources, chemistry, and pharmacology. In: Hall S, Strichartz G (eds) Marine toxins: origins, structure, and molecular pharmacology. American Chemical Society, Washington, DC
38. Hille B (1992) Ionic channels of excitable membranes, 2nd edn. Sinauer, Sunderland
39. Ritchie JM, Rogart RB (1977) The binding of saxitoxin and tetrodotoxin to excitable tissue. Rev Physiol Biochem Pharmacol 79:42–50
40. Kao CY et al (1967) Vasomotor and respiratory depressant actions of tetrodotoxin and saxitoxin. Arch Int Pharmacodyn Ther 165:438–450
41. Kao CY (1993) Paralytic shellfish poisoning. In: Falconer IR (ed) Algal toxins in seafood and drinking water. Academic, London, pp 75–86
42. Doucette G et al (2006) Phycotoxin pathways in aquatic food webs: transfer, accumulation, and degradation. In: Graneli E, Turner JT (eds) Ecology of harmful algae. Springer, Berlin/Heidelberg
43. Anderson DM, White AW (1989) Toxic dinoflagellates and marine mammal mortalities. Woods Hole Oceanographic Institution Technical Report. WHOI-89-36 (CRC-89-6). Woods Hole Oceanographic Institution, Woods Hole
44. Geraci JR (1989) Clinical investigations of the 1987–1988 mass mortality of bottlenose dolphins along the US central and south Atlantic coast. In: Final report to the national marine fisheries service. U. S. Navy, Office of Naval Research and Marine Mammal Commission, Ontario Veterinary College, University of Guelph, Guelph, pp 1–63
45. Geraci JR et al (1989) Humpback whales (*Megaptera novaeangliae*) fatally poisoned by dinoflagellate toxin. Can J Fish Aquat Sci 46:1895–1898
46. Shumway SE (1995) Phycotoxin-related shellfish poisoning: bivalve molluscs are not the only vectors. Rev Fish Sci 3:1–31
47. White AW (1981) Marine zooplankton can accumulate and retain dinoflagellate toxins and cause fish kills. Limnol Oceanogr 26:103–109
48. Doucette GJ et al (2006) PSP toxins in North Atlantic right whales (*Eubalaena glacialis*) and their zooplankton prey in the Bay of Fundy. Canada Mar Ecol Prog Ser 306:303–313
49. Deeds JR et al (2008) Non-traditional vectors for paralytic shellfish poisoning. Mar Drugs 6:308–348
50. Shumway SE (1990) A review of the effects of algal blooms on shellfish and aquaculture. J World Aquacul Soc 21:65–104
51. RaLonde R (1996) Paralytic shellfish poisoning: the Alaska problem. In: Alaska's marine resources. Marine Advisory Program, Anchorage
52. Bricelj VM, Lee JH, Cembella AD (1991) Influence of dinoflagellate cell toxicity on uptake and loss of paralytic shellfish toxins in the northern quahog *Mercenaria mercenaria*. Mar Ecol Prog Ser 74:33–46

53. Reboreda A et al (2010) Decrease in marine toxin content in bivalves by industrial processes. Toxicon 55:235–243
54. Trainer VL (2002) Harmful algal blooms on the U.S. west coast. In: Taylor FJ, Trainer VL (eds) Harmful algal blooms in the PICES region of the North Pacific. PICES Scientific Report No. 23. North Pacific Marine Science Organization, Sidney, pp 89–118
55. Nishitani L, Chew KK (1988) PSP toxins in the Pacific coast states: monitoring programs and effects on bivalve industries. J Shellfish Res 7:653–669
56. Anderson DM et al (2000) Estimated annual economic impacts from harmful algal blooms (HABs) in the US. Woods Hole Oceanographic Institute, Woods Hole
57. Anderson DM et al (2005) Initial observation of the 2005 *Alexandrium fundyense* bloom in southern New England: general patterns and mechanisms. Deep Sea Res II 52:2856–2876
58. Jin D, Thunberg E, Hoagland P (2008) Economic impact of the 2005 red tide event on commercial shellfish fisheries in New England. Ocean Coast Manag 51(5):420–429
59. Adachi M, Sako Y, Ishida Y (1996) Analysis of *Alexandrium* (Dinophyceae) species using sequence of the 5.8S ribosomal DNA and internal transcribed spacer regions. J Phycol 32:424–432
60. John U, Fensome RA, Medlin LK (2003) The application of a molecular clock based on molecular sequences and fossil record to explain biogeographic distributions within the *Alexandrium tamarense* "species complex" (Dinophyceae). Mol Biol Evol 20:1015–1027
61. Medlin L et al (1998) Sequence comparison links toxic European isolates of *Alexandrium tamarense* from the Orkney Islands to toxic North American stocks. Eur J Protistol 34:329–335
62. Hollingworth T, Wekell MM (1990) Paralytic shellfish poison biological method. Final action. In: Hellrich K (ed) Official methods of analysis. AOAC, Arlington, pp 881–882
63. AOAC (2000) AOAC official method 959.08 paralytic shellfish poison. In: Horwitz W (ed) Official methods of analysis of AOAC international. AOAC International, Arlington
64. Wekell JC, Hurst J, Lefebvre KA (2004) The origin of the regulatory limits for PSP and ASP toxins in shellfish. J Shellfish Res 23:927–930
65. Food and Agriculture Organization of the United Nations (2004) Marine biotoxins. Paper 80. FAO Food and Nutrition, Rome, pp 5–52
66. Jellet JF et al (1992) Paralytic shellfish poison (saxitoxin family) bioassays: automated endpoint determination and standardization of the in vitro tissue culture bioassay and comparison with the standard mouse bioassay. Toxicon 30:1143–1156
67. Kogure K et al (1988) A tissue culture assay for the tetrodotoxin, saxitoxin, and related toxins. Toxicon 26:191–197
68. Shimojo RY, Iwaoka WT (2000) A rapid hemolysis assay for the detection of sodium channel-specific marine toxins. Toxicology 154:1–7
69. Davio SR, Fontelo PA (1984) A competitive displacement assay to detect saxitoxin and tetrodotoxin. Anal Biochem 141:199–204
70. Doucette GJ et al (1997) Development and preliminary validation of a microtiter plate-based receptor binding assay for paralytic shellfish poisoning toxins. Toxicon 35(5):625–636
71. Powell CL, Doucette GJ (1999) A receptor binding assay for paralytic shellfish poisoning toxins: recent advances and applications. Nat Toxins 7:393–400
72. Chu FS, Huang X, Wei RD (1990) Enzyme-linked immunosorbent assay for microcystins in blue-green algal blooms. J Assoc Off Anal Chem 73(3):451–456
73. Usleber E, Schnieder E, Terplan G (1991) Direct enzyme immunoassay in microtitration plate and test strip format for the detection of saxitoxin in shellfish. Lett Appl Microbiol 13:275
74. Bergantin JH, Sevilla F III (2010) Quartz crystal microbalance biosensor for saxitoxin based on immobilized sodium channel receptors. Anal Lett 43(3):476–486
75. Metfies K et al (2005) Electrochemical detection of the toxic dinoflagellate *Alexandrium ostenfeldii* with a DNA-biosensor. Biosens Bioelectron 20(7):1349–1357

76. Diercks S, Metfies K, Medlin LK (2008) Development and adaptation of a multiprobe biosensor for the use in a semi-automated device for the detection of toxic algae. Biosens Bioelectron 23:1527–1533
77. Fonfria ES et al (2007) Paralytic shellfish poisoning detection by surface plasmon resonance-based biosensors in shellfish matrixes. Anal Chem 79(16):6303–6311
78. Chu FS et al (1996) Screening of paralytic shellfish poisoning toxins in naturally occurring samples with three different direct competitive enzyme-linked immunosorbent assays. J Agric Food Chem 44(12):4043–4047
79. Garthwaite I et al (2001) Integrated enzyme-linked immunosorbent assay screening system for amnesic, neurotoxic, diarrhetic, and paralytic shellfish poisoning toxins found in New Zealand. J AOAC Int 84:1643–1648
80. Kawatsu K et al (2002) Development and application of an enzyme immunoassay based on a monoclonal antibody against gonyautoxin components of paralytic shellfish poisoning toxins. J Food Prot 65(8):1304–1308
81. Usleber E et al (2001) Immunoassay methods for paralytic shellfish poisoning toxins. J AOAC Int 84:1649–1656
82. Usleber E, Donald M, Straka M (1997) Comparison of enzyme immunoassay and mouse bioassay for determining paralytic shellfish poisoning toxins in shellfish. Food Addit Contam 14(2):193–198
83. Gas F et al (2010) One step immunochromatographic assay for the rapid detection of *Alexandrium minutum*. Biosens Bioelectron 25:1235–1239
84. Jellet JF et al (2002) Detection of paralytic shellfish poisoning (PSP) toxins in shellfish tissue using MIST Alert, a new rapid test, in parallel with the regulatory AOAC mouse bioassay. Toxicon 40(10):1407–1425
85. Costa PR et al (2009) Comparative determination of paralytic shellfish toxins (PSTs) using five different toxin detection methods in shellfish species collected in the Aleutian islands, Alaska. Toxicon 54:313–320
86. Oshiro M et al (2006) Paralytic shellfish poisoning surveillance in California using the Jellett Rapid PSP test. Harmful Algae 5:69–73
87. Sullivan JJ (1988) Methods of analysis for DSP and PSP toxins in shellfish: a review. J Shellfish Res 7:587–595
88. Dell'Aversano C, Hess P, Quilliam MA (2005) Hydrophilic interaction liquid chromatography-mass spectrometry analysis of paralytic shellfish poisoning (PSP) toxins. J Chromatogr A1081:190
89. Oshima Y (1995) Post-column derivation HPLC methods for paralytic shellfish poisons. In: Hallegraeff GM, Anderson DM, Cembella AD (eds) Manual on harmful marine microalgae. IOC of UNESCO, Paris, pp 81–94
90. Rourke WA et al (2008) Rapid postcolumn methodology for determination of paralytic shellfish toxins in shellfish tissue. J AOAC Int 91(3):589–597
91. Thomas K et al (2006) Analysis of PSP toxins by liquid chromatography with postcolumn oxidation and fluorescence detection. In: Henshilwood K et al (eds) Molluscan shellfish safety. The Marine Institute, Galway, pp 132–138
92. Lawrence JF, Menard C (1991) Liquid chromatographic determination of paralytic shellfish poisons in shellfish after prechromatographic oxidation. J Assoc Off Anal Chem 74:1006–1012
93. Lawrence JF et al (1991) A study of ten toxins associated with paralytic shellfish poison using prechromatographic oxidation and liquid chromatography with fluorescence detection. J Assoc Off Anal Chem 74:419–425
94. Lawrence JF, Niedzwiadek B, Menard C (2004) Quantitative determination of paralytic shellfish poisoning toxins in shellfish using prechromatographic oxidation and liquid chromatography with fluorescence detection: interlaboratory study. J AOAC Int 87(1):83–100

95. Quilliam MA, Janecek M, Lawerence JF (1993) Characterization of the oxidation products of paralytic shellfish poisoning toxins by liquid chromatography/mass spectrometry. Rapid Commun Mass Spectrom 6:14–24
96. Laycock MV et al (1994) Isolation and purification procedures for the preparation of paralytic shellfish poisoning toxin standards. Nat Toxins 2:175–183
97. NRC (2003) National Research Council Canada (NRC) – Institute for Marine Biosciences. http://www.nrc-cnrc.gc.ca/eng/ibp/imb.html [cited]
98. Van De Riet, J et al (2011) Liquid chromatography post-column oxidation (PCOX) method for the determination of paralytic shellfish toxins in mussels, clams, oysters, and scallops: collaborative study. J AOAC International 94:1154–1176
99. Anderson DM (1997) Bloom dynamics of toxic *Alexandrium* species in the northeastern US. Limnol Oceanogr 42(5 Pt 2):1009–1022
100. Hurst JW (1975) History of paralytic shellfish poisoning on the Maine coast 1958–1974. In: First international conference on toxic dinoflagellate blooms. Massachusetts Science and Technology Foundation, Wakefield
101. Anderson DM, Glibert PM, Burkholder JM (2002) Harmful algal blooms and eutrophication: nutrient sources, composition, and consequences. Estuaries 25:704–726
102. Boesch DF et al (1997) Harmful algal blooms in coastal waters: options for prevention, control, and mitigation. NOAA Coastal Ocean Program Decision Analysis Series No.10. NOAA Coastal Ocean Office, Silver Spring, 46pp + appendix
103. Mulligan HF (1975) Oceanographic factors associated with New England red tide blooms. In: First international conference on toxic dinoflagellate blooms. Massachusetts Science and Technology Foundation, Wakefield
104. He R et al (2008) Historic 2005 toxic bloom of *Alexandrium fundyense* in the western Gulf of Maine: 2. coupled biophysical numerical modeling. J Geophys Res Oceans 113: C07040–C07042
105. McGillicuddy DJ Jr et al (2005) Mechanisms regulating large-scale seasonal fluctuations in *Alexandrium fundyense* populations in the Gulf of Maine: results from a physical-biological model. Deep Sea Res II 52:2698–2714
106. McGillicuddy DJ Jr et al (2003) A mechanism for offshore initiation of harmful algal blooms in the coastal Gulf of Maine. J Plankton Res 25(9):1131–1138
107. Anderson DM, Taylor CD, Armbrust EV (1987) The effects of darkness and anaerobiosis on dinoflagellate cyst germination. Limnol Oceanogr 32:340–351
108. Anderson DM et al (1982) Vertical and horizontal distributions of dinoflagellates cyst in sediments. Limnol Oceanogr 27:757–765
109. Sinclair GA et al (2006) Nitrate uptake by *Karenia brevis*. I. Influences of prior environmental exposure and biochemical state on diel uptake of nitrate. Mar Ecol Prog Ser 328:117–124
110. Sinclair GA et al (2006) Nitrate uptake by *Karenia brevis*. II. Behavior and uptake physiology in a nitrate-depleted mesocosm with a bottom nutrient source. Mar Ecol Prog Ser 328:125–131
111. Walsh JJ, Steidinger KA (2001) Saharan dust and Florida red tides: the cyanophyte connection. J Geophy Res Oceans 106(C6):11597–11612
112. Tester PA, Steidinger KA (1997) *Gymnodinium breve* red tide blooms: initiation, transport, and consequences of surface circulation. Limnol Oceanogr 42(5):1039–1051
113. Vargo GA et al (2008) Nutrient availability in support of *Karenia brevis* blooms on the central West Florida shelf: what keeps *Karenia* blooming? Cont Shelf Res 28(1):73–98
114. Baden DG, Mende TJ (1982) Toxicity of 2 toxins from the Florida red tide marine dinoflagellate *Ptychodiscus brevis*. Toxicon 20(2):457–461
115. Flewelling LJ et al (2005) Red tides and marine mammal mortalities. Nature 435(7043):755–756
116. Davis CC (1948) *Gymnodinium brevis* sp. nov a cause of discolored water and animal mortality in the Gulf of Mexico. Bot Gaz 109(3):358–360

117. Fire SE et al (2008) Prevalence of brevetoxins in prey fish of bottlenose dolphins in Sarasota Bay, Florida. Mar Ecol Prog Ser 368:283–294
118. Kirkpatrick B et al (2010) Inland transport of aerosolized Florida red tide toxins. Harmful Algae 9(2):186–189
119. Kirkpatrick B et al (2006) Environmental exposures to Florida red tides: effects on emergency room respiratory diagnoses admissions. Harmful Algae 5(5):526–533
120. Zaias J et al (2011) Repeated exposure to aerosolized brevetoxin-3 induces prolonged airway hyperresponsiveness and lung inflammation in sheep. Inhal Toxicol 23(4):205–211
121. Hetland RD, Campbell L (2007) Convergent blooms of *Karenia brevis* along the Texas coast. Geophys Res Lett 34(19):5
122. Flewelling LJ et al (2010) Brevetoxins in sharks and rays (Chondrichthyes, Elasmobranchii) from Florida coastal waters. Mar Biol 157(9):1937–1953
123. Gannon DP et al (2009) Effects of *Karenia brevis* harmful algal blooms on nearshore fish communities in southwest Florida. Mar Ecol Prog Ser 378:171–186
124. Kirkpatrick B et al (2004) Literature review of Florida red tide: implications for human health effects. Harmful Algae 3(2):99–115
125. Backer LC et al (2003) Recreational exposure to aerosolized brevetoxins during Florida red tide events. Harmful Algae 2(1):19–28
126. Fleming LE et al (2005) Initial evaluation of the effects of aerosolized Florida red tide toxins (Brevetoxins) in persons with asthma. Environ Health Perspect 113(5):650–657
127. Steidinger KA et al (1999) Harmful algal blooms in Florida. Unpublished technical report submitted to the Florida Harmful Algal Bloom Task Force. Florida Marine Research Institute, St. Petersburg, p 63
128. Hoagland P et al (2009) The costs of respiratory illnesses arising from Florida gulf coast *Karenia brevis* blooms. Environ Health Perspect 117(8):1239–1243
129. Morgan KL, Larkin SL, Adams CM (2010) Red tides and participation in marine-based activities: estimating the response of Southwest Florida residents. Harmful Algae 9(3):333–341
130. Morgan KL, Larkin SL, Adams CM (2009) Firm-level economic effects of HABS: a tool for business loss assessment. Harmful Algae 8(2):212–218
131. Plakas SM, Dickey RW (2010) Advances in monitoring and toxicity assessment of brevetoxins in molluscan shellfish. Toxicon 56(2):137–149
132. Stumpf R et al (2008) Hydrodynamic accumulation of *Karenia* on the west coast of Florida. Cont Shelf Res 28:189–213
133. Sinclair G (2008) Environmental, behavioral, and physiological constraints on the persistence of near-bottom populations of *Karenia brevis*. Department of marine, earth, and atmospheric sciences, North Carolina State University, Raleigh, p 181
134. Schofield O et al (1999) Optical monitoring and forecasting systems for harmful algal blooms: possibility or pipe dream? J Phycol 35(6):1477–1496
135. Janowitz GS, Kamykowski D (2006) Modeled *Karenia brevis* accumulation in the vicinity of a coastal nutrient front. Mar Ecol Prog Ser 314:49–59
136. Liu G, Janowitz G, Kamykowski D (2001) Influence of environmental nutrient conditions on *Gymnodinium breve* (Dinophyceae) population dynamics: a numerical study. Mar Ecol Prog Ser 213:13–37
137. Janowitz GS, Kamykowski D (1999) An expanded eularian model of phytoplankton environmental response. Ecol Model 118:237–247
138. Van Dolah FM, Leighfield TA (1999) Diel phasing of the cell-cycle in the Florida red tide dinoflagellate, *Gymnodinium breve*. J Phycol 35(6):1404–1411
139. Kamykowski D, Milligan EJ, Reed RE (1998) Biochemical relationships with the orientation of the autotrophic dinoflagellate *Gymnodinium breve* under nutrient replete conditions. Mar Ecol Prog Ser 167:105–117
140. Kamykowski D, Yamazaki H (1997) A study of metabolism-influence orientation in the diel vertical migration of marine dinoflagellates. Limnol Oceanogr 42(5):1189–1202

141. Dragovich A, Finucane J, Mays B (1961) Counts of red tide organisms, *Gymnodinium breve* and associated oceanographic data from Florida west coast, 1957–1959. In: U.S. Fish and Wildlife Service, Special. Science Report on Fish, Washington, DC, 175p
142. Steidinger KA et al (1998) Bloom dynamics and physiology of *Gymnodinium breve* with emphasis on the Gulf of Mexico. In: Anderson DM, Cembella AD, Hallegraeff GM (eds) Physiological ecology of harmful algal blooms. Springer, Berlin, pp 133–153
143. Masserini RT, Fanning KA (2000) A sensor package for the simultaneous determination of nanomolar concentrations of nitrite, nitrate, and ammonia in seawater by fluorescence detection. Mar Chem 68(4):323–333
144. Hecky RE, Kilham P (1988) Nutrient Limitation of phytoplankton in freshwater and marine environments: a review on the effects of enrichment. Limnol Oceanogr 33(4):796–822
145. Walsh J, Steidinger K (2001) Saharan dust and Florida red tides: the cyanophyte connection. J Geophys Res 106(C6):11,597–11,612
146. Mulholland MR et al (2006) Nitrogen fixation and release of fixed nitrogen by *Trichodesmium* spp. in the Gulf of Mexico. Limnol Oceanogr 51(4):1762–1776
147. Lester KM et al (2008) Zooplankton and *Karenia brevis* in the Gulf of Mexico. Cont Shelf Res 28(1):99–111
148. Sinclair GA, Kamykowski D (2008) Benthic-pelagic coupling in sediment-associated populations of *Karenia brevis*. J Plankton Res 30(7):829–838
149. Janowitz GS, Kamykowski D, Liu G (2008) A three-dimensional wind and behaviorally driven population dynamics model for *Karenia brevis*. Cont Shelf Res 28(1):177–188
150. Hasle GR (2002) Are most of the domoic acid-producing species of the diatom genus *Pseudo-nitzschia* cosmopolites? Harmful Algae 1:137–146
151. Smetacek V et al (2002) Mesoscale distribution of dominant diatom species relative to the hydrographical field along the Antarctic polar front. Deep-Sea Res II 49:3835–3848
152. Fryxell GA, Villac MC, Shapiro LP (1997) The occurrence of the toxic diatom genus *Pseudo-nitzschia* (Bacillariophyceae) on the West Coast of the USA, 1920–1996: a review. Phycologia 36:419–437
153. Qi Y, Wang J, Zheng L (1994) The taxonomy and bloom ecology of Pseudo-nitzschia on the coasts of China. In: IOC-WESTPAC third international scientific symposium, Bali, pp 88–95
154. Zou JZ, Zhou MJ, Zhang C (1993) Ecological features of toxic *Nitzschia pungens* Grunow in Chinese coastal waters. In: Smayda TJ, Shimizu Y (eds) Toxic phytoplankton blooms in the sea. Elsevier, Amsterdam
155. Evans KM et al (2004) Microsatellite marker development and genetic variation in the toxic marine diatom *Pseudo-nitzschia multiseries* (Bacillariophyceae). J Phycol 40(5):911–920
156. Evans KM, Hayes PK (2004) Microsatellite markers for the cosmopolitan marine diatom *Pseudo-nitzschia pungens*. Mol Ecol Notes 4(1):125–126
157. Evans KM, Kuhn SF, Hayes PK (2005) High levels of genetic diversity and low levels of genetic differentiation in North Sea *Pseudo-nitzschia pungens* (Bacillariophyceae) populations. J Phycol 41(3):506–514
158. Hubbard KA, Rocap G, Armbrust EV (2008) Inter- and intraspecific community structure within the diatom genus *Pseudo-nitzschia* (Bacillariophyceae). J Phycol 44:637–649
159. Orsini L et al (2004) Multiple rDNA ITS-types within the diatom *Pseudo-nitzschia delicatissima* (Bacillariophyceae) and their relative abundances across a spring bloom in the Gulf of Naples. Mar Ecol Prog Ser 271:87–98
160. Thessen AE, Bowers HA, Stoecker DK (2009) Intra- and interspecies differences in growth and toxicity of Pseudo-nitzschia while using different nitrogen sources. Harmful Algae 8:792–810
161. Bates SS (2000) Domoic-acid-producing diatoms: another genus added. J Phycol 36:978–985
162. Bates SS et al (1989) Pennate diatom *Nizschia pungens* as the primary source of domoic acid, a toxin in shellfish from eastern Prince Edward Island, Canada. Can J Fish Aquatic Sci 46:1203–1215

163. Lefebvre KA, Robertson A (2010) Domoic acid and human exposure risks: a review. Toxicon 56:218–230
164. Teitelbaum J, Carpenter S, Cashman NR (1990) Neurologic sequelae after ingestion of mussels contaminated with domoic acid. N Engl J Med 323:1632–1633
165. Peng YG et al (1994) Neuroexcitatory and the neurotoxic actions of the amnesic shellfish poison, domoic acid. Neuropharmacol Neurotoxicol 5:981–985
166. Gulland FMD et al (2002) Domoic acid toxicity in California sea lions (*Zalophus californianus*): clinical signs, treatment and survival. Vet Rec 150:475–480
167. Perl TM et al (1990) An outbreak of toxic encephalopathy caused by eating mussels contaminated with domoic acid. N Engl J Med 322:1775–1780
168. Todd ECD (1993) Domoic acid and amnesic shellfish poisoning: a review. J Food Prot 56:69–83
169. Wright JLC et al (1989) Identification of domoic acid, a neuroexcitatory amino acid, in toxic mussels from eastern Prince Edward Island. Can J Chem 67:481–490
170. Garrison DL et al (1992) Confirmation of domoic acid production by *Pseudo-nitzschia australis* (Bacillariophyceae) cultures. J Phycol 28:604–607
171. Lefebvre KA et al (2002) Domoic acid in planktivorous fish in relation to toxic *Pseudo-nitzschia* cell densities. Mar Biol 140:625–631
172. Scholin CA et al (2000) Mortality of sea lions along the central California coast linked to a toxic diatom bloom. Nature 403:80–84
173. Wekell JC et al (1994) Occurrence of domoic acid in Washington state razor clams (*Siliqua patula*) during 1991–1993. Nat Toxins 2:197–205
174. Fritz L et al (1992) An outbreak of domoic acid poisoning attributed to the pennate diatom *Pseudonitzschia australis*. J Phycol 28:439–442
175. Work TM et al (1993) Epidemiology of DA poisoning in brown pelicans (*Pelecanus occidentalis*) and Brandt's cormorants (*Phalacrocarax penicillatus*) in California. J Zoo Wildl Med 24:54–62
176. Sierra-Beltran AP et al (1997) Sea bird mortality at Cabo San Lucas, Mexico: evidence that toxic diatom blooms are spreading. Toxicon 35:447–453
177. Sierra-Beltran AP et al (1998) An overview of the marine food poisoning in Mexico. Toxicon 36:1493–1502
178. Lefebvre KA et al (1999) Detection of domoic acid in northern anchovies and California sea lions associated with an unusual mortality event. Nat Toxins 7(3):85–92
179. Goldstein T et al (2008) Novel symptomatology and changing epidemiology of domoic acid toxicosis in California sea lions (*Zalophus californianus*): an increasing risk to marine mammal health. Proc R Soc Biol Sci Ser B 275:267–276
180. Maucher JM, Ramsdell JS (2005) Domoic acid transfer to milk: evaluation of a potential route of neonatal exposure. Environ Health Perspect 115:1743–1746
181. Grattan LM et al (2007) Domoic acid neurotoxicity in native Americans in the pacific northwest: human health project methods and update. In: Fourth symposium on harmful algae in the US, Woods Hole
182. Trainer VL et al (2007) Recent domoic acid closures of shellfish harvest areas in Washington State inland waters. Harmful Algae 6:449–459
183. Trainer VL, Hickey BM, Horner RA (2002) Biological and physical dynamics of domoic acid production off the Washington coast. Limnol Oceanogr 47:1438–1446
184. Dyson K, Huppert DD (2010) Regional economic impacts of razor clam beach closures due to harmful algal blooms (HABs) on the Pacific coast of Washington. Harmful Algae 9:264–271
185. Bill BD et al (2006) The first closure of shellfish harvesting due to domoic acid in Puget Sound, Washington, USA. Afr J Mar Sci 28:437–442
186. Ayers D, Reed H (2004) Managing important recreational and commercial shellfish fisheries around harmful algal blooms. In: 2003 Georgia Basin/Puget sound research conference, Olympia

187. Hasle GR, Syvertsen EE (1997) *Marine Diatoms*. In: Tomas CR (ed) Identifying marine diatoms and dinoflagellates. Academic, San Diego
188. Hasle GR, Lundholm N (2005) *Pseudo-nitzschia seriata* f. *obtusa* (Bacillariophyceae) raised in rank based on morphological, phylogenetic and distributional data. Phycologia 44:608–619
189. Lundholm N et al (2003) A study of the *Pseudo-nitzschia pseudodelicatissima/cuspidata* complex (Bacillariophyceae): what is *P. pseudodelicatissima*? J Phycol 39:797–813
190. Lundholm N et al (2006) Inter- and intraspecific variation of the *Pseudo-nitzschia delicatissima* complex (Bacillariophyceae) illustrated by rRNA probes, morphological data and phylogenetic analyses. J Phycol 42(2):464–481
191. Fritz L (1992) The use of cellular probes in studying marine phytoplankton. Korean J Phycol 7:319–324
192. Rhodes LL (1998) Identification of potentially toxic *Pseudo-nitzschia* (Bacillariophyceae) in New Zealand coastal waters, using lectins. N Z J Mar Freshw Res 32:537–544
193. Rhodes LL et al (1998) Domoic acid producing *Pseudo-nitzschia* species educed by whole cell DNA probe-based and immunochemical assays. In: Reguera B et al (eds) Harmful algae. Xunta de Galicia and IOC of UNESCO, Santiago de Compostela, pp 274–277
194. Cho ES et al (1999) The rapid differentiation of toxic *Alexandrium* and *Pseudo-nitzschia* species using fluorescent lectin probes. J Korean Soc Oceanogr 35:167–171
195. Fraga S et al (1998) *Pseudo-nitzschia* species isolated from Galician waters: toxicity, DNA content and lectin binding assay. In: Harmful algae: eighth international conference on harmful algae, Vigo, pp 270–273
196. Rhodes L, Scholin C, Garthwaite I (1998) *Pseudo-nitzschia* in New Zealand and the role of DNA probes and immunoassays in refining marine biotoxin monitoring programmes. Nat Toxins 6:105–111
197. Andree KB et al (2011) Quantitative PCR coupled with melt curve analysis for detection of selected *Pseudo-nitzschia* spp. (Bacillariophyceae) from the northwestern Mediterranean Sea. Appl Environ Microbiol 77(5):1651–1659
198. Cho ES, Kodaki Y, Park JG (2001) The comparison of two strains of toxic *Pseudo-nitzschia multiseries* (Hasle) Hasle and non-toxic *Pseudo-nitzschia pungens* (Grunow) Hasle isolated from Chinhae Bay, Korea. Algae 16:275–285
199. Lundholm N et al (2002) Morphology, phylogeny and taxonomy of species within *Pseudo-nitzschia americana* complex (Bacillariophyceae) with descriptions of two new species, *Pseudo-nitzschia brasiliana* and *Pseudo-nitzschia linea*. Phycologia 41:480–497
200. Manhart JR et al (1995) *Pseudo-nitzschia pungens* and *P. multiseries* (Bacillariophyceae): nuclear ribosomal DNA's and species differences. J Phycol 31:421–427
201. McDonald SM, Sarno D, Zingone A (2007) Identifying *Pseudo-nitzschia* species in natural samples using genus-specific PCR primers and clone libraries. Harmful Algae 6(8):849–860
202. Miller P, Scholin CA (1996) Identification of cultured *Pseudo-nitzschia* (Bacillariophyceae) using species specific LSU rRNA-targeted probes. J Phycol 32:646–655
203. Miller P, Scholin CA (1998) Identification and enumeration of cultured and wild *Pseudo-nitzschia* (Bacillariophyceae) using species specific LSU rRNA-targeted fluorescent probes and filter-based whole cell hybridization. J Phycol 34:371–382
204. Scholin CA et al (1996) Identification of *Pseudo-nitzschia australis* (Bacillariophyceae) using rRNA-targeted probes in whole cell and sandwich hybridization formats. Phycologia 35:190–197
205. Scholin CA et al (1999) DNA probes and a receptor-binding assay for detection of *Pseudo-nitzschia* (Bacillariophyceae) species and domoic acid activity in cultured and natural samples. J Phycol 35:1356–1367
206. Scholin CA et al (1997) Detection and quantification of *Pseudo-nitzschia australis* in cultured and natural populations using LSU rRNA-targeted probes. Limnol Oceanogr 42:1265–1272
207. Scholin CA et al (1994) Ribosomal DNA sequences discriminate among toxic and non-toxic *Pseudo-nitzschia* species. Nat Toxins 2:152–165

208. Cho ES et al (2002) Monthly monitoring of domoic acid producer *Pseudo-nitzschia multiseries* (Hasle) Hasle using species-specific DNA probes and WGA lectins and abundance of *Pseudo-nitzschia* species (Bacillariophyceae) from Chinhae Bay, Korea. Bot Mar 45:364–372
209. Orsini L et al (2002) Toxic *Pseudo-nitzschia multistriata* (Bacillariophyceae) from the Gulf of Naples: morphology, toxin analysis and phylogenetic relationships with other *Pseudo-nitzschia* species. Eur J Phycol 37:247–257
210. Parsons ML et al (1999) *Pseudo-nitzschia* species (Bacillariophyceae) in Louisiana coastal waters: molecular probes field trials, genetic variability, and domoic acid analyses. J Phycol 35:1368–1378
211. Vrieling EG et al (1996) Identification of a domoic acid-producing *Pseudo-nitzschia* species (Bacillariophyceae) in the Dutch Wadden sea with electron microscopy and molecular probes. Eur J Phycol 31:333–340
212. Greenfield D et al (2008) Field applications of the second-generation environmental sample processor (ESP) for remote detection of harmful algae: 2006–2007. Limnol Oceanogr Method 6:667–679
213. Greenfield DI et al (2006) Application of the environmental sample processor (ESP) methodology for quantifying *Pseudo-nitzschia australis* using ribosomal RNA-targeted probes in sandwich and fluorescent in situ hybridization. Limnol Oceanogr Method 4:426–435
214. Scholin C et al (2009) Remote detection of marine microbes, small invertebrates, harmful algae and biotoxins using the environmental sample processor (ESP). Oceanography 22:158–167
215. AOAC (1990) Paralytic shellfish poison. Biological method. Final action. In: Hellrich K (ed) Official method of analysis. Association of Official Analytical Chemists (AOAC), Arlington, pp 881–882
216. Van Dolah FM et al (1997) A microplate receptor assay for the amnesic shellfish poisoning toxin, domoic acid, utilizing a cloned glutamate receptor. Anal Biochem 245:102–105
217. Garthwaite I et al (1998) An immunoassay for determination of domoic acid in shellfish and sea water. In: Reguera B et al (eds) Harmful algae. Xunta de Galicia and IOC of UNESCO, Santiago de Compostela, pp 559–562
218. Kawatsu K, Hamano Y, Noguchi T (1999) Production and characterization of a monoclonal antibody against domoic acid and its application to enzyme immunoassay. Toxicon 37:1579–1589
219. Smith DS, Kitts DD (1994) A competitive enzyme-linked immunoassay for domoic acid determination in human body fluids. Food Chem Toxicol 32(12):1147–1154
220. Smith DS, Kitts DD (1995) Enzyme Immunoassay for the determination of domoic acid in mussel extracts. J Agric Food Chem 43:367–371
221. Fernandez ML, Cembella AD (1995) Part B. Mammalian bioassays. In: Hallegraeff GM, Anderson DM, Cembella AD (eds) Manual on harmful marine microalgae. UNESCO, Paris, pp 213–228
222. Lawrence JF et al (1989) Liquid chromatographic determination of domoic acid in shellfish products using the paralytic shellfish extraction procedure of the association of official analytical chemists. J Chromatogr 462:349–356
223. Lawrence JF, Charbonneau CF, Menard C (1991) Liquid chromatographic determination of domoic acid in mussels, using AOAC paralytic shellfish poison extraction procedures: collaborative study. J Assoc Off Anal Chem 74(1):68–72
224. Ciminiello P et al (2005) Hydrophilic interaction liquid chromatography/mass spectrometry for determination of domoic acid in Adriatic shellfish. Rapid Commun Mass Spectrom 19(4):2030–2038
225. Hummert C, Reichelt M, Luckas B (1997) Automatic HPLC-UV determination of domoic acid in mussels and algae. Chromatographia 45:284–288

226. Quilliam MA et al (1989) High-performance liquid-chromatography of domoic acid, a marine neurotoxin, with application to shellfish and plankton. Int J Environ Anal Chem 36:139–154
227. European Union Reference Laboratory for Marine Biotoxins (ed) (2010) Standard operating procedure for determination of domoic acid (ADP toxins) in molluscs by UPLC-MS. European Union Reference Laboratory for Marine Biotoxins/Agencia Espanola de Seguridad Alimentaria y Nutricion, Vigo
228. Rafuse C et al (2004) Rapid monitoring of toxic phytoplankton and zooplankton with a lateral-flow immunochromatographic assay for ASP and PSP toxins. In: Steidinger KA (ed) Harmful algae. Florida Fish and Wildlife Conservation Commission and Intergovernmental Oceanographic Commission of UNESCO, St. Petersburg
229. Turrell E et al (2008) Detection of *Pseudo-nitzschia* (Bacillariophyceae) species and amnesic shellfish toxins in Scottish coastal waters using oligonucleotide probes and the Jellet rapid test. Harmful Algae 7(4):443–458
230. Parsons ML, Dortch Q, Turner RE (2002) Sedimentological evidence of an increase in *Pseudo-nitzschia* (Bacillariophyceae) abundance in response to coastal eutrophication. Limnol Oceanogr 47(2):551–558
231. Dortch Q et al (2000) *Pseudo-nitzschia* spp. in the northern Gulf of Mexico: overview and response to increasing eutrophication. In: Symposium on harmful marine algae in the U.S. Marine Biological Laboratory, Woods Hole, p 27
232. McFadyen A, Hickey BM, Foreman MGG (2005) Transport of surface waters from the Juan de Fuca eddy region to the Washington coast. Cont Shelf Res 25:2008–2021
233. Bates SS et al (1991) Controls on domoic acid production by the diatom *Nitzschia pungens* f. *multiseries* in culture: nutrient and irradiance. Can J Fish Aquatic Sci 48:1136–1144
234. Pan Y, Subba Rao DV, Mann KH (1996) Changes in domoic acid production and cellular chemical composition of the toxigenic diatom *Pseudo-nitzschia multiseries* under phosphate limitation. J Phycol 32:371–381
235. Pan Y et al (1996) Effects of silicate limitation on production of domoic acid, a neurotoxin, by the diatom *Pseudo-nitzschia multiseries*. I. Batch culture studies. Mar Ecol Prog Ser 131:225–233
236. Pan Y et al (1996) Effects of silicate limitation on production of domoic acid, a neurotoxin, by the diatom *Pseudo-nitzschia multiseries*. II. Continuous culture studies. Mar Ecol Prog Ser 131:235–243
237. Van Apeldoorn ME, van Egmond HP, Speijers GJA (1999) Amnesic shellfish poisoning: a review. In: RIVM report 388802 019. National Institute of Public Health and the Environment, the Netherlands
238. Maldonado MT, Hughes MP, Rue EL (2002) The effects of Fe and Cu on growth and domoic acid production by *Pseudo-nitzschia multiseries* and *Pseudo-nitzschia australis*. Limnol Oceanogr 47:515–526
239. de Baar HJW et al (2005) Synthesis of iron fertilization experiments: from the iron age in the age of enlightenment. J Geophys Res Oceans 110:1–24
240. Marchetti A et al (2008) Identification and assessment of domoic acid production in oceanic *Pseudo-nitzschia* (Bacillariophyceae) from iron-limited waters in the northeast subartic Pacific. J Phycol 44(3):650–661
241. Bates SS et al (1995) Enhancement of domoic acid production by reintroducing bacteria to axenic cultures of the diatom *Pseudo-nitzschia multiseries*. Nat Toxins 3:429–435
242. Bates SS et al (2004) Interaction between bacteria and the domoic-acid-producing diatom *Pseudo-nitzschia multiseries* (Hasle) Hasle; can bacteria produce domoic acid autonomously? Harmful Algae 3:11–20
243. Kaczmarska I et al (2005) Diversity and distribution of epibiotic bacteria on *Pseudo-nitzschia multiseries* (Bacillariophyceae) in culture, and comparison with those on diatoms in native seawater. Harmful Algae 4:725–741

244. Brock TD (1973) Evolutionary and ecological aspects of the cyanophytes. In: Carr NG, Whitton BA (eds) The biology of blue-green algae. University of California Press, Berkeley, pp 487–500
245. Whitton BA, Potts M (eds) (2002) The ecology of cyanobacteria: their diversity in time and space. Academic, New York
246. Schofp JW (2002) The fossil record: tracing the roots of the cyanobacterial lineage. In: Whitton BA, Potts M (eds) The ecology of cyanobacteria: their diversity in time and space. Academic, New York, pp 13–35
247. Francis G (1878) Poisonous Australian lake. Nature 18:11–12
248. Codd GA, Bell SG, Brooks WP (1989) Cyanobacterial toxins in water. Water Sci Technol 21:1–13
249. Pilotto LS et al (1997) Health effects of exposure to cyanobacteria (blue-green algae) due to recreational water-related activities. Aust N Z J Public Health 21:562–566
250. Stewart I (2008) Cyanobacterial poisoning in livestock, wild animals, and birds–an overview. In: Hudnell HK (ed) Cyanobacterial harmful algal blooms: state of the science and research needs. Springer, New York, p 500
251. Teixera MGLC et al (1993) Gastroenteritis epidemic in the area of the Itaparica Dam, Bahia, Brazil. Bull Pan Am Health Organ 27:244–253
252. Tisdale ES (1931) Epidemic of intestinal disorders in Charleston, W.Va., occurring simultaneously with unprecedented water supply conditions. Am J Public Health 21:198–200
253. Turner PC et al (1990) Pneumonia associated with cyanobacteria. Br Med J 300:1440–1441
254. Metcalf JS, Codd GA (2004) Cyanobacterial toxins in the water environment. A review of current knowledge. Foundation for Water Research, Marlow
255. Chorus I, Bartram J (eds) (1999) Toxic cyanobacteria in water. A guide to their public health consequences, monitoring, and management. E & FN Spon, London (on behalf of the World Health Organization)
256. Falconer IR (1993) Measurement of toxins from blue-green algae in water and foodstuffs. In: Falconer IR (ed) Algal toxins in seafood and drinking water. Academic, New York, pp 165–176
257. Cheng YS et al (2007) Characterization of aerosols containing microcystin. Mar Drugs 5:136–150
258. Falconer IR, Buckley TH (1989) Tumour promotion by *Microcystis* sp., a blue-green algae occurring in water supplies. Med J Australia 150:351–352
259. Falconer IR, Humpage AR (1996) Tumour promotion by cyanobacterial toxins. Phycologia 35(Suppl 6):74–79
260. Fujiki H, Sueoka E, Suganuma M (1996) Carcinogenesis of microcystins. In: Watanabe MF et al (eds) Toxic microcystis. CRC Press, Boca Raton, pp 203–232
261. Humpage AR et al (2000) Microcystins (cyanobacterial toxins) in drinking water enhance the growth of aberrant crypt foci in the mouse colon. J Toxicol Environ Health A 61:155–165
262. Zhou L, Yu H, Chen K (2002) Relationship between microcystin in drinking water and colorectal cancer. Biomed Environ Sci 15(2):166–171
263. World Health Organization (2008) Guidelines for drinking-water quality: incorporating the 1st and 2nd addenda. World Health Organization, Geneva
264. Burns J (2008) Toxic cyanobacteria in Florida waters. In: Hudnell HK (ed) Cyanobacterial harmful algal blooms: state of the science and research needs. Springer, New York, p 500
265. Westrick JA (2008) Cyanobacterial toxin removal in drinking water treatment processes and recreational water. In: Hudnell HK (ed) Cyanobacterial harmful algal blooms: state of the science and research needs. Springer, New York, p 500
266. Carmichael WW (2001) Assessment of blue-green algal toxins in raw and finished drinking water. AWWA Research Foundation and American Water Works Association, Denver
267. Chorus I, Fastner F (2001) Recreational exposure to cyanotoxins. In: Chorus I (ed) Cyanotoxins, occurrence, causes, consequences. Springer, Heidelberg, pp 190–199

268. Miller MA et al (2010) Evidence for a novel marine harmful algal bloom: cyanotoxin (microcystin) transfer from land to sea otters. PLoS One 5(9):e12576
269. Wiegand C, Pflugmacher S (2005) Ecotoxicological effects of selected cyanobacterial secondary metabolites a short review. Toxicol Appl Pharmacol 203:201–218
270. Funari E, Testai E (2007) Human health risk assessment related to cyanotoxins exposure. Crit Rev Toxicol 38(2):97–125
271. Henricksen P et al (1997) Detection of an anatoxin-a(s)-like anticholinesterase in natural blooms and cultures of cyanobacteria/blue-green algae from Danish lakes and in the stomach contents of poisoned birds. Toxicon 35(1111):901–913
272. Matsunaga S et al (1989) Anatoxin-a(s), a potent anticholinesterase from *Anabaena flos-aquae*. J Am Chem Soc 111(494):8021–8023
273. Mahmood NA, Carmichael WW (1986) Paralytic shellfish poison produced by the freshwater cyanobacterium *Aphanizomeno flos-aquae* NH-5. Toxicon 24:175–186
274. Stewart I, Schluter PJ, Shaw GR (2006) Cyanobacterial lipopolysaccharides and human health – a review. Environ Health A Global Access Sci Source 5:7
275. Smith JL, Boyer GL, Zimba PV (2008) A review of cyanobacterial odorous and bioactive metabolites: impacts and management alternatives in aquaculture. Aquaculture 280:5–20
276. Fujiki H et al (1985) A blue-green alga from Okinawa contains aplysiatoxins, the third class of tumour promoters. Jpn J Cancer Res 76:257–259
277. Shimizu Y (1996) Microalgal metabolites: a new perspective. Annu Rev Microbiol 50:431–465
278. Fujiki H et al (1990) New tumour promoters from marine natural products. In: Hall S, Strichartz G (eds) Marine toxins. Origin, structure and molecular pharmacology. American Chemical Society, Washington, DC, pp 232–240
279. Gorham PR, Carmichael WW (1988) Hazards of freshwater blue-green algae (cyanobacteria). In: Lembi CA, Waaland JR (eds) Algae and human affairs. Cambridge University Press, New York, pp 404–431
280. Aguirre AA et al (2006) Hazards associated with the consumption of sea turtle meat and eggs: a review for health care workers and the general public. EcoHealth 3(3):141–153
281. Yasumoto Y (1998) Fish poisoning due to toxins of microalgal origins in the Pacific. Toxicon 36:1515–1518
282. Papapetropoulos S (2007) Is there a role for naturally occurring cyanobacterial toxins in neurodegeneration? The beta-*N*-methylamino-*L*-alanine (BMAA) paradigm. Neurochem Int 50:998–1003
283. Wilson JM et al (2002) Behavioral and neurological correlates of ALS-parkinsonian dementia complex in adult mice fed washed cycad flour. Neuromol Med 1:207–221
284. Rao SD et al (2006) BMAA selectively injures motor neurons via AMP/kainate receptor activation. Exp Neurol 201:244–252
285. Lobner D et al (2007) Beta-*N*-methyl-amino-*L*-alanine enhances neurotoxicity through multiple mechanisms. Neurobiol Dis 25(2):360–365
286. Buenz EJ, Howe CL (2007) Beta-methylamino-alanine (BMAA) injures hippocampal neurons in vivo. Neurotoxicology 28(3):702–704
287. Liu XQ et al (2010) Selective death of cholinergic neurons induced by beta-methylamino-L-alanine. Neuroreport 21(1):55–58
288. Karlsson O et al (2009) Selective brain uptake and behavioral effects of the cyanobacterial toxin BMAA (β-*N*-methylamino-*L*-alanine) following neonatal administration to rodents. Toxicol Sci 109(2):286–295
289. Wilde SB et al (2005) Avian vacuolar myelinopathy linked to exotic aquatic plants and a novel cyanobacterial species. Environ Toxicol 20(3):348–353
290. Bidigare RR et al (2009) Cyanobacteria and BMAA: possible linkage with avian vacuolar myelinopathy (AVM) in the southeastern United States. Amyotroph Lateral Scler 10(Suppl 2):71–73

291. Fischer JR, Lewis-Weis LA, Tate CM (2003) Experimental vacuolar myelinopathy in red-tailed hawks. J Wildl Dis 39(2):400–406
292. Birrenkot AH et al (2004) Establishing a food-chain link between aquatic plant material and avian vacuolar myelinopathy in mallards (*Anas plytyrhynchos*). J Wildl Dis 40(3):485–494
293. Murch SJ, Cox PA, Banack SA (2004) A mechanism for slow release of biomagnified cyanobacterial neurotoxins and neurodegenerative disease in Guam. Proc Natl Acad Sci U S A 101:12228–12231
294. Esterhuizen M, Downing TG (2008) Beta-*N*-methylamino-*L*-alanine (BMAA) in novel South African cyanobacterial isolates. Ecotoxicol Environ Saf 71:309–313
295. Johnson HE et al (2008) Cyanobacteria (*Nostoc commune*) used as a dietary item in the Peruvian highlands produces the neurotoxic amino acid BMAA. J Ethnopharmacol 118:159–165
296. Li A et al (2010) Detection of the neurotoxin BMAA within cyanobacteria isolated from freshwater in China. Toxicon 55:947–953
297. Metcalf JS et al (2008) Co-occurrence of beta-*N*-methylamino-*L*-alanine, a neurotoxic amino acid with other cyanobacterial toxins in British waterbodies, 1990–2004. Environ Microbiol 10(3):702–708
298. Cox PA et al (2005) Diverse taxa of cyanobacteria produce beta-*N*-methylamino-*L*-alanine, a neurotoxic amino acid. Proc Natl Acad Sci U S A 102:5074–5078
299. Caller TA et al (2009) A cluster of amyotrophic lateral sclerosis in New Hampshire: a possible role for toxic cyanobacteria blooms. Amyotroph Lateral Scler 10(Suppl 2):101–108
300. Tucker CS (2000) Off-flavor problems in aquaculture. Rev Fish Sci 8(1):45–88
301. Dodds WK et al (2009) Eutrophication of U.S. freshwaters: analysis of potential economic damages. Environ Sci Technol 43:12–19
302. World Health Organization (2003) Guidelines for safe recreational water environments. Volume 1: coastal and fresh waters. World Health Organization, Geneva
303. Steffensen DA (2008) Economic cost of cyanobacterial blooms. In: Hudnell HK (ed) Cyanobacterial harmful algal blooms: state of the science and research needs, vol 619, Advances in experimental medicine and biology. Springer, New York, 500pp
304. Kenefick SL et al (1992) Odorous substances and cyanobacterial toxins in praire drinking water sources. Water Sci Technol 25:147–154
305. Paerl HW (2008) Nutrient and other environmental controls of harmful cyanobacterial blooms along the freshwater-marine continuum. In: Hudnell HK (ed) Cyanobacterial harmful algal blooms: state of the science and research needs, vol 619, Advances in experimental medicine and biology. Springer, New York, 500pp
306. Garcia-Villada L et al (2004) Occurrence of copper resistant mutants in the toxic cyanobacteria *Microcystis aeruginosa*: characterization and future implications in the use of copper sulphate as algaecide. Water Res 38:2207–2213
307. Hitzfeld BC, Hoeger SJ, Dietrich DR (2000) Cyanobacterial Toxins: removal during drinking water treatment, and human risk assessment. Environ Health Perspect 108(6):113–122
308. Hoeger SJ, Hitzfeld BC, Dietrich DR (2005) Occurrence and elimination of cyanobacterial toxins in drinking water treatment plants. Toxicol Appl Pharmacol 203:231–242
309. Hoeger SJ et al (2004) Occurrence and elimination of cyanobacterial toxins in two Australian drinking water treatment plants. Toxicon 43(14248):639–649
310. Jones G, Gurney S, Rocan D (1996) Water quality/toxic algae study interim report: summary of the 1995 field season results. Manitoba Environment, 4 June 1996
311. Lambert TW, Holmes CFB, Hrudey SE (1996) Adsorption of microcystin-LR by activating carbon and removal in full-scale water treatment. Water Res 30(6):1411–1422
312. Wannemacher JRW et al (1993) Treatment for removal of biotoxins from drinking water. US Army Biomedical Research and Development Laboratory, Fort Detrick, Frederick
313. James H, Lloyd A (2002) Blue-green algae and their toxins – Great Britain's perspective. In: Proceedings of the health effects of exposure to cyanobacteria toxins: state of the science, Sarasota

314. Kim S-C, Lee D-K (2005) Inactivation of algal blooms in eutrophic water of drinking water supplies with the photocatalysis of TiO2 thin film on hollow glass beads. Water Sci Technol 52(9):145–152
315. Ueno Y et al (1996) Detection of microcystins, a blue-green algal hepatotoxin, in drinking water sampled in Haimen and Fusui, endemic areas of primary liver cancer in China, by highly sensitive immunoassay. Carcinogenesis 17:1317
316. Koskenniemi K et al (2007) Quantitative real-time PCR detection of toxic *Nodularia* cyanobacteria in the Baltic Sea. Appl Environ Microbiol 73(7):2173–2179
317. Rinta-Kanto JM et al (2005) Quantification of toxic *Microcystis* spp. during the 2003 and 2004 blooms in western Lake Erie using quantitative real-time PCR. Environ Sci Technol 39(11):4198–4205
318. Codd GA et al (2001) Analysis of cyanobacterial toxins by physiochemical and biochemical methods. J AOAC Int 84:1625–1635
319. Dorr FA et al (2010) Methods for detection of anatoxin-a(s) by liquid chromatography coupled to electrospray ionization-tandem mass spectrometry. Toxicon 55:92–99
320. Fischer WJ et al (2001) Congener-independent immunoassay for microcystins and nodularins. Environ Sci Technol 35:4849–4856
321. Metcalf JS, Bell SG, Codd GA (2001) Colorimetric immuno-protein phophatase inhibition assay for specific detection of microcystins and nodularins of cyanobacteria. Appl Environ Microbiol 67:904–909
322. Anderson WB, Slawson RM, Mayfield CI (2002) A review of drinking-water-associated endotoxin, including potential routes of human exposure. Can J Microbiol 48:567–587
323. Carmichael WW (1992) Occurrence of toxic cyanobacteria. US EPA. Agency, Cincinnati, OH, pp 15–26
324. Oberholster PJ, Botha A-M, Grobbelaar JU (2004) Microcystis aeruginosa: source of toxic microcystins in drinking water. African J Biotech 3:159–168
325. Carmichael WW (2008) A world overview one-hundred, twenty-seven years of research on toxic cyanobacteria – where do we go from here? In: Hudnell HK (ed) Cyanobacterial harmful algal blooms: state of the science and research needs, vol 619, Advances in experimental medicine and biology. Springer, New York, 500 pp
326. Burns J, Williams C, Chapman A (2002) Cyanobacteria and their toxins in Florida surface waters. In: Proceedings of the health effects of exposure to cyanobacteria toxins: state of the science, Sarasota
327. Giddings M et al (2002) Cyanobacterial toxins: the development and evaluation of method to determine microcystin levels in Canadian water supplies. In: Proceedings of the health effects of exposure to cyanobacteria toxins: state of the science, Sarasota
328. Boyer GL (2008) Cyanobacterial toxins in New York and the lower great lakes ecosystems. In: Hudnell HK (ed) Cyanobacterial harmful algal blooms: state of the science and research needs. Springer, New York, p 500
329. Fristachi A et al (2007) A preliminary exposure assessment of microcystins from consumption of drinking water in the United States. Lake Reserv Manag 23(2):203–210
330. Osswald J et al (2007) Toxicology and detection methods of the alkaloid neurotoxin produced by cyanobacteria, anatoxin-a. Environ Int 33:1070–1089
331. Falconer IR (2005) Cyanobacterial toxins of drinking water supplies: cylindrospermopsins and microcystins. CRC Press, Florida
332. Walsby AE, Schanz F, Schmid M (2006) The Burgundy-blood phenomenon: a model of buoyancy change explains autumnal waterblooms by *Planktothrix rubescens* in Lake Zurich. New Phytol 169:109–122
333. Metting B, Pyne JW (1986) Biologically-active compounds from microalgae. Enzyme Microb Technol 8:386–394
334. Sellner KG (1997) Physiology, ecology, and toxic sproperties of marine cyanobacteria blooms. Limnol Oceanogr 42(5 Pt 2):1089–1104

Books and Reviews

Anderson DM (2009) Approaches to monitoring, control and management of harmful algal blooms (HABs). Ocean Coast Manag 52(7):342–347

Babin M et al (2005) New approaches and technologies for observing harmful algal blooms. Oceanography 18(2):210–227

Bauer M et al (2010) The importance of human dimensions research in managing harmful algal blooms. Front Ecol Environ 8(2):75–83

Fleming LE et al (2011) Review of Florida red tide and human health effects. Harmful Algae 10(2):224–233

Graneli E, Turner JT (eds) (2006) Ecology of harmful algae. Springer, Berlin/Heidelberg

Hoagland P et al (2002) The economic effects of harmful algal blooms in the United States: estimates, assessment issues, and information needs. Estuaries Coast 25(4):819–837

Hudnell HK (ed) (2008) Advances in experimental medicine and biology, vol 619. Springer, New York

Hudnell HK (2010) Harmful algal blooms and natural toxins in fresh and marine waters – exposure, occurrence, detection, toxicity, control, management and policy. Toxicon 55(5):1024–1034

Hudnell HK et al (2010) Freshwater harmful algal bloom (FHAB) suppression with solar powered circulation (SPC). Harmful Algae 9:208–217

Landsberg JH (2002) The effects of harmful algal blooms on aquatic organisms. Rev Fish Sci 10(2):113–390

Lewis RJ, Poli M (2010) Toxins in seafood. Toxicon 56(2):107–258 (Special Issue)

Secher S (2009) Measures to control harmful algal blooms. The Plymouth Student Scientist 2(1):212–227

Shumway S, Rodrick G (eds) (2009) Shellfish safety and quality. Woodhead, Cambridge

Valiela I et al (1997) Macroalgal blooms in shallow estuaries: controls and ecophysiological and ecosystem consequences. Limnol Oceanogr 42(5 Pt 2):1105–1118

Chapter 14
Microbial Risk Assessment of Pathogens in Water

Gertjan Medema

Glossary

Dose-response assessment	The determination of the relationship between the magnitude of exposure (dose) to a microbiological agent and the severity and/or frequency of the associated adverse health effects (response).
Exposure assessment	Qualitative and/or quantitative evaluation of the likely intake of microbial hazard via all relevant sources or a specific source.
Exposure	Concentration or amount of an infectious microorganism that reaches the target population, or organism usually expressed in numerical terms of substance, concentration, duration, and frequency.
HACCP: Hazard Analysis Critical Control Point	A system that identifies, evaluates, and controls hazards that are significant for water safety.
Hazard	A biological agent with the potential to cause an adverse health effect.
Hazard identification	The identification of microbiological and biological agents capable of causing adverse health effects that may be present in water.

This chapter was originally published as part of the Encyclopedia of Sustainability Science and Technology edited by Robert A. Meyers. DOI:10.1007/978-1-4419-0851-3

G. Medema (✉)
Water Quality & Health, KWR Watercycle Research Institute, Nieuwegein 3433 PE, The Netherlands

Sanitary Engineering, Delft University of Technology, Delft 2628 CN, The Netherlands
e-mail: gertjan.medema@kwrwater.nl; g.medema@tudelft.nl

E.A. Laws (ed.), *Environmental Toxicology: Selected Entries from the Encyclopedia of Sustainability Science and Technology*, DOI 10.1007/978-1-4614-5764-0_14,

Hazardous event	An event that may lead to the presence of a hazard in drinking water.
Health effects	Changes in morphology, physiology growth, development or life span of an organism, which results in impairment of functional capacity or impairment of capacity to compensate for additional stress or increase in susceptibility to the harmful effects or other environmental influences.
Infection	Colonization of a human (tissue) by a microorganism.
Infectious disease	Colonization by a pathogenic microorganism leading to overt symptoms of disease.
Pathogen	A microorganism capable of causing disease.
QMRA	Quantitative Microbial Risk Assessment.
Risk assessment	A scientifically based process consisting of the following steps: (1) hazard identification, (2) exposure assessment, (3) effect assessment, and (4) risk characterization.
Risk characterization	The qualitative and quantitative estimation, including attendant uncertainties of the probability of occurrence and severity of known or potential adverse health effects in a given population based on hazard identification, hazard characterization, and exposure assessment.
Risk	The likelihood of occurrence of an adverse health effect consequent to a hazard in drinking water.
Uncertainty	Lack of knowledge about specific factors, parameters, or models. Uncertainty includes parameter uncertainty (measurement errors, sampling errors, systematic errors), model uncertainty (uncertainty due to necessary simplification of real-world processes, mis-specification of the model structure, model misuse, use of inappropriate surrogate variables), and scenario uncertainty (descriptive errors, aggregation errors, errors in professional judgment, incomplete analysis).
Variability	Intrinsic heterogeneity in a population, process, or parameter.
Water Safety Plan (WSP)	A management plan developed to address all aspects of water supply that are under the direct control of the water supplier focused on the control of water production, treatment, and distribution to deliver drinking water.

Definition of the Subject

Water can transmit infectious diseases. Water can be transport vehicle. A range of pathogenic microorganisms is shed into the water cycle by infected hosts (man or animal) and transported to new hosts by the water cycle. Water can also be a niche for (opportunistic) pathogens. These pathogens grow in water ecosystems (natural or man-made) and may infect humans that come into contact with this water. Management of the risk of waterborne disease transmission requires knowledge about the nature of the pathogens, their potential growth, fate and transport in the water cycle, the routes of exposure to humans and the health effects that may result from this exposure in the human population, as well as the effect of potential mitigation measures. The challenge is to combine all this knowledge into information that risk managers can use. Quantitative Microbial Risk Assessment (QMRA) has developed as a new scientific discipline over the last 2 decades as a transparent, science-based approach that allows the risk manager to use the best available scientific evidence as basis for risk management decisions.

Introduction

We run risks. We always have. From being eaten by lions, being slaughtered by a rivaling tribe, to being hit by a car. A principal objective of decision making has always been to reduce risks. From avoiding lions, building walls around cities to regulating traffic. To make wise decisions, it is important to have good information about risks. Risk assessment aims to aid decision makers by collating and evaluating this type of information. Risk assessment is increasingly applied in our society, for a wide range of activities: economy, finance, insurances, traffic, infrastructure, health, and environment. What all these activities have in common is that we want to reduce our risk and need to spend resources on mitigation measures. Our resources are limited, so we need to allocate them wisely and proportionally. Risk assessment helps to keep proportions. Risk assessment as a formal discipline has emerged after World War II, paralleling the developments in air and road traffic, the nuclear power and chemical industries and the need to improve the safety of these activities. The process of risk assessment tries to determine the probability that a hazardous event will occur and the probable magnitude of the adverse effects that such an event will have. In the Netherlands, where a substantial part of the country lies below sea level and is protected against flooding by dikes, the height and strength of the dikes are based on assessing the probability of a storm event and the probable magnitude of the adverse effects of flooding part of the country.

In the health and environment arena, risk assessment science has developed over the last few decades. In environmental health, scientists try to establish the probability of exposure of humans to toxic chemicals or pathogens and the probable magnitude of the health effects of this exposure. Risk assessment has become a dominant tool in environmental policy-making, For chemical risks, this is well established (although not without debate [52]). Regulatory agencies are using chemical risk assessment to set standards for toxic chemicals in water. For risks of pathogenic microbes via water, the use of risk assessment was first proposed in the early 1990s [60]. The World Health Organization has been instrumental in the introduction of microbial risk assessment as a basis for safety management of the water we use for drinking, recreation, and food crop irrigation [73, 74].

The Safe Water Framework

An international group of experts, assembled by the World Health Organization, discussed the approach to assess and manage the health risk of pathogenic microorganisms in drinking water, recreational water, and wastewater reuse [7]. This group agreed that future guidelines for safe water and sanitation should integrate risk assessment and risk management into a single framework, the Safe Water Framework. The simplest form of the framework is shown in Fig. 14.1.

The risk that is assessed and managed in this approach is a health risk. It is clearly an iterative cycle in which risk assessment is a basis for decision making in risk management. The four steps of the cycle are described in the next paragraphs, using drinking water safety as an example. In the World Health Organization (WHO) guidelines for the safe use of wastewater, excreta, and grey water [74], these same steps are used for assessing and managing the risk of these water systems.

Health Targets

Health targets are benchmarks for water suppliers, set by the regulator as part of their health policy. Health targets for drinking water are traditionally strict because of the large impact of contaminated tap water and the basic need for safe drinking water.

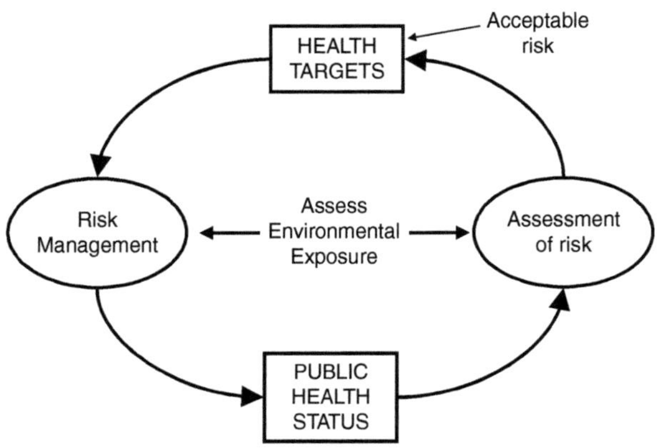

Fig. 14.1 Safe Water Framework for integrated risk assessment and risk management

That leads to the question of what level of health risk through drinking water could be tolerated, given the overall health status of the consumer population and the contribution of drinking water to the overall health risk of this population in relation to other routes of exposure, such as food, person-to-person or animal contact, recreational water, etc. This is a question that typically needs answering on the level of the regulator, who can translate this information into a health target for drinking water, considering other factors such as relative contribution of drinking water–transmitted disease to the overall health burden and the economic climate.

The health target is the level of tolerable risk for drinking water, which could be expressed as the tolerable risk of infection through drinking water (i.e., risk of infection $<10^{-4}$ per person per year [61]) or the tolerable amount of disease burden (i.e., $<10^{-6}$ disability adjusted life years per person per year [31, 73]). The health target could be translated into water quality targets for pathogens (analogous to the toxic chemicals). In the latter case, rather than producing a standard and monitoring requirement for all pathogens that could be transmitted through drinking water, the use of a suite of "index pathogens" is advisable. Establishment of adequate control against this suite of pathogens should offer protection against the other known (and even unknown) pathogens.

It is emphasized that the health targets may be different in different health status situations. The question of what is a tolerable level of risk is a judgment in which the society as a whole has a role to play; the decision on the cost-benefit is for each country to decide [71, 73]. It is important that health-based targets, defined by the relevant health authority, are realistic under local operating conditions and are set to protect and improve public health. Health-based targets underpin development of Water Safety Plans [73] and provide information with which to evaluate the adequacy of existing installations, and assist in identifying the level and type of inspection and analytical verifications appropriate.

Risk Management

Managing the safety of drinking water has been the core business of water supply companies for more than a century. Over this period, risk management has evolved into a culture, with codes and specifications of good practice. In the last few decades, quality management systems have been used in the water industry to formalize these practices. Currently, water suppliers in several European Union (EU) countries are using a Hazard Analysis and Critical Control Points (HACCP) based approach for management of (microbiological and other) risks. The basic principles of HACCP are to understand the system and the hazards/hazardous events that may challenge the system and their (health) priority and to ensure that control measures are in place and functioning. HACCP-based systems typically focus on good practice and even more specifically on ensuring that good practice is maintained at all times. HACCP fits within existing quality management systems (i.e., ISO 9001 c.s.). HACCP is the risk management tool that is used in food safety.

The Codex Alimentarius (FAO/WHO code for food safety) defines HACCP as a system that identifies, evaluates, and controls hazards that are significant for food safety [10]. The HACCP system is well established in the food industry.

Although there are many aspects of drinking water that are similar to food, there are also differences. Based on experiences of water suppliers with HACCP, the HACCP system has been refined and tailored for application in drinking water abstraction, treatment, and distribution in WHO's Water Safety Plan. The Water Safety Plan is described in the third revision of the Guidelines for Drinking Water Quality [73].

The principal components of the Water Safety Plan are:

System assessment to determine whether the water supply chain (from source through treatment to the point of consumption) as a whole can deliver water of a quality that meets the above targets.

Operational monitoring of the control measures in the supply chain that are of particular importance in securing drinking water safety.

Management plans documenting the system assessment and monitoring, and describing actions to be taken in normal operation and incident conditions, including upgrade and improvement documentation and communication.

In the Water Safety Plan, the risk assessment question: "Do we meet the health target?" is answered in the *System Assessment* and the risk management questions "How do we ensure and demonstrate that we always meet the target?" and "How do we respond to incidents?" are answered in the *Operational monitoring of control measures* and the *Management plans*.

For an overview of the Water Safety Plan and its context, the reader is referred to the WHO GDWQ and the Water Safety Plan guidance documents that are published on the website of WHO Water, Sanitation, and Health.

Public Health Status

The primary objective of drinking water safety management is the adequate protection of public health. The incidence of waterborne illness in the population or the occurrences of waterborne outbreaks are direct triggers for curative risk management. A more preventative incentive for assessing the water-related health risks and the installation of risk management is to demonstrate that the water supply is providing an adequate level of protection of public health.

The installation of health targets in national legislation and the risk management actions of water utilities should result in an improvement of the status of public health. Without addressing this, it is impossible to see if the health targets set and risk management actions taken are effective and if money spent for improving water supply results in a relevant health gain. This step in the process is the place where the health risk of drinking water can be compared to other routes of exposure and to other health risks. It allows comparison of the effort and resources put into the provision of safe drinking water versus resources allocated to manage other health risks.

The risk assessment and management framework is a circular process that can be run in an iterative manner. This fits well with the incremental nature of health decision-making, the efficient use of scarce resources, and the increase of information each time the circle is completed.

Risk Assessment

Risk assessment is used to answer the question: "Is my system able to produce and deliver drinking water that meets the health targets?" The risk assessment process requires quantitative information about the exposure of drinking water consumers to pathogens. This is provided by exposure assessment, one of the components of risk assessment. Quantitative information about pathogens in water sources, their removal by treatment and protection of the distribution network and drinking water consumption is collected and translated into an estimate of the exposure of consumers to pathogens through drinking water. To complete the risk assessment, the potential effect (the risk) of pathogen exposure is estimated through known dose-response models. As will be indicated later, the exposure assessment also provides valuable information to aid risk management in the prioritization of control measures.

An important question in risk management, especially in settings with an already high standard of drinking water safety, is "How far do we need to go with control measures?" This is an optimization that weighs the safety of the consumer against the costs of drinking water.

Quantitative microbial risk assessment (QMRA) can provide an objective and scientific basis for risk management decisions. Water utilities can use QMRA to assess whether they meet the health targets with their water treatment, storage, and distribution systems. This also provides the information to set the critical limits in the Water Safety Plans to ensure good performance. Good performance can now be based on a quantitative assessment of the contribution of the Critical Point (such as a disinfection or filtration process) to the overall safety, and limits can be set to ensure that the multiple barrier chain of water collection, treatment, and distribution as a whole does meet the target.

Risk assessment and risk management should not be regarded as two separate steps in the harmonized framework. To answer the question "Which control measures should be put in place to meet the target?" both the HACCP-based system and quantitative risk assessment provide valuable input: the hazardous events, the most important barriers in the system, the contribution of each of the barriers, target levels for control, the occurrence of weak elements in the chain, the quality of the available information, etc.

Quantitative Microbial Risk Assessment

Quantitative Microbial Risk Assessment (QMRA) is derived from the chemical risk assessment paradigm that encompasses four basic elements:

- A characterization of the problem, including the hazard
- Exposure assessment
- Effect assessment (dose-response)
- Risk characterization

Several QMRA frameworks have been published, such as the generic International Life Sciences Institute (ILSI) framework [8]. Here, most attention is given to exposure assessment and risk characterization of pathogens in drinking water. Therefore, the generic ILSI QMRA framework is expanded to highlight the elements that are important for exposure assessment and risk characterization in drinking water, and put in the overall WHO Safe Water Framework (Fig. 14.2).

Element 1. Problem Formulation and Hazard Identification

This is the initializing phase of QMRA to establish which specific questions need to be addressed. The scope and the boundaries of the QMRA process are determined

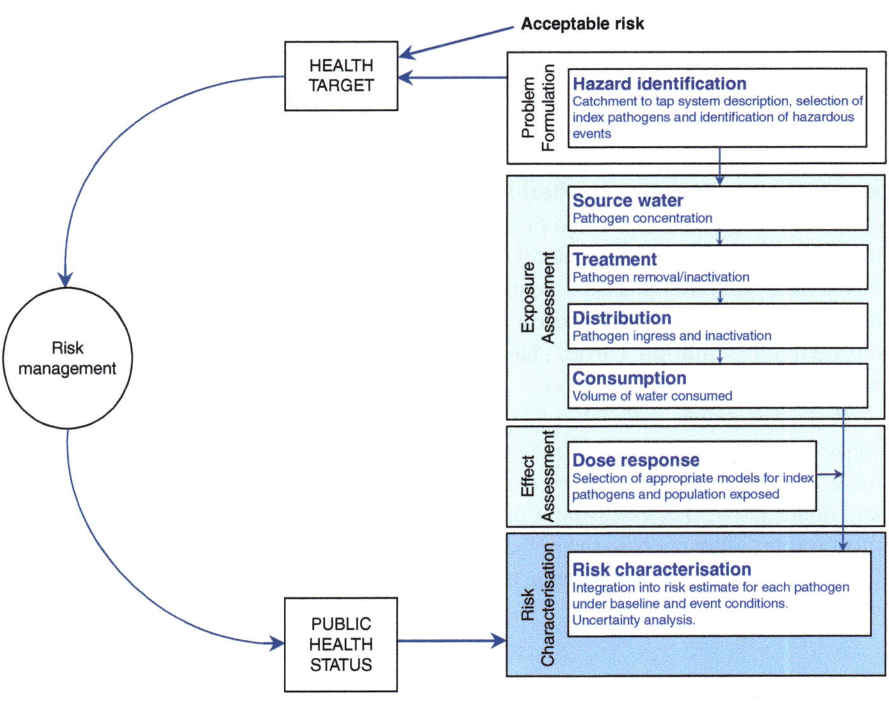

Fig. 14.2 The steps of quantitative microbial risk assessment in the Safe Water Framework

in this phase. This requires communication between the risk managers (regulators, public health agencies, water utilities) and the risk assessors. The basic question to QMRA is: "Is my system able to meet the health targets?"

To conduct a QMRA, a good description of the system under evaluation is necessary and the hazards and hazardous events need to be identified.

Step 1. Description of the System from Source to Tap

The system for water treatment from catchment to tap is described, identifying the principal control elements and strategies.

Step 2. Hazard Identification

Hazard identification is the identification of the microorganisms within the system boundaries that cause human illness, the processes by which each microorganism causes illness and the type of illness(es) caused, and the identification of possible transmission routes and the significance of these routes [26]. QMRA is usually focused on a specific transmission route, in this example drinking water from a surface water source.

The ideal QMRA does not focus on a single pathogen only, but on a suite of "index pathogens" that cover the range of health risks and control challenges for the particular water supply system defined. Adequate control of these index pathogens implies that the health risk of other known pathogens is also adequately controlled by the system and that the system also offers protection against unknown pathogens.

Hazard identification consists of the following steps:

Description of the characteristics of the pathogens, especially those related to waterborne transmission (survival in water, resistance to treatment, etc.).

Description of what is known about the transmission routes of these pathogens and specifically what is known about waterborne transmission, the causes of waterborne outbreaks, and the relative significance of waterborne transmission compared to other routes.

Description of the illness (type, duration, incubation time, etc.) caused by the pathogens in the risk assessment, and available information about sequellae.

Description of what is known about protective immunity and secondary transmission.

Step 3. Description of Hazardous Events

In many cases, the majority of the risk is not determined during the normal (baseline) situation, but during hazardous events, such as rainfall leading to a high load of pathogens in source waters, or treatment failure or distribution network failure (or combinations thereof). It is therefore important to ensure that these hazardous events are incorporated in the QMRA, or that a separate QMRA is conducted to determine the (health) significance of the event.

Element 2. Exposure Assessment

Exposure assessment is the quantitative assessment of the probability that drinking water consumers ingest pathogens. A QMRA of drinking water usually requires the assessment of the levels of pathogens in source water and the changes to these levels by treatment, storage, and distribution, and finally the volume of water consumed.

Step 4. Assess Pathogen Occurrence in Source Water

Collect information about the occurrence of pathogens in source water. This is preferably based on a catchment survey, identifying the principal sources of contamination of the catchment and the conditions that may lead to peak events in source water, such as heavy rainfall or resuspension of sediments. Pathogen monitoring in source water can be carried out, using the information of the catchment survey, which needs to include assessment of peak events. The pathogen detection methods are ideally targeted to viable and infectious pathogens. The performance characteristics of the available detection methods for pathogens can have implications for the applicability of the data in risk assessment. These should be identified and evaluated in (the early stages of) the risk assessment process.

Step 5. Assess the Elimination of Pathogens During Treatment

Collect information about the removal or inactivation of pathogens during drinking water treatment processes. Ideally, data on removal of pathogens at full scale are used. In practice, however, several other sources of data have to be used to estimate pathogen removal, such as pathogen data of pilot or lab scale systems or data on model parameters (indicator bacteria, phages, spores, particles, etc.) on full, pilot, or lab scale.

The efficacy of treatment processes may vary, depending on feed water composition, operational control, temperature, etc. Moments or periods of poor or suboptimal performance are hazardous events and hence most significant for risk assessment.

Step 6. Assess the Changes in Water Quality During Storage and Distribution

Determine the likelihood of recontamination of stored and distributed water (e.g., by the *E. coli* monitoring of water in these reservoirs and pipes or loss of disinfectant residual) and the significance of these contamination events. In well-maintained piped supplies, recontamination events are rare and could be regarded as a result of a hazardous event (heavy rainfall, cross-connection, poor hygiene during repairs, etc.). In other piped and non-piped settings, recontamination events are common and may dominate the health risk.

Step 7. Consumption of Drinking Water

The other component of exposure assessment is the volume of water consumed by the population. Not only the average volume of water consumed is important, also the person-to-person variation in consumption behavior and especially consumption behavior of risk groups (in terms of sensitivity to infection or high level of consumption) is relevant. The available data suggest there is considerable difference between drinking water consumption within the population. This variation needs to be captured and incorporated in the risk assessment. Household treatment/point-of-use devices affect the exposure. Hence, consumption data should be on consumption of drinking water without further treatment, such as heating or filters and include water that is drunk directly, but also cold tap water used for food preparation, ice, etc.

Step 8. Dose (Exposure) Estimation

Dose (or exposure) is the number of pathogens consumed per unit time. The information obtained in all steps of the exposure assessment needs to be combined into an estimate of the ingested dose. This is preferably a stochastic estimation, including the variability and uncertainty in all steps of the exposure assessment.

Element 3. Effect Assessment

The effect assessment is the determination of the health outcomes associated with the (level of) exposure to waterborne pathogens.

Step 9. Dose-Response Data

Dose-response characterizes the relation between dose magnitude, infectivity, and quantitative health effects to an exposed population. The microbial dose-response analysis records the incidence of a particular effect against dose of the agent. In most cases, this particular effect is infection, rather than symptoms of illness. For *Cryptosporidium parvum* for instance, there is a clear relation between ingested dose and the probability of infection, but not between dose and symptoms of intestinal illness.

Although the dataset is increasing, the number of dose-response studies with human volunteers is limited. Of most pathogens, only one or a few strains are tested in healthy adult volunteers. Information about strain-to-strain variability and the influence of the immune response of the hosts is still limited.

There are several dose-response models available and the type of model can have a very significant impact on the response that is attributed to exposure to low doses. The models and their limitations should be well understood when applying these in QMRA. Synergistic effects between pathogens are not incorporated in the current models.

Step 10. Host Characterization

For infectious diseases, the host susceptibility plays an important role in the health outcome of exposure. Exposure of persons with protective immunity will result in lower health outcomes than exposure of risk groups. During "Host Characterization" the characteristics of the potentially exposed populations that are suspected for susceptibility to a particular pathogen are evaluated.

Step 11. Health Outcome

Until now quantitative microbial risk assessment has been primarily focused on estimating the risk of infection. The relation between ingested dose and infection is relatively well defined, while the relation between dose and other health outcomes (illness, sequellae) is not available or less clear. This is one of the reasons why it is difficult to establish a direct relation between QMRA (on probability of infection) and epidemiological data (on symptoms of disease). The use of the risk (or probability) of infection is justified by the degree of conservatism in using infection as an endpoint and the inability to quantify the risk of more susceptible subpopulations [43].

However, waterborne diseases differ in nature, severity, and duration. A metric that takes into account the overall health burden of waterborne diseases is necessary. Ideally, this metric can also be used to describe the burden of the disease of chemical compounds, such as carcinogens, so all health risks can be weighed on the same scale.

In the new WHO guidelines for Drinking-Water Quality (GDWQ), the concept of Disability Adjusted Life Years (DALY) [31] is introduced as burden of disease metric in the drinking water guidelines.

The basic principle of the DALY approach is to weigh each health effect for its severity with (usually) death as the most severe outcome, multiply this weight with the duration of the health effect ("duration" of death being the remaining group life expectancy), and with the number of people in a population affected by the particular outcome. Summarizing all the health outcomes caused by a certain agent results in an estimate of the burden of disease attributable to this agent.

To be able to use DALYs in the QMRA, ideally the relation between exposure (dose) and different health outcomes is known. In the absence of sufficient data (which is usually the case), the dose-response relation for infection (as the first step of the disease process) can be combined with data on the fraction of the exposed population falling ill from exposure (for instance, from attack rates in waterborne outbreaks) and data on the fraction of the ill population that contract more severe health outcomes (from health surveillance data).

Element 4. Risk Characterization

In the process of risk characterization, the information obtained in the exposure assessment and the effect assessment are integrated to obtain a risk estimate. This can be done as a point estimation: a point estimate of exposure can be entered into the dose-response relation to compute a point estimate of the risk of infection. The point estimate can be the "best" estimate, to obtain a measure of central tendency of the risk. In the case of computing various risk scenarios, the computed point estimates give a quantitative estimate of the consequences of the circumstances that produce a risk scenario.

An approach that allows the incorporation of the variability and uncertainty in the steps of the risk assessment chain is promoted by [23, 66]. This encompasses the characterization of the distribution of all data used for risk assessment and to combine these distributions into a distribution of the computed risk, for instance, by Monte Carlo analysis. This approach not only provides the risk manager with important information about the (un)certainty of the risk estimate, but also with the relative contribution of the uncertainty and variability in all steps of the risk assessment. It therefore guides the risk manager to the most appropriate options for efficiently minimizing the risk and the most significant research items to reduce the overall uncertainty of the risk estimate.

With high-level water supply, the baseline risk is usually very low. Under such conditions, hazardous events, such as peak contamination in the source water, treatment failure and especially the combination thereof and contamination events in the distribution network, are responsible for the majority of the risk. Most waterborne outbreaks have been traced to a combination of hazardous events [35] and it is likely that many events result in the presence of pathogens in tap water and hence the transmission of disease. Wherever possible, identify and evaluate these events separately in QMRA to understand the significance of these events. Analysis of events also brings forward opportunities for optimization of the system to prevent these events from occurring or reduce their impact on health.

Tiered Approach

Risk assessment is well suited for a tiered approach and this is also commonly used in risk assessment practice, both in human health risk assessment and in ecological risk assessment. The tiered approach allows an effective interaction between risk assessment and risk management, starting with a crude risk assessment, usually based on limited information to determine the urgency of the perceived problem, to prioritize the risk of different water supply sites or scenarios, and to determine the need of a more detailed study for a particular situation. This allows the effective allocation of resources to the sites or situations that give rise to the highest risk. There is no strict definition of the tiers, only that the initial QMRA is usually generic and simple and the specificity and complexity increase in subsequent tiers.

The most basic (but also most important) QMRA is a screening-level study. Starting with whatever information is available, a crude first evaluation is made. Usually, the available information is not specific to the system that is studied, but has to be extrapolated from the available scientific literature. So, in its simplest form, a QMRA can be performed with only a generic description of the water supply system.

The screening-level assessment may show that the risks are negligible, without much scientific doubt. In that case, the screening-level risk assessment can be used to demonstrate the safety of the system. Setting up a more detailed study is not warranted. Or the screening-level risk assessment may highlight that the risk is unacceptably high, again without much scientific doubt. Such a screening-level risk assessment is also very useful in comparing different scenarios for risk management, for example, different water treatment options.

If the outcome of the screening-level risk assessment is that there may be a health risk that is not negligible, there is an incentive for a next iteration of the risk assessment, the collection of site-specific data, for instance, on the presence of *Cryptosporidium* in the source water or catchment. The QMRA is repeated with the new, site-specific information. The options for the outcome of this second-level QMRA are the same as for the first iteration. In general, a result of any risk assessment is the identification of which information is missing and the prioritization of research needs [21].

The screening-level risk assessments usually work with point estimates of risk. The tendency is to use conservative or worst-case estimates, to "be on the safe side." But worst-case estimates, by nature, may overestimate the risk and it is not clear to the risk manager what the uncertainty of the calculated risk is, only that the uncertainty will be toward the lower risk values (the nature of a worst-case assumption). More helpful for the risk manager is to provide a range of risks (interval estimate) that denote the variability and uncertainty in the risk estimate. In the case of the screening-level risk assessment, this can be achieved by using an average, worst, and best case, to illustrate the range of the risk that can be deduced from the available information and the level of certainty that is embedded in the QMRA.

Interval estimates require information about variability and uncertainty. Variability is the result of intrinsic heterogeneity in the input of the risk assessment, such as the variation in *Cryptosporidium* concentration in source water over time, or the variation in the removal of particles by a filtration process over time. Variability can be characterized if sufficient data points are collected. Uncertainty is the result of unknown errors in inputs of the risk assessment, such as errors in the measurement of *Cryptosporidium* or the assumption that certain indicator organisms can be used to describe the removal of *Cryptosporidium* by filtration. Uncertainty can be characterized by specific research activities, for example, to determine the recovery efficiency of the *Cryptosporidium* enumeration method or to compare the removal of *Cryptosporidium* to indicator organisms by filtration.

When sufficient data are available, a probabilistic risk assessment can be performed, where the input is described by statistical distribution functions to describe the confidence interval of the input itself and of the calculated risk.

Good QMRA Practice

Food safety has a longer history of employing microbial risk assessment to facilitate risk management. Several international bodies have produced guidance on good microbial risk assessment practice [13, 72]. The principles of good QMRA practice are also applicable to water safety. General principles are:

- Risk assessment should be clearly separated from risk management.
- Risk assessment should be soundly based on science.
- Risk assessment should be transparent: clear, understandable, and reproducible. It should follow a harmonized procedure based on the accepted standards of best practice.
- The scope and objectives of the risk assessment should be clearly defined and stated at the onset, in collaboration with the risk manager who is going to apply the results.
- The data used are evaluated to determine their quality and relevance to the assessment (taking into account their overall weight in the risk and uncertainty). If data are judged irrelevant or of too low quality, this should be justified. All data that are used are referenced.
- If data are variable, the variability should be documented and taken into account in the risk assessment, preferably in a probabilistic manner.
- All assumptions are documented and explained. Where alternative assumptions could have been made, they can be evaluated together with other uncertainties.
- The risk assessment should include a description of the uncertainties encountered in the risk assessment process. Their relative influence on the risk assessment outcome should be described, preferably in a quantitative (probabilistic) manner. Where point estimates are used for uncertain (or variable) quantities,

the selected values should be justified and their influence on the assessment included in the uncertainty analysis.
- Conclusions should reflect the objectives and scope of the risk assessment, and include uncertainties and data gaps.

Uncertainty Analysis

Uncertainty is inherent in risk assessment [54]. Many (if not all) data have a degree of uncertainty. Sources of uncertainty in QMRA include:

- Extrapolation from dose-response data (though, unlike with toxic chemicals, many dose-response data are from human exposure)
- Limitations of pathogen detection methods
- Estimates of exposure

It is important to include the uncertainties in all steps of the risk characterization. The uncertainties in the estimates of exposure are usually dominant. Two approaches are used to determine how the uncertainty in the information in individual steps of the risk assessment affect the uncertainty of the overall risk estimate: sensitivity analysis and Monte Carlo simulation. In sensitivity analysis, the value of each parameter in the risk assessment is varied, one at a time, along the uncertainty range of that parameter (e.g., Average and maximum concentration of a pathogen in water) to determine the effect on the final risk estimate. This procedure generates (1) the range of possible values of the final risk estimate and (2) the uncertainty in which of the parameters contribute most to the uncertainty of the final risk estimate. Sensitivity analysis is typically done in screening-level risk assessments. In probabilistic risk assessments, Monte Carlo simulation is the most widely applied method. Monte Carlo simulation needs a deterministic model for the risk assessment. The uncertainty (and variability) in each of the parameters in the risk assessment is expressed in a probability distribution. The simulation computes a final risk estimate by randomly selecting a value for each parameter in the model from the probability distribution for each parameter. This is repeated many (1,000–10,000) times, each time using a different set of random values from the probability functions. Monte Carlo simulation produces distributions of possible outcome values for the final risk estimate and the shape of the distribution identifies both the general tendency of the risk and the uncertainty of the risk estimate. Also here, the procedure gives information about the contribution of the uncertainty in individual parameters to the uncertainty in the overall risk estimate. While sensitivity analysis evaluates the impact of the uncertainty in each parameter separately and uses few values in the range of possible values of each parameter, Monte Carlo simulation evaluates the impact of the uncertainty in each parameter in combination with all other parameters and uses all possible values and the probability that they occur in the range of each parameter. Burmaster and Anderson [9] published principles of good practice for the use of Monte Carlo simulation in health risk assessments.

Applications of QMRA

The first quantitative microbial risk assessment studies on drinking water were conducted on viruses and *Giardia* [60]. Since the dose-response data from the first human volunteer study on *Cryptosporidium* [12] became available, several authors have performed QMRA for *Cryptosporidium* in water supply (Table 14.1). This makes the health risk of *Cryptosporidium* through drinking water the most intensively studied object in QMRA studies to date. The overview of QMRA studies for *Cryptosporidium* in water supply illustrates several issues:

1. QMRA studies were conducted to:

   - Evaluate the health risk of *Cryptosporidium* in specific water supply systems or water supply scenarios.
   - Balance the health risk of *Cryptosporidium* in ozonated drinking water to the health risk of bromate formation by ozone [30]. For the assessment of exposure to *Cryptosporidium*, they used raw water monitoring data on *Cryptosporidium*, data on the removal of anaerobic spores by conventional treatment and an ozone disinfection model (the Hom model published by [17]) and a bromate formation model. The ingested dose of oocysts and bromate ions was translated to DALYs to allow comparison of the microbiological and chemical health risk. In their scenario, the health benefits of microorganism inactivation by ozonation outweighed the health losses by bromate formation.
   - Demonstrate the need for additional treatment with UV [1]. They used monitoring data of *Cryptosporidium* in treated water, using a cell-culture-PCR technique to determine the concentration of infectious oocysts in treated water.
   - Demonstrate the need for treatment optimization [46, 48].
   - Illustrate the value of QMRA [47, 48, 59, 66, 68] and relation of QMRA to the Water Safety Plan [49, 65].
   - Evaluate the risk of cryptosporidiosis in different water supply and sanitation scenarios [69].
   - Evaluate the impact of failures in treatment and distribution on the health risk [70]. Failure reports were collected from operational logs/interviews. These failures were translated into an estimate of *Cryptosporidium* (and other pathogen) occurrence (which was the most uncertain step in this QMRA). They indicated that in this system, the health risk associated with normal operation was higher than from the very infrequent and short lasting reported incidents.
   - Prioritize research needs [21], which illustrates how QMRA can be used to determine the relative significance of major, well-controlled and minor, less well-controlled routes of exposure and the impact of moments of reduced treatment performance.

Table 14.1 QMRA studies on the risk of *Cryptosporidium* in public water supply

Authors	Exposure assessment	Effect assessment	Outcome	Type	Probability of infection average/ 95%–range
Medema et al. [47]	*Cryptosporidium* in source water, recovery data [39], viability data [39], removal of oocysts by full scale conventional treatment systems, [39], tap water consumption data [63]	Volunteer study with the Iowa strain [12]	Probability of infection	Probabilistic	3.6×10^{-5} [a] ($3.5 \times 10^{-7} - 1.8 \times 10^{-3}$)
Rose et al. [62]	*Cryptosporidium* in treated water [39]	Volunteer study with the Iowa strain [12]	Probability of infection	Point estimates	5.0×10^{-2} ($4.4 \times 10^{-3} - 1$)
Rose et al. [62]	*Cryptosporidium* in ice prepared from tap water at the time of an outbreak, the latter corrected for the effect of freezing/thawing (90% loss of detectable oocysts) and for the recovery	Volunteer study with the Iowa strain [12]	Probability of infection	Point estimates and comparison of observed and expected illness cases	–
Havelaar et al. [29]	*Cryptosporidium* in source water, recovery data, removal of anaerobic spores by conventional treatment, NL cold tap water consumption data	Volunteer study with the Iowa strain [12]	Probability of infection	Probabilistic	1.3×10^{-4} [a] ($10^{-5} - 10^{-3}$)
Teunis et al. [66]	*Cryptosporidium* in source water, recovery data, viability data [39], removal of anaerobic spores by conventional treatment, NL cold tap water consumption data	Volunteer study with the Iowa strain [12]	Probability of infection	Probabilistic	1.3×10^{-4} [a] ($4 \times 10^{-5} - 4 \times 10^{-4}$)
Teunis and Havelaar [68]	*Cryptosporidium* concentration in source water [5], recovery data [41], viable type morphology [39], removal by storage [66], removal of anaerobic spores by conventional treatment, NL cold tap water consumption data	Volunteer study with the Iowa strain [12]	Probability of inf ection, illness and DALYs	Probabilistic	No treatment failure: 2.0×10^{-12} 95%: 2.8×10^{-10} Treatment failure: 1.5×10^{-8} 95%: 2.1×10^{-6}

(continued)

Table 14.1 (continued)

Authors	Exposure assessment	Effect assessment	Outcome	Type	Probability of infection average/ 95%–range
Perz et al. [56]	Assumed concentration of *Cryptosporidium* in tap water, consumption of tap water [63], reduced by 40% for cold tap water consumption and by a further reduction of 33% for AIDS patients	Volunteer study with the Iowa strain [12], assumed threefold higher infectivity for AIDS patients	Probability of infection and illness (probability of illness 0.5 for general population and 1.0 for AIDS patients). Estimated reported cases in general and AIDS population	Point estimates, using two assumed concentrations of *Cryptosporidium* in tap water	$1.0 \times 10^{-3/-2}$ in general population $2.1 \times 10^{-3/-2}$ in AIDS population
Havelaar et al. [30] Gale [20]	*Cryptosporidium* in source water, recovery data, viability data [39], removal of anaerobic spores by conventional treatment, Hom model ozone inactivation [17], NL cold tap water consumption data. The exposure was compared to the exposure to bromate that was formed in the ozonation	Volunteer study with the Iowa strain [12]	DALY	Probabilistic, comparing *Cryptosporidium* to bromate burden of disease	1.0×10^{-3} [a] ($7.6 \times 10^{-4} - 1.5 \times 10^{-3}$)
Haas et al. [24] Haas [26]	*Cryptosporidium* concentration in ice manufactured from tap water during an outbreak, estimation of the inactivation by freezing and thawing, estimation of the duration of the contamination (on onset of cases), attack rate during the outbreak, tap water consumption data [63]	Volunteer study with the Iowa strain [12]	Probability of infection	Point estimate, comparing expected and observed illness	1.1×10^{-2} [b]
Haas et al. [26]	*Cryptosporidium* concentration in distributed water during an outbreak, estimation of the duration of the contamination (on onset of cases), attack rate during	Volunteer study with the Iowa strain [12]	Probability of infection	Point estimate, comparing expected and observed illness	3.6×10^{-4} [b]

(continued)

Table 14.1 (continued)

Authors	Exposure assessment	Effect assessment	Outcome	Type	Probability of infection average/ 95%–range
	the outbreak, assumed 1 L tap water consumption				
Gale [19, 20]	*Cryptosporidium* in source water [37] and removal of oocysts by full scale conventional treatment systems, [40], data on heterogeneity	Volunteer study with the Iowa strain, including immunity	Probability of infection		1.5×10^{-3} [b]
Haas and Eisenberg [27]	*Cryptosporidium* in different source watersheds, unfiltered system with chlorination, so removal/ inactivation by treatment assumed as 0, tap water consumption data [63]	Volunteer study with the Iowa strain [12]	Probability of infection	Point estimate and probabilistic	1.2×10^{-2} 1.2×10^{-3} ($1.2 \times 10^{-4} - 7.7 \times 10^{-2}$)
Medema et al. [48]	*Cryptosporidium* in source water, recovery data, removal of anaerobic spores by conventional treatment, NL cold tap water consumption data	Volunteer study with the Iowa strain [12]	Probability of infection	Point estimate	$1.1 \times 10^{-3} - 3.5 \times 10^{-2}$
	Cryptosporidium in source water, recovery data, removal of bacteriophages by soil passage and of *Cryptosporidium* in soil column studies, NL cold tap water consumption data	Volunteer study with the Iowa strain [12]	Probability of infection	Point estimate	0
	Cryptosporidium in source water, recovery data, viability and genotype data, removal of anaerobic spores by conventional treatment, NL cold tap water consumption data	Volunteer study with the Iowa strain [12]	Probability of infection	Probabilistic	$<1.0 \times 10^{-4}$ with 91% certainty

(continued)

Table 14.1 (continued)

Authors	Exposure assessment	Effect assessment	Outcome	Type	Probability of infection average/ 95%–range
Westrell et al. [70]	*Cryptosporidium* in source water, removal of particles by conventional treatment, inactivation by disinfection [18, 38], removal of oocysts by membrane filtration [2, 33]	Volunteer study with the Iowa strain [12]	Probability of infection	Probabilistic	Normal operation: 6.0×10^{-4} [a] (6×10^{-6}–4×10^{-2}) Filtration error: 4.0×10^{-5} [a] (6×10^{-7}–2×10^{-3})
	Cryptosporidium in sewage, reports of the water supply on treatment failure and contamination incidents in the distribution network	Volunteer study with the Iowa strain [12]	Probability of infection	Probabilistic	Reservoir contamination: 7×10^{-7} [a] (2×10^{-8}–2×10^{-6})
Masago et al. [46]	*Cryptosporidium* in source water [28], effect of rainfall, viability data [39], failure model for removal by conventional treatment, NL cold tap water consumption data	Volunteer study with the Iowa strain [12]	Probability of infection	Probabilistic	2.0×10^{-4} [a] (2.5×10^{-5} [c] – 2.5×10^{-3})
Gale [21]	Theoretical assumptions in scenario studies of treatment by-pass or failure	Volunteer study with the Iowa strain [12]	Probability of infection	Probabilistic	–
Pouillot et al. [59]	Assumed concentration in distributed water, recovery data, viability data (expert knowledge), French cold tap water consumption	Volunteer study with the Iowa strain for both infection and illness [12], immunodeficient mouse model [75]	Probability of infection and of illness for immunocompetent and immunodeficient persons	Probabilistic	At 2 oocysts/100 L: 1.8×10^{-2} 95%: 5.4×10^{-2}
Pouillot et al. [59]	*Cryptosporidium* in distributed water, recovery data, viability data (expert knowledge),	Volunteer study with the Iowa strain for both infection and illness [12],	Probability of infection and of illness for immunocompetent and immunodeficient persons	Probabilistic	2.1×10^{-2} 95%: 6.7×10^{-2}

(continued)

Table 14.1 (continued)

Authors	Exposure assessment	Effect assessment	Outcome	Type	Probability of infection average/ 95%–range
	French cold tap water consumption	immunodeficient mouse model [75]			
Havelaar et al. [30]	*Cryptosporidium* in source water, recovery data, *Cryptosporidium* challenge study of conventional treatment	–	Quality score of exposure assessment factors	Uncertainty analysis	–
Haas et al. [25] JAWWA 88:131	Calculation of a *Cryptosporidium* concentration that corresponds with the 10^{-4} probability of infection (3.27×10^{-5}) oocysts L^{-1} (95% CI: 1.8–6.4 × 10)	Volunteer study with the Iowa strain [12]	Probability of infection	Probabilistic	(1×10^{-4})
Aboytes et al. [1]	*Cryptosporidium* in filtered drinking water, recovery data, infectivity data (cell-culture PCR)	Volunteer studies with the Iowa, UCP and TAMU with Bayesian data-analysis [51]	Probability of infection	Point estimate with confidence interval	8.2×10^{-3} 95%: 1.2×10^{-2}
EPA [15]	*Cryptosporidium* monitoring data (ICR and beyond), recovery data, infectivity fraction, treatment performance credits, USDA consumption data	Volunteer studies with Iowa, TAMU, UCP, using different models	Probability of infection, illness, death and cost	Probabilistic with sensitivity analysis	Scenario evaluation Pre-LT2 filtered: 8×10^{-5} ($<10^{-6}$ – 0.02); unfiltered 0.02 (0.002 to ~0.5)

[a]Median
[b]Average daily risk of infection during the outbreak
[c]Minimum annual risk

- Perform a cost-benefit analysis of *Cryptosporidium* regulation that requires additional drinking water treatment for systems with relatively high levels of *Cryptosporidium* in source water [15].

2. Exposure assessment is in many studies hampered by incomplete "site-specific" data. The gaps in the site-specific data are filled by using data from the scientific literature. This is particularly true for the studies in the 1990s. As the use of QMRA progressed, more authors have collected site-specific information about most if not all steps in the exposure assessment.
3. Most studies used the dose-response data of the Iowa strain of *C. parvum* as published by DuPont and coworkers [12]. Over the years, the dose-response relationships of more *C. parvum* strains have been published. One recent study on the risk of *Cryptosporidium* to fire fighters using recycled water used the dose-response data of the TAMU strain of *C. parvum* as this was the most infective strain [11]. Medema [50] present an approach for the use of a *C. parvum* dose-response relation, that combines the dose-response data that are published for four different isolates of *C. parvum* (Iowa, TAMU, UCP and Moredun).
4. The most frequently used health outcome is the probability of infection; a few studies also determined the probability of illness of the general population and the immunodeficient population [56, 59]. Two studies calculated the DALY resulting from the waterborne transmission of *Cryptosporidium* [67, 30].
5. Using the data of the Milwaukee outbreak [44], the calculated probability of infection/illness with QMRA was compared to the observed probability of illness in the outbreak as observed in the epidemiological investigations [24, 26]. The authors concluded that the results of QMRA and epidemiological investigation were consistent. The analysis of the exposure of the Milwaukee residents to *Cryptosporidium* via tap water was hampered by the lack of timely measurements of *Cryptosporidium* in the contaminated water. Unfortunately, this is the rule rather than the exception in waterborne outbreaks. The concentration had to be inferred from oocyst concentrations found in samples of ice that was prepared at the time of the water supply contamination and was corrected for the expected loss of detectable oocysts after freezing/thawing. The exposure assessment was therefore not very certain. In addition, the reported magnitude of the Milwaukee outbreak has been criticized by [36]. They claim that the background prevalence of gastrointestinal illness in the USA is much higher (1.2–1.4 episodes per person per year, or 0.10–0.12 per person per month) than the prevalence used by [44] (0.005 per person per month). Use of higher background prevalence would drastically reduce the estimated size of the Milwaukee outbreak.
6. The setup of the QMRAs sometimes used point estimates, but more generally a probabilistic approach is used to be able to estimate the level of uncertainty of the calculated probability of infection or illness.
7. Between the different studies, the calculated probability of infection can differ considerably see (Table 14.1). Within studies, the uncertainty of the risk

estimate toward the higher health risk (illustrated by the difference between the average or median risk and the 95% confidence limit) is limited to around a factor of 10.

In general, it can be seen from these examples that QMRA has become an established tool to evaluate health risks of *Cryptosporidium* in (piped) drinking water supplies. QMRA requires input from data on exposure and dose-response and can be done in different levels of complexity. The next paragraphs give examples of the application of QMRA in water and illustrate the stepwise (tiered) approach that can be taken in QMRA and that QMRA can be conducted and be valuable in the absence of site-specific data and in developing countries.

QMRA to Assess the Safety of a Drinking Water Supply

Suppose that a water utility wants to evaluate if its surface water supply is at risk of significantly transmitting *Cryptosporidium* to its consumers, but has no specific information about *Cryptosporidium* in its source water or removal by its water treatment processes. A first exercise to get an idea of the level of risk could be a screening-level QMRA. The information on *Cryptosporidium* levels in source water can be derived from watershed use (see [50]), and for the water treatment processes default log-credits for the removal or inactivation of *Cryptosporidium* are available [49]. For instance, if the water supply system uses a watershed that can be characterized as moderately polluted and treats this source water with off-stream storage reservoirs and a conventional (coagulation/filtration/chlorination) water treatment system, using the scientific database, the expected concentration of *Cryptosporidium* in source water can be estimated at 0.1/L and the removal by the subsequent water treatment processes can be estimated at 0.5 + 2.5 = 3.0 logs removal. Hence, the estimated concentration of *Cryptosporidium* in drinking water is 1×10^{-4}/L. With a conservative best estimate of consumption of cold tap water of 0.78 L/day (3.49 glasses of 0.25 L, [53]), the average probability of exposure to *Cryptosporidium* is 8.7×10^{-5} per person per day. With the combined dose-response relation of the four *C. parvum* strains, the probability of infection is estimated at 3.8×10^{-5} per person per day, which amounts to 1.4×10^{-2} (=1.4%) per person per year. This is a first estimate of the health risk related to *Cryptosporidium* in this specific water supply system. Similarly, such an exercise can be used to evaluate different scenarios of risk management to reduce this risk (if required) such as measures to improve the catchment or install additional treatment processes. An example of a practical application of such a screening-level risk assessment is given in Medema [50], where a large water supply company uses the screening-level QMRA to prioritize risk management of its water supply systems.

Comparing Water Supply Scenarios with QMRA

Piped and non-piped water supply in Uganda [34].

In Kampala, 72% of the population uses piped water supplies. 20% of the population uses piped water through household connections; the rest collects water at standpipes and stores it in-house. The piped water is produced from Lake Victoria water through (coagulation/settling) rapid sand filtration followed by chlorination. The rest of the population (28%) uses protected springs for their water supply.

Data on thermotolerant coliforms were available from Lake Victoria and from the protected springs and the household containers. Using an estimate of the percentage of *E. coli* within the thermotolerant coliforms and an estimate of the percentage of pathogenic *E. coli* within *E. coli*, the thermotolerant coliform concentration data were translated to pathogenic *E. coli* concentrations. For the removal of (pathogenic) *E. coli* by the water treatment processes, the authors used a 3-log credit for the physical removal processes and an additional 2-log credit for the chlorination. This was used to calculate the concentration of pathogenic *E. coli* in drinking water. With data or estimates on consumption of unheated drinking water, dose-response for infection, probability of illness when infected, and disease burden (DALY), the concentration of pathogenic *E. coli* in drinking water was translated into the estimated disease burden by exposure (Table 14.2).

Similar assessments were made for *Cryptosporidium* and Rotavirus exposure for the population using piped water supply. For *Cryptosporidium*, they showed that treatment failure would result in a very significant increase of the disease burden (from 10^{-4} to 4 DALYs per person per year). The authors have compared the calculated levels of disease burden to the WHO reference level of risk (10^{-6} DALY). Upgrading the treatment would be necessary to achieve this health target, but the authors argue that, given the low level of access to piped water in the home and the disease burden associated with the use of alternative (more contaminated) sources, this would not be cost effective. Improving access to piped water supply in homes, sanitation and hygiene would be more effective in reducing the disease burden.

This example illustrates that QMRA is feasible also in settings with limited data. The authors discuss limitations and assumptions used in their study, but illustrate the value of system assessment to inform risk management of the area where control measures will be most effective.

QMRA to Evaluate the Health Risk of Hazardous Events

Many outbreaks of intestinal illness caused by consumption of contaminated drinking water in affluent nations have been associated with hazardous events, such as heavy rainfall (both for surface and groundwater systems), failures in

Table 14.2 Assessment of disease burden for pathogenic *E. coli* from different water types (adapted from [34])

	Piped water following treatment	Piped water in distribution	Household storage water	Protected spring water
Raw water quality thermotolerant coliforms/L	150		30	140
Raw water quality *E. coli*/L	143		28.5	133
Raw water pathogenic *E. coli*/L	11.5		2.3	10.6
Treatment effect (log)	5		0	0
Drinking water quality (/L)	1.15×10^{-4}	0.18	2.3	10.6
Consumption of unheated drinking water (L)	1			
Exposure (pathogens/day)	1.15×10^{-4}	0.18	2.3	1.06×10^{1}
Dose-response parameter (exponential)	0.001			
Risk of infection (day)	1.15×10^{-7}	1.80×10^{-4}	2.30×10^{-3}	1.06×10^{-2}
Risk of infection (year)	4.20×10^{-5}	6.57×10^{-2}	8.40×10^{-1}	3.87×10^{0}
Risk of diarrheal disease given infection	0.25			
Risk of diarrheal disease	1.05×10^{-5}	1.64×10^{-2}	2.10×10^{-1}	9.67×10^{-1}
Exposed fraction	0.31	0.1	0.42	0.28
Disease burden (DALYs)	1.04×10^{-6}	5.26×10^{-4}	2.82×10^{-2}	8.67×10^{-2}

a treatment process, failures in the integrity of the infrastructure (wells, distribution network), cross-connections in the distribution network, etc. For an overview, see [35]. Additional hazardous events can be identified for non-piped supplies, especially contamination of the water in storage containers. Also, events that lead to a stop in supply of drinking water (due to power or treatment failure, or indeed absence of sufficient quantities of source water) are hazardous events in themselves, since water is essential for life and hygiene.

Water quality testing can help to identify peak events. Often, peak events can be indicated by simple parameters, such as rainfall, river flow, turbidity, etc., and hence their detection does not require advanced equipment or expertise. It does require knowledge of the water supply system, including its catchment. In Microrisk, a European study on microbial risk assessment of drinking water, information was needed about pathogen occurrence in source (surface) water of the water supply systems under study [49]. Knowing the potential importance of peak events, catchment surveys were conducted to identify contamination sources and to identify events that could lead to peak pathogen contamination of the source water. One system used bank filtration and subsequent treatment to produce drinking water from a large river. Historical (50 years) data on the water level of the river showed that an increase of $\geq$3 m within 5 days occurred 1.1% of the time (3.9 days per year on average). This river level rise was used as a criterion to trigger peak

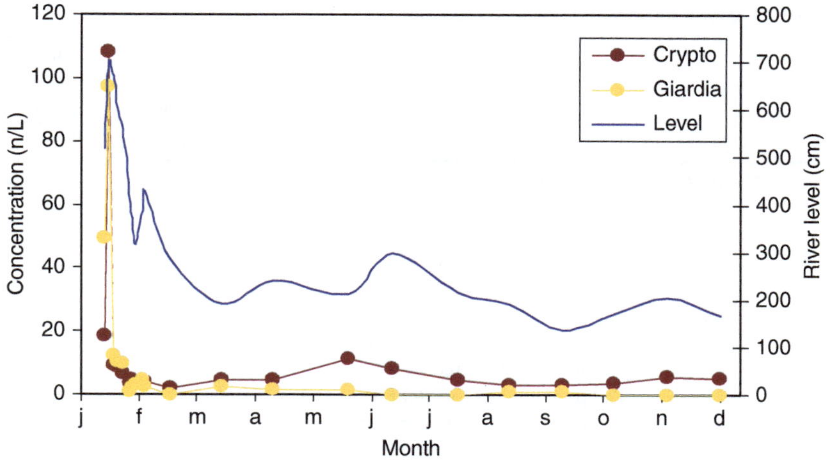

Fig. 14.3 *Cryptosporidium* and *Giardia* in river water during a peak event (Data from [49])

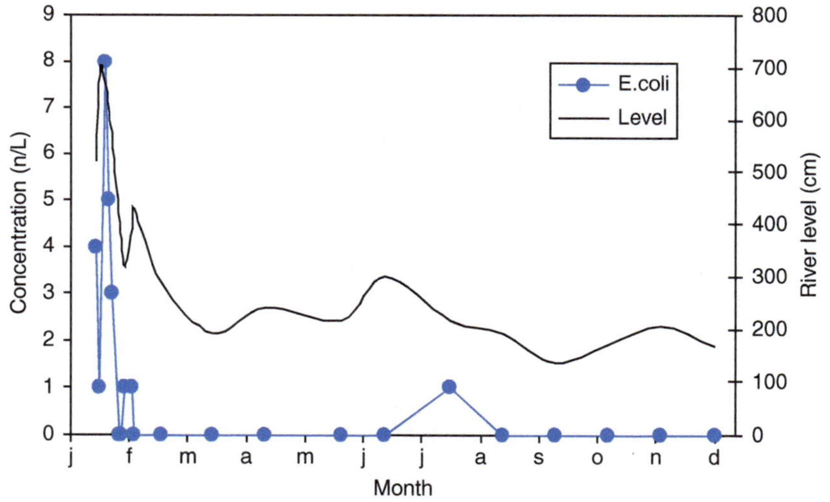

Fig. 14.4 *E. coli* breakthrough of bank filtration during a peak event in the river see (Fig. 14.3, data from [49])

event sampling. A dry weather flow sampling scheme was also in place, with monthly pathogen samples. During monitoring, one peak event was encountered and peak event samples were taken, showing a sharp increase in the concentration of *Cryptosporidium* and *Giardia* concentration in the river (Fig. 14.3). Event samples were also taken from the bank filtrate. The *E. coli* were detected in the bank filtrate only at the time of the peak event (Fig. 14.4).

Similarly hazardous events may occur in water treatment (i.e., disinfection failure) or distribution (cross-connection, ingress during main breaks, no pressure

Table 14.3 Hazardous event impacts on risk

CTS	Pathogen	Hazardous event	Total duration of event	Baseline risk (person^{-1} year^{-1})	Baseline + hazardous event risk (person^{-1} year^{-1})
1	*Cryptosporidium*	Loss of filtration due to petroleum spill necessitating cleanup. Only remaining treatment is chlorination	7 days	1.4×10^{-5}	1.7×10^{-2}
5	*Norovirus*	No intake closures leading to periodic high concentration of virus in source water	57 days	$<5.8 \times 10^{-4}$	2.7×10^{-2}
		Delay in intake closure of 4 h for each of 29 events of high virus concentration in source water per year	4.75 days		3.4×10^{-3}
6	*Campylobacter*	Loss of disinfection capacity: total suboptimal chlorination periods based on analysis of SCADA data – worst case of total loss of disinfection assumed	1.5 h	2.5×10^{-6}	3.2×10^{-6}
8	*Campylobacter*	Short-circuiting leads to reduced (1 log) removal in storage reservoir for 24 h. Nine short-circuiting events occur per year	9 days	1.7×10^{-5}	3.4×10^{-5}
		Short-circuiting leads to reduced (1 log) removal in storage reservoir for 24 h. Nine short-circuiting events occur per year. During one of these periods chlorination loss occurs due to power failure for 2.4 h (0.1 days)	0.1 days		1.8×10^{-4}

The risk estimates in brackets are based on upper 95th percentile uncertainty and are derived from upper limit inputs rather than typical source water concentrations

period or repair). A QMRA to determine the health effect of ingress of fecal contamination in municipal piped distribution networks is given in [42]. In the Microrisk project, the health risk associated with several source and treatment hazardous event scenarios in the different water supply systems (called Catchment-to-Tap Systems or CTS) studied was determined and compared to the baseline health risk in these systems in a Monte Carlo simulation [64].

Hazardous events were identified in discussions with local water suppliers and from SCADA data. Of these, five were selected and evaluated with QMRA (Table 14.3).

In the case of the CTS 1 (a surface water supply) the local managers were concerned about the prospect of a motorway fuel spill and its potential impact on the treatment plant. It was speculated that even small quantities could foul major filters (Rapid Sand Filter and Granular Activated Carbon filters) and reactors (ozone contact tanks) and necessitate cleaning. This led us to simulate a cleanup period of 7 days during which protection was provided by chlorination alone and hence the system was vulnerable to *Cryptosporidium* contamination because of its resistance to chlorine.

It can be seen that the annual risk of infection by *Cryptosporidium* rises by a factor of 1,000 and the estimated probability of infection is much higher than 10^{-4} per person per year. Further, even if the repair period could be reduced to 1–2 days, the additional risk would still be great and hence other action such as a boiled water alert on top of chlorination would need to be considered.

CTS 5 is a surface water supply system with the option to close water intake. If no intake management were in place the average annual risk would have been at least 19 times higher. The impact of a delay in closing the intake was also substantial. This highlighted the need for timely warning of event onset where source extraction is being managed.

CTS 6 included extensive diary and SCADA (Supervisory Control and Data Acquisition) data detailing performance of the chlorination. This information allowed determination whether chlorination failure was occurring. Analysis of the in-line chlorine monitoring data indicated that at worst chlorine dosing failed for a total time of 1.5 h over a 12-month period. The impact of simulated worst-case failure on *Campylobacter* showed a detectable but only small increase in health risk.

The final scenario considered was that of multiple concurrent hazardous events. A concern for CTS 8 and CTS 6 type systems, which draw their supply from a reservoir, is that during high run-off events there can be concurrent polluted input and short-circuiting [32]. Further, storms frequently cause power failures, which could affect treatment plant equipment such as dosing pumps. Two scenarios were considered with these events in mind. Concurrent contamination of runoff and short-circuiting of the reservoirs were estimated to double health risk for *Campylobacter*. With the combination of a short duration power failure leading to chlorination loss during a storm could increase annualized risk 11-fold in a short time, confirming the need for avoiding or actively managing periods of concurrent hazardous events.

The value of the hazardous event analyses illustrated lies not only in the actual estimates presented. They also demonstrate how QMRA can be used to evaluate events and other hazardous scenarios to produce risk estimates useful for management. In the case of CTS 1, it was clear that filtration shut down even for a short period posed high risks because of the contamination levels in the source water. Selective water intake at CTS 5 is a beneficial management activity. At CTS 6, chlorine dosing was shown to be maintained at a level sufficient to reduce risks arising from plant failure. The CTS 8 analysis showed that baseline operating conditions provide sufficient barrier protection to mitigate a run-off and short-circuiting event, but with a concurrent event (chlorination failure) pose a significant threat.

QMRA for Water Reuse

In (semi) arid conditions, there is (increasing) water scarcity and competition between agriculture and urban uses of this scarce resource. Wastewater is in most cases a reliable (in terms of quantity) source of water and valuable source of nutrients for agriculture. Wastewater reuse in agriculture is a form of water and nutrient recycling that is practiced worldwide, especially in arid and semi-arid areas. Also the (re)use of gray water in urban areas for applications such as toilet flushing in homes, gardening, etc., is becoming more common.

The new WHO Guidelines for safe use of wastewater, excreta, and gray water are based on the Safe Water Framework (Fig. 14.1). QMRA is presented in these WHO guidelines as useful tool to estimate the health risks associated with wastewater reuse in different scenarios and for different pathogens. The guidelines contain several references to the application of QMRA in wastewater reuse. In the next paragraphs, three examples of the use of QMRA in water reuse are given.

Comparing Risks Between Different Uses of Reclaimed Wastewater (California)

The first QMRA to estimate the disease risk associated with the reuse of (treated) wastewater was [3]. They evaluated the risk of an infection with enteric viruses (Poliovirus 1 and 3 and Echovirus 12) when chlorinated or unchlorinated tertiary effluent was used for:

- Irrigation of a golf course
  The exposure scenario was a golf course with night time irrigation with tertiary treated wastewater effluent and person golfing twice a week. Each day this person would be exposed to 1 mL of reclaimed water during handling and cleaning of golf balls. The pathogen concentration in this reclaimed water was calculated from data on enteric viruses in chlorinated and unchlorinated effluent and virus decay on the golf field.
- Spray irrigation of food crops
  After spray irrigation, it was assumed that 10 mL of reclaimed water was left on each portion of crops eaten raw. The spray irrigation was stopped 14 days before harvesting and the virus die-off due to desiccation and sunlight exposure was included in the calculation.
- Swimming in recreational water
  This recreational water was assumed to be an impoundment that was, during summer, completely made up out of reclaimed water. No dilution or die-off was assumed. A swimmer was assumed to ingest 100 mL each swimming day and to swim 40 days in a year.
- Groundwater recharge near domestic wells
  This exposure scenario was based on the proposed Californian groundwater recharge regulations. The nearest domestic well was assumed to receive 50%

Table 14.4 Annual risk of exposure to viruses for different applications of reclaimed water

Exposure scenario	Echovirus 12	Poliovirus 1	Poliovirus 3
Irrigation of golf course	1.0×10^{-3}	3.5 E-5	2.5 E-2
Spray irrigation food crops	4.5×10^{-6}	1.5 E-7	1.1 E-4
Recreational impoundment	7.4×10^{-2}	2.6 E-3	8.4 E-1
Groundwater recharge	5.9×10^{-8}	5.4 E-9	2.3 E-8

reclaimed water that had been passing through 3 m of unsaturated soil beneath the recharge basin during a period of 6 months. The people drinking from this well were assumed to consume 2 L/day.

The input data were:

- Concentration of culturable enteric viruses in unchlorinated secondary effluent: 5–734/L (90% and maximum, respectively)
- Concentration of culturable enteric viruses in chlorinated tertiary effluent: 0.01–1.1/L
- Removal of enteric viruses by full tertiary treatment (flocculation, clarification, filtration, chlorination): 5 logs
- Virus decay rate: 0.69/day (first order die-off kinetics)
- Fraction of virus remaining after percolation through the unsaturated soil $c/c_0 = 10^{-0.007L}$, where L is the depth of the unsaturated zone in centimeters
- Dose-response parameters for echovirus 12 and poliovirus 1 and 3

The concentration of viruses in reclaimed water was taken from data from surveys of secondary and tertiary effluent. They calculated the exposure to the viruses in the different exposure scenarios. Annual risks were calculated from the maximum concentration found in chlorinated tertiary effluent (1.1 culturable virus unit L^{-1}) and exposure in the different applications (Table 14.4).

This QMRA showed that the virus risk was highest when reclaimed wastewater was used in recreational impoundments and golf course irrigation. This maximum concentration was found in only 0.1% of the samples (with 99% of the samples with virus concentrations below the detection limit), so they also calculated the risk with a virus concentration of 1/100 L, which were approximately 100-fold (2 logs) lower.

The value of the QMRA was that it provided a comparative basis for addressing the treatment and fate of enteric viruses in wastewater reuse and showed that the risk can further be mitigated by controlling exposure to reclaimed water.

Health Risk of Reuse for Crop Irrigation (Australia; Probabilistic)

In the previous example, the available data and assumptions were used to generate point estimates. This example shows how the variability in available data can be used to determine the uncertainty that is associated with each of the components in a QMRA.

Table 14.5 Best and extreme estimates for parameters of exposure to viruses in wastewater reused for irrigation of lettuce

Model component	"Best" estimate	"Extreme" estimate
Virus occurrence	2.6 (virus units L^{-1})	470,000 (virus units L^{-1})
Virus attachment (*f*)	0.024	0.071
Virus inactivation: S(t)	h1 = 2.5 day^{-1}	h1 = 2.0 day^{-1}
Bi-phasic inactivation	h2 = 0.5 day^{-1}	h2 = 0.3 day^{-1}
$Ct = aC_0 * h1 + (1 - a)C_0h2$	a = 0.12%	a = 0.96%
Consumption per event *q*	100 g	300 g

Sources: Californian dataset used by [3, 76]. All other data were derived from [57, 58].

The use of wastewater for irrigation of food crops that are eaten raw is common practice in many arid and semi-arid regions [57] constructed a QMRA-model for evaluating the risks associated with the consumption of wastewater irrigated lettuce crops. The exposure assessment in this model consisted of four process steps:

Exposure to viruses was calculated as:

$$\text{Exposure} = N \times f \times S(t) \times q$$

where

N is the number of viruses in the irrigation water applied to the crop
f is the fraction of those viruses that survive the irrigation process and attach to the lettuce plant
S(*t*) is the fraction of viruses remaining infectious at consumption
q is the quantity of crop consumed

For each step a best estimate and an extreme estimate were selected (Table 14.5). This allowed analysis of the sensitivity of the QMRA to each of the model parameters.

The authors calculated the Factor Sensitivity (FS = log ($N_{extreme}/N_{best}$), with *N* being the number of viruses in the extreme or best estimate) for each of the components. Already obvious from the table above is the high impact of the estimate for the virus concentration in wastewater (FS = 5.49). Less obvious from the table is the high impact of the estimate of virus inactivation (FS = 2.2). This is of course time dependent; the authors used 14 days as the time between final irrigation and consumption. A shorter interval reduces the impact of virus inactivation, since the inactivation is less. The uncertainty associated with virus attachment (FS = 0.45) and consumption (FS = 0.48) was considerably less.

This simple mathematical approach yielded not only the risk estimates associated with wastewater reuse for food crop irrigation, but also the (un)certainty associated with each of the components in the exposure of crop consumers to viruses that remain on the crops at the time of consumption.

Guidelines for Safe Reuse (Australia)

QMRA can be used to estimate health risks from exposure to pathogens via wastewater reuse in agriculture, as illustrated in the above examples. In the National guidelines for water recycling in [6], QMRA is used for a different purpose: to calculate health-based performance targets for recycled water systems. In these guidelines, the Australians use a health-based target as a benchmark for safety that has to be met by each water reuse system. They use the health-based target that WHO has defined in their GDWQ: 10^{-6} disability adjusted life years per year (DALY, see Box 1 for more information about this disease burden metric) as their tolerable level of risk.

This health-based target is translated to performance targets for the reuse system with respect to microbial hazards. The concentration of pathogens in the source water for the reuse system (raw/treated sewage, gray water, etc.) and the level of exposure of people to the recycled water (via crops, aerosols, ingestion) determine how much reduction of pathogen exposure is required to meet the 10^{-6} DALY/year target.

In formula:

$$PT = \log(C \times E \times N/DALYd)$$

in which

PT is the performance target (required log reduction)
C is the concentration of pathogens in source water (in these guidelines: 95th percentile of concentration data)
E is the exposure (volume (*L*))
N is the average frequency of exposure (number/person/year)
DALYd is the pathogen dose that is equivalent to a DALY of 10^{-6} per year, a translation of the 10^{-6} DALY target to a pathogen dose target, taking into account the pathogen's dose-response relation and the fraction of persons that contract illness when infected.

Since sewage and gray water may contain a wide range of pathogens and it is not feasible to do this QMRA for all, it is more practical to select reference pathogens, pathogens that represent a major group of pathogens. The philosophy is that when risk management is aimed at these reference pathogens, the other pathogens from these groups will also be adequately controlled. For protozoa and helminth eggs, *Cryptosporidium* is selected as reference pathogen because it is reasonably infective and more difficult to control by chlorination and filtration than other protozoa or helminth eggs (DALYd is 1.6×10^{-2}, 95th percentile in sewage: 2,000/L). For bacteria, *Campylobacter* is selected because of its infectivity and high prevalence (DALYd is 3.7×10^{-2}, 95th percentile in sewage: 7,000/L). For viruses, rotavirus is selected because of its high infectivity and the availability of dose-response data. Since no data on rotavirus in sewage were available, but data on adenoviruses occurrence were available, these latter data are used and combined with the rotavirus dose-response data (DALYd is 2.5×10^{-3}, 95th percentile in sewage: 8,000/L).

So with concentration C in source water as known and the DALYd as a constant per reference pathogen, the level and frequency of exposure are needed to determine the performance target for the reuse system.

For a range of intended uses of recycled water the associated level and frequency of exposure was (point) estimated from available scientific and statistic data. For example, for exposure by consumption of commercial food crops irrigated with recycled water the level of exposure was estimated at 5 mL for a service of lettuce and 1 mL for a service of other raw produce, with an annual frequency of 70 and 140 services, respectively. Similar exposure estimates were determined for garden irrigation, municipal irrigation, fire fighting, toilet flushing, washing machine use, and cross-connections.

Now the performance target for the use of recycled wastewater for commercial crop irrigation can be calculated:

Exposure for lettuce is 0.005×70, for other raw produce 0.001×140; this totals to 0.49 L/year

$$\begin{aligned} PT_{\text{Cryptosporidium}} &= 2,000 \times 0.49/\left(1.6 \times 10^{-2}\right) \\ &= 4.8 \text{ log} \\ PT_{\text{Campylobacter}} &= 7,000 \times 0.49/\left(3.7 \times 10^{-2}\right) \\ &= 5.0 \text{ log} \\ PT_{\text{Rotavirus}} &= 8,000 \times 0.49/\left(2.5 \times 10^{-3}\right) \\ &= 6.1 \text{ log} \end{aligned}$$

There are different ways to manage the risk associated with water recycling: prevent pathogens from entering recycled water, remove pathogens from recycled water by treatment processes, and reduce exposure by using restrictions or preventive on-site measures: restricted access, withholding periods before harvesting, controlled application (drip or subsurface irrigation). The Australian guidelines have assigned default performance credits to a range of treatment processes and on-site preventive measures and give examples of how the combination of these two types of risk management options can be used to achieve safe water recycling.

Box 14.1 DALY

Disability Adjusted Life Years (DALYs) is as a metric for translating the risk of disease burden a general health burden per case of illness. The DALY accounts for the years lived with a disability (YLD) plus the years of life lost (YLL) due to the hazard (compared to the average expected age of death in a community). One DALY per million people a year roughly equates to one cancer death per 100,000 in a 70-year lifetime [73]. The DALY is calculated as the product of the probability of each illness outcome with a severity factor and the duration (years). Calculation of the DALY contribution per infection is undertaken using:

$$\mathrm{DALY} = \sum_{i=1}^{n} P(\mathrm{ill}|\,\mathrm{inf}) \times P(\mathrm{outcome}_i|\mathrm{ill}) \times \mathrm{Duration}_i \times \mathrm{Severity}_i$$

where n is the total number of outcomes considered

P(ill|inf) is the probability of illness given infection

P(outcome|ill) is the probability of outcome i given illness

$\mathrm{Duration}_i$ is the duration (years) of outcome i

$\mathrm{Severity}_i$ is the severity weighting for outcome i.

The advantage of using DALYs over an infection risk end point is that it not only reflects the effects of acute end points (e.g., diarrheal illness) but also the likelihood and severity of more serious disease outcomes (e.g., Guillain-Barré syndrome associated with *Campylobacter*). Disease burden per case varies widely, but can be focused on a locality. For example, the disease burden per 1,000 cases of rotavirus diarrhea is 480 DALYs in low-income regions, where child mortality frequently occurs. However, it is only 14 DALYs per 1,000 cases in high-income regions, where hospital facilities are accessible to the great majority of the population. Disease burden estimates for different drinking water contaminants is summarized in Table 14.B1.

Table 14.B1 Summary of disease burden estimates for different drinking water contaminants

	Disease burden per 1,000 cases		
	YLD	YLL	DALY
Cryptosporidium parvum	1.34	0.13	1.47
Campylobacter spp	3.2	1.4	4.6
STEC O157	13.8	40.9	54.7
Rotavirus			
High-income countries	2.0	12	14
Low-income countries	2.2	480	482
Hepatitis-A virus			
High-income countries, 15–49 years	5	250	255
Low-income countries	3	74	77

Source: Reproduced from [31].

Future Directions

The examples given in the previous paragraphs illustrate how QMRA can be applied to assess microbial health risks associated with systems where people may be exposed to pathogens through the use of water. QMRA is used to evaluate individual systems (against health-based targets), compare different systems or scenarios and to evaluate the significance of hazardous events and system failures in municipal piped water supply, but also non-piped water supply, and for waste-water and gray water reuse. Others have also demonstrated the use in recreational waters [4].

Risk assessment also allows comparison of the effort and resources put into the provision of safe water systems and resources allocated to manage other health risks. However, given the current state of the art and especially the lack of available quantitative data, QMRA has to rely partly on assumptions. Given the current level of uncertainty in quantitative risk assessments of water systems, the outcome should be regarded as an indication of the level of safety, rather than an absolute assessment of health risk. The outcome can be used to guide the risk management direction to pathogen control and to select the most appropriate control measures.

The benefit of risk assessment is that it gives a better understanding/breakdown of the problems and of important data. Additionally, the risk concept allows us to focus and prioritize research on the areas where important pieces of information are missing.

Improving the Technique of QMRA

The science of risk assessment is increasingly complex; most of the current QMRA work uses the probability of infection as end point. Infection is the first step in the disease process, but does not reflect the severity of the disease, including potential serious health effects that may arise in a particular subpopulation. Some studies have been using burden-of-disease and cost-of-illness measures [45]. This improves the assessment of the magnitude of the adverse effect of pathogens exposure via water and allows balancing pathogen risks with other risks. The dynamics of infectious diseases with secondary transmission and the effect of immunity and sensitive subpopulations have been largely neglected. Several studies are exploring ways to incorporate these disease dynamics into account [14].

The large variability of pathogens in water and the limited availability of data (especially in relation to peak events) and the variability in treatment efficacy are very important issues to take into consideration in QMRA. More data need to be collected, and monitoring programs of water suppliers should be targeted more toward the provision of information for QMRA. Pathogens to be selected for QMRA should be detectable in the water systems with reliable analytical techniques. The use of reference pathogens, pathogens that are critical for the control measures taken in water supply, is recommended. The variability and limited data available will cause uncertainty in the risk assessment, but compared to chemical risk assessment with large uncertainty factors, this is not inhibitive for the implementation of microbial risk assessment.

Improving the Utility of QMRA

QMRA can be done at different levels of sophistication. Sophisticated QMRA can take considerable amounts of time and resources. The level of detail in the QMRA and the extent of the uncertainty analysis that is needed to address a particular problem has to be appropriate only to the extent that is needed to help risk managers decide. QMRA lends itself well for a tiered approach, where the sophistication increases only if the risk manager requires better information to make a decision.

The National Research Council in the USA has advised USEPA to adopt a framework for risk-based decision making to make risk assessments more useful for risk management decisions [55]. In this framework, improved stakeholder involvement should also help to improve the acceptance and utility of risk assessment.

QMRA is a process that requires input from several disciplines. Researchers that are trained in a specific discipline have to learn to combine their data and knowledge with data and knowledge from other disciplines in a (probabilistic) risk assessment framework. And risk assessment is being extended to address broader questions in environment and health: risk-risk trade-offs and cost-benefit analysis. Development of guidance and training on QMRA is needed to strengthen the capacity of QMRA researchers.

Assessing the microbial risks of water systems is a relatively young field of science. It has the capacity to further professionalize safety management in water by providing science-based, objective, credible and proportionate information to help risk managers make informed decisions.

Bibliography

1. Aboytes R, DiGiovanni G, Abrams F, Rheinecker C, McElroy W, Shaw N, LeChevallier MW (2004) Detection of infectious *Cryptosporidium* in filtered drinking water. J Am Water Works Assoc 96(9):88–98
2. Adham SS, Trussel RR, Gagliardo PF, Olivieri AW (1998) Membranes: a barrier to microorganisms. Water Supply 16(1–2):336–340
3. Asano T, Leong LYC, Rigby MG, Sakaji RH (1992) Evaluation of the California wastewater reclamation criteria using enteric virus monitoring data. Water Sci Technol 26(7–8): 1513–1524

4. Ashbolt NJ, Bruno M (2003) Application and refinement of the WHO risk framework for recreational waters in Sydney, Australia. J Water Health 1(3):125–131
5. Atherholt TB, LeChevallier MW, Norton WD, Rosen JS (1998) Effect of rainfall on *Giardia* and *Cryptosporidium*. J Am Water Works Assoc 90:66–80
6. Australia (2005) Australian guidelines for water recycling: managing health and environmental risks (phase 1). Natural Resource Management Ministerial Council, Environment Protection and Heritage Council, Australian Health Ministers' Conference, Australia
7. Bartram J, Fewtrell L, Stenström TA (2001) Harmonised assessment of risk and risk management for water-related infectious disease: an overview. In: Fewtrell L, Bartram J (eds) Water quality: guidelines, standards and health. IWA Publishing, London, pp 1–16
8. Benford D (2001) Principals of risk assessment of food and drinking water related to human health, ILSI Europe concise monograph series. International Life Science Institute, Belgium, pp 1–43
9. Burmaster DE, Anderson PD (1994) Prnciples of good practice for the use of Monte Carlo techniques in human health and ecological risk assessment. Risk Anal 14(4):477–481
10. CODEX (1997) Codex Alimentarius Commission: procedure manual. Joint FAO/WHO Food Standards Programme. FAO, Rome
11. Deere D, Davison A (2004) Health risk assessment of fire fighting from recycled water. Occasional Paper 11. Water Services of Australia, Sydney
12. DuPont HL, Chappell CL, Sterling CR, Okhuysen PC, Rose JB, Jakubowski W (1995) The infectivity of *Cryptosporidium parvum* in healthy volunteers. N Engl J Med 332(13):855–859
13. EFSA (2009) Transparancy in risk assessment – scientific aspects. Scientific opinion. EFSA J 1051:1–22
14. Eisenberg JNS, Soller JA, Scott J, Eisenberg DM, Colford JM (2004) A dynamic model to assess microbial health risks associated with beneficial uses of biosolids. Risk Anal 24(1):221–236
15. EPA (Environmental Protection Agency) (2005) Economic analysis for the final long term 2 enhanced surface water treatment rule. EPA, Office of Water report EPA 815-R-06-001. EPA, Washington
16. Fewtrell L, Bartram J (2001) Water quality: guidelines, standards and health. Risk assessment and management for water related infectious diseases. IWA Publishing, London
17. Finch GR, Black EK, Gyurek L, Belosevic M (1993) Ozone inactivation of *Cryptosporidium parvum* in demand-free phosphate buffer determined by in vitro excystation and animal infectivity. Appl Environ Microbiol 59(12):4203–4210
18. Finch GR, Gyurek LL, Liyanage LRJ, Belosevic M (1997) Effect of various disinfecion methods on the inactivation of *Cryptosporidium*. American Water Works Association Research Foundation, Denver
19. Gale P (1996) Developments in microbiological risk assessment models for drinking water – a short review. J Appl Bacteriol 81:403–410
20. Gale P (2000) Risk assessment model for a waterborne outbreak of cryptosporidiosis. Water Sci Technol 41(7):1–7
21. Gale P (2002) Using risk assessment to identify future research requirements. J Am Water Works Assoc 94(9):30–42
22. Gale P, Stanfield G (2000) *Cryptosporidium* during a simulated outbreak. J Am Water Works Assoc 92(9):105
23. Haas CN (1997) Importance of distributional form in characterizing inputs to Monte Carlo risk assessments. Risk Anal 17(1):107–113
24. Haas CN (2000) Epidemiology, Microbiology, and Risk Assessment of Waterborne Pathogens Including *Cryptosporidium*. J Food Prot 63(6):827–831
25. Haas CN, Crockett CS, Rose JB, Gerba CP, Fazil AM (1996) Assessing the Risk Posed by Oocysts in Drinking Water. J Am Water Works Assoc 88(9):131–136
26. Haas CN, Rose JB, Gerba CP (1999) Quantitative microbial risk assessment. Wiley, New York
27. Haas CN, Eisenberg JNS (2001) Risk assessment. In: Fewtrell L, Bartram J (eds) Water quality: guidelines, standards and health. IWA Publishing, London, pp 161–183

28. Hashimoto A, Hirata T (1999) Occurrence of *Cryptosporidium* oocysts and *Giardia* cysts in Sagami river, Japan. In: Proceedings of the Asian water quallity 1999; 7th reginal IAWQ conference. Taipei, Taiwan, vol 2, pp 956–961
29. Havelaar AH, Teunis PFM, Medema GJ (1996) Risk assessment of waterborne pathogens. Presented at the symposium waterborne pathogens, Bonn, 22–25 May
30. Havelaar AH, den Hollander AEM, Teunis PFM, Evers EG, van Kranen HJ, Versteegh JFM, van Koten JEM, Slob W (2000) Balancing the risks and benefits of drinking water disinfection: disability adjusted-life years on the scale. Environ Health Perspect 108(4):315–321
31. Havelaar AH, Melse JM (2003) Quantifying health risks in the WHO guidelines for drinking water quality. A burden of disease approach. Report 734301022, RIVM, Bilthoven
32. Hipsey MR, Brookes JD, Antenucci JP, Burch MD, Regel R (2004) A three-dimensional model of Cryptosporidium dynamics in lakes and reservoirs: a new model for risk management. Int J River Basin Manag 2(3):181–197
33. Hirata T, Hashimoto A (1998) Experimental assessment of the efficacy of microfiltration and ultrafiltration for *Cryptosporidium* removal. Water Sci Technol 41(7):103–107
34. Howard G, Pedley S, Tibatemwa S (2006) Quantitative microbial risk assessment to estimate health risks attributable to water supply: can the technique be applied in developing countries with limited data? J Water Health 4(1):49–66
35. Hrudey S, Hrudey E (2004) Safe drinking water, Lessons from recent outbreaks in affluent nations. IWA Publishing, London, 486 pp
36. Hunter PR, Syed Q (2001) Community surveys of self-reported diarrhoea can dramatically overestimate the size of outbreaks of waterborne cryptosporidiosis. Water Sci Technol 43(12): 27–30
37. Hutton P, Ashbolt N, Vesey G, Walker J, Ongerth J (1995) Cryptosporidium and Giardia in the aquatic environment of Sydney, Australia. In: Betts WB et al (eds) Protozoan parasites and water. Royal Society of Chemistry, Cambridge
38. Korich DG, Mead JR, Madore MS, Sinclair NA, Sterling CR (1990) Effects of ozone, chlorine dioxide, chlorine, and monochloramine on *Cryptosporidium parvum* oocyst viability. Appl Environ Microbiol 56(5):1423–1428
39. LeChevallier MW, Norton WD, Lee RG (1991) Occurrence of *Giardia* and *Cryptosporidium* spp. in surface water supplies. Appl Environ Microbiol 56:2610–2616
40. LeChevallier MW, Norton WD (1995) *Giardia* and *Cryptosporidium* in raw and finished water. J Am Water Works Assoc 87(9):54–68
41. LeChevalier MW, Norton W, Abbaszadegan M, Atherholt T, Rosen J (1998) Development of a monitoring strategy to determine variations in *Giardia* and *Cryptosporidium* levels in a watershed. In: Proceeding of source water protection international 1998, Dallas
42. van Lieverloo JHML, Blokker EJ, Medema GJ (2007) Quantitative microbial risk assessment of distributed drinking water using faecal indicator incidence and concentrations. J Water Health 5(Suppl 1):131–149
43. Macler BA, Regli S (1993) Use of microbial risk assessment in setting United-States drinking water standards. Int J Food Microbiol 18:245–256
44. MacKenzie WR, Hoxie NJ, Proctor ME, Gradus MS, Blair KA, Peterson DE, Kazmierczak JJ, Addiss DG, Fox KR, Rose JB, David JP (1994) A massive outbreak in Milwaukee of *Cryptosporidium* infection transmitted through the public water supply. N Engl J Med 331(3):161–167
45. Mangen MJJ, Havelaar AH, Bernsen RAJAM, Van Koningsveld R, De Wit GA (2005) The costs of human Campylobacter infections and sequelae in the Netherlands: a DALY and cost-of-illness approach. Food Econ Acta Agric Scand 2(1):35–51
46. Masago Y, Katayama H, Hashimoto A, Hirata T, Ohgaki S (2002) Assessment of risk of infection due to *Cryptosporidium parvum* in drinking water. Water Sci Technol 46(11/12):319–324

47. Medema GJ, Teunis PFM, Gornik V, Havelaar AH, Exner M (1995) Estimation of the *Cryptosporidium* infection risk via drinking water. In: Betts WB et al (eds) Protozoan parasites and water. Royal Society of Chemistry, Cambridge, pp 53–56
48. Medema GJ, Hoogenboezem W, van der Veer AJ, Ketelaars HAM, Hijnen WAM, Nobel PJ (2003) Quantitative risk assessment of Cryptosporidium in surface water treatment. Water Sci Technol 47:241–247
49. Medema GJ, Loret JF, Stenström TA, Ashbolt NJ (2006) Quantitative microbial risk assessment in the water safety plan. Report Microrisk, Kiwa Water Research, Nieuwegein. www.microrisk.com
50. Medema GJ (2009) Risk assessment of *Cryptosporidium* in drinking water. World Health Organization, Geneva, Switzerland
51. Messner MJ, Chappell CL, Okhuysen PC (2001) Risk assessment for *Cryptosporidium*: a hierarchical Bayesian analysis of human dose response data. Water Res 35(16):3934–3940
52. Michaels D (2008) Doubt is their product: how industry's assault on science threathens your health. Oxford University Press, New York
53. Mons MN, van der Wielen JHML, Blokker EJM, Sinclair MI, Hulshof KFAM, Dangendorf F, Hunter PR, Medema GJ (2007) Estimation of the consumption of cold tap water for microbiological risk assessment: an overview of studies and statistical analysis of data. J Water Health 5(Suppl 1):151–170
54. Morgan MG, Henrion M (1990) Uncertainty. A guide to dealing with uncertainty in quantitative risk and policy analysis. Cambridge University Press, New York, 332 p
55. NRC (2009) Science and decisions, Advancing risk assessment. National Research Council. The national Academies Press, Washington, DC
56. Perz JF, Ennever FK, Le Blancq SM (1998) Cryptosporidium in tap water: comparison of predicted risks with observed levels of disease. Am J Epidemiol 147(3):289–301
57. Petterson SR, Ashbolt NJ, Sharma A (2001) Microbial risks from wastewater irrigation of salad crops: a screening-level risk assessment. Water Environ Res 73(6):667–672
58. Petterson SA, Ashbolt NJ (2005) WHO guidelines for the safe use of wastewater and excreta in agriculture. Microbial Risk Assessment Section. www.who.int/wsh
59. Pouillot R, Beaudeau P, Denis J-B, Derouin F (2004) A quantitative risk assessment of waterborne cryptosporidiosis in France using second-order Monte Carlo simulation. Risk Anal 24(1):1–17
60. Regli S, Rose JB, Haas CN, Gerba CP (1991) Modeling the risk from Giardia and viruses in drinking water. J Am Water Works Assoc 83:76–84
61. Regli S, Odom R, Cromwell J, Lustic M, Blank V (1999) Benefits and costs of the IESWTR. J Am Water Works Assoc 91(4):148–158
62. Rose JB, Lisle JT, Haas CN (1995) Risk assessment methods for *Cryptosporidium* and *Giardia* in contaminated water. In: Betts WB et al (eds) Protozoan parasites and water. Royal Society of Chemistry, Cambridge, pp 238–242
63. Roseberry AM, Burmaster DE (1992) Lognormal distributions for water intake by children and adults. Risk Anal 12:99–104
64. Roser D, Petterson S, Signor R, Ashbolt N, Nilsson P, Thorwaldsdotter R (2006) How to implement QMRA to estimate baseline and hazardous event risks with management end uses in mind. Report Microrisk, Kiwa Water Research, Nieuwegein. www.microrisk.com
65. Smeets P (2008) Stochastic modelling of drinking water treatment in quantitative microbiological risk assessment. PhD thesis, Delft University, The Netherlands
66. Teunis PFM, Medema GJ, Kruidenier L, Havelaar AH (1997) Assessment of the risk of infection by *Cryptosporidium* and *Giardia* in drinking water from a surface water source. Water Res 31(6):1333–1346
67. Teunis PFM, Havelaar AH (1997) *Cryptosporidium* in drinking water. Evaluation of the ILSI/RSI quantitative risk assessment framework. RIVM, Bilthoven, The Netherlands
68. Teunis PFM, Havelaar AH (1999) *Cryptosporidium* in drinking water. Evaluation of the ILSI/RSI quantitative risk assessment framework. Report no. 284 550 006 RIVM, Bilthoven

69. Westrell T (2004) Microbial risk assessment and its implications for risk management in urban water systems. Thesis Linköping University, Linköping
70. Westrell T, Bergstedt O, Stenstrom TA, Ashbolt NJ (2003) A theroretical approach to assess microbial risks due to failures in drinking water systems. Int J Environ Health Res 13:181–197
71. WHO (1993) Guidelines for drinking water quality, 2nd edn. WHO, Geneva
72. WHO (1999) Principles and guidelines for the conduct of microbiological risk assessment. WHO, Geneva, CAC/GL-30
73. WHO (2004) Guidelines for drinking water quality, 3rd edn. WHO, Geneva
74. WHO (2006) Guidelines for the safe use of wastewater, excreta and grey water. WHO, Geneva
75. Yang SS, Benson SK, Du C, Healey MC (2000) Infection of immunosuppressed C57BL/6N adult mice with a single oocyst of *Cryptosporidium parvum.* J Parasitol 86:884–887
76. Yates MV, Gerba CP (1998) Microbial considerations in wastewater reclamation and reuse. In: Asano T (ed) wastewater reclamation and reuse, vol 10, Water Quality Management Library. Technomic Publishing, Lancaster, pp 437–488

Chapter 15
Pathogen and Nutrient Transfer Through and Across Agricultural Soils

David M. Oliver and Louise A. Heathwaite

Glossary

Hydrological connectivity	The linkage of spatial locations through different hydrological flow paths (surface and subsurface) within the catchment drainage network.
Farmyard manure	Feces and urine mixed with bedding material (such as straw) used for housed livestock, and recycled back to land as an organic fertilizer.
Fecal indicator bacteria (FIB)	Nonpathogenic microbial parameters that can be used as surrogate measures of infection risk to humans.
Leaching	The movement and loss of soluble elements and colloids from soil via drainage water to both surface water and ground water environments.
Matrix flow	The slow percolation of water through the soil pore system.

This chapter was originally published as part of the Encyclopedia of Sustainability Science and Technology edited by Robert A. Meyers. DOI:10.1007/978-1-4419-0851-3

D.M. Oliver (✉)
Biological & Environmental Sciences, School of Natural Sciences, University of Stirling, Stirling, UK

Centre for Sustainable Water Management, Lancaster Environment Centre, Lancaster University, Lancaster LA1 4YQ, UK
e-mail: d.m.oliver@lancs.ac.uk

L.A. Heathwaite
Centre for Sustainable Water Management, Lancaster Environment Centre, Lancaster University, Lancaster LA1 4YQ, UK

E.A. Laws (ed.), *Environmental Toxicology: Selected Entries from the Encyclopedia of Sustainability Science and Technology*, DOI 10.1007/978-1-4614-5764-0_15,

Mobilization	Term used – in the context of this paper – to describe the initiation of contaminant transfer and the process by which those contaminants begin movement from soil.
Nonpoint source pollution	Comprises contamination and pollution arising from many dispersed sources.
Pathogens	Microorganisms capable of causing disease or illness in a host and used here to refer to bacteria and protozoa originating from fecal material.
Preferential flow	Rapid movement of water and contaminants through the soil architecture. Much of the flow is focused in regions of enhanced flux, such as earthworm burrows or larger soil pores (macropore flow).
Slurry	A liquid mix of feces and urine produced by housed livestock combined with water during management, and usually incorporating some bedding material to give dry matter content of 1–10%.
Surface runoff	Flow generated from rainfall and other water sources that facilitates the transfer of contaminants across the soil surface due to saturation excess or infiltration excess conditions.
Transfer	A term used here to describe the movement of pollutants through soil-water systems.

Definition of the Subject

Human activity can place heavy stress on agricultural soils across the world. Soil systems are continually manipulated in order to support the increase in crop yields and accommodate more intensive livestock production and thus provide the planet's ever-growing population with a diverse array of ecosystem services, among which food production features highly. The recycling of livestock manures to land provides a sustainable solution to support the ecosystem services that soils provide and a host of benefits both in terms of improving soil structure and also soil fertility. However, livestock manures and feces may contain a high number of fecal microorganisms that pose a threat to human well-being and potentially large concentrations of nutrients harmful to the ecology of freshwater systems that the soils often buffer. To manage water quality in agricultural catchments value should be placed on adopting appropriate land, livestock, and manure management. To do this effectively requires a comprehensive understanding of the role of soil in facilitating contaminant transfer via a suite of hydrological pathways, and its ability to filter or absorb key pathogens or limiting nutrients during their passage through the soil network. Uncertainties in quantifying episodic transfers of contaminants through spatially and temporally dynamic pathways in soils at different scales ensures that multidisciplinary teams of scientists will not be short of challenges posed by soil complexity for decades to come.

Introduction

Soils are central to the functioning of agricultural systems and provide the planet's ever-growing population with a diverse array of ecosystem services, among which food production features highly. If managed effectively they can provide a host of benefits to society by, among other things, buffering flooding and hindering the transfer of numerous pollutants [1]. Conversely, if managed inappropriately the consequences can be far reaching, moving beyond the farm boundary to impact on local, regional and national health, livelihoods, food security, and downstream environmental quality. In terms of water quality, this can mean contamination of the ground and surface water supplies with the potential for chronic and acute impacts on human health [2, 3].

In the past nonpoint source pollution from agriculture has been synonymous with nutrient and sediment impacts on water quality. However, there is now a growing recognition that microbial pollution is a significant contributor to water quality impairment across the world [4]. Indeed, pathogen contamination of receiving waters has implications not only for water quality but also human health [5] with pathogens, or indicators of their presence, now topping the list of leading contaminants causing watercourse impairment across the USA [6]. Contamination of surface waters with microbial pollutants and nutrients is therefore potentially problematic both in an ecological and human health context. However, to put this in perspective, while the consequences of large-scale waterborne disease outbreaks linked to agriculture can be fatal, their occurrence is relatively rare. Nonetheless, coastal waters and often lakes and rivers too can be used for bathing, and their deterioration in water quality with fecally derived microbes from agricultural lands provides a key route of recreational exposure of potential pathogens to the human population. The reader is referred to Bioaccumulation/Biomagnifications in Food Chains for a substantial discussion surrounding further impacts of microbial transfer from land to water on food chain contamination. For nutrients, eutrophication is a clear consequence of overloading surface waters with supplies of nitrogen (N) and phosphorus (P). While being aesthetically displeasing, its toxic impacts on freshwater ecology include oxygen depletion leading to fish kills, and cyanobacterial blooms that can prove harmful to humans and animals (e.g., dogs and sheep) [7]. The degree to which eutrophication is considered a problem throughout the world depends on the place and people concerned, but it is an issue for a large number of water quality regulators across many nations.

Pathogen-contaminated groundwater has caused much waterborne disease worldwide [8]. Similarly, nitrate contamination of groundwater supplies is often linked to nonpoint source pollution in agricultural catchments, and it too can cause adverse effects on human health [9]. Waterborne outbreaks of any variety often have a large impact on society, but the actual disease burden attributed from such incidents in Europe is difficult to approximate and most likely underestimated [10]. An example of an outbreak of illness linked to microbial contamination of fresh water supplies is the Walkerton *Escherichia coli* O157:H7 Outbreak, Ontario, Canada in 2000 whereby over 2,500 people became ill through a drinking water well contaminated with farm runoff, and at least seven died directly as a result of drinking the contaminated water. The health implications (both anecdotal and case report data) of exposure to freshwater cyanobacteria are perhaps more contested and debated and have been reviewed recently by Stewart et al. [11]. Protection of surface and groundwater supplies from both nutrients and microbial contaminants thus requires a need to understand the differing transport behavior of these pollutants in agricultural systems, and critically this means understanding the role of soil in providing a potential buffer between agricultural activity and receiving waters. Thus, there is considerable value in appreciating that the quality of approaches to land management should be reflected in the quality of watercourses that drain catchments. This chapter considers the movement of pathogens through and across soil systems in agricultural settings and draws on nutrient studies to provide a comparative analysis of the differences in the importance of pathway functioning for different contaminant typologies, which ultimately is a key requirement for making measured assessments of the potential for pollution swapping linked to management shifts on-farm. The reader is referred to Microbial Risk Assessment of Pathogens in Water for a discussion of chemicals and pathogens in aquatic systems, and therefore this current chapter does not consider in-stream dynamics of nutrients or pathogens and their response to aquatic conditions when delivered into receiving waters. However, the movement of these biological and chemical contaminants through hillslopes and associated soil matrices is critical in linking on-farm activity to wider downstream impacts at both riverine and coastal environments, but also groundwater supplies too as identified in Fig. 15.1 (see Microbial Risk Assessment of Pathogens in Water).

Agriculture, Livestock, Manures, and Contaminants

Pasture that receives manure, through either land application in varying forms (e.g., solid, liquid) or via direct deposition by livestock, will accommodate some degree of risk for contributing fecal microorganisms and nutrients to watercourses, as will arable soils (though direct defecation is not an important input for these systems). Many jurisdictions mandate how livestock wastes are managed to protect adjacent water quality from microbial and chemical contaminants that pose an environmental and human health challenge [12]. The likelihood of microbial or nutrient loss from land to water may vary in relative magnitude from negligible through to very high risk potential. This will depend on the combination of a number of site-specific source, transfer and connectivity drivers (see Fig. 15.2) thought to moderate the risk of contaminant loss from land to water. Landscape features and management

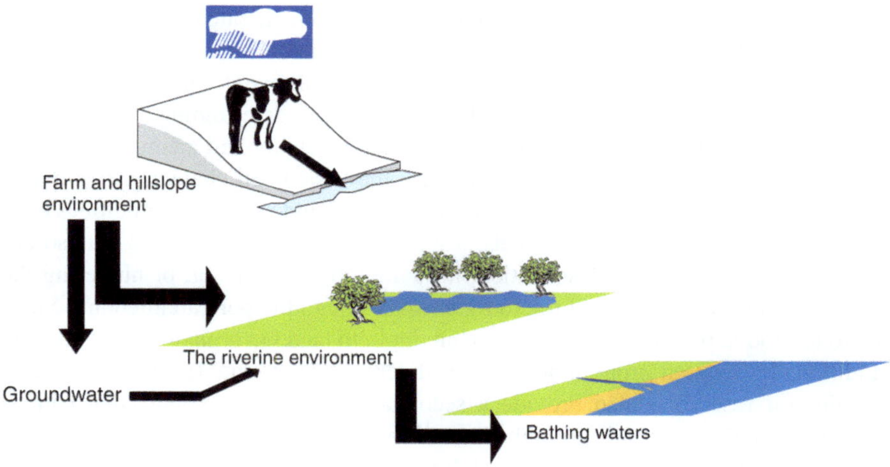

Fig. 15.1 "Farm gate to the bathing zone": diverse subsectors of catchments requiring investigation of nutrient and fecally derived microorganism fate and transfer

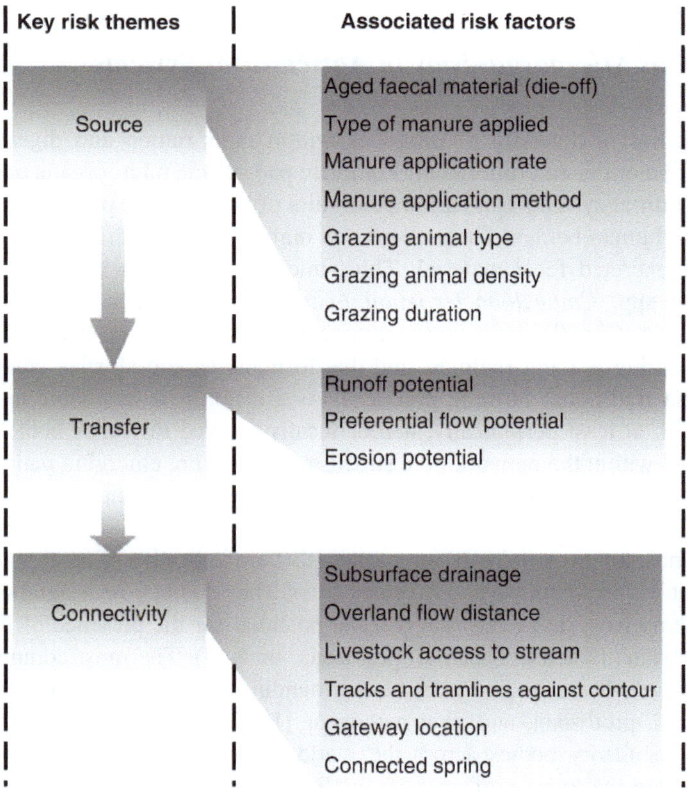

Fig. 15.2 Risk factors associated with source, transfer, and connectivity of pathogens and nutrients in the farm environment. The in-stream and in-river impacts are considered in Microbial Risk Assessment of Pathogens in Water

practices help determine this risk potential [13]. In effect, catchment contaminant dynamics depend on a complex interaction between spatial patterns of land use and management, soils, antecedent conditions, and rainfall/event characteristics [14]. For example, different soil types can play a large role in determining the magnitude of transfer of fecal microbes, delivered to land via applied manures, through different soil hydrological pathways. In turn, the varied soil characteristics linked to different soil types can impact on the survivability and viability of microbes once within the soil habitat (either as freely suspended entities, or associated with soil or organic matter). Soils are also critical in facilitating, enhancing, or hindering the transfer of nutrients and fecal microorganisms through the pore architecture. Similar to microbial pollutants, nutrients too are input to land via organic wastes, but are additionally sourced from inorganic fertilizers. Background levels of nutrients within soil reservoirs also provide a secondary source of nutrients that can be mobilized and transferred through the soil system, with a proportion ultimately delivered to receiving waters. Clearly, how land use patterns affect the transport, source, and relative longevity of pathogens (and nutrients) in the environment is a significant issue for policy formulation [15, 16].

Dangerous Microorganisms in Agricultural Systems

Livestock host a diversity of microbes within their rumen and digestive tract. A proportion of this microflora may comprise pathogenic microorganisms meaning that agricultural systems can harbor reservoirs of bacteria and protozoa potentially harmful to human beings. These reservoirs may be found both within animals but also their excreted fecal material. Thus, microorganisms such as *E. coli* O157, *Salmonella* spp., *Campylobacter jejuni*, *Listeria monocytogenes*, *Cryptosporidium parvum*, and *Giardia intestinalis* may contaminate pasture because of feces excreted during grazing regimes, and due to manures and slurries being applied to land via traditional farming practices. As a result, it is inevitable that the soil system will, at least periodically, harbor fecally derived microbes both at the soil surface and within the network pore structure [17]. Future emerging pathogens will likely arise from existing ones through adaptation and via unpredictable changes linked to biotic and abiotic mechanisms [18].

Fecal indicator bacteria (FIB) are bacteria that are indicative of fecal pollution and are present within all human and livestock feces. They do not pose a significant health hazard themselves; rather they suggest the potential for the presence of pathogenic microorganisms (whether bacterial, protozoan, or viral). The most commonly used FIB are *E. coli*. There are lively debates surrounding the validity of FIB as surrogates for bacterial, protozoan, and viral pathogens [19, 20], but currently FIB are used by many regulatory bodies across the world to monitor microbial water quality. In considering the transfer of microbial pollutants through and across soil there are a number of critical research questions that have driven the continued investigation of this research area. These are summarized in Fig. 15.3. Clearly there is a need

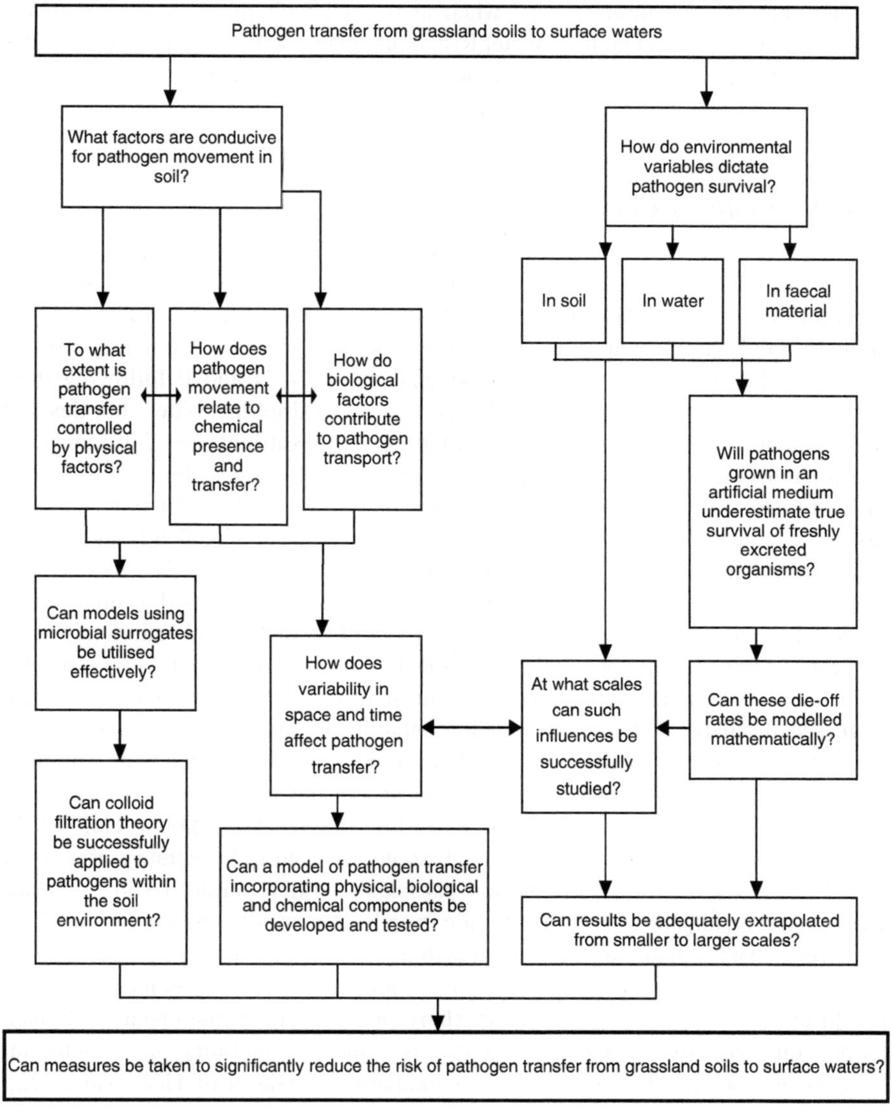

Fig. 15.3 A questioning framework of key research needs for pathogen transfer studies in the environment (Reproduced from Oliver et al. [17])

to account not only for microbial transfer, but simultaneously appreciate their survivability when excreted into an environment outside of the animal gut. If survivability is so unlikely under the environmental conditions typical of agricultural settings then the likelihood for mobilization and delivery to watercourses and downstream impact is much reduced. That said, many microbial pathogens and associated indicators are able to survive for lengthy periods outside of the gut, meaning that there is a key need to understand how they transfer in addition to how their population profiles may change through time when associated with different environmental matrices. The entry Recreational Water Risk: Pathogens and Fecal Indicators provides a comprehensive overview of monitoring approaches used in the detection of pathogens and FIB in the environment.

The Nutrients of Concern

The primary nutrients of concern considered in this chapter with regard to movement through soils are phosphorus (P) and nitrogen (N), specifically nitrate (NO_3^-). Both P and N are key limiting nutrients in aquatic systems. Contamination of surface and groundwater with anthropogenic inputs of N and P following their passage through and across soils risks the quality of drinking water supplies and may lead to excessive harmful algal blooms [21]. Nutrients are introduced as a comparative contaminant within this Chapter to highlight differences in terms of their transfer dynamics in relation to microbial transfers. The FIB and pathogens discussed within this chapter generally range in size from 1 to 5 μm and can be classified as biological "particles." In contrast, phosphorus can be divided into a particulate and soluble fraction (soluble in this instance being based on an artificial laboratory subdivision of 0.45 μm), and NO_3^- is highly soluble. This intuitively leads to clear differences in modes of transfer linked to physical mobilization by water for particulate contaminant typologies, in contrast to dilution effects associated with increased discharge volume for soluble contaminants and therefore suggests that there are differences in transfer linked to contaminant functional type. Another important property of P that can impact on its transfer dynamics is related to its high capacity for sorption to soil relative to NO_3^- and its association as colloidal P. Indeed, in soils NO_3^- remains free in soil solution because the nitrate anion is not adsorbed onto soil surfaces due to soil particles having a partial negative surface charge. As a result, NO_3^- is highly mobile in soils. In contrast, ammonium (NH_4^+) is positively charged and can therefore attach and associate with soil particles to be retained within the soil body. However, this chapter only provides an introduction to the differing properties of NO_3^- and P relative to microbial contaminants and much more comprehensive detail specifically on nutrient mobility in agricultural systems can be found within recent comprehensive reviews of Heathwaite [22], Edmeades [23], and Haygarth et al. [24].

Sources of Pathogens, Indicators, and Nutrients

The presence of pathogens, FIB and nutrients within the farm environment can be largely attributed to either point or nonpoint sources. Point sources are readily identifiable inputs that can be traced to a point of origin. They represent "end-of-pipe" locations and can include agricultural ditches, leaking septic tanks and manure stores, and farm hard standings. Nonpoint sources relate to inputs that occur over a large area and are attributed to land use. Nonpoint sources can include dung deposited to pasture by grazing animals [25, 26] and liquid and solid manures spread to land [27]. In addition, treated human wastes can be recycled back to land (biosolids) and still may contain a number of fecal microorganisms of concern. Untreated human waste (i.e., waste from septic tanks) may sometimes be spread back to land, though in the UK this is against regulatory policy. However, anecdotal evidence suggests this activity does take place, though published data on such practice are notably scarce. Wildlife can also serve as a source of microbial pollutants and indeed additional P and N inputs to land via their excretions. However, the significance and quantification of the wildlife source component is an often overlooked contributor to pasture-based budgets [28].

The soil matrix can sustain fecal microorganism survival, especially if microbes are incorporated in association with fecal material. Indeed some studies have detailed bacterial cell survival in excess of several months [29]. This can be due to protective microsites and protection from desiccation and UV radiation. Clearly then, it is not only fresh additions of manures to land that may be capable of impacting the microbial quality of watercourses, and aged fecal remnants can still contribute bacterial numbers to runoff following rainfall [30]. Slurries, solid manures, and feces all harbor different numbers of fecal microbes, and all impact on differential fecal microbe die-off patterns [31]. Furthermore, the consistency of each manure type (e.g., dry matter content) allows for differing degrees of mobility of fecal microbes accommodated within these manure source locations [32].

Nutrient sources in agricultural systems are somewhat different in that they do not depend on a fecal store to convey signal strength in receiving waters. For example, natural sources of N and P exist in the environment, and for N there are atmospheric sources too. The soil can therefore serve as a reservoir for N and P, meaning that even without continued anthropogenic inputs to land there will still be considerable losses of NO_3^- and P from land to water following rainfall events. This is a critical difference between the fecally derived bacterial and protozoan contaminant typologies and nutrient contaminant typologies. While microbial contaminants can persist for lengthy periods, there are not significant soil stores of fecal pathogens and their indicators *independent* of manure applications or livestock activity, though some studies report the potential for naturalized *E. coli* populations in the soil environment [33–35]. In contrast, liver fluke and intestinal worms and similar livestock pathogens do have significant soil stores that are important for their cycling, but they are not considered in this chapter under the definition of pathogens (e.g., fecal bacteria or fecal protozoa). Therefore, lag time between cessation of contaminant inputs to land and a measured reduction of pollutant losses at the edge-of-field scale varies by the pollutant type and depends strongly on the behavior of the pollution source [36]. Circumstances where farming practices have led to excessive soil P levels can be particularly problematic because even if nutrient management leads to a reduction of P inputs to levels below crop removal rates, the timescale needed to exhaust the P from the soil to the point where dissolved P in runoff is effectively reduced may well be years to decades [36]. Of course, soil with a high P index (high P content) does not always result in high

P transport as this is dependent on the way that the P is "locked" into the soil and its likelihood for subsequent mobilization.

Manure and slurries are applied to land via a range of techniques and equipment. Broadcast-applied manures (i.e., those spread to land using a splash plate and distributed onto the surface of blades of grass) are likely to be considered more risky in terms of contaminant loss than injected or plowed manures because of the opportunity for incidental contaminant losses from land to water [37]. However, broadcast applied slurry is likely to lead to a more rapid destruction of associated bacteria through UV radiation and desiccation [38]. Livestock grazing is equally, if not more important than manure applications in loading the landscape with fecal microbes [39]. The key difference between land applied manures and livestock excretions is the resulting spatial patterning of contaminants across the landscape. Similarly for nutrients, in this case NO_3^-, McGechan and Topp [40] have found higher levels of NO_3^- pollution in tile drains following grazing compared to fields receiving slurry and have suggested that high levels of NO_3^- pollution could be attributed to various factors, including the fact that cows tend to congregate in certain areas of a field at a localized stocking rate much higher than the overall stocking rate, and due to deposition of N at times when grass cannot utilize it as a plant nutrient. Animal type is a critical factor to consider with regard to the source strength of contaminants in farmed environments. This is because age and variety of livestock governs the number of organisms per gram of feces excreted onto pasture (e.g., lambs shed higher numbers of *E. coli* than beef cattle) but also dictates the daily excretion rates of feces (and therefore nutrients and microbes contained within) to farmed land and impacts on the differential release of FIB from various types of livestock feces at different times of the year. Logically, grazing density bears an impact on land-based loading of fecal microorganisms and nutrients too, and it has been shown that the concentration of FIB in streams in sub-catchments with high stocking densities can be four to eight times higher compared to sub-catchments accommodating low stocking densities [41], and grazing duration is potentially important for both nutrient and microbial contaminants based on the accumulating reservoirs of fecal excretion to land through time.

The Transfer of Pathogens and Nutrients Through and Across Soils

The transfer of water across and through the soil environment, and contaminants dissolved or entrained within, is subject to a series of spatial and temporal controls that dictate this potential for transfer and which vary both spatially and temporally [42]. Temporal controls on microbial and nutrient transfers include meteorological inputs that provide an energy source and driving force to initiate transfer processes. Additionally, hydrological pathways are susceptible to change through time, for example, pore networks can collapse, drain pathways can silt up, and extreme events can remove existing pathways and reconfigure hydrological conduits. The significance of a particular pathway in facilitating microbial and chemical transfers from land to receiving waters is controlled by properties of the pollutant of concern and also the form of the contaminant, for example, whether it be particulate, particle associated, or soluble (i.e., the contaminant functional type) [43]. Climate and weather too are key dictating factors controlling the magnitude of transferred loads. Whether considering individual storm events or interannual variability, these climatic drivers can have a significant effect on the magnitude of multiple pollutants over many timescales. Spatial factors impacting on transfer include the particular soil type that microbes or nutrients are negotiating their passage through or across. Soil types dictate hydrological pathways, some being more likely to hinder transfer via filtration, whereas others may be more likely to permit rapid transfers via large cracks or pores in the soil (macropore flow). Research has demonstrated the importance of water flow velocity and soil particle size distributions, for example, on *E. coli* transport through soil [44].

For transfers across the soil surface, it would appear logical to assume that slope gradient and slope shape can be useful indicators for transfer potential, and many models accommodate a slope term as a control over particle transfer, with increased transfer associated with increased slopes. However, some studies suggest the need for caution when dealing with what appears to be a relatively straightforward assumption. Such relationships tend to be based on data collected over a wide range of slopes and using relatively small soil flumes. Recent work by Armstrong and Quinton [45] used laboratory rainfall simulation on a large soil flume to investigate interrill soil erosion of a silt loam under a rainfall intensity of 47 mm h^{-1} on 3%, 6%, and 9% slopes. Results from replicate experiments showed wide variations in runoff and sediment concentration that were explained by the complexities in interrill soil erosion processes. Critically, the data also demonstrated that at low slope (on arable soils) processes related to surface area connectivity, soil saturation, flow patterns, and water depth may dominate over those related to gravity, and Armstrong and Quinton [45] therefore query the validity of risk assessments and soil erosion models with a dominant slope term when assessing soil erosion from agricultural land at low slopes. However, it is important to acknowledge the scale at which this reported study was conducted, and caveats should be applied in upscaling such laboratory-based findings to field situations. For more topographically diverse landscapes, alternative approaches use landscape wetness as a metric of delivery potential by integrating the small-scale spatial variation in runoff generation and probability of hydrological connection linked to topography [46, 47].

Physical, geochemical, and biological processes can be used to classify the key transport mechanisms responsible for microorganism and nutrient movement within soils. The physical processes include advection, whereby pathogens or FIB and nutrients are carried in bulk water and move according to the water velocity, and dispersion, which can involve the spreading of microorganisms and nutrients as they move along the water path. Geochemical processes act to delay microbial and particulate transfer through the soil matrix and consist of filtration, sorption, and sedimentation mechanisms [48, 49]. Finally, biological processes, such as growth and chemotactic responses, may influence microbial transfer through the soil habitat, albeit to a lesser extent [50, 51]. This ability to self-propel through the soil system because of their biological attributes (e.g., pili, fimbriae) differentiates microbes from their nutrient counterparts. While at the pore scale this transfer mechanism is important, in the grander scheme of the catchment or hillslope, the role of biological transfer is likely to be minimal in contrast to the physical mobilization facilitated by rainfall and the resulting water flows.

The transfer of agriculturally sourced contaminants from soils to ground and surface waters is largely driven by rainfall and resulting surface and subsurface runoff. Conceptually, transfers can be classified as low energy or high energy. For example, slow flow microbial and nutrient transfers may operate between storm events and are thought to be associated with the steady percolation of precipitation inputs through the soil profile. This contrasts with overland flows and bypass flows resulting from high-energy precipitation (or storm) events, which enable the physical movement of soils, manures, and potential pathogens into streams, creating a more rapid and direct transfer route. However, it is interesting to draw attention to findings from Sharpley et al.'s [52] recent study that showed that small storms were more important than large storms in delivering P to watercourses over a 10-year research period. These two contrasting energy levels of transfer will be discussed in later sections of this chapter and are illustrated in Fig. 15.4. Nitrate (and some forms of P), being soluble, follow a different trend than that of microorganisms and particulate P, and instead can be diluted under heavy water flow [53]. Table 15.1 is provided as a point of reference to direct the reader to scale-appropriate studies of the transfer of a variety of agricultural contaminants through soil systems. Current understanding of the transfer capability of different hydrological pathways in space and time is growing for microbial pollutants, but this remains a complex and underexploited interdisciplinary research topic necessitating the collaboration of soil hydrologists and environmental microbiologists. While understanding the spatial and temporal intricacies of environmental transfer of fecal microorganisms lags some way behind current knowledge surrounding N and P dynamics in agricultural systems, that does not mean the knowledge of P and N dynamics is satisfactory. There remains for these nutrients need to further the understanding of the relative importance of different hydrological pathways in providing *delivery* (in quantifiable terms) from land to water [95]. Some have argued for the critical development of appropriate methods to quantify the delivery process and thus new monitoring tools capable of providing a framework for understanding contaminant transfer and delivery at a range of scales in agricultural catchments [88]. Deasy et al. [88] highlighted the dominant role of a field drain in transferring P from land to water and argued that overland flow inputs, despite being directly connected to the stream and containing higher P concentrations, contributed less to the stream P flux. Such data are key reminders that the obvious and visible routes of transfer do not always equate into dominant and critical flow pathways upon which mitigation efforts must be focused to protect our water resources. The P transfer continuum concept has been outlined by Haygarth et al. [96] as a framework within which the factors that

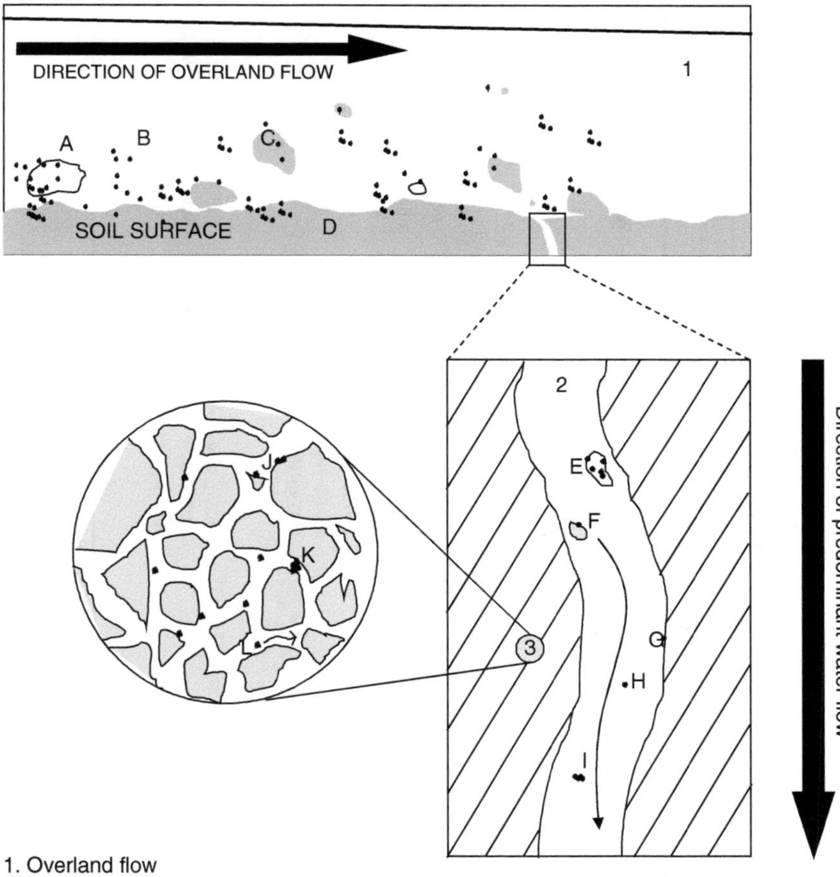

Fig. 15.4 Natural transfer pathways available to fecally derived microorganisms applied to soil surfaces (Reproduced from Oliver et al. [17])

contribute to nonpoint P fluxes from catchments can be operationally divided into *source*, *mobilization*, and *delivery*. These divisions are a convenient way of grouping processes that determine P behavior and accommodate transferability in terms of conceptualizing other contaminants in agricultural systems (cf. Fig. 15.2).

Table 15.1 A "look-up" table of research studies investigating microbial and nutrient transfer through and across soils at a range of different scales

	Contaminant type			
Scale of study	Bacterial	Protozoan	P	N
Laboratory soil columns/boxes	Artz et al. [54] Donnison and Ross [58] Garbrecht et al. [44] Horswell et al. [64] Rosa et al. [65]	Mawdsley et al. [55] Boyer et al. [59] Harter et al. [62]	Matula [56] Tarkalson and Leytem [60] Brock et al. [63]	Entry et al. [57] Miller et al. [61]
Vertical Lysimeters	Bech et al. [66] Brennan et al. [34] Guzman et al. [69]		Turner and Haygarth [67]	Di et al. [68]
Soil plot : small	Muirhead et al. [70] Abu Ashour and Lee [72] Soupir et al. [75] Mishra et al. [77] Thiagarajan et al. [79] Oliver et al. [82] Kouznetsov et al. [84] Collins et al. [86]	Ferguson et al. [71] Ramirez et al. [80]	Preedy et al. [37] Tunney et al. [73] Quinton et al. [74] Withers et al. [76] Heathwaite et al. [78] Haygarth et al. [81]	Alfaro et al. [83] Olson et al. [85]
Hillslope Sub-catchment - Catchment	Kay et al., [87] Close et al. [90] Davies-Colley et al. [93] McKergow and Davies-Colley [14]	Keeley and Faulconer, [15]	Deasy et al. [42, 88] Heathwaite and Johnes [91] Rothwell et al. [94]	Botter et al. [89] Dougherty et al. [92] Heathwaite and Johnes [91]

Soil Characteristics Impacting on Transfer

Rainfall and resulting runoff and drainage from farmed land is critical for the transfer of agricultural pollutants such as phosphorus, NO_3^- and sediments from land to water (e.g., [37, 97]), and there are a series of key soil characteristics that impact on the transfer of pathogens and FIB through lysimeters, hillslopes, and catchments too. Soil water content, soil structure, and soil texture are among those that shall be discussed within this chapter. The physiochemical properties of soils and the resulting interactions with potential pollutants means that the navigation of pathogens and nutrients through the soil architecture is extremely complex, and particular attention will be given to these rapid versus slow transfers as dictated by soil properties in the following sections of the chapter. Particulate-type contaminants experience more rapid travel times in coarser textured soils with larger pore spaces as opposed to finer textured soils, with the soil matrix operating as a filtration system. In soils where matrix flow dominates it is therefore generally accepted that water-induced particle transport is strongly correlated with particle size.

In addition to the physical make-up of soil and the associated network of conduits, pores, and microhabitats, it is pertinent to appreciate the variability in sorption properties attributed to different soil types. This property of soil and associated colloidal material can have a marked influence on microbial and P transfer [98, 99]. The major soil components affecting sorption of bacteria and P are clay and organic matter, and particle and colloid facilitated TP delivery from soils to water via different hydrological pathways has been shown to be dominated by the transfer of TP associated with clay and colloidal fractions [81]. For *Cryptosporidium*, hydrophobicity and zeta potential have been highlighted to exert a significant influence in the adhesion mechanisms of the oocysts [100], and more recently the presence of manure in solution has been shown to enhance the extent of adhesion of these protozoa to soil particles. However, the role of manure in facilitating attachment of *Cryptosporidium* to soil particles is complex, with indications that an optimal concentration of a "facilitating" component of the manure exists (at somewhere between 0% and 1.0%) [101]. Clearly, the sorption of microorganisms and nutrients to soil surfaces cannot be attributed to a single factor; instead it is important to appreciate an array of forces interacting to govern microbial retention. Table 15.2 provides a definition of the common mechanisms through which FIB and bacterial pathogens associate with soil particles.

From a conventional soil science perspective the concept of sorption is normally applied to chemical contaminants and the derivation of their associated adsorption isotherm. Elements of sorption theory can be considered transferable for the determination of bacterial association with soil. Adsorption is an important process that partly governs the aqueous concentration of contaminants in soil environments, and therefore their mobility. The common approach is to use a batch equilibration method to obtain adsorption coefficients. This is based on the assumption that partitioning of the contaminant is a thermodynamically motivated distribution between two readily separable, homogeneous phases [102]. Many other studies

Table 15.2 Summary table of mechanisms of bacterial sorption/attachment to soil particles

Potential mechanism of sorption/attachment	Description
Van der Waal forces of attraction	Cells overcome electrostatic repulsion and are held at a finite distance from a particle surface
Charge attraction of opposite signs	Metal oxide coatings on soil particles confer a positive charge to the particle surface and this results in a much tighter adhesion of the cell to the surface
Hydrophobic forces	Hydrophobic bacteria have the tendency to adsorb to surfaces in water due to repulsion from the polar water molecule
Positive chemotaxis (cellular mobility)	Higher nutrient content of a particle surface may cause motile bacteria to move toward surface as a result of nutrient gradient
Irreversible anchoring via pili and fimbriae	Biological mechanism of attachment where contact is made between cell and surface via hair-like appendages. The presence of a localized positive charge at the tip of fimbriae can aid attachment to a negatively charged soil particle
Charge fluctuations	Heterogeneities in surface charge provide locally favorable regions for attachment

have investigated the adsorption of a broad range of potential contaminants, including heavy metals, pesticides, herbicides, and fertilizers.

The adsorption isotherms of nutrients such as P often fit models such as those described by the single and double Langmuir, Freundlich, and Tempkin equations whose profiles suggest that a maximum loading of P with the soil occurs, after which no more sorbate may be adsorbed. The linear isotherm is essentially a special case of the Freundlich isotherm, and this suggests that no point is reached whereby the solid fraction has had all its available sorption sites exhausted. Oliver [103] observed a linear isotherm when investigating the affinity of *E. coli* to soil particles suggesting that sites for cell interactions remain available. However, the linearity of the data may in fact reflect multilayer cellular associations with particles whereby cell consortia have associated with a contact point of the particle. Koopmans et al. [104] discuss isotherm linearity with respect to P adsorption and propose that a linear profile is often either: (1) a function of a narrow range of concentrations being tested; or (2) linked to the use of a wide soil to solution ratio. In the *E. coli* isotherm described by Oliver [103] neither of the two factors discussed by Koopmans et al. [104] were thought to apply to the data because the range of concentrations used was large in terms of a realistic minima and maxima, and an appropriate soil to water ratio was determined in one of the preliminary stages of designing the isotherm experiment. Of course, if the concentration of cells was increased further then the *E. coli* isotherm may reach a plateau at a potential retention maximum, but the applicability of such results to field situations is likely to be limited. While some studies have determined that over 99% of cells adsorb to soil [105], the numbers do not necessarily equate to real-world observations whereby a greater proportion than the remaining 1% of total applied cells loaded onto the soil surface via manure and feces may emerge via runoff and subsurface drainage. The reasons for this are numerous. For starters, hydrological factors play a role at larger scales and are unaccounted for in laboratory batch-scale experiments. Essentially, water flux may prevent cell interaction with the soil because preferential and rapid overland flow pathways can allow bacteria to bypass the soil matrix and those soil horizons that adsorb contaminants strongest. Additionally, bacteria entering the soil system under field conditions are not necessarily free-living cells as introduced to the soil-water system under experimental conditions. Fecal bacteria may already be associated with organic matter derived from excrement, and this may impact on soil–cell interactions. The extent to which laboratory findings can be extrapolated to the environment is, therefore, a little unclear. However, such comparative studies can provide a mechanistic understanding of an important process and as long as their limitations are acknowledged such experiments are important tools in the development of the understanding of soil-contaminant interactions.

Colloid and Particle Associated Transfer

In addition to retention of microbial and nutrient contaminants within the soil architecture there also exist the potential for these contaminants to associate with colloidal and particulate soil material, which may subsequently provide a vehicle for contaminant transfer through the pore network. There are two particularly important properties associated with colloids that enable them to function as important contaminant carriers. The first is that colloids have a very large specific surface area, in excess of $10\ m^2\ g^{-1}$ [106]. Second, these colloids remain stable in suspension for significant periods, and if bacteria attach or nutrients such as P adsorb to the large available surface area, their dispersal through the soil may be aided considerably. Those colloids that remain more in the center stream of the flow path are likely to remain uncaptured and thus migrate along these faster, more permeable flow paths.

The strong positive relationships between suspended sediments and P in drainage waters reported throughout the literature demonstrate the clear linkage between this nutrient and soil particles. Much work has considered colloidal P transfer through agricultural systems, and it is an accepted mechanism by which a significant fraction of P loss from agricultural land can occur. Colloidal material falling within the notional "dissolved" fraction (i.e., $<0.45\ \mu m$) has thus been shown to serve as an important vehicle for the transfer of P from soils [78]. In fact, a whole range of particle and colloid size fractions have been shown to function as P carriers with linear regression highlighting strong significant relationships ($r^2 = 0.86$) between water extractable supernatant turbidity and colloidal P release across particle sizes ranging between 2 and 0.0003 μm for a suite of different soil types [78]. The association of P with these colloidal fractions allows for a relatively rapid transfer of sorbed material because of the exclusion of colloidal size fractions from soil micropores and restricts their passage to the more rapid flow routes via macropores.

Association of FIB with soil particles has been discussed too (e.g., [72, 107–109]), though there remains a large degree of uncertainty regarding “vehicles” for cell transfer under environmental settings and a clear need for continued research to consolidate understanding of processes governing cell partitioning with different sedimentary fractions and the importance of such interactions in freshwater systems [5]. Importantly, the size of *E. coli* itself is generally regarded as being approximately 2 μm in length and 0.5 μm wide. In one of the few studies examining the preferential attachment of *E. coli* to different particle size fractions of an agricultural grassland soil [108] it was shown that 35% of introduced *E. coli* cells were associated with soil particulates >2 μm diameter. Of this 35%, most of the *E. coli* (14%) were found to be associated with the size fraction 15–4 μm. This was attributed to the larger number of particles within this size range and its consequently greater surface area available for attachment. When results were *normalized* with respect to estimates of the surface area available for bacterial cell attachment to each size fraction, it was found that *E. coli* preferentially attached to those soil particles within the size range 30–16 μm. For soil particles >2 μm, *E. coli* showed at least 3.9 times more preference to associate with the 30–16 μm than any other fraction apart, from the <2 μm grouping. Undoubtedly the range of methods used to arrive at estimates of bacterial attachment to suspended sediments can in many respects be responsible for the discrepancies observed in different studies [110]. Within emerging drainage water, a measure of turbidity should be correlated with FIB concentrations if the majority of cells are truly associated with suspended solids and, in addition, such association is unaffected by other water quality parameters. However, both strong and very weak relationships have been observed by various authors between *E. coli* concentrations and turbidity, and between *E. coli* concentrations and total suspended solids concentrations [110]. This is perhaps not too surprising given that the microbial consortia being measured reflect a combination of life-cycle stages. For example, turbid water can often be found draining pasture that has not been grazed for several months. The length of time between the removal of livestock and the collection of runoff water can mean that the majority of FIB may have perished as a result of UV radiation and desiccation and so the relationship between turbidity and FIB concentrations is weak. In contrast, a storm event that generates highly turbid runoff from pasture grazed by a large number of cattle will likely result in a strong relationship between the suspended particle concentration and FIB numbers because the population of FIB present are abundant within freshly deposited material. Tracking the actual emergence of FIB, pathogens, and turbidity peaks with storm hydrographs provides one interesting means of analyzing the strength of association of microbial contaminants with the soil fraction. Such observation as reported by Oliver et al. [103] identified a similar but delayed emergence pattern of *E. coli* and turbidity peaks in subsurface runoff via artificial drainage from grassland plots. Thus, the peaks for turbidity, flow, and *E. coli* concentration were not identical, but occurred in respective order as a function of increasing time. The results suggested that *E. coli* may have been associated with hydrologically energized soil particles of a particular size fraction.

Rapid Pathways of Contaminant Transfer

Intensity of water flow through soils is undisputedly one of the most critical factors responsible for driving the transfer of fecal microbes and nutrients from land to receiving waterbodies. However, when entering receiving waters that may have a high discharge there is the potential for high dilution capacity. Surface runoff processes that generate overland flow are often perceived to be the dominant transfer pathway for pathogens and microbial contaminants. The growing evidence base detailing pathogen and FIB transfers through agricultural systems is beginning to add weight to such hypotheses. If antecedent soil conditions are conducive to generate overland flow and heavy rains occur shortly after slurry application or an intensive grazing season, then there is potential for significant runoff of fecal microorganisms and particulate P following their entrainment into a variety of flow pathways. Overland flow that directly connects with a watercourse has been shown to deliver substantial loads of FIB to the stream network [86], though the numbers may be diluted if entering a receiving water with high discharge. It is likely that a greater *uninterrupted* overland flow distance and contributing area will enhance the potential for accumulated carriage and delivery of FIB to the watercourse. However, there is little proof that overland flow, once started, actually delivers contaminants to streams and primary water systems in a single event. It may instead provide a pulsing mechanism of transfer or be interrupted by buffers before having impact on receiving waters. Alternatively, particulate contaminants (e.g., cells or particulate P) that are transferred via the surface runoff pathway may be deposited and resuspended numerous times before being finally delivered to a watercourse. The role of high rainfall and storm events and the resulting flow signatures in dictating microbial transfer from soil to water has been documented by a growing number of studies. Recent examples include: McKergow and Davies-Colley [14], Sinclair et al. [111], Wilkes et al. [20], Davies-Colley et al. [93], and Oliver et al. [82]. Similarly for P and N a large amount of literature is available on the response of these nutrients to hydrological drivers (e.g., [53, 78, 89, 112]). Some recent studies on P delivery from land to water have however challenged prior assumptions of the importance of large storm events [52, 113]. Sharlpley et al. [88] compared the surface runoff contributing area and stream flow and total P response for 248 storms over a 10-year period from 1997 through to 2006 and found that 93% of storm flows had a return period of <1 year and delivered 63% of the flow and 47% of the total P load. Similarly Jordan et al. [113] were able to demonstrate the importance of high resolution sampling frequency on detecting P delivery to water from diffuse sources independent of storm events and others have recognized the need to adopt appropriate time intervals and lengthy data records to observe useful diffuse pollution trends across scales [22].

Coarser textured soils tend to allow for rapid vertical transfer of NO_3^- relative to fine textured soils, though it must be acknowledged that rapid NO_3^- transfers can occur within finer textured soils via cracks or fissures, particularly if NO_3^- is heavily loaded onto the soil surface. Overall, NO_3^- losses via rapid overland flow pathways are relatively minor compared with NO_3^- lost via leaching and drainage. For a soluble contaminant such as NO_3^- it is interesting to observe response trends over both seasonal and storm-event timescales. For example, NO_3^- emergence generally reduces during the growing season and increases during the winter owing to uptake from plants during the summer. However, superimposed on this seasonal pattern of increasing and decreasing concentrations are dilution effects attributed to storm events whereby heavy rainfall effectively causes a reduction in concentration through increased water volume. Downscaling from seasonal timeframes to individual storm-events, the relationship between flow and NO_3^- concentration is in fact much more complex and not a straightforward dilution for every rain event.

For microbial transfer, Tyrrel and Quinton [114] summarize overland flow transport scenarios as (1) incorporation of free microbes into overland flow; (2) mobilization of soil or waste particles into overland flow carrying attached microbes; and (3) detachment of microbes from soil surfaces arising from shearing forces of raindrop or flow action. However, there remains much work to be done in quantifying the efficiency of overland flow in facilitating the wash-in of fecal material from pasture to stream. The complexity of predicting contaminant delivery to water courses cannot be understated, and while wash-in of fecal matter has long been recognized as a consequence of overland flows generated within the contributing areas of a catchment [115], within large and complex watersheds the bacteriological and chemical quality of a stream is the resultant effect of a variety of indistinguishable sources, and so determination of the loading capability of a particular transfer route is difficult. Poaching and pugging, which are terms used to refer to the compaction and breakup of soil due to trampling by livestock, can lead to a lower soil infiltration rate [116] and can enhance the initiation of overland flow. Poaching and pugging are therefore of critical importance in localized areas and hence may be more important at such scales within surface water dominated catchments compared to groundwater dominated catchments, where their effects are likely to be more hydrologically isolated [117]. Despite the apparent complexity in understanding contaminant transfers through soil systems at catchment scales some studies suggest that complex behavior patterns can be reduced to surprisingly low variability in model outputs [118].

Rapid routes of water transfer other than overland flow do exist and include macropores and artificial drainage systems, and a suite of rapid flow pathways are illustrated in Fig. 15.5. Such bypass routes within the soil matrix facilitate a rapid transfer of water and the soluble and particulate contaminants carried within. The significance of large pores and voids in facilitating the movement of water and colloidal material through the soil has long been acknowledged [119–121]. Macropores may be formed naturally or through soil fauna activity, plant root presence, or soil shrinkage. Preferential flow in soil has both environmental and human health implications since it favors contaminant transport to groundwater without interaction with the chemically and biologically reactive upper layer of soil [122]. The interconnected pore network therefore allows for a slow transfer of particulate-type contaminants via infiltration into the soil matrix coupled with a rapid transfer of microbes and particles (and solutes) within larger sized pores

Fig. 15.5 Rapid transfer pathways across and through soils: (**a**) overland flow conduits across a clay loam soil; (**b**) soil cracking on exposed grassland soil surface for a clay loam; (**c**) drain flow exported from mole and tile drains on clay loam grassland plots; (**d**) creation of macropores within soil by earthworms

capable of increased soil water velocities. In the absence of macropores the distribution of microbes may be mostly confined to the uppermost zones of the soil profile [54]. However, for FIB, after a dry spell, the rapid entry of cells into soil following rainfall does not guarantee rapid downward transport because of the potential for cellular retention in the upper dry topsoil aggregates [123]. Rather, as soils wet up over time the interaction with wet soil particles reduces and cells are able to transfer more freely. Thus, Guber et al. [123] showed that FIB were able to associate with dry soil aggregates in increased number compared to wet aggregates, leading the authors to propose an interesting interplay between soil water content prior to a rainfall event and the subsequent transfer of manure-borne bacteria with infiltrating rainfall. Similarly it has been shown that fecal bacteria sticking efficiency to quartz sand particles decreases with distance traveled [124].

Although macropores often make up only a small volume of the soil body they can serve as vertical and lateral routes of relatively rapid water flow and allow microorganisms, among other colloids and contaminants, to successfully bypass the sieving and constraining architecture of the soil matrix (see Fig. 15.4). Attempts have been made to develop relationships between soil type classes in New Zealand and breakthrough curves [125] and to therefore generate datasets of regionalized potential for microbial bypass flow based on soil classifications that become useful for larger scale modeling scenarios. The role of continuous macropores within silt loam and loam soils has been well documented by Abu-Ashour et al. [120]. Their experimental protocol was designed to create artificial macropores (vertical and straight) through sieved and packed soil. A comparison was made between columns (175 mm high × 89 mm diameter) accommodating artificial macropores (and columns without) in their ability to transfer a slug of cell suspension to depth. Irrigation was applied 24 h after inoculation at a rate of 60 mL h^{-1} for 2 h. For all packed soils without macropores no biotracer was ever detected in the effluent, regardless of initial water content, rainfall rate, or soil type. The study concluded that initial soil moisture appeared to have a notable effect on bacterial movement through soils especially in the presence of a macropore. The effect of the macropore was not substantial when the soil was dry. Aislabie et al. [126] compared four contrasting soil types for macroporous transfer of bacteria and showed the importance of macropore flow in breakthrough curves of clayey soils. This complements the studies of Paterson et al. [127] and Mawdsley et al. [65], who also demonstrated greater recoveries of introduced microbes in leachates from clay loam compared with loamy sand cores. The study of McLeod et al. [128] observed FIB numbers emerging from saturated intact columns (500 × 500 mm) of 2 contrasting soil types following dairy shed effluent application and simulated rainfall (5 mm h^{-1}). The patterns of microbial emergence were indicative of bypass flow and led the authors to stress the need to consider the nature of soil hydraulics in addition to conventional theory of direct entry into mole drains. Critically, studies reporting on the role of macropore flow have provided strong evidence to suggest that very different conclusions can be reached regarding the efficiency of soils as bacterial filters (or particulate filters in general) depending on whether soil cores used within experiments are disturbed or intact.

The action of disrupting macropore continuity can therefore prove effective in limiting contaminant transfer via these rapid flow conduits. For example, on arable soils there can be increased water transfer within no till soils relative to tilled soils. This is thought to be a direct influence of the increased macropore continuity within the no till soils. The action of tillage physically disrupts the soil structure and significantly reduces the extent of macropore connectivity within the soil system. Such a finding has been shown to hold true for reducing *Cryptosporidium* transfer to tile drains [80]. Tillage therefore represents a useful management approach for limiting rapid translocation of microbial pollutants (and particulate associated P) in arable soils and indicates that alternative methods of disrupting the soil structure on a no-till soil would be beneficial for reducing pathogen and FIB loss from land to water. This provides a useful management option for arable land, but for livestock systems and grazed grasslands such approaches are obviously not possible without destruction of pasture.

The presence of subsurface drains (e.g., mole and tile drains) effectively increases subsurface connectivity provided that the age and condition of the features is such that the pathway still functions efficiently as a hydrological conduit. Field drains have been reported to be a rapid route of nutrient export from

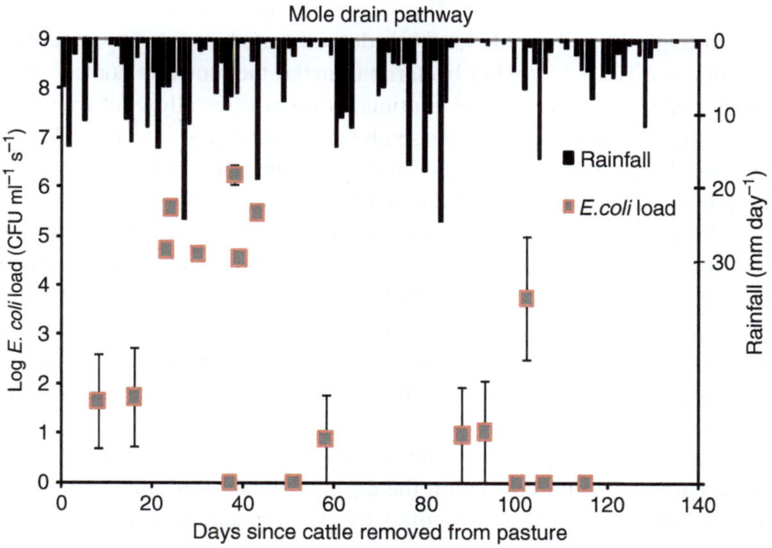

Fig. 15.6 *Escherichia coli* loads exported from mole and tile drains of grassland plots in relation to daily rainfall. X-axis represents day of sampling since cattle removed. Error bars represent 1 S. E. of logarithmic mean (Modified from Oliver et al. [82])

agricultural land [129] and have also been shown to export *E. coli* from land to water during storm events at similar loads to that of undrained fields using replicated 1 ha plots [82]. The use of natural fluorescence as a tracer has also demonstrated the importance of drain flow in contributing toward nonpoint pollution impacts [130]. Figure 15.6 shows the relationship between *E. coli* load exported via a mole drain pathway on replicated 1 ha grassland plots (clay loam) in response to rainfall events following the removal of livestock (in this case four beef steers) at the end of a 6-month grazing season. So while the installation of mole and tile drains to lower the water table may be seen as advantageous in limiting the load of potential pollutants transferred rapidly by surface runoff or near surface flow pathways it must also be appreciated that contaminants may be re-routed via different subsurface pathways [82]. Figure 15.6 shows clearly that significant loads of *E. coli* can be exported from drained plots, even after livestock have been removed, and that these bacteria must have transferred through the soil profile, probably via fissures and cracks in the clay soil to reach drain pathways.

Slow Pathways of Contaminant Transfer

Much research has focused on the infiltration and vertical transfer of pathogens, FIB, and nutrients in soil leachate, and for particulate contaminants there has been considerable assessment of FIB and pathogen movement in relation to colloid filtration theory (e.g., [49, 67, 130]). Vertical displacement of microbes and nutrients through the soil profile has been demonstrated in a variety of soil column experiments (see Table 15.1). For bacterial transfer, the moisture content of the soil determines cell movement because continuous water films allow for transfer that is essentially limited to the aqueous phase and the solid-liquid interface. Thus, it has been proposed that appreciable bacterial movement in soil can only occur if there are enough water filled pores of the required diameter to enable a continuous pathway [131]. Nitrate is undoubtedly the main form of N leached from agricultural systems, and greater concentrations will be detected in leachate from soils where N inputs on the land surface produce NO_3^- in excess of amounts required by plant uptake. The soluble and highly mobile nature of NO_3^- means that those soils in climatic zones typical of wet weather regimes will deliver potentially high concentrations of NO_3^-, though because of dilution the concentrations detected will be less.

Under conditions that promote water movement through soil pores, the carriage and translocation of microbes within the drainage water through the soil structure will depend on sieving effects imposed by pore openings. Under conditions of limited pore clogging, bacterial transfer is possible to considerable depths below the soil surface layers following the initial application of a surface applied manure source [132]. However, the transfer of microbes and particulates through the soil architecture can lead to pore clogging that may subsequently restrict such contaminant transfer by the physical blocking of a pore entrance. Bioclogging of the pore network is clearly a function of the particle size of the porous medium and the diameter of the microbial consortia that transfer through the soil system. Bacteria, protozoa, and particulate phosphorus suspended in flow can be effectively strained by the soil matrix and then accumulate within soil passages when pore openings are too small to permit their continued transfer. These pores then become immobile regions that may exist in the filter matrix in the form of ineffective micropores, resulting in the trapping of microbes and sediments in dead end pores. Recent work by Donnison and Ross [58] investigated the effect of soil type on transfer of zoonotic bacteria to rural streams using intact cores and turfs of a gley soil and a sandy loam. Farm dairy effluent, containing laboratory grown *E. coli* O157:H7 and *Campylobacter*, was added to cores and turfs that were stored at 10°C. It was found that the relative timing of application of zoonotic bacteria to soil and that of subsequent rainfall determine whether significant numbers of these bacteria are likely to reach streams. This timing varied for different soil types. For the sandy loam soil there was little transport of cells to drainage after 14 days but for the gley soil there was little change in the proportion of surviving *E. coli* O157:H7 that was washed out over 35 days, and it was found that *E. coli* O157 could be mobilized relatively easily from this soil. Donnison and Ross [58] highlighted that although far fewer bacteria are washed out than are retained, even immediately after application, the actual numbers can be substantial and could contribute to waterborne disease.

Under conditions of low rainfall, the soil may act as a more efficient filter matrix; this is because the slow percolation of FIB, pathogens, and P into the soil associated with light rainfall will increase microbial and P exposure time to soil surfaces in contrast to rapid water flows, and allow for an increased potential of contact with soil. This does not follow for NO_3^-, which does not interact or associate with the surrounding soil. In well-structured soils N will be retained to a high degree when protected within the bulk of the soil and will therefore only transfer through the soil profile with the slow percolation of mobile water. For microbes and P, implications of forming a cell-particle or P-particle composite may have impact once delivered to surface waters. The significant proportion of *E. coli*, for example, that remains unattached or associated with particles ≤ 2 μm are likely to remain in the water column and cause potential contamination problems further downstream. In contrast, the proportion of *E. coli* associated with larger soil particles are more likely to sediment out into the stream bed and therefore pose a delayed threat to water quality upon their resuspension. Further affinity studies of FIB are required, using different soil types, to test for the effects of an array of environmental variables such as pH, soil mineralogy, and temperature on the affinity of *E. coli* for soil.

Interesting observations have been made regarding the effect of cattle manure and slurry application on the percolation of the pathogens *E. coli* O157:H7 and *S. enterica* serovar Typhimurium. In the study of Semenov et al. [133] a greater number of cells were found to percolate to greater soil depths after slurry application compared to cells applied via manure application. Such results suggest that surface application of solid manures rather than their liquid counterparts may decrease the risk of contamination of groundwater supplies. Phosphate sorption (and release) in soils has also been investigated in relation to fertilizer sources. Such sorption is affected by reactions that take place at the solution-soil surface interface and recent work has demonstrated that phosphate binding strength can be up to 50% less in manured soils than in soils fertilized with inorganic triple superphosphate [134]. Such findings imply manure applications to some soil types result in an increased net negative surface charge and therefore a reduced soil phosphate adsorption capacity leading to increased losses from the soil system. These slower vertical transfers may eventually allow for connection with groundwater supplies, which are covered in the next section of this chapter.

The Role of Groundwater

Microbial and nutrient transport to surface water can occur by deposition of manure directly in the water or by wash-off in surface and subsurface runoff. However, transport to groundwater is a somewhat lesser concern in that it requires that the contaminant of study move through soil and bedrock to reach the water table. Nevertheless, groundwater systems are the predominant reservoir and strategic reserve of global freshwater storage at ca. 30% of the global water total and 98% of freshwater in liquid form [22], and protection of the microbial and chemical quality of this precious freshwater resource is therefore essential to safeguard human health and maintain sustainable freshwater resources. While a number of contaminants may interact with groundwater supplies, it is generally accepted that the most widely studied contaminant of groundwater supplies is NO_3^-. However, it cannot be disputed that the groundwater pollutants that most concern human health are microbiological, causing disease and sometimes death [135].

Manure application to land is regarded as the primary source of bacterial groundwater contamination in agricultural settings [136]. However, most pathogens are suspected to have life cycles far shorter (i.e., high die-off or inactivation rates outside of the animal gut) than typical groundwater travel times, except in karstic aquifers because of their dual porosity or where the source is very close to the point of water abstraction. Even so, continued investigation on microbial persistence within a range of agricultural environments is needed, and published persistence profiles of robust pathogens such as *C. parvum* propose that this pathogen can remain viable well in excess of hundreds of days [17], suggesting potential implications for groundwater quality upon emergence. The larger size of this protozoan relative to bacterial cells means it is more likely to be filtered by the overlying soil, but direct wash-in of such pathogens into private water supplies is not unheard of and can cause significant human illness. Of course, in addition to nitrates, bacteria, and protozoa, there remain other contaminants of groundwater, their presence dictated by their mobility and reactivity with the aquifer matrix and its overlying soil [135].

A recent case study of contamination of groundwater with microbial pollutants is illustrated by a small dryland watershed in central Chile [137]. This study suggested that concentrations of FIB were temporally dynamic with levels varying between seasons with higher concentrations in winter. As discussed in earlier sections of this chapter, causes of contamination could be linked to the easy access of domestic animals to the water source (i.e., wells in this case) and to the permeable well casing material. Local precipitation runoff would also have a direct influence on the bacterial concentrations found in such wells, with seasonality influencing runoff volumes and climatic characteristics important for driving transfer. Undoubtedly, the climatic conditions typical of winter can also favor survival of pathogens and FIB, with cooler temperatures often considered advantageous for lengthy persistence of these microbes. Studies of *Cryptosporidium* through undisturbed, macroporous karst soil have demonstrated that leaching is an important mechanism of oocyst transport in karst soils but have also provided evidence for the significant role played by macropores (detailed earlier in this chapter) in the soil physical structure. Thus oocysts leaching from soils into the epikarst could accumulate and remain viable for months until hydrological conditions are right for flushing the oocysts into the conduit flow system [59].

Studies have also shown that high concentrations of nitrate in groundwater samples do not relate well to concentrations of bacterial contaminants. A Vietnamese study took measurements of NO_3^- and FIB in groundwater samples taken from both dug wells and bores and found that a significant number (18%) of samples had NO_3^- concentrations in excess of the WHO Guideline value for drinking water of 50 mg L^{-1} (11.3 mg L^{-1} as nitrate-nitrogen). High concentrations of FIB were found in many of the dug wells and even in the deeper drilled bores, but there was no correlation between NO_3^- concentration and bacterial content [138], probably reflecting the different transfer pathways and mobilities of these contaminant typologies through and across the surrounding landscape. Attempts to curb groundwater contamination from nitrate and FIB include novel approaches to mitigation and management such as tree planting within agricultural systems to allow for a more efficient use of resources, since the rooting system of the trees captures nutrients that are not captured by crops, for example. In particular, intercropping systems have been shown to offer potential mitigating effects on *E coli* movement to the groundwater [92].

In contrast to NO_3^-, phosphate is mostly immobile in the subsurface, with the flux of P to groundwater being controlled by the degree of attenuation in soil. It has therefore generally been viewed as presenting minimal threat to groundwater quality. However, recent contrary views have been published contesting the long-held belief that adsorption and metal complex formation retain the majority of potentially mobile phosphorus. The relative contributions of potential sources for these elevated concentrations of P in groundwater are currently unclear but there is evidence to suggest that they are probably partly anthropogenic. Recent findings therefore suggest that groundwater P concentrations are such that they may be a more important contributor to surface water phosphorus, contrary to prior thinking [139]. Fracture flow is critically important because it allows turbulent flow to transport colloidal- and particulate-attached P to groundwater via a pathway where there is low rock–water interaction (see earlier discussion of macropore flow). Similar to concepts governing high levels of emergence of contaminants via macropore flow, the connectivity between fractures rather than fracture presence per se provide major controls on water residence time distribution from the surface system to the receiving surface water body.

Groundwater–surface water interactions are a final point to note for the upwelling of nutrient sources into streams and rivers. Spatial and temporal distributions of NO_3^- can be observed along the upwelling flow path from groundwater to surface water along stream reaches. A UK study on a Cumbrian stream reach identified that upwelling flows dominated the exchange between groundwater and surface water throughout a period of investigation. In particular, for this stream reach, NO_3^- concentrations along upwelling flow paths appeared to follow two opposite trends, with both decreasing and increasing nitrate concentrations being observed at different points in the experimental reach. The magnitude of variation in NO_3^- concentration along the upwelling flow path to the streambed was thought to be governed by the sediment structure and characteristics in the two contrasting field sites [140]. Research studies detailing microbial pollution linked to groundwater–surface water interactions are scarce, though it is an interesting aspect for future consideration. Under baseflow conditions, the general assumption is that no resuspension of FIB occurs under steady state flow conditions, and only during storm event flows do stream sediments become resuspended in the overlying water column [141]. However, there may be potential for uncontaminated (FIB-free) groundwater to become contaminated as it travels through upwelling flow paths of the hyporheic zone and the sediment associated store of FIB, possibly adding to the baseflow load of FIB. This is clearly an underexplored research agenda for FIB at current time.

Future Directions

Continued research conducted across a range of scales to understand better the transfer of pathogens and chemicals through and over soils is paramount. Different scales of study investigating mechanistic processes through to catchment scale real-world responses of landscape systems all serve key roles in enhancing understanding of pollutant dynamics in the soil-water continuum. Ultimately, management of livestock and their manure must be undertaken with a view to ensure the sustainability of key ecosystem services, such as the provision of clean and safe recreational and drinking water [142] and targeting of mitigation options to protect watercourses needs to be grounded on sound science. The database of such scientific results continues to grow and provides an expanding (though this is immature for FIB and pathogens relative to P and N) body of empirical science that can form the "evidence base" for good regulatory practice [4]. A growing number of initiatives continue to emerge that advocate the drive for integrated catchment management and the need for greater stakeholder engagement to help reduce contaminant transfers. For successful mitigation of pollutant transfers through and across soil that reduce impact on water quality there is a clear need to tailor mitigation options to reap the most "bang for the buck," and therefore management interventions and changes must be considered in terms of both cost and efficiency [143]. Making informed decisions at the field scale is therefore crucial because agricultural land is heterogeneous, and inherent spatial variability in soils and hydrological flow pathways influences the loss of pollutants from land to water, often at subfield scales [144]. Clearly, land management options such as manure incorporation into soil and reduced application rates to pasture can help avoid scenarios likely to enhance transport and serve as viable approaches to minimize nonpoint-source contamination while ensuring the least public health risk [64]. However, mitigation strategies are often much more complex to implement in practice and accommodate site specific constraints. The lag time between implementation of management practices on the land and water quality response is an unfortunate reality in watershed management and adds an extra layer of complexity to matters [36].

While efforts continue in an attempt to limit contaminant transfers through and across soils, there are parallel challenges for the research and policy community. It is critical that soil and water scientists make progress on quantifying losses from (and delivery via) spatially and temporally dynamic pathways rather than only appreciating the existence of such pathways for facilitating pollutant transfers. In particular, quantifying exports of pathogens remains elusive at farm and catchment scales. However, future research agendas for pathogen transport through soil systems must accommodate the need to characterize microbial behavior across a broad range of environmentally relevant conditions. Critically, the transfer pathways of both pathogens and chemicals will be impacted by the likely changes in climatic conditions [43], which may also impact indirectly on contaminant transfers because of changes to transport and movement of animals, intensity of livestock farming, and habitat change [145, 146]. Flooding too represents a transfer mechanism by which pathogens, FIB, and nutrients can be catastrophically exported from agricultural land in considerable numbers and result in much greater exposure of the human population to pollutants [3, 147]. Thus flooding represents an example of a very different route of transfer, which at present remains poorly understood [148] but which is sure to warrant much attention as an exposure pathway of increasing significance.

What this chapter has attempted to show, in part, is that various pollutant typologies (soluble, particulate, inert, biological etc.) behave and respond to environmental drivers and conditions in different ways. Pathways through and across the soil are used to a different extent by different pollutant types and are influenced by spatial and temporal characteristics of rainfall and soil types. The warning message for mitigation options implemented to reduce pollutant transfer through soil systems is that different strategies will impact on pollutants in different ways. A mitigation strategy designed to impede particulate transfer may, in fact, enhance the movement of soluble contaminants, or even impact on greenhouse gas emissions. The need for holistic understanding of multi-pollutants is now essential to reflect the potential for such "pollution swapping" risks [149, 150]. In turn, approaches for modeling agricultural systems and the flows and transfers of associated contaminants within the soil-water continuum now require frameworks that can explore the effectiveness of nonpoint pollutant mitigation options for multiple pollutants (including those designed for pathogens). Undoubtedly, such tools are high on the agenda for policy-makers who are required to identify which mitigation options are likely to reduce the target pollutant without increasing impact from others [151]. Indeed, integrated water and agricultural management is needed so that decision makers can recognize interdependencies in environmental systems and prioritize management responsibilities for optimal protection of environmental resources [152, 153].

Finally, new technologies continue to push forward the boundaries of science. Higher resolution and continuous sampling is highlighting trends in diffuse pollutants previously undetected and substantiating our knowledge of catchment response to environmental drivers. For detection of microbial pollutants and associated transfer routes through the environment, the future is likely to embrace molecular approaches. Studies are beginning to emerge that investigate molecular versus culture-based approaches for the detection of FIB, and such initiatives are key in examining relationships between quantitative PCR (qPCR) and culture-based FIB counts in an attempt to optimize and standardize methods for current indicators. However, currently, very few studies have attempted to quantify the uncertainty in quantitative qPCR data, though this is likely to be rectified over the coming years. Indeed, such tools are continuing to be investigated, developed, and tested on "end-point receptor" waters (e.g., coastal waters) (e.g., [154]). With continued evaluation of these approaches against standard culture-based techniques (for bacterial studies), there is much potential for these methods to be developed further to determine potential sources and pathways of pollution linked to both surface and groundwater contamination.

Bibliography

Primary Literature

1. Haygarth PM, Ritz K (2009) The future of soils and land use in the UK: soil systems for the provision of land-based ecosystem services. Land Use Policy 26S:S187–S197
2. Pepper IL, Gerba CP, Newby DT, Rice CW (2009) Soil: a public health threat or savior? Crit Rev Environ Sci Technol 39:416–432
3. Bridge JW, Oliver DM, Chadwick D, Godfray C, Heathwaite AL, Kay D, Maheswaran R, McGonigle D, Nichols G, Pickup R, Porter J, Wastling J, Banwart SA (under review). Minimising the waterborne disease burden in affluent nations: a key role for environmental sciences. Bull World Health Organ
4. Kay D, Crowther J, Fewtrell L, Francis CA, Hopkins M, Kay C, McDonald AT, Stapleton CM, Watkins J, Wyer MD (2008) Quantification and control of microbial pollution from agriculture: a new policy challenge? Environ Sci Policy 11:171–184
5. Kay D, Edwards AC, Ferrier RC, Francis C, Kay C, Rushby L, Watkins J, McDonald AT, Wyer M, Crowther J, Wilkinson J (2007) Catchment microbial dynamics: the emergence of a research agenda. Prog Phys Geogr 31:59–76
6. USEPA (2010) National summary of impaired waters and TMDL information. U.S. Environmental Protection Agency, Washington. Available at: http://iaspub.epa.gov/waters10/attains_nation_cy.control?p_report_type=T. Accessed 4 Feb 2010
7. Hunter PR (1992) Cyanobacteria and human health. J Med Microbiol 36:301–302
8. Pang L, Nowostawska U, Ryan JN, Williamson WM, Walshe G, Hunter KA (2009) Modifying the surface charge of pathogen-sized microspheres for studying pathogen transport in groundwater. J Environ Qual 38:2210–2217
9. Infascelli R, Pelorosso R, Boccia L (2009) Spatial assessment of animal manure spreading and groundwater nitrate pollution. Geospat Health 4:27–38
10. Bartram J, Thyssen N, Gowers A, Pond K, Lack T (2002) Water and health in Europe. A joint report from the European Environment Agency and the WHO regional office foe Europe. WHO Regional Publications, Copenhagan
11. Stewart I, Webb PM, Schluter PJ, Shaw GR (2006) Recreational and occupational field exposure to freshwater cyanobacteria – a review of anecdotal and case reports, epidemiological studies and the challenges for epidemiologic assessment. Environ Health 5:6
12. Topp E, Scott A, Lapen DR, Lyautey AR, Duriez P (2009) Livestock waste treatment systems for reducing environmental exposure to hazardous enteric pathogens: some considerations. Bioresour Technol 100:5395–5398
13. Monaghan RM, Carey PL, Wilcock RJ, Drewry JJ, Houlbrooke DJ, Quinn JM, Thorrold BS (2009) Linkages between land management activities and stream water quality in a border dyke-irrigated pastoral catchment. Agric Ecosyst Environ 129:201–211
14. McKergow LA, Davies-Colley RJ (2009) Stormflow dynamics and loads of *Escherichia coli* in a large mixed land use catchment. Hydrol Process 24:276–289
15. Kay D, Anthony S, Crowther J, Chambers BJ, Nicholson FA, Chadwick D, Stapleton CM, Wyer MD (2010) Microbial water pollution: a screening tool for initial catchment-scale assessment and source apportionment. Sci Total Environ (in press)
16. Arinaminpathy N, McLean AR, Godfray HCJ (2009) Future UK land use policy and the risk of infectious disease in humans, livestock and wild animals. Land use Policy s6S:S124–S133
17. Oliver DM, Clegg CD, Haygarth PM, Heathwaite AL (2005a) Assessing the potential for pathogen transfer from grassland soils to surface waters. Adv Agron 85:125–180
18. Sobsey MD, Pillai SD (2009) Where future emerging pathogens will come from and what approaches can be used to find them, besides VFARs. J Water Health 7:S75–S93

19. St.-Pierre K, Levesque S, Frost E, Carrier N, Arbeit RD, Michaud S (2009) Thermotolerant coliforms are not a good surrogate for *Campylobacter* spp. In environmental waters. Appl Environ Microbiol 75:6736–6744
20. Wilkes G, Edge T, Gannon V, Jokinen C, Lyautey E, Medeiros D, Neumann N, Ruecker N, Topp E, Lapen D (2009) Seasonal relationships among indicator bacteria, pathogenic bacteria, *Cryptosporidium* oocysts, *Giardia* cysts and hydrological indices for surface waters within an agricultural landscape. Water Res 43:2209–2223
21. Conley DJ, Paerl HW, Howarth HW, Boesch DF, Seitzinger SP, Havens KE, Lancelot C, Likens GE (2009) Controlling eutrophication: nitrogen and phosphorus. Science 323:1014–1015
22. Heathwaite AL (2010) Multiple stressors on water availability at global to catchment scales: understanding human impact on nutrient cycles to protect water quality and water availability in the long term. Freshw Biol 55:241–257
23. Edmeades DC (2003) The long-term effects of manures and fertilizers on soil productivity and quality: a review. Nutr Cycl Agroecosyst 66:165–180
24. Haygarth PM, Heathwaite AL, Jarvis SC, Harrod TR (2000) Hydrological factors for phosphorus transfer from agricultural soils. Adv Agron 69:153–178
25. Muirhead RW (2009) Soil and fecal material reservoirs of *Escherichia coli* in a grazed pasture. NZ J Agric Res 52:1–8
26. Oliver DM, Page T, Heathwaite AL, Haygarth PM (2010a) Reshaping models of *E. coli* population dynamics in livestock feces: increased bacterial risk to humans? Environ Int 36:1–7
27. Chadwick DR, Fish RD, Oliver DM, Heathwaite AL, Hodgson CJ, Winter M (2008) Management of livestock and their manure to reduce the risk of microbial transfers to water – the case for an interdisciplinary approach. Trends Food Sci Technol 19:240–247
28. Oliver DM, Heathwaite AL, Fish RD, Chadwick DR, Hodgson CJ, Winter M, Butler AJ (2009) Scale appropriate modeling of diffuse microbial pollution from agriculture. Prog Phys Geogr 33:358–377
29. Unc A, Goss MJ (2006) Culturable *Escherichia coli* in soil mixed with two types of manure. Soil Sci Soc Am J 70:763–769
30. Oliver DM, Page T, Hodgson CJ, Heathwaite AL, Chadwick DR, Fish RD, Winter M (2010b) Development and testing of a risk indexing framework to determine fields-scale critical source areas of faecal bacteria on grassland. Environ Modell Softw 25:503–512
31. Oliver DM, Haygarth PM, Clegg CD, Heathwaite AL (2006) Differential *E. coli* die off patterns associated with agricultural matrices. Environ Sci Technol 40:5710–5716
32. Hodgson CJ, Bulmer N, Chadwick DR, Oliver DM, Heathwaite AL, Fish RD, Winter M (2009) Establishing relative release kinetics of fecal indicator organisms from different fecal matrices. Lett Appl Microbiol 49:124–130
33. Brennan FP, O'Flaherty V, Kramers G, Grant J, Richards KG (2010) Long-term persistence and leaching of *E. coli* in temperate maritime soils. Appl Environ Microbiol 76(5):1449–1455
34. Lyautey E, Zexun L, Lapen DR, Wilkes G, Scott A, Berkers T, Edge TA, Topp E (2010) Distribution and diversity of *Escherichia coli* populations in the South Nation River Drainage Basin, Eastern Ontario, Canada. Appl Environ Microbiol 76(5):1486–1496
35. Nautiyal CS, Rehman A, Chauhan PS (2010) Environmental *Escherichia coli* occur as natural plant growth promoting soil bacterium. Arch Microbiol 192:185–193
36. Meals DW, Dressing SA, Davenport TE (2009) Lag time in water quality response to best management practices: a review. J Environ Qual 39:85–96
37. Preedy N, McTiernan K, Matthews R, Heathwaite AL, Haygarth PM (2001) Rapid incidental phosphorus transfers from grassland. J Environ Qual 30:2105–2122
38. Hutchison ML, Walters LD, Moore A, Crookes KM, Avery SM (2004) Effect of length of time before incorporation on survival of pathogenic bacteria present in livestock wastes applied to agricultural soil. Appl Environ Microbiol 70:5111–5118

39. Oliver DM, Heathwaite AL, Hodgson CJ, Chadwick DR (2007a) Mitigation and current management attempts to limit pathogen survival and movement within farmed grasslands. Adv Agron 93:95–152
40. McGechan MB, Topp CFE (2004) Modeling environmental impacts of deposition of excreted nitrogen by dairy cows. Agric Ecosyst Environ 103:149–164
41. Aitken MN (2003) Impact of agricultural practices and river catchment characteristics on river and bathing water quality. Water Sci Technol 48:217–224
42. Deasy C, Brazier RE, Heathwaite AL, Hodgkinson R (2009) Pathways of runoff and sediment transfer in small agricultural catchments. Hydrol Process 23:1349–1358
43. Boxall ABA, Hardy A, Beulke S, Boucard T, Burgin L, Falloon PD, Haygarth PM, Hutchinson T, Kovats RS, Leonardi G, Levy LS, Nichols G, Parsons SA, Potts L, Stone D, Topp E, Turley DB, Walsh K, Wellington EMH, Williams RJ (2009) Impacts of climate change on indirect human exposure to pathogens and chemicals from agriculture. Environ Health Perspect 117:508–514
44. Garbrecht K, Fox GA, Guzman JA, Alexander D (2009) *E. coli* transport through soil columns: implications for bioretention cell removal efficiency. Trans ASABE 52:481–486
45. Armstrong A, Quinton JN (2010) Variability of interrill erosion at low slopes. Earth Surf Process Landforms
46. Lane SN, Brookes CJ, Kirkby AJ, Holden J (2004) A network-indexbased version of TOPMODEL for use with high-resolution digital topographic data. Hydrol Process 18:191–201
47. Reaney SM, Lane SN, Heathwaite AL, Dugdale LJ (2010) Risk-based modeling of diffuse land use impacts from rural landscapes upon salmonid fry abundance. Ecol Model (in press)
48. Yaron B, Dror I, Berkorwitz B (2010) Contaminant geochemistry – a new perspective. Naturwissenschaften 97:1–17
49. Bridge JW, Banwart SA, Heathwaite AL (2006) Noninvasive quantitative measurement of colloid transport in mesoscale porour media using time lapse fluorescence imaging. Environ Sci Technol 40:5930–5936
50. Zonia L, Bray D (2009) Swimming patterns and dynamics of simulated *Escherichia coli* bacteria. J R Soc Interface 6:1035–1046
51. Tu YH, Shimizu TS, Berg HC (2008) Modelling the chemotactic response of *Escherichia coli* to time-varying stimuli. Proc Natl Acad Sci USA 105:14855–14860
52. Sharpley AN, Kleinman PJA, Heathwaite AL, Bburek WJ, Folmar GJ, Schmidt JR (2008) Phosphorus loss from an agricultural watershed as a function of storm size. J Environ Qual 37:362–368
53. Granger SJ, Hawkins JMB, Bol R, White SM, nadan P, Old G, Bilotta GS, Brazier RE, Macleod CJA, Haygarth PM (2010) High temporal resolution monitoring of multiple pollutant responses in drainage from an intensively managed grassland catchment caused by a summer storm. Water Air Soil Pollut 205:377–393
54. Artz RRE, Townend J, Brown K, Towers W, Killham K (2005) Soil macropores and compaction control the leaching potential of *Escherichia coli* O157:H7. Environ Microbiol 7:241–248
55. Mawdsley JL, Brooks AE, Merry RJ (1996) Movement of the protozoan pathogen *Cryptosporidium parvum* through three contrasting soil types. Biol Fertil Soils 21:30–36
56. Matula J (2009) Possible phosphorus losses from the top layer of agricultural soils by rainfall simulations in relation to multi-nutrient soil tests. Plant Soil Environ 55:511–518
57. Entry JA, Sojka RE, Hicks BJ (2010) Matrix-based fertilizers reduce nutrient and bacterial leaching after manure application in a greenhouse column study. J Environ Qual 39:384–392
58. Donnison A, Ross C (2009) Survival and retention of *Escherichia coli* O157:H7 and *Campylobacter* in contrasting soils from the Toenepi catchment. NZ J Agric Res 52:133–144
59. Boyer DG, Kuczynska E, Fayer R (2009) Transport, fate, and infectivity of *Cryptosporidium parvum* oocysts released from manure and leached through macroporous soil. Environ Geol 58:1011–1019

60. Tarkalson DD, Leytem AB (2009) Phosphorus mobility in soil columns treated with dairy manures and commercial fertilizer. Soil Sci 174:73–80
61. Miller JJ, Beasley BW, Chanasyk DS, Larney FJ, Olson BM (2008) Short-term nitrogen leaching potential of fresh and composted beef cattle manure applied to disturbed soil cores. Compost Sci Util 16:12–19
62. Harter T, Atwill ER, Hou L, Karle BM, Tate KW (2008) Developing risk models of *Cryptosporidium* transport in soils from vegetated, tilted, soilbox experiments. J Environ Qual 37:245–258
63. Brock EH, Ketterings QM, Kleinman PJA (2007) Measuring and predicting the phosphorus sorption capacity of manure-amended soils. Soil Sci 172:266–278
64. Horswell J, Hewitt J, Prosser J, Van Schaik A, Croucher D, Macdonald C, Burford P, Susarla P, Bickers P, Speir T (2010) Mobility and survival of *Salmonella Typhimurium* and human adenovirus from spiked sewage sludge applied to soil columns. J Appl Microbiol 108:104–114
65. Rosa BA, Yim MI, Burdenuk L, Kjartanson B, Leung KT (2010) The transport of *Escherichia coli* through freeze-fractured clay soil. Water Air Soil Pollut 210:243–254
66. Bech TB, Johnsen K, Dalsgaard A, Laegdsmand M, Jacobsen OH, Jacobsen CS (2010) Transport and distribution *of Salmonella enterica* serovar Typhimurium in loamy and sandy soil monoliths with applied liquid manure. Appl Environ Microbiol 76:710–714
67. Turner BL, Haygarth PM (2000) Phosphorus forms and concentrations in leachate under four grassland soil types. Soil Sci Soc Am J 64:1090–1099
68. Di HJ, Cameron KC, Shen JP, He JZ, Winefield JS (2009) A lysimeter study of nitrate leaching from grazed grassland as affected by a nitrification inhibitor, dicyandiamide, and relationships with ammonia oxidizing bacteria and archaea. Soil Use Manage 25:454–451
69. Guzman JA, Fox GA, Malone RW, Kanwar RS (2009) *Escherichia coli* transport from surface-applied manure to subsurface drains through artificial biopores. J Environ Qual 38: 2412–2421
70. Muirhead RW, Collins RP, Bremer PJ (2006) Interaction of *Escherichia coli* and soil particles in runoff. Appl Environ Microbiol 72:3406–3411
71. Ferguson CM, Davies CM, Kaucner C, Krogh M, Rodehutskors J, Deere DA, Ashbolt NJ (2007) Field scale quantification of microbial transport from bovine feces under simulated rainfall events. J Water Health 5:83–95
72. Abu-Ashour J, Lee H (2000) Transport of bacteria on sloping soil surfaces by runoff. Environ Toxicol 15:149–153
73. Tunney H, Kurz I, Bourke D, O'Reilly C (2007) RTDI programme 2000–2006 eutrophication from agricultural sources: the impact of the grazing animal on phosphorous loss from grazed pasture. Teagasc, Wexford
74. Quinton JN, Catt JA, Hess TM (2001) The selective removal of phosphorus from soil: is event size important? J Environ Qual 30:538–545
75. Soupir ML, Mostaghimi S, Yagow ER, Hagedorn C, Vaughan DH (2006) Transport of fecal bacteria from poultry litter and cattle manures applied to pastureland. Water Air Soil Pollut 169:125–136
76. Withers PJA, Hodgkinson RA, Bates A, Withers CL (2007) Soil cultivation effects on sediment and phosphorus mobilization in surface runoff from three contrasting soil types in England. Soil Tillage Res 93:438–451
77. Mishra A, Benham BL, Mostaghimi S (2008) Bacterial transport from agricultural lands fertilized with animal manure. Water Air Soil Pollut 189:127–134
78. Heathwaite AL, Haygarth PM, Matthews R, Preedy N, Butler P (2005) Evaluating colloidal phosphorus delivery to surface waters from non-point agricultural sources. J Environ Qual 34:287–298
79. Thiagarajan A, Gordon R, Madani A, Stratton GW (2007) Discharge of *Escherichia coli* from agricultural surface and subsurface drainage water: tillage effects. Water Air Soil Pollut 182:3–12

80. Ramirez NE, Wang P, Lejeune J, Shipitalo MJ, Ward LA, Sreevatsan S, Dick WA (2009) Efffect of tillage and rainfall on transport of manure-applied *Cryptosporidium* parvum oocysts through soil. J Environ Qual 38:2394–2401
81. Haygarth PM, Hepworth L, Jarvis SC (1998) Forms of phosphorus transfer in hydrological pathways from soil under grazed grassland. Eur J Soil Sci 49:65–72
82. Oliver DM, Heathwaite AL, Haygarth PM, Clegg CD (2005) Transfer of *Escherichia coli* to water from drained and undrained grassland after grazing. J Environ Qual 34:918–925
83. Alfaro M, Salazar F, Iraira S, Teuber N, Villarroel D, Ramirez L (2008) Nitrogen, phosphorus and potassium losses in a grazing system with different stocking rates in a volcanic soil. Chil J Agric Res 68:146–155
84. Kouznetsov MY, Roodsari R, Pachepsky YA, Shelton DR, Sadeghi AM, Shirmohammadi A, Starr JL (2007) Modeling manure-borne bromide and fecal coliform transport with runoff and infiltration at a hillslope. J Environ Manage 84:336–346
85. Olson BM, Bennett DR, McKenzie RH, Ormann TD, Atkins RP (2009) Nitrate leaching in two irrigated soils with different rates of cattle manure. J Environ Qual 38:2218–2228
86. Collins R, Elliot S, Adams R (2005) Overland flow delivery of fecal bacteria to a headwater pastoral stream. J Appl Microbiol 99:126–132
87. Kay D, Aitken M, Crowther J, Dickinson I, Edwards AC, Francis C, Hopkins M, Jeffrey W, Kay C, McDonald AT, McDonald D, Stapleton CM, Watkins J, Wilkinson J, Wyer M (2007) Reducing fluxes of fecal indicator compliance parameters to bathing waters from non-point agricultural sources: the Brighouse Bay study. Scotland Environ Pollut 147:138–149
88. Deasy C, Heathwaite AL, brazier RE (2008) A field methodology for quantifying phosphorus transfer and delivery to streams in first order agricultural catchments. J Hydrol 350:329–338
89. Botter G, Milan E, Bertuzzo E, Zanardo S, Marani M, Rinaldo A (2009) Inferences from catchment-scale tracer circulation experiments. J Hydrol 369:368–380
90. Close M, Dann R, Ball A, Pirie R, Savill M, Smith Z (2008) Microbial groundwater quality and its health implications for a border-strip irrigated dairy farm catchment, South Island, New Zealand. J Water Health 6:83–98
91. Heathwaite AL, Johnes PJ (1996) Contribution of nitrogen species and phosphorus fractions to stream water quality in agricultural catchments. Hydrol Process 10:971–983
92. Dougherty MC, Thevathasan NV, Gordon AM, Lee H, Kort J (2009) Nitrate and *Escherichia coli* NAR analysis in tile drain effluent from a mixed tree intercrop and monocrop system. Agric Ecosyst Environ 131:77–84
93. Davies-Colley R, Lydiard E, Nagels J (2008) Stormflow-dominated loads of fecal pollution from an intensively dairy-farmed catchment. Water Sci Technol 57:1519–1523
94. Rothwell JJ, Dise NB, Taylor KG, Allott TEH, Scholefield P, Davies H, Neal C (2010) A spatial and seasonal assessment of river water chemistry across North West England. Sci Total Environ 408:841–855
95. Beven K, Heathwaite AL, Haygarth PM, Walling D, Brazier R, Withers P (2005) On the concept of delivery of sediment and nutrients to stream channels. Hydrol Process 19:551–556
96. Haygarth PM, Condron LM, Heathwaite AL, Turner BL, Harris GP (2005) The phosphorus transfer continuum: linking source to impact with an interdisciplinary and multi-scaled approach. Sci Total Environ 344:5–14
97. Armstrong A, Quinton JN (2009) Pumped rainfall simulators: the impact of rain pulses on sediment concentration and size. Earth Surf Process Land 34:1310–1314
98. Mankin KR, Wang L, Hutchinson SL, Marchin GL (2007) *Escherichia coli* sorption to sand and silt loam soil. Trans ASABE 50:1159–1165
99. Penn CJ, Mullins GL, Zelanzny LW (2005) Mineralogy in relation to phosphorus sorption and dissolved phorphorus losses in runoff. Soil Sci Soc Am J 69:1532–1540
100. Drozd C, Schwartzbrod J (1996) Hydrophobic and electrostatic cell surface properties of Cryptosporidium parvum. Appl Environ Microbiol 62:1227–1232
101. Kuczynska E, Shelton DR, Pachepsky Y (2005) Effect of bovine manure on *Cryptosporidium parvum* oocyst attachment to soil. Appl Environ Microbiol 71:6394–6397

102. Voice TC, Weber WJ (1985) Sorbent concentration effects in liquid/solid partitioning. Environ Sci Technol 19:789–796
103. Oliver DM (2005) Hydrological pathways and processes of *Escherichia coli* transfer from grassland soils to surface waters. Ph.D. thesis, University of Sheffield
104. Koopmans GF, McDowell RW, Chardon WJ, Oenema O, Dolfing J (2002) Soil phosphorus quantity-intensity relationships to predict increased soil phosphorus loss to overland and subsurface flow. Chemosphere 48:679–687
105. Ling TY, Achberger EC, Drapcho CM, Bengtson RL (2002) Quantifying adsorption of an indicator bacteria in a soil-water system. Trans ASAE 45:669–674
106. Kretzschmar R, Borkovec M, Grolimund D, Elimelech M (1999) Mobile subsurface colloids and their role in contaminant transport. Adv Agron 66:121–193
107. Soupir ML, Mostaghimi S, Dillaha T (2010) Attachment of *Escherichia coli* and Enterococci to particles in runoff from bare soils. J Environ Qual (in press)
108. Oliver DM, Clegg CD, Heathwaite AL, Haygarth PM (2007b) Preferential attachment of *Escherichia coli* to different particle size fractions of an agricultural grassland soil. Water Air Soil Pollut 185:369–375
109. Muirhead RW, Collins RP, Bremer PJ (2006) The association of *E. coli* and soil particles in overland flow. Water Sci Technol 54:153–159
110. Pachepsky YA, Shelton DR (2011) *Escherichia coli* and fecal coliforms in freshwater and estuarine sediments. Crit Rev Environ Sci Technol (in press)
111. Sinclair A, Hebb D, Jamieson R, Gordon R, Benedict K, Fuller K, Stratton GW, Madani A (2009) Growing season surface water loading of fecal indicator organisms within a rural watershed. Water Res 43:1199–1206
112. Haygarth PM, Wood FL, Heathwaite AL, Butler P (2005) Phosphorus dynamics observed through increasing scales in a nested headwater-to-river channel study. Sci Total Environ 344:83–106
113. Jordan P, Arnscheidt J, McGrogan H, McCormick S (2007) Characterising phosphorus transfers in rural catchments using a continuous bank-side analyser. Hydrol Earth Syst Sci 11:372–381
114. Tyrrel SF, Quinton JN (2003) Overland flow transport of pathogens from agricultural land receiving fecal wastes. J Appl Microbiol 94:87S–93S
115. McDonald A, Kay D (1981) Enteric bacterial concentrations in reservoir feeder streams – baseflow characteristics and response to hydrograph events. Water Res 15:961–968
116. Heathwaite AL, Burt TP, Trudgill ST (1990) The effect of land use on nitrogen, phosphorus and suspended sediment delivery to streams in a small catchment in southwest England. In: Boardman J, Foster LDL, Dearing JA (eds) Soil erosion on agricultural land. Wiley, Chichester/New York/Brisbane/Toronto/Singapore, pp 161–177
117. Page T (2008) Uncertainty assessment of phosphorus risk to surface waters. Environment Agency Science Report – SC050035
118. Kirchner JW, Feng XH, Neal C (2000) Fractal stream chemistry and its implications for contaminant transport in catchments. Nature 403:524–527
119. Beven K, Germann P (1982) Macropores and water-flow in soils. Water Resour Res 18:1311–1325
120. Abu-Ashour J, Joy DM, Lee H, Whiteley HR, Zelin S (1998) Movement of bacteria in unsaturated soil columns with macropores. Trans ASAE 41:1043–1050
121. Geohring LD, McHugh OV, Walter MT, Steenhuis TS, Akhtar MS (2001) Phosphorus transport into subsurface drains by macrpores after manure applications: implications for best manure management practices. Soil Sci 166:896–909
122. Allaire SE, Roulier S, Cessna AJ (2009) Quantifying preferential flow in soils: a review of different techniques. J Hydrol 378:179–204
123. Guber AK, Pachepsky YA, Shelton DR, Yu O (2009) Association of fecal coliforms with soil aggregates: effect of water content and bovine manure application. Soil Sci 174:543–548

124. Lutterodt G, Basnet M, Foppen JWA, Uhlenbrook S (2009) The effect of surface characteristics on the transport of multiple *Escherichia coli* isolates in large scale columns of quartz sand. Water Res 43:595–604
125. McLeod M, Aislabie J, Ryburn J, McGill A (2008) Regionalizing potential for microbial bypass flow through New Zealand soils. J Environ Qual 37:1959–1967
126. Aislabie J, Smith JJ, Fraser R, McLeod M (2001) Leaching of bacterial indicators of faecal contamination through four New Zealand soils. Aust J Soil Res 39:1397–1406
127. Paterson E, Kemp JS, Gammack SM, Fitzpatrick EA, Cresser MS, Mullins CE, Killham K (1993) Leaching of genetically-modified pseudomonas fluorescens through intact soil microcosms – influence of soil type. Biol Fertil Soils 15:308–314
128. McLeod M, Aislabie J, Ryburn J, McGilla A, taylor M (2003) Microbial and chemical tracer movement through two southland soils, New Zealand. Aust J Soil Res 41:1163–1169
129. Heathwaite AL, Burke SP, Bolton L (2006) Field drains as a route of rapid nutrient export from agricultural land receiving biosolids. Sci Total Environ 365:33–46
130. Mosaddeghi MR, Mahboubi AA, Zandsalimi S, Unc A (2009) Influence of organic waste type and soil structure on the bacterial filtration rates in unsaturated intact soil columns. J Environ Manage 90:730–739
131. Bowen GD, Rovira AD (1999) The rhizosphere and its management to improve plant growth. Adv Agron 66:1–102
132. Gagliardi JV, Karns JS (2000) Leaching of *Escherichia coli* O157:H7 in diverse soils under various agricultural management practices. Appl Environ Microbiol 66:877–883
133. Semenov AV, van Overbeek L, van Brugger AHC (2009) Percolation and survival of *Escherichia coli* O157:H7 and *Salmonella enterica* Serovar Typhimurium in soil amended with contaminated dairy or slurry. Appl Environ Microbiol 75:3206–3215
134. Jiao Y, Whalen JK, Hendershot WH (2007) Phosphate sorption and release in a sandy-loam soil as influenced by fertilizer sources. Soil Sci Soc Am J 71:118–124
135. Lerner D, Harris B (2009) The relationship between land use and groundwater resources and quality. Land Use Policy 26S:S265–S273
136. Pachepsky YA, Sadeghi AM, Bradford SA, Shelton DR, Guber AK, Dao T (2006) Transport and fate of manure-borne pathogens: modeling perspective. Agric Water Manage 86:81–92
137. Valenzuela M, Lagos B, Claret M, Mondaca MA, Perez C, Parra O (2009) Fecal contamination of groundwater in a small rural dryland watershed in rural Chile. Chil J Agric Res 69:235–243
138. Cam PD, Lan NTP, Smith GD, Verma N (2008) Nitrate and bacterial contamination in well waters in Vinh Phuc province, Vietnam. J Water Health 6:275–279
139. Holman IP, Whelan MJ, Howden NJK, Bellamy PH, Willby NJ, Rivas-Casado M, McConvey P (2008) Phosphorus in groundwater-an overlooked contributor to eutrophication? Hydrol Process 22:5121–5126
140. Krause S, Heathwaite L, Binley A, Keenan P (2009) Nitrate concentration changes at the groundwater-surface water interface of a small Cumbrian river. Hydrol Process 23:2195–2211
141. Rehmann CR, Soupir ML (2009) Importance of interactions between the water column and the sediment for microbial concentrations in streams. Water Res 43:4579–4589
142. Pretty J (2008) Agricultural sustainability: concepts, principles and evidence. Philos Trans R Soc B Biol Sci 363:447–465
143. Haygarth PM, ApSimon H, Betson M, Harris D, Hodgkinson R, Withers PJA (2009) Mitigating non-point phosphorus transfer from agriculture according to cost and efficiency. J Environ Qual 38:2012–2022
144. Hewett CJM, Quinn PF, Whitehead PG, Heathwaite AL, Flynn NJ (2004) Towards a nutrient export risk matrix approach to managing agricultural pollution at source. Hydrol Earth Syst Sci 8:834–845

145. Gale P, Drew T, Phipps LP, David G, Wooldridge M (2009) The effect of climate change on the occurrence and prevalence of livestock diseases in Great Britain: a review. J Appl Microbiol 106:1409–1423
146. Semenza JC, Menne B (2009) Climate change and infectious diseases in Europe. Lancet Infect Dis 9:365–375
147. Nagels JW, Davies-Colley RJ, Donnison AM, Muirhead RW (2002) Faecal contamination over flood events in a pastoral agricultural stream in New Zealand. Water Sci Technol 12:45–52
148. Wheater H, Evans E (2009) Land use, water management and future flood risk. Land Use Policy 26S:S251–S264
149. Stevens CJ, Quinton JN (2009a) Policy implications of pollution swapping. Phys Chem Earth 34:589–594
150. Stevens CJ, Quinton JN (2009b) Non-point pollution swapping in arable agricultural systems. Crit Rev Environ Sci Technol 39:478–520
151. Cuttle SP, Macleod CJA, Chadwick DR, Scholefield D, Haygarth PM, Newell-Price P, Harris D, Shepherd MA, Chambers BJ, Humphrey R (2007) An inventory of methods to control non-point water pollution from agriculture (DWPA) user manual. Defra, London, Defra project code ES0203, 113p
152. Fish RD, Ioris AAR, Watson NM (2010) Integrating water and agricultural management: collaborative governance for a complex policy problem. Sci Total Environ (in press)
153. Macleod CJA, Scholefield D, Haygarth PM (2007) Integration for sustainable catchment management. Sci Total Environ 373:591–602
154. Fremaux B, Gritzfeld J, Boa T, Yost CK (2009) Evaluation of host specific bacteriodales 16S rRNA gene markers as complementary tool for detecting fecal pollution in a prairie watershed. Water Res 43:4838–4849
155. Naden PS, Old GH, Eliot-Laize C, Granger SJ, Hawkins JMB, Bol R, Haygarth P (2010) Assessment of natural fluorescence as a tracer of non-point agricultural pollution from slurry spreading on intensely-farmed grasslands. Water Res 44(6):1701–1712

Books and Reviews

Burton CH, Turner C (2003) Manure management: treatment strategies for sustainable agriculture, 2nd edn. Silsoe Research Institute, Bedford

Ferguson C, Husman AMD, Altavilla N, Deere DA, Ashbolt N (2003) Fate and transport of surface water pathogens in watersheds. Crit Rev Environ Sci Technol 33:299–361

Haygarth PM, Jarvis SC (2002) Agriculture, hydrology and water quality. CABI, Wallingford

McDowell RW, Houlbrooke DJ, Muirhead RW, Mueller K, Shepherd M, Cuttle SP (2008) Grazed pastures and surface water quality. Nova Science Publishers, Hauppauge, p 238

Unc A, Goss MJ (2004) Transport of bacteria from manure and protection of water resources. Appl Soil Ecol 25:1–18

Whitehead PG, Wilby RL, Batterbee RW, Kernan M, Wade AJ (2009) A review of the potential impacts of climate change on surface water quality. Hydrol Sci J 54:101–123

Chapter 16
Recreational Water Risk: Pathogens and Fecal Indicators

Alexandria B. Boehm and Jeffrey A. Soller

Glossary

Microbial pollution	Pathogens.
Quantitative microbial risk assessment	Assessment of risk from exposure to pathogens.
Epidemiology	The study of human health in response to a treatment.
Recreational waterborne illness	Illness resulting from exposure to microbial pollution, includes gastroenteritis, respiratory illness, ear, nose, eye, and throat ailments, and skin rash.
Bacterial pathogens	Pathogens that are bacterial.
Viral pathogens	Pathogens that are viruses.
Protozoan pathogens	Pathogens that are protozoa.

This chapter was originally published as part of the Encyclopedia of Sustainability Science and Technology edited by Robert A. Meyers. DOI:10.1007/978-1-4419-0851-3

A.B. Boehm (✉)
Environmental and Water Studies, Department of Civil and Environmental Engineering, Stanford University, Stanford, CA 94305, USA
e-mail: aboehm@stanford.edu

J.A. Soller
Soller Environmental, LLC, 3022 King St, Berkeley, CA 94703, USA
e-mail: jsoller@sollerenvironmental.com

E.A. Laws (ed.), *Environmental Toxicology: Selected Entries from the Encyclopedia of Sustainability Science and Technology*, DOI 10.1007/978-1-4614-5764-0_16,

Definition of the Subject and Its Importance

Pathogens can enter recreational waterbodies (lakes, rivers, ocean beaches) from a number of different sources. The illness acquired by swimmers after exposure to pathogen-polluted recreational waters is termed recreational waterborne illness (RWI). Since many RWI go unreported to health care agencies, the true number of RWI each year can only be estimated; however, numbers are believed to be high – over one million in southern California alone. The risk of RWI from exposure to recreational waters can be measured using epidemiological studies and estimated using quantitative microbial risk assessments.

Introduction

Exposure to microbially polluted recreational waters can cause a variety of adverse health effects in humans including neurological infections, skin infections, earaches, eye infections, gastrointestinal illnesses, and respiratory infections [1]. Microbial pollution refers to the presence of organisms that cause illness in humans either through the production of toxins or their colonization of the human body.

It is estimated that globally, exposure to coastal waters polluted with wastewater results in an excess 120 million gastrointestinal and 50 million severe respiratory illnesses per year [2], including illnesses acquired through consumption of contaminated shellfish. In southern California, there are an estimated 1.5 million cases of gastrointestinal illnesses each year due to recreational exposure to polluted waters [3]. Moreover, there were 259 recreational water outbreaks that occurred in the USA between 1970 and 2000 [4].

Estimating the number of individuals acquiring illness through exposure to recreational waters is challenging. Typically, individuals who acquire recreational waterborne illness (RWI) do not seek medical attention because most of the illnesses tend to be mild and self-limiting. In addition, most RWIs are not reportable, so incidence levels are highly uncertain. The estimates mentioned previously were obtained using a variety of assumptions about the contamination of water and exposures. Epidemiology studies and quantitative microbial risk assessments (QMRA) are two scientifically rigorous methods that are used to estimate rates of RWI as a function of water quality.

This chapter provides a brief overview of the pathogens present in coastal waters. Readers are referred to other references for more information on these topics [5–7]. The two most common methods of assessing risk of exposure to pathogens in coastal waters, epidemiology studies, and QMRA, are also described. Example applications of these methods to assess risk of illness from exposure to pathogen-polluted coastal waters are presented. Critical research gaps are identified and summarized.

Pathogens in Coastal Waters

Pathogens present in coastal waters can be characterized into two broad groups. The first group consists of autochthonous pathogens that have an ecological niche in recreational waters. The other group is composed of allochthonous pathogens that come from human and animal wastes that have been discharged into these waterbodies.

Autochthonous pathogens include harmful algae, organisms in the genus *Vibrio*, and some protozoa such as *Naeglaria fowleri*. Harmful diatoms and dinoflagellates can cause a variety of ailments in humans from their production of toxins [8, 9]. *Vibrio vulnificus* and *V. parahaemolyticus* can infect open wounds and cause gastroenteritis if ingested from seawater. Other *Vibrio* species (such as *V. cholerae* and *V. mimicus*) can be pathogenic and cause similar ailments. *N. fowleri* is primarily found in freshwater, and when it enters the human body, can cause a rare but serious brain infection. While these organisms are extremely important from a human health perspective, this chapter will not discuss risks associated with autochthonous pathogens.

The focus of this chapter is allochthonous pathogens (Table 16.1) including the eight pathogens that cause a large proportion (>95%) of all non-foodborne illnesses in the USA [10]: the viruses norovirus, rotavirus, and adenovirus; the bacteria *Campylobacter*, *Salmonella* and pathogenic *Escherichia coli*; and the protozoans *Cryptosporidium* spp. and *Giardia lamblia*. Other pathogens that are important etiologies of RWI include enteroviruses, *Staphylococcus aureus,* and methicillin-resistant *S. aureus* (MRSA). All these pathogens, with the exception of *S. aureus* and MRSA, cause gastrointestinal illness. *Staphylococcus* causes skin infections. Adenoviruses can cause gastrointestinal illness, as well as eye and respiratory infections.

The detection of pathogens in environmental matrices is methodologically challenging. Allochthonous pathogens are rare microbes in the environment. In seawater there are on the order of one million autochthonous bacteria and 10 million autochthonous viruses in a milliliter. Whereas allochthonous pathogens may be present at levels of 1 per 10 liters l or lower. Thus, enumerating the allochthonous pathogens is particularly difficult because the presence of all the other organisms in the sample can interfere with the detection of the rare target. Because the field of pathogen detection is still evolving and because allochthonous pathogens densities are typically low, there are limited data on pathogen concentrations and occurrence in recreational waters. In many cases, authors only provide data on the presence or absence of pathogens in recreational waters and do not provide concentrations. Some example concentrations are provided in Table 16.1.

Treated and untreated wastewater, human and other animal feces, stormwater and urban runoff, and agricultural runoff can all contain microbes that are pathogenic to humans [11, 12]. When discharged to coastal waters, concentrations of pathogens may be high and can pose high levels of human health risks.

Table 16.1 Allothchonous human pathogens detected in coastal waters

	Concentration/Occurrence	Reference
Viruses		
Enteroviruses	Present in 9 of 72 1-liter samples using RT-PCR at Avalon Beach, CA[a]	[81]
Adenoviruses	Present in 15 of 30 250-liter samples using PCR at Silver Beach, MI[a]	[63]
Hepatitis A	105–30,771 viral particles/l using Q-RTPCR at Imperial Beach, CA[a]	[82]
Norovirus	2 of 19 samples in 110-liters using RT-PCR at Key West sites (FL)[a]	[83]
Rotavirus (reovirus)	2 of 19 sites with 2–5 MPN/L at Italian coastline	[84]
Bacteria		
Campylobacter	Detected in 25 of 192 100 – 1,000 mL Spanish marine recreational water samples using culture-based methods	[85]
Salmonella	Detected in 70–100% of samples from a lagoon in Brazil using culture-based methods, volume assayed not reported	[86]
Staphylococcus	60–70% of approx. 100 mL seawater samples from Doheny and Avalon Beach, CA using culture-based methods	[13]
Pathogenic *Escherichia coli*	2 of 377 *E. coli* isolates from North Carolina and Southern California coastal waters using combined culture and PCR methods	[87]
Shigella	100% of algal mat samples from Lake Michigan near Burns Ditch by PCR	[18]
Protozoa		
Cryptosporidium	13.7 ± 1.7 oocysts/L on weekends at Chesapeake Bay beach, MD	[88]
Giardia	9.1 ± 1.1 cysts/L on weekends at Chesapeake Bay beach, MD	[88]

[a]Volumes reported do not account for the fact that a fraction of water sample was used during polymerase chain reaction (PCR), reverse-transcriptase (RT-)PCR, or quantitative (Q)PCR.

Pathogens are also found in environmental reservoirs that in some cases may serve as a source of pathogens to recreational waters. Beach sands can harbor *Campylobacter, Salmonella*, and *S. aureus* [13–16]. Aquatic sediments can accumulate bacteria, protozoa, and viruses [17]. Marine and lacustrine kelp species may also harbor bacterial pathogens including pathogenic *E. coli*, *Salmonella*, and *Campylobacter* [18, 19].

Assessing Risk: Epidemiology and Indicator Organisms

The monitoring of recreational waters for all RWI pathogens to assess the safety of swimming is not scientifically or economically feasible [20]. Microbial indicator organisms have been used for centuries as indicators of the presence of human pathogens. Internationally, many countries use fecal and total coliforms as a basis for their recreational water quality criteria, standards, or guidelines [21]. Other countries rely on measurements of enterococci (or fecal streptococci), *E. coli*, or both for their recreational waters, most based on guidelines provided by World Health Organization [22] and/or the US Environmental Protection Agency (EPA). These organisms were chosen as indicators because their concentrations are high in human wastewater and feces, they are relatively simple to measure, and their presence in coastal waters is correlated to adverse health outcomes in swimmers through epidemiology studies conducted in wastewater-impacted waters [23–26]. The epidemiology studies that correlate indicator concentration to adverse health outcomes are key to the use of indicators to assess risk.

Epidemiology studies evaluate illness resulting from exposure to a particular contaminant or activity. When applied to RWI, epidemiology studies evaluate the illness rates in swimmers versus non-swimmers, and characterize illness rates as a function of indicator organism concentration. The studies involve the collection of health and behavior data contemporaneously with concentrations of indicator organisms. RWI epidemiology studies are either case-control randomized trial or prospective cohort designs. In case-control studies, swimmer and non-swimmer activities are prescribed by randomization at the onset of the study. In these studies, exposures are well controlled. Subject recruitment is done in advance of the study. In prospective cohort studies, subjects are recruited at the study site and are enrolled when they arrive at the shoreline. Exposures are self-prescribed by subjects and behavioral data on exposure is collected using self-reports at the end of the day. In the prospective cohort design, there is less control over exposure, but the exposures are more realistic as they are not prescribed by the study design. Additional types of studies that have been employed to study RWI include cross-sectional studies and event studies. The former is similar in design to a cohort study; the latter takes advantage of a sporting event, for example, for data collection.

Since the 1950s, numerous epidemiological studies have been conducted throughout the world to evaluate the association between recreational water quality and RWI (including GI symptoms; eye infections; skin complaints; ear, nose, and throat infections; and respiratory illness) [23–26]. Most of these studies investigated wastewater effluent-impacted marine and estuarine waters alone or in combination with freshwater. Several investigated freshwater recreational environments or non-wastewater effluent-impacted waters. These studies indicate that the rates of some adverse health outcomes are higher in swimmers compared with non-swimmers [23].

Taken as a whole, the weight of evidence from these studies indicates that fecal indicator bacteria (fecal streptococcus/*Enterococcus*, in particular) are able to predict GI and in some cases, respiratory illnesses from exposure to recreational waters [23, 25, 26]. This broad base of information stems from studies conducted throughout much of the developed world (Table 16.2, Adapted from [27]).

Several meta-analyses and/or systematic reviews have summarized the available recreational water epidemiology studies [23, 25, 26]. Pruss et al. [23] conducted a systematic review to initiate development of new WHO guidelines for

Table 16.2 Recreational water epidemiology studies included in reviews by Prüss [23], Wade et al. [25], and Zmirou et al. [26]. Location refers to geographic location of study. Water Type refers to whether the study was conducted at a marine or fresh water. Study design denotes whether to study was a cohort, randomized trial, cross section, or event study

Reference	Location	Water type	Study design	Review article
Alexander et al. [89]	UK	Marine	Cohort	Wade, Zmirou
Bandaranayake [90]	New Zealand	Marine	Cohort	Prüss
Brown et al. [49]	UK	Marine	Cohort	Zmirou
Cabelli [38]	USA	Marine	Cohort	Wade, Prüss, Zmirou
Cabelli [44]	Egypt	Marine	Cohort	Wade, Prüss
Calderon et al. [31]	USA	Fresh	Cohort	Wade
Cheung et al. [36]	Hong Kong	Marine	Cohort	Wade, Prüss, Zmirou
Corbett et al. [52]	Australia	Marine	Cohort	Wade, Prüss, Zmirou
Dufour [39]	USA	Fresh	Cohort	Wade, Prüss, Zmirou
Fattal et al. [42]	Israel	Marine	Cohort	Wade, Prüss, Zmirou
Ferley et al. [47]	France	Fresh	Cohort	Wade, Prüss, Zmirou
Fewtrell et al. [91]	UK	Fresh	Event	Wade, Zmirou
Fewtrell et al. [92]	UK	Marine	Cohort	Wade, Zmirou
Kay et al. [71]	UK	Marine	Randomized trial	Wade
Fleisher et al. [93]	UK	Marine	Randomized trial	Prüss
Foulon et al. [48]	France	Marine	Cross-sectional	Wade
Haile et al. [34, 94]	USA	Marine	Cohort	Wade, Prüss, Zmirou
Kay et al. [71]	UK	Marine	Randomized trial	Wade, Prüss, Zmirou
Kueh et al. [95]	Hong Kong	Marine	Cohort	Wade, Prüss
Lee et al. [96]	UK	Fresh	Event	Wade
Lightfoot [97]	Canada	Fresh	Cohort	Wade, Prüss
Marino et al. [98]	Spain	Marine	Cohort	Wade
McBride et al. [33]	New Zealand	Marine	Cohort	Wade
Medema et al. [99]	The Netherlands	Fresh	Event	Wade
CSIR [100]	South Africa	Marine	Cohort	Prüss
Mujeriego et al. [46]	Spain	Marine	Cohort	Prüss
Philipp et al. [101]	UK	Marine	Event	Wade, Zmirou
Pike [102]	UK	Marine	Cohort	Wade, Prüss, Zmirou
Prieto et al. [103]	Spain	Marine	Cohort	Wade
Seyfried et al. [40]	Canada	Fresh	Cohort	Wade, Prüss, Zmirou
Stevenson [104]	USA	Fresh	Cohort	Wade, Prüss, Zmirou
UNEP/WHO [105]	Israel	Marine	Cohort	Prüss
UNEP/WHO [106]	Spain	Marine	Cohort	Prüss
Van Asperen et al. [107]	The Netherlands	Fresh	Event	Wade, Zmirou
Van Dijk et al. [108]	UK	Marine	Cohort	Prüss
Von Schirnding et al. [109]	South Africa	Marine	Cohort	Wade, Zmirou

recreational use of the water environment. The comprehensive review of 22 published studies on sewage pollution of recreational water and health outcomes concluded that the epidemiological basis had been laid to develop WHO guidelines on fecal pollution based on a causal association between GI illness symptoms and increased concentrations of bacterial indicators (i.e., enterococci for marine, enterococci and *E. coli* for fresh) in recreational waters.

Zmirou et al. [26] examined 18 published studies to provide a scientific basis for establishing new standards for the microbial quality of marine and fresh recreational waters to replace the 30 year-old European Union bathing water quality guidelines [28]. The researchers provided four major results: (1) increased concentrations of fecal coliforms or *E. coli* and enterococci in both fresh and marine recreational waters are associated with increased risks of acute GI illness, with enterococci eliciting four to eight times greater excess risks than fecal coliforms or *E. coli* at the same concentrations; (2) GI illness risks associated with enterococci occur at lower indicator concentrations in marine versus fresh recreational waters; (3) increased concentrations of total coliforms have little or no association with GI illness risk; and (4) no evidence exists of a concentration threshold of indicator microorganisms below which there would be no GI illness risk to bathers.

Wade et al. [25] conducted a systematic review and meta-analysis of 27 published studies to evaluate the evidence linking specific microbial indicators of recreational water quality to specific health outcomes under non-outbreak (endemic) conditions. The study was conducted at the request of the United States National Academy of Sciences. Secondary goals included identifying and describing critical study design issues and evaluating the potential for health effects at or below the current regulatory criteria [29]. They concluded that (1) enterococci and, to a lesser extent, *E. coli* are adequate indicators (predictors) of GI illness in marine recreational waters, but fecal coliforms are not; (2) the risk of GI illness is considerably lower in studies with enterococci and *E. coli* densities below those established by EPA [29], thus providing support for their regulatory use; (3) *E. coli* is a more reliable and consistent predictor of GI illness than enterococci or other indicators in fresh recreational waters; and (4) studies that used a non-swimming control group and that focused on children found elevated GI illness risks.

Based on these meta-analyses, the weight of evidence indicates that there is a relationship between levels of specific indicator bacteria and RWI in coastal waters impacted by wastewater. However, as discussed earlier there are many sources of indicator bacteria to coastal waters, and many of these sources contain different pathogens with diverse health risks. Along coastlines with good sewage infrastructure and regulated anthropogenic discharges, wastewater is unlikely to be contributing substantial amounts of indicator organisms to the swimming areas on a regular basis [27]. Important sources of indicator organisms and pathogens are probably nonpoint in nature, emanating from soils, animal feces, urban runoff, stormwater runoff, or agricultural runoff.

There are only a few epidemiology studies that examine the link between RWI and fecal indicator organisms in recreational waters polluted with sources other than wastewater. Review of these studies suggests the relationship between indicator concentration and RWI risks are equivocal. On one hand, Colford et al. [30] found that the incidence of swimmer illness was greater than the incidence of non-swimmer illness, but swimmer illness was not associated with any of the bacterial indicator organisms at a marine beach where bacterial contamination was not attributable to wastewater discharges. Similarly, Calderon et al. [31] found no statistically significant association between swimmers' illness risk and indicator concentrations in a freshwater pond where agricultural runoff was the source of contamination. McBride [32] suggested that if more swimmers had been included in the Calderon et al. [31] study, achieving statistically significant results would have been possible, however. At a marine bathing study in New Zealand, McBride et al. [33] indicated that RWI risks posed by animal versus human fecal material were not substantially different; however, the study's limited range of indicator organisms' concentrations precluded the development of a detailed statistical model of health risks versus indicator density.

In the first study to be conducted in waters directly impacted by urban runoff, Haile et al. [34] reported associations between swimmer health and indicator densities. However, this nonpoint runoff source was known to contain human sources of fecal contamination, based on the presence of human enteric viruses. Dwight and colleagues [35] found that surfers exposed to Southern California urban runoff had higher illness rates than surfers exposed to Northern California rural runoff. The results from a Hong Kong marine water study [36] and a German freshwater study [37] are more difficult to interpret regarding risks from human versus nonhuman sources because in both studies, the analyses combined the results from sites with different predominant contamination sources.

An additional meta-analysis examined the differential risks associated with exposure to human and nonhuman animal fecal material [24]. Illness risk associated with bathing in water polluted primarily with human fecal material was reviewed based on studies from the USA [38, 39], Canada [40, 41], Israel [42, 43], Egypt [44, 45], Spain [46], France [47, 48], the UK [49, 50], Hong Kong [51], and Australia [52, 53]. Most of these studies showed a positive correlation between GI illness and fecal indicator density; there was little equivalent evidence from waters polluted primarily with animal feces. The only study specifically designed to address swimming-associated illness in animal-impacted waters was that of Calderon et al. [31] who found no statistically significant association between GI illness and fecal indicator bacteria densities. Based on this observation, Sinton and colleagues [24] concluded that reliable epidemiological evidence was lacking, and that other sources of information were needed to identify and apportion human and nonhuman fecal inputs to natural waters.

Given these studies, there are not sufficient epidemiological data to conclude that concentration of indicator organisms in coastal waters not impacted by wastewater are predictive of health risks in all cases. More epidemiology data may help address the lack of information. However, fecal contamination in coastal waters not impacted by wastewater is likely to be highly variable and emanate from a complex mix of sources. Other approaches may be more useful for addressing the relationship between indicators and health risk along these shorelines. Despite the lack of epidemiological evidence for a relationship between indicator organism concentration and health risk, it is well established in the outbreak literature that water contaminated with animal feces can cause illness in humans [4].

Assessing Risk: Quantitative Microbial Risk Assessment (QMRA) Modeling

QMRA is a health risk modeling approach that translates microbial exposures into infection or illness risk estimates. For RWI, the dose received by individuals is derived from estimates of the volume of water ingested during an exposure event and concentration of pathogen(s) in that volume of water. Once a dose (number of pathogens per exposure) is determined then a risk of infection or illness is derived from applying published dose-response models for specific pathogens [54], which are derived from human feeding trials or outbreak data [55, 56].

Estimations of key model parameters (such as ingested volume of water and pathogen concentration) are generally described as probability density functions (PDFs) (i.e., distributions to account for the stochastic nature of the parameter) to help account for inherent variability as well as various methods and model uncertainties. The characterized risk is best described as a distribution as well, to capture variability and uncertainty as best as possible. This characterized risk can then be compared to "tolerable risk" or a site benchmark for the recreational water (e.g., 8 or 19 gastrointestinal illnesses per 1,000 bathers as used in the 1986 EPA criteria for freshwater and marine recreation, respectively).

An initial screening-level risk assessment for a site may start by only using point estimates to describe model parameters (e.g., WHO, 2004). Then, to reduce uncertainties in the risk estimate, more complexity is built into a QMRA model using PDFs to better represent a specific site, in what is called a "static" model. An alternative approach known as "dynamic QMRA" takes into account secondary infections to the broader community, as well as addressing the susceptibility versus nonsusceptibility of individuals to infection and illness [57–59].

QMRA can be useful for a number of purposes. First, it can be used to look at hypothetical risks under different scenarios of pathogen sources and/or recreational activities and exposure routes [60, 61]. This can provide a human health interpretation of environmental pathogen concentrations, or it can provide guidance for decision-making with respect to alternative management options. Second, QMRA can be used to augment the understanding of recreational water epidemiology studies [62]. A number of studies have evaluated pathogen risks in recreational waters using QMRA [60, 61, 63–70]. A handful of these are reviewed below.

Diallo et al. [69] examined the risk associated with recreational exposure to canals in Thailand containing *Giardia*, *Cryptosporidum*, and diarrhegenic *E. coli*. They found the predicted risk of illness from the protozoa was two orders of magnitude higher than the actual protozoan infection rate in the region of Thailand, while the rate of gastrointestinal illness predicted from diarrhegenic *E. coli* matched actual observed rates of the disease in the region. The authors suggest that the illness rates predicted for the protozoa are much higher due to immunity in the community, which was not considered in the QMRA modeling. Another possibility is massive under-reporting of protozoa illness rates, which is a documented issue for diarrheal diseases.

Ashbolt and Bruno [65] used QMRA in conjunction with the published relationship between enterococci concentrations and probability of gastrointestinal illness and acute respiratory illness observed during a UK beach epidemiology study [71]. The authors showed that by assuming that the etiology of illnesses was viral, a fixed ratio between enterococci and viruses of 1–175, a volume of water ingested of 50 mL, and 50% illness rate of those infected with virus, they were able to model the observed illness rates (at exposures greater than 60 CFU/100 mL enterococci) assuming viruses had an exponential dose-response curve. Further, they were able to model the observed acute respiratory illness using the dose-response curve of adenovirus.

Gerba et al. [70] estimated risk of exposure to rotavirus in bathing waters using the few previously reported data available at the time on rotavirus concentrations in bathing waters. They found the risk of illness to be 1/10–1/100 with a one time bathing event. These authors chose to use rotavirus as a model pathogen because it is one of the most infectious viruses for which a dose-response model is available. A major limitation to their study is the lack of data on concentrations of rotavirus in bathing waters. Surface water concentrations were estimated to be between 0.24/L and 29/L. Ingested volumes used were 100 mL for recreational exposure.

Schoen and Ashbolt [67] explored the relative risk associated with exposure to seagull feces, poorly treated sewage, and a mixture of both sources in seawater with enterococci at levels of 35 CFU/100 mL (a USEPA standard). Authors assumed that seagull feces contained *Campylobacter* and *Salmonella*, while sewage contained norovirus, *G. intestinalis*, *Cryptosporidium* spp., and *Salmonella*. A distribution of ratios of enterococci to pathogen concentrations were considered to reflect the uncertainty in this parameter. The authors showed that when enterococci came exclusively from seagull feces, the risk of illness is less than the benchmark of 0.01 on which US standards are based.

Soller et al. [61] examined the risk of viral gastroenteritis associated with recreational and non-recreational use of a river downstream of a wastewater treatment plant discharge. Two wastewater treatment scenarios were compared with the goal of evaluating the public health benefit of increased treatment of the effluent. The authors employed a health protective approach by assuming that the etiology of illness would be a virus with clinical features identical to those of rotavirus. They also assumed that removal of the virus in treatment facilities would mirror that of coliphage and that coliphage and the virus occurred in a ratio that varied from 0.001 to 1. Exposures were informed by a hydrologic model of the area and observations of swimmer behavior. Unlike the other models discussed above, this model incorporated secondary transmission by allowing illness to be passed from person to person.

Steyn et al. [68] compared the risks in recreational surfaces waters in South Africa expected from measured *E. coli* concentrations (applying epidemiology study-derived dose-response curves) and measured *Salmonella* concentrations (applying the QMRA method and the *Salmonella* dose-response curve assuming ingestion of 100 mL of water for exposure). The researchers found that risks derived from the *Salmonella* QMRA model were higher than those derived from the *E. coli* concentrations. Their results suggest using *E. coli* to assess risk from exposure to *Salmonella* may be inadequate.

Wong et al. [63] estimated risk from exposure to adenoviruses during recreation contact with water at Great Lakes beaches impacted by point sources of treated wastewater. The authors measured adenoviruses using an MPN culture-dependent assay at their study site, assumed an adenovirus dose-response model, and an ingestion rate of 100 mL of water. The authors found that 0.24–2.4 illnesses per 1,000 swimmers were likely to have occurred from adenoviruses, a range of frequencies below the EPA guideline of 8 illnesses per 1,000 swimmers. Although an epidemiology study was done at the same time as their study, the epidemiology data were not available for comparison of actual illness rates.

Soller et al. [62] used QMRA to understand more fully the reported epidemiologic results from studies conducted on the Great Lakes in the US during 2003 and 2004 by identifying pathogens that could have caused the observed illnesses in those studies. The reference pathogens used for this analysis were *Norovirus*, rotavirus, adenovirus, *Cryptosporidium* spp., *G. lamblia, Campylobacter jejuni, Salmonella enterica, and E. coli* O157:H7. Two QMRA-based approaches were used to estimate the pathogen combinations that would be consistent with observed illness rates: in the first, swimming-associated gastrointestinal (GI) illnesses were assumed to occur in the same proportion as known illnesses in the US due to all non-foodborne sources, and in the second, pathogens were assumed to occur in the recreational waters in the same proportion as they occur in disinfected secondary effluent. The results indicated that human enteric viruses and in particular, norovirus could have caused the vast majority of the observed swimming-associated GI illnesses during the 2003/2004 water epidemiology studies [72, 73]. Evaluation of the time to onset of illness strongly supports the principal finding, and sensitivity analyses support the overall trends of the analyses even given their substantial uncertainties. These results are notable because little is known about the specific microbial agents that are responsible for the observed illnesses in swimmers. While several studies have attempted to collect biological specimens (blood or stool) as part of epidemiologic research at beach sites, these efforts have to date been largely unsuccessful in identifying the agents responsible for the observed increase in GI symptoms among swimmers [50].

Soller et al. [74] conducted a QMRA investigation to determine whether estimated risks following exposure to recreational waters impacted by gull, poultry, pig, or cattle fecal contamination are substantially different from those associated with waters impacted by human sources such as treated wastewater. Previously published QMRA methods were employed and extended to meet these objectives [67]. Health outcomes used in the analyses were infection via reference pathogens by water ingestion during recreation and subsequent GI illness. Illness risks from the reference pathogens were calculated for exposure to contaminated recreational water with fecal indicator bacterial densities at the U.S. regulatory limits: 35 CFU/100 mL enterococci and 126 CFU/100 mL *E. coli*. The probabilities of GI illness from reference pathogens were calculated using dose-response relationships from the literature and Monte Carlo simulations. The primary findings from the analysis were that: (1) GI illness risks associated with human exposure to recreational waters directly impacted by fresh cattle feces are not substantially different from those impacted by human sources; and (2) the risks associated with human exposure to recreational waters impacted by fresh gull, poultry, or pig feces are substantially lower than those impacted by human sources.

Several observations may be drawn from the QMRA studies summarized above. First, the assembled studies focused on a small subset of the pathogens potentially important in waterborne exposure during recreation. The pathogen analyzed most frequently is rotavirus, primarily due to its high infectivity and the availability of dose-response data. It is likely that future QMRAs will also focus on norovirus [62, 67, 74] now that a published dose-response relationship is available [75].

Second, modeling variability in pathogen source density appears to be hampered by scarcity of both data and analysis techniques. Two common methods for accounting for source variability among studies are (1) use of empirical distributions for pathogen density based on relatively short time series, and (2) assumption of log-normal distribution of pathogen densities. There are advantages and drawbacks to both of these approaches. Drawbacks to the use of empirical distributions are inconsistency in sampling strategies used to develop databases, frequent non-detects, constraint of pathogen densities to those observed in a limited number of samples, and lack of availability of pathogen concentration data. Use of distributions to describe pathogen density in sources overcomes some of the constraints associated with use of empirical distributions, but choosing a distributional family can be problematic. Among the studies reviewed, many studies employed point estimates for pathogen density. Among studies using distributions to describe pathogen variability, the following distributions were employed: normal, triangular, log-normal, negative binomial, uniform, and Poisson.

Third, most of the studies reviewed do not account for variability and uncertainty in dose-response model parameters. As with variability in exposure, this observation likely indicates that high quality and diverse dose-response model data are not available. The use of a small number of dose-response models may indicate that some QMRA modelers chose the pathogens to model based on the availability of dose-response models. Lack of dose-response models for many pathogens of concern and for differing routes of exposure (i.e., cutaneous exposure to *S. aureus*) is a major data gap. The need for dose-response models corresponding to different exposure routes (i.e., ingestion, inhalation, etc.) arises from the ability of some waterborne pathogens such as adenovirus to initiate infection via multiple routes.

Finally, secondary transmission and immunity are often neglected in risk estimation. Studies that have included secondary transmission and waterborne illnesses [57, 58, 60, 76–78] have demonstrated that consideration of secondary transmission and immunity can influence overall risk associated with exposure to pathogens significantly and in unintuitive ways.

Future Directions

Health data collected from recreational swimmers confirm measurable health effects from exposure to contaminated coastal waters. To adequately protect the health of swimmers and others who recreate in coastal waters or consume fish and shellfish from coastal waters, it is essential to understand the microbial hazards present and the risks they represent to human health. While there are data available on microbial hazards and risks, many aspects remain poorly understood and characterized.

Research is needed to further characterize microbial hazards in recreational waters. Pathogen detection techniques that allow detection of infectious pathogens rapidly in environmental waters are needed. There are a number of pathogen and indicator detection technologies that are in development or have been recently developed [79, 80]. However, many of these have not been applied to natural waters, or are just in the "proof of concept" stage. Work is needed to transit new detection technologies that work at the bench-scale to the field scale – to detect pathogens in environmental waters. Once this is accomplished, they should be applied to a wide array of waterbodies to fully understand pathogen and indicator occurrence, concentrations, fate, and transport.

A better understanding of the risks of exposure to pathogens from different sources is needed. Some QMRA studies have addressed this issue, but further research is needed to ground truth the QMRA results with epidemiology data, or data on infection rates and etiologies, as determined through analysis of bodily fluid from individuals with RWI. Gastroenteritis is the most well-studied RWI; more work is needed to understand the importance of other RWI including skin infections and respiratory ailments. More dose-response data for a wider array of pathogens are needed to provide more refined estimates of risk using QMRA. The importance of secondary infections and immunity for RWI should also be further characterized.

Finally, better surveillance systems are needed for RWI so that prediction systems can be developed. As global climate changes in the coming decades, the scientific community needs to be able to anticipate how this might change the burden of RWI. Microbial pollution of coastal waters is expected to change as temperatures, runoff frequency and volumes, and rainfall pattern change. A thorough understanding of the occurrence of RWI, and the distribution, fate, and transport of waterborne pathogens will enable us to better anticipate the effects of climate change.

Bibliography

1. CDC http://www.cdc.gov/healthywater/swimming/rwi.html (5 January)
2. Shuval H (2003) Estimating the global burden of thalassogenic diseases: human infectious diseases caused by wastewater pollution in the environment. J Water Health 1:53–64

3. Given S, Pendleton LH, Boehm AB (2006) Regional public health cost estimates of contaminated coastal waters: a case study of gastroenteritis at Southern California beaches. Environ Sci Technol 40:4851–4858
4. Craun GF (2003) Causes of waterborne outbreaks in the United States. Wat Sci Tech 24:17
5. Santo-Domingo JW, Hansel J (2008) Waterborne diseases and microbial water quality monitoring for recreational water bodies using regulatory methods. In: Walsh PJ, Smith SL, Fleming LE, Solo-Gabriele HM, Gerwick WH (eds) Oceans and human health. Academic, San Diego
6. Griffin DW, Donaldson KA, Paul JH, Rose JB (2003) Pathogenic human viruses in coastal waters. Clin Microbiol Rev 16(1):129–143
7. Maier RM, Pepper IL, Gerba CP (2000) Environmental microbiology. Academic, San Diego
8. Boehm AB, Bischel HN (2010) Oceans and human health. In: Nriagu J (ed) Encyclopedia of environmental health. Elsevier, New York, NY
9. Walsh PJ, Smith SL, Fleming LE, Solo-Gabriele HM, Gerwick WH (2008) Oceans and human health. Academic, San Diego
10. Mead PS, Slutsker L, Dietz V, McCaig LF, Bresee JS, Shapiro C, Griffin PM, Tauxe RV (1999) Food related illness and death in the United States. Emerg Infect Dis 5(5):607–625
11. Olivieri AW, Boehm AB, Sommers CA, Soller JA, Eisenberg JNS, Danielson R (2007) Development of a protocol for risk assessment of separate stormwater system microorganisms; Water Environment Research Foundation. Project 03-SW-2, Final Project Report
12. U.S. EPA (2009) Review of published studies to characterize relative risks from different sources of fecal contamination in recreational water, EPA 822-R-09-001. In Lett Appl Microbiol, Office of Science and Technology, Washington
13. Goodwin KD, Pobuda M (2009) Performance of CHROMagar™ Staph aureus and CHROMagar™ MRSA for detection of Staphylococcus aureus in seawater and beach sand – Comparison of culture, agglutination, and molecular analyses. Water Res 43(19):4802–4811
14. Bolton F, Surman SB, Martin K, Wareing DR, Humphrey TJ (1999) Presence of Campylobacter and Salmonella in sand from bathing beaches. Epidemiol Infect 122:7–13
15. Elmanama AA, Fahd MI, Afifi S, Abdallah S, Bahr S (2004) Microbiological beach sand quality in Gaza Strip in comparison to seawater quality. Environ Res 99:1–10
16. Obiri-Danso K, Jones K (2000) Intertidal sediments as reservoirs for hippurate negative campylobacters, salmonellae and faecal indicators in three EU recognized bathing waters in North West England. Water Res 34:519–527
17. LaBelle RL, Gerba CP, Goyal SM, Melnick JL, Cech I, Bogdan GF (1980) Relationships between environmental factors, bacterial indicators, and the occurrence of enteric viruses in estuarine sediments. Appl Environ Microbiol 39(3):588–596
18. Ishii S, Yan T, Shivley DA, Byappanahalli MN, Whitman RL, Sadowsky MJ (2006) Cladophora (Chlorophyta) spp. harbor human bacterial pathogens in nearshore water of Lake Michigan. Appl Environ Microbiol 72:4545–4553
19. Byappanahalli MN, Sawdey R, Ishii S, Shively DA, Ferguson JA, Whitman RL, Sadowsky MJ (2009) Seasonal stability of Cladophora-associated Salmonella in Lake Michigan watersheds. Water Res 43:806–814
20. U.S. EPA (2009) Review of published studies to characterize relative risks from different sources of fecal contamination in recreational water, EPA 822-R-09-001. In Office of Science and Technology, Washington
21. WHO (1999) Health-based monitoring of recreational waters: the feasibility of a new approach (The 'Annapolis Protocol'); WHO/SDE/WSH/99.1, World Health Organization, 1999
22. WHO (2003) Guidelines for safe recreational-water environments, coastal and fresh-waters, vol 1. World Health Organization, Geneva

23. Prüss A (1998) Review of epidemiological studies on health effects from exposure to recreational water. Int J Epidemiol 27(1):1–9
24. Sinton LW, Finlay RK, Hnnah DJ (1998) Distinguishing human from animal faecal contamination in water: a review. NZ J Mar Freshw Res 32:323–348
25. Wade TJ, Pai N, Eisenberg J, Colford JM (2003) Do US EPA water quality guidelines for recreational waters prevent gastrointestinal illness? A systematic review and meta-analysis. Environ Health Perspect 111(8):1102–1109
26. Zmirou D, Pena L, Ledrans M, Letertre A (2003) Risks associated with the microbiological quality of bodies of fresh and marine water used for recreational purposes: summary estimates based on published epidemiological studies. Arch Environ Health 58(11):703–711
27. Boehm AB, Ashbolt NJ, Colford JM, Dunbar J, Fleming LE, Gold MA, Hansel JA, Hunter PR, Ichida AM, McGee CD, Soller JA, Weisberg SB (2009) A sea change ahead for recreational water quality criteria. J Water Health 7:9–20
28. EEC (1976) Bathing water quality: Directive 76/160/EEC, 1–6
29. U.S. EPA (1986) Ambient water quality criteria for bacteria, EPA440/5-84-002, Office of Water, Washington
30. Colford JM Jr, Wade TJ, Schiff KC, Wright CC, Griffith JF, Sandhu SK, Burns S, Sobsey M, Lovelace G, Weisberg SB (2007) Water quality indicators and the risk of illness at beaches with nonpoint sources of fecal contamination. Epidemiology 18(1):27–35
31. Calderon R, Mood E, Dufour A (1991) Health effects of swimmers and nonpoint sources of contaminated water. Int J Environ Health Res 1:21–31
32. McBride GB, Calderon RL, Mood EW, Dufour AP (1991) Health effects of swimmers and non-point sources of contaminated water. Int J Environ Health Res 1:21–31
33. McBride GB, Salmond CE, Bandaranayake DR, Turner SJ, Lewis GD, Till DG (1998) Health effects of marine bathing in New Zealand. Int J Environ Health Res 8(3):173–189
34. Haile RW, Witte JS, Gold M, Cressey R, McGee C, Millikan RC, Glasser A, Harawa N, Ervin C, Harmon P, Harper J, Dermand J, Alamillo J, Barrett K, Nides M, Wang GY (1999) The health effects of swimming in ocean water contaminated by storm drain runoff. Epidemiology 10(4):355–363
35. Dwight RH, Baker DB, Semenza JC, Olson BH (2004) Health effects associated with recreational coastal water use: urban versus rural California. Am J Public Health 94(4):565–567
36. Cheung WH, Chang KC, Hung RP, Kleevens JW (1990) Health effects of beach water pollution in Hong Kong. Epidemiol Infect 105(1):139–162
37. Wiedenmann A, Kruger P, Dietz K, Lopez-Pila JM, Szewzyk R, Botzenhart K (2006) A randomized controlled trial assessing infectious disease risks from bathing in fresh recreational waters in relation to the concentration of Escherichia coli, intestinal enterococci, Clostridium perfringens, and somatic coliphages. Environ Health Perspect 114(2):228–236
38. Cabelli VJ (1983) Health effects criteria for marine recreational waters, EPA-600/1-80-031. USEPA, Cincinnati
39. Dufour A (1984) Health effects criteria for fresh recreational waters; EPA-600 1-84-004. US EPA, Cincinnati
40. Seyfried PL, Tobin RS, Brown NE, Ness PF (1985) A prospective study of swimming-related illness. I. Swimming-associated health risk. Am J Public Health 75(9):1068–1070
41. Seyfried PL, Tobin RS, Brown NE, Ness PF (1985) A prospective study of swimming-related illness. II. Morbidity and the microbiological quality of water. Am J Public Health 75(9):1071–1075
42. Fattal B, Peleg-Olevsky E, Yoshpe-Purer Y, Shuval H (1986) Association between morbidity among bathers and microbial quality of seawater. Water Sci Technol 18(11):59–69
43. Fattal DB, Peleg-Olevsky E, Agursky T, Shuval PHI (1987) The association between seawater pollution as measured by bacterial indicators and morbidity among bathers at Mediterranean bathing beaches of Israel. Chemosphere 16(2/3):565–570

44. Cabelli V (1983) Public health and water quality significance of viral diseases transmitted by drinking water and recreational water. Water Sci Technol 15:1–15
45. El-Sharkawi HF (1979) The relation between the state of pollution in Alexandria swimming beaches and the occurrence of typhoid among bathers. Bull High Inst Public Health 9:337–351
46. Mujeriego R, Bravo J, Feliu M (1982) Recreation in coastal waters: public health implications. Vier Journee Etudes Pollutions, 585–594
47. Ferley JP, Zmirou D, Balducci F, Baleux B, Fera P, Larbaigt G, Jacq E, Moissonnier B, Blineau A, Boudot J (1989) Epidemiological significance of microbiological pollution criteria for river recreational waters. Int J Epidemiol 18(1):198–205
48. Foulon G, Maurin J, Quoi N, Martin-Boyer G (1983) Relationship between the microbial quality of water and health effects. Revue des Sciences de L'Eau 2:127–143
49. Brown JM, Campbell EA, Rickards AD, Wheeler D (1987) Sewage pollution of bathing water. Lancet 2(8569):1208–1209
50. Jones F, Kay D, Stanwell-Smith R, Wyer M (1991) Results of the first pilot-scale controlled cohort epidemiological investigation into the possible health effects of bathing in seawater at Langland Bay, Swansea. Water Environ J 5(1):91–98
51. Holmes P (1989) Research into health risks at bathing beaches in Hong Kong. J Inst Water Environ Manage 3:488–495
52. Corbett SJ, Rubin GL, Curry GK, Kleinbaum DG (1993) The health effects of swimming at Sydney beaches. The Sydney beach users study advisory group. Am J Public Health 83(12):1701–1706
53. Harrington JF, Wilcox DN, Giles PS, Ashbolt NJ, Evans JC, Kirton HC (1993) The health of Sydney surfers: an epidemiological study. Water Sci Technol 27:175–181
54. Haas CN (1983) Estimation of risk due to low doses of microorganisms: a comparison of alternative methodologies. Am J Epidemiol 55:573–582
55. Haas CN, Rose JB, Gerba CP (1999) Quantitative microbial risk assessment. Wiley, New York
56. McBride G, Till D, Ryan T, Balll A, Lewis G, Palmer S, Weinstein P (2002) Pathogen occurrence and human health risk assessment analysis. In: Freshwater Microbiology Research Programme, Ministry for the Environment, Ministry of Health, New Zealand
57. Soller JA (2009) Potential implications of person-to-person transmission of viral infection to US EPA's Groundwater Rule. J Water Health 7(2):208–223
58. Soller JA, Eisenberg JNS (2008) An evaluation of parsimony for microbial risk assessment models. Environmetrics 19(1):61–78
59. Eisenberg JNS, Seto EYW, Olivieri AW, Spear RC (1996) Quantifying water pathogen risk in an epidemiological framework. Risk Anal 16(4):549–563
60. Soller JA, Eisenberg J, DeGeorge J, Cooper R, Tchobanoglous G, Olivieri A (2006) A public health evaluation of recreational water impairment. J Water Health 4(1):1–19
61. Soller JA, Olivieri AW, Crook J, Cooper RC, Tchobanoglous G, Parkin RT, Spear RC, Eisenberg JN (2003) Risk-based approach to evaluate the public health benefit of additional wastewater treatment. Environ Sci Technol 37(9):1882
62. Soller JA, Bartrand T, Ashbolt NJ, Ravenscroft J, Wade TJ (2010) Estimating the primary aetiologic agents in recreational freshwaters impacted by human sources of faecal contamination. Water Res 44(16):4736–4747
63. Wong M, Kumar L, Jenkins TM, Xagoraraki I, Phanikumar MS, Rose JB (2009) Evaluation of public health risks at recreational beaches in Lake Michigan via detection of enteric viruses and a human-specific bacteriological marker. Water Res 43(4):1137–1149
64. Ashbolt A, Reidy C, Haas CN (1997) In Microbial health risk at Sydney's coastal bathing beaches. In: 17th Australian Water and Wastewater Association meeting, AWWA, Melbourne, 1997, pp 104–111
65. Ashbolt N, Bruno M (2003) Application and refinement of the WHO risk framework for recreational waters in Sydney, Australia. J Water Health 1(3):125–131

66. Roser D, Ashbolt N (2007) Source water quality assessment and the management of pathogens in surface catchments and aquifers. Research report 29, Bolivar
67. Schoen ME, Ashbolt NJ (2010) Assessing pathogen risk to swimmers at non-sewage impacted recreational beaches. Environ Sci Technol 44(7):2286–2291
68. Steyn M, Jagals P, Genthe B (2004) Assessment of microbial infection risks posed by ingestion of water during domestic water use and full-contact recreation in a mid-southern African region. Water Sci Technol 50(1):301–308
69. Diallo MB, Anceno AJ, Tawatsupa B, Houpt ER, Wangsuphachart V, Shipin OV (2008) Infection risk assessment of diarrhea-related pathogens in a tropical canal network. Sci Total Environ 407(1):223–232
70. Gerba CP, Rose JB, Haas CN, Crabtree KD (1996) Waterborne rotavirus: a risk assessment. Water Res 30(12):2929–2940
71. Kay D, Fleisher JM, Salmon RL, Jones F, Wyer WD, Godfree AF, Zelenauch-Jacquotte Z, Shore R (1994) Predicting likelihood of gastroenteritis from sea bathing: results from randomised exposure. Lancet 344:905–909
72. Wade TJ, Calderon RL, Brenner KP, Sams E, Beach M, Haugland R, Wymer L, Dufour AP (2008) High sensitivity of children to swimming-associated gastrointestinal illness: results using a rapid assay of recreational water quality. Epidemiology 19(3):375–383
73. Wade TJ, Calderon RL, Sams E, Beach M, Brenner KP, Williams AH, Dufour AP (2006) Rapidly measured indicators of recreational water quality are predictive of swimming-associated gastrointestinal illness. Environ Health Perspect 114(1):24–28
74. Soller JA, Schoen ME, Bartrand T, Ravenscroft J, Ashbolt NJ (2010) Estimated Human Health Risks from Exposure to Recreational Waters Impacted by Human and Non-human Sources of Faecal Contamination. Water Res 44(10):4674–4691
75. Teunis PF, Moe CL, Liu PSEM, Lindesmith L, Baric RS, Le Pendu J, Calderon RL (2008) Norwalk virus: how infectious is it? J Med Virol 80(8):1468–1476
76. Eisenberg JNS, Soller JA, Scott J, Eisenberg DM, Colford JMJ (2004) A dynamic model to assess microbial health risks associated with beneficial uses of biosolids. Risk Anal 24(1):221–236
77. Eisenberg JNS, Moore K, Soller JA, Eisenberg D, Colford JMJ (2008) Microbial risk assessment framework for exposure to amended sludge projects. Environ Health Perspect 116(6):727–733
78. Eisenberg JNS, Seto EYW, Colford JM, Olivieri AW, Spear RC (1998) An analysis of the Milwaukee cryptosporidiosis outbreak based on a dynamic model of the infection process. Epidemiol 9(3):255–263
79. Vikesland PJ, Wigginton KR (2010) Nanomaterial enabled biosensors for pathogen monitoring - A review. Environ Sci Technol 44(10):3656–3669
80. Goodwin KD, Litaker RW (2008) Emerging Technologies fir monitoring recreational waters for bacteria and viruses. In: Walsh PJ, Smith SL, Fleming LE, Solo-Gabriele HM, Gerwick WH (eds) Oceans and human health. Academic, San Diego
81. Boehm AB, Yamahara KM, Love DC, Peterson BM, McNeill K, Nelson KL (2009) Covariation and photoinactivation of traditional and novel indicator organisms and human viruses at a sewage – impacted marine beach. Environ Sci Technol 43:8046–8052
82. Gersberg RM, Rose MA, Robles-Sikisaka R, Dhar AK (2006) Quantitative detection of hepatitis A virus and enteroviruses near the United States-Mexico border and correlation with levels of fecal indicator bacteria. Appl Environ Microbiol 72(12):7438–7444
83. Griffin DW, Gibson CJ, Lipp EK, Riley K, Paul JH, Rose JB (1999) Detection of viral pathogens by reverse transcriptase PCR and of microbial indicators by standard methods in the canals of the Florida Keys. Appl Environ Microbiol 65(9):4118–4125
84. Aulicino FA, Orsini P, Carere M, Mastrantonio A (2001) Bacteriological and virological quality of seawater bathing areas along the Tyrrhenian coast. Int J Environ Health Res 11:5–11

85. Alonso JL, Alonso MA (1993) Presence of Campylobacter in marine waters of Valencia. Spain Water Res 27(10):1559–1562
86. Gonzalez AM, Paranhos R, Lutterbach MS (2010) Relationships between fecal indicators and pathogenic microorganisms in a tropical lagoon in Rio de Janeiro, Brazil. Environ Monit Assess 164(1–4):207–219
87. Lang AL, Tsai Y-L, Mayer CL, Patton KC, Palmer CJ (1994) Multiplex PCR for Detection of the Heat-Labile Toxin Gene and Shiga-Like Toxin I and II Genes in Escherichia coli Isolated from Natural Waters. Appl Environ Microbiol 60(9):3145–3149
88. Graczyk TK, Sunderland D, Tamang L, Lucy FE, Breysse PN (2007) Bather density and levels of Cryptosporidium, Giardia, and pathogenic microsporidian spores in recreational bathing water. Parasitol Res 101(6):1729–1731
89. Alexander LM, Heaven A, Tennant A, Morris R (1992) Symptomatology of children in contact with sea water contaminated with sewage. J Epidemiol Community Health 46(4):340–344
90. Bandaranayake DR, Turner SJ, McBride GB, Lewis G, Till D (1995) Health effects of bathing at selected New Zealand marine beaches. New Zealand Department of Health, Wellington
91. Fewtrell L, Godfree AF, Jones F, Kay D, Salmon RL, Wyer MD (1992) Health effects of white-water canoeing. Lancet 339:1587–1589
92. Fewtrell L, Kay D, Salmon R, Wyer M, Newman G, Bowering G (1994) The health effects of low-contact water activities in fresh and estuarine waters. J Inst Water Environ Manag 8:97–101
93. Fleisher JM, Kay D, Salmon RL, Jones F, Wyer MD, Godfree AF (1996) Marine waters contaminated with domestic sewage: nonenteric illnesses associated with bather exposure in the United Kingdom. Am J Public Health 86(9):1228–1234
94. Haile RW, Witte JS, Gold M, Cressey R, McGee C, Millikan RC, Glasser A, Harawa N, Ervin C, Harmon P, Harper J, Dermand J, Alamillo J, Barrett K, Nides M, Wang G.-Y (1996) An epidemiological study of possible adverse health effects of swimming in Santa Monica Bay. Santa Monica Bay Restoration Project, Santa Monica, CA, 70pp
95. Kueh CSW, Tam TY, Lee T, Wong SL, Lloyd OL, Yu ITS, Wong TW, Tam JS, Bassett DCJ (1995) Epidemiological study of swimming-associated illnesses relating to bathing-beach water quality. Water Sci Technol 31:1–4
96. Lee JV, Dawson SR, Ward S, Surman SB, Neal KR (1997) Bacteriophages are a better indicator of illness rates than bacteria amongst users of a white water course fed by a lowland river. Water Sci Technol 35:165–170
97. Lightfoot N (1989) A prospective study of swimming related illness at six freshwater beaches in southern Ontario. University of Toronto, Toronto
98. Marino F, Morinigo M, Martinez-Manzanares E, Borrego J (1995) Microbiological-epidemiological study of selected marine beaches in Malaga (Spain). Water Sci Technol 31:5–9
99. Medema G, Van Asperen I, Klokman-Houweling J, Nooitgedagt A, Van de Laar M (1995) The relationship between health effects in triathletes and microbiological quality of freshwater. Water Sci Technol 31:19–26
100. Medical Research Council & Council for Scientific and Industrial Research (CSIR) (1995) Pathogenic microorganisms/epidemiological microbiological study. Final report (1991–1995), South Africa
101. Philipp R, Evans EJ, Hughes AO, Grisdale SK, Enticott RG, Jephcott AE (1985) Health risks of snorkel swimming in untreated water. Int J Epidemiol 14(4):624–627
102. Pike E (1994) Health effects of sea bathing (WMI 9021), phase III—final report to the department of the environment. Water Research Centre, Medmenham
103. Prieto M, Lopez B, Juanes J, Revilla J, Llorca J, Delgado-Rodriguez M (2001) Recreation in coastal waters: health risks associated with bathing in sea water. J Epidemiol Community Health 55(6):442–447

104. Stevenson AH (1953) Studies of bathing water quality and health. Am J Public Health Nations Health 43:529–538
105. United Nations Environment Programme/World Health Organization (1991) Epidemiological studies related to environmental quality criteria for bathing waters. Shellfish-growing waters and edible marine organisms (Activity D). Final report on project on relationship between microbial quality of coastal seawater and rotavirus-Induced gastroenteritis among bathers (1986–88). United Nations Environment Programme, Athens, Greece
106. United Nations Environment Programme/World Health Organization (1991) Epidemiological studies related to environmental quality criteria for bathing waters, Shellfish-growing waters and edible marine organisms (Activity D). Final report on epidemiological study on bathers from selected beaches in Malaga, Spain (1988–1989). United Nations Environment Programme, Athens, Greece
107. van Asperen IA, Medema G, Borgdorff MW, Sprenger MJ, Havelaar AH (1998) Risk of gastroenteritis among triathletes in relation to faecal pollution of fresh waters. Int J Epidemiol 27(2):309–315
108. van Dijk PAH, Lacey RF, Pike EB (1996) Health effects of sea bathing—further analysis of data from UK beach surveys. Department of the Environment, WRc pic, Medmenham
109. von Schirnding YE, Kfir R, Cabelli V, Franklin L, Joubert G (1992) Morbidity among bathers exposed to polluted seawater. A prospective epidemiological study. S Afr Med J 81(11):543–546

Chapter 17
Science, Policy, and Risk Management: Case of Seafood Safety

Damaris A.F. Meujo and Mark T. Hamann

Glossary

Action levels	"Action levels and tolerances represent limits at or above which the FDA will take legal action to remove products from the market" – FDA.
An outbreak	Involves two or more ill people – CSPI.
Biological contaminants	In the context of this document, these are pathogenic microorganisms (bacteria, viruses, and parasites) found in seafood.
Chemical contaminants	In the context of this entry, are regrouped under this denomination, all nonbiological contaminants (deleterious chemicals) traceable to seafood.

This chapter was originally published as part of the Encyclopedia of Sustainability Science and Technology edited by Robert A. Meyers. DOI:10.1007/978-1-4419-0851-3

D.A.F. Meujo (✉)
Department of Pharmacognosy, The University of Mississippi, P.O. Box 1848, University, MS 38677, USA
e-mail: dmeujofo@olemiss.edu

M.T. Hamann
Department of Pharmacognosy, The University of Mississippi, P.O. Box 1848, University, MS 38677, USA

Department of Chemistry and Biochemistry, The University of Mississippi, University, MS 38677, USA

Department of Pharmacology, The University of Mississippi, University, MS 38677, USA

Institute of Marine and Environmental Technology, University of Maryland Center for Environmental Science, Baltimore, MD 21202, USA
e-mail: mthamann@olemiss.edu

E.A. Laws (ed.), *Environmental Toxicology: Selected Entries from the Encyclopedia of Sustainability Science and Technology*, DOI 10.1007/978-1-4614-5764-0_17,

Environmental pollutants	Seafood-associated deleterious substances traceable to the environment such as heavy metals and persistent organic pollutants.
Etiological agent	A microorganism responsible for a given disease.
Food safety hazards	According to the Seafood HACCP Regulation, a "food safety hazard" is "any biological, chemical, or physical property that may cause food to be unsafe for human consumption."
Seafood	Edible marine plants and animals (fish and shellfish) are usually grouped under the denomination of seafood in some contexts, these are referred to as "fish and fishery products" [1]. This same term is often given a broader meaning: all edible aquatic plants and animals.
Seafood-associated toxins	Harmful chemical substances produced either by seafood-associated bacterial contaminants, cyanobacteria, or toxic microscopic algae (dinoflagellates and diatoms) on which seafood feed.
Tolerance threshold	Maximum allowable amount of ubiquitous deleterious substance in seafood.

Definition of the Subject

In order to function properly, the human body needs a wide range of essential nutrients, which it gets from food that is ingested on a daily basis. Unfortunately, food also represents a vector for harmful creatures (bacterial, viral, protozoan pathogens) and chemical substances (organic toxins as well as toxic metals and various environmental contaminants). According to the most recent surveys of the Center for Science for Public Interest (CSPI), for more than a decade now, seafood has ranked first as the most likely source of foodborne disease outbreaks of established origin [2, 3]. Based on these surveys, seafood-associated hazards that have caused the largest number of outbreaks are toxins (especially scombrotoxin and ciguatera), followed by bacteria, the most problematic of which are *Vibrio* spp, and finally viruses (especially norovirus). Though food safety is primarily the responsibility of regulatory agencies, several other groups are involved. These include industries, consumers, and the scientific community upon which rests the responsibility of developing cutting-edge technologies capable of eliminating seafood-associated biological and chemical contaminants. The international community also relies on science for the development of revolutionary technologies for a faster, cheaper, easier, and more accurate detection of seafood-associated health hazards; tools without which enforcing laws and regulations set forth by regulatory agencies is virtually impossible. In this entry, different categories of seafood-associated health hazards as well as a few relevant regulatory and scientific efforts dedicated to reduce the incidence of seafood-borne illnesses are reviewed.

Introduction

Seafood constitutes a significant portion of the world's food supply and is renowned for its delightful taste. It is a critical component of the human diet because of its unique nutritional properties. Fish, for instance, is a good source of protein as its major components are proteins and lipids. All essential amino acids can be derived from fish consumption. Approximately 40% of the lipids found in fish are comprised of highly unsaturated long-chain fatty acids. Other outstanding nutritional qualities are reduced saturated fats and carbohydrates and plentiful essential nutrients. Some fish species are a valuable source of important nutrients such as vitamins A and D, phosphorus, iron, calcium, magnesium, selenium, and iodine [4, 5].

There are numerous reports of health benefits associated with the consumption of seafood. Several of these health benefits have been attributed to seafood's high content of vital nutrients, such as n-3 polyunsaturated fatty acids (PUFAs), specifically eicosapentaenoic acid (EPA), and docosahexaenoic (DHA). These health benefits include a reduced risk of developing serious diseases such as depression, [6] myocardial infarction, Alzheimer's [7], dementia [8], and weight loss [9]. A number of reports have associated seafood consumption with a reduced risk of mortality among individuals suffering from coronary heart disease [10] and a reduced risk of developing diseases such as ischemic and thrombotic strokes, colon and intestinal cancers, as well as others [9, 11–13]. These positive effects are counted among the factors that have driven current market trends. There has been a steady increase in the world's per capita fish and fishery products consumption for several decades now [14]. According to the December 2009 Food Outlook Report of the Food and Agriculture Organization (FAO), the annual per capita fish consumption in the world during the years 2007–2009 was estimated at ~17.1 kg. It is important to note that in the 1970s, 1980s, and 1990s, these values were 11.5, 12.8, and 16.4 kg per capita, respectively [15, 16].

As indicated by recent estimates, there has been a net increase in the demand for seafood in countries around the world [17]. In the UK for instance, the seafood retail market has experienced a considerable increase between the years 2003 and 2007, increasing from £2.4 billion (retail price) to an estimated £3.25 billion in 2007 [14]. A significant increase in seafood demand in developing countries has been observed, as well [16]. Millions of tons of seafood are caught each year worldwide to sustain the current demand. There has been a steady increase in the total world fish production since the 1950s, from 19.3 million tons to about 134 million tons in 2002 [18]. According to a 2009 FAO report, the current world production (capture fisheries plus aquaculture) is estimated at 144.1 million tons, divided into 98.8 million from capture fisheries and 54.3 million from aquaculture [15, 16].

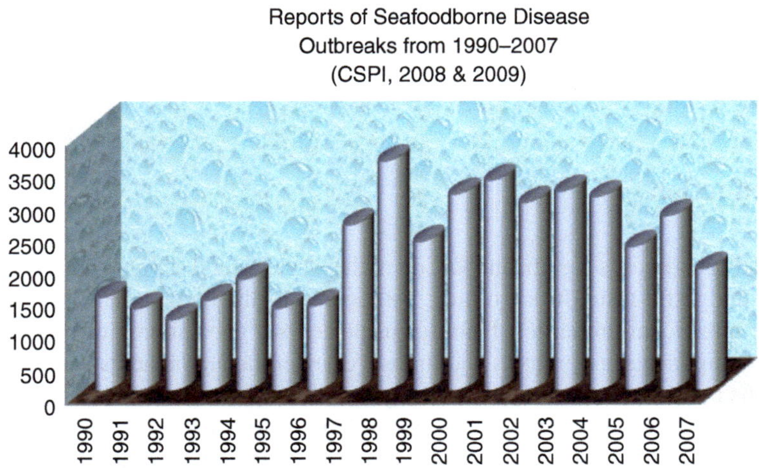

Fig. 17.1 Variation in the number of reported seafood-borne disease outbreaks since the 1990s in the USA

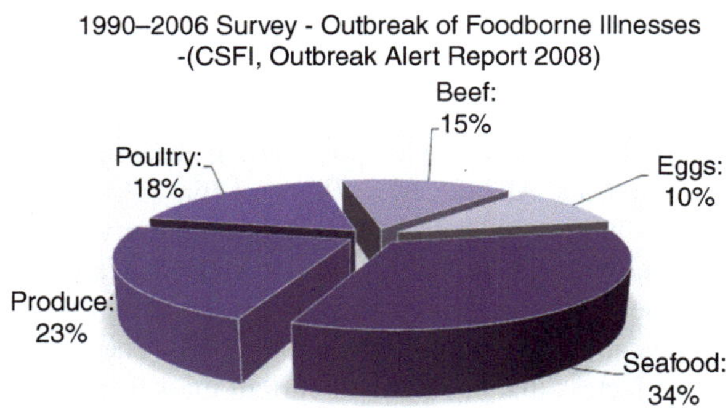

Fig. 17.2 Foodborne outbreaks reported during the years 1998–2007 by category of food in the USA

Unfortunately, seafood consumption is not without risks and food is an important vector of a wide range of health hazards (Fig. 17.1). Foodborne illnesses are a serious public health concern and according to the Centers for Disease Control and Prevention (CDC) roughly 76 million foodborne illnesses corresponding to about 325,000 hospitalizations and 5,000 deaths are recorded in the USA each year [19]. A recent survey conducted by the Center for Science for Public Interest (CSPI) [2] revealed that the food categories that were associated with the largest number of outbreaks in the USA during the period 1990–2006 were seafood, produce, poultry, beef, and eggs. Seafood was responsible for 1,140 out of 5,778 outbreaks and therefore was the most problematic food (Fig. 17.2). Also reported was an increase in the number of seafood-related outbreaks compared to the early 1990s (Fig. 17.1). It is important to mention that the number of reported cases of seafood-borne illness has remained constant over the years. Though it ranked second as far as number of outbreaks, produce caused the largest number of cases of illness during that same time frame [2].

"Food Safety Hazards" Associated with Seafood

Seafood-associated health hazards can be classified into two main categories: (1) biological contaminants (includes a long list of bacteria, viruses, parasites) and (2) chemical contaminants such as environmental pollutants (pesticides, heavy metals, approved or unapproved drug substances) and finally natural toxins from a variety of structural classes. According to the CSPI's 2008 report, the latter category was associated with the largest number of seafood-borne outbreaks from 1990 through 2006 (Fig. 17.3) [2]. Currently known health hazards associated with seafood are either naturally occurring or from various anthropogenic activities. Seafood becomes contaminated either as a result of feeding on poisonous phytoplankton species or in sewage-contaminated marine environments. Seafood contamination can also arise from inappropriate storage or accidental exposures during handling. Certain types of seafood are more likely vehicles of dangerous substances and pathogenic microorganisms than others.

Categories of seafood that have been associated with greater public safety risks are considered a high priority in sampling and surveillance efforts by the Food and Drug Administration (FDA) [20]. At the top of the FDA's seafood watch list are products such as molluskan shellfish from uncertified sources, refrigerated reduced oxygen packaged products, ready-to-eat seafood, seafood mixes containing cooked, raw, or partially cooked seafood components, as well as, scombrotoxin (histamine)-forming fish, aquaculture-derived seafood, and finally salt-cured or dried uneviscerated finfish [20]. According to the 2008 CSPI report, based on the number of reported outbreaks, finfish (such as tuna and grouper) was the most dangerous

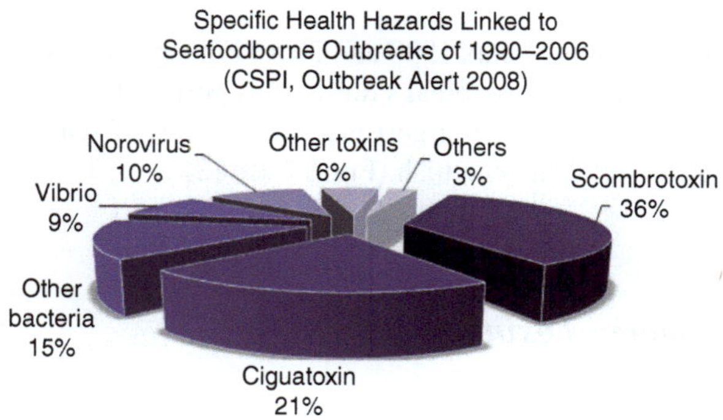

Fig. 17.3 Specific health hazards that caused seafood-borne outbreaks reported during the years 1990–2006 in the USA

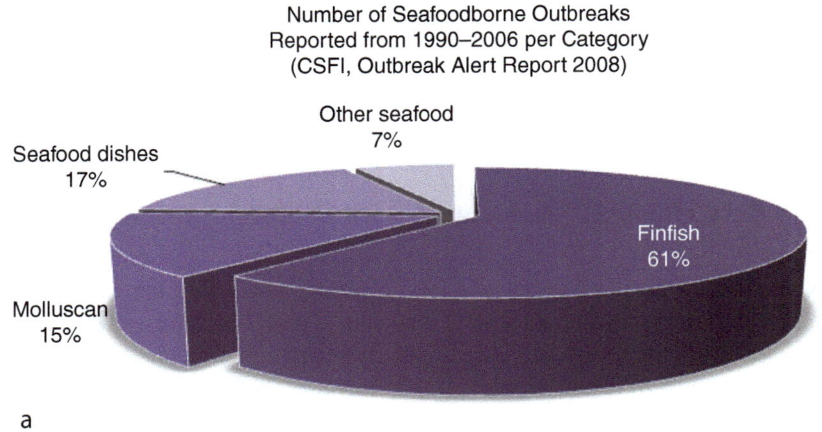

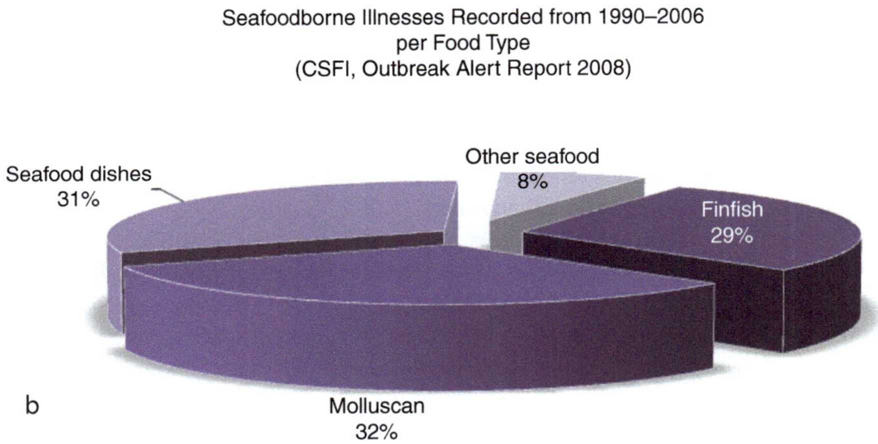

Fig. 17.4 Seafood-borne, outbreaks (**a**) and cases of illnesses (**b**) reported in the USA during the years 1990–2006 (by category of seafood)

type of seafood between the years 1990 and 2006. Finfish was responsible for 61% of all reported seafood-borne outbreaks during this period, followed by molluskan shellfish (15%) (Fig. 17.4a). The largest number of cases of seafood-borne illness was attributable to mollukcan shellfish (Fig. 17.4b) [2].

Seafood-Associated Toxins

Seafood-associated toxins, especially scombrotoxin and ciguatoxin, have been linked to the majority of seafood-borne outbreaks that occurred during the last decade and as such, could be viewed as the most dangerous seafood-associated "food safety hazard" [2, 3]. Scombrotoxin and ciguatoxin alone were responsible for 57% of all seafood-related outbreaks in the USA reported from 1990 to 2006 [2].

Seafood-associated toxins have been linked to a wide variety of intoxications. These include, poisoning associated with shellfish consumption, namely diarrheic shellfish poisoning (DSP), paralytic shellfish poisoning (PSP), amnesic shellfish poisoning (ASP), neurologic shellfish poisoning (NSP), aZaspiracid shellfish poisoning (AZP), and poisonings associated with fish consumption. The latter include ciguatera fish poisoning (CFP) and puffer fish poisoning [21]. It is necessary to point out that some seafood-associated biotoxins, namely ciguatoxin and toxins responsible for PSP, NSP, and ASP, can be lethal [22]. Seafood-associated toxins are generated either by bacterial contaminants that freely proliferate when seafood is improperly stored or by cyanobacteria and toxic microscopic algae (dinoflagellates and diatoms) on which the seafood feed. The blooms of these latter organisms, which occur from season to season, forming red tides (Harmful Algae Blooms [HAB]) are a subject of public health and environmental concerns, affecting the tourism and fishing industries [23].

Ciguatera Fish Poisoning

Ciguatera fish poisoning has been reported in countries around the world (Europe, Africa, America, Asia, and Oceania). This form of poisoning is frequent between the months of April through August and has been linked to the consumption of certain fish. Some examples are shark, barracuda, snapper, hogfish, horse-eye jack, red grouper, gray triggerfish, Spanish mackerel, narrowhead gray mullet, chinamanfish, swordfish, and amberjack. Ciguatera fish poisoning is caused by a set of heat-resistant polyether toxins known as ciguatoxins (Fig. 17.5), which are the product of in situ gambiertoxin biotransformation [23, 24]. Ciguatoxin, maitotoxin, palytoxin, and scaritoxin are members of this group. These toxins are produced by a variety of organisms; these include *Gambierdiscus toxicus*, *Gymnodinium sangieneum, G. polyedra, Ostreopsis lenticularis, Prorocentrum concavum, P. mexicanum,* and *P. rhathytum*, among others [21, 25]. Symptoms of ciguatera fish poisoning are mainly gastrointestinal and neurological. A few therapeutic approaches in case of poisoning include, but are not limited to, antihistamines, antiemetics (droperidol, prochlorperazin, metoclopramide), atropine, as well as intravenous hydration [23].

Scombrotoxic Fish Poisoning

Scombroid fish poisoning differs from other types of toxin-mediated seafood poisonings because the responsible toxin is not produced by a microalgae. Instead, it is generated under improper storage conditions (temperature $> 20°C$). This toxin is the result of a catalytic reaction involving the conversion of in situ histidine into histamine (Fig. 17.5). The enzyme responsible for this conversion, histidine decarboxylase, can be produced by several types of bacteria. These include various

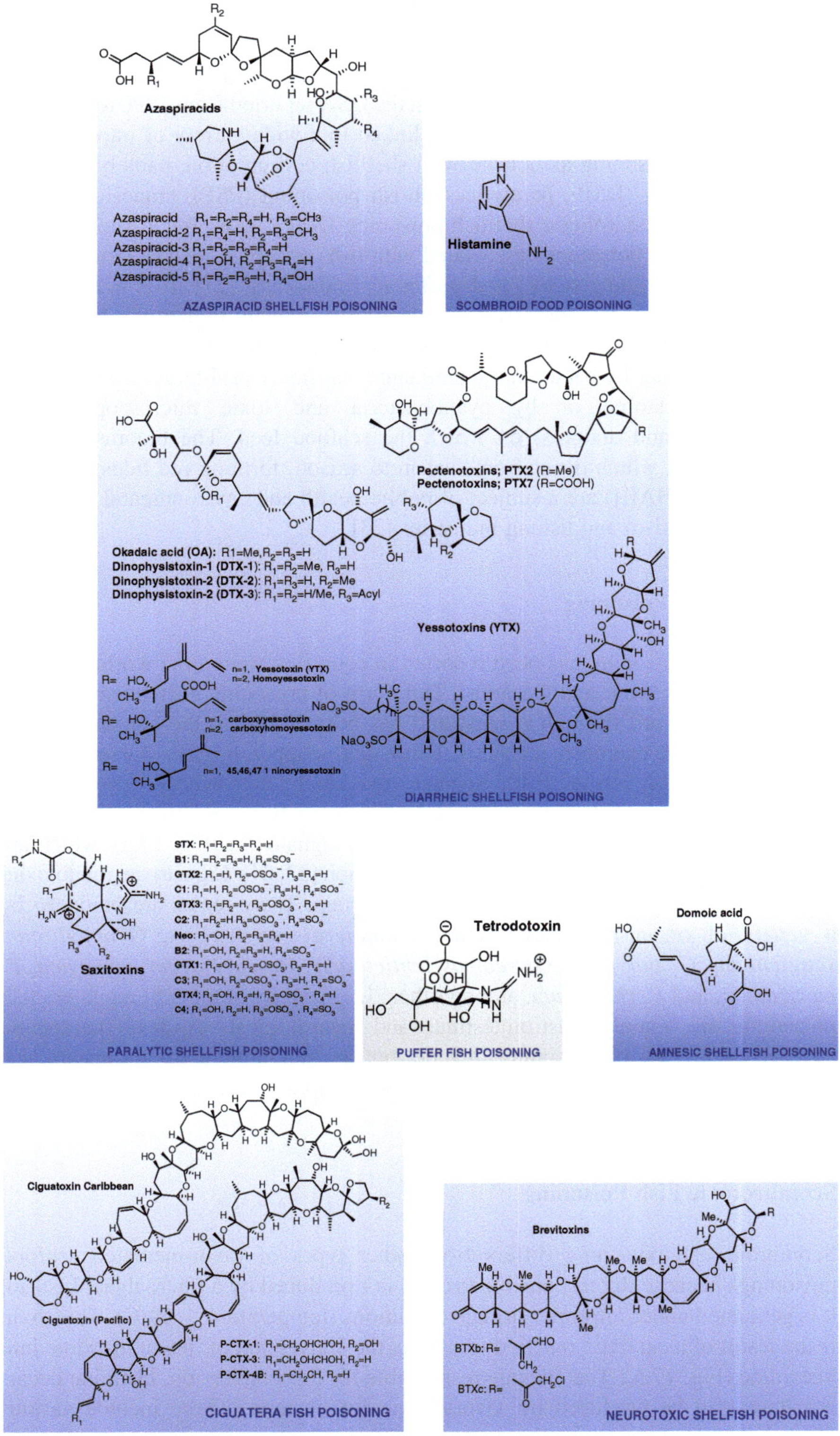

Fig. 17.5 Toxins involved in seafood-borne intoxications

Vibrio sp. *Clostridium, Enterobacteriaceae* (such as *Morganella morganii* and *Klebsiella pneumoniae* and *Hafnia alvei*) and *Lactobacillus* sp. [26]. Scombrotoxin is stable to both heat and cold conditions. Scombroid fish (fish containing a high level of free histidine) and non-scombroid fish have been implicated in this form of poisoning. These include fish such as amberjack, abalone, tunas, sardines, mackerel, bonito, and bluefish, just to name a few [25]; associated symptoms rank from mild and self-limiting to severe. Groups at risk for developing the severe form of this disease are people with respiratory and cardiac conditions or those on medication such as isoniazid and doxycycline that slow histamine degradation [23, 25, 27]. Symptoms of scombrotoxic fish poisoning include headache, abdominal cramps, nausea, diarrhea, and palpitations, among others. As far as pathophysiology, bioamines other than histamine are believed to play a critical role; a few examples are spermine, cadaverine, agmatine, and putrescine [25, 28]. There are several therapeutic approaches for this type of food poisoning. These include administration of activated charcoal, diphenhydramine, cimetidine, and famodine [27].

Other Major Seafood-Borne Poisonings

Paralytic shellfish poisoning (PSP) can be caused by a wide range of tetrahydropurine type toxins (carbamate, *N*-sulfo-carbamoyl, decarbamoyl, and deoxydecarbamoyl) [24] collectively called saxitoxins (Fig. 17.5). These are neurotoxins that act by blocking sodium channels, causing symptoms like numbness, paralysis, and disorientation. Saxitoxins are produced by dinoflagellates and blue-green algae. Dinoflagellates that have been linked to this form of poisoning include *Pyrodinium bahamense, Gymnodinium catenatum,* as well as several organisms belonging to the genus *Alexandrium* [21, 24, 29]. Various types of seafood can serve as vectors; these include clam, crabs, cockles, oysters, salmon, mackerels, scallops, and whales to name a few [21, 23]. PSP has been reported on all continents. Regrettably, there is no antidote for PSP, and therapeutic approaches are mostly supportive and include respiratory support in a life-threatening situation, gastric emptying, dialysis and enhancing renal clearance [23].

Diarrheic shellfish poisoning (DSP) has also been reported worldwide. Diarrhea, nausea, cramps, and vomiting are common signs of DSP. It is usually associated with consumption of contaminated clams, mussels, oysters, and scallops. This syndrome is caused by a group of acidic (okadaic acid and related dinophysistoxins) and neutral toxins (pectenotoxin). Yessotoxins have also been reported to cause this form of poisoning (Fig. 17.5) [24, 30]. These toxins are produced by a variety of marine dinoflagellates including *Dinophysis* spp. (*D. acuta, D. acuminate, D. caudate, D. mitra, D. norvegica*) as well as *Protoceratium* spp., *Prorocentrum* spp., *Gonyaulax* spp., and *Phalacroma* spp [21].

Amnesic shellfish poisoning (ASP) has been reported in various areas around the world including Europe, North, Central, and South America, Asia, and Oceania.

Several types of seafood, including anchovies, clams, crabs, oysters, mussels, mackerels, lobsters, scallops, and gastropods are potential vectors. ASP is caused by a marine biotoxin called domoic acid (Fig. 17.5). This toxin is produced by a red-brown marine diatom called *Pseudo-nitzschia pungens*. Diarrhea, nausea, and abdominal pain are examples of symptoms that indicate amnesic shellfish poisoning [21].

Neurotoxic shellfish poisoning (NSP) is caused by brevetoxins and its analogs (Fig. 17.5). Occurrences have been reported in countries around the world. This form of poisoning has been associated with the consumption of contaminated clams, mullets, mussels, oysters, tunas, and whelks. *Fibrocapsa japonica, Gymnodinium breve* (*Karenia brevis*), *Raphidophyceae* sp., and *Chattonella marina*, among others, are examples of organisms that produce these toxins [21, 24].

Puffer fish poisoning has been linked to the most potent and lethal marine neurotoxin: tedrodotoxin (Fig. 17.5). It is produced by a variety of animals including the California newt, trumpet shell, the blue ringed octopus, and puffer fish, especially a species known as fugu (present in the liver). Puffer fish is a delicacy in Japan and is the main vector for this form of poisoning. Once again, there is no antidote. The first case in Europe occurred in 2009 and involved an individual that had consumed trumpet shellfish (*Charonia sauliae*) harvested from the Atlantic Ocean in Southern Europe [31]. The main therapeutic approach upon poisoning is supportive and includes respiratory support (life-threatening circumstances). Activated charcoal, atropine, anticholinesterase agents, and alpha agonists, among others, are also recommended [32].

AZaspiracid shellfish poisoning (AZP). Mussels and oysters are known vectors of toxins responsible for azaspiracid shellfish poisoning. AZP has been reported in countries around Europe, namely Norway, Portugal, the UK, and Ireland. AZP is caused by marine toxins known as azaspiracids (Fig. 17.5), which are produced by *Protoceratum crassipes* and *Protoperidinium*. Nausea, vomiting, and diarrhea are a few symptoms of azaspiracid shellfish poisoning [21, 24].

Several other marine biotoxins not listed above have been reported. A few examples able to impair human health are gymnodimine, neosurugatoxin, prosurugatoxin, polycavernoside, and debromoaplysiatoxin [21]. It is also important to note that a number of biotoxin producers (dinoflagellates) have been associated with massive fish mortality and thus represent a major issue to the seafood and tourism industry worldwide. Examples in this case are *Pfiesteria piscicida* and *Karlodinium veneficum*; however, there is growing evidence that *P. piscicida* associated fish kills in the past may indeed have been *K. veneficum* derived. Blooms of *K. veneficum* have been linked to various episodes of massive fish kill around the world. In this particular case, a set of toxins believed to be the etiological agents and regrouped under the denomination karlotoxins or KmTxs has been isolated [33, 34]. Examples of such toxins include the karlotoxin-1 (KmTx-1), the 10-*O*-sulfo-KmTx-1, the KmTx-3, the 64-*E*-chloro-KmTx-3, the 10-*O*-sulfo-KmTx-3, the 65-*E*-chloro-KmTx-1, and, finally, the KmTx-2 (Fig. 17.6) [35], for which the relative and absolute configurations were assigned

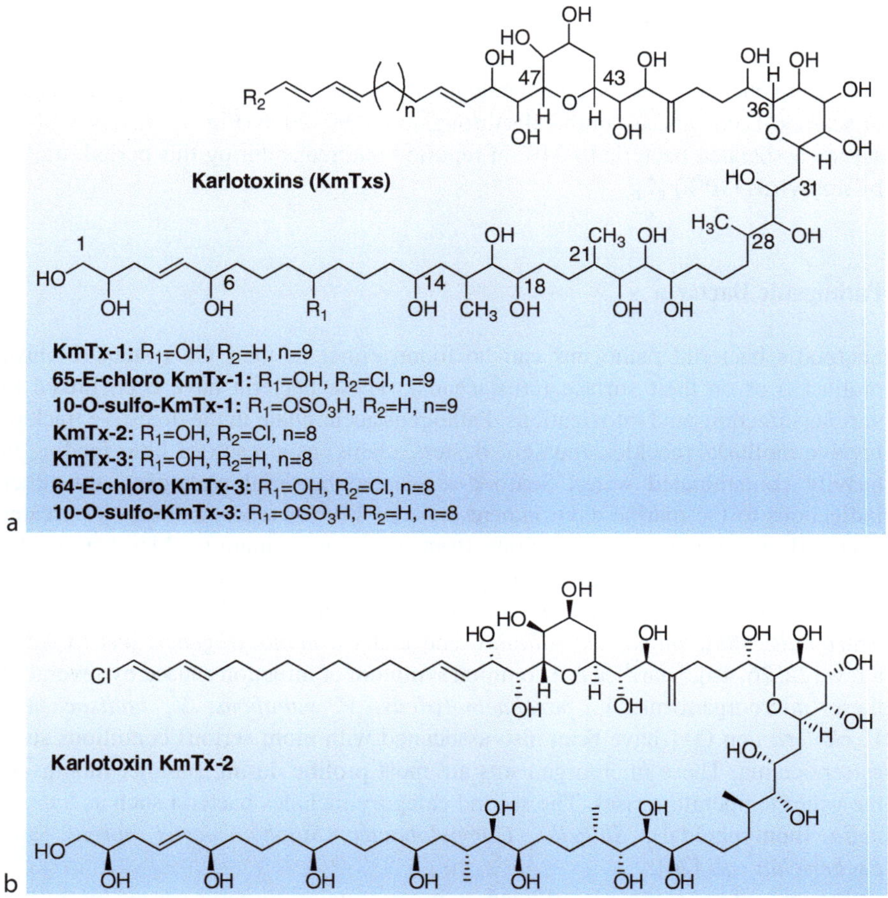

Fig. 17.6 (**a**) Structure of biotoxins produced by *Karlodinium veneficum*. (**b**) The absolute configuration of KmTx-2

only recently [36]. Ongoing work in this area actually began with research by Abbott and Ballantine in the 1940s and 1950s [37]. Karlotoxins kill fish by osmotic cell lysis, a result of the alteration of the ion transport system of the cell membrane. The fish dies as a result of damage to its vital gill epithelial tissues. These toxins are environmental pollutants and detectable in water during fish kill episodes. The ecological role of these toxins was investigated recently and because the toxins possess an allelopathic inhibitory effect on competitors as well as a prey immobilization it was established that the organism produce these toxins in order to facilitate feeding and control of competition during a bloom [38, 39].

Microbial Pathogens

Following chemical toxins, pathogenic microorganisms were the most likely cause of seafood-borne disease outbreaks throughout 1990–2006 (Fig. 17.3). The CSPI's survey associated bacteria to 24% of reported outbreaks during this period, trailed by norovirus (10%) [2].

Pathogenic Bacteria

Seafood's bacterial pathogens can be found either in their GI system (bivalve mollusks) or on their surface (crustaceans). These bacteria have been linked to various infections and intoxications. Pathogens accumulate in the digestive track of bivalve mollusks (cockles, mussels, oysters, clams) as a result of filter feeding in heavily contaminated water. Seafood-associated bacterial pathogens are either indigenous to the marine environment (case of *Vibrionaceae,* a source of greater concern) or nonindigenous (resulting from fecal contamination). Members of the first category are pathogenic *Vibrio* spp. such as *V. cholerae, V. parahaemolyticus, V. vulnificus, Clostridium botulinum* (non-proteolytic types B, E, F), *Aeromonas hydrophila, Plesiomonas shigelloides,* and *Listeria monocytogenes*, just to name a few [22, 26, 40]. Diarrhea is a common symptom of infection caused by several of these microorganisms. *V. parahaemolyticus, V. vulnificus, V. hollisae,* and *V. cholera* non O–1 have been also associated with more serious conditions such as septicemia. These microorganisms are most prolific during summer months as the water temperature rises. The second category includes bacteria such as *Salmonella* (nontyphoidal), *Shigella, Campylobacter, Staphylococcus aureus,* and *Escherichia coli* [26].

Proper seafood storage is sufficient to protect against diseases caused by several of these bacteria. It has been established that their concentration in seafood is normally low (below the minimum infective dose) and will remain so providing that the seafood is stored in conditions that are not conducive to bacterial growth, multiplication, or toxin production (refrigerated (4°C) and frozen (−18°C)). It is important to note that this does not apply to mollusks and their predators [28, 41]. While, in several cases, a low concentration in seafood is tolerated, FDA regulations become more stringent when it comes to bacteria such as *L. monocytogenes, V. vulnificus, Salmonella, C. botulinum*, and toxigenic 01 *V. cholera.* Current regulations require that these be undetectable in seafood requiring minimal cooking before consumption, for instance [1].

Vibrio spp

Vibrio spp., especially *V. parahaemolyticus* and *V. vulnificus*, are a cause of significant concern in the USA and several Asian countries (Japan, Taiwan, India, and China). *V. parahaemolyticus* (Vp), for instance, has been associated to the vast majority of seafood-borne gastroenteritis in the USA. This bacterium has been associated with fewer outbreaks in Europe [40, 41]. The pathophysiology of Vp is centered on several virulence factors; a few examples in this case are the thermostable direct hemolysin (TDH) and TDH-related hemolysin (TRH), which are encoded by tdh and trh genes, respectively [40]. Healthy individuals are also at risk of developing Vp associated infection [25]. The FDA can take legal action when seafood products are found to contain a Vp count $\geq 1 \times 10^4$/g [1]. Raw or improperly cooked fish and shellfish are potential vectors [25]. The minimal infective dose for this pathogen is $>10^6$/g [26]. *V. vulnificus* (Vv) infections can also result from consumption of raw or undercooked seafood. In addition, transmission can occur via wound infection. Though Vv is not a major issue in healthy individuals, in certain groups, Vv can cause serious infections or death. People suffering from alcoholic cirrhosis, hemochromatosis/cirrhosis, chronic hepatitis, postnecrotic cirrhosis, as well as diabetics and alcoholics are at higher risk. *V. vulnificus* is the second leading cause of seafood-related fatality in the USA [42–44]. Vv infections that occur as a result of consumption of contaminated seafood (especially raw oysters) are primarily septicemia and gastroenteritis. One therapeutic approach is the use of antimicrobial agents (tetracycline and intravenous doxycycline with ceftazidime) [43].

Other Seafood-Associated Bacteria

Clostridium botulinum – This bacterium is responsible for a condition known as botulism, the responsible agent being a toxin. Lightly preserved, semi-preserved, and fully preserved smoked, fermented, salted, and pickled fish products are likely vectors. Cold-smoked and fermented fish products are of greater risk. Chilling, autoclaving, and salting are approaches used to prevent botulism [26, 28]. *Listeria monocytogenes* – Infections caused by this other bacterium can, at worst, result in septicemia spreading to several organs and even, in the case of pregnant women, to the fetus. This type of infection can be fatal. Groups most at risk are pregnant women, neonates, fetuses, and immunocompromised patients. Shrimp is an important vector [26, 28, 45]. *S. aureus* has also been isolated from seafood. This bacterium is, as *C. botulinum,* a toxinogenic species. It produces toxins that are resistant both to enzymes degradation and heat. It is introduced in seafood as a result of environmental contamination or transferred from an infected worker involved in seafood handling [26, 28]. *Enterobacteriaceae* such as *Salmonella, Shigella*, and *E. coli* have also been reported in seafood. *Enterobacteriaceae* usually occurs in seafood as a result of fecal contamination. *Salmonella* is responsible for salmonellosis and is especially problematic for the shrimp industry [45]. Compared to other food categories, seafood is a less likely vector of *Salmonella* [25, 26]. The minimum infective dose for *Salmonella* sp has been estimated to be in the range of $<10^2$–$>10^6$. Non-bloody diarrhea, fever, abdominal pain and nausea, just to name a few, are indicators of infection by this

Table 17.1 Safety levels set by FDA for several seafood-associated bacteria

Hazards	FDA & EPA thresholds	Analytical approach	Targeted seafood	Higher risk populations
Salmonella sp.	Presence of organism[a]	Conventional culture methods	All fish	Severe in the elderly, infants, AIDS patients
E. coli	MPN of 230/100 g[b] APC – 500,000/g	Hemorrhagic colitis agar – direct plating method	Imported fresh and frozen clams and oysters	All people – most susceptible are young children and the elderly
S. aureus	Presence of staphylococcal enterotoxin, or a load $\geq 10^4$/g (MPN)[c]	Specific precipitation with antiserum	All fish	All people
C. botulinum	Presence of viable spores or vegetative cells or toxin[c]	Mouse neutralization test	All fish	All people

Compliance policy/programs
[a]Sec 555.300 Compliance Policy Guide
[b]Sec 560.600 Compliance Policy Guide
[c]Compliance Program 7303.842

pathogen. Symptoms of a *Shigella* infection, on the other hand, are bloody stools, severe abdominal cramps, fever, and dehydration. The minimum infective dose in this case has been estimated at 10^1–10^2. This is similar to what has been reported for *E. coli,* for which the minimal infective dose is 10^1–10^3 (Table 17.1) [26, 28].

Seafood-Associated Viruses

Seafood can also serve as a vector of viruses. Non-A, non-B enteral hepatitis viruses, hepatitis A virus (HAV), poliovirus, and norovirus, among others, have been associated with seafood-borne outbreaks [25, 26, 46, 47]. Viruses end up in seafood as a result of fecal contamination of the marine environment or when handled by an infected worker. Viruses are one of the most serious seafood-associated threats. So far, reports of seafood-borne infection outbreaks linked to viruses have emerged from countries around the world. HAV was associated with the largest seafood outbreak ever reported. This outbreak, which occurred in 1998, in the Chinese city of Shanghai, was linked to the consumption of contaminated clams and *over 292,000 cases were reported* [47, 48]. Another virus, namely norovirus, a single-stranded nonenveloped RNA virus, is currently responsible for roughly 50% of all foodborne outbreaks of gastroenteritis according to the CDC [49]. Norovirus was reported by CSPI as the most problematic seafood-associated virus during the period 1990–2006 (Fig. 17.3), as it caused 10% of all reported seafood outbreaks during that time [2]. It is important to point out that of the five genogroups of norovirus, GII has been linked to the majority of infections. Filter-feeding bivalve shellfish are important vectors [43]. Norovirus gastroenteritis-associated symptoms are vomiting, watery stools, non-bloody diarrhea with abdominal cramps, and nausea. It is a fairly resistant virus that can survive harsh conditions such as chlorine treatments (10 ppm), heating to 60°C (4 h), or freezing [49].

Seafood-Associated Parasites

Seafood (raw or undercooked) can also serve as a vector of pathogenic parasites. These include nematodes, cestodes, and trematodes. *Anisakis simplex, Pseudoterranova dicipiens, Gnathostoma* sp., *Capillaria* sp., and *Angiostrongylus* sp. are examples of seafood-associated nematodes. Examples of tapeworms that can be isolated from seafood include *Diphyllobothrium latum* and *D. pacificum; Clonorchis* sp., *Opisthorchis* sp., *Metagonimus yokagawai, Heterophyes* sp., *Paragonimus* sp., and *Echinostoma* sp., on the other hand, are a few examples of seafood-associated trematodes [26].

Several parasite-infected seafood dishes such as sushi, crab, sashimi, herring roe, and undercooked grilled fish have been associated with illnesses [1]. Compared to bacteria and viruses, however, parasites are of lesser concern. Trematodes (such as *Paragonimus westermani),* cestodes (such as *Diphyllobothrium latum, D. pacificum),* and nematodes (such as *Angiostrongylus cantonensis, Contraceacum osculatum)* have been associated with domestic fish and shellfish. Several other organisms have been associated with imported products instead. These include *Clonorchis sinensis, Heterophyes heterophyes, Metagonimus yokogawai, Opisthorchis felineus*, and *Gnathostoma spinigerum* [49]. Some nematodes, cestodes, and trematodes are of greatest concern as far as seafood safety. These include *Anisakis simplex, Pseudoterranova* spp*., Eustrongylides* spp*., Gnathostoma* spp*., Opisthorchis* spp., *Chlonorchis sinensis, and Paragonimus* spp., just to name a few [1, 49].

Toxic Heavy Metals

Heavy metals are a threat to the environment and public health and are problematic in regard to their long-term persistence. A variety of heavy metal contaminants has been reported in seafood. These elements originate from natural occurrences (marine volcanism, and geological and geothermal events) and anthropogenic activities. Several categories of anthropogenic activities threaten the marine environment. These include, activities that take place in tanneries, steel plants, battery industries, thermal power plants, and farms, especially those farms using heavy metal containing fertilizers and pesticides. Runoff from roadways has also been recently cited as an important source of contamination [4, 50, 51].

Heavy metals found in seafood include antimony, arsenic, cadmium, chromium, lead, mercury, nickel, copper, iron, manganese, selenium, zinc, aluminum, silver, strontium, thallium, and tin. Of these heavy metal contaminants, those of greatest concern are antimony, arsenic, cadmium, chromium, lead, mercury, and nickel. It is important to note that elements such as copper, selenium, iron, and zinc (known essential micronutrients) are toxic only at high concentrations [4].

Arsenic can be present under a variety of forms: toxic (inorganic) and nontoxic (organic). Arsenic is an extremely potent poison in its trivalent form and can cause a wide range of acute and chronic illnesses. A few examples include cancer, nephritis, hepatomegaly, peripheral symmetrical neuropathy, and palmar hyperkeratosis, among others. In seafood, arsenic is mainly present in a nonpoisonous form known as arsenobetaine or arsenocholine [46]. A compilation of data (about 100,000 results) received from 15 European countries revealed that seafood is among the food commodities with the highest arsenic levels [52].

Methyl mercury is a neurotoxic contaminant. Due to its potential effects on the fetus, it is one of the most regulated seafood-associated toxic metals. There are several related FDA recommendations to nursing/pregnant women, women of childbearing age, or children when it comes to seafood consumption. CH_3Hg^+ is present in nearly all seafood, but some types of fish such as shark, swordfish, king mackerel, and tilefish are believed to have a higher content. The FDA and the US Environmental Protection Agency (US EPA) have recommended being very selective of the type of fish consumed, limiting uptake to $\sim$12 oz/week as a healthy approach for groups at risk [5]. This has been supported by scientific data [12, 53, 54]. Several reports have shown that moderate to high consumption of fish species containing only a low amount of CH_3Hg^+ during pregnancy has a positive effect on fetal brain development. However, it is also important to note that several other studies, Myers et al. (2003) for instance, did not support the fact that exposure of pregnant women to methyl mercury through fish consumption could have deleterious effects on fetal development [13, 55].

Other toxic heavy metals include nickel, chromium, cadmium, selenium, and lead. Cadmium tends to be bioaccumulated by crustacea and bivalves. Clinical signs of cadmium poisoning include osteoporotic and osteomalacic disease, as well as kidney damage. Lead poisoning, on the other hand, has been associated with anemia, convulsions, paralysis, and proteinuria. This metal tends to accumulate in cortical and trabecular bone, kidney, lung, as well as the CNS. Edema, hepatitis, and hemorrhage are conditions that can result from selenium poisoning, which is also known, to result in congenital malformations and infertility. Arsenic (As), nickel (Ni), and chromium (Cr) are carcinogens [46].

Other Chemical Environmental Contaminants

Organochlorine compounds. Until recently, organochlorine compounds were widely used. A wide range of polychlorinated substances can be found in seafood. These include various insecticides, agrochemicals, industrial chemicals, and by-products. Examples of pesticides traceable to seafood include endrin, heptachlor, dieldrin, benzene hexachloride (BHC), chlordane, dichlorodiphenyltrichloroethane (DDT) and lindane. Polychlorinated biphenyls (PCBs) and dioxins, such as 2,3,7,8-tetrachlorodibenzo-*p*-dioxin, are example of industrial chemicals that contaminate seafood. Prior to 1970, PCBs were widely used industrially [56]. Because they are non-biodegradable, toxic in nature, and tend to bioaccumulate in seafood, organochlorine compounds now constitute a major environmental and public health concern. Several of these substances are known carcinogens (dioxins and polychlorinated biphenyls); but because seafood-associated health benefits outweigh potential risks, seafood consumption is recommended [13]. The level of these substances in fish is truly minimal. However, it is important to note that people whose diet is predominantly made of seafood are still at risk [56]. Reports on farm raised salmon (especially European farms) containing higher levels of these contaminants compared to wild-type salmon is believed to have had a serious impact on the consumer acceptance of this type of seafood [57, 58].

Antimicrobial drug contaminants of seafood. This is a problem mainly associated with aquaculture, as farmers rely more and more on various drugs and antimicrobial agents to deal with specific seafood diseases (bacterial, fungal, viral), to improve the quality of water, and to manage pest-related issues.

There are quite a few reasons why there are concerns over drug residues being present in seafood. Several antimicrobial agents used in aquaculture have been linked to various adverse health effects, such as hypersensitivity reactions. In vivo studies, in animals, have established several others as carcinogens; these include nitrofuran, malachite green, gentian violet, and fluoroquinolones. Fluoroquinolones have been also associated with antibiotic resistance. There are several drugs used to this end in the USA that have been approved by the Center for Veterinary Medicine (CVM) (Table 17.2). Substances such as ciprofloxacin, erythromycin, tetracyclines, chloramphenicol, nitrofurans, malachite green, gentian (crystal) violet, and fluoroquinolones have been frequently reported in seafood imported into the USA in recent years [20]. A short, recent survey by the FDA has shown that imports from China test positive for the presence of a variety of unapproved substances, such as gentian violet, malachite green, nitrofuran, and fluoroquinolones. Consequently, these products are the subject of great concern and the focus of regulatory surveillance [59].

Similar issues have been reported in Europe. In fact, during the years 1999–2002, the presence of residues of antimicrobial drugs, mainly nitrofuran and chloramphenicol, was one of the main reasons for detention or rejection of seafood imports into Europe (Fig. 17.7) [28]. A handful of aquaculture drugs approved in Europe are presented in Table 17.3.

What Are the Approaches Used to Assure Public Safety?

Regulatory authorities and the scientific community are two groups with complementary goals that play a key role in ensuring public safety against seafood-associated health hazards.

Table 17.2 Aquaculture drugs approved in the USA and action levels

FDA-approved aquaculture drugs			
Drug name	Tolerance level in the flesh	Type of seafood	Purpose
Unapproved drugs	No trace tolerated[a]	All fish	–
Chorionic gonadotropin[b]		Brood finfish	Reproductive
Formalin solution[c]		Salmon trout, catfish, largemouth bass, and bluegill	Antiparasitic and fungicidal/static
Tricaine methanesulfonate[d]		Catfish, salmon, and trout, pike and perch	–
Oxytetracycline	2.0 ppm[e]	Salmonids, catfish, and lobster	Disease control
Sulfamerazine[f]	Undetectable[g]	Trout	–
Sulfadimethoxine/ Ormetoprim combination[h]	0.1 ppm[i]	Salmonids and catfish.	–

Compliance policy/programs
[a]Sec 615.200 Compliance Policy Guide
[b]21 CFR 522.1081
[c]21 CFR 529.1030
[d]21 CFR 529.2503
[e]21 CFR 556.500
[f]21 CFR 558.582
[g]21 CFR 556.660
[h]21 CFR 558.575
[i]21 CFR 556.640

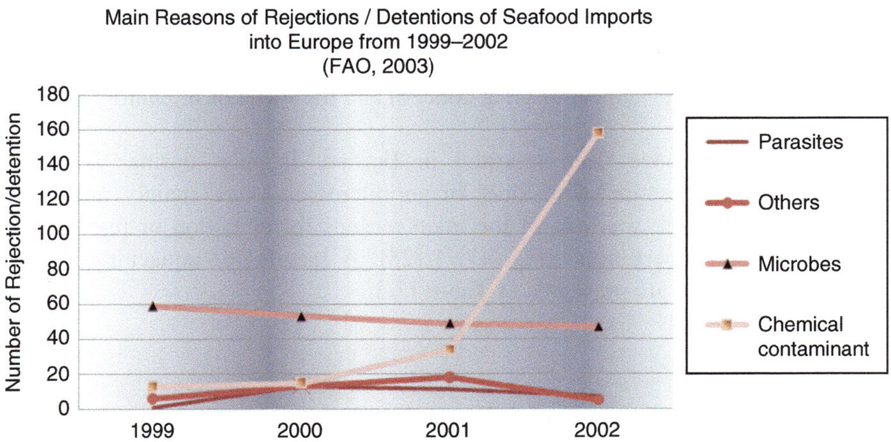

Fig. 17.7 Main causes of rejection or detention of seafood imports into Europe during the years 1999–2002

Table 17.3 Aquaculture drugs approved in Europe and action levels

The Council of the European Communities approved aquaculture drugs		
Drug name	Maximum Residue Limit (MRL) – muscle and skin in natural portion	Type of seafood
Trimethoprim	50 μg/kg[a]	Finfish
Flumequine	600 μg/kg[a]	Finfish
Oxolinic acid	100 μg/kg[a]	Finfish
Sarafloxacin	30 μg/kg[a]	Salmonidae
Tylosin A	100 μg/kg[a]	Finfish
Thiamphenicol	50 μg/kg[a]	Finfish
Colistin	150 μg/kg[a]	Finfish
Deltamethrin	10 μg/kg[a]	Finfish
Emamectin B1a	100 μg/kg[a]	Finfish
Oxolinic acid	300 μg/kg[a]	Finfish
Florfenicol	1000 μg/kg[a]	Fish
Thiamphenicol	50 μg/kg[a]	Finfish
Deltamethrin	10 μg/kg[a]	Finfish

Compliance policy/programs
[a]Council Regulation (EEC) No 2377/90 of 26 June 1990

Regulatory Agencies

Seafood regulatory agencies, via promulgation and enforcement of various laws and regulations, guidance, and recommendations, are usually at the heart of protecting the public from seafood-associated hazards. Numerous programs have been designed over the years to this end. In the USA, the primary responsibility of assuring seafood safety falls to the FDA, which is in charge of setting the maximum safe levels of unavoidable toxic substances in seafood [59]. The FDA has the authority to detain and even refuse import entries into the USA. A recent related event was the June 28, 2007, decision of the FDA to detain Chinese farm-raised catfish, basa, shrimp, and dace until they were cleared from containing unapproved drug substances [59]. As for domestic seafood, the FDA can recommend legal sanctions, which include "warning letters, seizure of products, injunction against further non-compliant practices, or prosecution of an individual or establishment" [20]. There are a few other regulatory agencies; these include the EPA (Environmental Protection Agency) in charge of chemical contaminants such as pesticides and the National Marine Fisheries Service for instance. In Europe, the EU parliament is at the heart of food safety control. This organization promulgates food safety-related laws, regulations, and directives, which are mandatory in all states of the European Union. The EU parliament works in close collaboration with the European Food Safety Authority, which was established by regulation (EC) No.178/2002. This latter organization, more science oriented, is in charge of risk assessment.

There is a long list of guidance and regulations, mostly to seafood industries that protect the public from dangers associated with seafood consumption. Several of these are preventive measures. A few examples in the USA are the National Shellfish Sanitation Program, the Salmon Control Plan, the Low-Acid Canned Food (LACF) Program, the Hazard Analysis & Critical Control Points (HACCP) Program, and the Good Manufacturing Practice regulation. The latter is intended to assure that recommended processing conditions were used. The Salmon Control Program, on the other hand, was designed to assure the safety of salmon consumers and is a cooperative approach involving the FDA, industries, and various associations. As part of the Shellfish Sanitation Program in the USA, the level of various pollutants in coastal water is to be monitored in order to classify each given area as suitable or unsuitable for shellfish harvest. As for the HACCP, it applies to domestic as well as imported seafood. HACCP requires both domestic and foreign processors of fish and fishery products to understand all concepts behind food safety hazards and through a system of precautionary control measures to prevent hazards from occurring [20, 59]. The EU parliaments, as well as regulatory agencies of countries around the world, have issued several similar regulations and directives that apply to domestic and imported seafood. A few examples of seafood-related regulations promulgated by the European Parliament include regulations (EC) No. 852/2004 and No. 853/2004, which established some key hygienic rules for food (including seafood) business operators and regulation (EC) No. 854/2004 in which key exigencies related to the organization of seafood official control programs are determined. Several of these regulations elaborate on the "tolerance threshold" for contaminants present in seafood. Examples of "tolerance threshold" for seafood-associated health hazards, set by regulatory agencies around the world are presented in Tables 17.4, 17.5, and 17.6.

The Scientific Community's Contribution in Ensuring Seafood Safety

Through the development of cutting-edge technologies to solve problems at hand, science has also played a critical role in ensuring public safety against dangers associated with seafood. A simple literature search with a keyword such as "seafood safety" in ScienceDirect shows a significant and steady increase in seafood safety-related research effort since 1991 (Fig. 17.8). The international community relies on science to develop cutting-edge technologies and techniques that can rid seafood from associated biological and chemical contaminants in order to bring a safer product to the market. The development of cutting-edge technologies for faster, cheaper, easier, and more accurate detection methods of seafood-associated health hazards is another way scientists have contributed in the protection of consumers against these dangers.

Table 17.4 Seafood-associated marine biotoxins and action level set by regulatory agencies around the world

Hazards	Detection method (analytical methods for regulatory purposes)	Tolerated thresholds	Targeted seafood
Paralytic shellfish poison	USA – mouse bioassay	0.8 ppm (80 μg/100 g) saxitoxin equivalent[a]	All fish
		800 ug/kg (live bivalve mollusks)	
	EU – mouse bioassay[b]	80 mg STX eq/100 g of meat[c,d]	Bivalve mollusks
	Africa – mouse bioassay[b]	80 μg STX eq/100 g	mollusks
	Canada – mouse bioassay[b]	<80 mg STX eq/100 g	Mollusks
	Asia – mouse bioassay[b]	400 MU/100 g	Shellfish
	Australia – mouse assay[b]	80 mg STX eq/100 g	Shellfish meat
Amnesic shellfish poison	USA – LC method	20 ppm domoic acid (in general)[a]	All fish
		30 ppm (in viscera of dungeness crab)[a]	
	Europe – LC method	20 mg/kg of domoic acid[c]	Live bivalve mollusks
	Canada – LC method	20 mg DA/kg	Mussel
	New Zealand – LC method	20 mg DA/kg	Shellfish
Neurotoxic shellfish poison		0.8 ppm (20 mouse units/ 100 g) (USA)	Clams, mussels, and oysters, fresh, frozen, or canned
Diarrheic shellfish poison	EU – mouse bioassay[b]	160 ug of okadaic acid/kg[c]	Mollusks, echinoderms, tunicates and marine gastropods
	Asia (Japan) mouse bioassay	5 MU/100 g whole meat	Shellfish
	Australia	16–20 μg OA eq/100 g	Shellfish
Azaspiracids	USA – mouse or rat bioassay	160 ug azaspiracid equivalents/kg[c]	Bivalve mollusks, echinoderms, tunicates, and marine gastropods
	Europe – mouse or rat bioassay	160 μg/kg	Live bivalves
Ciguatoxin		Presence[c]	Fishery products
Histamine	USA – extraction coupled to fluorescence spectroscopy	50 ppm	Tuna, mahi mahi, and related fish
	Europe	<200 ppm[e]	Scombridae,clupeidae, engraulidae, and coryphaenidae

Compliance policy/programs

[a]Compliance Program 7303.842 or Sec 540.250 Compliance Policy Guide

[b]FAO (2004) Marine Biotoxins Food and Agriculture Organization of the United Nations Rome, 2004 http://www.fao.org/docrep/007/y5486e/y5486e00.HTM available online, retrieved December 30, 2009

[c]Regulation (EC) No 853/2004 (European standards)

[d]EU Directive 91/492/EEC

[e]Council directives 91/493/EEC

Table 17.5 Seafood-associated toxic heavy metals and action level set by the FDA

Seafood Health Hazard	Tolerance threshold	Targeted seafood
Methyl mercury	1.0 ppm[a]	All fish
Arsenic	86 ppm (76 ppm for crustacea)[b]	Clams, oysters, and mussels
Cadmium	4 ppm (3 ppm for crustacean)[b]	Clams, oysters, and mussels
Chromium	13 ppm (12 ppm for crustacea)[b]	Clams, oysters, and mussels
Lead	1.7 ppm (1.5 ppm for crustacea)[b]	Clams, oysters, and mussels
Nickel	80 ppm (70 ppm for crustacean)[b]	Clams, oysters, and mussels

Compliance Policy/program
[a]Sec 540.600 Compliance Policy Guide
[b]Appendix 5 – FDA & EPA Safety Levels in Regulations and Guidance

Table 17.6 Seafood-associated environmental pollutants and action level set by the FDA

Seafood health hazard	Tolerance threshold	Targeted seafood
Polychlorinated biphenyls (PCBs)	2.0 ppm (edible portion)[a]	All fish
DDT, TDE and DDE	5.0 ppm (edible portion)[b]	All fish
Chlordane –	0.3 ppm (edible portion)[b]	All fish
Chlordecone –	0.3 ppm (0.4 ppm in crabmeat)[b]	All fish
Mirex	0.1 ppm[b]	All fish
Diquat	0.1 ppm[c]	All fish
Heptachlor and heptachlor epoxide	0.3 ppm[b]	All fish
Glyphosate	0.25 ppm[d] 3.0 ppm (for Shellfish)	Fin fish
Fluridone	0.5 ppm[e]	Fin fish and crayfish
Simazine	12 ppm[f]	Fin fish
Aldrin and dieldrin –	0.3 ppm[b]	Fin fish and shellfish

Compliance Policy/program
[a]21 CFR 109.30
[b]Sec 575.100 Compliance Policy Guide
[c]40 CFR 180.226
[d]40 CFR 180.364
[e]40 CFR 180.420
[f]40 CFR 180.213

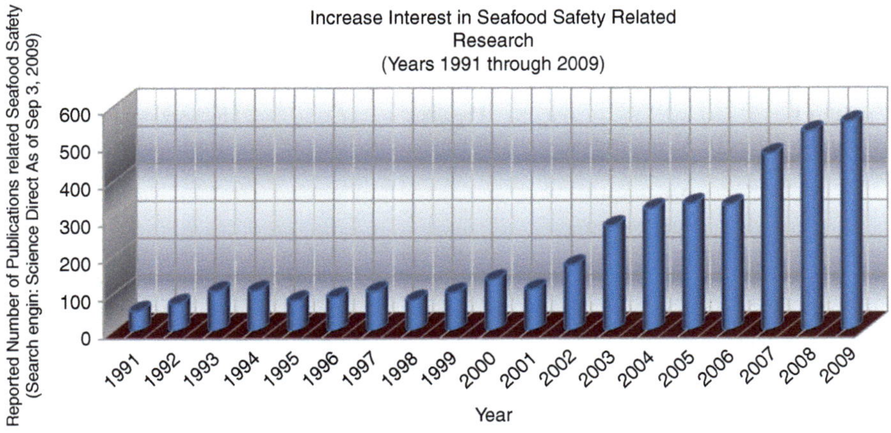

Fig. 17.8 Increased interest in seafood safety related research

Detection Tools for Seafood-Associated Health Hazards

Detection of biotoxins. Years of effort have yielded a wide range of approaches for toxin detection in seafood. These include various bioassays (in vivo and in vitro assays), biochemical techniques (immunoassays), and chemical techniques including fluorometric and colorimetric techniques, chromatographic techniques, electrophoretic techniques, mass spectrometry, and finally biosensor-based techniques (Table 17.7). In countries around the world, the mouse bioassay, despite its numerous shortcomings, and liquid chromatography are the two official methods recommended for biotoxin detection in seafood (Table 17.4).

Chromatographic techniques are at the heart of several effective analytical approaches to biotoxin separation and detection. Compared to animal assays, analytical techniques present the advantages of higher accuracy and sensitivity. While with animal bioassays, it is impossible to clearly determine the nature of contaminants, these techniques, coupled with the appropriate detection methods, will permit not only separation and accurate identification of incriminated toxins, but also, using a standard curve, their quantification. A variety of analytical approaches is currently available for the detection of marine toxins. These include Gas Chromatography (GC), Thin-Layer Chromatography (TLC), Liquid Chromatography (LC), Liquid Chromatography–Mass Spectrometry (LC-MS), and Capillary electrophoresis [21, 59]. In addition to advantages listed above, with chromatography-based detection methods, several toxins can be monitored simultaneously. A new liquid chromatography–tandem mass spectrometry method was developed recently [60]. With this method, up to 28 marine lipophilic toxins can be monitored at the same time. In this case, toxins are separated using a gradient of acetonitrile/water at alkaline pH on a new type of C18 column proven stable under these conditions: a Waters X-Bridge C_{18} (150 mm × 3 mm, 5 μm) [60].

Biosensor-based detection methods have also been investigated for application in seafood safety programs. They are not only economical, straightforward, and easy to use, but also offer the advantages of high sensitivities/low limit of detection, plus, these technologies are portable [21]. An example of a recent development in this field is a new planar interdigital sensor-based sensing system developed by Syaifudin et al. (2009). This approach involves simple monitoring of variations of reactive impedance of the planar interdigital sensors. Using this approach, as little as 12.6 μg/g of domoic acid in mussel meat, for instance, can be successfully detected [61].

ELISA has also been investigated widely for application in seafood biotoxin detection. An example of a recent development in this field was made by Zhou et al. (2010), who reported a reliable ELISA-based approach to monitor brevitoxin in mollusks with reduced interference from the matrices. Oysters and cockles were used in this experiment. With such a method, the limit of detection of brevitoxin is improved. The main advantage of immunoassays, which are based on antibody–antigen interactions, is high specificity [62].

Table 17.7 Proposed methods for biotoxin detection in seafood [21, 24, 63]

	PSP	DSP	ASP	Ciguatera	NSP	SFP
Bioassays						
In vivo assays	Mouse bioassay	Mouse bioassay	Mouse bioassay	Mouse bioassay	Mouse bioassay	
		Suckling mouse assay		Chicken assay	Fish bioassay	
		Rat bioassay		Mongoose and cat assay		
		Daphnia magna assay		Brine shrimp assay		
		Intestinal loop assays		Mosquito assay		
				Diptera larvae assay		
In vitro assays	In vitro hippocampal slice assay	Cytotoxicity assays (rat hepatocytes, KB cells, fibroblasts)	Receptor binding assays	Sodium channel binding assays for ciguatoxins	Neuroblastoma cell assay	
	Sodium channel blocking assay		Hippocampal slice preparations		Synaptosome binding assay	
					Hippocampal slice assay	
Biochemical techniques	Immunoassay (ELISA)	Immunoassay (RIA or ELISA)	Immunoassay (ELISA)	Immunoassay		(Radioimmunoassay
		Immunoassay (ELISA)			Acid phosphatase assay	Enzyme-linked immunosorbent assay (ELISA)
						Stick tests
						Immunoassays based on monoclonal antibodies
Solid-phase	immunobead assay)					
		TLC,GC, LC				

Chemical/ analytical techniques	Fluorometric and colorimetric technique, MS		Amino acid analysis	Chromatographic detection	Micellar electrokinetic capillary chromatography detection	Chromatography coupled with fluorimetric detection or derivatization techniques
	Chromatographic and	Electrophoretic techniques (TLC, GC,CE, HPLC)	MEKC, MS	TLC, LC, CE, MS	Nuclear magnetic resonance (NMR)/mass spectrometry (MS)	Electrospray LC/MS
					Ionspray LC/MS	
					Mass spectrometry	LC/MS/MS
Capillary zone				electrophoresis	Biosensors based techniques	
Sodium channels based biosensors	Immuno sensors	Optical immuno sensors				
	Enzyme inhibition based biosensors	Molecularly imprinted polymer based biosensors				

Drug residues in seafood. With the rise in demand of fish and shellfish, a large portion of seafood found in market places around the world comes from aquaculture, especially from China. As it is the case with any animal husbandry, veterinary drugs are often heavily used in aquaculture to control pests, infections, or to increase production. Unfortunately, drug residues, often molecules that have proven harmful to humans, are found in edible seafood tissues. As previously mentioned, lately in the USA as in Europe, there have been a significant number of alerts regarding seafood imports contaminated with unapproved drug residues (Fig. 17.7). Over the years, through the dedication of the scientific community, better and more improved detection methods have been made available. An example in this case is liquid chromatography (LC) coupled with triple-quadrupole mass-spectrometry (LC–QqQ-MS). Until recently, in a single run, this sort of method could only analyze related molecular entities. An upgraded and more robust version, a multicomponent quantitative HRLC–ToF-MS, was reported recently. This new approach has proven effective at simultaneously monitoring a wide range of unrelated drugs generally used in aquaculture or found in seafood tissues [64]. Smith et al. (2009) also developed an LC-ion trap mass spectrometry approach, effective at detecting several types of veterinary drugs in fish samples. In this case, the extraction of drug residues from seafood matrixes is completed in acetonitrile and hexane followed by HPLC separation on a phenyl column. In certain cases, imidazoles, macrolides, fluoroquinolones for instance, very low concentrations (0.01 ppm) could be detected using this technique. Other drugs involved in these studies included ionophores, macrolides, nitroimidazoles, benzimidazoles, anthelmintics, penicillins, quinolones, sulfonamides, tetracyclines, amphenicols, and tranquilizers, among others [64, 65].

Detection of microbial pathogens in seafood. The best approaches for the detection of microbes in seafood are usually a combination of culture-based and molecular-based techniques; the latter being often used to assist in bacterial strain identification, while culture is required for enrichment purpose. The kind of medium used to this end is dictated by the type of microorganisms targeted. Over the years, a tremendous amount of effort has been dedicated to the conception of superior media. Recommended Standard Operating Protocols for use in microorganism detection, in seafood, are described in detail in the FDA's 1998 Bacteriological Analytical Manual (BAM) [63]. Targets of quality control programs are numerous. These include various pathogenic bacteria such as *Vibrio*, *Salmonella, S. aureus, L. monocytogenes*, and *E. coli* as well as indicators of fecal contamination [26].

Several molecular and culture-based methods have been developed to assist in the detection of *V. parahaemolyticus* (Vp) and *V. vulnificus* (Vv), which are major issues as far as seafood safety is concerned. Commonly used culture-based approaches, which present the main disadvantages of being time consuming, laborious, and inaccurate, are the MPN and the ISO cultural methods [40]. Quite a few *Vibrio*-specific media has been developed to assist in the isolation of these bacteria from seafood matrixes; these include TCBS agar (for *V. cholera* and *V. parahaemolyticus)* and modified cellobiose polymyxin colistin (mCPC) and

CC agar for *V. vulnificus* [66]. The MPN method, coupled with various techniques to assist in the identification of suspect isolates, is recommended for detection of Vv and Vp in seafood. Examples of approaches used for these identifications are the establishment of a biochemical profile, DNA probes, or PCR. In the case of PCR for instance, DNA primers targeting tdh and *trh* genes are used to detect virulent strains of *V. parahaemolyticus*.

Until the advent of real-time PCR, this approach presented the main drawback of being limited to qualitative evaluation of food samples. Today, faster and quantitative assessment of *Vibrio*'s presence in food samples is possible [40] More sophisticated methods have been introduced. In 2009 for instance, Espiñeira et al. introduced a sequential multiplex PCR system for the detection of *Vibrio* sp. that have been involved in fish and shellfish poisoning, namely *V. cholerae*, *V. parahaemolyticus, V. vulnificus, V. alginoliticus*, and *V. mimicus*. This method, which has been validated, is not only able to detect problematic *Vibrio* species, but it can also, using a fragment analysis, confirm the viable/dead status of these microorganisms and most importantly, probe for the presence of important serogroups and virulence factors [67]. Genetic markers, a precious tool for PCR, have been identified for several other seafood-borne pathogens. These include cytotoxin-hemolysin (*V. vulnificus*), *ctxAB* (*V. cholerae*), *oriC*, chromosomal origin of replication (*Salmonella* spp.), listeriolysin O (*hly*), and the 16 S rRNA (*L. monocytogenes*), polymerase gene (hepatitis A virus), and polymerase gene for norovirus, just to name a few [66].

Examples of Recent Technological Breakthroughs in Efforts to Free Seafood from Associated Contaminants

In recent years, outstanding breakthrough techniques and cutting-edge technologies to aid in freeing seafood from associated contaminants have been developed. Molluskan shellfish and associated pathogens (especially *V. vulnificus* and *V. parahaemolyticus*), scombrotoxin, fish safety (especially of cold-smoked salmon), and bio-preservation of seafood are a few examples of highly investigated topics.

Molluskan shellfish and associated pathogens. Nowadays, seafood regulatory authorities face major issues with molluskan shellfish. For years now, molluskan shellfish has been classified as "high-risk" seafood by the FDA. They were responsible for the largest number of seafood-borne illnesses during 1990–2006 (Fig. 17.4). Because of their filter-feeding habits, they tend to accumulate a wide range of etiological agents (pathogenic bacteria, parasites, and viruses) as well as biotoxins. The emergence of innovative FDA-approved Post-Harvest Processing (PHP) technologies such as Individual Quick Freezing (IQF), Heat-Cool Pasteurization (HCP), and High Hydrostatic Pressure (HHP) has revolutionized the industry of seafood, particularly in regard to oysters. These technologies, which are currently commercially available, have made it possible to bring raw and healthy products to consumers. An end product of high quality (fresh taste and superior appearance) is the major advantage of IQF, HCP, and HHP. These processing techniques reduce the level of *Vibrio* bacteria to undetectable levels [67]. Though HHP is already approved by the FDA, several attempts to perfect this technology are currently on the way. In 2008 for instance, using a pressure-resistant strain of Vp (MLT 403) to picture the worst-case scenario, Kural et al. proposed better pasteurizing conditions. Under such conditions, a 5-log reduction in the load of the pressure-resistant Vp in live oysters could be achieved. These conditions were as follows: a 2-min treatment at pressure $\geq$350 MPa (1–35°C) and a 2-min treatment at 40°C (pressure $\geq$300 MPa) [68].

As far as seafood safety is concerned, HHP processing is an especially promising technology. Its inactivating effect on a variety of pathogenic agents isolable from seafood has been documented. These agents include viruses, parasites, and several other types of seafood-associated bacterial pathogens. The ability of HHP to inactivate oyster-associated viruses has been extensively studied [69–72]. A 5-min HHP treatment at pressure 400-MPa and 0°C was established as an effective approach to bring murine norovirus-1 to undetectable levels in oysters [71]. HAV can also be effectively inactivated from oyster tissues using HHP. Calci et al. (2005) could achieve a PFU reduction $>3 \log_{10}$ with a 1-min treatment at 400 MPa [69]. It is important to note that temperature, matrices' pH, and salinity have a great effect on the efficiency of pressure-mediated HAV inactivation [70].

Another recent development in the field of molluskan shellfish safety is the application of super critical CO_2 ($scCO_2$), a known antibacterial substance widely used in the food industry to reduce the load of oyster-associated bacteria. Two conditions, 100 bar and 37°C for 30 min and at 172 bar and 60°C for 60 min, were reported as able to induce 2-log and 3-log reductions in the oysters' aerobic plate count, respectively. No significant change in the physical appearance, smell, or texture was recorded (Fig. 17.9) [73].

As previously stated, *V. vulnificus* and *V. parahaemolyticus* are major seafood-associated health concerns. In recent years, there has been a tremendous amount of effort to design approaches to reduce their load, especially in molluskan shellfish. One of the latest innovations in this field is the introduction of a "weak acidic electrolyzed water"-based approach [74]. Quan et al. (2010) have demonstrated that weak acidic electrolyzed water possesses outstanding antibacterial potency against Vv and Vp, especially when compared to sodium hypochlorite (NaClO), a commercial sanitizer [74].

A chlorine dioxide (ClO_2)-based approach was also recently introduced. According to Wang et al. (2010), a 6-h treatment with 20 mg/L of ClO_2 is enough to disinfect oyster tissues contaminated with Vp. These authors particularly recommend this method because it is cost effective; it also has the potential of increasing seafood's shelf life (~12 days) [75]. In addition, *E. coli* O157:H7, *S. typhimurium*, and *L. monocytogenes*, other pathogens of importance regarding seafood safety, have proven sensitive to this type of treatment.

Scombrotoxin. Because of the large number of associated outbreaks, scombrotoxin can be viewed as the most dangerous seafood-associated health hazard [2, 3]. To store

Fig. 17.9 Appearance of oysters before and after a 40-min exposure to $scCO_2$

fish at low temperatures is generally considered sufficient to prevent the growth of causative bacteria. Unfortunately, such conditions cannot always be respected, especially during retail processes. Phuvasate and Su (2010) proposed an alternative to low temperature storage applicable under these circumstances, namely the use of Electrolyzed Oxidizing (EO) water and ice. According to these authors, the load of several histamine-producing bacteria on food surfaces and fish skin can be significantly reduced simply by using electrolyzed oxidizing water and ice. Conditions reported as effective using salmon and tuna's skin were, for EO water, 100 ppm chlorine for 120 min, and for EO ice, 100 ppm chlorine for 24 h, respectively. Experts now agree that electrolyzed oxidizing water possesses a great potential as far as seafood safety. Its use is recommended not only because it is environmentally friendly, safe, and cost effective, but also because its application is quite straightforward. Its antibacterial effects against several other seafood-associated bacteria have been reported. A few examples are *E. coli* O157:H7, *L. monocytogenes*, as well as *Salmonella enteritidis*, *Campylobacter jejuni*, *Enterobacter aerogenes*, and *S. aureus* [76].

Bio-preservation of seafood. To add various chemicals to seafood in order to reduce its bacterial load or inhibit the growth of unwanted bacteria is an option that many have proposed as a solution to some seafood safety issues. Regrettably, consumer acceptance of these products is not always guaranteed. As chemical/preservative-free, ready-to-eat seafood products are gaining in popularity alternatives to the use of chemicals have emerged, and an example is bio-preservation. Recently, there have been a few innovations in this field. An example in this case is the proposed use of *Carnobacterium divergens* M35 and divergicin M35 in an effort to rid seafood from one of its most persistent bacterial contaminants, namely *L. monocytogenes* [77]. Matamoros et al. (2009) characterized several strains of lactic acid bacteria that can be used to this end as well. These bacteria (*Lactobacillus fuchuensis, Leuconostoc gelidum, Lactococcus piscium,* and *Carnobacterium alterfunditum*) showed inhibitory potential against seafood spoiling and pathogenic bacteria [78]. Pinto et al. (2009) also reported two bacteriocins produced by lactic acid bacteria (*Enterococcus faecium* and *Pediococcus pentosaceus*) that can be used to this end [79].

Irradiation and other recent breakthroughs. Several other cutting-edge technologies designed to help deal with seafood-associated health issues have been introduced in recent years. A few examples are X-ray, gamma ray, and electron beam irradiation-based technologies. Gamma irradiation (0.5, 1, 2, and 5 kGy) and electron beam irradiation have recently proven an effective non thermal approach to reduce the load of *V. parahaemolyticus* as well as several other seafood-associated contaminants, namely *L. monocytogenes* and *S. aureus*, in a raw seafood dish (oyster *Jeotkal*). Organoleptic properties were not negatively affected by the irradiation. In this particular case, gamma irradiation appears a better alternative to electron beam irradiation [80].

The beneficial antibacterial effects of electron-beam irradiation applied to another seafood dish (salted and seasoned short-necked clam) were reported in 2009 by Kim et al. It is important to note that, in this case, no change in sensory qualities was observed. A significant microbial inactivation (coliform bacteria, aerobic bacteria, yeast and mold) was reported [81]. Similar results were achieved with cold-smoked salmon [82]. This technique appears to be a better alternative to HHP in regards to sanitization of cold-smoked salmon. In fact, while both approaches (irradiation at 2 kGy and HPP: 450 MPa for 5 min) yielded a safer final product, the visual aspect of HHP-treated, not irradiation-treated, salmon was negatively affected. The microbial load of both products did not exceed 6 $\log_{10}$ cfu/g after 35 days storage at 5°C [82].

X-ray irradiation has also been investigated for use in improving seafood microbiological quality. Several recent publications have demonstrated its beneficial effects on *V. parahaemolyticus*, *V. vulnificus* (shrimp contaminants). It is important to note that several other shrimp-associated bacterial contaminants, namely *E. coli*, *Salmonella enterica*, and *Shigella flexneri*, were inactivated as well [83, 84].

High Hydrostatic Pressure (HHP) was also reported as an excellent alternative to reduce the level of *Listeria* in fish. According to Gudbjornsdottir et al. (2010), a 700–900 MPa treatment of 10 s is sufficient to reduce the load of *L. innocua* to undetectable levels. This experiment was completed with cold-smoked salmon. In this case, though there was no lipid oxidation, some variations in color and microstructure of the final product were noted [85]. The potential of HHP to rid mackerel from parasites, such as the nematode *Anisakis simplex*, was also recently reported. In this particular case, a complete inactivation of the larvae in the fish tissue was achieved at 300 MPa after a 5-min exposure [86].

Another recent report has proposed using CO_2 in packaging fish. Schirmer et al. (2009) proposed using CO_2 combined with various organic acids: citric acid (3% w/w, pH 5) and acetic acid (1% w/w, pH 5) in packaging fresh fish, as an effective way to improve its quality and shelf life. This combination has proven efficient at completely inhibiting bacterial growth in naturally contaminated salmon stored at 4°C for 14 days. Monitored bacteria were *Enterobacteriaceae*, lactic acid bacteria, and sulphur reducing bacteria. Sensory analysis was not completed by these authors [87].

A better alternative to rid shrimp from *V. parahaemolyticus* was recently introduced. The use of chlorine is the current approach to reduce the level of shrimp-associated pathogenic bacteria. Unfortunately, health issues such as bronchitis and pulmonary edema in workers have been reported. Norhana et al. (2009) proposed an even simpler approach to deal with shrimp-associated acterial pathogens. These authors show that washing shrimp with acidic fruit juice, namely bilimbi (*Averrhoa bilimbi*) or tamarind (*Tamarindus indica* L.) was also an effective way to significantly reduce its load of bacteria [88]. In 2007, Chaiyakosa et al. reported another safer alternative to reducing the load of *V. parahaemolyticus* in shrimp, namely the use of Chitosan [89]. Another author reported this same substance as an effective means of bringing safer salmon to consumers. Packing cold-smoked salmon in chitosan-coated plastic films containing 4.5 mg/cm^2 sodium lactate or either a combination of 4.5 mg/cm^2 sodium lactate plus 0.6 mg/cm^2 potassium sorbate or 2.3 mg/cm^2 sodium lactate plus 500 IU/cm^2 nisin was found to be beneficial. It was established that for seafood conditioned using this approach and stored at low temperature (~4°C), *L. monocytogenes*' growth was inhibited for at least 6 weeks [90].

Future Directions and Conclusions

From pathogenic bacteria, parasites, and viruses of all sorts, to life threatening biotoxins, the range of chemical and biological contaminants present in seafood is broad in scope and challenging to manage. Despite years of efforts from regulatory authorities and the scientific community, the public at large is still at risk of dangers associated with seafood consumption. For several years now, seafood has ranked foremost as the most significant source of food-borne disease outbreaks of known origin. Contributing factors are numerous and represent key points upon which urgent action, from regulatory authorities and the scientific community is required. Special attention should be paid to deleterious agents that have been associated with the largest number of outbreaks in the past years. In the USA for instance, major threats were scombrotoxin, responsible for 36% of all reported seafood-borne outbreaks from 1990 to 2006, followed by ciguatera (responsible for 21% of outbreaks), bacteria (especially *Vibrio* spp) responsible for 24% and finally noroviruses, which cause about 10% of all outbreaks reported during this same period [2]. It is important to note that these same agents pose serious problems in other parts of the world, as well. Scombroid fish poisoning, for instance, is also a serious problem in countries like Japan and the UK [28] *Vibrio* spp, especially *V. parahaemolyticus,* have been reported as a major problem in several Asian countries [40]. A recent study, a 2005–2008 survey of shellfish (mussels, clams and oysters), showed that norovirus is a problem in Italy, as well [91].

The task of ensuring consumers protection against the dangers of seafood is shared by several groups. These include (1) regulatory authorities, which depend heavily on the scientific community, and are in charge of the promulgation and enforcement of laws and regulations, (2) establishments involved in the harvest, processing, storage, and retail of seafood, and (3) consumers themselves. Failures and flaws at various levels of the current safety management system explain why seafood has been a persistent issue for the past few years. A certain number of defects in the seafood regulatory system of the USA for instance are presented in a 2008 report of the Center for Science in the Public Interest (CSPI). Mentioned shortcomings are the "voluntary recall" approach currently in place, added to financial issues. CSPI sees in the limited budget allocated to the FDA (Fig. 17.10), a serious hindrance to its efficiency. This can be a serious hindrance, for instance, when it comes to law enforcement. CSPI reports an extremely low inspection rate of food processing companies by the FDA, a rate that is insignificant compared to the USDA's [2]. Currently, in the USA, there are approximately 13,400 seafood-processing establishments. The FDA reports having inspected only 3,066, 2,830, and 2,456 during the fiscal years 2004, 2005, and 2006 respectively.

Financial limitations are not the only obstacle to full efficiency of regulatory agencies. The unavailability of effective technological tools that could either help rid seafood from hazardous entities/substances, or assist regulatory authorities in effective risk assessment and management during various control programs is also critical. Without effectual detection methods, efficient law enforcement is nearly impossible. Though numerous approaches applicable to virtually all seafood-associated health hazards have been proposed so far, several gaps remain.

This is the case of biotoxins for instance, which alone were responsible for 63% of all reported seafood-borne diseases from 1990 to 2006. Because they are extremely resistant to various post-harvest processing techniques, as far as assuring seafood safety, preventive measures are a better option. More effective detection tools are thus critically needed in order to reduce the current incidence of biotoxin-linked seafood-borne illnesses. Presently, in this field, there is still a need for cost-effective,

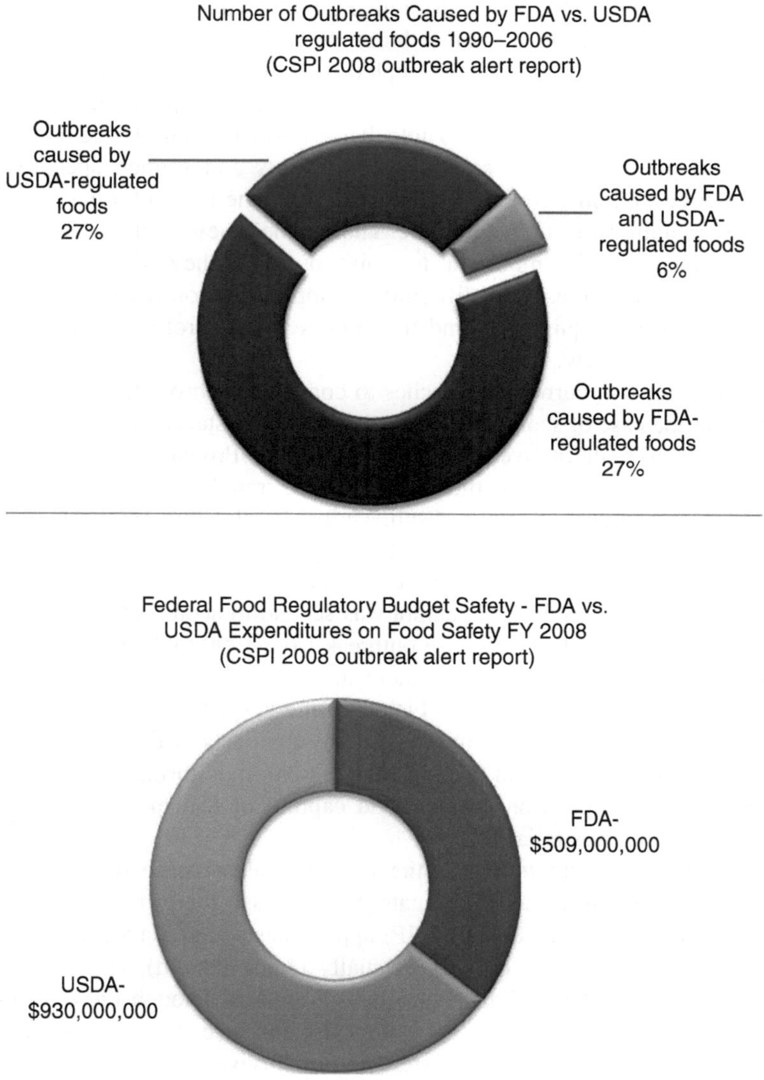

Fig. 17.10 Graphical representation of some limitations in the current federal food safety regulatory system with significant impact on seafood safety

rapid, sensitive, specific, and straightforward methods that can be operated by untrained personnel on a routine basis.

Though several approaches to toxin detection in seafood have been proposed so far, and even significant progress made recently (Table 17.7), animal-based assays and liquid chromatography, techniques that have their share of shortcomings, are still the official methods (Table 17.4). Animal bioassays for instance, are fit to assess the overall toxicity of a sample but cannot give insight on the nature of the toxin(s) involved. Moreover, these assays are time consuming, of limited accuracy and controversial (ethics). As stated previously, liquid chromatography, on the other hand, when coupled with effectual detection methods, is valuable at accurately identifying the nature and concentration of the incriminated toxin(s). However, it is important to note that, chromatography-based methods are lengthy and not cost-effective; heavy equipments and trained specialists are needed. Better methods have been proposed, but, none has so far been approved by regulatory authorities for routine use. Though these new methods present some clear advantages, it is important to point out that they also possess some weaknesses. Among these are difficulty of application on a routine basis, the necessity for heavy equipment, and the complexity of protocols and generated data, just to name a few.

There are flaws in current approaches to control scombroid fish poisoning and ciguatera. In the case of scombroid fish poisoning for instance, there is still a need for more effective alternatives to control Histamine-Producing Bacteria (HPB). Current recommendations of the FDA involve rapid cooling ($\leq$40°F) after visceration (for larger fish) upon death. Unfortunately, this approach possesses a few shortcomings. It is important to note that not all HPB are mesophiles. Several studies showed that histamine can be produced even at low temperature by physchrophilic HPB [28, 92]. Ciguatera was second to scombrotoxin as the most likely cause of seafood-borne disease outbreak from 1990 to 2006. Though ciguatera is such an issue in the USA, "there are neither standards, nor an official method" that applies to Ciguatera Fish Poisoning (CFP) in this country [24]. Innovations, regulation-wise, in this regard are obviously critical. Current efforts with respect to CFP prevention involve various toxin-monitoring programs, education, alongside with bans on the sale and capture of fish most likely to cause poisoning (Europe, Australia).

Bacteria were next to toxins as the most prevalent cause of seafood-borne illnesses during the past decade. Fortunately, in this case, there are currently several approved Post-Harvest-Processing (PHP) approaches aimed at reducing their load in seafood. These included IQF (Individually Quick Frozen), HCP (Heat Cold Pasteurization), and HHP (High Hydrostatic Pressure). Though these techniques have revolutionized the industry of oysters, for instance, there is still room for improvement, especially because oysters do not survive such processing [93]. In terms of shelf life, this can be an issue. To store, HHP-, IQF- and HCP-treated oysters at low temperatures is, unfortunately, not enough to solve the issue. Prapaiwong et al. [94], studying variations in the bacterial load (total aerobic bacterial counts) of HHP-treated oysters stored at 4°C for instance, determined that the bacterial count of processed products can rise quickly during storage and can even reach $\sim 10^7$ CFU/g in just 7 days [94]. Moreover, long-term storage presents the disadvantage of increasing the final production cost. Post-harvest-processing techniques non lethal to oysters would be, without the shadow of a doubt, a better alternative. Mahmoud and Burrage (2009c) reported X-ray irradiation as a better approach for reducing the load of oysters-associated *V. parahaemolyticus* because oysters are able to survive even extremely high X-ray doses [95]. Oysters have also been shown to be able to survive $scCO_2$ exposure, making it an attractive tool for further exploration [73].

Norovirus is another major seafood-associated health hazard. Major gaps in the current system are regulatory and scientific. Despite the clear threat posed by this virus, not much effort has apparently being exerted regulation-wise. As mentioned by Terio et al. [91], "there is no virological standard for bivalve shellfish in European legislation" [91]. Though several approaches to the detection of norovirus in seafood have been proposed so far, much still needs to be done; an area in need of improvement is the development of more efficient methods of viral RNA extraction. RNA extraction is a critical step in several virus detection protocols. RNeasy Kit was recently presented by Husman et al. (2009) as a most useful alternative for viral RNA extraction after comparing five such approaches side by side. A modified paramagnetic silica-based guanidium extraction based technique was also developed recently [96, 97].

Other challenges faced in norovirus detection are related to the virus' genetic variability and scarcity in samples. Interferences of seafood matrix, mainly, the presence of inhibitory substances, also represents a serious obstruction [97]. An effective TaqMan RT-PCR based approach for quantification of genogroups I and II norovirus, which presents the advantage of overcoming background inhibitory effects, was introduced recently by Gentry et al. (2009) [98]. An effective and more sensitive multiplex RT–PCR-based approach to norovirus and rotavirus detection in oysters was also introduced recently [99].

There is no doubt that the development of effective technologies able to surmount each and every one of these challenges will aid efforts to reduce outbreak incidents of seafood-borne infections attributable to viruses.

Compliance Policy/Regulation Related Citations

Compliance Program 7303.842
Sec 560.600 Compliance Policy Guide
Sec 555.300 Compliance Policy Guide
21 CFR 522.1081
21 CFR 529.1030
21 CFR 529.2503
21 CFR 556.500
21 CFR 558.450
21 CFR 558.582
21 CFR 556.660
21 CFR 558.575
21 CFR 556.640
Council Regulation (EEC) No 2377/90 of 26 June 1990
Sec 615.200 Compliance Policy Guide

21 CFR 556.660
Compliance Program 7303.842 or Sec 540.250 Compliance Policy Guide
Sec 540.600 Compliance Policy Guide
21 CFR 109.30
Sec 575.100 Compliance Policy Guide
40 CFR 180.226
Sec 615.200 Compliance Policy Guide
21 CFR 556.660
40 CFR 180.364
40 CFR 180.420
40 CFR 180.213
Sec 540.525
Regulation (EC) No 853/2004 (European standards)
EU Directive 91/492/EEC
Council directives 91/493/EEC

Acknowledgments We thank the NOAA (Mississippi-Alabama Sea Grant consortium, NIUST) Kraft Foods, CDC, and the NIH for financial support. We also thank Carol-Scott Defore, Mark Guy, and Derrick Donald for the precious assistance. Damaris A.F. Meujo is supported by a fellowship from the Medicines for Malaria Venture.

Bibliography

Primary Literature

1. Appendix 5 – FDA & EPA Safety Levels in Regulations and Guidance Fish and Fisheries Products Hazards and Controls Guidance
2. Outbreak Alert 2008, http://www.cspinet.org/new/pdf/outbreak_alert_2008_report_final.pdf
3. Outbreak Alert 2009, http://cspinet.org/new/pdf/outbreakalert report09.pdf
4. Fish and Seafood Utilization, http://www.fao.org/fishery/topic/424/en
5. What You Need to Know About Mercury in Fish and Shellfish March 2004, EPA-823-R-04-005, http://www.fda.gov/food/resourcesforyou/consumers/ucm110591.htm
6. Naliwaiko K, Araújo RL, da Fonseca RV, Castilho JC, Andreatini R, Bellissimo MI, Oliveira BH, Martins EF, Curi R, Fernandes LC, Ferraz AC (2004) Effects of fish oil on the central nervous system: a new potential antidepressant. Nutr Neurosci 7:91–99
7. Morris MC, Evans DA, Bienias JL, Tangney CC, Bennett DA, Wilson RS, Aggarwal N, Schneider J (2003) Consumption of fish and n-3 fatty acids and risk of incident Alzheimer disease. Arch Neurol 60:940–946
8. Barberger-Gateau P, Letenneur L, Deschamps V, Pérès K, Dartigues JF, Renaud S (2002) Fish, meat, and risk of dementia: cohort study. BMJ 325:932–933
9. Food is getting healthier and better, thanks to EU research, http://europa.eu/rapid/pressReleasesAction.do?reference=IP/06/1759&format=HTML&aged=0&language=EN&guiLanguage=en
10. Yokoyama M, Origasa H, Matsuzaki M, Matsuzawa Y, Saito Y, Ishikawa Y, Oikawa S, Sasaki J, Hishida H, Itakura H et al (2007) Effects of eicosapentaenoic acid on major coronary events in hypercholesterolaemic patients (JELIS): a randomized open-label, blinded endpoint analysis. Lancet 369:1090–1098

11. Cohen JT, Bellinger DC, Connor WE, Kris-Etherton PM, Lawrence RS (2005) A quantitative risk-benefit analysis of changes in population fish consumption. Am J Prev Med 29:325–334
12. Mozaffarian D, Longstreth WTJ, Lemaitre RN, Manolio TA, Kuller LH, Burke GL, Siscovick DS (2005) Fish consumption and stroke risk in elderly individuals: the cardiovascular health study. Arch Intern Med 165:200–206
13. Mozaffarian D, Rimm EB (2006) Fish intake, contaminants, and human health: evaluating the risks and the benefits. JAMA 296:1885–1899
14. Fish & Fish Products Market Report 2008, http://www.researchandmarkets.com/reportinfo.asp?cat_id=0&report_id= 597257&q=seafood market&p=1
15. Food Outlook – Global Market Analysis – Global Information and Early Warning System on Food and Agriculture, http://www.fao.org/docrep/012/ak341e/ak341e00.htm
16. Food Outlook – Global Market Analysis, http://www.fao.org/docrep/010/ai466e/ai466e00.HTM
17. Delgado CL, Wada N, Rosegrant MW, Meijer S, Ahmed M (2003) Outlook for fish to 2020: meeting global demand. International Food Policy Research Institute and The WorldFish Center, Washington, DC
18. Review of the State of World Marine Fishery Resources Food and Agriculture Organization of the United Nations, FAO FISHERIES TECHNICAL PAPER 457, http://www.fao.org/docrep/009/y5852e/Y5852E00.htm#TOC
19. Food-Related Illness and Death in the United States, http://www.cdc.gov/ncidod/EID/vol5no5/mead.htm
20. Report to Congress Food and Drug Administration Amendments Act of 2007 Public Law 110-85 Section 1006 – Enhanced Aquaculture and Seafood Inspection. Enhanced Aquaculture and Seafood Inspection – Report to Congress, http://www.fda.gov/Food/FoodSafety/Product-SpecificInformation/Seafood/SeafoodRegulatoryProgram/ucm150954.htm
21. Campas M, Beatriz PS, Jean-Louis M (2007) Biosensors to detect marine toxins: assessing seafood safety. Talanta 72:884–895
22. Fleming LE, Broad K, Clement A, Dewailly E, Elmir S, Knap A, Pomponi SA, Smith S, Gabriele HS, Walsh P (2006) Oceans and human health: emerging public health risks n the marine environment. Mar Pollut Bull 53:545–560
23. Marine Toxins, http://www.cdc.gov/ncidod/dbmd/disease info/marinetoxins_g.htm
24. Marine Biotoxins, http://www.fao.org/docrep/007/y5486e/y5486e00.HTM
25. BBB – Various Shellfish-Associated Toxins-Bad Bug Book: Foodborne Pathogenic Microorganisms and Natural Toxins Handbook Various Shellfish-Associated Toxins, http://www.fda.gov/Food/FoodSafety/FoodborneIllness/Food borneIllnessFoodbornePathogensNaturalToxins/BadBugBook/ ucm070008.htm
26. Assurance of Seafood Quality, http://www.fao.org/docrep/003/t1768e/T1768E00.htm#TOC
27. Clark RF, Williams SR, Nordt SP, Manoguerra AS (1999) A review of selected seafood poisonings. Undersea Hyperb Med 26:175–184
28. Assessment and Management of Seafood Safety and Quality, http://www.fao.org/docrep/006/y4743e/y4743e00.htm#Contents
29. Brief review of natural nonprotein neurotoxins, Applied science and analysis Inc., http://www.asanltr.com/newsletter/02-2/articles/Neurotoxins.htm
30. García C, Pereira P, Valle L, Lagos N (2003) Quantitation of diarrhetic shellfish poisoning toxins in Chilean Mussel using pyrenyldiazomethane as fluorescent labeling reagent. Biol Res 36:171–183
31. Fernández-Ortega JF, Morales-de los Santos JM, Herrera-Gutiérrez ME, Fernández-Sánchez V, Loureo PR, Rancaño AA, Téllez-Andrade A (2010) Seafood intoxication by tetrodotoxin: first case in Europe. J Emerg Med 39(5):612–617
32. Hwang DF, Noguchi T (2007) Tetrodotoxin poisoning. Adv Food Nutr Res 52:141–236
33. Bachvaroff TR, Adolf JE, Squier AH, Harvey HR, Place AR (2008) Characterization and quantification of karlotoxins by liquid chromatography-mass spectrometry. Harmful Algae 7:473–484

34. Van Wagoner RM, Deeds JR, Satake M, Ribeiro AA, Place AR, Wright JLC (2008) Isolation and characterization of karlotoxin 1, a new amphipathic toxin from *Karlodinium veneficum*. Tetrahedron Lett 49:6457–6461
35. Van Wagoner RM, Deeds JR, Tatters AO, Place AR, Tomas CR, Wright JLC (2010) Structure and relative potency of several Karlotoxins from *Karlodinium veneficum*. J Nat Prod 73(8):1360–1365
36. Peng J, Place AR, Yoshida W, Anklin C, Hamann MT (2010) Structure and absolute configuration of Karlotoxin-2, an Ichthyotoxin from the marine Dinoflagellate Karlodinium veneficum. J Am Chem Soc 132:3277–3279
37. Abbott BC, Ballantine D (1957) The toxin from *Gymnodinium veneflcum* Ballantine. J Mar Biol Assoc UK 36
38. Adolf JE, Bachvaroff TR, Krupatkina DN, Nonogaki H, Brown PJP, Lewitus AJ, Harvey HR, Place AR (2006) Species specificity and potential roles of Karlodinium micrum toxinAfr. J Mar Sci XI HAB Proc 28:177–180
39. Sheng J, Malkiel E, Katz J, Adolf JE, Place AR (2010) A dinoflagellate exploits toxins to immobilize prey prior to ingestion. PNAS 107:2082–2087
40. Su YC, Liu C (2007) *Vibrio parahaemolyticus*: a concern of seafood safety. Food Microbiol 24:549–558
41. Feldhusen F (2000) The role of seafood in bacterial foodborne diseases. Microb Infect 2:1651–1660
42. Andrews LS, DeBlanc S, Veal CD, Park DL (2003) Response of *Vibrio parahaemolyticus* O3: K6 to a hot water/cold shock pasteurization process. Food Add Contam 20:331–334
43. Interstate Shellfish Sanitation Conference Vibrio vulnificus fact sheet- Health care providers, http://www.issc.org/client_ resources/Education/VvFactSheet.pdf
44. Samir M, Haq BS, Hari HD (2005) Chronic liver disease and consumption of raw oysters: a potentially lethal combination – a review of *Vibrio vulnificus* septicemia. Am J Gastroenterol 100:1195–1199
45. Norhana MNW, Poole SE, Deeth HC, Dykes GA (2010) Prevalence, persistence and control of *Salmonella* and *Listeria* in shrimp and shrimp products: a review. Food Control 21:343–361
46. Committee on Evaluation of the Safety of Fishery Products. Seafood Safety (1991) Institute of Medicine (IOM). Food and Nutrition Board Institute of Medicine, http://www.nap.edu/openbook.php?record_id=1612&page=30
47. Potasman I, Paz A, Odeh M (2002) Infectious outbreaks associated with bivalve shellfish consumption: a worldwide perspective. Clin Infect Dis 35:921–928
48. Halliday ML, Kang LY, Zhou TK, Hu MD, Pan QC, Fu TY, Huang YS, Hu SLl (1991) An epidemic of hepatitis A attributable to the ingestion of raw clams in Shanghai, China. J Infect Dis 164:852–859
49. Norovirus: Technical Fact Sheet, http://www.cdc.gov/ncidod/dvrd/revb/gastro/norovirus-factsheet.htm
50. Kar D, Sur P, Mandal SK, Saha T, Kole RK (2008) Assessment of heavy metal pollution in surface water. Int J Environ Sci Tech 5:119–124
51. Heavy Metal Pollution – Heavy Metal Pollution is More Common Than You Think, http://www.fairfaxcounty.gov/nvswcd/newsletter/heavymetal.htm
52. Scientific Opinion on Arsenic in Food Question number: EFSA-Q 2008-425, http://www.efsa.europa.eu/EFSA/efsa_ locale-1178620753812_1211902959840.htm
53. Daniels JL, Longnecker MP, Rowland AS, Golding J (2004) Fish intake during pregnancy and early cognitive development of offspring. Epidemiology 15:394–402
54. Oken E, Wright RO, Kleinman KP, Bellinger D, Amarasiriwardena CJ, Hu H, Rich-Edwards JW, Gillman MW (2005) Maternal fish consumption, hair mercury, and infant cognition in a U. S. Cohort. Environ Health Perspect 113:1376–1380
55. Myers GJ, Davidson PW, Cox C, Shamlaye CF, Palumbo D, Cernichiari E, Sloane-Reeves JW GE, Kost J, Huang LS, Clarkson TW (2003) Prenatal methylmercury exposure from ocean fish consumption in the Seychelles child development study. Lancet 361:1686–1692

56. Smith AG, Gangolli SD (2002) Organochlorine chemicals in seafood: occurrence and health concerns. Food Chem Tox 40:767–779
57. Hites RA, Foran JA, Carpenter DO, Hamilton MC, Knuth BA, Schwager SJ (2004) Global assessment of organic contaminants in farmed salmon. Science 303:226–229
58. Willett WC (2005) Fish: balancing health risks and benefits. Am J Prev Med 29:320–321
59. How FDA Regulates Seafood: FDA Detains Imports of Farm-Raised Chinese Seafood, http://www.fda.gov/ForConsumers/ConsumerUpdates/ucm094558.htm
60. Gerssen A, Mulder PPJ, McElhinney MA, de Boer J (2009) Liquid chromatography-tandem mass spectrometry method for the detection of marine lipophilic toxins under alkaline conditions. J Chromatogr A 1216:1421–1430
61. Syaifudin ARM, Jayasundera KP, Mukhopadhyay SC (2009) A low cost novel sensing system for detection of dangerous marine biotoxins in seafood. Sens Actuators B 137:67–75
62. Zhou Y, Li YS, Pan FG, Zhang YY, Lu SY, Ren HL, Li ZH, Liu ZS, Zhang JH (2010) Development of a new monoclonal antibody based direct competitive enzyme-linked immunosorbent assay for detection of brevetoxins in food samples. Food Chem 118:467–471
63. Önal A (2007) Analytical, nutritional and clinical methods a review: current analytical methods for the determination of biogenic amines in foods. Food Chem 103:1475–1486
64. BY PRJB, Rutgers P, Stolker AAM, Nielen MWF (2009) Multi-residue screening of veterinary drugs in egg, fish and meat using high-resolution liquid chromatography accurate mass time-of-flight mass spectrometry. J Chromat A 1216:8206–8216
65. Smith SGC, Reimschuessel R, Decker C-S, Carson MC (2009) Simultaneous screening and confirmation of multiple classes of drug residues in fish by liquid chromatography-ion trap mass spectrometry. J Chromat A 1216:8224–8232
66. Bacteriological Analytical Manual (BAM) Bacteriological Analytical Manual – Revision A. Last Updated: 05/14/2009, http://www.fda.gov/Food/ScienceResearch/LaboratoryMethods/BacteriologicalAnalyticalManualBAM/default.htm
67. Espiñeira M, Atanassova M, Vieites JM, Santaclara FJ (2010) Validation of a method for the detection of five species, serogroups, biotypes and virulence factors of Vibrio by multiplex PCR in fish and seafood. Food Microbiol 27:122–131
68. Kural AG, Shearer AE, Kingsley DH, Chen H (2008) Conditions for high pressure inactivation of Vibrio parahaemolyticus in oysters. Int J Food Microbiol 127:1–5
69. Calci KR, Meade GK, Tezloff RC, Kingsley DH (2005) High-pressure inactivation of hepatitis A virus within oysters. Appl Environ Microbiol 71:339–343
70. Kingsley DH, Chen H (2009) Influence of pH, salt, and temperature on pressure inactivation of hepatitis A virus. Int J Food Microbiol 130:61–64
71. Li D, Tang Q, Wang J, Wang Y, Zhao Q, Xue C (2009) Effects of high-pressure processing on murine norovirus-1 in oysters (Crassostrea gigas) in situ. Food Control 20:992–996
72. Murchie LW, Kelly AL, Wiley M, Adair BM, Patterson M (2007) Inactivation of a calicivirus and enterovirus in shellfish by high pressure. Innovative Food Sci Emerg Technol 8:213–217
73. Meujo DAF, Kevin D, Peng J, Bowling JJ, Liu J, Hamann MT (2010) Reducing oyster-associated bacteria levels using supercritical fluid CO_2 as an agent of warm pasteurization. Int J Food Microbiol 138:63–70
74. Quan Y, Choi KD, Chung D, Shin IS (2010) Evaluation of bactericidal activity of weakly acidic electrolyzed water (WAEW) against *Vibrio vulnificus* and *Vibrio parahaemolyticus*. Int J Food Microbiol 136:255–260
75. Wang D, Zhang D, Chen W, Yu S, Shi X (2010) Retention of Vibrio parahaemolyticus in oyster tissues after chlorine dioxide treatment. Int J Food Microbiol 137:76–80
76. Huang YR, Hung YC, Hsu SY, Huang YW, Hwang DF (2008) Application of electrolyzed water in the food industry. Food Control 19:329–345
77. Tahiri I, Desbiens M, Kheadr E, Lacroix C, Fliss I (2009) Comparison of different application strategies of divergicin M35 for inactivation of Listeria monocytogenes in cold-smoked wild salmon. Food Microbiol 26:783–793

78. Matamoros S, Pilet MF, Gigout F, Prévost H, Leroi F (2009) Selection and evaluation of seafoodborne psychrotrophic lactic acid bacteria as inhibitors of pathogenic and spoilage bacteria. Food Microbiol 26:638–644
79. Pinto AL, Fernandes M, Pinto C, Albano H, Castilho F, Teixeira P, Gibbs PA (2009) Characterization of anti-*Listeria* bacteriocins isolated from shellfish: potential antimicrobials to control non-fermented seafood. Int J Food Microbiol 129:50–58
80. Song HP, Kim B, Jung S, Choe JH, Yun H, Kim YJ, Jo C (2009) Effect of gamma and electron beam irradiation on the survival of pathogens inoculated into salted, seasoned, and fermented oyster. LWT Food Sci Technol 42:1320–1324
81. Kim B, Song HP, Choe JH, Jung S, Jang A, Kim YJ, Jo C (2009) Application of electron-beam irradiation on the production of salted and seasoned short-necked clam, Tapes Pilippinarum, for safe distribution. Rad Phys Chem 78:585–587
82. Medina M, Cabeza MC, Bravo D, Cambero I, Montiel R, OrdóñezNuñez JAM, Hoz LA (2009) comparison between E-beam irradiation and high pressure treatment for cold-smoked salmon sanitation: microbiological aspects. Food Microbiol 26:224–227
83. Mahmoud BSM (2009) Effect of X-ray treatments on inoculated *Escherichia coli* O157: H7, *Salmonella enterica*, *Shigella flexneri* and *Vibrio parahaemolyticus* in ready-to-eat shrimp. Food Microbiol 26:860–864
84. Mahmoud BSM (2009) Reduction of *Vibrio vulnificus* in pure culture, half shell and whole shell oysters (*Crassostrea virginica*) by X-ray. Int J Food Microbiol 130:135–139
85. Gudbjornsdottir B, Jonsson A, Hafsteinsson H, Heinz V (2010) Effect of high-pressure processing on Listeria spp. and on the textural and microstructural properties of cold smoked salmon. LWT Food Sci Technol 43:366–374
86. Brutti A, Rovere P, Cavallero S, D'Amelio S, Danesi P, Arcangeli G (2010) Inactivation of Anisakis simplex larvae in raw fish using high hydrostatic pressure treatments. Food Control 21:331–333
87. Schirmer BC, Heiberg R, Eie T, Møretrø T, Maugesten T, Carlehøg M, Langsrud S (2009) A novel packaging method with a dissolving CO_2 headspace combined with organic acids prolongs the shelf life of fresh salmon. Int J Food Microbiol 133:154–160
88. Norhana MNW, Azman AMN, Poole SE, Deeth HC, Dykes GA (2009) Effects of bilimbi (*Averrhoa bilimbi* L.) and tamarind (*Tamarindus indica* L.) juice on *Listeria monocytogenes* Scott A and *Salmonella typhimurium* ATCC 14028 and the and the sensory properties of raw shrimps. Int J Food Microbiol 136:88–94
89. Chaiyakosa S, Charernjiratragul W, Umsakul K, Vuddhakul V (2007) Comparing the efficiency of chitosan with chlorine for reducing *Vibrio parahaemolyticus* in shrimp. Food Control 18:1031–1035
90. Ye M, Neetoo H, Chen H (2008) Effectiveness of chitosan-coated plastic films incorporating antimicrobials in inhibition of *Listeria monocytogenes* on cold-smoked salmon. Int J Food Microbiol 127:235–240
91. Terio V, Martella V, Moschidou P, Di Pinto P, Tantillo G, Buonavoglia C (2010) Food norovirus in retail shellfish microbiology. Food Microbiol 27:29–32
92. Bakar J, Yassoralipour A, Bakar FA, Rahman RA (2010) Biogenic amine changes in barramundi (*Lates calcarifer*) slices stored at 0°C and 4°C. Food Chem 119:467–470
93. Post-harvest Oyster Processing Technologies – Fact Sheet for Seafood Dealers and Processors, http://www.dmr.state.ms. us/Fisheries/Seafood-Technology/pdfs/fact-sheet-postharvest- oyster-processing.pdf
94. Prapaiwong N, Wallace RK, Arias CR (2009) Bacterial loads and microbial composition in high pressure treated oysters during storage. Int J Food Microbiol 131:145–150
95. Mahmoud BS, Burrage DD (2009) Inactivation of *Vibrio parahaemolyticus* in pure culture, whole live and half shell oysters (*Crassostrea virginica*) by X-ray. Lett Appl Microbiol 48:572–578

96. De Roda Husman AM, Lodder-Verschoor F, van den Berg HH, Le Guyader FS, van Pelt H, van der Poel WH, Rutjes SA (2007) Rapid virus detection procedure for molecular tracing of shellfish associated with disease outbreaks. J Food Prot 70:967–974
97. Le Guyader FS, Parnaudeau S, Schaeffer J, Bosch A, Loisy F, Pommepuy M, Atmar RL (2009) Detection and quantification of noroviruses in shellfish. Appl Environ Microbiol 75:618–624
98. Gentry J, Vinjé J, Lipp EK (2009) A rapid and efficient method for quantitation of genogroups I and II norovirus from oysters and application in other complex environmental samples. J Virol Meth 156:59–65
99. Xiaoxia K, Qingping W, Dapeng W, Jumei Z (2008) Simultaneous detection of norovirus and rotavirus in oysters by multiplex RT-PCR. Food Control 19:722–726

Books and Reviews

Balaban M, Odabaşi A, Damar S, Oliveira A (2007) Quality evaluation of seafood. In: Da-Wen Sun (ed) Computer vision technology for food quality evaluation. Academic, San Diego, pp 189–209

Ciminiello P, Dell' Aversano C, Fattorusso E, Forino M (2009) Recent developments in mediterranean harmful algal events. In: Fishbein JC (ed) Advances in molecular toxicology. Elsevier, New York, pp 1–41

Fung D (2009) Food spoilage, preservation and quality control. In: Encyclopedia of microbiology. 3rd edn. Elsevier, New York, pp 54–79

Halvorson H, Smolowitz R (2009) Aquaculture. In: Encyclopedia of microbiology, 3rd edn. Elsevier, New York, pp 17–22

Hwang D-F, Noguchi T (2007) Tetrodotoxin poisoning. In: Advances in food and nutrition research, vol 52. Elsevier, New York, pp 141–236

Jong E (2008) Toxic syndromes the travel and tropical medicine manual. In: Fish and shellfish poisoning, 4th edn. Saunders, Philadelphia, pp 474–480

Keener K (2007) Food regulations. In: Handbook of farm, dairy, and food machinery. William Andrew, Norwich, pp 15–43

Landrigan P, Kotelchuck D, Grandjean P (2007) Principles for prevention of the toxic effects of metals. In: Handbook on the toxicology of metals, 3rd edn, Academic, San Diego, pp 319–337

Le Guyader F, Atmar R (2007) Viruses in shellfish. In: Perspectives in Medical Virology, vol 17. Elsevier, Barcelona, pp 205–226

Ling KH, Nichols P, But P-H (2009) Fish-induced keriorrhea. In: Advances in food and nutrition research, vol 57. Elsevier, New York, pp 1–52

Niemira B, Zhang Q (2009) Advanced technologies for detection and elimination of pathogens. In: The produce contamination problem. Elsevier, New York, pp 425–443

Paulsen P, Luf W, Smulders F (2006) Different legislations on toxicants in foodstuffs. In: Food toxicants analysis. Elsevier, New York, pp 11–31

Rzeżutka A, Cook N (2009) Review of currently applied methodologies used for detection and typing of foodborne viruses. In: Global issues in food science and technology. Academic, San Diego, pp 229–246

Still K, Mohapatra A (2009) Biotoxins. In: Information resources in toxicology, 4th edn. Academic, San Diego, pp 91–102

Taylor S (2008) Molluscan shellfish allergy. In: Advances in food and nutrition research, vol 54. pp 139–177

Chapter 18
Sentinel Species in Oceans and Human Health

Lori H. Schwacke, Frances M. Gulland, and Susan White

Glossary

Bioaccumulation	The accumulation of a substance in an organism. Bioaccumulation occurs when a substance is absorbed by an organism at a more rapid rate than it is metabolized and/or excreted by the organism.
Bioavailable	Being in a state that can be readily absorbed by an animal.
Biomagnification	The increase in concentration of a chemical or toxin that occurs as it is passed up the food chain.
Confounding factor	A factor that correlates with both the exposure and response (i.e., independent and dependent variables in statistical terminology) so that it masks an actual association or falsely indicates an apparent association.
Epidemiology	The study of the distribution and determinants of health-related states in populations.
Exposure	Disease-causing factors, including infectious, toxic, nutritional, traumatic, genetic, degenerative, physiological, social, and behavioral.

This chapter was originally published as part of the Encyclopedia of Sustainability Science and Technology edited by Robert A. Meyers. DOI:10.1007/978-1-4419-0851-3

L.H. Schwacke (✉) • S. White
NOAA Center for Excellence in Oceans and Human Health, National Oceanic and Atmospheric Administration's Hollings Marine Laboratory, Charleston, SC, USA
e-mail: Lori.Schwacke@noaa.gov; Susan.White@noaa.gov

F.M. Gulland
The Marine Mammal Center, Sausalito, CA, USA
e-mail: gullandf@tmmc.org

E.A. Laws (ed.), *Environmental Toxicology: Selected Entries from the Encyclopedia of Sustainability Science and Technology*, DOI 10.1007/978-1-4614-5764-0_18,

HAB	Harmful algal bloom, a proliferation or aggregation of algae forming dense patches which are harmful to the environment, plants, or animals. The harmful algae may produce hazardous toxins or may harm other marine organisms by depleting oxygen and blocking sunlight.
Hazardous agent	Any chemical contaminant, biological toxin, or pathogen which presents a threat to human or animal health.
PAH	Polycyclic aromatic hydrocarbon, a class of environmental pollutants that occur in oil and coal and are produced as byproducts of fuel burning. The toxicity of PAHs depends on the structure of the specific compound but many are carcinogens and/or have been linked to congenital defects.
Pathogenicity	The ability of an agent to cause disease.
PCBs	Polychlorinated biphenyls, a class of persistent chemicals with a broad range of toxic effects. PCBs were widely used in industrial applications until the late 1970s when their manufacture was banned in the USA.
POPs	Persistent organic pollutants, organic compounds that persist in the environment because they are resistant to degradation and tend to bioaccumulate in animal tissues.
Zoonotic pathogen	A pathogen that can be transmitted between animals and humans.

Definition of the Subject and Its Importance

A sentinel marine species is one which can provide early warning of existing or emerging health hazards from the ocean environment. Sentinel species are generally considered in two categories: (1) those which are sensitive indicators of a chemical contaminant, biological toxin, or pathogen due to their ability to concentrate or integrate exposures within a food web or ecosystem, and (2) marine organisms with physiology and/or diet similar enough to humans such that they may provide early indication of potential adverse health effects and provide insight into toxic mechanisms of a given hazardous agent.

The sentinel concept has been extended to include marine habitats as well as species. Sentinel habitats generally encompass key ecosystem components (such as nursery grounds) and are particularly sensitive to exposure to a given hazardous agent or to other environmental perturbations which may increase or alter the distribution of a hazardous agent, thereby increasing the likelihood for human exposure.

Introduction

> "The concept of the combined health of people and wildlife – one health – when combined with the concept of the ocean as an essential modulator/caregiver of all life on Earth, inexorably leads to the prime tenet that the health of the "one ocean" is essential to the "one health" of all life on earth, including that of humans." NOAA Science Advisory Board Report (2010).

Few would argue the close connections among human health, animal health, and the health of the ecosystems of which they are a part. Recognizing these linkages, there has been a growing movement to expand collaborations among the medical community, veterinary and wildlife researchers, and environmental scientists to improve the understanding of how the environment, and changes in the environment, can affect human health. This holistic approach may be even more apropos when applied to the world's ocean, which is so critical for the elements and processes that sustain life on Earth. Oceans cover more than 70% of the Earth's surface, play an essential role in climate, support the life of nearly half the world's species, and provide a large portion of the protein necessary to support the Earth's human population. The recognition of the ocean's importance and the inextricable linkages between the health of people and the health of animals and ecosystems has led to the "One Ocean-One Health" research paradigm.

A natural component of the "One Ocean-One Health" paradigm is the use of marine animals as sentinels of potential emerging hazards. A sentinel by definition is one who is assigned to warn of danger, and the concept of using animals as sentinels to warn of environmental health hazards has long been recognized. As early as the 1960s, studies of Baltic fish and birds warned of the presence of polychlorinated biphenyls (PCBs) in the marine food chain [1], and in 1971, California sea lions alerted the scientific community to extraordinarily high levels of DDT off the coast of California and Baja California, Mexico [2]. A committee organized by the National Research Council described how animal sentinel systems, i.e., "systems in which data on animals exposed to contaminants in the environment are regularly and systematically collected and analyzed" could be used for identifying potential health hazards to other animals or humans [3]. Since that time, the use of sentinel animals has been promoted [4] and debated in the literature [5, 6] but has primarily been viewed in the context of chemical contaminant risk assessment.

Under the NOAA's Oceans and Human Health Initiative, the idea of using animal sentinels has been expanded to a more broad application to include marine toxins, particularly those associated with harmful algal blooms (HABs). HABs appear to be increasing in frequency and expanding in geographic distribution [7], heightening concerns for the human health impacts of HAB toxins which may include neurological, respiratory, or digestive distress or even death. While technology is advancing for the detection and prediction of HAB events, the degree to which a harmful algal bloom may represent a hazard for human health will depend not only on the occurrence of the bloom itself but also upon the physical processes and food web dynamics that drive the transport and persistence of the toxin, as well as the availability of potential vectors for human exposure. Monitoring of appropriate sentinels may assist not only in detecting the presence of hazardous toxins but also in elucidating the complex processes which drive toxin availability and the mechanisms by which the toxins exert their harmful effects. In fact, it has often been an increase in mortality or morbidity of marine wildlife that has signaled that a HAB toxin hazard is present [8, 9].

Sentinel marine species can also provide an early warning of threats from disease-causing pathogens which can affect seafood consumers, as well as for people who swim or recreate at beaches or in other coastal waters. Concern for infectious pathogens from the marine environment is perhaps greatest for food- or water-borne pathogens [10], and recurring human outbreaks of bacterial food poisoning resulting from contamination of shellfish with fecal microorganisms provide clear evidence that pathogens in marine ecosystems can impact human health [11]. However, nonfood marine wildlife may also act as an indicator or even reservoirs for zoonotic pathogens [12]. Some of these zoonotic pathogens may occur naturally in the marine environment, but others are being pumped into coastal waters through sewage or stormwater runoff. The potential for "spillover" across the marine animal–human interface increases as coastal development expands and a growing human population moves to our coast.

Regardless of the agent, whether chemical or biological, marine sentinels can signal presence in the environment and help differentiate between strict presence versus a presence in which humans are likely to be exposed. Additionally, a sentinel can provide indication of potential harmful effects, and if physiologically similar enough to humans, can offer insight into mechanisms of toxicity. The chapter starts by outlining the criteria that determine a good sentinel. The application of sentinels for surveillance or monitoring for potential exposures is discussed next; applications for chemical contaminants, marine toxins, and pathogens are discussed. The alternative use of sentinels for elucidating pathogenicity and to understand toxic pathways and modes of action is then presented. Finally, the concept of how a sentinel can be expanded to include sentinel habitats which can provide an early warning of ecosystem or community level effects that may increase or alter the distribution of hazardous agents, increasing the likelihood for human exposure, are discussed.

Desirable Characteristics for Marine Sentinels

In the ocean environment, sentinels to provide early warning of hazards may be even more critical in that many of the processes that drive the emergence of a hazard are complex and difficult to observe. Marine animals, which inhabit ocean waters 24/7 and/or consume a 100% seafood diet, receive the most intense and sustained exposures when ocean hazards are present. However, sentinel species and populations should be chosen carefully to ensure sensitivity to the exposure or associated health effect, as well as to ensure that the interpretation of data is meaningful and unambiguous. While the selection of an appropriate sentinel species will depend on the agent of concern and the desired region for surveillance, in general, the appropriateness of a given species as a sentinel can be judged based on the following criteria:

1. *Ease of sampling.* Tissue samples can be obtained with minimal effort and individuals can be randomly sampled from the population.
2. *Appropriate distribution.* The species should have a broad distribution to allow for analysis of geographic trends, or for localized studies have a distribution that overlaps with the agent of concern. The appropriate distribution will depend upon the agent of concern and the specific purpose of monitoring.
3. *Sensitivity for bioaccumulation and/or biomagnification.* The species should be particularly sensitive to accumulation of the agent in tissues or be sufficiently high on the food chain to be susceptible to biomagnification.
4. *Similar trophic position as humans or as a food animal.* Animals at a similar trophic position as humans and which have prey similar to those consumed by humans can provide an integrated index for potential exposures. Alternatively, animals consumed by humans provide a direct indication of potential dose.
5. *Indicator on appropriate spatial scale.* Some species will be better indicators of the immediate local environment while others may integrate across a geographic region. The sentinel should be chosen based on its ability to represent the appropriate spatial scale.
6. *Existing infrastructure and protocols for consistent collection, analysis, and data archiving.* Consistent sample collection and processing will ensure spatial and temporal comparability and is essential for long-term surveillance.

Meeting as many of these criteria as possible will aid in the design of statistically robust studies, facilitate the interpretation of results, and provide the best opportunity to obtain information relevant for human health. However, meeting all criteria is generally not possible, and certainly, the importance of meeting a given criterion must be weighed against the other benefits that a species may offer. For example, marine mammals are commonly selected sentinels despite the logistical difficulties in sampling (Table 18.1) and the legal and ethical constraints against experimental study and lethal sampling. However, physiological similarities to humans and high trophic position with diets that include many of the same types of seafood consumed by humans provide unique benefits that have prompted the inclusion of many marine mammal species as key elements of marine sentinel research. Due to the protected status of marine mammals in the USA [13], researchers rely heavily on epidemiological approaches for studying associations between exposures and disease, similar to approaches applied for human populations. In this case, it is often beneficial to supplement the research with experimental studies of other species (e.g., invertebrates, fish, rodents) to provide additional weight of evidence for establishing cause–effect relationships.

In addition to the above criteria for species selection, it is important to have existing infrastructure and protocols in place for consistent collection, analysis, and

Table 18.1 Examples of key sentinel species used in Oceans and Human Health Research and advantages and disadvantages of their use

Family/ Species	General		Chemical contaminants		Biological toxins		Zoonotic pathogens	
	Advantages	Disadvantages	Advantages	Disadvantages	Advantages	Disadvantages	Advantages	Disadvantages
Tursiops truncatus	Mammal; physiologically similar to humans so appropriate for examining pathogenicity	Ethical considerations of sampling; protected species under Marine Mammal Protection Act (MMPA); studies require permits understood	Integrator across food web	Some stocks may range across estuaries or even migrate across regions; requires understanding of stock structure	Integrator across food web	Difficult to obtain appropriate samples (e.g., urine, feces, stomach contents) from free-ranging animals	As a mammal, able to host many of the same pathogens as humans	Implications for human exposure potential not well
		Logistical considerations of sampling very large animals in water; adults generally range from 200 to 300 kg	Apex predator susceptible to biomagnification	Compounds may be preferentially metabolized, altering chemical profiles	Network in place for stranding response & mortality/morbidity investigation in U.S.			
		Long lifespan and reproductive cycle makes longitudinal studies difficult	Samples can be obtained via remote biopsy	Females offload lipophilic contaminants to young; tissue contaminants not representative of cumulative exposure				
Spend entire life in water		making it difficult to observe endpoints such as birth and reproduction	Broad coastal and oceanic distribution with discrete spatial stock structure allows for geographic comparisons					
Zalophus	*californianus*	Mammal; physiologically similar to humans so appropriate for examining pathogenicity	Ethical considerations of sampling; protected species under Marine Mammal Protection Act (MMPA); studies require permits	Distribution along U.S. Pacific coast		Distribution along U.S. Pacific coast and oceanic waters, overlapping with many algae of concern (e.g., *Pseudonitzchia* sp.)		As a mammal, able to host many of the same pathogens as humans
	Implications for human exposure	Periodically haul-out on shore, providing opportunity for	Logistical considerations ofsampling; age- and		Compounds may be preferentially	Integrator across food web		

(continued)

Table 18.1 (continued)

Family/ Species	General		Chemical contaminants		Biological toxins		Zoonotic pathogens	
	Advantages	Disadvantages	Advantages	Disadvantages	Advantages	Disadvantages	Advantages	Disadvantages
	potential not well understood	sampling and census of births	sex-classes are generally segregated	High trophic predator susceptible to biomagnificaiton	metabolized, altering chemical profiles			
Females offload lipophilic					contaminants to young; tissue contaminants not representative of cumulative exposure	Network in place for stranding response and mortality/morbidity investigation in U.S.		
Bivalves	Ease of sampling; not covered by Animal Welfare Act; greater social acceptance of use		Sessile, so good indicator of local environment	May not detect compunds which are transported via pathways other than water/sediments	Food source, so direct indication of potential dose for humans	Does not detect toxins that enter or persist in other parts of food web (e.g., through planktivorous or benthic fish)	Accumulates and concentrates many microbial pathogens	Not suitable host for many of human infectious pathogens
	High fecundity allows experiments on reproductin in reasonable time scale		Compounds generally not metabolically altered		Accumulates and concentrations toxins		Food source, so direct indication of potential dose for humans via ingestion route	
	Coastal distribution appropriate for monitoring pollutants being		Accumulates and concentrates many chemical contaminants					

(continued)

Table 18.1 (continued)

Family/ Species	General		Chemical contaminants		Biological toxins		Zoonotic pathogens	
	Advantages	Disadvantages	Advantages	Disadvantages	Advantages	Disadvantages	Advantages	Disadvantages
	input from terrestrial sources							
Salmonids	Not covered by Animal Welfare Act; greater social acceptance of use; can be lethally sampled Recreational and commercial fisheries offer opportunity to obtain samples	Non-mammal, so less similarity to human physiology	Internal organs/tissue (e.g., liver, bile) can be sampled to detect chemicals such as PAHs that do not persist in other tissues Food source, so direct indication of potential dose for humans if contaminants measured in edible tissue High trophic position; susceptible to biomagnification	Migrate during some life stages; requires an understanding of life history and range		Due to differences in physiology, mechanistic pathways may be dissimilar and many species are less sensitive to toxic effects		Not suitable host for many of human infectious pathogens

data archiving. This is imperative for consistent sample collection and processing which will ensure spatial and temporal comparability necessary for long-term surveillance.

Sentinel Applications for Monitoring Exposure

Chemical Contaminants

Traditionally, the use of sentinels has focused on chemical contaminants, and monitoring of sentinel animals as indicators of potential exposure offers several advantages over direct measurement in water, sediment, or other substrate. A primary advantage is that a measurement in a marine animal implies that the agent is bioavailable; that is, it is in a state that can be readily absorbed by an animal and, thus, is more likely to represent a human hazard.

Sentinels also may provide greater sensitivity for detection of an agent. For example, bivalves such as mussels and oysters filter and accumulate particles to concentrations much higher than in the ambient water and with little metabolic transformation, making them suitable sentinels for many chemical contaminants. Bivalves are sessile organisms and have been shown to be responsive to changes in ambient environmental levels of contaminants, making them particularly good indicators for local contamination [14]. In the USA, the Mussel Watch Program, sponsored by NOAA since 1986, monitors an array of legacy and emerging chemical contaminants in bivalves across the US coast [15]. No single species is common across all coastal regions, so the program samples a variety of bivalves: mussels (*Mytilus* spp.) from the North Atlantic and Pacific coasts, oysters (*Crassostrea virginica*) from the mid-Atlantic southward and in the Gulf of Mexico, and zebra mussels (*Dreissena* spp.) in the Great Lakes in order to provide a national perspective on chemical contaminant status and trends in estuarine and coastal waters.

Higher level marine predators may offer the greatest sensitivity for detection of lipophilic compounds such as polychlorinated biphenyls (PCBs), organochlorine pesticides, and polybrominated flame retardants. These persistent organic pollutants (POPs) accumulate in an animal's fatty tissues and biomagnify up the food chain. Upper trophic marine mammals, such as toothed whales, sea lions, some seal species, and polar bears are particularly susceptible to the biomagnification of POPs, often having tissue concentrations that are six to seven orders of magnitude higher than those found in water, sediments, or lower trophic biota. Some of the highest concentrations of polybrominated flame retardants, an emerging contaminant class of concern, have been reported in killer whales (*Orcinus orca*) which feed on other marine mammals and occupy an apex position in the Pacific marine food chain [16, 17]. Bottlenose dolphins (*Tursiops truncatus*) provided the first warning that legacy PCBs from a point source on the U.S. Atlantic coast had entered the coastal food web and spread far beyond the previously established source footprint [18]. In this case, the geographic extent of the contamination had not been recognized from monitoring of sediments and bivalves [19] because the contaminants were not elevated in these localized, less-mobile components of the ecosystem. This illustrates the utility of top predators as sentinels, not only because of their sensitivity to biomagnification, but as an integrative indicator across the broader coastal food web.

Historically, marine mammals were not considered useful species for biomonitoring due to the difficulty in obtaining tissue samples. However, after several large-scale mortality events of marine mammals, concern from the scientific community and the public prompted the establishment of a biomonitoring program to collect data to help elucidate temporal and geographic trends [20]. The National Marine Mammal Tissue Bank (NMMTB) was formally established in 1992 under the Marine Mammal Health and Stranding Response Act (Public Law 102–587) as a collaboration between NOAA Fisheries and the National Institute of Standards and Technology (NIST). Similar to NIST's Environmental Specimen Bank Program which has banked samples from Mussel Watch Program, the NMMTB focused on standardized collection, banking, and analysis of marine mammal tissues. Initially, samples were collected opportunistically from stranded and bycaught marine mammals, or those taken as part of authorized subsistence hunts. The reliance on stranded animals and carcasses somewhat hindered the interpretation of chemical analyses, as these animals may not be representative of the general population. The cause of stranding is often disease, and a diseased marine mammal is likely to carry abnormal levels of chemical contaminants [21]. Improved remote sampling techniques, including systems which use a crossbow, rifle or polespear with a dart to collect skin and blubber, now allow sampling of tissues from live marine mammals and have been applied for a variety of both pinniped and cetacean species [16, 22, 23]. The remote techniques allow for more robust sampling design, allowing studies to obtain a random sample of the free-ranging population. Methods for capture–release studies of some free-ranging marine mammals have also matured, and these types of studies, which allow sampling and banking of multiple tissue types, are conducted quite frequently across much of the US coast (e.g., [24–26]). Improved sampling techniques and standardization of protocols across studies have allowed for the synthesis of data to obtain an understanding of chemical contaminant trends in one common marine mammal species, the bottlenose dolphin, across the US southeast coast (Kucklick et al., submitted). The bottlenose dolphin is broadly distributed, and estuarine stocks generally maintain limited movements within a given region; therefore, monitoring of this species can offer a national perspective on chemical contamination of coastal waters.

Harmful Algal Bloom (HAB) Toxins

Sentinel species can warn of potential for exposure to known hazardous chemical agents but also can warn of previously unrecognized emerging hazards from algal toxins. While forecast systems using satellite data have been developed to detect and track some HABs, the forecast systems do not necessarily detect toxins which persist after a bloom or which accumulate in fish or invertebrate tissues, allowing the toxin to be transported through the food web. Because shellfish can accumulate algal toxins and are also harvested for human consumption, levels of algal toxins in shellfish are regularly monitored by states. However, upper trophic marine animals can often provide a broader perspective for detection of toxins in the marine food web. As an example, California sea lions (*Zalophus californianus)* are high-level predators that range from Baja California in the south to British Columbia in the north, feeding on species that often enter the human seafood market, such as anchovies, sardines, hake, rockfish, salmon, and market squid. The detection of domoic acid, an algal toxin produced by the marine diatom *Pseudo-nitzschia* spp., in California sea lions dying off the California coast in 1998 [9] raised public awareness of the presence of this algal toxin in a variety of seafood species.

Similarly, bottlenose dolphins in the Northern Gulf of Mexico provided the first indication that brevetoxin, a potent neurotoxin produced by the "red tide" dinoflagellate *Karenia brevis*, can accumulate in fish and remain in the food web long after a bloom has dissipated. An unusual mortality event, in which at least 107 bottlenose dolphins died, occurred on the coast of the Florida Panhandle in 2004 [8]. Although extensive sampling of water found only low levels of *K. brevis*, stomach contents of the dolphin carcasses showed high levels of brevetoxin and recent ingestion of menhaden, a planktivorous fish. Prior to this event, the common assumption, based on fish kills which occur during red tides, was that even low concentrations of brevetoxin would be lethal to fish and the toxin would therefore not accumulate in their tissues. However, sampling of fish that followed the dolphin mortalities showed high levels of brevetoxin that remained for several weeks. The event prompted long-term fish sampling and feeding studies which determined that because brevetoxins are sequestered in fish food, the fish may remain healthy while the toxin concentrations increase in their tissue, including the edible muscle tissue [27]. Brevetoxin accumulated to the highest levels in the liver and stomach of the fish and did not reach dangerous levels in the edible tissues. However, the findings raise awareness of the potential for brevetoxin exposure through consumption of fish and provide a warning of associated toxic effects in mammals. A follow-up capture–release health assessment of dolphins in the same region led to the finding that another marine toxin, domoic acid, was also present [25]. The low levels of domoic acid detected in urine and feces of dolphins during two separate capture–release sampling events and during one of the previous unusual mortality events, combined with the fact that domoic acid is rapidly eliminated from the body [28, 29], suggest that exposure to the toxin for dolphins in this area may be chronic.

Zoonotic Pathogens

Coastal waters have been historically monitored for presence of fecal indicator bacteria and advances in molecular technologies now allow for direct detection of some pathogens from environmental samples. However, despite technological advances, the difficulty in concentrating cells in order to extract the appropriate molecular markers generally limits the sensitivity of assays using environmental samples [30]. In some cases, marine animals may be more sensitive indicators for potential pathogenic hazards as they concentrate microbial pathogens in their tissues and may even act as pathogen reservoirs. Specifically, bivalves have been shown to accumulate bacterial as well as protozoan pathogens [31, 32]. Furthermore, a broad survey of seabirds and marine mammals along the northeast US coast revealed a number of human pathogens, most considered typically terrestrial, including *Giardia* spp., *Cryptosporidium* spp., influenza A, and avian influenza H3N8 virus [12]. A range of bacteria were isolated in the same study, and over 70% of these showed some level of antibiotic resistance. Similar reports of typically terrestrial pathogens occurring in free-ranging marine mammals have come from the Pacific US coast [33]. Many of the bacteria in marine mammals demonstrate resistance to multiple antibiotics, suggesting that their introduction to marine systems may be through human sewage or agricultural runoff [34]. The associated risk for human health is yet to be determined, but the presence and prevalence of these resistant pathogens in the coastal waters is a growing concern [35].

In some cases, surveillance of pathogens in marine mammals has demonstrated apparent shifts in distribution, potentially signaling an ecological response to climate change. An infection of a harbor porpoise (*Phocoena phocoena*) with *Cryptoccocus gatti* was the first of a cluster of human and animal cases that occurred on southern Vancouver Island, British Columbia between 1999 and 2002 [36]. Prior to this event, *C. gatti* was not recognized as a variant associated with disease in North America and was considered a tropical or semitropical pathogen. The emergence into more northern latitudes was potentially encouraged by a complex interaction of climatic variation, host susceptibility, and a creation of an ecological niche for the fungal pathogen [37, 38]. The range of the fungi may still be expanding as the first case in Hawaiian waters was recently reported in a spinner dolphin (*Stenella longirostris*) [39].

Another fungi, *Lacazia loboi*, also associated with tropical and semitropical climates, has emerged into more northern latitudes of North America where it was also first detected through surveillance of marine mammals. *L. loboi* infections have been reported in humans [40, 41] and multiple dolphin species [42, 43] and present as dermal and subcutaneous granulomas (wart-like lesions). While previously reported on the US coast in transitional tropical regions of the central Florida and Texas coasts [43, 44], a recent study reported the emergence as far north as the US mid-Atlantic coast [45].

Sentinels for Establishing Pathogenicity and Elucidating Mechanisms for Toxicity

Biological systems have the advantage that along with indicating potential for exposure, they can help to establish pathogenicity, provide insight into mechanisms of action, and contribute to the weight of evidence to establish causality of specific agents. In addition, the response of in situ sentinels to broad environmental changes provides insight to potential impacts on human populations. The mechanism of action of specific agents can be difficult to identify in natural systems due to the complex interactions among different exposures, as well as natural variability in response associated with different life history stages. Confirming that one specific factor is responsible for a physiological, anatomical, or health change in an animal or person can be difficult for logistical, ethical, or economical reasons. The careful selection of sentinel species can sometimes circumvent these constraints. Selection of a sentinel for understanding a mechanism of action of a hazard or establishing causality usually results in choice of an invertebrate or lower vertebrate such as a fish, so that lethal sampling and controlled exposures can be used. In contrast, use of sentinels to detect impacts of natural exposures, mixtures, and broad-scale ecological changes, such as those in response to changing ocean temperature or acidity, may focus on marine wildlife epidemiological studies. Below are examples of sentinel species that have increased the understanding of mode of action for chemical hazards, algal toxins, or pathogenic organisms.

Invertebrates

Invertebrates are useful sentinels for investigating the effects of pollutants and toxins on enzyme pathways, as these animals can be readily sampled and analyzed in laboratory situations. The phenomenon of "green oysters" was first documented in the late eighteenth century [46]; the intense green coloration was later demonstrated to be correlated with highly elevated copper concentrations [47]. Oysters and other bivalves detoxify accumulated copper and other metals by inducible metal-binding proteins (metallothioneins) and by sequestration in metal-rich granules, resulting in bioaccumulation [48]. Direct measurements of the effects of hazards on invertebrate enzyme pathways allow accumulation of evidence for harmful effects of an environmental hazard on animal systems that can contribute to risk assessments, despite the great difference between invertebrate and human physiology.

Amphibian, Fish, and Reptiles

Amphibians and reptiles are well-recognized sentinels of freshwater aquatic systems with the presence of developmental abnormalities in frogs and reproductive abnormalities in alligators associated epidemiologically with exposure to organochlorines and endocrine-disrupting chemicals in the environment [49]. Although their distribution is mostly limited to coastal and estuarine habitats, factors influencing health of freshwater systems, such as toxins like atrazine and PCBs, will ultimately make their way to marine waters.

Fish have become increasingly popular as sentinels of toxic effects as they contain most of the vertebrate genes of interest to human health but have fewer introns (sequences between genes) making sequencing studies more efficient. Although early studies using fish as sentinels focused on protein biomarkers of exposure, more recent studies have focused on alterations in genes that can be sequenced [50]. Fish are useful sentinels for estrogenic compounds in the environment, as the yolk protein precursor, vitellogenin (Vtg), is induced after exposure to estrogens and can be readily measured in individual fish [51]. Relationships have been demonstrated between Vtg or Vtg mRNA and skewed sex ratios, prevalence of intersexuality, and reduced fertilizing capacity of gametes, indicating that biochemical effects in an individual are likely to have population level impact on wildlife [52–54].

Evidence implicating environmental contaminants as causative agents of neoplasia in fish has accrued over the past five decades, with early reports focusing on epidemiological associations between contaminants and high prevalence of neoplasia and later reports showing direct induction of neoplasia following exposure of fish to certain compounds. In the 1960s, high incidences of papillomas in white croakers (*Genyonemus lineatus*) and Dover sole (*Paropyrys vetulus)* were observed at several sites in Southern California, while hepatic neoplasms were observed in white suckers (*Catostomus commersoni*) in Maryland [55, 56]. Studies of English sole, Pacific tomcod (*Microgadus proximus*), and rock sole (*Lepidopsetta bilineata*) in the heavily industrialized Puget Sound (Washington state) showed a tumor incidence that appeared to correlate with PCB levels [57, 58]. There was also a correlation between sediment polycyclic aromatic hydrocarbon (PAH) concentrations or bile fluorescent aromatic hydrocarbon concentrations and hepatic lesions. Laboratory studies confirming these observations were conducted by injecting English sole with an extract from either the contaminated or reference sediment or the model hepatocarcinogen benzo[a]pyrene [59]. The results demonstrated that injections with benzo[a]pyrene or the contaminated sediment extract increased the incidence of hepatotoxic lesions and preneoplastic lesions. The investigators formulated a model to determine prevalence associated with species of fish, type of exposure, location of capture, age, and gender. For example, an English sole from Elliot Bay, Washington, was 734 times more likely to have a neoplasm than a fish from a reference site, after controlling for both age and gender.

Effects and mechanisms of algal toxins have also been examined in fish, but fish generally have been found to be less sensitive to toxicity, particularly neurotoxicity, as compared to mammalian species [60]. Dosing studies of coho salmon (*Oncorhynchus kisutch*) with domoic acid showed that although the toxin was absorbed in the gut, the toxin was excreted via the kidneys and bile and did not reach sensitive nervous tissue [61].

Although not a true environmental sentinel, one freshwater fish species has been widely employed by marine researchers as an important model organism for understanding the effects of marine contaminants and toxins, particularly on embryonic development. The zebrafish (*Danio rerio*) is an ideal model species for understanding developmental processes and the potential impacts of toxins in that its embryos develop outside of the mother, and the egg as well as the embryo itself are transparent [62]. This feature allows for full observation of the developing organ systems. In addition, the species can be easily maintained, and there are now thousands of well-characterized mutant forms and behaviors which can be easily observed and tested. Zebrafish were initially used as a model to understand the health effects of PAHs, which derive from fossil fuels and were of concern following the Exxon Valdez oil spill, on marine fish. Additional sources of PAHs include vehicle exhaust and other forms of air pollution, which creates significant exposure potential for human populations in many areas. Research using the zebrafish model determined that PAHs can interrupt cardiac rhythm and contractility [63, 64], and this research may provide insight into the potential for PAH exposure to contribute to the incidence of cardiac disease in some human cohorts. The zebrafish model has also been applied to examine toxic mechanisms for algal toxins and microbial pathogens [64–66].

Marine Mammals

Domestic mammals have acted serendipitously as sentinels of human health when outbreaks of disease have been traced to an environmental factor that could impact humans. In Japan in the 1950s, domestic cats developed "dancing cat fever," which was eventually traced to high mercury concentrations in fish and shellfish in Minamata Bay [67]. The high concentrations were caused by effluents from a nearby factory using mercuric chloride in the production of vinyl chloride. More recently, attention has focused on marine mammals as sentinels of ocean change as they are charismatic and mortality events attract considerable public attention. Marine mammals have similar mammalian physiology to humans and are long lived so they may be effective indicators for chronic, latent, or slow-developing pathologies that are more difficult to detect in human populations exposed to lower levels of the same hazard (if it is in shared prey items or habitat). A variety of diseases and lesions have been observed in marine mammals worldwide that have heightened public awareness of their risk of environmental exposure to contaminants, biotoxins, and pathogens.

Since 1998, domoic acid producing algal blooms have become increasingly common along the California coast within the foraging range of sea lions. Accompanying this increase, there has been an increase in sea lions with epilepsy characterized by hippocampal atrophy, as well as cardiac lesions and reproductive failure [68–70]. Clinical examination, histology, and molecular diagnostics ruled out infectious causes for these health changes in sea lions, and epidemiological evidence indicated exposure to domoic acid was the factor most strongly associated with these problems. Thus sea lions have acted as sentinels for sublethal effects of domoic acid on mammalian health that can be further investigated using laboratory models such as rodents. A recent paper in the medical literature drew the public health community's attention to this potent marine toxin, hypothesizing that dietary exposure to doses of domoic acid that are subclinical in pregnant women may be sufficient to damage the fetal hippocampus and initiate epileptogenesis [71].

California sea lions have also drawn attention to the potential health impacts from high levels of POPs in the food chain off coastal California, as these compounds in sea lion blubber are associated with a high prevalence of cancer: 17% of stranded adult sea lions examined at necropsy have urogenital cancer [72]. This cancer is associated with a novel herpes virus and exposure to anthropogenic contaminants (such as PCBs and DDTs) [73]. More genetically inbred sea lions, and those with a specific major histocompatibility complex genotype, are more likely to develop cancer [74]. These data suggest that interactions among genes, toxic chemicals, and viruses result in cancer in this ubiquitous marine mammal that shares its coastal California environment with humans.

High levels of contaminants and cancer have also been documented in cetaceans, again raising concerns for health impacts of feeding in polluted waters [75]. A population of 450–500 belugas (*Delphinapterus leucas*) resides in the polluted estuary of the St. Lawrence River, Canada. Cancer has been identified in 50% of these belugas in a recent survey, and high concentrations of organochlorines, heavy metals, and benzo[a]pyrene (BaP) have been measured in tissues, which could act directly or indirectly to increase risk of cancer development [76]. Furthermore, lesions suggestive of endocrine disruption are common in this population of whales, including hyperplastic and degenerative changes of the adrenal gland, "adenomas" of the thyroid gland, hermaphroditism, and pseudohermaphroditism [77, 78]. Most recently, a high prevalence of thyroid hyperplasia was identified in these animals [79]. Polychlorinated biphenyls are one of the major recognized contaminants causing hyperplastic and neoplastic lesions of the thyroid gland [79]. Some polycyclic aromatic hydrocarbons, such as BaP, are among the most potent carcinogens, acting as initiators. Other compounds, such as PCBs, are recognized promoters for the induction of tumors from initiated cells. Thus, although the direct causal relationship between environmental contaminant exposure and either cancer or glandular lesions in the belugas is unknown, the existence of these lesions in wild marine mammals draws attention to the need for concern over increased risk to human health.

Marine mammals thus act as sentinels by attracting attention to naturally occurring diseases potentially linked to environmental exposure to chemical contaminants and other toxins. A limitation to their use as sentinels, however, is that experiments to confirm causality are usually ethically, legally, and logistically impossible. Thus teasing apart confounding variables in disease pathogenesis is difficult, unlike in fish (see above). For example, although California sea lions with cancer have higher levels of organochlorine compounds in blubber than control, non cancer animals, they are also thinner. Blubber contaminant residues are highly mobile, and two studies, one monitoring changes in blubber organochlorine levels during weight loss [80], the other using regression analyses to investigate relationships among blubber and plasma organochlorine levels [81] have demonstrated that weight loss resulting in thinner blubber levels must be considered as a confounding factor when investigating the relationship between cancer and blubber contaminants in sea lions.

Sentinel Habitats

The application of sentinels for early warning of health hazards in the marine environment cannot only be applied to marine organisms but also more broadly to marine habitats. Combining these two approaches to examine and monitor potential environmental impairments, particularly in the coastal environment, can provide a strong foundation investigating linkages among coastal land use, ecosystem condition, and risks to human and marine animal health. Certain sensitive coastal and marine habitats may exhibit ecological decline in quality and quantity that may be more rapidly detected prior to observable changes in the broader ecosystem. Monitoring of habitat level change often provides an early signal to researchers to look more closely at specific marine organisms for further understanding of any observed declining conditions and potential causality. Links between ecosystem condition and public health and well-being are poorly understood; however, working to better understand these linkages may provide additional insights into potential impacts on human health and well-being in the coastal environment.

National reports on the condition of coastal ecosystems within the United States note measurably diminished conditions [82, 83]. Major sources of impairment identified included chemical and microbial contamination, changes in freshwater inflows, nutrient enrichment, hypoxia, habitat modification and land use changes, wetland loss, increases in nonnative species, and overharvesting of fisheries. The cumulative and potentially synergistic effect of multiple stressors significantly contributes to diminished habitat quality and quantity. One of the earliest symptoms of broad-scale coastal ecosystem impairment has, historically, been decline in the amount and condition of habitats that are particularly sensitive to fluctuations in environmental conditions. The use of sentinel habitats, or first responders, for capturing the ecosystem level impacts of changing coastal conditions is growing.

Marine habitats that are strong candidates as sentinels often have the following attributes:

1. *Shallow water habitats with close proximity to land.* Shallow, nearshore habitats are immediate endpoints to terrestrial inputs and do not benefit from the dilution that occurs in deeper water offshore habitats. Terrestrial pollutants and/or changes in freshwater inflow volume directly impact these habitats and generally produce high cumulative exposure over time.

2. *Easily, and often, observed habitats.* Declining environmental conditions in marine habitats that the public regularly enjoy, due to their easy access, provide unambiguous signals that both the public and scientists understand such as a reduction in the extent or quality of that habitat.
3. *Habitats with multiple roles in sustaining overall ecosystem function.* Habitats that provide multiple key roles in marine environments (e.g., nurseries, refuges, feeding grounds, high biological productivity) provide a complexity of functions that can be assessed as healthy or in good ecological condition.
4. *Habitats with environmentally sensitive structural components.* Key organism sensitivities (e.g., narrow tolerances to environmental fluctuations in temperature, salinity, or light availability) to fluctuations in the natural range of environmental conditions provide indicators of impaired ecosystem condition rapidly.

Coral reefs, submerged aquatic vegetation (e.g., sea grass beds), oyster reefs, tidal creeks, and the estuarine wetlands associated with them are examples of sentinel habitats which generally decline in extent and condition years before system-wide impairment is documented by routine environmental monitoring [84–86]. Linking declines in environmental condition to specific causes (or multiple causes) is challenging but research continues to progress in this arena [87–90]. Coral reef and tidal creek habitats are highlighted below as examples of sentinels for broad-scale changes in environmental condition.

Coral Reefs

Tropical coral reefs exist within narrow ranges of environmental conditions (e.g., temperature and light) and tolerate only limited fluctuations in these conditions. Coral communities experiencing persistent environmental disturbances (e.g., coastal development, land-based pollution, warming temperatures) are one of the first coastal ocean ecosystem components to signal impairment including loss of diversity, increased incidence of disease, reduced growth, reduced reproduction, and mass mortality. These habitats are experiencing rapid declines from changing environmental conditions across the globe. Recent reports indicate that 58–70% of coral reefs worldwide are directly threatened by human-associated activities [88, 91, 92]. Coral reefs are accepted sentinels for the adverse effects of rapid global warming as they exhibit extreme sensitivity to episodes of prolonged thermal stress [93]; temperatures of just a few degrees above mean summer highs can cause bleaching involving significant losses of mutualistic symbiotic dinoflagellates (microalgae) from the tissues of corals. Such episodes of mass bleaching and mortality adversely affect the health, growth, and fitness of coral communities throughout the world, highlighting the global effects of climate change on ocean health. However, our understanding of the ecological and evolutionary response of corals to episodes of thermal stress remains inadequate, and studies on the mechanisms involved in coral death are few.

Tidal Creeks

Tidal creeks and estuarine salt marshes are the interface between the landscape and estuaries where freshwater from the land mixes with saline water, resulting in complex environments that are renowned for their ecological complexity, biological productivity, and seafood production [19, 86, 94–98]. Most of the US population lives in watersheds that drain into estuaries; as a result, estuaries are a repository for much of the pollution released into the environment [99–101]. The amount and timing of rainfall affects freshwater inflows into estuaries and the pollutants it contains. These processes are increasingly being affected by humans who in turn impact the health of both ecosystems and people. Billions of dollars have been spent to reduce anthropogenic impacts upon estuarine ecosystems with mixed results [99, 102, 103]. These shallow ecosystems can provide early warnings of ensuing harm for both marine organisms and human health [19].

Future Directions

Sentinel species data can be useful as additional weight of evidence in a risk assessment, in providing early warning of situations requiring further study, or for monitoring the course of remedial activities. The risks to aquatic organisms resulting from changes in the ocean serve as an early warning signal for humans because of the close linkage between human and animal health. Approximately 60% of all emerging infectious diseases in humans are zoonotic [104], and the molecular and metabolic processes of higher animals, including man, are very similar, so that aquatic organisms serve as effective sentinels for how ocean anomalies can affect human health. This underlying linkage is the basis for the "One Health" approach in which scientists have recognized that human health and animal health are inextricably linked (www.onehealthinitiative.com). The One Health initiative seeks to promote the health of all species by enhancing cooperation and collaboration among physicians, veterinarians, and other scientific health professionals. This paradigm of interdependence among human, wildlife, and ecosystem health has been well accepted for the terrestrial biosphere and now is being applied to the water-based biosphere, creating the "One Ocean-One Health" perspective. However, despite the increasing number of examples of health changes in sentinel marine species and habitats, the public health community is less engaged in concern for ocean "One Health" than the veterinary and ecological communities. There is now a real need to integrate findings from sentinel studies into the main stream public health community. Just as translational medicine grew from the need to effectively move basic research into medical practice and pharmaceutical applications, a call for "translational ecology" has been made to connect the outcomes of environmental research with public health experts and policy makers who can utilize the results and take appropriate action [105]. Soliciting the input from potential end users and connecting these end users with major funders of environmental research has been proposed as a mechanism to promote translational ecology, and this would be particularly appropriate for marine sentinel research. Funding opportunities geared toward joint eco-epidemiologic and human epidemiologic surveys so that data can be simultaneously collected and compared would go a long way toward promoting the appropriate interactions between ecological and human health researchers.

The data obtained from sentinel species need to be more readily available to the public health community. This could be achieved by use of tumor registries and reportable disease-tracking systems for animals so that geographic trends in disease prevalence and incidence can be detected and related to changes in human health. There is also a need for consistent surveillance and rapid-response teams of experts to investigate disease epidemics and morbidity and mortality in wildlife to determine if the cause was related to environmental changes. Making data on chemical and biological toxins measured in marine wildlife readily available to epidemiologists and other public health researchers would facilitate their use for generating hypotheses for human cohort studies. The Mussel Watch Program (http://ccma.nos.noaa.gov/about/coast/nsandt/musselwatch.aspx) provides a model for web-based data presentation.

Although the use of aquatic sentinel species is expanding, there are several areas of activity that need further attention. Specifically, there is a need to improve veterinary diagnostics and develop biomarkers of exposure and toxic effects that reflect similar biological events in both humans and sentinel species. In order for a sentinel to provide a true "early warning," diagnostics that observe early, potentially subclinical responses that occur shortly after exposure to a hazardous agent are required. These might include changes in hematologic or serum biochemical parameters but improved diagnostics that involve the use of novel "omics" approaches – genomics, proteinomics, and metabolomics – are needed. These techniques are being developed for clinical use in humans, but their development has been limited in the veterinary and ecological communities. Such techniques could help overcome some of the logistical difficulties facing researchers using species such as marine mammals as sentinels.

A further future direction for the field of sentinel species is the improvement of communication among government agencies, researchers, and public health officials regarding the availability, interpretation, and application of animal sentinel data and methods. Regulators need to be informed about the existence of sentinel animal databases, such as the Marine Mammal Health and Stranding Program in the USA (http://www.nmfs.noaa.gov/pr/health/) and reportable disease-tracking systems for domestic animals (http://www.oie.int/) and animal tumor registries.

Bibliography

Primary Literature

1. Jensen S (1966) Report of a new chemical hazard. New Science 32:612
2. Le Boeuf BJ, Bonnell ML (1971) DDT in California sea lions. Nature 234(5324):108–110
3. NRC (1991) Animals as sentinels of environmental health hazards. National Academy Press, Washington, DC
4. van der Schalie WH, Gardner HS Jr, Bantle JA, De Rosa CT, Finch RA, Reif JS, Reuter RH, Backer LC, Burger J, Folmar LC et al (1999) Animals as sentinels of human health hazards of environmental chemicals. Environ Health Perspect 107(4):309–315
5. Burkhart J, Gardner H (1997) Non-mammalian and environmental sentinels in human health: "back to the future?". Hum Ecol Risk Assess 3(3):309–328
6. Stahl R (1997) Can mammalian and non-mammalian "sentinel species" data be used to evaluate the human health implications of environmental contaminants? Hum Ecol Risk Assess 3(3):329–335
7. Van Dolah FM (2000) Marine algal toxins: origins, health effects, and their increased occurrence. Environ Health Perspect 108(Suppl 1):133–141
8. Flewelling LJ, Naar JP, Abbott JP, Baden DG, Barros NB, Bossart GD, Bottein MYD, Hammond DG, Haubold EM, Heil CA et al (2005) Red tides and marine mammal mortalities. Nature 435(7043):755–756
9. Scholin CA, Gulland F, Doucette GJ, Benson S, Busman M, Chavez FP, Cordaro J, DeLong R, De Vogelaere A, Harvey J et al (2000) Mortality of sea lions along the central California coast linked to a toxic diatom bloom. Nature 403(6765):80–84
10. Lipp EK, Rose JB (1997) The role of seafood in foodborne diseases in the United States of America. Rev Sci Tech 16(2):620–640
11. Potasman I, Paz A, Odeh M (2002) Infectious outbreaks associated with bivalve shellfish consumption: a worldwide perspective. Clin Infect Dis 35(8):921–928
12. Bogomolni AL, Gast RJ, Ellis JC, Dennett M, Pugliares KR, Lentell BJ, Moore MJ (2008) Victims or vectors: a survey of marine vertebrate zoonoses from coastal waters of the Northwest Atlantic. Dis Aquat Organ 81(1):13–38
13. Marine Mammal Protection Act (1972) 16 USC Chapter 31, vol 16, USC Chapter31, MMPA 1972
14. Roesijadi G, Young J, Drum A, Gurtisen J (1987) Behavior of trace metals in *Mytilus edulis* during a reciprocal transplant field experiment. Mar Ecol Prog Ser 15:155–170
15. Kimbrough K, Johnson W, Lauenstein G, Christensen J, Apeti D (2008) An assessment of two decades of contaminant monitoring in the Nation's Coastal Zone, NOAA Technical Memorandum NOS NCCOS 74. National Oceanic and Atmospheric Administration, National Ocean Service, National Centers for Coastal Ocean Science, Center for Coastal Monitoring and Assessment, Silver Spring, p 105
16. Ross PS, Ellis GM, Ikonomou MG, Barrett-Lennard LG, Addison RF (2000) High PCB concentrations in free-ranging Pacific killer whales, *Orcinus orca*: effects of age, sex and dietary preference. Mar Pollut Bull 40(6):504–515
17. Krahn MM, Hanson MB, Schorr GS, Emmons CK, Burrows DG, Bolton JL, Baird RW, Ylitalo GM (2009) Effects of age, sex and reproductive status on persistent organic pollutant concentrations in "southern resident" killer whales. Mar Pollut Bull 58(10):1522–1529
18. Balmer B (2011) Patterns of persistent organochlorine contaminants in relation to sighting patterns of bottlenose dolphins from coastal Georgia, USA. Sci Total Environ 409(11): 2094–2101
19. Sanger D, Blair A, DiDonato G, Washburn T, Jones S, Chapman R, Bergquist D, Riekerk G, Wirth E, Stewart J et al (2008) Support for integrated ecosystem assessments of NOAA's National Estuarine Research Reserves System (NERRS), Volume 1: the impacts of coastal

development on the ecology and human well-being of tidal creek ecosystems of the US Southeast, NOAA Technical Memorandum NOS NCCOS 82. National Oceanic and Atmospheric Administration, National Ocean Service, National Centers for Coastal Ocean Science, Hollings Marine Laboratory, Charleston, p 76

20. Pugh RS, Becker PR, Porter BJ, Ellisor MB, Moors AJ, Wise SA (2008) Design and applications of the National Institute of Standards and Technology's (NIST's) environmental specimen banking programs. Cell Preserv Technol 6(1):59–72
21. Aguilar A, Borrell A, Pastor T (1999) Biological factors affecting variability of persistent pollutant levels in cetaceans. In: Reijnders P, Aguilar A, Donavan G (eds) Chemical pollutants and cetaceans, vol Special Issue 1. International Whaling Commission, Cambridge, pp 83–116
22. Hansen LJ, Schwacke LH, Mitchum GB, Hohn AA, Wells RS, Zolman ES, Fair PA (2004) Geographic variation in polychorinated biphenyl and organochlorine pesticide concentrations in the blubber of bottlenose dolphins from the US Atlantic coast. Sci Total Environ 319(1–3):147–172
23. Whitehead H, Gordon J, Mathews E, Richard K (1990) Obtaining skin samples fromliving sperm whales. Mar Mamm Sci 6(4):316–326
24. Greig DJ, Gulland FM, Rios CA, Hall AJ (2010) Hematology and serum chemistry in stranded and wild-caught harbor seals in central California: reference intervals, predictors of survival, and parameters affecting blood variables. J Wildl Dis 46(4):1172–1184
25. Schwacke LH, Twiner MJ, De Guise S, Balmer BC, Wells RS, Townsend FI, Rotstein DC, Varela RA, Hansen LJ, Zolman ES et al (2010) Eosinophilia and biotoxin exposure in bottlenose dolphins (*Tursiops truncatus*) from a coastal area impacted by repeated mortality events. Environ Res 110(6):548–555
26. Wells RS, Rhinehart HL, Hansen LJ, Sweeney JC, Townsend FI, Stone R, Casper D, Scott MD, Hohn AA, Rowles TK (2004) Bottlenose dolphins as marine ecosystem sentinels: developing a health monitoring system. EcoHealth 1:246–254
27. Naar JP, Flewelling LJ, Lenzi A, Abbott JP, Granholm A, Jacocks HM, Gannon D, Henry M, Pierce R, Baden DG et al (2007) Brevetoxins, like ciguatoxins, are potent ichthyotoxic neurotoxins that accumulate in fish. Toxicon 50(5):707–723
28. Truelove J, Iverson F (1994) Serum domoic acid clearance and clinical observations in the cynomolgus monkey and Sprague-Dawley rat following a single i.v. dose. Bull Environ Contam Toxicol 52(4):479–486
29. Suzuki CA, Hierlihy SL (1993) Renal clearance of domoic acid in the rat. Food Chem Toxicol 31(10):701–706
30. Stewart JR, Gast RJ, Fujioka RS, Solo-Gabriele HM, Meschke JS, Amaral-Zettler LA, Del Castillo E, Polz MF, Collier TK, Strom MS et al (2008) The coastal environment and human health: microbial indicators, pathogens, sentinels and reservoirs. Environ Health 7(Suppl 2): S3
31. Kueh CS, Chan KY (1985) Bacteria in bivalve shellfish with special reference to the oyster. J Appl Bacteriol 59(1):41–47
32. Miller WA, Atwill ER, Gardner IA, Miller MA, Fritz HM, Hedrick RP, Melli AC, Barnes NM, Conrad PA (2005) Clams (*Corbicula fluminea*) as bioindicators of fecal contamination with *Cryptosporidium* and *Giardia* spp. in freshwater ecosystems in California. Int J Parasitol 35(6):673–684
33. Thornton SM, Nolan S, Gulland FM (1998) Bacterial isolates from California sea lions (*Zalophus californianus*), harbor seals (*Phoca vitulina*), and northern elephant seals (*Mirounga angustirostris*) admitted to a rehabilitation center along the central California coast, 1994–1995. J Zoo Wildl Med 29(2):171–176
34. Johnson SP, Nolan S, Gulland FM (1998) Antimicrobial susceptibility of bacteria isolated from pinnipeds stranded in central and northern California. J Zoo Wildl Med 29(3):288–294
35. Miranda CD, Zemelman R (2001) Antibiotic resistant bacteria in fish from the Concepcion Bay, Chile. Mar Pollut Bull 42(11):1096–1102

36. Stephen C, Lester S, Black W, Fyfe M, Raverty S (2002) Multispecies outbreak of cryptococcosis on southern Vancouver Island, British Columbia. Can Vet J 43(10):792–794
37. Datta K, Bartlett KH, Baer R, Byrnes E, Galanis E, Heitman J, Hoang L, Leslie MJ, MacDougall L, Magill SS et al (2009) Spread of *Cryptococcus gattii* into Pacific Northwest region of the United States. Emerg Infect Dis 15(8):1185–1191
38. Datta K, Bartlett KH, Marr KA (2009) *Cryptococcus gattii*: emergence in Western North America: exploitation of a Novel Ecological Niche. Interdiscip Perspect Infect Dis 2009:176532
39. Rotstein DS, West K, Levine G, Lockhart SR, Raverty S, Morshed MG, Rowles T (2010) *Cryptococcus gattiivgi* in a spinner dolphin (*Stenella longirostris*) from Hawaii. J Zoo Wildl Med 41(1):181–183
40. Rodriguez-Toro G (1993) Lobomycosis. Int J Dermatol 32(5):324–332
41. Rodriguez-Toro G, Tellez N (1992) Lobomycosis in Colombian Amer Indian patients. Mycopathologia 120(1):5–9
42. Van Bressem M, Van Waerebeek K, Reyes J, Felix F, Echegaray M, Siciliano S, Beneditto AD, Flach L, Viddi F, Avila I et al (2007) A preliminary overview of skin and skeletal diseases and traumata in small cetaceans from South American waters. Lat Am J Aquat Mamm 6(1):7–42
43. Cowan D (1993) Lobo's disease in a bottlenose dolphin (*Tursiops truncatus*) from Matagorda Bay, Texas. J Wildl Dis 29:488–489
44. Reif JS, Mazzoil MS, McCulloch SD, Varela RA, Goldstein JD, Fair PA, Bossart GD (2006) Lobomycosis in Atlantic bottlenose dolphins from the Indian River Lagoon, Florida. J Am Vet Med Assoc 228(1):104–108
45. Rotstein DS, Burdett LG, McLellan W, Schwacke L, Rowles T, Terio KA, Schultz S, Pabst A (2009) Lobomycosis in offshore bottlenose dolphins (*Tursiops truncatus*), North Carolina. Emerg Infect Dis 15(4):588–590
46. O'Shaughnessy A (1866) On green oysters. Annual Magazine of Natural History 18:221–228
47. Orton J (1923) Summary of an account of investigations into the cause or causes of the unusual mortality among oyster in English oyster beds during 1920 and 1921. J Mar Biol Assoc UK 13:1–23
48. Langston W, Bebianno M, Burt G (1998) Metal handling strategies in molluscs. In: Langston W, Bebianno M (eds) Metal matabolism in aquatic environments. Chapman and Hall, London, pp 219–283
49. Crain DA, Guillette LJ Jr (1998) Reptiles as models of contaminant-induced endocrine disruption. Anim Reprod Sci 53(1–4):77–86
50. Cossins AR, Crawford DL (2005) Fish as models for environmental genomics. Nat Rev Genet 6(4):324–333
51. Sumpter JP, Jobling S (1995) Vitellogenesis as a biomarker for estrogenic contamination of the aquatic environment. Environ Health Perspect 103(Suppl 7):173–178
52. Jobling S, Coey S, Whitmore JG, Kime DE, Van Look KJ, McAllister BG, Beresford N, Henshaw AC, Brighty G, Tyler CR et al (2002) Wild intersex roach (*Rutilus rutilus*) have reduced fertility. Biol Reprod 67(2):515–524
53. Jobling S, Nolan M, Tyler C (1998) Widespread sexual distribution in wild fish. Environ Sci Technol 32(2):498–506
54. Grosell M, Walsh P (2006) Sentinel species and animal models of human health. Oceanography 19:126–133
55. Black JJ, Baumann PC (1991) Carcinogens and cancers in freshwater fishes. Environ Health Perspect 90:27–33
56. Mix M (1986) Cancerous diseases in aquatic animals and their association with environmental pollutants: a critical review of the literature. Mar Environ Res 20:1–141
57. McCain BB, Pierce KV, Wellings SR, Miller BS (1977) Hepatomas in marine fish from an urban estuary. Bull Environ Contam Toxicol 18(1):1–2

58. Pierce KV, McCain BB, Wellings SR (1978) Pathology of hepatomas and other liver abnormalities in English sole (*Parophrys vetulus*) from the Duwamish River estuary, Seattle, Washington. J Natl Cancer Inst 60(6):1445–1453
59. Schwiewe M, Weber D, Myers M, Jacques F, Reichert W, Krone C, Malins D, McCain B, Chan S, Varanasi U (1991) Induction of cellular alteration and other hepatic lesions in English sole (*Parophrys vetulus*) exposed to an extract of an urban marine sediment. Can J Fish Aquat Sci 48:1750–1760
60. Nogueira I, Lobo-da-Cunha A, Afonso A, Rivera S, Azevedo J, Monteiro R, Cervantes R, Gago-Martinez A, Vasconcelos V (2010) Toxic effects of domoic acid in the Seabream *Sparus aurata*. Mar Drugs 8(10):2721–2732
61. Lefebvre KA, Noren DP, Schultz IR, Bogard SM, Wilson J, Eberhart BT (2007) Uptake, tissue distribution and excretion of domoic acid after oral exposure in coho salmon (*Oncorhynchus kisutch*). Aquat Toxicol 81(3):266–274
62. Dahm R (2006) The zebrafish exposed: 'see-through' mutants may hold the key to unraveling the mysteries of embryonic development. Am Sci 94:446–453
63. Incardona JP, Day HL, Collier TK, Scholz NL (2006) Developmental toxicity of 4-ring polycyclic aromatic hydrocarbons in zebrafish is differentially dependent on AH receptor isoforms and hepatic cytochrome P4501A metabolism. Toxicol Appl Pharmacol 217(3):308–321
64. Billiard SM, Timme-Laragy AR, Wassenberg DM, Cockman C, Di Giulio RT (2006) The role of the aryl hydrocarbon receptor pathway in mediating synergistic developmental toxicity of polycyclic aromatic hydrocarbons to zebrafish. Toxicol Sci 92(2):526–536
65. Saslowsky DE, Cho JA, Chinnapen H, Massol RH, Chinnapen DJ, Wagner JS, De Luca HE, Kam W, Paw BH, Lencer WI (2010) Intoxication of zebrafish and mammalian cells by cholera toxin depends on the flotillin/reggie proteins but not Derlin-1 or −2. J Clin Invest 120:4399–4409
66. Lefebvre KA, Tilton SC, Bammler TK, Beyer RP, Srinouanprachan S, Stapleton PL, Farin FM, Gallagher EP (2009) Gene expression profiles in zebrafish brain after acute exposure to domoic acid at symptomatic and asymptomatic doses. Toxicol Sci 107(1):65–77
67. Eto K (1997) Pathology of minamata disease. Toxicol Pathol 25(6):614–623
68. Goldstein T, Zabka TS, Delong RL, Wheeler EA, Ylitalo G, Bargu S, Silver M, Leighfield T, Van Dolah F, Langlois G et al (2009) The role of domoic acid in abortion and premature parturition of California sea lions (*Zalophus californianus*) on San Miguel Island, California. J Wildl Dis 45(1):91–108
69. Zabka TS, Goldstein T, Cross C, Mueller RW, Kreuder-Johnson C, Gill S, Gulland FM (2009) Characterization of a degenerative cardiomyopathy associated with domoic acid toxicity in California sea lions (*Zalophus californianus*). Vet Pathol 46(1):105–119
70. Goldstein T, Mazet JA, Zabka TS, Langlois G, Colegrove KM, Silver M, Bargu S, Van Dolah F, Leighfield T, Conrad PA et al (2008) Novel symptomatology and changing epidemiology of domoic acid toxicosis in California sea lions (*Zalophus californianus*): an increasing risk to marine mammal health. Proc Biol Sci 275(1632):267–276
71. Stewart I (2010) Environmental risk factors for temporal lobe epilepsy–is prenatal exposure to the marine algal neurotoxin domoic acid a potentially preventable cause? Med Hypotheses 74(3):466–481
72. Gulland FM, Trupkiewicz JG, Spraker TR, Lowenstine LJ (1996) Metastatic carcinoma of probable transitional cell origin in 66 free-living California sea lions (*Zalophus californianus*), 1979 to 1994. J Wildl Dis 32(2):250–258
73. Ylitalo GM, Stein JE, Hom T, Johnson LL, Tilbury KL, Hall AJ, Rowles T, Greig D, Lowenstine LJ, Gulland FM (2005) The role of organochlorines in cancer-associated mortality in California sea lions (*Zalophus californianus*). Mar Pollut Bull 50(1):30–39
74. Acevedo-Whitehouse K, Gulland F, Greig D, Amos W (2003) Inbreeding: disease susceptibility in California sea lions. Nature 422(6927):35

75. Martineau D, Lagace A, Beland P, Higgins R, Armstrong D, Shugart LR (1988) Pathology of stranded beluga whales (*Delphinapterus leucas*) from the St. Lawrence Estuary, Quebec, Canada. J Comp Pathol 98(3):287–311
76. De Guise S, Lagace A, Beland P (1994) Tumors in St. Lawrence beluga whales (*Delphinapterus leucas*). Vet Pathol 31(4):444–449
77. Girard C, Lagace A, Higgins R, Beland P (1991) Adenocarcinoma of the salivary gland in a beluga whale (*Delphinapterus leucas*). J Vet Diagn Invest 3(3):264–265
78. De Guise S, Lagace A, Beland P (1994) True hermaphroditism in a St. Lawrence beluga whale (*Delphinapterus leucas*). J Wildl Dis 30(2):287–290
79. Mikaelian I, Labelle P, Kopal M, De Guise S, Martineau D (2003) Adenomatous hyperplasia of the thyroid gland in beluga whales (*Delphinapterus leucas*) from the St. Lawrence Estuary and Hudson Bay, Quebec, Canada. Vet Pathol 40(6):698–703
80. Hall AJ, Gulland FM, Ylitalo GM, Greig DJ, Lowenstine L (2008) Changes in blubber contaminant concentrations in California sea lions (*Zalophus californianus*) associated with weight loss and gain during rehabilitation. Environ Sci Technol 42(11):4181–4187
81. Yordy JE, Wells RS, Balmer BC, Schwacke LH, Rowles TK, Kucklick JR (2010) Partitioning of persistent organic pollutants between blubber and blood of wild bottlenose dolphins: implications for biomonitoring and health. Environ Sci Technol 44(12):4789–4795
82. Pew Oceans Commission (2003) America's living oceans: charting a course for sea change. A report to the nation, May 2003. Pew Oceans Commission, Arlington
83. US Commission on Ocean Policy (2004) An ocean blueprint for the 21st Century. Final report, Washington, DC. ISBN#0-9759462-0-X
84. Bayley S, Stotts VD, Springer PF (1978) Changes in submerged aquatic macrophyte populations at the head of Chesapeake Bay, 1958–1975. Estuaries 1(3):171–182
85. Dustan P, Halas JC (1987) Changes in the reef-coral community of Carysfort Reef, Key Largo, Florida: 1974–1982. Coral Reefs 6:91–106
86. Holland AF, Sanger DM, Gawle CP, Lerberg SB, Santiago MS, Riekerk GHM, Zimmerman LE, Scott GI (2004) Linkages between tidal creek ecosystems and the landscape and demographic attributes of their watersheds. J Exp Mar Biol Ecol 298:151–178
87. Kemp WM, Twilley RR, Stevenson JC, Boynton WR, Means JC (1983) The decline of submerged vascular plants in upper Chesapeake Bay: summary of results concerning possible causes. Mar Technol Soc J 17:78–89
88. Hoegh-Guldberg O (1999) Coral bleaching, climate change and the future of the world's coral reefs. Review. Mar Freshw Res 50:839–866
89. Porter JW, Tougas JI (2001) Reef ecosystems: threats to their biodiversity. In: Levin S (ed) Encyclopedia of biodiversity, vol 5. Academic, New York, pp 73–95
90. Turgeon DD et al (2002) The state of coral reef ecosystems of the United States and Pacific Freely Associated States: 2002. National Oceanic and Atmospheric Administration/National Ocean Service/National Centers for Coastal Ocean Science, Silver Spring, p 265
91. Goreau T, McClanahan T, Hayes R, Strong A (2000) Conservation of coral reefs after the 1998 Global Bleaching Event. Conserv Biol 14(1):5–15
92. Knowlton N (2001) The future of coral reefs. Proc Natl Acad Sci USA 98(10):5419–5425
93. Coles SL, Brown BE (2003) Coral bleaching–capacity for acclimatization and adaptation. Adv Mar Biol 46:183–223
94. Kneib RT (1997) The role of tidal marshes in the ecology of estuarine nekton. Oceanogr Mar Biol Ann Rev 35:163–220
95. Mallin MA, Burkholder JM, Cahoon LB, Posey MH (2000) North and South Carolina coasts. Mar Pollut Bull 41:56–75
96. Lerberg SB, Holland AF, Sanger DM (2000) Responses of tidal creek macrobenthic communities to the effects of watershed development. Estuaries 23(6):838–853
97. Sanger DM, Holland AF, Scott GI (1999) Tidal creek and salt marsh sediments in South Carolina coastal estuaries: I. Distribution of trace metals. Arch Environ Contam Toxicol 37:445–457

98. Sanger DM, Holland AF, Scott GI (1999) Tidal creek and salt marsh sediments in South Carolina coastal estuaries: II. Distribution of organic contaminants. Arch Environ Contam Toxicol 37:458–471
99. Boesch DF, Brinsfield RB, Magnien RE (2001) Chesapeake Bay eutrophication: scientific understanding, ecosystem restoration, and challenges for agriculture. J Environ Qual 30(2):303–320
100. Beach D (2002) Coastal sprawl: the effects of urban design on aquatic ecosystems in the United States. Pew Oceans Commission, Arlington
101. DiDonato GT, Stewart JR, Sanger DM, Robinson BJ, Thompson BC, Holland AF, Van Dolah RF (2009) Effects of changing land use on the microbial water quality of tidal creeks. Mar Pollut Bull 58(1):97–106
102. U S Environmental Protection Agency (2001) National coastal condition report, EPA-620/R-01/005. Office of Research and Development, Office of Water, Washington, DC
103. H. John Heinz III Center for Science EatE (2008) The state of the nation's ecosystems: measuring the lands, waters, and living resources of the United States. Island, Washington, DC
104. Jones KE, Patel NG, Levy MA, Storeygard A, Balk D, Gittleman JL, Daszak P (2008) Global trends in emerging infectious diseases. Nature 451(7181):990–993
105. Schlesinger WH (2010) Translational ecology. Science 329(5992):609

Books and Reviews

Bossart G (2006) Marine mammals as sentinel species for oceans and human health. Oceanography 19:134–137

Boyd IL, Bowen WD, Iverson SJ (eds) (2010) Marine mammal ecology and conservation, a handbook of techniques. Oxford University Press, New York

Gulland FM (1999) Stranded seals: important sentinels. J Am Vet Med Assoc 214:1191

Scotch ML, Odofin PR (2009) Linkages between animal and human health sentinel data. BMC Vet Res 5:15

Chapter 19
Solar Radiation and Human Health

Gunther Seckmeyer, Armin Zittermann, Richard McKenzie, and Ruediger Greinert

Glossary

Action spectrum	Weighting function describing the wavelength dependence of the biological response. Usually, it is normalized to 1 at a specific wavelength. In the UV, action spectra need to be known accurately over several orders of magnitude.
Direct spectral irradiance $E_{\lambda,D}$	Radiant energy dQ arriving from the disk of the sun per time interval dt, per wavelength interval $d\lambda$, and per area dA on a surface normal to the solar beam.

This chapter was originally published as part of the Encyclopedia of Sustainability Science and Technology edited by Robert A. Meyers. DOI:10.1007/978-1-4419-0851-3

G. Seckmeyer (✉)
Institute of Meteorology and Climatology, Leibniz University of Hannover,
Herrenhaeuser Str. 2, 30419 Hannover, Germany
e-mail: seckmeyer@muk.uni-hannover.de

A. Zittermann
Clinic for Thoracic and Cardiovascular Surgery, Heart Center North Rhine-Westphalia,
Ruhr University Bochum, Bad Oeynhausen, Germany
e-mail: azittermann@hdz-nrw.de

R. McKenzie
National Institute for Water and Atmospheric Research (NIWA), Lauder, New Zealand
e-mail: r.mckenzie@niwa.co.nz

R. Greinert
Centre of Dermatology, Elbe Kliniken Buxtehude, Am Krankenhaus 1,
21614 Buxtehude, Germany
e-mail: grein25@gmx.de

E.A. Laws (ed.), *Environmental Toxicology: Selected Entries from the Encyclopedia of Sustainability Science and Technology*, DOI 10.1007/978-1-4614-5764-0_19,

$$E_{\lambda,D} = \frac{dQ}{dt\,dA\,d\lambda}$$

The angular field of view of an instrument measuring direct normal spectral irradiance must be sufficiently small to reduce uncertainties caused by circumscolar radiation. Recommendations for view-limiting geometries can be found in WMO [166].

Erythemally weighted irradiance E_{CIE}

Global spectral irradiance $E_G(\lambda)$ multiplied with the action spectrum for erythema, $C(\lambda)$, proposed by CIE [1] and integrated over wavelengths λ:

$$E_{\mathrm{CIE}} = \int_{250\mathrm{nm}}^{400\,\mathrm{nm}} E_G(\lambda)\cdot C(\lambda)\,d\lambda$$

Exposure

The spectral exposure Ex_λ is the radiance L_λ integrated over the relevant areas dA of the human body. In this context, the spectral radiance originates from the Sun's direct beam and any scattered components.

$$Ex_\lambda = \int_{t_1}^{t_2}\left(\oint_{(A)} L_\lambda(\epsilon,\varphi,t,\lambda)\cdot dA\cos\epsilon\right)dt$$

where $\mathrm{T} = t_2 - t_1$ is the exposure time. $Ex_\lambda(\lambda)$ may be weighted with a biological action spectrum and integrated over the wavelength to assess its biological impact. In this case the exposure is no longer a function of the wavelength and has the unit J.

Global spectral irradiance $E_{\lambda,G}$

Radiant energy dQ arriving per time interval dt, per wavelength interval $d\lambda$, and per area dA on a horizontally oriented surface from all parts of the sky above the horizontal, including the disk of the sun itself:

$$E_{\lambda,G} = \frac{dQ}{dt\,dA\,d\lambda} = E_{\lambda,D}\cdot\cos(\vartheta) + E_{\lambda,S}$$

where ϑ is the solar zenith angle.

Spectral radiance $\mathrm{L}_{\lambda,}$

This can be defined in terms of emitted or received radiation. Here the latter applies. Radiant energy

dQ per time interval dt, per wavelength interval $d\lambda$, per area dA, and per solid angle $d\Omega$ on a receiver oriented normal to the source.

$$L_\lambda = \frac{dQ}{dt\, dA\, d\lambda d\Omega}$$

UV index

A measure of solar UV radiation at the Earth's surface that is used for public information. According to [2], the UV index is calculated considering the following items:

1. Calculation of the erythemally weighted irradiance E_{CIE} (see above) by utilization of the CIE action spectrum [1] normalized to 1.0 at 298 nm.
2. A minimum requirement is to report the daily maximum UV index.
3. The index is expressed by multiplying the weighted irradiance in W m^{-2} by 40.0 (this leads to an open-ended index which is normally between 0 and 16 at sea level, but with larger values possible at high altitudes).

Remarks:

(a) The definition of the UV index given above may be revised in the future.
(b) According to the alternative definition given in [3], the UV index is calculated as the daily maximum erythemally weighted irradiance in W m^{-2}, averaged over a duration of between 10 and 30 min and multiplied by 40.

UV-A radiation

Electromagnetic radiation between 315 and 400 nm [4]. UV-A radiation is a summarizing term only and, unlike UVA irradiance, not a physical quantity.

UV-B radiation

Electromagnetic radiation between 280 and 315 nm [4]. UV-B radiation is a summarizing term only and, unlike UVB irradiance, not a physical quantity.

Vitamin D

Vitamin D is produced photochemically by UV exposure and conversion of 7-dehydrocholesterol into previtamin D_3, which is rapidly converted to vitamin D_3. The active form of vitamin D_3, 1,25-dihydroxyvitamin D_3, is a hormone.

Definition of the Subject

Solar radiation has both direct and indirect impacts on human health. Only direct effects are described here. (Space is too limited to describe the indirect effects, which are numerous, complex and imbedded into important feedback mechanisms; the most important indirect effects for human health are on the availability and quality of food, effects on aquatic and terrestrial plants and ecosystems, deterioration in air quality, damage to materials, and energy-related issues that drive the world economy.)

While solar energy has been stored in the form of oil and gas for millions of years, it has become evident that these resources are limited and that harnessing renewable energies will be necessary in the future. Closely related are the effects on the environment for food production, which relies on solar energy.

The sun's spectrum extends over a much-broader range of wavelengths than what is detected by the human eye, which responds only from 380 (violet light) to 780 nm (red light), with a peak response near 550 nm (green light). About 50% of the total solar irradiance originates from the near-infrared region. A relatively small but nevertheless important component is contained in the ultraviolet (UV).

Usually, any effect of solar radiation on biological organisms is wavelength dependent. A frequent concept is to describe these effects by means of biological weighting functions, often called action spectra, which quantify the wavelength dependence of effects introduced by electromagnetic radiation on biological matter. Depending on the effect and the organism involved, different biological weighting functions $W(\lambda)$ are used. The biologically effective irradiance, E_{weighted}, is calculated by multiplying the global spectral irradiance, $E_G(\lambda)$, with the action spectrum $W(\lambda)$, and integrating over wavelengths λ:

$$E_{\text{weighted}} = \int E_G(\lambda) \times W(\lambda)\, d\lambda$$

Note that the equation above is applicable for additive effects only (the so-called Bunsen–Roscoe law). Whether the concept is applicable or not must be proven or judged for each individual case. An important weighting function is the action spectrum for erythema (sunburn), as proposed by CIE [1], which describes the wavelength dependence of the reddening of human skin by UV radiation (see also below "erythemally weighted irradiance" E_{CIE}).

Introduction

Humans have long been aware of a close relationship between human health and solar radiation. This is reflected by the fact that in many religions, the sun plays an important role. In many ancient cultures, the sun itself was considered as a god (e.g., in Egypt). The systematic study of the effects of solar radiation progressed in parallel with the evolution of science and medicine. For an assessment of the spectral dependencies of solar radiation, a separation of the different spectra was necessary. A major step was probably the analysis by Joseph von Fraunhofer in the early nineteenth century, who separated the different components of sunlight with high precision using his newly constructed prisms. Later on, the investigations carried out by Carl Dorno, and his publication about "light and air in mountainous regions" that he published in 1911, were other important milestones. For a long time, the UVB radiation was actually called "Dorno radiation."

In this chapter, the present knowledge of the relationship between solar radiation and human health is described. Historical aspects, which are course interesting in their own right, are not considered further. This chapter is separated into the knowledge of spectral effects (UV, visible, and infrared), followed by a short section that deals with effects for which the spectral dependence is not yet known. Solar radiation has both negative and positive effects on human health. The negative effects of UV radiation have been emphasized in recent decades – on the background that skin cancer has been the cancer with the highest rate of increase. In recent years, however, more attention is being given to the positive side of solar radiation. At the end of this chapter, the benefits and dangers of solar radiation are summarized.

UV Radiation and Health Effects

The atmosphere is largely transparent in the visible region. Conversely, gaseous absorption is important in the UV. The latter is subdivided into three spectral bands: UV-A (315–400 nm) radiation, UV-B (280–315 nm) radiation, and UV-C (100–280 nm) radiation. (some authors still use 320 nm as the boundary between UV-A and UV-B radiation.) This separation has been originally introduced due to its relation with human effects. Nowadays, it is better understood that these effects cannot be strictly separated and, thus, only the general term "UV" is often used. UV-A radiation is largely unaffected by gases in the atmosphere. Conversely, only a small amount of UV-B radiation proceeds to the Earth's surface because it is strongly absorbed by atmospheric ozone, whose absorption cross section increases rapidly toward shorter wavelengths in this spectral region. UV-C comprises less than 0.6% of the incident solar spectrum at the top of the atmosphere, but none of it reaches the surface. Radiation below 240 nm is absorbed by molecular oxygen (O_2, which comprises 20% of the atmosphere). This radiation is capable of photo-dissociating molecules to form two oxygen atoms. The atoms so formed then recombine with another oxygen molecule to form ozone (O_3). Under the terms of the Montreal Protocol on Protection of the Ozone Layer, assessments of the knowledge about the science of Ozone Depletion and UV radiation [5, 6] and assessments of the Environmental Impacts of Ozone Depletion [7] are produced regularly.

If human eyes were sensitive to UV radiation rather than to visible radiation, perceptions of the world would be very different. The sky appears blue because air molecules scatter blue light more strongly than red light. This dependence on the wavelength (λ) of light is very strong. Rayleigh scattering theory shows that it varies as λ^{-4} so, at 300 nm, the strength of this scattering is about 32 times greater than at 600 nm. That is why photographers choose to use blue-blocking filters (i.e., red filters) if they want to obtain dramatic images of cloudy skies in black and white pictures since such filters can greatly enhance the contrast between the bright cloud and dark background.

Because of this wavelength-dependent scattering by air molecules, the UV radiation field is much more diffuse than visible radiation (light). An eye sensitive to UV radiation would therefore perceive a much smaller contrast between clouds and sky, and a smaller contrast between sky and sun. Under clear skies, the diffuse UV component would normally dominate over the direct beam component, whereas in the visible, the diffuse component is typically only 5–15% of the direct. When the sun is low in the sky, Rayleigh extinction becomes so large that the solar disk in the UV region would not even be detectable from the surrounding skylight. The sphere of consciousness would be greatly reduced because objects more distant than even a few kilometers would be completely lost in the gloom of diffuse light. It would be like living in a light fog. Shadows would be less distinct. When the sky is overcast, places normally in the shade would receive more light than under clear skies. One practical implication of this more diffuse field is that any reduction in UV radiation in a shaded area is smaller than the reduction in visible light, as perceived with the eye. Therefore, the protection from the sun offered by shade may be less complete that anticipated from the standpoint of UV effects. The brightness of objects would appear very different too. Most natural surfaces reflect a significant fraction of the incoming visible radiation. But in the UV region, very few surfaces are strongly reflecting. Most natural surfaces would appear quite dark because they reflect less than 5% of the incoming UV-B radiation. Snow and clouds are two of the few exceptions to this rule.

Another aspect of UV radiation is the ratio of diffuse to direct irradiance is much higher in the UV compared to visible wavelengths. Therefore, under clear skies, hazy skies, or very lightly cloudy skies, the diffuse UV irradiance comes from all possible directions, rather than preferentially from around the sun disk (Fig. 19.1). This further decreases the effectiveness of partial shading structures.

The variability in UV irradiance between common sources is described in Table 19.1.

Vitamin D

Vitamin D Metabolism

Solar UV-B irradiation (280–315 nm) of the skin is the major source of vitamin D3 for humans, whereas dietary intake of vitamin D2 or vitamin D3 is a second, less important source. The dietary contribution to vitamin D supply usually does not

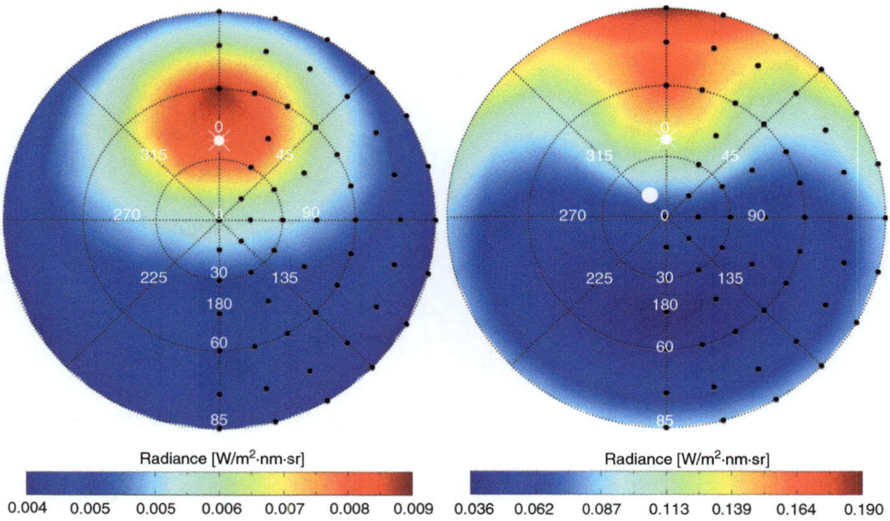

Fig. 19.1 Radiance distribution measured during clear skies in Hannover during noon. The position of the sun – south at noon – is given by the white symbol. *Left* panel 305 nm, *right* panel 500 nm. It can be recognized that the spatial distribution is quite different at these two wavelengths and that the distribution of sky radiance is more uniform in the UV, which can be understood as a result of double scattering by Rayleigh scattering

Table 19.1 Relationships between various weightings for a typical dermatological UV-A tanning salon compared with natural sunlight (noon sun in summer and winter at 45° S (Lauder New Zealand) and a powerful 1,000 W quartz halogen lamp at a distance of 0.5 m. UV_Ery and UV-VitD are erythemally weighted and vitamin D-weighted UV radiation, respectively

Quantity	Unit	UV-A tanning salon	Winter sun	Summer sun	1,000 W FEL at 0.5 m
UV-A	(μW cm^{-2})	17,598	1797	6161.5	106.2
UV-B	(μW cm^{-2})	208.3	17.9	206.6	6.6
UV_Ery	(μW cm^{-2})	51.5	2.6	28.2	7.0
UV_VitD	(μW cm^{-2})	63.8	3.2	56.7	3.8
UVI		20.6	1	11.3	2.8
T(2SED*)	(min)	6.5	128.2	11.8	47.4
Vit. D/Ery		1.2	1.2	2	0.5
UV-B / UV-A		0.012	0.010	0.034	0.062

exceed 10–20% in free-living persons. Only few fatty salt water fish such as herring, salmon, and sardines are good sources of vitamin D. Skin synthesis of vitamin D3 is very effective. The relevant quantity is the exposure, as defined above (in units of W). Some authors consider summer exposure of arms and legs for 5–30 min (depending on latitude and skin pigmentation) between the hours of 10 a.m. and 3 p.m. twice a week to be adequate. However, it has been found that the vitamin D action spectrum is more uncertain than previously thought. Consequently, in

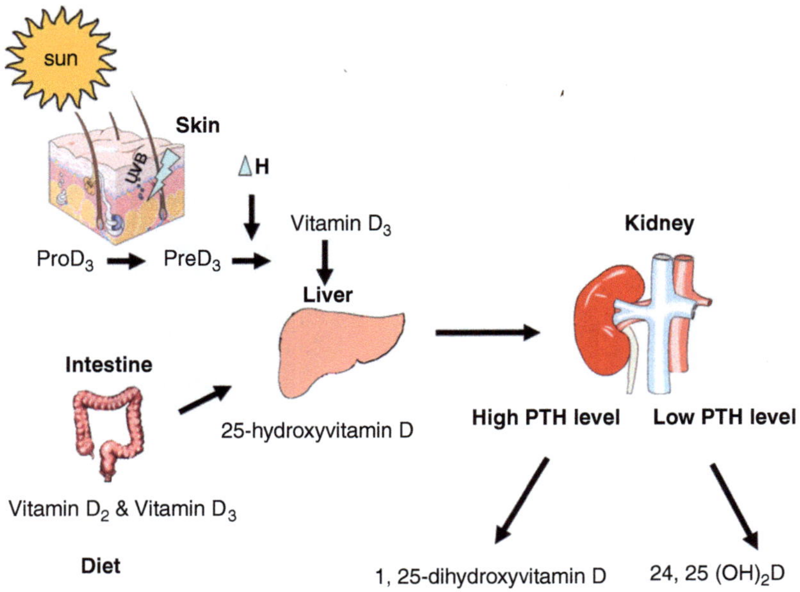

Fig. 19.2 Vitamin D metabolism

winter, it may not be possible to gain enough vitamin D by casual exposure to sunlight. Exposure of skin to sunlight at regular intervals results in the photochemical conversion of 7-dehydrocholesterol into previtamin D_3, which is rapidly converted to vitamin D_3. In the midsummer months (e.g., June and July in the northern hemisphere), the amount of vitamin D_3 increases to a maximum of approximately 12% of the amount of 7-dehydrocholesterol at latitudes of 34° and below, to 9% in one hour in Boston (42° N) and 11% in 3 h in Edmonton (52° N). Analyses after exposure of the 7-dehydrocholesterol solution at several places in the southern hemisphere showed concordant results [8]. The sunlight-generated vitamin D_3 is a precursor of the active hormonal form of vitamin D_3. Because any excess in previtamin D3 or vitamin D3 is destroyed by sunlight, excessive exposure to sunlight does not cause vitamin D intoxication. It has been reported that exposure to one minimal erythemal dose while wearing only a bathing suit is equivalent to ingestion of approximately 500 μg of vitamin D [9]. Because of its skin synthesis, vitamin D_3 is not really a vitamin for humans. However, if skin exposure to UV-B radiation is negligible (e.g., in winter) or if people live exclusively indoors (institutionalized people), then there is indeed an absolute regular requirement for the fat soluble vitamin D, which must be met through proper dietary intake.

Once in circulation through the body, vitamin D is metabolized by a hepatic hydroxylase into 25-hydroxyvitamin D (25[OH]D) and by a renal 1α-hydroxylase into the vitamin D hormone 1,25-dihydroxyvitamin D (1,25-dihydroxyvitamin D) (Fig. 19.2). The latter step is under the control of parathyroid hormone (PTH) and serum phosphate. However, in case of vitamin D deficiency/insufficiency, renal synthesis of 1,25-dihydroxyvitamin D becomes substrate dependent, i.e., dependent on the circulating 25(OH)D concentration [10].

The active form of vitamin D, 1,25-dihydroxyvitamin D, is known as a regulator of systemic calcium homeostasis. Together with the parathyroid hormone (PTH) and calcitonin, 1,25-dihydroxyvitamin D carries out the policing job of regulating serum calcium levels in highly evolved mammals. This regulating mechanism is done with astounding efficiency, indicating the importance of serum calcium homeostasis. Serum calcium homeostasis is essential for blood coagulation, activation of various intracellular processes, and bone health. Therefore, serum calcium level of 2.5 mmol/L is maintained in all vertebrate life forms (both aquatic and terrestrial). The regulatory system of calciotropic hormones appears to have developed after transition of life from water (calcium–phosphorus rich environment) to land (environment poor in calcium and phosphorus). This assumption is supported by the fact that, in fish, PTH is missing and calcitonin is inactive [11]. Together with PTH, vitamin D is responsible for maintaining serum calcium levels by increasing intestinal calcium absorption, renal calcium reabsorption, and calcium resorption from bone. Without vitamin D, only 10–15% of dietary calcium and about 60% of phosphorus can be absorbed. vitamin D increases the efficiency of intestinal calcium absorption to 30–40% and phosphorus absorption to approximately 80% [11].

1,25-dihydroxyvitamin D also plays a pivotal role in the intracellular calcium homeostasis and has pleiotropic effects in various tissues. The biological actions of 1,25-dihydroxyvitamin D are mediated through the vitamin D receptor (VDR). In this context, 1,25-dihydroxyvitamin D functions as a steroid hormone that binds to a cytosolic VDR, resulting in a selective demasking of the genome of the nucleolus. The vitamin D receptor is nearly ubiquitously expressed, and almost all cells respond to 1,25-dihydroxyvitamin D exposure. About 3% of the human genome is regulated, directly and/or indirectly, by the vitamin D endocrine system [12]. Apart from the kidney, 1,25-dihydroxyvitamin D can be locally produced in several tissues. These tissues include monocytes, dendritic cells, B lymphocytes, colonocytes, vascular smooth muscle cells, and endothelial cells. Consequently, for 1,25-dihydroxyvitamin D, a paracrine role separate from its calcium-regulating function has been proposed.

Reference Values

There is general agreement that the measurement of 25(OH)D is the appropriate tool for the assessment of vitamin D status. Nevertheless, there is currently no consensus on adequate/optimal circulating 25(OH)D concentrations. Cutoff values range between 50 nmol/L [13], 90–100 nmol/L [14], and more than 100 nmol/L [15]. This inconsistency is in part due to different criteria of defining inadequacy. Many researchers do not differentiate between different stages of vitamin D status. However, similar to other vitamins, it is possible to categorize the stages of vitamin D status into deficiency, insufficiency, hypovitaminosis, adequacy, and intoxication. In the deficiency range, severe vitamin D-specific clinical symptoms such as rickets, osteomalacia, calcium malabsorption, severe hyperparathyroidism, low 1,25-dihydroxyvitamin D concentrations, impaired immune system, and ultimately death can occur. In the insufficiency range, reduced bone mineral density, impaired muscle function, low intestinal calcium absorption rates, elevated PTH levels, and slightly reduced 1,25-dihydroxyvitamin D concentration are encountered. In the stage of hypovitaminosis, the body's vitamin reserves are already low, and slightly enhanced PTH levels may be present, although the corresponding concentrations are usually still in the reference range. In the stage of adequacy, no disturbances in vitamin D-dependent body functions do occur. Only excessive oral intake can lead to vitamin D intoxication, resulting in intestinal calcium hyperabsorption, hypercalcemia, soft tissue calcification, and death. On an individual basis, the consequences of an insufficient vitamin D status may be mild, but the consequences on a population scale may be more important because of the large number of people who are affected. People with long-lasting vitamin D insufficiency may have the highest risk to develop vitamin-D-related chronic diseases.

Unfortunately, insufficient vitamin D status is prevalent around the world. Over the six regions Asia, Europe, Middle East and Africa, Latin America, North America, and Oceania, it has been found that between 50% and more than 90% of people have 25(OH)D concentrations below 50 nmol/L [16]. Low vitamin D status is most common in South Asia and the Middle East. Insufficient vitamin D status and even vitamin D deficiency is widespread, and is actually reemerging as a major health problem globally. Urbanization is an important risk factor for vitamin D insufficiency/deficiency in large parts of the adult population. Other factors include short daylight periods, long cold or hot seasons – or strict cultural traditions – imposing extensive clothing, modern and also traditional lifestyles such as indoor working, predominantly indoor leisure time activities, and the aging factor (institutionalization of elder persons). In a British study in the 1958 birth cohort, for example, 25-hydroxyvitamin D concentrations below 40 nmol/L were more prevalent in Scotland than in the South of England [17]. In highly urbanized or polluted areas, individual daily sun exposure is usually too low to achieve a 25(OH)D level of 75 nmol/L. Without food supplementation, the diet is then not able to close the gap in vitamin D supply. There have been studies linking vitamin D deficiency to latitude [18–21], but there are also studies that could not confirm such correlations [22].

Musculoskeletal Disease

Urbanization and industrialization (with their associated air pollution problems) have long been known as a major cause of childhood rickets in western countries. Rickets is now on the increase in many developing countries and is also reemerging as an important health problem in countries with strong sun avoidance policies, or in cultures requiring strictly modest dressing codes. In adults, severe vitamin D deficiency causes osteomalacia, a disease resulting in bone demineralization and muscle weakness.

Due to low calcium absorption rates, vitamin D insufficiency can also contribute to the bone disease osteoporosis. It is estimated that up to 50% of women and more than 20% of men 50 years of age or older will sustain an osteoporotic fracture in their remaining lifetime. A meta-analysis of randomized clinical trials that evaluated the effect of vitamin D supplementation came to the conclusion that the fracture preventing effect of vitamin D is approximately 20% when serum 25(OH) D levels of 75–80 nmol/L are achieved [23]. Another meta-analysis of randomized controlled trials (RCTs) came to the conclusion that daily doses of 17.5–20 μg supplemental vitamin D are able to prevent elderly adults to fall down by improving muscle function [24]. The relative risk of falls was reduced by approximately 20% when serum 25(OH)D concentrations of 60 nmol/L or more could be maintained. In contrast to "high-dose" supplemental vitamin D, low-dose daily supplemental vitamin D (5–15 μg) is not able to prevent falls.

Cancer

Since vitamin D is a key regulator of various cellular metabolic pathways, it is important for cellular maturation, differentiation, and apoptosis [25]. In 2008, the WHO published a report from the International Agency for Research on cancer [26] that came to the conclusion that there is (1) consistent epidemiological evidence for an inverse association between 25(OH)D and colorectal cancer and colorectal adenomas, (2) suggested epidemiological evidence for an inverse association between 25(OH)D and breast cancer, (3) insufficient evidence for an inverse association between 25(OH)D and other types of cancer, and (4) the need for new RCTs. One such RCT has recently been published [27]: In a 4-year, population-based study, where the primary outcome was fracture incidence and the principal secondary outcome was cancer incidence, 1,179 community-dwelling women were randomly assigned to receive 1,500 mg supplemental calcium/day alone (Ca-only), supplemental calcium plus 27.5 μg vitamin D/day (Ca + D), or placebo. The cancer incidence was 60–77% lower in the Ca + D women and 43% lower in the Ca-only group than in the placebo control subjects ($P < 0.03$).

Diabetes Mellitus

In vitro and in vivo studies suggest that vitamin D can prevent pancreatic beta-cell destruction and reduce the incidence of autoimmune diabetes. This may be due, at least in part, to a suppression of proinflammatory cytokines such as the tumor necrosis factor (TNF)-α. Recently, the relationship between UVB irradiance and age-standardized incidence rates of type-1 diabetes mellitus in children aged 14 years or less was analyzed according to 51 regions of the world [28]. Incidence rates were generally greater at higher latitudes and were inversely associated with UV-B irradiance. As early as 2001, Hyppönen et al. [29] demonstrated, in a birth cohort study, that vitamin D supplementation was associated with a decreased frequency of type-1 diabetes. In contrast, children suspected of having rickets during the first year of life had a three times higher relative risk compared with those without such a suspicion. A meta-analysis of four case-control studies has shown that the risk of type-1 diabetes is reduced by 29% in infants who are supplemented with vitamin D compared to those who are not supplemented [30]. There is also some evidence of a dose-response effect, with those using higher amounts of vitamin D being at lower risk of developing type-1 diabetes.

Observational studies, e.g., Pittas et al. [31], show a relatively consistent association between low vitamin D status and prevalent type-2 diabetes, with an odds ratio of 0.36 for highest vs. lowest 25-hydroxyvitamin D among nonblacks. Evidence from RCTs with vitamin D and/or calcium supplementation suggests that combined vitamin D and calcium supplementation may have a role in the prevention of type-2 diabetes only in populations at high risk (i.e., individuals with glucose intolerance). Whereas vitamin D supplementation did not improve glycemic control in diabetic subjects with initial serum 25(OH)D levels above 50 nmol/L [32], administration of 100 μg vitamin D3 improved insulin sensitivity in vitamin D deficient and insulin-resistant South Asian women [33]. Insulin resistance was most improved when endpoint serum 25(OH)D reached at least 80 nmol/L. Optimal vitamin D concentrations for reducing insulin resistance were shown to be 80–119 nmol/L.

Cardiovascular Disease

Globally, cardiovascular disease (CVD) is the number one cause of death. In 2005, it was responsible for approximately 30% of deaths worldwide. CVD includes various illnesses such as coronary heart disease, peripheral arterial disease, cerebrovascular disease such as stroke, and congestive heart failure. There is accumulating evidence that the vitamin D hormone 1,25-dihydroxyvitamin D exerts important physiological effects in cardiomyocytes, vascular smooth muscle cells, and the vascular endothelium. The mechanisms have been reviewed in detail elsewhere [34]. Hypertension is a key risk factor for CVD. A recently published systematic review and meta-analysis came to the conclusion that vitamin D produces a fall in systolic blood pressure of −6.18 mmHg and a nonsignificant fall in diastolic blood pressure of −2.56 mmHg in hypertensive patients. No reduction in blood pressure is seen in studies examining patients who are normotensive at baseline [35].

Several large prospective observational or cohort studies have demonstrated that a higher vitamin D status is associated with approximately 50% lower cardiovascular morbidity and mortality risk compared with low vitamin D status (Table 19.2).

Table 19.2 Evidence for association of circulating 25-hydroxyvitamin D level with cardiovascular morbidity and mortality

Study	Design	Population	Comparator	Odds/hazard ratio or Relative risk (95% CI)
Fatal stroke [36]	Prospective cohort study with coronary angiography	3,258	Per z value of 25(OH)D	OR 0.58 (0.43–0.78)
Cardiovascular morbidity [37]	Prospective observational study	1,739	25(OH)D > 37.5 nmol/L vs. < 25 nmol/L	HR 0.55 (0.32–0.97)
Myocardial infarction [38]	Nested case control study	1,354	25(OH)D > 75 nmol/L vs < 37.5 nmol/L	RR 0.48 (0.28–0.81)
Cardiovascular mortality [39]	Prospective cohort study with coronary angiography	3,258	Median 25(OH)D 70 nmol/L vs 19 nmol/L	HR 0.45 (0.32–0.64)
Cardiovascular mortality [36]	Prospective observational study in individuals 50–75 years	614	Three highest vs. lowest 25 (OH) D quartile	HR 0.19 (0.07–0.50)
Cardiovascular mortality [40]	Prospective observational study in individuals > 65 years.	3,408	25(OH)D > 100 nmol/L vs < 25 nmol	HR 0.42 (0.21–0.86)

Mortality

In 2007, a meta-analysis was published, including RCTs on vitamin D and mortality that were not primarily designed to assess mortality [29]. The authors concluded that vitamin D supplementation is linked to lower all-cause mortality rates in middle-aged and elderly patients with low serum concentrations of 25(OH)D than in unsupplemented subjects. Daily doses of vitamin D ranged between 10 and 50 μg. Risk reduction was 7% during a mean follow-up of 5.7 years. Several large prospective cohort studies published in 2008 and 2009 have provided further evidence for association of low circulating 25(OH)D levels with enhanced all-cause mortality (Table 19.3). However, there is some concern that vitamin D has a biphasic effect on mortality with an enhanced risk at deficient 25(OH)D levels but also at levels above 125 nmol/L [41]. The scientific background of this point of view has recently been questioned [42].

Sunburn (Erythema)

Since UV radiation is not detected by the human eye, or by usual light meters, a different standard is needed to quantify how much of it is present at any time. The erythemally weighted irradiance is defined as above, in units of W m^{-2}. For health

Table 19.3 Evidence for association of circulating 25-hydroxyvitamin D level or vitamin D supplementation with all-cause mortality

Study	Design	Population	Comparator	Hazard ratio or relative risk (95% CI)
[43]	Meta-analysis of 18 vitamin D supplementation studies	57,311	Supplemented vs unsupplemented	RR 0.93 (0.87–0.99)
[39]	Prospective cohort study with coronary angiography	3,258	Median 25(OH)D 70 nmol/L vs 19 nmol/L	HR 0.48 (0.37–0.63)
[44]	Prospective observational study in postmenopausal women	1,232	≥ 50 nmol/L vs. < 50 nmol/L	HR 0.46 (0.27–0.79)
[45]	Prospective cohort study in patients with colorectal cancer	304	Mean 41 nmol/L vs. 100 nmol/L	HR 0.52 (0.29–0.94)
[40]	Prospective observational study in individuals > 65 years.	3,408	25(OH)D > 100 nmol/L vs < 25 nmol/L	HR 0.55 (0.34–0.88)
[36]	Prospective observational study in individuals 50–75 years	614	Three highest quartiles vs. lowest quartile	HR 0.51 (0.28–0.93)

purposes, the erythemally weighted UV irradiance is generally reported to the public in terms of a dimensionless number called the UV Index (i.e., UVI, – see below). Rather than referring to the response of the human eye (i.e., the "photopic" response) for light, it refers to the response of human skin to sunburn (i.e., the "erythemal" response), which has no contribution from visible radiation but increases rapidly toward shorter wavelengths within the UV range.

Although UVI was developed to represent damage to human skin, it may be applied to other processes because many biological UV effects have similar action spectra. Since UVI is based on the erythemal action spectrum, its sensitivity to ozone change is the same as for erythema. For a 1% reduction in ozone, UVI increases by approximately 1.1%. The UVI is an open-ended scale. In the British Isles, it peaks at ~6, whereas at northern midlatitudes, it peaks at ~10. Peak values are higher in the southern hemisphere because of the lower ozone amounts, closer summertime Sun–Earth separation, and generally cleaner air. For example, the peak UVI in New Zealand is approximately 40–50% larger than at corresponding northern latitudes, and it often exceeds 12 [46]. In the tropics, it is even much higher. The highest values on Earth occur in the Peruvian Andes at latitudes 10–15° S, where UVI can reach 25 [47]. Outside the Earth's atmosphere, the UVI would be ~300.

The relatively high UV levels in New Zealand and Australia [46, 48] are a contributing factor to the high rates of skin cancer in those countries. There are other factors, however. For those of European extraction, skin types are better adapted to the lower UV levels in Europe, especially in the north [48]. Outdoor lifestyle choices can also be important. In temperate climates, it is quite comfortable to spend long periods of time in direct sunlight. The folly of that behavior manifests itself only later – when the discomfort of sunburn is felt, and much later when skin cancer is experienced.

UV Index

Sunburning UV is often reported to the public in terms of UVI. The upper panel in Fig. 19.3 shows the spectral UV irradiance for two sun angles and an ozone column of 300 Dobson Units (DU) (Note the log scale for the y-axis). The erythemal (i.e., the "skin-reddening," or "sunburning") weighting function is shown on the right axis. The lower panel shows the corresponding spectra of erythemally weighted UV irradiance (UV_{Ery}).

The internationally agreed UVI is defined in terms of UV_{Ery}. The weighting function involves an arbitrary normalization to unity at wavelengths shorter than 298 nm, so UV_{Ery} is not strictly an SI unit. Furthermore, when UV information was first provided to the public, another normalization was applied to give a maximum UVI of ~10 in Canada, where it was first used. The UVI is therefore a unitless number corresponding to the integral under the curves in the lower panels, multiplied by 40 m^2/W. Mathematically,

$$\mathrm{UVI} = 40\int \mathrm{I}(\lambda)\mathrm{w}(\lambda)d\lambda,$$

where

λ is the wavelength in nm
$\mathrm{I}(\lambda)$ is the irradiance in W m^{-2} nm^{-1} and
$\mathrm{w}(\lambda)$ is the erythemal weighting function, defined as:

$\mathrm{w}(\lambda) = 1.0$ for $250 < \lambda \leq 298$ nm
$\mathrm{w}(\lambda) = 10^{0.094(298 - \lambda)}$ for $298 < \lambda \leq 328$ nm
$\mathrm{w}(\lambda) = 10^{0.015(139 - \lambda)}$ for $328 < \lambda \leq 400$ nm
$\mathrm{w}(\lambda) = 0.0$ for $\lambda > 400$ nm

Exposure Times to Optimize Health Effects of UV Radiation

Attempts have been made to estimate the exposure times from sunlight to optimize health [49]. Results are given in terms of UVI (Fig. 19.4). The exposure times also depend on skin type and possible application of any sunscreens. In the case of vitamin D production, the optimal exposure time also depends on the area of skin exposed. There is a huge variation between the optimal exposure times in summer

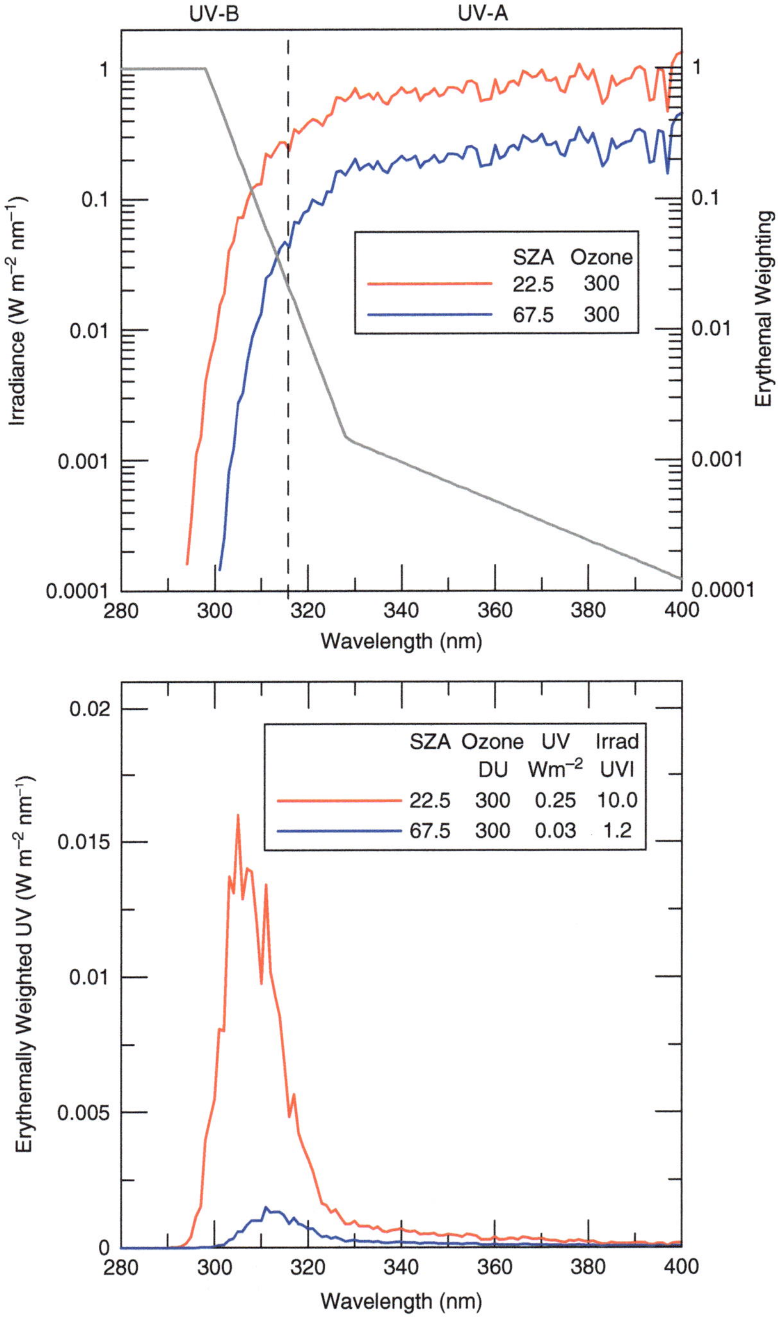

Fig. 19.3 *Upper panel*: UV ir radiance spectra for noon sunlight at midlatitude (45° S) in summer and winter. The *grey* line shows the erythemal action spectrum. Note the logarithmic scale on the y-axis. *Lower panel*: The same solar irradiance spectra weighted by the erythemal action spectrum

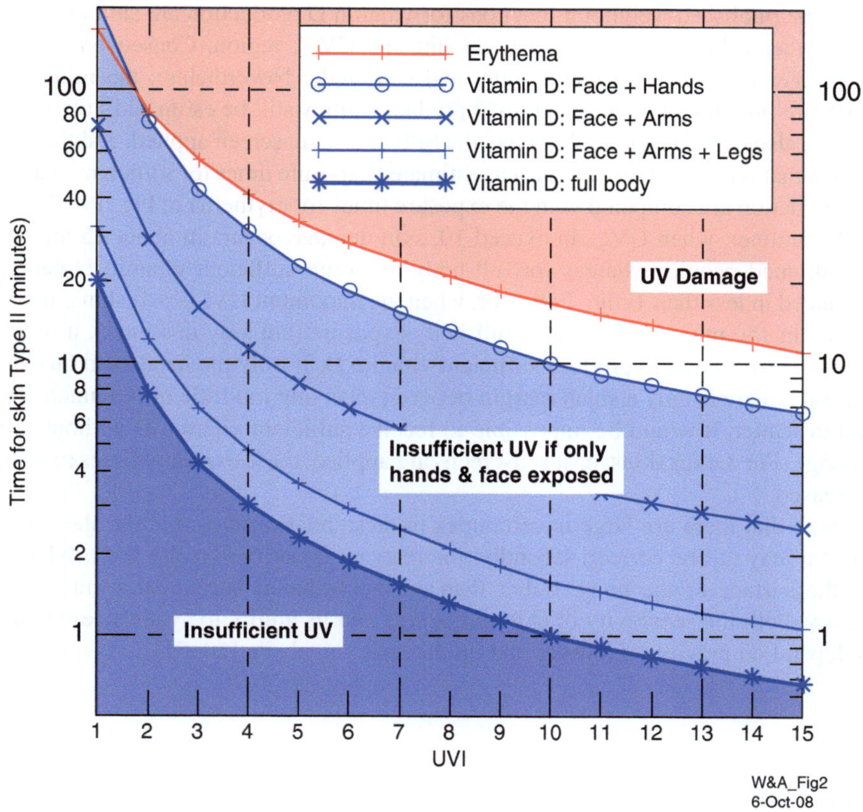

Fig. 19.4 Approximate exposure times leading to erythema or to sustained adequate vitamin D production, as a function of UVI. Exposure times corresponding to the *red*-shaded area represent too much UV, leading to erythema (skin reddening- and sunburn). Exposures within the dark *blue* region represents too little UV to maintain an intake of 1,000 IU with full-body exposure. The *unshaded* area represents the range of times for optimal exposure, assuming only the hands and face are exposed. The other *blue* curves give the lower exposure limits for adequate vitamin D for different areas of skin exposed. All times shown are for fair skin (skin type II). For brown skin or black skin (skin types IV and VI), all exposure times should be multiplied by 2 and 5, respectively

and winter at midlatitude sites. There is no simple relationship between UVI and visible radiation. In the case of sunlight, it is true that as light levels increase, so does UV radiation, but the relationship is complex and nonlinear.

The large seasonal variations are also a health risk. In winter, the UV irradiation may not be high enough to maintain adequate vitamin D. The problem is exacerbated because the skin becomes tanned to protect the body from further damage in summer, and this tan persists into the winter, therefore inhibiting the ability to absorb further UV radiation. Those with darker skins (e.g., Africans, many Maoris, Pacific Islanders, and Asians) are at greater risk.

Some published weighting functions for vitamin D production are similar to that for erythema but do not extend as far into the UV-A region. Consequently, it is a stronger function of ozone amount and sun angle. Nevertheless, the minimum exposure time for sufficient vitamin D production may still be estimated in terms of UVI. It also depends on the skin type, the SPF of any sunscreen applied, and the area of skin that is exposed to sunlight. The estimated exposure times for sufficient vitamin D production are compared with the exposure times for erythema in Fig. 19.2 [49].

In summer, when UVI can exceed 10, skin damage occurs in about 15 min for fair-skinned people, whereas for full-body exposure sufficient vitamin D can be produced in less than 1 min. In winter, when the maximum UVI is ~1, skin damage occurs in 150 min, and even with full-body exposure (unlikely in winter), it would take at least 20 min to produce sufficient vitamin D. If only the hands and face are exposed, there is only a small margin between receiving too little or too much UV; and in winter, it would be impossible to receive sufficient vitamin D without skin damage. For darker skins, or if sunscreens are applied, these exposure times must be increased.

Note that there are large uncertainties in these relationships. Firstly, the action spectra may not be correct; secondly, the relevant quantity is really the irradiance on the surface of the body, rather than on the assumed horizontal surface; and thirdly, both the sensitivity of skin to damage and its ability to synthesize vitamin D depend on previous exposure and on the anatomical site [50, 51].

Skin Cancer

Three types of skin cancer, basal cell carcinoma (BCC), squamous cell carcinoma (SCC), and malignant melanoma (MM), sum up to the most frequent type of cancer in the white population worldwide. For MM, the incidence increases more rapidly than for any other type of cancer. Data for the non-melanocytic skin cancers, NMSCs (BCC and SCC), for e.g., USA and Germany show incidences in the range of 1,000,000 new cases per year (USA, 2005) [52] and about 100,000 new cases per year (Germany, 2003) [53]; 80% of these are found on sun-exposed areas of the human body [54].

Corresponding numbers for MM (the most deadly skin cancer, 20–25% of diagnosed patients die) are about 62,000 new cases per year (USA, 2006) [55] and about 15,000 (Germany, 2004, Robert-Koch-Institut: http://www.rki.de). Most of melanoma cases occur on sun-exposed areas of the body. The estimated 10-year survival rates for patients without evident metastases range from 88% for those patients with tumors smaller than 1mm without ulceration, to 32% for those patients with tumors larger than 4 mm with ulceration [56].

The skin cancer incidence is still increasing in most countries [57, 58], although some stabilization has been observed for parts of western Europe and Scandinavia [59, 60]. Nevertheless, being the most frequent cancer, skin cancer represents a huge problem for the public. The personal, medical, clinical and, last but not least, the final burden of skin cancer has therefore to be reduced by means of primary prevention and early detection.

Environmental Risk Factor(s)

Although environmental arsenic exposure [60] or certain types of human papillomaviruses (as cofactors in association with UV) [61], as well as ionizing radiation [62] have been considered to play a role during the pathogenesis of NMSC, the overwhelming number of epidemiological and experimental investigations recognize UV-radiation as the main environmental risk factor [63–68]. The role of UV-radiation in melanogenesis has been discussed in past decades because the MM etiology strongly depends on genetic predispositions (e.g., allelic variances in the melanocortin 1 receptor [69]). However, recent epidemiological and experimental investigations, which are mainly dealing with UV-associated induction of benign nevi and the UV-induced mutation patterns, e.g., in the *BRAF* – gene (in nevi), clearly underline once more the important role of UV-radiation as a risk factor for malignant melanoma [63, 70–74]. It is furthermore important to consider that different types of skin cancer have a different dependence on UV-exposure patterns. Whereas MMs are mainly induced by intermittent UV-exposure (e.g., sunburns) [63, 71, 72], SCC-induction is highly dependent on cumulative UV-exposure, and BCC-induction depends on both cumulative and intermittent exposure patterns [63].

Because the main environmental risk factor in the etiology of skin cancer – UV-radiation (regardless of natural (sun) or artificial (sunbeds) origin) – is known, skin cancer is one of the key cancers that can be prevented by means of primary prevention (i.e., avoiding or reducing risks).

The importance of UV-radiation as *the* environmental (and artificial) risk factor for the induction of skin cancer has recently been underscored by the International Agency for the Research of Cancer (IARC) which characterized UV-radiation from the sun (and from sunbeds) as a group-1 carcinogen ("carcinogenic to humans") [75]. IARC has also been able to show in a meta-analysis that regular use of sunbeds before the age of 35 increases melanoma risk later in life by 75% [76].

Biological Damage Leading to Skin Cancer

UV-photon absorption of DNA leads to photochemical conversion of absorbed energy into photodimerization between adjacent pyrimidine bases. According to an action spectrum, UVB is about 1,000 times more effective than UVA in inducing this kind of photodimerization [77]. UVB also induces, at a much lower frequency, inter- and intra-strand crosslinks, protein-DNA crosslinks, DNA strand-breaks, and rare base adducts [78–80]. Photodimerizations between adjacent pyrimidine bases are by far the most prevalent photoreactions resulting from the direct action of

UV-radiation. Two major photoproducts are formed via this reaction pathway: the cyclobutyl pyrimidine dimer (CPD) and the pyrimidine(6–4)pyrimidone photoproduct (6-4PP). CPDs are formed between adjacent pyrimidines linked by a cyclobutyl ring between the five and six carbons of adjacent thymine (Thy) and/or cytosine (Cyt) bases. The 6-4PP links the 6 carbons on a 3' Cyt or Thy with the four carbons of a 5' Cyt. However, using the same wavelength of excitation in the UVB, 6-4PPs are induced only at 15–30% the rate of CPDs [81]. Using fluorescently labeled monoclonal antibodies against CPD and a calibration with a radioimmunoassay (RIA) [82], it was shown that a UVB dose of 300 Jm^{-2} is able to induce several hundred thousands of CPDs per genome per cell [83]. Both induction of CPDs and repair of these lesions via nucleotide-excision repair (NER) are UV-dose dependent. Increasing both the UVB dose and the amount of premutagenic CPDs leads to a increase of the time constant for NER to remove CDPs [83]. Furthermore, deficiencies in NER of CPDs have been linked with increased risk of cutaneous malignant melanoma and non-melanoma skin cancers [84–86].

When CPDs are not repaired, these DNA lesions can lead to mutations in the DNA sequence. Mutations are in the form of C→T and CC→TT transitions, known as "UV-signature mutations" [87–90]. These mutations have long been believed to be a specific signature for UVB-radiation. However, recent work suggests that UV-A also induces CPDs that lead to UV-signature mutations [91–93]. This view is supported by results showing that UV-A induces mainly CPDs in human skin [94]. A so called "A-rule" has been proposed to explain how these signature mutations arise [95]. According to the A-rule, DNA polymerase (pol η) inserts, in a default-mechanism, adenine (A) residues,opposite to those lesions it cannot interpret. A mutation is then created upon DNA replication of the strand containing base-pair changes. At a CC cyclobutane dimer, therefore, a CC → TT transition occurs because two A-residues are placed opposite the dimer by default, in place of two guanine (G)-residues. A similar default mechanism for 6–4 PPs might be responsible for C→T transitions [95]. Recent findings show that defects in efficiency of translesion polymerase η (despite other polymerases) seem to play an important role in UV-induced mutagenesis [96, 97].

UV-signature mutations have been detected in a number of tumor suppressor genes and oncogenes (including, e.g., *patched, p16, ras* and *p53*) in human SCC, BCC, and malignant melanoma [95]. Because the important biological function of p53 results from transcriptional activation of a large number of genes involved in fundamental cellular processes like cell cycle control, apoptosis, DNA replication, repair, genome instability, and senescence, it is not surprising that genetic alterations of the p53 gene have most frequently been found in human tumors [98]: especially in skin cancers like SCC, BCC, and precancerous actinic keratoses [99–103]. Convincing models for the initial steps of UV-carcinogenesis, especially for SCC, already exist [104, 105] based on findings that p53 mutations in non-melanoma skin cancers are detected at higher frequency (50–90%) than those in internal malignancy, and that the predominant alterations are C→T and CC→TT transitions at dipyrimidine sites [103]. Very recent investigations, exploring the melanoma genome, furthermore show that UV-induced signature mutations play the primary role in melanoma induction in humans [106].

Etiology of Skin Cancer

In recent years, a huge amount of information has been obtained in the fields of genetics, molecular pathways, and cellular changes, which are important players in skin cancer induction, promotion, progression, and metastasis. Excellent reviews are now available [107–112]. Very briefly, there is convincing evidence that the BCC etiology is highly dependent on dysregulation of the hedgehog signaling pathway, whereas for the SCC etiology, p53-regulated pathways are of outstanding importance. For malignant melanoma, at least three molecular pathways have been found to be nearly invariably disregulated in melanocytic tumors, including the RAS-RAF-MEK-ERK pathway (through mutation of BRAF, NRAS or KIT), the p16 INK4A-CDK4-RB pathway (through mutation of INK4A or CDK4), and the ARF-p53 pathway (through mutation of ARF or TP53). Less frequently targeted pathways include the PI3K-AKT pathway (through mutation of NRAS, PTEN, or PIK3CA) and the canonical Wnt signaling pathway (through mutation of CTNNB1 or APC) [113].

It is to be expected that future research dealing with biological markers will give further insight into the etiology of skin cancer. Furthermore, very recent findings in the field of stem-cell research very clearly show that epidermal stem cells and their regulation on the genetic and epigenetic levels are the main cell targets involved in skin carcinogenesis [113–120]. Investigating the genetic and epigenetic regulation of (cancer-) stem cell fate, therefore, will enduringly increase knowledge about the etiology, as well as about (bio-) markers indicative for progression, staging, and metastasis of skin cancers.

Immunosuppression

It is now generally accepted that chronic and/or intermittent UV-exposure from the sun or from artificial sources (e.g., sunbeds) can initiate and promote skin cancer development through two major mechanisms: (1) induction of UV-dependent mutations; and (2) UV-induced immunosuppression, which might impair recognition of, e.g., UV-induced tumors (as an antigen source) by immunocompetent cells in the skin [121]. The “skin immune system” (SIS) is composed of several different cell types: keratinocytes (KC), melanocytes (MC), fibroblasts, monocytes, epidermal homing T cells, dermal macrophages, and Langerhans cells. These cells interact in a complex network via a number of soluble mediators like cytokines, interleukines, and prostaglandins and build up immune response and immunosurveillance of the skin [122].

UV-induced immunosuppression has been demonstrated initially by in vivo experiments with mice. These studies were able to show that (UV-induced) skin tumors are not rejected when transplanted to previously UV-irradiated mice. Later on, it was shown that UV-radiation mostly impaired the cellular immune response, leaving humoral immunological pathways almost untouched. The cell-mediated immune response was then studied in a large number of investigations with contact or delayed hypersensitivity reactions (CHS or DTH) as a biological endpoint in human skin [122, 123]. These investigations show that antigenes applied to the skin are taken up by antigen presenting cells (APC), where they are processed and finally presented to T lymphocytes to induce the complex immune response to eliminate the antigen.

UV-induced immunosuppression works both locally and systemically. Locally, the site of hapten application corresponds to the UV-irradiated area of skin. Systemic immunosuppression, on the other hand, induces the effect far away from hapten application. Local immunosuppression is mediated by direct UV-induced alterations in APC function, whereas systemic immunosuppresion needs mediators like e.g., Interleukin-1, 10, 12, tumor necrosis factor (TNF-α) or tumor-growth factor β (TGF-β). Furthermore, dose-rate effects of UV-irradiation seem to have an influence on whether immunosuppresion is induced locally or systemically [124].

UV-radiation needs photoreceptors that are able to "translate" UV-interaction into immunomodulatory effects at the cellular level. According to absorption- (and action-) spectra, the most important chromophores involved in immunosuppression are: urocanic acid (UCA) and DNA [123].

UCA is one of the major UV absorbing components in the *stratum corneum* of human skin and undergoes *trans* to *cis* isomerization after UVB irradiation [125]. Cis-UCA then modulates the action of several cytokines, including TNF-α, IL-6, IL-8, IL-12, and others [126, 127], in this way influencing the complex reaction pathways of SIS. It is well accepted now that UV-B-radiation-induced UCA changes, which are involved in immunosuppressive pathways, both work locally and systemically in mice and humans [122, 125, 128], although there exists some evidence that UV-induced production of cis-UCA might be insufficient to suppress CHS [129].

Another, possibly more important molecular target to interact with UV-radiation and to induce immunosuppresion, is DNA [130]. DNA is able to absorb UV-B irradiation directly, which creates cyclobutane pyrimidine dimers (CPD) and (with lower yield) 6–4 photoproducts [90]. These are known to be pre-cancerogenic DNA lesions, giving rise to UV-signature mutations, prominantely involved in the skin cancer etiology. However, these lesions (if not repaired) are also responsible for direct activation of genes involved in immune reactions [131]. CPD play an important role in immunosuppression through alterations of APC function [132], cytokine production, e.g., IL-10, and inhibition of transcription factors [133]. An increase of enzymatic repair activity for CPDs (via support of repair enzymes T4 endonuclease V) restores CHS and DTH response after UV-irradiation [134].

There is convincing evidence that cis-UCA and CPDs mediate their immunosuppressive properties through the impairment of immunocompetent cells populations like APCs, especially Langehans cells [135, 136]. One of the major actions of UV radiation on LC is that it makes them unable to prime Th1 lymphocytes, therefore inducing some tolerance against antigenes [137]. This leads to the hypothesis that UV induced cytokine might affect the critical balance between Th1 and Th2 cells in favor of an (immunosuppressive) Th2 response [138, 139]. There is a convergence of evidence now supporting this hypothesis.

Most of the immunosuppressive effects of UV-radiation have been documented after experimental UV-B irradiation on subjects [122]. However, there is increasing evidence that UVA-radiation (315–400 nm) might also be capable of inducing immunosuppression in human skin [140–143]. From a mechanistic point of view, this seems reasonable because it has recently been shown that UVA (apart from other premutagenic DNA-lesions) predominantly induces CPDs [94] (known to be involved in immunosuppression, see above) in human skin.

Skin Treatments

During the past decade, UV-radiation has been used extensively to cure certain types of skin diseases, especially psoriasis, which is shortly discussed here as an example.

Psoriasis is one of the most frequent inflammatory skin disorders. Its prevalence is estimated to be 2% in the Caucasian population, and it may develop at any age [144]. Immunological mechanisms play an important role, and it is now recognized that psoriasis is the most important T-cell mediated inflammatory disease in humans [145]. The primary immune defect in psoriasis appears to be an increase in cell signaling via chemokines and cytokines that act on upregulation of gene expression and cause hyperproliferation of keratinocytes [146]. There exists, however, strong evidence for an equally important role of polygenetic inheritance in complex genetics of psoriasis [147]. At least eight genetic loci (PSORS1–PSORS8) have been identified so far [147, 148].

Classic psoriarsis treatments use UV radiation. Broadband UV-B, narrowband UV-B (311 nm), as well as psoralen plus UV-A (PUVA) treatments have been used in the past [149]. Among other things, UV radiation reduces the number of antigen presenting cells and affects cell signaling pathways responsible for a decrease of hyperproliferating keratinocytes. UV-B-radiation seems to be less effective than PUVA-therapy. However, the latter carries an elevated risk of skin cancer induction [150–153]. For instance, high-dose PUVA patients (with more than 337 PUVA treatments) carry a 68-fold increase in overall risk of SCC (fourfold increased risk of BCC) [154].

Eye Damage

With the exception of snow- and ice-covered surfaces, most surfaces have a very low UV reflectivity. However, over ice- or snow-covered areas, reflections can directly increase the eye exposure. In some case, this natural reflectance can be as high as 100% [155]. These enhanced radiation fields can contribute to eye damage. Some impacts on the eye are described below. For more details, see [7].

Photoconjunctivitis

Photoconjunctivitis is an inflammation of the conjunctiva of the eye caused by UV radiation. The action spectrum is similar to DNA damaging radiation. The threshold irradiation for photoconjunctivitis is 50 J m^{-2} [156].

Photokeratitis

Photokeratitis is an inflammation of the cornea's epithelial layer caused by UV radiation. The action spectrum is again similar to DNA damaging radiation. The threshold irradiation is 50 J m^{-2} [156].

Cataract

Cataracts are irreversible turbidities of the eye that can be caused by infrared (IR) or UV radiation. IR cataracts are caused by direct absorption of IR radiation or, indirectly, by heat transfer from the iris to the lens. They are usually caused by long-term (many years) exposure to large infrared sources and are therefore rare as a result of sun exposure only. UV cataracts are quite common. However, there is no known action spectrum for IR or UV cataracts [156].

Visible Radiation and Health Effects

Circadian Effects

It is well known that visible radiation has an impact on human health by affecting the eyes. The major task for the eyes is to gather and further process incoming visible radiation. Basically, the eye acts like a camera. In this "picture," the retina is the film or, more appropriately, the sensitive array that converts incoming photons to electrical signals that are further processed in the brain. For 150 years, it has been known that there are two types of sensors in the retina: the rods and the cones. There is some evidence that visual perception has an indirect impact on human health through psychological effects. However, this is not discussed further here.

In recent years, another photoreceptor, called melanopsin, has been found. It is likely that this photoreceptor has been developed early in the evolution of man. Melanopsin has a lower sensitivity and a coarser spatial resolution compared to the rods and cones. It is "designed" to help synchronizing the inner clock with the solar day.

The inner clock is thought to be essential to human health. The concentrations of many hormones are regulated by the inner clock, and it is known that a long-term disturbance of the inner clock has negative impact on humans [157]. For example, it is known that night-shift workers have a higher incidence of cancer.

The suppression of melatonin is closely linked with melanopsin concentration. The known action spectrum of melatonin suppression has its maximum in the blue part of the spectrum, at about 460 nm and almost no sensitivity in the red part [156]. Since the spectral transmission of the aging human eye progressively shifts toward longer wavelengths, it is assumed that elderly people have a reduced ability to synchronize the inner clock from blue light.

Since the solar spectrum changes appreciably during the day – certainly much more than the visual impression, which is regulated by the brain to recognize objects regardless of the incoming solar spectrum – it can be assumed that the red-shift of the solar spectrum during mornings and evenings may also have an impact on the melatonin suppression as well. This is an example of impacts of the solar spectrum on human health that is being understood, and that should be regarded when artificial lighting is used.

Damage of the Retina

Radiation (including infrared) that is absorbed by the retina can cause thermal damage. UV does not play a role in this case because it has a relatively low thermal contribution and is mostly absorbed before it reaches the retina. What is relevant to the damage of the retina is not the irradiance but its thermal impact [156].

Infrared Effects

For a long time, infrared radiation was considered as biologically irrelevant for human health. Consequently, the study of the impact of infrared radiation only began relatively recently [158]. Infrared radiation produces a high concentration of reactive oxygen species (ROS), i.e., free radicals, that can damage the skin. It was shown that the generation of free radicals depends not only on the incident irradiance but also on the temperature of the skin, which is increased by infrared radiation [159]. A doubling of the irradiance from 400 to 800 W/m^2 can increase the temperature by ~10°C. Such a temperature change enables biochemical reactions in the skin. There are nonlinear dose-effect relationships. The production of free radicals starts at about 37°C and becomes much stronger at about 41°C. However, further increase in temperature to 43°C causes little additional damage. At low latitude sites or in a likely warmer climate of the near future, increased infrared irradiation will lead to an increased formation of ROS, which is potentially dangerous for human health [159]. As a consequence, infrared protection may be beneficial, in as much as the threshold skin temperature of 41°C is not reached. For example, at a latitude of 43.4° N (corresponding to Monaco), these experiments value can be reached after only 20 min of sun exposure. However, these experiments were made with artificial light (Sullux 500 W), which emits a nearly continuous spectrum, whereas the solar irradiance has absorption bands, especially from water vapor. It is not known to what extent such effects can play a role when evaluating the potential danger of infrared radiation for humans. In addition, the spectral transmittance of infrared radiation must be considered. In the near-infrared (IR-A), the spectral transmittance of the skin varies between 2% and 10%. The maximum transmittance is reached at about 1,150 nm.

Several other studies of infrared affects have been undertaken [160, 161] Schieke 2002 [155]; Jantschitsch 2009 [163]. They use a special infrared light source (called "hydrasun") that includes a water filter to minimize heating from long-wave radiation. Experiments with this source show that there is a time- and dose-dependent metalloproteinase-1 (MMP-1) that leads to increased skin aging [162], similar to that from UV exposure. The necessary dose may be reached after 2.5 h of exposure at midlatitudes in summer. However, another study [160] found that even with a high irradiance (380 W/cm^2), there is no systematic induction of MMP-1, which is clearly in contradiction with the findings in [162]. Another study [161] found that, without increased temperature, there is a protection of infrared radiation to UV exposure – a result confirmed by another study [163].

In summary: there are clear indications for effects of infrared radiation for human health, but the complete process is far from being well understood.

Effects for Which the Spectral Responsivity Is Not Yet Known

Depression

Depression can occur as a result of a lack of light in winter. This is often described as seasonal affective disorder (SAD), also known as winter depression. The U.S. national library of medicine notes that some people experience a serious mood change when the seasons change. They may sleep too much, have little energy, and crave sweets and starchy foods. They may also feel depressed. Though symptoms can be severe, they usually clear up naturally. It has been estimated that 1.5–9% of adults in the U.S. experience SAD [164]. There are many different treatments for classic (*winter*-based) SAD, usually using artificial bright lights, or a temporary stay at a southern, sunny location. However, the spectral sensitivity of SAD is not known yet. It may be cured just by exposing the eyes with bright light, but it may be also connected to a deficiency in vitamin D.

Human Behavior in Response to Solar Radiation

Humans change their behavior in response to changes in solar radiation. Possible reactions include: avoiding too intense solar irradiation by seeking shadow, or seeking high level of solar irradiance for cultural or cosmetic reasons. As a consequence, human health is not only influenced by changes in solar irradiation (geographically, daily, seasonally), but to a large extent also by human behavior. Of course, such reactions depend on many factors, both socioeconomic and individual, and it is therefore hard to give general statements on the advantages or disadvantages of changes induced by human behavior in response to solar radiation.

Future Directions

The descriptions above clearly show that solar radiation has both positive and negative effects on human health. Positive effects are the synchronization of the inner clock by natural sunlight and avoidance of winter depression, which affects many people at higher latitudes. Another positive effect is the vitamin D production, for which an overdose by natural sunlight is not known. There is clear evidence that high vitamin D levels are very beneficial to humans. The list of the resulting positive effects has increased considerably in recent years.

On the other hand, there is also clear evidence that UV radiation is mainly responsible for skin cancer, for which the incidence rates are rapidly increasing, probably as a result of changed behavior. Also, negative effects of UV radiation on the eyes are well known. Strong indications of negative impacts of infrared radiation have been found recently.

The positive aspects just mentioned have already led to changes in the "sun-smart" recommendations given by some national cancer agencies. In the past, the most frequent recommendation may have been along the lines "avoid the sun as much as possible." This is now altered to "avoid being sunburnt." Other suggestions are "avoid the sun in summer, seek it in winter." Despite the great number of studies and publications on these topics, there seems to be still insufficient scientific knowledge to convey simple messages to the public about the optimum solar exposure. It is clear that overexposure leads to severe negative effects, whereas underexposure has a negative impact on human health. It is concluded that more research is needed to assess under which conditions either overexposure or underexposure occur and must be considered. This is a rather difficult question for which simple answers are not to be expected soon due to the complexity of human health and of varying physical and meteorological conditions [165]. In addition, the relevant quantity to consider in vitamin D studies is not the irradiance on a horizontal surface but the actual exposure, which requires knowledge of the spectral radiance. This quantity, however, is far from being as well known as the widely used irradiance. More studies are required since the solar radiation regime is expected to be modified as a consequence of climate change.

Bibliography

1. McKinlay AF, Diffey BL (1987) A reference action spectrum for ultra-violet induced erythema in human skin. In: Passchier WF, Bosnajakovic BFM (eds) Human exposure to ultraviolet radiation: risks and regulations. Elsevier, Amsterdam, pp 83–87
2. Kerr JB, Seckmeyer G et al Surface ultraviolet radiation: past, present and future. In: Chapter 7 of Scientific assessment of ozone depletion: 2006, Global ozone research and monitoring project, report no. 50. World Meteorological Organization, Geneva
3. ICNRP/UNEP (1995) Global solar UV Index. Oberschleissheim, Germany
4. CIE (1970) Internationales Wörterbuch der Lichttechnik, Commission International de l'Éclairage, Paris
5. WMO (2007) Scientific Assessment of Ozone Depletion: 2006, vol Global ozone research and monitoring project – report no. 50. (World Meteorological Organization), Geneva
6. Bais AF, Lubin D, Coauthors: Arola A, Bernhard G, Blumthaler M, Chubarova N, Erlick C, Gies HP, Krotkov N, Lantz K, Mayer B, McKenzie RL, Piacentini RD, Seckmeyer G, Slusser JR, Zerefos CS (2007) Surface ultraviolet radiation: past, present and future. In: Albritton DL, Ajavon AN, Watson RT (eds) Chapter 7 of WMO scientific assessment of ozone depletion: 2006, Global ozone research and monitoring project, report no. 55. World Meteorological Organization, Geneva, p 58
7. UNEP (2007) Environmental effects of ozone depletion and its interactions with climate change: 2006 assessment. Photochem Photobiol Sci 6:201–332
8. Holick MF (1994) McCollum award lecture, vitamin D: new horizons for the 21st cen-tury. Am J Clin Nutr 60:619–630
9. Holick MF (2007) Vitamin D deficiency. N Engl J Med 357:266–281
10. Hewison M, Burke F, Evans KN, Lammas DA (2007) Extra-renal 25-hydroxyvitamin D3-1alpha-hydroxylase in human health and disease. J Steroid Biochem Mol Biol 103:316–321
11. Rao D, Raghuramulu N (1999) Is vitamin D redundant in an aquatic habitat? J Nutr Sci Vitaminol 45:1–8
12. Bouillon R, Carmeliet G, Daci E, Segaert S, Verstuyf A (1998) Vitamin D metabolism and action. Osteoporos Int 8:S13–S19
13. Hintzpeter B, Mensink G, Thierfelder W, Müller M, Scheidt-Nave C (2008) Vitamin D status and health correlates among German adults. Eur J Clin Nutr 62:1079–1089
14. Bischoff-Ferrari HA, Giovannucci E, Willett WC, Dietrich T, Dawson-Hughes B (2006) Estimation of optimal serum concentrations of 25-hydroxyvitamin D for multiple health outcomes. Am J Clin Nutr 84:18–28

15. Gorham E, Mohr S, Garland F, Garland C (2009) Vitamin D for cancer prevention and survival. Clin Rev Bone Miner Metab 7:159–175
16. Mithal A, Wahl D, Bonjour J, Burckhardt P, Dawson-Hughes B, Eisman J, El-Hajj Fuleihan G, Josse R, Lips P, Morales-Torres J, IOF Committee of Scientific Advisors (CSA) Nutrition Working Group (2009) Global vitamin D status and determinants of hypovitaminosis D. Osteoporos Int 20:1807–1820
17. Chris Power C, Hyppönen E (2007) Hypovitaminosis D in British adults at age 45 y: nationwide cohort study of dietary and life-style predictors. Am J Clin Nutr 85:860–868
18. Engelsen O, Brustad M, Aksnes L, Lund E (2005) Daily duration of vitamin D synthesis in human skin with relation to latitude, total ozone, altitude, ground cover, aerosols and cloud thickness. Photochem Photobiol 81:1287–1290
19. Genuis SJ, Schwalfenberg GK, Hiltz MN, Vaselenak SA (2009) Vitamin d status of clinical practice populations at higher latitudes: analysis and applications. Int J Environ Res Public Health 6:151–173
20. Grant W (2004) Geographic variation of prostate cancer mortality rates in the United States: Implications for prostate cancer risk related to vitamin D. Int J Cancer 111:470–471
21. Kimlin MG (2008) Geographic location and vitamin D synthesis. Mol Aspects Med 29(6):453–461
22. van der Mei IAF, Ponsonby A-L, Engelsen O, Pasco JA, McGrath JJ, Eyles DW, Blizzard L, Dwyer T, Lucas R, Jones G (2007) The high prevalence of vitamin D insufficiency across Australian populations is only partly explained by season and latitude. Environ Health Perspect 115:1132–1139
23. Bischoff-Ferrari H, Willett W, Wong J, Stuck A, Staehelin H, Orav E, Thoma A, Kiel D, Henschkowski J (2009) Prevention of nonvertebral fractures with oral vitamin D and dose dependency: a meta-analysis of randomized controlled trials. Arch Intern Med 169:551–561
24. Bischoff-Ferrari H, Dawson-Hughes B, Staehelin H, Orav J, Stuck A, Theiler R, Wong J, Egli A, Kiel D, Henschkowski J (2009) Fall prevention with supplemental and active forms of vitamin D: a meta-analysis of randomised controlled trials. BMJ 339:b3692, <javascript: AL_get(this,%20'jour',%20'BMJ.');>
25. Zittermann A (2003) Vitamin D in preventive medicine: are we ignoring the evidence? Br J Nutr 89:552–572
26. WHO (2008) International agency for research on cancer. Vitamin D and cancer. IARC working group reports, vol 5. WHO Press, Geneva, p 148
27. Lappe JM, Travers-Gustafson D, Davies KM, Recker RR, Heaney RP (2007) Vitamin D and calcium supplementation reduces cancer risk: results of a randomized trial. Am J Clin Nutr 85:1586–1591
28. Mohr S, Garland C, Gorham E, Garland F (2008) The association between ultraviolet B irradiance, vitamin D status and incidence rates of type 1 diabetes in 51 regions worldwide. Diabetologia 51:1391–1398
29. Hyppönen E (2001) Intake of vitamin D and risk of type 1 diabetes: a birth-cohort study. Lancet 358:1500–1503
30. Zipitis C, Akobeng A (2008) Vitamin D supplementation in early childhood and risk of type 1 diabetes: a systematic review and meta-analysis. Arch Dis Child 93:512–517
31. Pittas A, Lau J, Hu F, Dawson-Hughes B (2007) The role of vitamin D and calcium in type 2 diabetes. A systematic review and meta-analysis. J Clin Endocrinol Metab 92:2017–2029
32. Jorde R, Figenschau Y (2009) Supplementation with cholecalciferol does not improve glycemic control in diabtetic subjects with normal serum 25-hydroxyvitamin D levels. Eur J Nutr 48:349–354
33. von Hurst P, Stonehouse W, Coad J (2009) Vitamin D supplementation reduces insulin resistance in South Asian women living in New Zealand who are insulin resistant and vitamin D deficient – a randomised, placebo-controlled trial. Br J Nutr 103(4):549–555
34. Zittermann A, Koerfer R (2008) Vitamin D in the prevention and treatment of coronary heart disease. Curr Opin Clin Nutr Metab Care 11:752–757

35. Witham M, Nadir M, Struthers A (2009) Effect of vitamin D on blood pressure: a systematic review and meta-analysis. J Hypertens 27:1948–1954
36. Pilz S, Dobnig H, Nijpels G, Heine R, Stehouwer C, Snijder M, van Dam R, Dekker J (2009) Vitamin D and mortality in older men and women. Clin Endocrinol 71:666–672
37. Wang T, Pencina M, Booth S, Jacques P, Ingelsson E, Lanier K, Benjamin E, D'Agostino R, Wolf M, Vasan R (2008) Vitamin D deficiency and risk of cardiovascular disease. Circulation 117:503–511
38. Giovannucci E, Liu Y, Hollis B, Rimm E (2008) 25-hydroxyvitamin D and risk of myocardial infarction in men: a prospective study. Arch Intern Med 168:1174–1180
39. Dobnig H, Pilz S, Scharnagl H, Renner W, Seelhorst U, Wellnitz B, Kinkeldei J, Boehm B, Weihrauch G, Maerz W (2008) Independent association of low serum 25-hydroxyvitamin D and 1,25-dihydroxyvitamin D levels with all-cause and cardiovascular mortality. Arch Intern Med 168:1340–1349
40. Ginde A, Scragg R, Schwartz R, Camargo C (2009) Prospective study of serum 25-Hydroxyvitamin D level, cardiovascular disease mortality, and all-cause mortality in older U.S. adults. J Am Geriatr Soc 57:1595–1603
41. Melamed M, Michos E, Post W, Astor B (2008) 25-hydroxyvitamin D levels and the risk of mortality in the general population. Arch Intern Med 168:1629–1637
42. Zittermann A, Grant W (2009) 25-hydroxyvitamin D levels and all-cause mortality. Arch Intern Med 169:1075–1076
43. Autier P, Gandini S (2007) Vitamin D supplementation and total mortality: a meta-analysis of randomized controlled trials. Arch Intern Med 167:1730–1737
44. Kuroda T, Shiraki M, Tanaka S, Ohta H (2009) Contributions of 25-hydroxyvitamin D, co-morbidities and bone mass to mortality in Japanese postmenopausal women. Bone 44:168–172
45. Ng K, Meyerhardt J, Wu K, Feskanich D, Hollis B, Giovannucci E, Fuchs C (2008) Circulating 25-hydroxyvitamin d levels and survival in patients with colorectal cancer. J Clin Oncol 26:2984–2991
46. McKenzie RL, Bodeker GE, Scott G, Slusser J (2006) Geographical differences in erythemally-weighted UV measured at mid-latitude USDA sites. Photochem Photobiol Sci 5:343–352
47. Liley JB, McKenzie RL (2006) Where on Earth has the highest UV? In: UV Radiation and its effects: an update, 2006, pp. 26–37. http://www.niwa.co.nz/our-services/online-services/uv-and-ozone/workshops/2006/Liley_2.pdf
48. Seckmeyer G, Glandorf M, Wichers C, McKenzie RL, Henriques D, Carvalho F, Webb AR, Siani AM, Bais A, Kjeldstad B, Brogniez C, Werle P, Koskela T, Lakkala K, Lenoble J, Groebner J, Slaper H, denOuter PN, Feister U (2008) Europe's darker atmosphere in the UV-B. Photochem Photobiol Sci 7:925–930. doi:10.1039/b804109a
49. McKenzie RL, Liley JB, Björn LO (2009) UV radiation: balancing risks and benefits. Photochem Photobiol 85:88–98
50. Knuschke P, Unverricht I, Ott G (2007) "Baseline reference of solar UV exposures" to assess individual UV exposure levels in the population. In: Grobner J (ed) One century of UV radiation research. PMOD, Davos pp 127–128
51. Knuschke P, Unverricht I, Ott G, Jansen M (2007) Personenbezogene messung der UV-Exposition von Arbeitnehmern im freien, Bundesanstalt fur Arbeitsschutz und Arbeitsmedian (BAUA) report no. Dortmund/Berlin/Dresden, p 195. Available from: http://www.baua.de/cae/servlet/contentblob/699494/publicationFile/46848/F1777.pdf
52. Jemal A, Ward E, Hao Y, Thun M (2005) Trends in the leading causes of death in the United States, 1970–2002. JAMA 294:1255–1259
53. Katalinic A, Kunze U, Schafer T (2003) Epidemiology of cutaneous melanoma and non-melanoma skin cancer in Schleswig-Holstein, Germany: incidence, clinical subtypes, tumour stages and localization (epidemiology of skin cancer). Br J Dermatol 149:1200–1206

54. Diepgen TL, Mahler V (2002) The epidemiology of skin cancer. Br J Dermatol 146(Suppl 61):1–6
55. Jemal A, Siegel R, Ward E, Murray T, Xu J, Smigal C, Thun MJ (2006) Cancer statistics, 2006. CA Cancer J Clin 56:106–130
56. Balch CM, Buzaid AC, Soong SJ, Atkins MB, Cascinelli N, Coit DG, Fleming ID, Gershenwald JE, Houghton A Jr, Kirkwood JM, McMasters KM, Mihm MF, Morton DL, Reintgen DS, Ross MI, Sober A, Thompson JA, Thompson JF (2001) Final version of the American joint committee on cancer staging system for cutaneous melanoma. J Clin Oncol 19:3635–3648
57. Boyle P, Dore JF, Autier P, Ringborg U (2004) Cancer of the skin: a forgotten problem in Europe. Ann Oncol 15:5–6
58. de Vries E, Bray FI, Coebergh JW, Parkin DM (2003) Changing epidemiology of malignant cutaneous melanoma in Europe 1953–1997: rising trends in incidence and mortality but recent stabilizations in western Europe and decreases in Scandinavia. Int J Cancer 107:119–126
59. de Vries E, Coebergh JW (2004) Cutaneous malignant melanoma in Europe. Eur J Cancer 40:2355–2366
60. Pesch B, Ranft U, Jakubis P, Nieuwenhuijsen MJ, Hergemoller A, Unfried K, Jakubis M, Miskovic P, Keegan T (2002) Environmental arsenic exposure from a coal-burning power plant as a potential risk factor for nonmelanoma skin carcinoma: results from a case-control study in the district of Prievidza, Slovakia. Am J Epidemiol 155:798–809
61. Nindl I, Gottschling M, Stockfleth E (2007) Human papillomaviruses and non-melanoma skin cancer: basic virology and clinical manifestations. Dis Markers 23:247–259
62. Lichter MD, Karagas MR, Mott LA, Spencer SK, Stukel TA, Greenberg ER, The New Hampshire Skin Cancer Study Group (2000) Therapeutic ionizing radiation and the incidence of basal cell carcinoma and squamous cell carcinoma. Arch Dermatol 136:1007–1011
63. Armstrong BK, Kricker A (2001) The epidemiology of UV induced skin cancer. J Photochem Photobiol B 63:8–18
64. Cleaver JE, Crowley E (2002) UV damage, DNA repair and skin carcinogenesis. Front Biosci 7:d1024–d1043
65. de Gruijl FR, van Kranen HJ, Mullenders LH (2001) UV-induced DNA damage, repair, mutations and oncogenic pathways in skin cancer. J Photochem Photobiol B 63:9–27
66. Madan V, Hoban P, Strange RC, Fryer AA, Lear JT (2006) Genetics and risk factors for basal cell carcinoma. Br J Dermatol 154(Suppl 1):5–7
67. Norval M, Cullen AP, de Gruijl FR, Longstreth J, Takizawa Y, Lucas RM, Noonan FP, van der Leun JC (2007) The effects on human health from stratospheric ozone depletion and its interactions with climate change. Photochem Photobiol Sci 6:232–251
68. Ramos J, Villa J, Ruiz A, Armstrong R, Matta J (2004) UV dose determines key characteristics of nonmelanoma skin cancer. Cancer Epidemiol Biomark Prev 13:2006–2011
69. Abdel-Malek ZA, Knittel J, Kadekaro AL, Swope VB, Starner R (2008) The melanocortin 1 receptor and the UV response of human melanocytes–a shift in paradigm. Photochem Photobiol 84:501–508
70. Besaratinia A, Pfeifer GP (2008) Sunlight ultraviolet irradiation and BRAF V600 mutagenesis in human melanoma. Hum Mutat 29:983–991
71. Dulon M, Weichenthal M, Blettner M, Breitbart M, Hetzer M, Greinert R, Baumgardt-Elms C, Breitbart EW (2002) Sun exposure and number of nevi in 5- to 6-year-old European children. J Clin Epidemiol 55:1075–1081
72. Harrison SL, MacLennan R, Buettner PG (2008) Sun exposure and the incidence of melanocytic nevi in young Australian children. Cancer Epidemiol Biomark Prev 17: 2318–2324
73. Leiter U, Garbe C (2008) Epidemiology of melanoma and nonmelanoma skin cancer–the role of sunlight. Adv Exp Med Biol 624:89–103

74. Thomas NE, Edmiston SN, Alexander A, Millikan RC, Groben PA, Hao H, Tolbert D, Berwick M, Busam K, Begg CB, Mattingly D, Ollila DW, Tse CK, Hummer A, Lee-Taylor J, Conway K (2007) Number of nevi and early-life ambient UV exposure are associated with BRAF-mutant melanoma. Cancer Epidemiol Biomark Prev 16:991–997
75. El Ghissassi F, Baan R, Straif K, Grosse Y, Secretan B, Bouvard V, Benbrahim-Tallaa L, Guha N, Freeman C, Galichet L, Cogliano V (2009) A review of human carcinogens–part D: radiation. Lancet Oncol 10:751–752
76. International Agency for Research on Cancer Working Group (2007) The association of use of sunbeds with cutaneous malignant melanoma and other skin cancers: a systematic review. Int J Cancer 120:116–1122
77. Rosenstein BS, Mitchell DL (1987) Action spectra for the induction of pyrimidine(6–4) pyrimidone photoproducts and cyclobutane pyrimidine dimers in normal human skin fibroblasts. Photochem Photobiol 45:775–780
78. Cadet J, Anselmino C, Douki T, Voituriez L (1992) Photochemistry of nucleic acids in cells. J Photochem Photobiol B 15:277–298
79. Cadet J, Sage E, Douki T (2005) Ultraviolet radiation-mediated damage to cellular DNA. Mutat Res 571:3–17
80. Douki T, Court M, Sauvaigo S, Odin F, Cadet J (2000) Formation of the main UV-induced thymine dimeric lesions within isolated and cellular DNA as measured by high performance liquid chromatography-tandem mass spectrometry. J Biol Chem 275:11678–11685
81. Mitchell DL, Brash DE, Nairn RS (1990) Rapid repair kinetics of pyrimidine(6–4) pyrimidone photoproducts in human cells are due to excision rather than conformational change. Nucleic Acids Res 18:963–971
82. Mitchell DL (2006) Quantification of photoproducts in mammalian cell DNA using radioimmunoassay. Method Mol Biol 314:239–249
83. Greinert R, Boguhn O, Harder D, Breitbart EW, Mitchell DL, Volkmer B (2000) The dose dependence of cyclobutane dimer induction and repair in UVB-irradiated human keratinocytes. Photochem Photobiol 72:701–708
84. Wang LE, Li C, Strom SS, Goldberg LH, Brewster A, Guo Z, Qiao Y, Clayman GL, Lee JJ, El Naggar AK, Prieto VG, Duvic M, Lippman SM, Weber RS, Kripke ML, Wei Q (2007) Repair capacity for UV light induced DNA damage associated with risk of nonmelanoma skin cancer and tumor progression. Clin Cancer Res 13:6532–6539
85. Wang LE, Xiong P, Strom SS, Goldberg LH, Lee JE, Ross MI, Mansfield PF, Gershenwald JE, Prieto VG, Cormier JN, Duvic M, Clayman GL, Weber RS, Lippman SM, Amos CI, Spitz MR, Wei Q (2005) In vitro sensitivity to ultraviolet B light and skin cancer risk: a case-control analysis. J Natl Cancer Inst 97:1822–1831
86. Wei Q, Lee JE, Gershenwald JE, Ross MI, Mansfield PF, Strom SS, Wang LE, Guo Z, Qiao Y, Amos CI, Spitz MR, Duvic M (2003) Repair of UV light-induced DNA damage and risk of cutaneous malignant melanoma. J Natl Cancer Inst 95:308–315
87. Dumaz N, Drougard C, Sarasin A, Daya-Grosjean L (1993) Specific UV-induced mutation spectrum in the p53 gene of skin tumors from DNA-repair-deficient xeroderma pigmentosum patients. Proc Natl Acad Sci U S A 90:10529–10533
88. Nakazawa H, English D, Randell PL, Nakazawa K, Martel N, Armstrong BK, Yamasaki H (1994) UV and skin cancer: specific p53 gene mutation in normal skin as a biologically relevant exposure measurement. Proc Natl Acad Sci U S A 91:360–364
89. Ravanat JL, Douki T, Cadet J (2001) Direct and indirect effects of UV radiation on DNA and its components. J Photochem Photobiol 63:88–102
90. Sage E, Lamolet B, Brulay E, Moustacchi E, Chteauneuf A, Drobetsky EA (1996) Mutagenic specificity of solar UV light in nucleotide excision repair-deficient rodent cells. Proc Natl Acad Sci U S A 93:176–180
91. Besaratinia A, Kim SI, Pfeifer GP (2008) Rapid repair of UVA-induced oxidized purines and persistence of UVB-induced dipyrimidine lesions determine the mutagenicity of sunlight in mouse cells. FASEB J 22:2379–2392

92. Ikehata H, Kudo H, Masuda T, Ono T (2003) UVA induces C–>T transitions at methyl-CpG-associated dipyrimidine sites in mouse skin epidermis more frequently than UVB. Mutagenesis 18:511–519
93. Runger TM, Kappes UP (2008) Mechanisms of mutation formation with long-wave ultraviolet light (UVA). Photodermatol Photoimmunol Photomed 24:2–10
94. Mouret S, Baudouin C, Charveron M, Favier A, Cadet J, Douki T (2006) Cyclobutane pyrimidine dimers are predominant DNA lesions in whole human skin exposed to UVA radiation. Proc Natl Acad Sci U S A 103:13765–13770
95. Matsumura Y, Ananthaswamy HN (2002) Molecular mechanisms of photocarcinogenesis. Front Biosci 7:d765–d783
96. Cruet-Hennequart S, Gallagher K, Sokol AM, Villalan S, Prendergast AM, Carty MP (2010) DNA Polymerase eta, a key protein in translesion synthesis in human cells. Subcell Biochem 50:189–209
97. Hendel A, Ziv O, Gueranger Q, Geacintov N, Livneh Z (2008) Reduced efficiency and increased mutagenicity of translesion DNA synthesis across a TT cyclobutane pyrimidine dimer, but not a TT 6–4 photoproduct, in human cells lacking DNA polymerase eta. DNA Repair Amst 7:1636–1646
98. Hussain SP, Hollstein MH, Harris CC (2000) p53 tumor suppressor gene: at the crossroads of molecular carcinogenesis, molecular epidemiology, and human risk assessment. Ann NY Acad Sci 919:79–85
99. Benjamin CL, Ullrich SE, Kripke ML, Ananthaswamy HN (2008) p53 tumor suppressor gene: a critical molecular target for UV induction and prevention of skin cancer. Photochem Photobiol 84:55–62
100. Brash DE (2006) Roles of the transcription factor p53 in keratinocyte carcinomas. Br J Dermatol 154(Suppl 1):8–10
101. Brash DE, Ziegler A, Jonason AS, Simon JA, Kunala S, Leffell DJ (1996) Sunlight and sunburn in human skin cancer: p53, apoptosis, and tumor promotion. J Investig Dermatol Symp Proc 1:136–142
102. Nelson MA, Einspahr JG, Alberts DS, Balfour CA, Wymer JA, Welch KL, Salasche SJ, Bangert JL, Grogan TM, Bozzo PO (1994) Analysis of the p53 gene in human precancerous actinic keratosis lesions and squamous cell cancers. Cancer Lett 85:23–29
103. Ziegler A, Leffell DJ, Kunala S, Sharma HW, Gailani M, Simon JA, Halperin AJ, Baden HP, Shapiro PE, Bale AE, Brash DE (1993) Mutation hotspots due to sunlight in the p53 gene of nonmelanoma skin cancers. Proc Natl Acad Sci U S A 90:4216–4220
104. Brash DE (1997) Sunlight and the onset of skin cancer. Trends Genet 13:410–414
105. Fusenig NE, Boukamp P (1998) Multiple stages and genetic alterations in immortalization, malignant transformation, and tumor progression of human skin keratinocytes. Mol Carcinog 23:144–158
106. Pleasance ED, Cheetham RK, Stephens PJ, McBride DJ, Humphray SJ, Greenman CD, Varela I, Lin ML, Ordonez GR, Bignell GR, Ye K, Alipaz J, Bauer MJ, Beare D, Butler A, Carter RJ, Chen L, Cox AJ, Edkins S, Kokko-Gonzales PI, Gormley NA, Grocock RJ, Haudenschild CD, Hims MM, James T, Jia M, Kingsbury Z, Leroy C, Marshall J, Menzies A, Mudie LJ, Ning Z, Royce T, Schulz-Trieglaff OB, Spiridou A, Stebbings LA, Szajkowski L, Teague J, Williamson D, Chin L, Ross MT, Campbell PJ, Bentley DR, Futreal PA, Stratton MR (2010) A comprehensive catalogue of somatic mutations from a human cancer genome. Nature 463:191–196
107. Benjamin CL, Melnikova VO, Ananthaswamy HN (2008) P53 protein and pathogenesis of melanoma and nonmelanoma skin cancer. Adv Exp Med Biol 624:265–282
108. Bosserhoff AK (2006) Novel biomarkers in malignant melanoma. Clin Chim Acta 367:28–35
109. Daya-Grosjean L, Couve-Privat S (2005) Sonic hedgehog signaling in basal cell carcinomas. Cancer Lett 225:181–192
110. Hussein MR (2005) Ultraviolet radiation and skin cancer: molecular mechanisms. J Cutan Pathol 32:191–205

111. Lupi O (2007) Correlations between the sonic Hedgehog pathway and basal cell carcinoma. Int J Dermatol 46:1113–1117
112. Sekulic A, Haluska P Jr, Miller AJ, Genebriera DL, Ejadi S, Pulido JS, Salomao DR, Thorland EC, Vile RG, Swanson DL, Pockaj BA, Laman SD, Pittelkow MR, Markovic SN (2008) Malignant melanoma in the 21st century: the emerging molecular landscape. Mayo Clin Proc 83:825–846
113. Dahl C, Guldberg P (2007) The genome and epigenome of malignant melanoma. APMIS 115:1161–1176
114. Faurschou A, Haedersdal M, Poulsen T, Wulf HC (2007) Squamous cell carcinoma induced by ultraviolet radiation originates from cells of the hair follicle in mice. Exp Dermatol 16:485–489
115. Gerdes MJ, Yuspa SH (2005) The contribution of epidermal stem cells to skin cancer. Stem Cell Rev 1:225–231
116. Greinert R (2009) Skin cancer: new markers for better prevention. Pathobiology 76:64–81
117. Grichnik JM (2008) Melanoma, nevogenesis, and stem cell biology. J Invest Dermatol 128:2365–2380
118. Schwabe M, Lubbert M (2007) Epigenetic lesions in malignant melanoma. Curr Pharm Biotechnol 8:382–387
119. Yan X, Owens DM (2008) The skin: a home to multiple classes of epithelial progenitor cells. Stem Cell Rev 4:113–118
120. Zabierowski SE, Herlyn M (2008) Melanoma stem cells: the dark seed of melanoma. J Clin Oncol 26:2890–2894
121. Muller HK, Malley RC, McGee HM, Scott DK, Wozniak T, Woods GM (2008) Effect of UV radiation on the neonatal skin immune system- implications for melanoma. Photochem Photobiol 84:47–54
122. Amerio P, Carbone A, Auriemma M, Varrati S, Tulli A (2009) UV induced skin immunosuppression. Anti-Infamm Anti-Allergy Agents Med Chem 8:3–13
123. Norval M (2000) The impact of ultraviolet radiation on immune response. Radiat Prot Dosim 91:51–56
124. Shimizu T, Streilein JW (1994) Local and systemic consequences of acute, low-dose ultraviolet B radiation are mediated by different immune regulatory mechanisms. Eur J Immunol 24:1765–1770
125. Noonan FP, De Fabo EC (1992) Immunosuppression by ultraviolet B radiation: initiation by urocanic acid. Immunol Today 13:250–254
126. Kaneko K, Smetana-Just U, Matsui M, Young AR, John S, Norval M, Walker SL (2008) cis-Urocanic acid initiates gene transcription in primary human keratinocytes. J Immunol 181:217–224
127. Kurimoto I, Streilein JW (1992) cis-urocanic acid suppression of contact hypersensitivity induction is mediated via tumor necrosis factor-alpha. J Immunol 148:3072–3078
128. Kondo S, Sauder DN, McKenzie RC, Fujisawa H, Shivji GM, El Ghorr A, Norval M (1995) The role of cis-urocanic acid in UVB-induced suppression of contact hypersensitivity. Immunol Lett 48:181–186
129. el Ghorr AA, Norval M (1995) A monoclonal antibody to cis-urocanic acid prevents the ultraviolet-induced changes in Langerhans cells and delayed hypersensitivity responses in mice, although not preventing dendritic cell accumulation in lymph nodes draining the site of irradiation and contact hypersensitivity responses. J Invest Dermatol 105:264–268
130. Kuchel JM, Barnetson RS, Halliday GM (2005) Cyclobutane pyrimidine dimer formation is a molecular trigger for solar-simulated ultraviolet radiation-induced suppression of memory immunity in humans. Photochem Photobiol Sci 4:577–582
131. Yarosh DB (2002) Enhanced DNA repair of cyclobutane pyrimidine dimers changes the biological response to UV-B radiation. Mutat Res 509:221–226
132. Vink AA, Strickland FM, Bucana C, Cox PA, Roza L, Yarosh DB, Kripke ML (1996) Localization of DNA damage and its role in altered antigen-presenting cell function in ultraviolet-irradiated mice. J Exp Med 183:1491–1500

133. Tommasi S, Swiderski PM, Tu Y, Kaplan BE, Pfeifer GP (1996) Inhibition of transcription factor binding by ultraviolet-induced pyrimidine dimers. Biochemistry 35:15693–15703
134. Kripke ML, Cox PA, Alas LG, Yarosh DB (1992) Pyrimidine dimers in DNA initiate systemic immunosuppression in UV-irradiated mice. Proc Natl Acad Sci U S A 89:7516–7520
135. Donawho CK, Muller HK, Bucana CD, Kripke ML (1996) Enhanced growth of murine melanoma in ultraviolet-irradiated skin is associated with local inhibition of immune effector mechanisms. J Immunol 157:781–786
136. Ullrich SE (1995) Modulation of immunity by ultraviolet radiation: key effects on antigen presentation. J Invest Dermatol 105:30S–36S
137. Simon JC, Tigelaar RE, Bergstresser PR, Edelbaum D, Cruz PD (1991) Ultraviolet B radiation converts Langerhans cells from immunogenic to tolerogenic antigen-presenting cells. Induction of specific clonal anergy in CD4+ T helper 1 cells. J Immunol 146:485–491
138. Shreedhar V, Giese T, Sung VW, Ullrich SE (1998) A cytokine cascade including prostaglandin E2, IL-4, and IL-10 is responsible for UV-induced systemic immune suppression. J Immunol 160:3783–3789
139. Ullrich SE (1996) Does exposure to UV radiation induce a shift to a Th-2-like immune reaction? Photochem Photobiol 64:254–258
140. Clement-Lacroix P, Michel L, Moysan A, Morliere P, Dubertret L (1996) UVA-induced immune suppression in human skin: protective effect of vitamin E in human epidermal cells in vitro. Br J Dermatol 134:77–84
141. Rana S, Rogers LJ, Halliday GM (2010) Immunosuppressive ultraviolet-A radiation inhibits the development of skin memory CD8 T cells. Photochem Photobiol Sci 9:25–30
142. Seite S, Zucchi H, Moyal D, Tison S, Compan D, Christiaens F, Gueniche A, Fourtanier A (2003) Alterations in human epidermal Langerhans cells by ultraviolet radiation: quantitative and morphological study. Br J Dermatol 148:291–299
143. Stoebner PE, Poosti R, Djoukelfit K, Martinez J, Meunier L (2007) Decreased human epidermal antigen-presenting cell activity after ultraviolet A exposure: dose-response effects and protection by sunscreens. Br J Dermatol 156:1315–1320
144. Christophers E (2001) Psoriasis–epidemiology and clinical spectrum. Clin Exp Dermatol 26:314–320
145. Krueger JG (2002) The immunologic basis for the treatment of psoriasis with new biologic agents. J Am Acad Dermatol 46:1–23
146. Traub M, Marshall K (2007) Psoriasis–pathophysiology, conventional, and alternative approaches to treatment. Altern Med Rev 12:319–330
147. Barker JN (2001) Genetic aspects of psoriasis. Clin Exp Dermatol 26:321–325
148. Valdimarsson H (2007) The genetic basis of psoriasis. Clin Dermatol 25:563–567
149. Gambichler T, Breuckmann F, Boms S, Altmeyer P, Kreuter A (2005) Narrowband UVB phototherapy in skin conditions beyond psoriasis. J Am Acad Dermatol 52:660–670
150. Lim JL, Stern RS (2005) High levels of ultraviolet B exposure increase the risk of non-melanoma skin cancer in psoralen and ultraviolet A-treated patients. J Invest Dermatol 124: 505–513
151. Nijsten TE, Stern RS (2003) The increased risk of skin cancer is persistent after discontinuation of psoralen+ultraviolet A: a cohort study. J Invest Dermatol 121:252–258
152. Stern RS, Bolshakov S, Nataraj AJ, Ananthaswamy HN (2002) p53 mutation in nonmelanoma skin cancers occurring in psoralen ultraviolet a-treated patients: evidence for heterogeneity and field cancerization. J Invest Dermatol 119:522–526
153. Stern RS, Liebman EJ, Vakeva L (1998) Oral psoralen and ultraviolet-A light (PUVA) treatment of psoriasis and persistent risk of nonmelanoma skin cancer. PUVA follow-up study. J Natl Cancer Inst 90:1278–1284
154. Peritz AE, Gasparro FP (1999) Psoriasis, PUVA, and skin cancer–molecular epidemiology: the curious question of T–>A transversions. J Investig Dermatol Symp Proc 4:11–16
155. Wuttke S, Seckmeyer G, Konig-Langlo G (2006) Measurements of spectral snow albedo at Neumayer, Antarctica. Annles Geophysicae 24:7–21

156. DIN (2008) German Institute for Normalisation Report – Several papers, ISBN 978-3-410-16938. In: *2. DIN Expertenforum, Wirkung des Lichts auf Menschen*, Beuth, Berlin, p 43
157. Schmidt C, Collette F, Leclercq Y, Sterpenich V, Vandewalle G, Berthomier P, Berthomier C, Phillips C, Tinguely G, Darsaud A, Gais S, Schabus M, Desseilles M, Dang-Vu TT, Salmon E, Balteau E, Degueldre C, Luxen A, Maquet P, Cajochen C, Peigneux P (2009) Homeostatic sleep pressure and responses to sustained attention in the suprachiasmatic area Science 465:516–519
158. Kligman LH (1982) Intensification of ultraviolet-induced dermal damage by infrared radiation. Arch Dermatol Res 272:229–267
159. Zastrow L, Groth N (2008) The missing link – light-induced (280–1600 nm) free radical formation in human skin. Skin Pharmacol Physiol 22:31–44
160. Gebbers N, Hirt-Burri N (2007) Water-filtered infrared-a radiation (wira) is not implicated in cellular degeneration of human skin. GMS Ger Med Sci 5:1–14
161. Menendez S, Coulomb B (1998) Non-coherent near infrared radiation protects normal human dermal fibroplasts from solar ultraviolet toxicity. J Investig Dermatol 111:629–633
162. Schieke SM, Stege H (2002) Infrared-a radiation-induced matrix metalloproteinase 1 expression is mediated through exracellular signal-regulated kinase 1/2 activation in human dermal fibroblasts. J Investig Dermatol 119:1323–1329
163. Jantschitsch C, Majewski S (2009) Infrared radiation confers resistanse to uv-induces apoptosus via reduction of DNA damage and upregulation of antiapoptotic proteins. Soc Investig Dermatol 129:1271–1279
164. Modell J, Rosenthal N, Harriett A, Krishen A, Asgharian A, Foster V, Metz A, Rockett C, Wightman D (2005) Seasonal affective disorder and its prevention by anticipatory treatment with bupropion XL biological psychiatry. Biol Psychiatry 58:658–667
165. Pissulla D, Seckmeyer G, Cordero RR, Blumthaler M, Schallhart B, Webb A, Kifte ASR, Bais AF, Kouremeti N, Cede A, Herman J, Kowalewski M (2009) Comparison of different calibration methods to derive spectral radiance as a function of incident and azimuth angle. Photochem Photobiol Sci 8:516–527
166. WMO (1983) Guide to meteorological instruments and methods of observation (Fifth edn), World Meteorological Organization, Report No. 8, Geneva

Chapter 20
Toxic Chemical Risks

Edward A. Laws

Glossary

Alkaloid	A colorless, complex, and bitter organic base containing nitrogen and usually oxygen that is often physiologically active.
Analgesic	Painkiller.
Apoptosis	A genetically directed process of cell self-destruction that eliminates DNA-damaged, superfluous, or unwanted cells.
Cold turkey	Cold flashes with goose bumps, one of the symptoms associated with withdrawal from a drug to which a person has become addicted.
Dopamine	A neurotransmitter in the brain that plays an important role, inter alia, in behavior and reward.
Opioid	A chemical that binds to and activates the opioid receptors in the nervous system and gastrointestinal tract; includes both natural opiates (derived from opium) and synthetic drugs possessing narcotic properties similar to opiates but not derived from opium.
Secondhand smoke	Smoke inhaled by a nonsmoker and consisting of mainstream smoke (smoke exhaled by a smoker) and sidestream smoke (smoke from the end of a lighted cigarette, pipe, or cigar).
Teratogenic	Causing developmental malformations.

This chapter was originally published as part of the Encyclopedia of Sustainability Science and Technology edited by Robert A. Meyers. DOI:10.1007/978-1-4419-0851-3

E.A. Laws (✉)
Louisiana State University, School of the Coast and Environment, 3141 Energy, Coast and Environment Building, Baton Rouge, LA 70803, USA
e-mail: edlaws@lsu.edu

E.A. Laws (ed.), *Environmental Toxicology: Selected Entries from the Encyclopedia of Sustainability Science and Technology*, DOI 10.1007/978-1-4614-5764-0_20,

Definition of the Subject

At the present time there are approximately 100,000 chemicals in the environment, with an additional 500–1,000 added each year. Although not all these chemicals are toxic, merely keeping up-to-date information on the possible toxicity of so many compounds is a prodigious task. Federal agencies such as the Environmental Protection Agency (EPA), Food and Drug Administration (FDA), and Consumer Product Safety Commission (CPSC) are responsible for establishing rules and guidelines that ensure, insofar as possible, that risks associated with exposure to toxic chemicals are within acceptable limits. Legislation such as the Clean Air Act, originally passed in 1963, and Clean Water Act, a 1977 amendment to the 1972 Federal Water Pollution Control Act, have done much to advance the cause of environmental safety. Ironically perhaps, in the United States the primary sources of exposure to toxic chemicals and poisons are related to lifestyle issues, including in particular smoking, drinking (alcohol), use/abuse of pharmaceutical and recreational drugs, and obesity. Most incidents of poisoning occur in the home, where unintentional exposure to toxic chemicals is often the result of ignorance and/or carelessness. Education and public awareness have therefore become important components of the effort to address the issue of exposure to toxic chemicals in the United States.

Introduction

On the first day of class students in an honors section of an introductory environmental science class at Louisiana State University were asked to answer a series of multiple-choice questions to determine how much they knew about the subject before taking the course. One of the questions asked (Dr. Vince Wilson, personal communication):

The statement, "Water can be a poison," is an example of

1. No agent has a single effect
2. The water quality of the Baton Rouge Public Water System
3. Dose makes the poison
4. A sensitive population
5. Toxicology is an empirical science

The correct answer is "dose makes the poison," the translation of a statement made in German by the sixteenth century physician Paracelsus. It underscores the fact that many substances commonly considered to be toxic are presumably harmless if the dose or degree of exposure is sufficiently small. This line of reasoning underlies the concept of an acceptable daily intake (ADI), for example, of food additives and contaminants [1, p. 2]. The ADI is a dose of the substance that would be expected to be associated with no appreciable adverse health effects if administered over the lifetime of a healthy person. On the other hand, consumption of too much of an otherwise benign substance can have serious health consequences. Water intoxication, for example, is a potentially fatal condition that occurs when consumption of too much water causes an abnormal balance of electrolytes in the body.

In judging the risk associated with exposure to toxic chemicals, one must consider the duration of exposure as well as the dose. In toxicology the distinction is typically made between short-term or acute exposure and long-term or chronic exposure. Experimentally acute and chronic exposures are defined to be exposures lasting up to 14 days and greater than 1 year, respectively, with "intermediate exposure" used to characterize exposures lasting 14 days to 1 year [2]. These definitions are arbitrary, and in practice exposure times may vary continuously from seconds to an entire lifetime.

Alcohol consumption is a good example of an issue linked to exposure time. In most states in the United States, a person is considered to be intoxicated if his/her blood alcohol content exceeds 0.08%. Consumption of alcohol can have fatal consequences if the blood alcohol content approaches ~0.40%. Death results from alcohol's depressant effect on vital areas of the brain that control consciousness, respiration, and heart rate. To provide some context to these numbers, a person weighing ~100 pounds (45 kg) who consumed the contents of two 12-ounce cans of beer (5% alcohol content) over a short timeframe (~30–40 min) would likely have a blood alcohol content of ~0.10% (i.e., above the legal limit for driving) and a blood alcohol content of ~0.40% (i.e., potentially lethal) if he/she consumed eight 12-ounce cans of beer over the same timeframe. On the other hand, a 150-pound (68 kg) person who consumed the same quantities of beer over the same timeframe would likely have blood alcohol levels of only ~0.06% (i.e., below the legal limit for driving) and 0.25% (intoxicated and perhaps experiencing serious side effects, but probably not life-threatening), respectively. Incidentally, the tendency of people to vomit when they have had too much to drink is a natural defense mechanism and allows the body to rid itself of alcohol that has not already been absorbed into the bloodstream. Unfortunately this defense mechanism is not infallible. When some people have had too much to drink they lose consciousness and never recover, either because of the effect of the alcohol on their brain or because they choke on their own vomit and are unable to clear their air passages.

The effects of consuming 12-ounce cans of beer are very different if the timeframe is longer. If a 100-pound person consumed one 12-ounce can of beer per day for 8 days, for example, his/her blood alcohol level would probably never rise to more than 0.05%, below the legal limit for driving and far below the level associated with potentially fatal effects.

Perhaps surprisingly, most (~90%) exposures to toxic chemicals occur in the home and involve everyday household items. Not surprisingly, most exposures (85–90%) are accidents, although some are intentional (e.g., suicides). With the notable exception of cigarette smoke, about 75% of cases in the United States involve ingestion of a poison (as opposed, e.g., to inhalation of a toxin or exposure via dermal contact), and, with the exception of alcohol consumption, over half the cases involve children under the age of six. Interestingly, the greatest public health problems associated with exposure to toxic chemicals involve lifestyle decisions as opposed, for example, to living downwind or downstream from an obvious source of pollution.

Lifestyle-Associated Exposure to Toxic Chemicals

Smoking

In the United States and many other countries, by far the single greatest source of exposure to toxic chemicals is tobacco smoke, primarily the result of cigarette smoking. Cigarette smoke contains several hundred toxins, of which 44, the so-called Hoffmann analytes [3–5], are believed to be most relevant to smoking-related health effects [6]. These include the following:

- Ammonia (NH_3): Nicotine, which is the addictive component of cigarettes, exists in two forms, bound and free. The free form of nicotine vaporizes more readily, and from the vapor phase is rapidly absorbed by the lungs and transported by the bloodstream to the brain. Adding ammonia to cigarettes facilitates the conversion of bound to free nicotine and thereby enhances the pleasurable effects associated with secretion of dopamine by brain neurons that have been stimulated either directly or indirectly by nicotine. Ammonia is also a toxic gas that can damage the respiratory system when it combines with water to form ammonium hydroxide:

  $$NH_3 + H_2O \rightarrow NH_4OH$$

  NH_4OH is highly alkaline and corrosive. Exposure to NH_4OH is associated with coughing, sore throat, shortness of breath, and a burning sensation, all of which may be associated with smoking.
- Arsenic (As): Arsenic is a well-known toxin that has been used in pesticides for more than 1,000 years. The most commonly used formulation has been lead arsenate, which was used extensively for control of insect pests until the end of World War II, when more effective synthetic organic insecticides became available. However, it is estimated that smokers breathe in between 0.8 and 2.4 μg of As per pack of cigarettes [7] due to continued use of arsenic-containing pesticides by tobacco growers. The toxicity of As is related to its disruptive effects on a wide variety of metabolic processes leading to, inter alia, oxidative stress, lactic acidosis, anemia, central nervous system dysfunction, and apparent thiamine deficiency.
- Benzene (C_6H_6): Benzene, a human carcinogen and known cause of leukemia, is found in cigarette smoke, which is believed to account for about half of human exposure. Chronic exposure to benzene can result in a variety of health problems including anemia, leukemia, and adverse effects on the immune system [8, 9].
- Benzo[a]pyrene ($C_{20}H_{12}$): Benzo[a]pyrene is a five-ring polycyclic aromatic hydrocarbon formed by the fusion of a benzene ring with pyrene. It is found in all smoke associated with the combustion of organic matter, including tobacco leaves. Metabolites of benzo[a]pyrene have been shown to be mutagenic and are highly carcinogenic [10]. Amounts of benzo[a]pyrene in cigarettes range from ~3 to 30 ng per cigarette and are roughly in proportion to declared tar values [11].

The carcinogenic effects of benzo[a]pyrene are associated with binding of its diol epoxide metabolites to the tumor suppressor protein P53, which normally functions as a cell cycle regulator. Damage to P53 is related to more than half of all human cancers and to an even higher percentage of lung cancers [12, 13].

- Cadmium (Cd): In general leafy vegetables such as lettuce and spinach contain high levels of cadmium, which they accumulate from the soil. This is a natural phenomenon, and tobacco leaves are no exception. A cigarette smoker will typically take in about 3.0 μg of Cd per pack, of which 25% will be absorbed [14]. Chronic exposure to Cd is associated with kidney and liver damage, and in severe cases to softening of the bones [15]. Smokers typically have about twice the body burden of Cd as nonsmokers [16, 17].
- Carbon monoxide (CO): Carbon monoxide is a colorless, odorless gas that results from the combustion of organic matter with inadequate amounts of oxygen. The concentration of CO in the air inhaled with cigarette smoke is about 2%, of which ~85% is absorbed [18, 19]. A person smoking 20 cigarettes per day (the minimum number of cigarettes per pack in the United States is 20) would be expected to absorb ~200 mg of CO [20], considerably higher than the maximum of 24 mg d^{-1} that can be safely tolerated [18]. The adverse effects of CO exposure are typically associated with its tendency to bind to hemoglobin and hence to interfere with the ability of hemoglobin to absorb and transport oxygen. Ironically, the affinity between CO and hemoglobin is about 230 times greater than the affinity between O_2 and hemoglobin [21, 22]. Additional adverse effects may be caused by the tendency of CO to bind to myoglobin, a protein that binds iron and oxygen in muscle tissues, and to cytochrome oxidase, an enzyme in the respiratory electron transport chain.
- Formaldehyde (CH_2O): Formaldehyde is a colorless but pungent gas found in cigarette smoke as a result of the pyrolysis of saccharides used as tobacco ingredients [23–25]. It is carcinogenic [26–29] and causes damage to the lungs when inhaled [30, 31]. The amount of formaldehyde inhaled in cigarette smoke is closely correlated with the tar content of the cigarette and ranges as high as 40–50 μg per cigarette [20]. Estimated exposure levels in the oral cavity during smoking range as high as 115 parts per million (ppm), which is considerably higher than the Occupational Safety and Health Administration (OSHA) short-term (15 min) exposure limit (STEL) of 2 ppm.
- Hydrogen cyanide (HCN): Hydrogen cyanide is a colorless, poisonous gas that was used as a genocidal agent by the Nazis during World War II. A byproduct of burning tobacco, HCN is found in cigarette smoke in amounts sufficient to produce exposures of 144–351 μg per cigarette [6, 32]. To put this number in context, the STEL for HCN is 4.7 ppm in the air one breathes. An average person inhales about 11 m^3 of air per day or 115 L in 15 min. If the air contained 4.7 ppm HCN vapor, the person would be inhaling about 0.54 mL or 22 μmol of HCN vapor in 15 min (this calculation assumes that HCN vapor behaves like an ideal gas and that the temperature is 25 °C or 77 °F). Twenty-two micromoles of HCN have a mass of 594 μg. Thus smoking 2–4 cigarettes in 15 min would very

likely lead to the inhalation of a quantity of HCN greater than the OHSA short-term exposure limit for this toxic gas. Headache, nausea, and vertigo are symptoms associated with chronic exposure to HCN [33].

- Nicotine ($C_{10}H_{14}N_2$): Nicotine is an alkaloid that is responsible for the addictive nature of tobacco use. However, it has also been implicated in DNA damage, oxidative stress, and inhibition of apoptosis [34, 35]. The latter effect could be a factor in the etiology of smoking-associated cancer of the lungs and oral cavity, that is, inhibition of programmed cell death may allow the multiplication of "chemically initiated cells" [35]. Doses of nicotine associated with cigarette smoking are highly correlated with the tar content of cigarettes and range as high as 1.0–1.2 mg per cigarette [20].
- TSNAs: This acronym stands for tobacco-specific N-nitrosamines, a category that includes four compounds linked to smoking-related cancers: (1) N′-nitrosonornicotine (NNN), (2) 4-methyl-N-nitrosamino-1-(3-pyridyl)-1-butanone (NNK), (3) N-nitrosoanatabine (NAT), and (4) N-nitrosoanabasine (NAB). Together with the polycyclic aromatic hydrocarbons (e.g., benzo[a]pyrene) these are among the most potent cancer-causing agents in cigarette smoke [36–38]. NNN, NNK, and NNAL, for example, cause cancer in mice, rats, and hamsters [36]. The TSNAs are produced during tobacco processing and smoking and are derived from nicotine and minor tobacco alkaloids. A variety of experimental studies have demonstrated that TSNAs are in fact procarcinogens, that is, agents that require metabolic activation before they become carcinogenic. Their activation involves α-hydroxylation to form unstable α-hydroxynitrosamines, which then decompose to diazohydroxides that react with cellular components, including DNA and hemoglobin. There is good evidence that TSNAs contribute to the increased risk for cancer (1) of the upper digestive tract among tobacco chewers and (2) of lung cancer, especially pulmonary adenocarcinoma, in smokers [36].

Considering the large number of toxic compounds in cigarette smoke, including in particular those implicated in causing cancer, it should come as no surprise that smoking is associated with a considerable health risk. Approximately 440,000 people die each year in the United States as a result of cigarette smoking, including approximately 50,000 who die from exposure to secondhand smoke. The combined death toll accounts for almost 20% of all deaths in the United States annually (the death rate in the United States is about 2.4 million people per year) and exceeds all deaths associated with alcohol, AIDS, car accidents, illegal drugs, murders, and suicides combined [39]. The lifespan of smokers is about 14 years shorter than that of nonsmokers [40]. Statistics compiled by the Centers for Disease Control (CDC) indicate that persons who smoke increase their risk of dying from bronchitis and emphysema roughly tenfold. Men and women who smoke increase their risk of dying from lung cancer by more than a factor of 22 and by nearly a factor of 12, respectively [39]. In the United States, cigarette smoking is estimated to be responsible for $193 billion in annual health-related economic losses (this figure includes $96 billion in direct medical costs and another ~$97 billion in lost productivity),

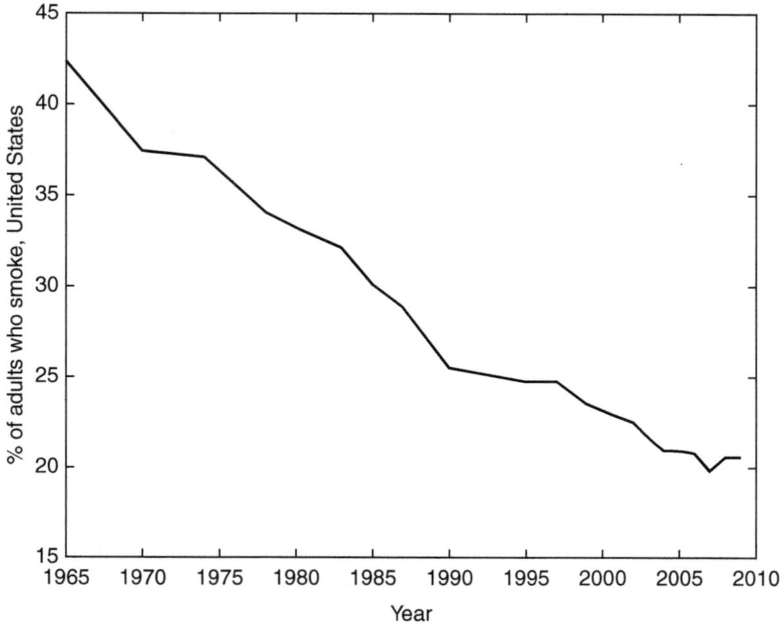

Fig. 20.1 Percentage of adults in the United States who smoke [42]

a figure that translates to about $10.47 in economic costs per pack of cigarettes sold [41].

The good news is that smoking in the United States is on the decline. The percentage of adults who smoke has declined more-or-less monotonically from a high of 42.4% in 1965 to 20.6% in 2008–2009 (Fig. 20.1). The percentage of high school students who smoke has declined even more rapidly, from a high of 36% in 1997 to 19.5% in 2009 [42].

The pattern evident in Fig. 20.1 reflects increasing public awareness of the health risks associated with cigarette smoking and, in recent years, an effort by the federal government to reduce tobacco-related diseases and deaths. The National Tobacco Control Program (NTCP), created by the CDC in 1999, provides funding and technical support to state and territorial health departments with the following goals:

- Prevent initiation of smoking by youth.
- Promote quitting among both adults and youth.
- Eliminate exposure to secondhand smoke.
- Identify and eliminate disparities among population groups.

More information about the program is available at the NTCP Web site:
http://www.cdc.gov/tobacco/tobacco_control_programs/ntcp/index.htm

Alcohol Consumption

Alcohol is probably the oldest drug used by human beings, with commercial production of wine traceable to as early as 1,500 B.C. [43]. The alcohol in question is ethanol (ethyl alcohol), which is produced by the fermentation of sugar via the overall reaction:

$$C_6H_{12}O_6 \rightarrow 2C_2H_5OH + 2CO_2$$
$$\text{sugar} \rightarrow \text{ethanol} + \text{carbon dioxide}$$

The pleasurable effects of ethanol are associated with its effects on the brain. From the digestive tract ethanol rapidly passes to the blood stream and from there to all parts of the body, including the brain. Ethanol depresses the membrane responsiveness of all neurons in the body, which explains the slower reaction time and sedative effects associated with ethanol consumption. Ethanol also helps to increase the release of dopamine by brain neurons, which results in artificial feelings of pleasure. This interference with normal brain functionality eventually causes the brain to reduce dopamine activity, leading over time to a greatly reduced ability to experience pleasure by normal means. This in turn leads to alcohol addiction.

In addition to being addictive, ethanol is also toxic. In fact, it is routinely used as an antiseptic because it kills bacterial cells. At a sufficiently high concentration it will kill virtually any cell. Even doses typical of recreational consumption of alcoholic beverages lead to a loss of a small number of cells in the brain and liver. If repeated often enough, this can lead to health problems, for example, hepatic cirrhosis and cancers. The fact that the ethanol content of wine is about 10%, for example, is no accident. The yeasts that typically mediate the fermentation process in winemaking die if the ethanol content of the product exceeds about 12%.

The organ primarily responsible for eliminating ethanol from the human body is the liver, where ethanol is converted to acetaldehyde via the enzyme alcohol dehydrogenase:

$$2C_2H_5OH + O_2 \rightarrow 2C_2H_4O + 2H_2O$$
$$\text{ethanol} + O_2 \rightarrow \text{acetaldehyde} + H_2O$$

This does not solve the toxicity problem, however, because acetaldehyde itself is toxic and is responsible for many of the symptoms associated with hangover. In the liver, acetaldehyde is converted to acetic acid ($C_2H_4O_2$) via acetaldehyde dehydrogenase. Simplistically:

$$2C_2H_4O + O_2 \rightarrow 2C_2H_4O_2$$
$$\text{acetaldehyde} + O_2 \rightarrow \text{acetic acid}$$

Problem solved? The answer is yes if a person consumes ethanol infrequently enough, but as noted in the introduction to this chapter, the dose makes the poison. Serious health problems can arise if a person consumes ethanol too frequently over a long period of time. Health problems associated with such long-term ethanol consumption include the following [44]:

- Liver damage: Alcoholic hepatitis, cirrhosis of the liver, and cancer of the liver can all result from chronic ethanol consumption. Alcoholic hepatitis is an inflammation of the liver caused by excessive ethanol consumption. Cirrhosis of the liver is caused by a gradual deterioration of the functionality of the liver and is characterized by replacement of normal liver tissue by scar tissue.
- Brain damage: There are a variety of symptoms here, including impaired judgment, vision, and coordination (leading in some cases to accidents), cell death (leading to dementia and in some cases to paralysis), and in cases of acute overdoses, coma and death.
- Impacts on the reproductive system: Here again, there are many symptoms, including impotence due to hormone imbalances, birth defects caused by mutations of egg or sperm cells, fetal alcohol syndrome (damage to the fetus associated with the fact that ethanol passes from the mother's bloodstream across the placenta to the fetus), teratogenic effects, and breast cancer.
- Damage to the digestive tract: Impacts include inflammation and cancer of the throat, esophagus, stomach, pancreas, small intestine, and colon.

The greatest risks from ethanol consumption are associated with the development of a condition called alcoholism. Simplistically, this is a condition that develops when a person's brain has been sufficiently altered by ethanol consumption that the person must drink to feel normal. The result is compulsive drinking. Failure to drink leads to serious withdrawal symptoms that include nausea, sweating, shakiness, anxiety, and in a small percentage of cases, to a condition called delirium tremens, which can be fatal. Most people who become alcoholics and subsequently stop drinking do so only with assistance.

How many people are alcoholics? About 10% of Americans are estimated to have "an alcohol problem" and 7.7% "abuse alcohol or are alcoholic" [45]. Thirty-eight percent of American families report a "history of alcoholism" [46].

The costs associated with alcoholism in terms of lives, human health, and societal impacts are very large. In the United States, more than 100,000 people die annually as a result of excessive alcohol consumption. This is about 4% of the annual death rate in the United States. The specific causes of death include, inter alia, drunken driving, cirrhosis of the liver, cancer, and stroke. In the case of drunken driving, many victims are not themselves alcoholics but instead are innocent bystanders. Roughly half of all homicides, 40% of assaults, and one-third of all fatal car accidents, suicides, and hospital admissions in the United States are alcohol related. The economic costs associated with alcohol abuse and alcoholism in the United States are roughly $230 billion per year, almost 80% more than the costs related to all other addictive drugs combined [45, 47].

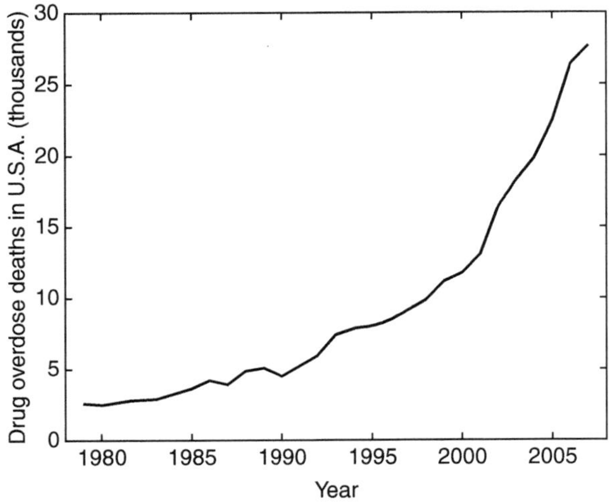

Fig. 20.2 Number of deaths per year in the United States associated with unintentional drug overdoses [52]

Most people who drink are not alcoholics. In the United States, roughly 44% of adults (18 years and older) consume an average of at least one drink per month, including 60% of persons 21–34 years of age, 46% for adults aged 60–64, and 33% of persons 65 or older [48]. Clearly only a small fraction of persons who consume ethanol become alcoholics. The risk of becoming an alcoholic is to some extent a function of genetics. Alcoholism runs in families [49, 50]. But the risk of becoming an alcoholic is also a function of one's environment, including peer pressure, the influence of family and friends, and the ease with which one can obtain alcohol [48].

Drug Abuse/Misuse

In the last 30 years, the number of deaths associated with unintentional drug overdoses has increased dramatically in the United States (Fig. 20.2). Most of the increase has not been due to the use of illegal drugs such as cocaine and heroin but rather due to misuse/abuse of prescription painkillers (analgesics) that are opioids (Fig. 20.3). Opiates are compounds derived from opium poppies and include morphine and codeine. Opioid analgesics are synthetic or semisynthetic analogues of opiates and include oxycodone (e.g., OxyContin®), hydrocodone (e.g., Vicodin®), propoxyphene (e.g., Darvon®), and hydromorphone (e.g., Dilaudid®) [51].

Naively one might assume that deaths associated with opioid analgesics are primarily linked to little children getting into medicine cabinets or old people mixing up their pills, but that is not the case. According to Paulozzi [53], the trend in opioid-related deaths apparent in Fig. 20.3 is due to the increasing use of "prescription drugs, especially opioid painkillers, among people during the working years of life." A high percentage of the people who die have a history of drug abuse,

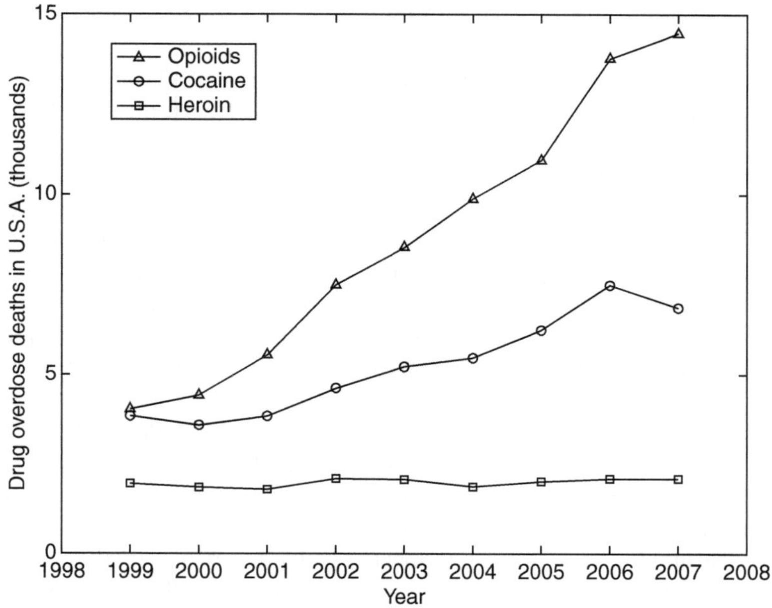

Fig. 20.3 Number of deaths per year in the United States associated with unintentional overdoses of opioids, cocaine, and heroin [54, 55]

and many have no prescription for the drug. Mixing of prescription drugs with illegal drugs is common, and "some alter the prescription drugs by crushing and snorting them or dissolving and injecting them" [53].

The problems associated with the use of opioid analgesics are an excellent example of the adage that the dose makes the poison. When used as prescribed, opioid analgesics can be very effective painkillers. They bind to specific proteins (opioid receptors) in the brain, spinal cord, and gastrointestinal tract, and in this way they can block the transmission of pain signals to the brain.

There are side effects associated with opioid/opiate use. One is suppression of coughing, which is why codeine, for example, is incorporated into a number of cough syrups and cold remedies. Another is constipation, which explains why loperamide, an opioid derived from piperdine, is used to treat diarrhea. More insidious side effects include drowsiness and a feeling of euphoria. The former is associated with the lethal effects of an opioid overdose, which leads to a suppression of breathing and may ultimately cause respiratory failure. The latter effect is responsible for the abuse and compulsive use of opioids. Persons addicted to opioids experience withdrawal symptoms if use of the drugs is reduced or stopped. These symptoms include muscle and bone pain, insomnia, diarrhea, vomiting, involuntary leg movements, and cold flashes with goose bumps (i.e., cold turkey). Interestingly, the synthetic opioid methadone blocks the effects of other opioids (including heroin), eliminates withdrawal symptoms, relieves drug craving, and has been used for decades to treat opioid addicts [56]. Fortunately, properly supervised use of opioid analgesics seldom results in clinical addiction.

Cocaine is an illegal drug and one of the most addictive of all stimulants. Derived from the coca bush, cocaine is marketed "on the street" in two forms: a water-soluble hydrochloride salt and a water-insoluble cocaine base. A cocaine user can either inject the water-soluble form with a hypodermic needle or snort (i.e., inhale) the hydrochloride salt powder through the nose, where it is readily absorbed through nasal tissues into the bloodstream. The base form is processed to produce a substance known as crack that can be smoked. The term "crack" comes from the crackling sound heard when the mixture is smoked. Once it has entered the blood stream, cocaine travels to the brain, where it produces a euphoric effect that is responsible for its popularity and addictive properties. In a nutshell, cocaine causes dopamine to accumulate at nerve synapses (small gaps between two neurons) in the brain and thereby heightens the pleasurable sensations associated with dopamine secretion. Rather than stimulating the secretion of dopamine by a neuron, the mechanism involves blockage of the natural process by which dopamine is recycled from the synapse to the neuron by specialized proteins called dopamine transporters.

The pleasurable effects of cocaine appear rapidly but are of short duration. When cocaine is absorbed rapidly via either injection or smoking, the effects are intense but dissipate after 5–10 min. Snorting is associated with a delayed response that is less intense but persists for 15–30 min [57].

In addition to the general euphoric sensation associated with cocaine use, there are other psychological effects. Users feel mentally alert, talkative, and energetic and temporarily may feel less need for sleeping and eating. Physiological effects include constricted blood vessels and an increase in body temperature, pulse, and blood pressure. Medical complications arising from the physiological effects of cocaine include the following [57]:

- Cardiovascular effects: arrhythmias (heart rhythm disturbances), heart attacks
- Neurological effects: strokes, seizures, headaches, coma
- Gastrointestinal complications: abdominal pain, nausea

Deaths associated with cocaine use are typically associated respiratory failures preceded by cardiac arrest or seizures. In addition, there is an even greater danger associated with the use of cocaine during alcohol consumption. Alcohol and cocaine together are the most common two-drug combination responsible for drug-related deaths [57].

Clearly there is a very significant risk associated with cocaine abuse, and with the exception of its very limited use as a topical anesthetic in nasal and lacrimal duct surgery, it is associated with no beneficial applications. Moreover, there are presently no medications approved by the Food and Drug Administration (FDA) to treat cocaine addiction. In 2007, cocaine abuse accounted for 13% of all admissions to drug abuse treatment programs in the United States, and a majority of those cocaine addicts were abusing multiple drugs [57]. Treatment of cocaine addiction is a complex process, not only because of the frequent problem of multiple-drug abuse but also because of the complex biological (e.g., changes in the brain), social,

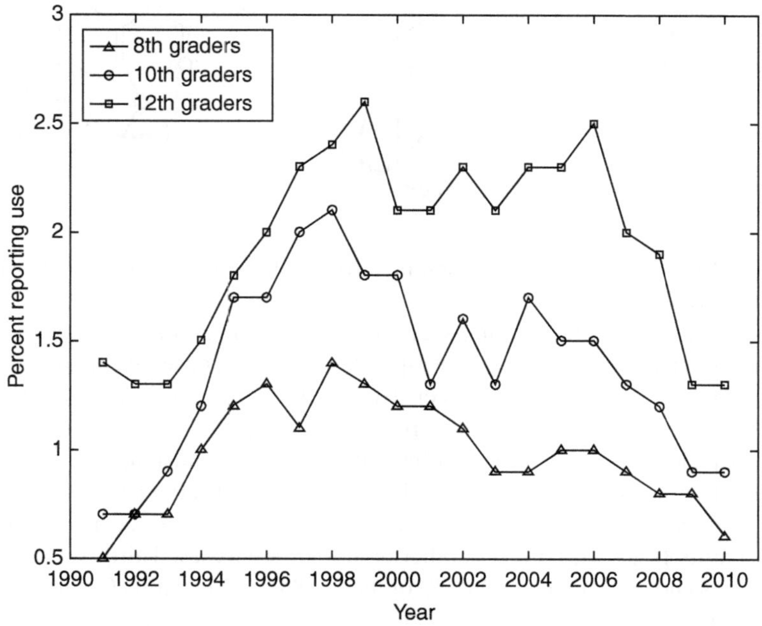

Fig. 20.4 Trends in a 30-day prevalence of cocaine abuse among 8th, 10th, and 12th graders [59]

familial, and environmental problems that must typically be addressed in the treatment process. Flynn et al. [58] examined the factors that contributed to long-term recovery from cocaine dependence of 708 patients from 45 treatment programs in the United States and concluded that only 33% of the sample were "highly favorable" at follow-up. The major reasons cited for improvements were "motivations to change, positive influences of family, strength from religion and spirituality, and help from drug treatment."

Given the highly addictive nature of cocaine, the absence of approved medications for treatment of addiction, and the mixed success of treatment programs, Benjamin Franklin's observation that "an ounce of prevention is worth a pound of cure" applies (Franklin's comment was advice related to fighting fires; he founded Philadelphia's Union Fire Company). The good news is that use of cocaine by teenagers has been declining in the United States since roughly 1997 (Fig. 20.4). This decline, however, has followed a period of increase during the 1990s. It is apparent from Fig. 20.4 that the percentage of US teenagers who use cocaine is little different now than in 1990.

Heroin is the most abused and the most rapidly acting of the natural opiates. It is obtained by the acetylation of the two hydroxyl groups of morphine with acetyl chloride, that is, the two –OH groups are replaced with $-OCOCH_3$ groups (Fig. 20.5). This modification makes heroin much less soluble in water than morphine but allows it to pass the blood-brain barrier much more rapidly than morphine. Interestingly, once inside the brain, heroin is converted to morphine and

Fig. 20.5 Structures of morphine (*left*) [61] and heroin (*right*) [62]

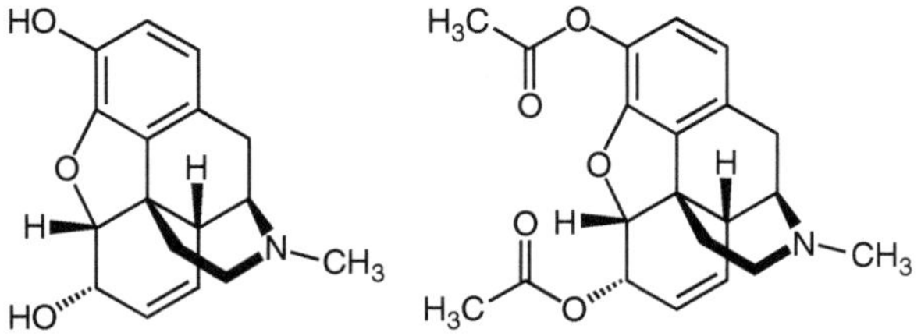

then rapidly binds to opioid receptors [60], leading to a surge of pleasurable sensation known as a "rush."

Heroin may be injected, sniffed/snorted, or smoked. Intravenous injection is very common because it provides the most rapid (less than 10 s) onset of euphoria. It is not uncommon for a heroin abuser to inject up to four times per day. Feelings of euphoria follow intramuscular injection and sniffing/snorting by 5–8 and 10–15 min, respectively [60]. Heroin is addictive regardless of how it is administered.

The risks associated with heroin addiction are numerous. For persons who inject, sharing of needles can lead to some of the most serious consequences: infection with hepatitis B and C, HIV, and other blood-borne viruses. Thirty-six percent of AIDS cases and an estimated 70–80% of the 35,000 new hepatitis C infections that occur in the United States each year are attributable to injection drug use [62, 63].

Then there are the effects of the heroin itself. After the initial rush, heroin users typically feel drowsy for several hours. Cardiac function and breathing slow, and in cases of overdose may slow to the point of death. On the street heroin is typically "cut" with other drugs or with substances such as sugar, starch, or powdered milk. As a result, users of street heroin are often unsure of the purity of the drug they are using. This uncertainty can easily lead to overdoses.

Chronic heroin injection can lead to a variety of medical problems, including scarred and/or collapsed veins, bacterial infections, and liver or kidney disease [62]. In addition to these physiological effects, heroin addicts gradually spend more and more of their time and energy obtaining and using heroin. "Once they are addicted, the heroin abuser's primary purpose in life becomes seeking and using drugs" [62]. Withdrawal symptoms include restlessness, muscle and bone pain, insomnia, diarrhea, vomiting, involuntary leg movements, and the aforementioned cold turkey effects. These typically begin to manifest themselves within 1–2 days after the last dose of heroin and subside within 1 week, although in some cases they last for many months [62].

The good news is that there are a variety of effective treatments available for heroin addiction. As already noted, methadone effectively blocks the effects of heroin and has been used successfully to treat heroin addicts for decades. When properly used methadone is not intoxicating or sedating, it may be used safely for as long as 10 years or more. It is taken orally and suppresses withdrawal symptoms for 24–36 h. Very importantly, it relieves the craving for heroin, which is a major cause of relapse. A more recent development has been the use of buprenorphine, a semisynthetic opioid, for the same purpose. Use of buprenorphine is associated with less risk of addiction and can be prescribed in the privacy of a doctor's office.

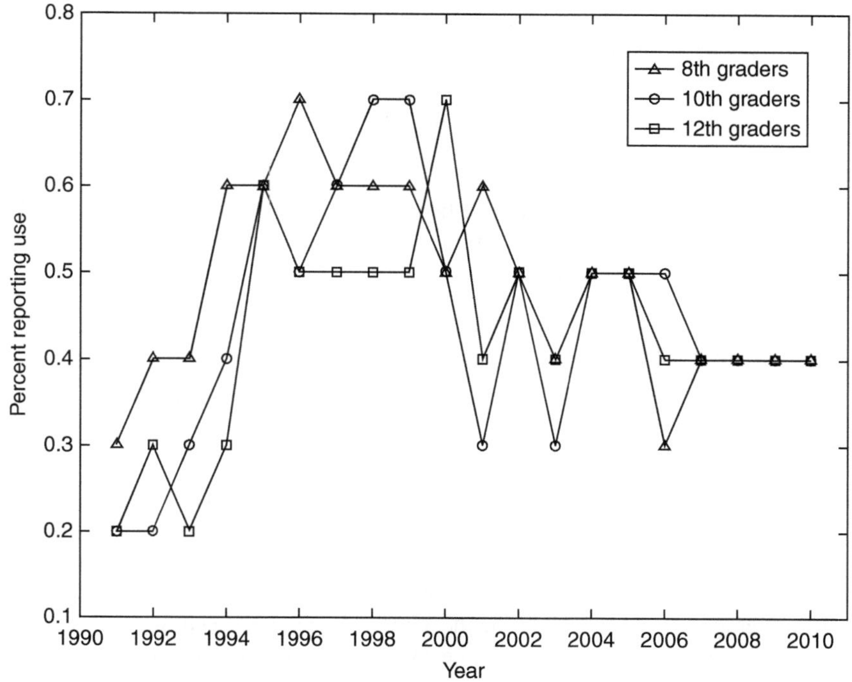

Fig. 20.6 Trends in 30-day prevalence of heroin abuse among 8th, 10th, and 12th graders [59]

Another treatment, a combination of buprenorphine and naloxone (Suboxone®), is designed to minimize the possibility of abuse [62].

According to a 2003 survey of drug use in the United States [64], roughly 3.7 million people had used heroin at least once in their lives, 119,000 had used heroin within the previous month, and 314,000 had used heroin within the previous year. The number of deaths attributable to heroin use in the United States has been constant for the last 20 years (Fig. 20.3) and has averaged about 2,000 per year. The fraction of teenagers who abuse heroin has been constant since ~2000 (Fig. 20.6) and is higher now than in the early 1990s. Thus heroin abuse and addiction continue to be serious problems in the United States.

Obesity

Food is perhaps the quintessential example of a substance for which the dose makes the poison. Most people do not think of food as a poison, but obesity is a serious problem in the United States and a disease formally recognized by the Centers for Disease Control and Prevention.

For a number of years public health agencies have used a metric called the body mass index (BMI) to determine whether a person's weight is normal or not.

If a person's weight and height are quantified in pounds and inches, respectively, the BMI is determined from the following equation:

$$\mathrm{BMI} = 703\frac{\mathrm{weight}}{\mathrm{height}^2}$$

Thus, for example, a person whose height is 70.3 in. and whose weight is 140.6 pounds would have a BMI of 20. As a rule of thumb, a person is categorized as follows based on his/her BMI:

- Underweight: <18.5
- Normal: 18.5–25.0
- Overweight: 25.0–30.0
- Obese: 30.0–40.0
- Extremely obese: >40.0

These are somewhat arbitrary definitions and in individual cases should be interpreted with the understanding that what is normal for one person may not be normal for another. With that caveat, the BMI index has proven very useful for assessing the general health of a population relative to the issue of obesity.

The average weight of adults in the United States has been slowly increasing since 1960 (Fig. 20.7, upper left). The average weight of both men and women increased by ~25 pounds between 1960 and 2000. Interestingly, there was no change in the percentage of adults in the overweight category (Fig. 20.7, upper right), but there were rather dramatic increases in the percentages of adults characterized as obese (Fig. 20.7, lower left) and extremely obese (Fig. 20.7, lower right).

At the present time roughly 100 million adults in the United States are overweight or obese. Obesity substantially increases a person's risk of the following [67]:

- Morbidity from hypertension
- Dyslipidemia (abnormal amounts of lipids such as cholesterol and fat in the blood)
- Type II diabetes (adult-onset diabetes)
- Coronary heart disease
- Stroke
- Gallbladder disease
- Osteoarthritis (degenerative joint disease)
- Sleep apnea (abnormal pauses or abnormally low breathing during sleep)
- Respiratory problems
- Endometrial (lining of the uterus), breast, prostate, and colon cancers
- Social stigmatization and discrimination

While there is no question that obesity significantly increases a person's risk of developing a number of potentially serious health problems, there is less agreement about what to do about obesity. The number of diet/exercise books in a typical bookstore is a good indication of the uncertainty associated with treatment of this

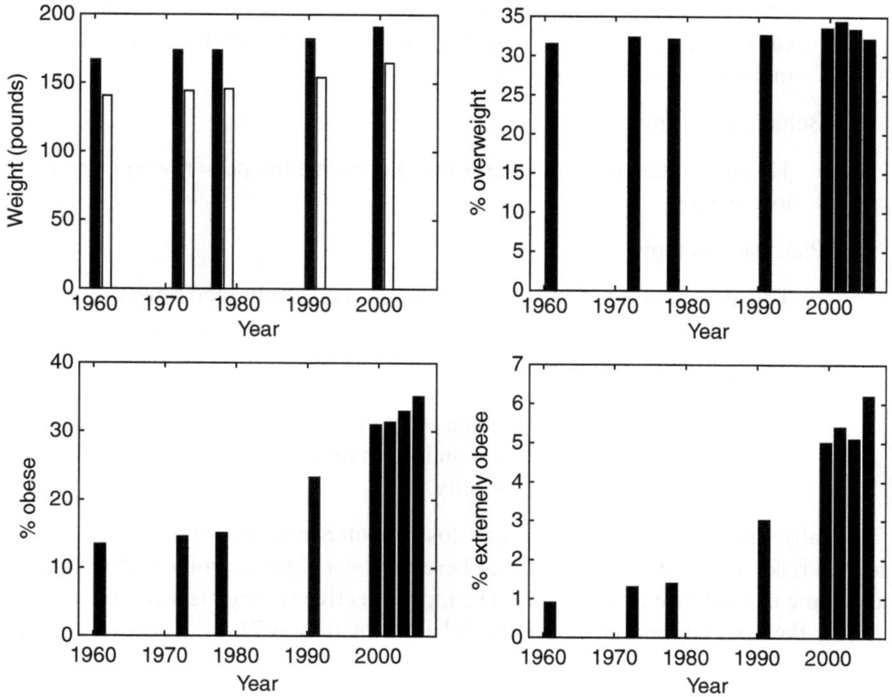

Fig. 20.7 Metrics of US adults 20–74 years old with respect to weight. *Upper left*: Average weight of adult males (▪) and females (□). *Upper right*: Percentage of adults with BMIs of 25–30. *Lower left*: Percentage of adults with BMIs of 30–40. *Lower right*: Percentage of adults with BMIs greater than 40 [65, 66]

disease. Guidelines recommended by the National Heart, Lung, and Blood Institute include the following:

- Goals
  - Initial target of weight loss should be 10%. If successful, further weight loss may be attempted.
  - Initial rate of weight loss should be 1–2 pounds per week for up to 6 months.
- Methods
  - Dietary therapy
    - Low-calorie diets with reduced fat content.
    - Reducing fat without reducing calories is futile.
    - Target caloric reduction should be 500–1,000 kcal per day.
  - Physical activity
    - The target here is at least 30 min of moderate-intensity physical activity on most, and preferably all, days of the week. Although physical

activity will not be the primary determinant of weight loss, it will increase cardiorespiratory fitness and may decrease abdominal fat and help with maintenance of weight loss.

- Behavior therapy
  - This translates to identifying ways to motivate the person who is trying to lose weight.
- Pharmacotherapy
  - This is the use of FDA-approved drugs and is only recommended if their use is supervised and their efficacy assessed by a physician.
- Weight loss surgery
  - This is only recommended in cases of clinically severe obesity when less invasive methods have failed and the patient is at high risk for obesity-associated morbidity or mortality.

Finally, there is the issue of weight loss maintenance. As with other lifestyle-related risks, the tendency to become obese can be a chronic problem that requires a lifetime of vigilance to overcome. The most effective strategy is a combination of dietary therapy, physical activity, and behavior therapy [67].

Exposure to Toxic Chemicals in the Home

Lead

Of all the toxic chemicals to which humans are exposed, lead has perhaps the longest history of exposure and today impacts the health of the greatest number of people. Lead is mined primarily from deposits of the mineral galena, or lead sulfide (PbS). Since metallic lead can be separated from PbS by heating at low temperatures easily achieved by burning wood or charcoal, it was not difficult for early civilizations to extract lead. The overall reaction is:

$$PbS + O_2 \rightarrow SO_2 + Pb$$

From written and archeological evidence, it is known that lead was used by the Egyptians as long ago as 1500 BC. The Romans used lead extensively to line their aqueducts and water mains, and both the Greeks and Romans used lead to line cooking vessels, since bronze pots tend to give food a bitter taste [68]. At least one author has suggested that endemic lead poisoning caused by the consumption of contaminated food and drink contributed significantly to the fall of the Roman

Empire [69]. The use of lead or lead salts to sweeten wine was common in Europe after the ninth century, and associated outbreaks of lead poisoning were not infrequent. In colonial America lead poisoning was also a problem, mainly due to the use of lead condensing tubes for distilling rum and the use of earthenware with a high lead content [68].

For obvious reasons, the principal uses of lead today are of an industrial rather than culinary nature. By far the major use, accounting for 87% of total lead consumption in the United States, is the production of batteries. However, lead batteries have never been a major source of human exposure to lead. During much of the twentieth century by far the most important source of human exposure was the use of lead in gasoline. Although leaded gasoline is no longer sold in the United States, lead was used extensively in the United States as a gasoline additive for roughly 60 years. This practice dates back to 1923, when tetraethyl lead was first introduced as an antiknock additive. This use was suspended during part of 1925 and 1926 pending the establishment of safety standards for its manufacture and handling, but after the spring of 1926 lead additives were commonly used in most gasoline sold in the Unites States and throughout the world. Beginning in 1960, organic alkyl lead compounds were blended into gasoline along with tetraethyl lead to further improve antiknock characteristics, and the additives were subsequently referred to as lead alkyls rather than just tetraethyl lead [68]. As of 1973, the lead content of gasoline sold in the United States averaged about 0.58 g/L.

Because of concern over pollution of the environment with lead, in 1974 the Environmental Protection Agency (EPA) requested a phasedown in the use of lead alkyls in gasoline to 0.45 g/L on January 1, 1975, and to 0.13 g/L as of January 1, 1979. In December 1974, however, the US Circuit Court of Appeals in Washington ruled against the EPA, stating that evidence did not support the EPA contention that automobile emissions contributed significantly to blood lead levels, and that the EPA regulation was arbitrary and capricious.

Since that time the legal status of gasoline lead additives has changed significantly. This change has resulted partly from the requirement for exhaust emission control devices on automobiles. The devices function poorly if at all when lead is present in the exhaust fumes, since the lead poisons the catalyst, which is designed to help oxidize unburned hydrocarbons. Furthermore, there is increasing awareness of the danger of environmental lead pollution due to the use of leaded gasoline. Since July 1, 1977, all new cars sold in the United States have been required to run on unleaded gas, defined as gasoline containing less than 0.013 g/L lead. Finally, the government required that the lead content of leaded gasoline be reduced to 0.13 g/L by July 1, 1985, and to 0.10 g/L by January 1, 1986. The result of these regulations was a tenfold decline in the use of lead in gasoline in the United States between the early 1970s and 1986. Leaded gasoline is now completely phased out in the United States. European countries began slowly phasing out leaded gasoline in the 1980s and mandated elimination in the 1990s. Leaded gasoline was phased out in sub-Saharan Africa by 2006 [70]. China banned the sale of leaded gasoline in 2000, although production by clandestine factories for use in older vehicles continued until at least 2004 [71].

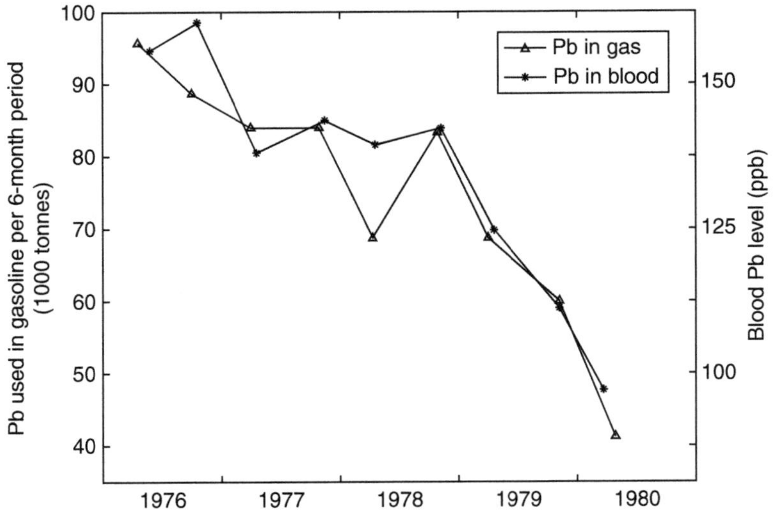

Fig. 20.8 Change in blood lead levels during phaseout of leaded gasoline in the United States [72]

The decline in the blood lead levels of people in the United States during the phaseout of leaded gasoline (Fig. 20.8) is provocative considering the Circuit Court of Appeals 1974 ruling that evidence did not support the EPA's contention that automobile emissions contributed significantly to blood lead levels. During the period 1976–1980, blood lead levels declined in almost direct proportion to the use of leaded gasoline, suggesting that automobile emissions not only were contributing significantly to blood lead levels, but were in fact the dominant contributors.

Why all the concern about lead? With respect to toxicology, lead is a general metabolic poison. The pathological effects are varied, but in general reflect the tendency of lead to interact with proteins and hence to damage tissue and interfere with the proper functioning of enzymes [68]. Lead is known to inhibit active transport mechanisms involving ATP, to depress the activity of the enzyme cholinesterase, to suppress cellular oxidation–reduction reactions, and to inhibit protein synthesis [68]. Lead poisoning is also associated with the following problems:

- Anemia: Lead is known to disrupt several enzymes involved in the production of heme, a constituent of hemoglobin and various other respiratory pigments, and has been shown to interfere with the uptake of iron by red blood cells. Anemia associated with lead poisoning is undoubtedly caused in part by these effects. There is also evidence, however, that lead causes a shortening of the life of red blood cells, probably due to disruption of the red cell membrane [68].
- Damage to the Central Nervous System (CNS): Damage to both the peripheral and central nervous system, including the brain, may be caused by a degeneration of nerve fibers and interference with the permeability of capillaries in the brain due to lead intoxication. The exact mechanism responsible for these effects is not known, and indeed several factors, including enzyme inhibition and tissue

damage, may be involved [68]. The CNS effects mainly impact children under 5 years of age, since lead does not pass the blood-brain barrier as easily in adults. Asymptomatic exposure to low doses of lead in young children (toddlers) can lead to permanent decreases in IQ.

- Kidney Damage: Kidney damage characterized by atrophy of the renal tubules is a well-established effect of lead poisoning. The damage is associated with elevated levels of amino acids, sugar, and phosphate in the urine [68]. Gout caused by kidney damage has also been associated with lead poisoning, the consumption of illicitly distilled whiskey (moonshine) being a common cause. Concentrations of lead in moonshine often exceed 10 ppm due to the use of lead in the distillation apparatus [73]. Waldron and Stöfen [68] cite one example of a hospital in which 37 of 43 cases of gout admitted in 1967 were caused by the consumption of lead-contaminated liquor.
- Effects on Children and Reproduction: From the standpoint of human health, the greatest danger of lead poisoning is undoubtedly to young children, particularly those living in urban areas. Neurological damage caused by the poisoning of such children may be permanent and can result in impaired physical as well as mental development. Even the unborn fetus is not protected from the effects of lead poisoning. There is evidence that the brain of the fetus is much more sensitive to lead poisoning than the brain of the infant or young child, and lead has repeatedly been shown to cause birth defects in experimental animals [73]. Although there is little information to indicate that lead has a teratogenic effect on humans, exposure of pregnant women to lead can induce miscarriages and stillbirths [68].

Most of the body burden of lead is found in the bones, and although the lead content of bones can now be assayed using noninvasive x-ray fluorescence techniques [74], historically it has been much more practical to relate health effects on living persons to the level of lead in the blood, PbB. Table 20.1 summarizes the lowest PbB levels associated with various health effects as summarized by the EPA [73] and CDC [75].

PbB levels as low as 100 $\mu g\ L^{-1}$ do not cause distinctive symptoms, but are associated with decreased intelligence and impaired neurobehavioral development, decreased stature or growth, decreased hearing acuity, and decreased ability to maintain a steady posture. The activity of δ-aminolaevulinic acid dehydratase (ALA-d), an enzyme involved in heme metabolism, is also adversely affected at a PbB level of 100 $\mu g\ L^{-1}$. Interference with vitamin D metabolism occurs at PbB levels of 100–150 $\mu g\ L^{-1}$, and increased concentrations of protoporphyrin, a substrate needed for heme synthesis, and ALA begin to appear in red blood cells and urine, respectively, at slightly higher PbB levels. Actual decreases in hemoglobin, somewhat loosely referred to in Table 20.1 as anemia, occur at PbB levels of about 400–500 $\mu g\ L^{-1}$. Based on this evidence, the Centers for Disease Control have concluded that children's PbB levels should not exceed 100 $\mu g\ L^{-1}$ [75].

Unfortunately a great many children as well as adults in the United States have PbB levels greater than 100 $\mu g\ L^{-1}$. For example, a study of urban children in seven

Table 20.1 Lowest PbB levels associated with health effects in humans

Lead in blood ($\mu g\ L^{-1}$)	Effect	Population group
100	ALA-d inhibition	Children and adults
	Adverse effects on intelligence, hearing, and growth	Children
100–150	Interference with vitamin D metabolism	Children
150–200	Protoporphyrin elevation in red blood cells	Women and children
250–300	Protoporphyrin elevation in red blood cells	Adult males
400	Increased urinary ALA excretion	Children and adults
400	Anemia	Children
400	Coproporphyrin elevation	Adults and children
500	Anemia	Adults
500–600	Cognitive deficits	Children
500–600	Peripheral neuropathies	Adults and children
800–1,000	Encephalopathic symptoms	Children
1,000–1,200	Encephalopathic symptoms	Adults

different US cities between 1967 and 1970 revealed that about 29% had blood levels over 400 $\mu g\ L^{-1}$, and almost 9% had blood lead levels over 500 $\mu g\ L^{-1}$ [68]. A blood lead level of 500–800 $\mu g\ L^{-1}$ is considered to be the threshold level of classical lead poisoning [76]. At the present time in the United States ~250,000 children less than 6 years old have blood lead levels exceeding 100 $\mu g\ L^{-1}$ [77].

Now that leaded gasoline has been phased out, the principal sources of exposure to lead are deteriorating lead-based paint, lead-contaminated dust, and lead-contaminated residential soil [78]. Of these three, the historical use of lead-based paint has proven to be the most problematic. Lead was used for many years in paints as a pigment, and in oil paints lead naphthenate is used as a drying agent. In 1918, 40% of all painters were estimated to have symptoms of lead poisoning because of their contact with such paint [79]. The most serious problem associated with the use of lead in paints, however, has been the poisoning of small children who eat paint chips or teethe on windowsills or other surfaces that have been painted with lead-based paints. Dust from deteriorating paint may also contaminate windowsills, carpets, furniture, and toys and be ingested by children who touch the dust and subsequently put their fingers in their mouths. Between 1954 and 1967, a total of 2,038 children were treated for lead poisoning in New York City alone, and of these 128 died [79]. On January 1, 1973, the Food and Drug Administration (FDA) banned the use of paint containing over 0.5% lead on residential surfaces accessible to children [68], and effective February 27, 1978, paint containing over 0.06% lead was banned as hazardous for use in residences, schools, hospitals, parks, playgrounds, and public buildings under the aegis of the Consumer Product Safety Act. The only exceptions specifically written into the 1977 legislation were paints and coatings for motor vehicles and boats. The problem of poisoning from lead-based paints remains a serious public health problem, however, because many older buildings contain lead-based paints. A 1988 Public Health Service report revealed that 52% of American homes had layers of lead-based paint on their walls and woodwork

[74]. Today lead poisoning is the number one environmental health risk for children in the United States, and 80% of lead poisonings are caused by lead-based paint in homes and apartments built before 1978 [80].

Naively one might assume that simply painting over lead-based paint would solve the problem, but according to the EPA, that is "not enough" [78].

> To permanently remove lead hazards, you must hire a certified lead "abatement" contractor. Abatement (or permanent hazard elimination) methods include removing, sealing, or enclosing lead-based paint with special materials.... Always hire a person with special training for correcting lead problems – someone who knows how to do this work safely and has the proper equipment to clean up thoroughly [78].

Short of having the lead-based paint removed by a certified contractor, people who find themselves living in housing units painted with lead-based paint are advised to take the following steps [78]:

- If you rent, notify your landlord of peeling or chipping paint.
- Clean up paint chips immediately.
- Clean floors, window frames, windowsills, and other surfaces weekly. Use a mop, sponge, or paper towel with warm water and a general all-purpose cleaner or a cleaner made specifically for lead. Thoroughly rinse sponges and mop heads after cleaning dirty or dusty areas.
- Wash children's hands often, especially before they eat and before nap time and bedtime.
- Keep play areas clean. Wash bottles, pacifiers, toys, and stuffed animals regularly.
- Keep children from chewing windowsills or other painted surfaces.
- Clean or remove shoes before entering your home to avoid tracking in lead from soil.

Other Chemicals

Almost 90% of all poison exposures occur in the home, and after lead there is a rather long list of toxic chemicals to which people have been exposed, in most (~87%) cases unintentionally. Seventy-five percent of poison exposures involve ingestion of a toxic substance, and children less than 6 years old are the victims in roughly half the incidents, although they account for only 2% of the poison fatalities. In almost 40% of the cases involving children, cosmetics and personal care products, cleaning compounds, analgesics, and plants (e.g., holly berries and mistletoe berries) are the source of the toxin(s). Although adults account for a small percentage of poison exposures, they account for more than 75% of all fatalities related to poison exposure [81]. The following examples illustrate several facets of the poison exposure problem.

1. Antibacterials found in many cosmetic and personal care products: Triclosan is a chlorinated organic bactericide used in soaps, deodorants, toothpastes, shaving

creams, mouthwashes, and cleaning supplies and is infused in consumer products such as kitchen utensils, toys, bedding, socks, and trash bags [82]. Its use for 40 years has been regulated by three different federal agencies, the FDA, EPA, and Consumer Product Safety Commission. Remarkably, the FDA has been working on the development of rules for the use of triclosan for much of that time but has yet to promulgate any [83]. Japan and Canada, on the other hand, have banned triclosan in consumer products.

What is the problem? Concerns about the effects of triclosan on human health are related to several issues. First, several studies have linked triclosan to allergic reactions in children [84–87]. Second, triclosan reacts with chlorine in water to produce chloroform, which the EPA classifies as a probable human carcinogen, and smaller amounts of other chlorinated compounds that, in the presence of sunlight, are converted to dioxin [88–91], some forms of which are highly toxic and very potent endocrine disruptors. So widespread is the use of triclosan that it has been found in the urine of 75% of people tested by the CDC [83] and even in human breast milk [92, 93].

Given this information, one might wonder why the US government has not taken some action to restrict the use of triclosan, but there is another side to the story. In commenting on the presence of triclosan in breast milk, Dayan [92], for example, concludes, "There is no evidence to indicate that the presence of a miniscule amount of triclosan in breast milk presents a risk to babies," and while it is true that triclosan reacts with chlorine in tap water to produce chloroform, the concentrations of chloroform reported in the study by Rule et al. [94] were smaller than the amounts often present in chlorinated drinking water [82]. With respect to allergic reactions, Campbell and Zirwas [86] acknowledge that, "Several cases of contact dermatitis secondary to triclosan have been reported" but that, "Allergy to triclosan-based products is uncommon," and Schena et al. [85], in a study of patients with chronic eczema, concluded, "Triclosan is well tolerated and has a very low sensitizing potential even in high-risk patients affected by eczema."

There is no question that very large numbers of people (arguably almost everyone) in the United States are exposed to triclosan, but is there a health risk associated with exposure to this chemical? It seems that the jury is still out on this question. And would the health risk of bacterial infection be heightened if use of triclosan were banned? Remarkably, a study carried out by the University of Michigan concluded, "Washing hands with an antibacterial soap was no more effective in preventing infectious illness than plain soap. Moreover, antibacterial soaps at formulations sold to the public do not remove any more bacteria from the hands during washing than plain soaps" [95].

2. Dangerous mixtures of household chemicals: Without realizing the danger, people sometimes mix bleach with ammonia. The resulting chemical reactions can produce a real witches' brew of toxic gases. Bleach is just a dilute (typically 6%) aqueous solution of sodium hypochlorite ($NaOCl$). Ammonia (NH_3) is a gas, and when dissolved in water exists as a mixture of the dissolved gas and ammonium

ion (NH_4^+). When roughly equal amounts of bleach and ammonia are combined the following reaction occurs:

$$2NaOCl + 2NH_3 \rightarrow 2NaONH_3 + Cl_2$$

Chlorine gas, Cl_2, is a highly corrosive gas that can do serious damage to the lungs, nasal passages, and trachea when inhaled. It was used as a chemical weapon during World War I and by Nazi Germany in World War II. When more bleach is added than ammonia the following reaction occurs:

$$3NaOCl + NH_3 \rightarrow 3NaOH + NCl_3$$

NaOH is a very strong base and caustic to soft tissue. NCl_3, nitrogen trichloride, is highly toxic and sufficiently volatile that it is likely to explode in the face of the unsuspecting chemical mixer. And finally, if ammonia is added in excess of the amount of bleach, the following series of reactions may occur:

$$NH_3 + NaOCl \rightarrow +NaOH + NH_2Cl$$
$$NH_3 + NH_2Cl + NaOH \rightarrow N_2H_4 + NaCl + H_2O$$
$$2NH_2Cl + N_2H_4 \rightarrow 2NH_4Cl + N_2$$

The third reaction is very exothermic (releases a large amount of heat) and typically leads to an explosion.
Another common mistake is to mix bleach with either toilet bowl cleaners or vinegar. People presumably do this because they think mixing the two will produce a better cleaner and disinfectant. In both cases, one of the products is Cl_2, which, as already noted, is a very toxic gas. Mixing bleach and toilet bowl cleaner, which contains hydrochloric acid (HCl), leads to the following reaction:

$$NaOCl + HCl \rightarrow Cl_2 + NaOH$$

Mixing bleach with vinegar, which contains acetic acid (CH_3CO_2H), leads to the following reaction:

$$2NaOCl + 2CH_3CO_2H \rightarrow Cl_2 + 2H_2O + 2CH_3CO_2^- + 2Na^+$$

In either case, the result could be very painful and possibly fatal for the unwary mixer of chemicals.

Future Directions

Given the fact that most exposures to toxic chemicals are associated with lifestyle issues, it is provocative to ask what lies ahead. Clearly the prevalence of smoking has been on the decline in the Unites States, and if the trend apparent in Fig. 20.1 is extrapolated another 50 years, only a small percentage of adults will be smoking by that time. The risks associated with smoking are well established. The challenges to reducing the societal costs associated with smoking now lie in the realms of education and social science. How far the percentage of smokers can be reduced through education and public awareness efforts is unclear at this time.

The risks associated with alcohol consumption are also well defined, and because alcohol consumption is much more socially acceptable than smoking, it is unlikely that the costs associated with drinking will be substantially reduced in the years ahead.

Abuse of drugs is a major public health problem whose solution, if there is one, will lie in the realm of education and social science. Despite the best efforts of law enforcement agencies, it is apparent that if people want illegal drugs, they will get them, and if they want prescription drugs without a prescription, that can happen too. The solution will require reducing the desire/need for the artificial euphoria associated with use of drugs such as opioids, cocaine, and heroin. Drug abuse is a difficult and complex problem, and particularly in the case of opioids (Fig. 20.3) with no apparent resolution in sight.

Obesity is a serious problem in the United States, and awareness of the magnitude of the problem has been slow to emerge. Trends in recent years (Fig. 20.7) are not encouraging. Education and public awareness of the costs of obesity will help, but much of the underlying problem is the simple fact that people like to eat.

Lead has been historically and remains today a major environmental pollutant. The phasing out of leaded gasoline has been a very positive step, but now there is the legacy of the many homes and apartments that were painted with lead-based paint prior to 1978. Eventually these will be taken out of service, but the process has been slow because of the cost and hazards associated with renovating interiors painted with lead-based paint.

Numerous unintentional exposures to toxic chemicals occur in and around the home. Improved education and public awareness of the hazards of certain chemicals can help reduce the frequency of these exposures. The presence of potential toxins in consumer products is an issue that must be addressed by government agencies such as the FDA, EPA, and Consumer Product Safety Commission. In some cases a judgment must be made between the benefits and the risks. In the United States, 35,000–40,000 people die in automobile accidents each year, but there is no movement to take cars off the roads. While the benefits and costs associated with automobile use are arguably well defined, such is not the case with triclosan, for which both the benefits and risks are in dispute.

Bibliography

Primary Literature

1. WHO (1987) Principles for the safety assessment of food additives and contaminants in food. WHO Press, Geneva
2. CDC (2009) Glossary of terms. http://www.atsdr.cdc.gov/glossary.html. Accessed 16 Mar, 2011
3. Hoffmann D, Hoffmann I (2001) The changing cigarette: chemical studies and bioassays. In: Burns DM, Benowitz NL (eds) Risks associated with smoking cigarettes with low machine-measured yields of tar and nicotine. U.S. Department of Health and Human Services, Public Health Service, National Institutes of Health, National Cancer Institute, Washington, DC, pp 159–191
4. Hoffmann D, Hoffmann I (1998) Tobacco smoke components. Beitr Tabakforsch Int 18:49–52
5. Hoffmann D, Hoffmann I, El-Bayoumy K (2001) The less harmful cigarette: a controversial issue. A tribute to Ernst L. Wynder. Chem Res Toxicol 14(7):767–790
6. Baker RR, da Silva JRP, Smith G (2004) The effect of tobacco ingredients on smoke chemistry. Part I: Flavourings and additives. Food Chem Toxicol 42:S3–S37
7. Staff (1990) Proposed identification of inorganic arsenic as a toxic air contaminant. California Air Resources Board and Department of Health Services, Sacramento
8. Korte JE, Hertz-Picciotto I, Schulz MR, Ball LM, Duell EJ (2000) The contribution of benzene to smoking-induced leukemia. Environ Health Perspect 108(4):333–339
9. Staff (2004) The health consequences of smoking: a report of the surgeon general. Centers for Disease Control and Prevention, Atlanta, GA
10. Denissenko MF, Pao A, Tang MS, Pfeifer GP (1996) Preferential formation of benzo a pyrene adducts at lung cancer mutational hotspots in P53. Science 274(5286):430–432
11. Kaiserman MJ, Rickert WS (1992) Carcinogens in tobacco-smoke – benzo a pyrene from Canadian cigarettes and cigarette tobacco. Am J Public Health 82(7):1023–1026
12. Greenblatt MS, Bennett WP, Hollstein M, Harris CC (1994) Mutations in the P53 tumor-suppressor gene – clues to cancer etiology and molecular pathogenesis. Cancer Res 54(18):4855–4878
13. Lane DP (1992) Cancer – P53, guardian of the genome. Nature 358(6381):15–16
14. EPA (1980) Ambient water quality for cadmium. EPA 440/5-80-025, Washington, DC
15. Kasuya M et al (1992) Water pollution by cadmium and the onset of Itai-Itai disease. Water Sci Technol 25(11):149–156
16. Olsson IM et al (2002) Cadmium in blood and urine – Impact of sex, age, dietary intake, iron status, and former smoking – Association of renal effects. Environ Health Perspect 110(12):1185–1190
17. Paschal DC et al (2000) Exposure of the US population aged 6 years and older to cadmium: 1988–1994. Arch Environ Contam Toxicol 38(3):377–383
18. Bokhoven C, Niessen HJ (1961) Amounts of oxides of nitrogen and carbon monoxide In cigarette smoke, with and without inhalation. Nature 192(480):458
19. Kaufman DW, Helmrich SP, Rosenberg L, Miettinen OS, Shapiro S (1983) Nicotine and carbon-monoxide content of cigarette-smoke and the risk of myocardial-infarction in young men. N Engl J Med 308(8):409–413
20. Counts ME, Hsu FS, Laffoon SW, Dwyer RW, Cox RH (2004) Mainstream smoke constituent yields and predicting relationships from a worldwide market sample of cigarette brands: ISO smoking conditions. Regul Toxicol Pharmacol 39(2):111–134
21. Bateman DN (2003) Carbon monoxide. Medicine 31(10):41–42
22. Townsend CL, Maynard RL (2002) Effects on health of prolonged exposure to low concentrations of carbon monoxide. Occup Environ Med 59(10):708–711
23. Baker RR (2006) The generation of formaldehyde in cigarettes – overview and recent experiments. Food Chem Toxicol 44(11):1799–1822

24. Baker RR, Coburn S, Liu C (2006) The pyrolytic formation of formaldehyde from sugars and tobacco. J Anal Appl Pyrol 77(1):12–21
25. Li S, Banyasz JL, Parrish ME, Lyons-Hart J, Shafer KH (2002) Formaldehyde in the gas phase of mainstream cigarette smoke. J Anal Appl Pyrol 65(2):137–145
26. Albert RE et al (1982) Gaseous formaldehyde and hydrogen-chloride induction of nasal cancer in the rat. J Natl Cancer Inst 68(4):597–603
27. Blair A, Saracci R, Stewart PA, Hayes RB, Shy C (1990) Epidemiologic evidence on the relationship between formaldehyde exposure and cancer. Scand J Work Environ Health 16(6): 381–393
28. Casanova M, Morgan KT, Gross EA, Moss OR, Heck HD (1994) DNA-protein cross-links and cell replication at specific sites in the nose of F344 rats exposed subchronically to formaldehyde. Fundam Appl Toxicol 23(4):525–536
29. Olsen JH et al (1984) Occupational formaldehyde exposure and increased nasal cancer risk in man. Int J Cancer 34(5):639–644
30. Horvath EP, Anderson H, Pierce WE, Hanrahan L, Wendlick JD (1988) Effects of formaldehyde on the mucous-membranes and lungs – a study of an industrial-population. JAMA 259(5):701–707
31. Liteplo RG, Meek ME (2003) Inhaled formaldehyde: exposure estimation, hazard characterization, and exposure-response analysis. J Toxicol Environ Health B Crit Rev 6(1):85–114
32. Rodgman A, Perfetti TA (2008) The chemical components of tobacco and tobacco smoke. CRC Press, Boca Raton, 1840 pp
33. Hathaway GJ, Proctor NH, Hughes JP, Fischman ML (1991) Proctor and Hughes' chemical hazards of the workplace. Van Nostrand Reinhold, New York, NY
34. Argentin G, Cicchetti R (2004) Genotoxic and antiapoptotic effect of nicotine on human gingival fibroblasts. Toxicol Sci 79(1):75–81
35. Campain JA (2004) Nicotine: potentially a multifunctional carcinogen? Toxicol Sci 79(1):1–3
36. Hoffmann D, Brunnemann KD, Prokopczyk B, Djordjevic MV (1994) Tobacco-specific N-nitrosamines and areca-derived N-nitrosamines – chemistry, biochemistry, carcinogenicity, and relevance to humans. J Toxicol Environ Health 41(1):1–52
37. Pfeifer GP et al (2002) Tobacco smoke carcinogens, DNA damage and p53 mutations in smoking-associated cancers. Oncogene 21(48):7435–7451
38. Hoffmann D, Hoffmann I (1997) The changing cigarette, 1950–1995. J Toxicol Environ Health 50(4):307–364
39. CDC (2011) Toll of tobacco in the United States. University of Pennsylvania Center for Interdisciplinary Research on Nicotine Addiction, Philadelphia, PA. http://www.med.upenn.edu/cirna/pdf/USA_Figures.pdf. Accessed 26 Apr 2011
40. CDC (2011) Annual smoking-attributable mortality, years of potential life lost, and economic costs – United States, 1995–1999. http://www.cdc.gov/tobacco/data_statistics/mmwrs/byyear/2002/mm5114a2/highlights.htm. Accessed 6 Mar 2011
41. CDC (2011) Economic facts about U.S. tobacco production and use. http://www.cdc.gov/tobacco/data_statistics/fact_sheets/economics/econ_facts/#costs. Accessed 7 Mar 2011
42. CDC (2011) Trends in current cigarette smoking among high school students and adults, united states, 1965–2009. Centers for Disease Control and Prevention, Atlanta, GA. http://www.cdc.gov/tobacco/data_statistics/tables/trends/cig_smoking/index.htm. Accessed 26 Apr 2011
43. Courtwright DT (2001) Forces of habit: drugs and the making of the modern world. Harvard University Press, Cambridge, MA
44. Lord JL et al (1987) Coming to grips with alcoholism. US News World Rep 103(22):57–62
45. Anonymous (2011) Alcohol abuse statistics. http://www.about-alcohol-abuse.com/Alcohol_Abuse_Statistics.html. Accessed 8 Mar 2011
46. Harford TC (1992) Family history of alcoholism in the United-States – prevalence and demographic characteristics. Br J Addict 87(6):931–935
47. Anonymous (2010) The party's over. Nature 468(7323):475

48. Anonymous (2011) Alcoholism. http://www.addiction-help-line.com/alcohol_addiction.html. Accessed 9 Mar 2011
49. Anthenelli RM, Schuckit MA (1993) Affective and anxiety disorders and alcohol and drug dependence: diagnosis and treatment. J Addict Disord 12:73–87
50. Goodwin DW (1985) Alcoholism and genetics – the sins of the fathers. Arch Gen Psychiatry 42(2):171–174
51. Paulozzi LJ, Ryan GW (2006) Opioid analgesics and rates of fatal drug poisoning in the United States. Am J Prev Med 31(6):506–511
52. CDC (2011) CDC WONDER. http://wonder.cdc.gov/. Accessed 11 Mar 2011
53. Paulozzi LJ (2008) Trends in unintentional drug overdose deaths. http://www.hhs.gov/asl/testify/2008/03/t20080312b.html. Accessed 11 Mar 2011
54. Warner ME, Chen LH, Makuc DM (2009) Increase in fatal poisonings involving opioid analgesics in the United States, 1999–2006. Centers for Disease Control and Prevention, Atlanta, GA
55. CDC (2011) Unintentional drug poisoning in the United States. http://www.cdc.gov/HomeandRecreationalSafety/pdf/poison-issue-brief.pdf. Accessed 11 Mar 2011
56. Anonymous (2011) Opioids. http://www.drug-addiction.com/opioids.htm. Accessed 11 Mar 2011
57. NIDA (2010) Cocaine: abuse and addiction. U.S. Department of Health and Human Services, National Institutes of Health, Research Triangle Park, NC
58. Flynn PM, Joe GW, Broome KM, Simpson DD, Brown BS (2003) Looking back on cocaine dependence: reasons for recovery. Am J Addict 12(5):398–411
59. Johnston LD, O'Malley PM, Bachman JG, Schulenberg JE (2010) Marijuana use is rising; ecstasy use is beginning to rise; and alcohol use is declining among U.S. teens. http://www.monitoringthefuture.org. Accessed 11 Mar 2011
60. NIDA (2005) Heroin abuse and addiction. U.S. Department of Health and Human Services, National Institutes of Health, Research Triangle Park, NC
61. Anoonymous (2011) Morphine. http://en.wikipedia.org/wiki/Morphine. Accessed 12 Mar 2011
62. Anonymous (2011) Heroin. http://en.wikipedia.org/wiki/Heroin. Accessed 12 Mar 2011
63. CDC (2002) Drug-associated HIV transmission continues in the United States. Centers for Disease Control and Prevention, Atlanta, GA
64. Anonymous (2004) Results from the 2003 national survey on drug use and health: national findings. Office of Applied Studies, Research Triangle Park, NC
65. NCHS (2008) Prevalence of overweight, obesity and extreme obesity among adults: United States, trends 1976-80 through 2005–2006. National Center for Health Statistics, Hyattsville, MD
66. NCHS (2004) Mean body weight, height, and body mass index, United States 1960–2002. National Center for Health Statistics, Hyattsville, MD
67. NHLBI (1998) Clinical guidelines on the identification, evaluation, and treatment of overweight and obesity in adults. National Heart, Lung, and Blood Institute, Research Triangle Park, NC
68. Waldron HA, Stöfen D (1974) Sub-clinical lead poisoning. Academic, New York
69. Gilfillan SC (1965) Lead poisoning and the fall of Rome. J Occup Environ Med 7(2):53–60
70. IPIECA (2010) Leaded gasoline phase-out. http://www.ipieca.org/topic/operations-and-fuels/leaded-gasoline-phase-out. Accessed 12 Mar 2011
71. Mcdonald J (2007) China's lead problems go beyond toys. http://www.usatoday.com/money/economy/2007-08-15-3066711946_x.htm. Accessed 12 Mar 2011
72. Annest JL (1983) Trends in the blood lead levels of the U.S. population. In: Rutter M, Jones RR (eds) Lead versus health. John Wiley and Sons, Chichester and New York, pp 33–58
73. EPA (1980) Ambient water quality for lead. EPA-440/4-80-057. Environmental Protection Agency, Washington, DC
74. Blackman A (1991) Controlling a childhood menace. Time 137(8):68–69
75. CDC (1991) Preventing lead poisoning in young children. Centers for Disease Control and Prevention, Atlanta, GA

76. Patterson CC (1965) Contaminated and natural lead environments and man. Arch Environ Health 11:344–360
77. CDC (2011) Lead. http://www.cdc.gov/nceh/lead/. Accessed 13 Mar 2011
78. EPA (2011) Lead in paint, dust, and soil. http://www.epa.gov/lead/. Accessed 12 Mar 2011
79. Craig PP, Berlin E (1971) The air of poverty. Environment 13(5):2–9
80. Anonymous (2011) Frequently asked questions. http://www.leadpro.com/faq.html. Accessed 12 Mar 2011
81. Litovitz TL et al (2001) 2000 annual report of the American association of poison control centers toxic exposure surveillance system. Am J Emerg Med 19(5):337–395
82. Anonymous (2011) Triclosan. http://en.wikipedia.org/wiki/Triclosan. Accessed 13 Mar 2011
83. Layton L (2010) FDA says studies on triclosan, used in sanitizers and soaps, raise concerns. http://www.washingtonpost.com/wp-dyn/content/article/2010/04/07/AR2010040704621.html. Accessed 13 Mar 2011
84. Wong CSM, Beck MH (2001) Allergic contact dermatitis from triclosan in antibacterial handwashes. Contact Dermat 45(5):307–307
85. Schena D, Papagrigoraki A, Girolomoni G (2008) Sensitizing potential of triclosan and triclosan-based skin care products in patients with chronic eczema. Dermatol Ther 21: S35–S38
86. Campbell L, Zirwas MJ (2006) Triclosan. Dermatitis 17(4):204–207
87. Bhutani T, Jacob SE (2009) Triclosan: a potential allergen in suture-line allergic contact dermatitis. Dermatol Surg 35(5):888–889
88. Mezcua M et al (2004) Evidence of 2,7/2,8-dibenzodichloro-p-dioxin as a photodegradation product of triclosan in water and wastewater samples. Anal Chim Acta 524(1–2):241–247
89. Latch DE et al (2005) Aqueous photochemistry of triclosan: Formation of 2,4-dichlorophenol, 2,8-dichlorodibenzo-p-dioxin, and oligomerization products. Environ Toxicol Chem 24(3):517–525
90. Latch DE, Packer JL, Arnold WA, McNeill K (2003) Photochemical conversion of triclosan to 2,8-dichlorodibenzo-p-dioxin in aqueous solution. J Photochem Photobiol A Chem 158(1): 63–66
91. Aranami K, Readman JW (2007) Photolytic degradation of triclosan in freshwater and seawater. Chemosphere 66(6):1052–1056
92. Dayan AD (2007) Risk assessment of triclosan Irgasan (R) in human breast milk. Food Chem Toxicol 45(1):125–129
93. Allmyr M, Adolfsson-Erici M, McLachlan MS, Sandborgh-Englund G (2006) Triclosan in plasma and milk from Swedish nursing mothers and their exposure via personal care products. Sci Total Environ 372(1):87–93
94. Rule KL, Ebbett VR, Vikesland PJ (2005) Formation of chloroform and chlorinated organics by free-chlorine-mediated oxidation of triclosan. Environ Sci Technol 39(9):3176–3185
95. Anonymous (2007) Plain soap as effective as antibacterial but without the risk. http://www.physorg.com/news106418144.html. Accessed 14 Mar 2011

Books and Reviews

Higgins TE, Sachdev JA, Engleman SA (2010) Toxic chemicals: risk prevention through use reduction. CRC Press, Boca Raton

Nielsen E, Ostergaard G, Larsen JC (2008) Toxicological risk assessment of chemicals: a practical guide. Informa Healthcare, London

Penningroth S (2010) Essentials of toxic chemical risk: science and society. CRC Press, Boca Raton

Richardson ML (ed) (1996) Risk reduction: chemicals and energy into the 21st century. Taylor and Francis, London

Chapter 21
Ultraviolet Radiation: Distribution and Variability

Julian Gröbner

Glossary

Effective surface albedo	The effective surface albedo represents the reflectivity of a homogeneous surface which produces the same downwelling UV irradiance as in the presence of an inhomogeneous surface.
Erythemal UV radiation	UV radiation weighted with the nominal action spectrum for the spectral sensitivity of Human skin to sunburn (erythema).
Solar zenith angle	Angle between the sun and the normal to the Earth's surface. An SZA of 0° corresponds to an overhead sun (solar elevation 90°), while a SZA of 90° corresponds to the sun at the horizon.
Spectroradiometer	Instrument using a grating or prism to split the incoming radiation in its spectral components so as to measure the intensity of the solar spectrum at individual wavelengths.
Stratosphere	Part of the atmosphere between approximately 10 and 40 km which is characterized by increasing atmospheric temperatures with height due to the photo dissociation of ozone molecules by solar UV radiation.
Irradiance	Radiation flux through a unit surface area.

This chapter was originally published as part of the Encyclopedia of Sustainability Science and Technology edited by Robert A. Meyers. DOI:10.1007/978-1-4419-0851-3

J. Gröbner (✉)
European Ultraviolet Calibration Center, Physikalisch-Meteorologisches Observatorium Davos, World Radiation Center, Dorfstrasse 33, 7260 Davos Dorf, Switzerland
e-mail: julian.groebner@pmodwrc.ch

E.A. Laws (ed.), *Environmental Toxicology: Selected Entries from the Encyclopedia of Sustainability Science and Technology*, DOI 10.1007/978-1-4614-5764-0_21,

Radiation dose	Amount of radiation reaching a subject during a defined time interval. A daily dose corresponds to the total radiation reaching the surface during one whole day.
UV Radiation	Radiation in the wavelength range 100–400 nm. It is subdivided into three regions called UV-C (100–280 nm), UV-B (280–315 nm), and UV-A (315–400 nm).

Definition of the Subject and Its Importance

Solar UV radiation is the most energetic radiation that is able to penetrate the atmosphere and reach the Earth's surface. Due to the very low intensities of the radiation in this wavelength range, its investigation started comparably late with respect to the other spectral regions of the solar spectrum. Dorno was the first to investigate systematically the seasonal and spectral variability of solar UV radiation at Davos, Switzerland. Starting in 1907, his measurements allowed to determine the distinct differences of radiation in this wavelength range, which were later explained by the absorption of atmospheric ozone (mainly located in the stratosphere). Using a specially designed spectroradiometer, he discovered the relationship between the cut-on of the solar spectrum in the UV-B and the path through the atmosphere. In view of his achievements, UV radiation at the time was also called Dorno radiation. In the 1960s, Paul Bener, also from Davos, continued the investigations initiated by Dorno, by building specially designed spectroradiometers that were unique in his time. Systematic measurements under different atmospheric and environmental conditions provided for the first time a benchmark dataset to use as reference for validating radiative transfer models (RTMs), which were emerging at the time. Thus, the dependence of solar UV radiation on clouds, aerosols, surface reflectivity, and altitude were investigated systematically. Moreover, polarization measurements of the sky radiance were obtained at that time, and are still among the most precise measurements available. Recently, the interest for UV radiation has been linked to its biological effects on humans and on the biosphere in general. The discovery of decreasing ozone levels due to anthropogenic chemicals released in the atmosphere in the early 1980 s has led to intense research activity in the field of solar UV measurements to determine the effect of declining ozone levels on UV radiation and its implications for the environment.

In this entry, the spatial distribution of solar UV radiation at the surface will be discussed, as well as the most important parameters responsible for variations in intensity of the surface solar UV radiation.

Introduction

Solar ultraviolet radiation falling on the Earth's surface originates from the sun. On its passage through the atmosphere, the radiation is scattered and absorbed depending on a variety of factors. Even though the mechanisms are qualitatively similar to what occurs at longer wavelengths, several important differences exist: Ultraviolet radiation is strongly absorbed by atmospheric ozone, located high up in the atmosphere. Furthermore, ultraviolet radiation is strongly scattered by air molecules.

Ultraviolet radiation is classified into three wavelength regions: Radiation at wavelengths just shorter than visible light is called UV-A and extends from 315 to 400 nm. The radiation at shorter wavelengths is more energetic and is classified as UV-B radiation, from 280 nm to 315 nm. At even shorter wavelengths between 100 and 280 nm, the radiation is termed UV-C and is fully absorbed by atmospheric trace gases before reaching the Earth's surface [1]. Due to the spectral characteristics of the ozone absorption cross sections, UV radiation shows a dramatic decrease in intensity – over several orders of magnitude – between approximately 330 and 290 nm. This short wavelength cutoff at around 290 nm is therefore mainly influenced by ozone absorption.

Solar ultraviolet radiation contains only a small part of the total energy of the solar radiation reaching the Earth's surface. Nevertheless, radiation at these short wavelengths is energetic enough to affect biological tissue, such as breaking DNA bonds [2]. Ultraviolet radiation has detrimental effects on plants, human beings, and underwater life forms such as plankton, fish, larvae, and algae.

Information on the level of ultraviolet radiation reaching the Earth's surface can be obtained from ground-based measurements, radiative transfer modeling, and more recently can be inferred from measurements by satellite sensors.

The most precise and accurate method is to measure the incoming UV radiation from the surface using spectroradiometers [3]. These instruments measure the intensity of radiation in very narrow wavelength bands and can thereby reconstruct the solar spectrum with high fidelity. Since the UV spectrum shows such important intensity changes with wavelength, the knowledge of the solar spectrum gives the most complete knowledge on the surface UV radiation.

Narrowband multifilter radiometers measure the solar spectrum in several narrow UV wavelength bands with resolutions of the order of 2–10 nm. This type of instrument is more compact and robust than a spectroradiometer, but requires a spectroradiometer for its calibration [4].

The most common type of instrument measuring solar UV radiation is a broadband filter radiometer. This instrument uses filters to be sensitive to a broad wavelength interval, and returns a single value representing the solar UV spectrum weighted with the instrument spectral sensitivity. Various types of such broadband instruments exist, depending on what spectral range they are sensitive: UV-B, UV-A, or the total UV range. While these instruments are easy to operate and require very little maintenance, their calibration is complex and necessitates experienced and well-equipped laboratories [5]. Furthermore, due to the gradual deterioration of their filters, these instruments need frequent recalibrations if long-term measurements are required.

Radiative transfer models (RTMs) solve the radiation equation to retrieve the surface UV radiation from a prescribed atmospheric state [6, 7]. While current RTMs solve the radiation equation with sufficient precision, the results are practically limited by the knowledge of the model parameters that are input to the model. These model parameters (essentially the composition of the atmosphere and the vertical distribution of their components) are usually not very well known and induce corresponding large uncertainties in the computed UV radiation [8].

Knowledge of the spatial distribution of UV radiation is readily obtained from remote-sensing instruments located on satellite platforms. These instruments determine UV radiation from the backscattered solar radiation that is reflected by the atmospheric layers and the Earth's surface [9]. The surface UV radiation is then obtained from elaborate model calculations. Due to the difficulty inherent in such an approach, the accuracy of the retrieved UV radiation, although determined over large areas with high spatial resolution, has much larger uncertainties than the corresponding surface point-like measurements. For example, a typical problem faced by satellite sensors is the distinction between snow and clouds, which both have a high reflectivity as seen from space. In the first case, surface UV radiation is enhanced, whereas in the second case it is strongly reduced, depending on cloud type and thickness [10]. Additionally, comparisons between surface UV measurements and different satellite sensors have shown that the radiation absorbed by tropospheric aerosols is not well taken into account by satellite sensors and retrieval algorithms. Thus, UV radiation in polluted areas can be overestimated by up to 30% [11, 12].

Factors Affecting Surface Ultraviolet Radiation

The map shown in Fig. 21.1 displays the daily erythemal UV dose reaching the Earth's surface, averaged over the whole year 2005. The map was obtained from measurements by the Ozone Monitoring Instrument (OMI) [13]. Each pixel has a size of $1 \times 1°$. As can be seen in the figure, the highest UV radiation is found close to the equator, where the solar elevation is the highest. In addition, the Andes and the Himalayas mountain ranges are clearly visible due to the much higher UV radiation with respect to the surrounding areas. Less obvious is the reduced UV radiation over large parts of the Amazonian rainforest, due to the predominant cloud cover over this area compared to the Atlantic and Pacific Ocean at the same latitude.

In what follows, the parameters affecting surface UV radiation will be described in detail.

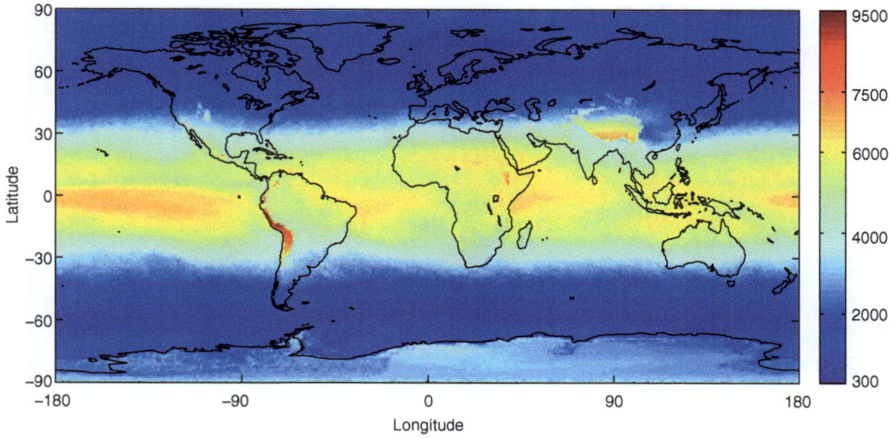

Fig. 21.1 Average daily dose of surface solar erythemal UV radiation in J/m^2 from the OMI Satellite for the year 2005

Earth to Sun Distance

Due to the point-like size of the sun as seen from Earth, the radiation intensity coming from it decreases as the inverse square of the distance between it and the Earth. This variability is of geometric nature and is therefore independent of wavelength, that is, radiation at all wavelengths is affected equally. Since the Earth path around the Sun follows an ellipse in which the sun is at one focus, slight changes in the distance during one annual rotation are evident. The Earth is closest to the sun in early January and farthest from it in early July. The relative difference between the minimum and maximum intensity is 6.8%. This radiation variability has consequences on the geographical distribution of UV radiation since the maximum occurs during the winter in the Northern hemisphere and correspondingly in (austral) summer in the Southern hemisphere and vice versa. Thus, purely from geometrical considerations, the radiation intensity is higher in summer in the Southern hemisphere than in the Northern hemisphere.

Solar Zenith Angle

The largest variation of surface UV radiation is due to the elevation of the sun above the horizon. The dependence of solar UV radiation on solar zenith angle (SZA) is responsible for the diurnal variation of cloud-free solar UV radiation, which normally has a maximum at local noon. The solar zenith angle is also responsible for the latitudinal dependence of surface UV radiation, as shown in Fig. 21.1. The overall latitudinal gradient dominates the UV variability on a global scale.

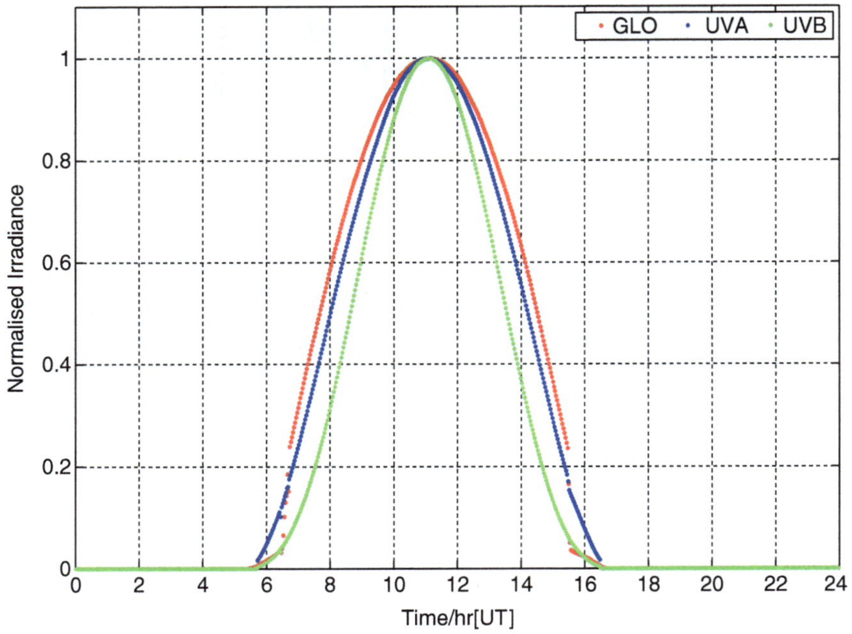

Fig. 21.2 Typical diurnal variation of global solar radiation (0.3–3 μm), red curve, UV-A radiation, blue curve, and UV-B radiation, green curve, during a clear day at Davos. The discontinuity in global radiation around both 7 and 16 UT is due to surrounding mountains blocking the path of the direct radiation from the sun

This pronounced variability is much higher than at longer wavelengths and is due to the absorption of UV radiation by atmospheric ozone, as explained below.

When the solar zenith angle is low, that is, the sun is high in the sky, the path through the atmosphere is short, so that both the absorption and scattering processes are minimized with respect to high SZA when the path through the atmosphere is much larger. This dependence on SZA is mainly due to the atmospheric ozone contained in the atmosphere, since it strongly absorbs UV radiation on its passage through the atmosphere. As is shown in Fig. 21.2, the diurnal shape of UV-B radiation is much steeper compared with the variation at longer wavelengths, which are not affected by atmospheric ozone.

Clouds

After SZA, clouds represent the most important factor responsible for surface UV variability [14]. Due to their high temporal and spatial variability, clouds are the main reason why it is so difficult to accurately predict UV radiation. Surface UV radiation can be attenuated by up to 99% under certain specific thick cloud types, such as cumulonimbus clouds. In the presence of scattered cumulus clouds, UV

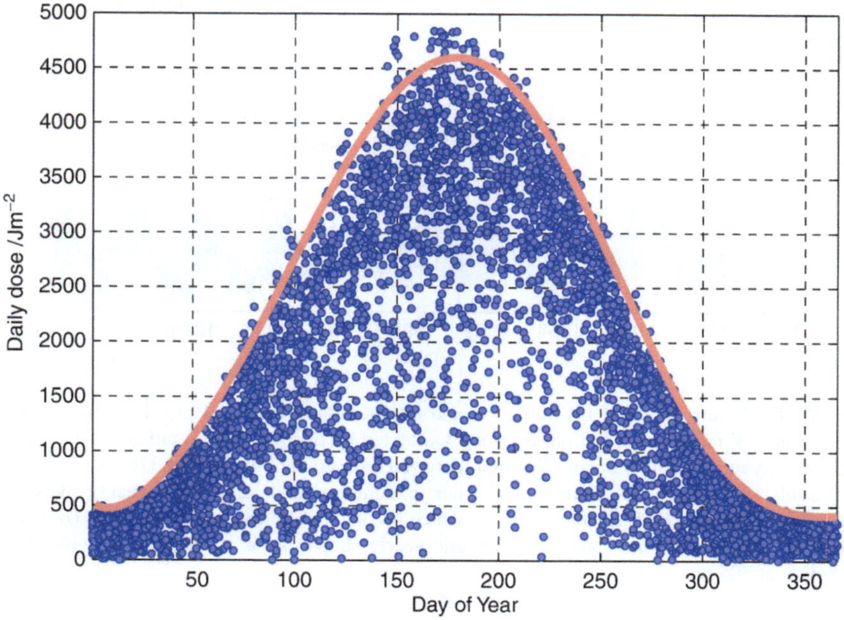

Fig. 21.3 Daily erythemally weighted UV radiation dose at Ispra, Italy for the years 1992–2004. Each point represents the radiation dose reaching the Earth's surface during a single day. The red curve represents the maximum radiation dose expected on a cloud-free day. The large scatter around this curve is due to the effect of clouds, which can reduce UV radiation by up to 99%. Instances of UV radiation higher than the theoretical cloud-free maximum are due to enhanced UV radiation caused by the lensing (or funnel) effect from the sides of cumulus clouds. This effect can locally and temporarily increase UV radiation by up to 20%

radiation is frequently enhanced by the lensing effect from tall adjacent vertical cloud sides [15]. This enhancement can be up to 20% and occurs usually only during short periods of time for particular solar and cloud situations. Figure 21.3 displays the UV radiation measured at Ispra, Italy between 1992 and 2004. The red curve represents the radiation level possible on a cloud-free day, while the individual points are actual measurements. The large scatter is predominantly due to attenuation by clouds, while the overall yearly variability is due to seasonal changes in solar elevation.

Molecular Scattering

Air molecules in the atmosphere scatter incoming radiation according to its wavelength. As shown by Lord Rayleigh in 1871, the scattering efficiency scales with the inverse fourth power of wavelength ($1/\lambda^4$), thereby scattering short wavelengths much more than longer ones [16]. This relationship, for example, leads to the blue

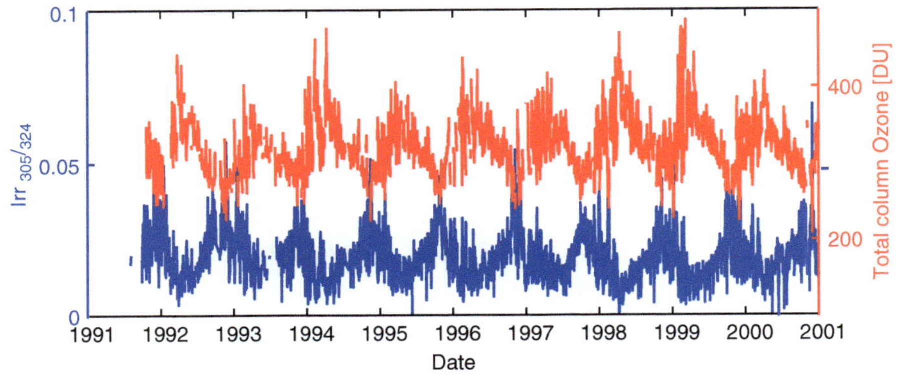

Fig. 21.4 Measurements of UV radiation and total column ozone at Ispra, Italy from 1991 to 2001. The UV radiation at 305 nm is strongly affected by ozone, whereas radiation at 324 nm is only weakly attenuated by ozone. The ratio between radiation at 305 nm and 324 nm (*blue curve*) is largely insensitive to clouds and other parameters that are only weakly depending on wavelength, thus highlighting the dependence on ozone. The total column ozone (*red curve*) measured at the same site shows a systematic yearly shape that is anti-correlated with the UV radiation at 305 nm

color of the sky in the visible radiation spectrum. For UV wavelengths, the effect is even stronger and is the reason why a large fraction of the incoming solar UV radiation is scattered and reaches the Earth as diffuse radiation. Indeed, even at low SZA, as much as half of the surface UV radiation flux is scattered solar radiation.

Ozone

The inverse correlation between UV Radiation and atmospheric ozone is well documented, and has been thoroughly discussed in the popular and scientific literature. As has been shown, a 1% decrease in total atmospheric ozone leads to an approximately 1% increase in UV-B radiation [17]. The relationship between UV radiation and ozone is strongly dependent on wavelength because the efficiency of ozone in absorbing UV radiation increases exponentially with decreasing wavelength.

Due to this inverse relationship, a global decrease in total column ozone would raise UV levels to a level that would become deleterious to many life forms living on the Earth.

The inverse correlation between UV radiation and ozone is generally masked by the much larger variability due to SZA and clouds. In Fig. 21.4, the correlation between UV-B radiation and ozone is shown by normalizing the irradiance at 305 nm by that at 324 nm. The resulting ratio remains very sensitive to ozone, while becoming insensitive to parameters such as clouds or aerosols, which both affect radiation at 305 and 324 nm nearly equally.

Stratospheric Ozone Hole

A dramatic ozone decrease has been occurring in the spring period over the Antarctic continent since the early 1980s [18]. Man-made chemical species accumulate in the winter period over the Antarctic, due to atmospheric conditions specific to the Antarctic continent during winter. In the presence of sunlight, these chemical species react with ozone molecules in the stratosphere and destroy a large fraction of the atmospheric ozone. As much as two thirds of the total ozone contained in the atmosphere can thus be destroyed, which leads in turn to important enhancements in the UV radiation reaching the Earth's surface.

Due to the dynamical nature of the polar vortex, ozone-poor air occasionally reaches also Punta Arenas and Ushuaia at the southern extreme of the South American continent. For instance, during periods of significant ozone reductions, biologically effective UV radiation at Punta Arenas in Chile were more than a factor of 2 higher than during normal conditions [19].

Aerosols

Aerosols in the atmosphere both scatter and absorb radiation. The amount and nature of these processes depend directly on the microphysical properties of the aerosols. Due to the large spatial and temporal variability in aerosols created by anthropogenic or natural processes, the quantification of these processes remains very challenging.

Two main parameters are used to quantify the effect of aerosols on UV radiation. The amount of aerosols, termed the aerosol optical depth (AOD) is a measure of the total aerosol load of the atmosphere. Incoming direct solar UV radiation is attenuated due to scattering and absorbing processes, which are a direct function of AOD. The relative efficiency of aerosols in either scattering or absorbing radiation is described by a second parameter, called the single-scattering albedo (SSA). This parameter can be obtained from microphysical properties of the aerosols. In practice, SSA is very difficult to quantify for a particular atmospheric situation due to the large variability of the aerosol composition. UV radiation can be attenuated by 30% or more in presence of heavy aerosol loads [20].

Surface Reflectivity

Radiation reflected from the surface is backscattered by the atmosphere and thus increases diffuse radiation. This enhancement of diffuse radiation occurs from the scattering that occurs on the molecules and aerosols present in the lower atmospheric layers. Due to the large molecular scattering effect in the UV wavelength

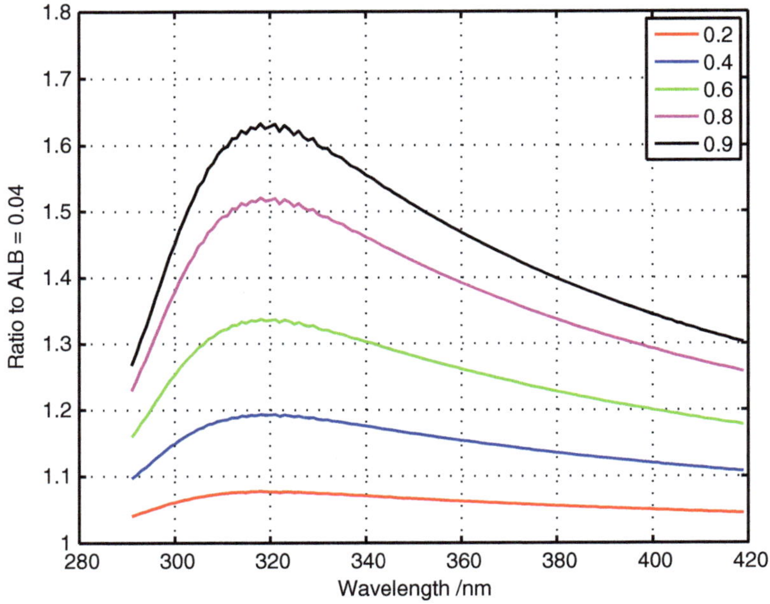

Fig. 21.5 Enhanced UV radiation for different surface albedo values with respect to the case with a UV albedo of 0.04. The gradual increase with decreasing wavelength is due to the increasing scattering efficiency of air molecules, while the decrease below about 320 nm is caused by tropospheric ozone absorption

range, a significant fraction of the reflected radiation is scattered back to the Earth surface, thereby enhancing the downwelling solar UV radiation. As can be seen in Fig. 21.5, this enhancement increases toward shorter wavelengths as is expected from the increased molecular scattering efficiency.

The largest enhancement occurs in presence of snow, which has a reflectivity of up to 90% [21]. Studies have shown that in regions with a complete snow cover UV radiation can be enhanced by up to 50% comparatively to a snow-free situation [22]. Radiative transfer models are generally used to infer the effect of surface albedo on UV radiation. As is shown in Fig. 21.4, the enhancement due to surface albedo has a specific spectral signature, which is mainly due to the interplay between two parameters: At wavelengths above $\approx$ 330 nm, where ozone absorption is negligible, the enhancement is due to Rayleigh scattering. Around 320 nm, a distinct maximum is found followed by a gradual decrease toward shorter wavelengths, which is due to the increasing effectiveness of tropospheric ozone in absorbing UV radiation. Since the reflected radiation is scattered several times before reaching the surface, the likelihood of being absorbed by tropospheric ozone is enhanced.

The enhancement of surface UV radiation due to reflected radiation from the surface that is backscattered to the surface is influenced by the surrounding area around the measurement site. Indeed, model calculations demonstrate that radiation

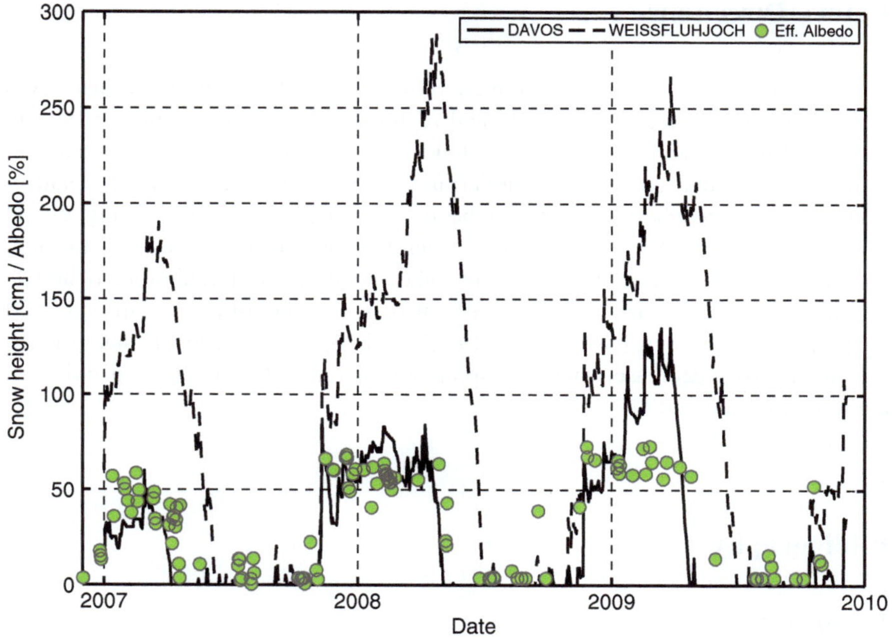

Fig. 21.6 Effective surface albedo in percent, as determined from UV measurements during cloud-free periods at Davos (elevation 1,590 m). The black and dotted lines represent the actual snow height measured in Davos and Weissfluhjoch (elevation 2,540 m, horizontal distance from Davos, 2 km), respectively. During winter, the effective albedo is as high as 70%, enhancing UV radiation by up to 30% with respect to snow-free conditions. During summer, the effective albedo drops to values close to 2–3%

reflected as far away as 30 km can influence the locally incident radiation [23, 24]. State-of-the art three-dimensional radiative transfer models can calculate the actual radiation for an arbitrary surface environment composed of a complex patchwork of albedo conditions, and be only limited by the spatial resolution of the model. In contrast, most common radiative transfer models can only prescribe a single albedo value for the whole surface, which is then called an effective surface albedo [25]. This effective surface albedo represents the overall reflection characteristics of a usually inhomogeneous surface over a circular area with a radius of up to 30 km.

Figure 21.6 shows the effective albedo determined for Davos from 2007 to 2009 by combining measurements and radiative transfer calculations. As can be seen in the figure, during the winter period when Davos and the surrounding areas are fully covered by snow, the effective albedo has an average value of about 0.6, whereas during the summer months the surface albedo decreases to values around 0.03. The corresponding radiation increase with respect to the snow-free summer period is as large as 30%. It should be noted however that, due to the lower solar elevation in winter than summer, the daily radiation dose is much higher in summer than winter.

Future Directions

The intensity of solar UV radiation in the future will depend on possible changes in several factors, in relation to global and regional changes in the climate system. These factors include ozone, aerosols, clouds, and surface reflectivity, which are all expected to change on both a regional and global scale due to climate change. Changes in stratospheric ozone can be modeled quite accurately using global chemistry models. Their results indicate that the total ozone levels will not only recover as a consequence of the measures taken under the Montreal protocol and its amendments, but that a "super-recovery" of ozone is likely to occur around 2050 [26]. This super-recovery will manifest itself with higher ozone levels than prior to the global ozone decrease, with correspondingly reduced UV levels, depending on the region considered [27].

Bibliography

Primary Literature

1. Coblentz WW (1932) The Copenhagen meeting of the second international congress on light. Science 76:412–417
2. Diffey BL (1991) Solar ultraviolet radiation effects on biological systems. Phys Med Biol 36(3):299–328
3. Gröbner J (2001) Characterisation of spectrophotometers used for spectral solar ultraviolet radiation measurements. Rad Prot Dosimetry 97:415–418
4. Dahlback A (1996) Measurements of biologically effective UV doses, total ozone abundances, and cloud effects with multichannel, moderate bandwidth filter instruments. Appl Opt 35:6514–6521
5. Hülsen G (2007) Characterization and calibration of ultraviolet broadband radiometers measuring erythemally weighted irradiance. Appl Opt 46:5877–5886
6. Chandrasekhar S (1960) Radiative transfer. Dover Press, New York. ISBN 0-486-60590-6
7. Stamnes K et al (1988) Numerically stable algorithm for discrete-ordinate-method radiative transfer in multiple scattering and emitting layered media. Appl Opt 27:2502–2509
8. Weihs P et al (1997) Accuracy of spectral UV model calculations 1. Consideration of uncertainties in input parameters. J Geophys Res 102(D1):1541–1550
9. Herman J et al (1999) Distribution of UV radiation at the Earth's surface from TOMS-measured UV-backscattered radiances. J Geophys Res 104(D10):12059–12076
10. Arola A et al (2003) A new approach to estimating the albedo for snow-covered surfaces in the satellite UV method. J Geophys Res 108:4531. doi:10.1029/2003JD003492
11. McKenzie R et al (2001) Satellite retrievals of erythemal UV dose compared with ground-based measurements at northern and southern midlatitudes. J Geophys Res 106:D20. doi:10.1029/2001JD000545
12. Arola A et al (2005) Assessment of TOMS UV bias due to absorbing aerosols. J Geophys Res 110:D23211. doi:10.1029/2005JD005913
13. Tanskanen A et al (2007) Validation of daily erythemal doses from Ozone Monitoring Instrument with ground-based UV measurement data. J Geophys Res 112:D24S44. doi:10.1029/2007JD008830

14. Bener P (1964) Investigation on the influence of clouds on the ultraviolet sky radiation. Contract AF 61(052)-618, Tech. Note 3, Davos-Platz, Switzerland
15. Mims FM III, Frederick JE (1994) Cumulus clouds and UV-B. Nature 371:291
16. Rayleigh (1871) On the scattering of light by small particles. Phil Mag 41, 107, 274
17. Madronich S et al (1998) Changes in biologically active ultraviolet radiation reaching the Earth's surface. J Photochem Photobiol B Biol 46(1–3):5–19
18. Farman JC et al (1985) Large losses of total ozone in Antartica reveal seasonal ClOx/NOx interaction. Nature 315:207–210
19. Kirchhoff V et al (1997) UV-B enhancements at Punta Arenas, Chile. J Photochem Photobiol B Biol 38:174–177
20. Kylling A et al (1998) Effect of aerosols on solar UV irradiances during the photochemical activity and solar ultraviolet radiation campaign. J Geophys Res 103:26051–26060
21. Blumthaler M et al (1988) Solar UVB albedo of various surfaces. Photochem Photobiol 48(1):85–88
22. Bernhard G et al (2007) Ultraviolet and visible radiation at Barrow, Alaska: climatology and influencing factors on the basis of version 2 National Science Foundation network data. J Geophys Res 112:D09101. doi:10.1029/2006JD007865
23. Degünther M et al (1998) Case study on the influence of inhomogeneous surface albedo on UV irradiance. Geophys Res Lett 25:3587–3590
24. Mayer B, Degünther M (2000) Comment on "Measurements of erythemal irradiance near Davis station, Antarctica: effect of inhomogeneous surface Albedo". Geophys Res Lett 27: 3480–3490
25. Kylling A et al (2000) Determination of an effective spectral surface albedo from ground-based global and direct UV irradiance measurements. J Geophys Res 105(D4):4949–4959
26. Austin J et al (2006) Ensemble simulation of the decline and recovery of stratospheric ozone. J Geophys Res 111:D16314. doi:10.1029/2005JD006907
27. WMO (2007) Scientific assessment of ozone depletion: 2006. Global ozone research and monitoring project report No. 50. World Meteorological Organization, Geneva

Books and Reviews

Seckmeyer et al (2001) Instruments to measure solar ultraviolet radiation, part 1: spectral instruments (WMO TD No. 1066), available at http://www.wmo.ch/pages/prog/arep/gaw/gaw-reports.html

Seckmeyer et al., Instruments to measure solar ultraviolet radiation – Part 2: Broadband instruments measuring erythemally weighted solar irradiance (WMO TD No. 1289), 55 p, July 2008, available at http://www.wmo.ch/pages/prog/arep/gaw/gaw-reports.html

Webb et al (2003) Quality assurance in monitoring solar utraviolet radiation: The state of the art (WMO TD No. 1180), available at http://www.wmo.ch/pages/prog/arep/gaw/gaw-reports.html

Zerefos C (1997) Solar ultraviolet radiation: modelling, measurements and effects (NATO ASI Series/Global Environmental Change). ISBN 3-540-62711-1

Chapter 22
UV Effects on Living Organisms

Philipp Weihs, Alois W. Schmalwieser, and Günther Schauberger

Glossary

Abiotic	An abiotic factor is a feature or process that is not derived from, or caused by, living organisms but that might affect living organisms in their environment.
Biologically effective UV radiation	The biologically effective UV radiation is obtained by integrating the biologically effective spectral irradiance over the whole wavelength range.
Biotic	A biotic factor is a feature or process derived from, or caused by, living organisms, which might affect other living organisms in their environment.
Chloroplast	Chloroplast is the site of photosynthesis. It is distinguished by its green color, caused by the presence of chlorophyll.
Chromophore	A chemical group capable of selective light absorption resulting in the coloration of certain organic compounds. In biological molecules that serve to capture or detect light energy, the chromophore is the moiety that causes a conformational change of the molecule when hit by light.
Ecotherm	An organism that regulates its body temperature largely by exchanging heat with its surroundings (e.g., fish, reptile, . . .).

This chapter was originally published as part of the Encyclopedia of Sustainability Science and Technology edited by Robert A. Meyers. DOI:10.1007/978-1-4419-0851-3

P. Weihs (✉)
University of Natural Resources and Applied Life Sciences, A 1190 Vienna, Austria
e-mail: philipp.weihs@boku.ac.at; weihs@boku.ac.at

A.W. Schmalwieser • G. Schauberger
WG Environmental Health, University of Veterinary Medicine, Veterinaerplatz 1,
A 1210 Vienna, Austria
e-mail: Alois.Schmalwieser@vetmeduni.ac.at; Gunther.Schauberger@vetmeduni.ac.at

E.A. Laws (ed.), *Environmental Toxicology: Selected Entries from the Encyclopedia of Sustainability Science and Technology*, DOI 10.1007/978-1-4614-5764-0_22,

Eye damage	This includes the keratoconjunctivitis (an inflammation of the cornea and conjunctiva, snow blindness), an acute effect, and chronic damage of the eye lens, caused by absorbing UVA radiation.
Flavonoids	An umbrella term that designates a wide spectrum of pigments (including anthocyanins, among others).
Genetic damage	DNA absorbs UVB radiation and the absorbed energy can break bonds in the DNA. Most of the DNA breakages are repaired by proteins present in the cell nucleus, but unrepaired genetic damage of the DNA can lead to skin cancers.
Health effects	In humans and animals, prolonged exposure to solar UV radiation may result in acute and chronic health effects on the skin, eye, and immune system. UVB exposure induces the production of vitamin D in the skin. The majority of positive health effects are related to this vitamin.
Minimal erythemal dose (MED)	The amount of UV radiation exposure required to cause a just perceptible reddening of the skin.
Mitochondria	"Powerhouse" of the cell, which acts as the site for the production of high-energy compounds. These high-energy compounds are the vital energy source for several cellular processes.
Monomer	A molecule that can combine with others to form a polymer.
Morphogenesis	Biological process of growth, structure, form, or shape.
Phototaxis	A response to a directional light stimulus that involves movement of a whole organism. This is advantageous for phototrophic organisms as they can orient themselves most efficiently to receive light for photosynthesis. Phototaxis is called positive if the movement is in the direction of increasing light intensity and negative if the direction is opposite.
Plant growth	The rate of increase in weight and size of plant organs, such as leaf, stem, or root.
Skin damage	Beside the sunburn (erythema), a well-known acute effect, chronic skin damage is also observed. This includes elastoses (damage of collagen fibers) and, thereby, aging of the skin and skin cancer. (The three most common skin cancers are basal cell cancer, squamous cell cancer, and melanoma, each of which is named after the type of skin cell from which it arises. Melanoma is one of the less common types of skin cancer but causes the majority (75%) of skin cancer–related deaths.)
Spectral weighting function	Wavelength dependency of biologic efficiency of UV radiation. The multiplication of spectral irradiance and a spectral weighting function delivers the biologically effective spectral irradiance. The spectral weighting function is unique for each photobiological process.

UV (ultraviolet) radiation — Electromagnetic energy in the wavelength range between 200 and 400 nm. In the solar spectrum on Earth's surface the wavelength range is limited to 280 and 400 nm. The spectral range is divided in UVC (200–280 nm), UVB (280–315 nm), and UVA (315–400 nm).

Yield — Economic product harvested from plants, for example, seed from pods of soybean, roots from carrots (*Daucus carota*), grain from wheat (*Triticum aestivum*), or seed and lint from cotton (*Gossypium hursutum*).

Definition of the Subject

Since the energy of an electromagnetic photon increases with decreasing wavelength, the biological effectiveness – defined as the capacity of radiation to produce a specific biological effect – also increases with decreasing wavelength. The largest biological effects are therefore generated in the shortest wavelength range (280–400 nm) of the incident solar radiation spectrum, which is referred to as ultraviolet (UV) radiation. Living organisms (i.e., living systems such as animals, plants, and microorganisms) are influenced by solar radiation and especially also by UV radiation. Three important aspects have to be taken into account in an investigation regarding the UV impact on living organisms: First, the damage potential of UV has to be studied; second, the faculty of living organisms to protect themselves and to adapt has to be examined; and third, the environment and its impact on the UV radiation field has to be taken into account.

Introduction

The decrease in the ozone layer, which was discovered in the early 1980s, has raised big concerns among the scientific community and among the decision makers. The ozone layer shields the Earth from the very harmful UV radiation that belongs to the part of the solar spectrum with the largest photon energy. Before the ozone hole discovery, it was well known that UV radiation may lead to sunburn on humans and animals, and it was also known that UV may have a negative impact on the eyes. It was, however, not clear what impact UV would have on the biosphere if the ozone layer was to diminish more. Would the biosphere be able to adapt? Or, conversely, would the expected increase in UV eradicate all life on Earth? A lot of money was therefore invested in research on the impacts of UV on the biosphere: the UV effects on humans, as well as on microorganisms, plants and animals, were investigated. In this article, the focus is on microorganisms, plants and animals.

An overview of the research results obtained in these topics during the last 30 years is given, in the form of a detailed review of the findings on the damage potential of UV, and on the adaptation potential and the UV defense mechanisms of microorganisms, plants and animals.

Biological Effectiveness of UV Radiation

The main concept that is used throughout this article is that of the "dose," which describes the biologically effective UV radiation reaching the skin, considered as a relevant receiving surface, and taking into account the Grotthus–Draper law (named after the chemists Christian J.D.T. von Grotthus and John W. Draper). This law states that a tissue effect can only occur if radiation is absorbed. However, absorbed radiation does not necessarily cause a biochemical reaction. Absorbed radiation may simply be converted into heat, or it may be re-emitted as light at a different wavelength. The latter phenomenon is called fluorescence.

The spectral sensitivity of these biochemical effects is taken into account by a spectral weighting function, whose convolution with the incident solar spectrum results in biologically effective radiation (it would be good to introduce the MED here or in next section).

Biologically Effective Irradiance and Dose

The biologically effective irradiance E_{biol} is calculated by a dimensionless biological weighting function σ_λ and the incident spectral irradiance E_λ from:

$$E_{biol} = \int_{280\text{ nm}}^{400\text{ nm}} E_\lambda \, \sigma_\lambda \, \mathrm{d}\lambda. \tag{22.1}$$

The corresponding biologically effective dose (irradiation) is obtained as the product of the biologically effective irradiance E_{biol} and the time of exposure t:

$$D_{biol} = E_{biol} \, t \tag{22.2}$$

if it can be assumed that E_{biol} is constant over the period. Otherwise, a summation over time would be necessary.

According to the rule of Bunsen and Roscoe [1] see (Fig. 22.1), also called reciprocity law, a photochemical reaction is directly proportional to the dose D_{biol}, irrespective of the time t over which this dose is delivered. This often implicit assumption is a simple working hypothesis even though it has been shown that at a constant total dose, the irradiance (intensity) is a major factor that determines both

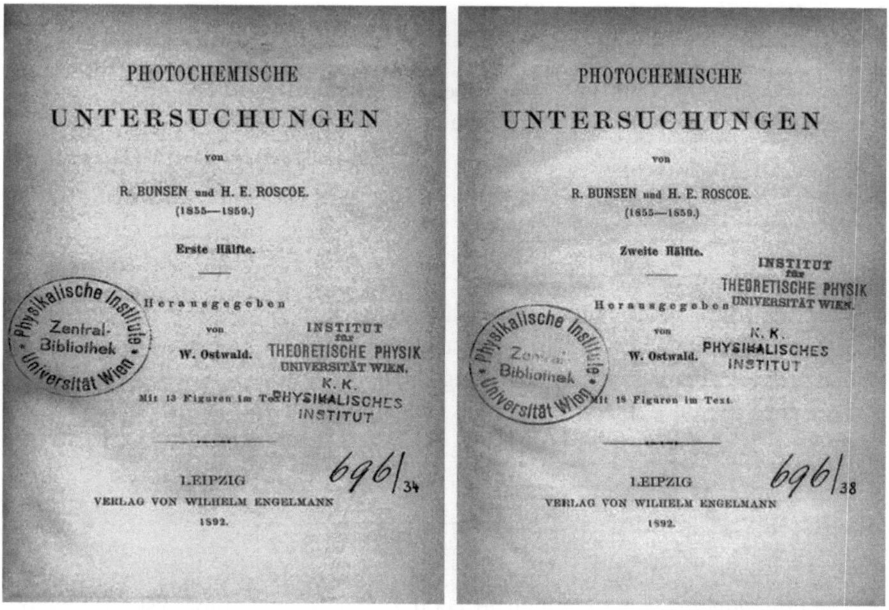
PHOTOCHEMISCHE
UNTERSUCHUNGEN
von
R. BUNSEN und H. E. ROSCOE.
(1855—1859.)
Erste Hälfte.
Herausgegeben
von
W. Ostwald.
Mit 13 Figuren im Text.
LEIPZIG
VERLAG VON WILHELM ENGELMANN
1892.

Physikalische Institute Zentral-Bibliothek Universität Wien
INSTITUT für THEORETISCHE PHYSIK UNIVERSITÄT WIEN.
K. K. PHYSIKALISCHES INSTITUT
696/34

PHOTOCHEMISCHE
UNTERSUCHUNGEN
von
R. BUNSEN und H. E. ROSCOE.
(1855—1859.)
Zweite Hälfte.
Herausgegeben
von
W. Ostwald.
Mit 18 Figuren im Text.
LEIPZIG
VERLAG VON WILHELM ENGELMANN
1892.

INSTITUT für THEORETISCHE PHYSIK UNIVERSITÄT WIEN.
K. K. PHYSIKALISCHES INSTITUT
Physikalische Institute Zentral-Bibliothek Universität Wien
696/38

Fig. 22.1 Cover page of a re-print of the paper by Bunsen and Roscoe [1], published by Wilhelm Engelmann in the series "Ostwalds Klassiker der modernen Wissenschaft," Leipzig 1892 [2]

the quality and quantity of the response [2, 3], in the case of UV-induced cell death, photocarcinogenesis, psoralen photochemistry, and effects of low-level laser radiation.

Reciprocity law failures were commonly observed in experiments conducted at either very low or very high radiant fluxes. To account for these failures, Schwarzschild, an astronomer, proposed a modification of the reciprocity law that resulted in a better fit to his low intensity, stellar data. This empirical model later became known as Schwarzschild's law [4] and is given by either [5]

$$D_{biol} = E^{p}{}_{biol}\, t \tag{22.3}$$

or

$$D_{biol} = E_{biol}\, t^{p} \tag{22.4}$$

where p is the Schwarzschild coefficient. When $p = 1$, then the Schwarzschild and the reciprocity law are identical. Martin et al. [5] found in a review that, in 11 experiments with erythema, only one deviated from the reciprocity law by $p = 1.2$. For many other biological effects the Schwarzschild coefficient was found to be $p = 1$ for the majority of experiments.

Dose–Response Relationship

A dose-response function (DRF) describes the influence of a biologically effective UV dose, D, on an observed biological effect.

Erythema

The erythema is caused by ultraviolet radiation. It can be observed on animals in the same way as on humans. In particular, the dose-response functions and the time course of the reddening of the skin of pigs were investigated in detail.

The dose-response function of skin reddening can be described by a logit function,

$$\Delta EI = \frac{\Delta EI_{\max}}{1 + e^{\frac{\log D - \log D_{0.5}}{-k}}} \tag{22.5}$$

The erythema index ΔEI is defined via the international color system, L*a*b*. The ΔEI value is the difference of the red color value a* before and after UV exposure, the UV Dose D (J/m^2), the plateau of the dose-response function (maximum value of the erythema index) $\Delta EI_{\max}$, the dose of the midpoint of the function $D_{0,5}$ and the slope of the logit function at the midpoint k (Fig. 22.2). The measurements of the redness of the skin can be done by a color meter (e.g., Minolta Chroma Meter CR 300) that uses the international color system L*a*b*.

Most photobiological effects do not appear immediately after or during exposure but after a certain latent period. For certain effects, the latent period may be in the order of decades (e.g., skin cancer in human), while for other effects, the period is rather short (e.g., erythema).

The visibility of an effect may depend also on time. In a first phase the visibility of an effect increases until reaching a maximum. Afterward the visibility decreases. If an effect is reversible, the visibility of the effect goes back to zero; if the effect is irreversible, it stays visible. Repair mechanisms (photorepair) of the organisms are crucial to make the effect disappear. Sustained irradiation can be additive and enhances the remaining effect (cumulative effect).

The temporal course of an effect may depend on dose and wavelength.

The dose-response function, as well as the time course of the erythema in pigs, were investigated by Reischl et al. [6] (Fig. 22.3). In Fig. 22.4, the normalized dose-response function of ten pigs (crossbreeding of "Edelschwein" and "Landrasse") show similar behavior. The differences of the plateau and the slope of the function can be explained by individual sensitivity, similar to the four skin types of humans. The median of the MED was calculated as 165 J/m^2, the lower quartile 143 J/m^2, and the upper quartile 184 J/m^2. In Table 22.1 the minimal erythema dose (MED) is summarized for cattle, horse and pig.

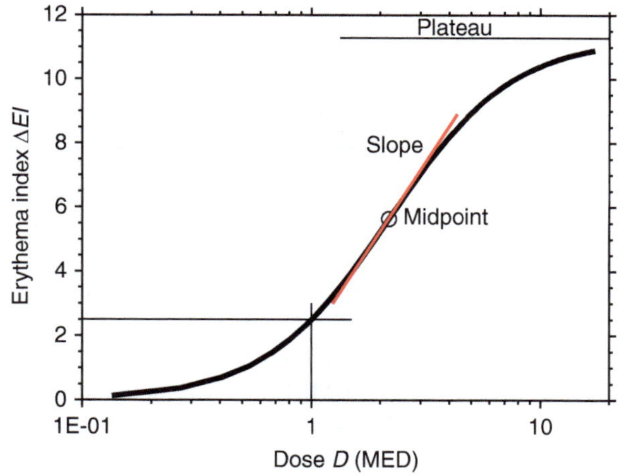

Fig. 22.2 UVB-caused dose-response curve, 24 h after exposure, for a pig with a minimal erythema dose of MED = 148 J/m^2. In this specific case, the plateau $\Delta EI_{\max}$ equals 11.28, the midpoint $D_{0.5}$ is 2.75 MED, and the slope k is 0.269

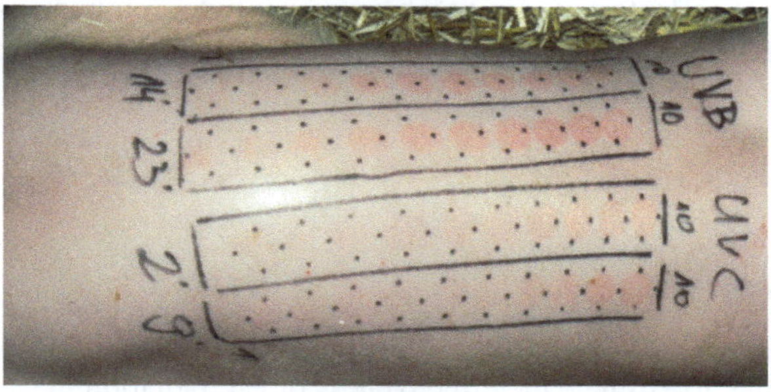

Fig. 22.3 Erythema (on pig skin?) caused by UVB and UVC radiation [6]

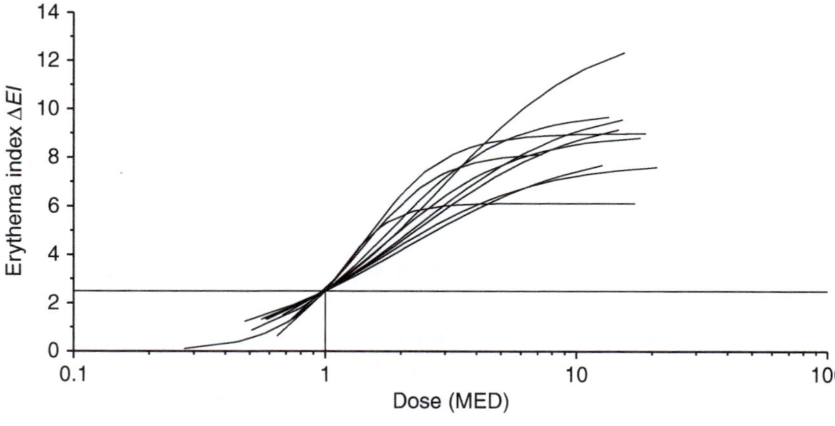

Fig. 22.4 Dose-response function for pigs, normalized by the individual minimal erythema dose (MED). The MED was determined by an erythema index $\Delta EI = 2.5$, which is the limit of the visual detection 24 h after a UVB exposure [6]

Table 22.1 Minimal erythema dose (MED) for several species

Specie	Minimal erythema dose (MED) (J/m^2)	Author
Cattle	100	Mehlhorn and Steiger [8][a]
Horse	450	Kasper [9][b]
Pig	165	Reischl et al. [6][b]

[a]Spectral weighting function not clearly defined
[b]Reference erythema weighting function by MacKinley and Diffey [10]

To describe the time course of the erythema index ΔEI, the doses were divided into three groups: low doses if below 1 MED (sub-erythemal doses), medium doses between 1 and 7 MED, and high doses above 7 MED. The time course of the artificial UVB and UVC exposure is depicted in Fig. 22.5.

The time course of erythema was bimodal after UVB and UVC exposure. Both spectral ranges show a peak for low doses within about 12 h. Higher doses showed a greater intensity of redness and later fading by UVB than by UVC. Two effects could be distinguished. The first effect caused an early maximum of erythema by low doses, whereas the second effect caused a later, higher maximum by higher doses. These two effects can also be observed in studies on humans [7]. For both spectral ranges, the time course of erythema was primarily related to the UV dose and not to the spectral range.

Action Spectra

If one would like to estimate the effectiveness of UV radiation in causing a certain effect, the action spectrum (or biological weighting function) has to be known to weigh the incident spectral irradiance. An action spectrum (AS) describes the spectral efficiency of light in initiating a photochemical or photobiological process.

Conventionally, an AS is determined by exposing a sample to $n(\lambda)$ photons at wavelength λ to produce the effect. Then the wavelength and photon number are varied in such a way that the same effect is produced at all wavelengths. The AS s is then

$$s(\lambda) = C/n(\lambda) \tag{22.6}$$

where C is a constant.

The number of photons is called radiant exposure or dose. It is the product of irradiance and the duration of irradiation.

If only one chromophore is present and if this chromophore is in an unbound and monomeric state, the action spectrum will have a shape identical to that of the absorption spectrum of the chromophore. In samples with many absorbing molecules, action spectroscopy is a powerful tool used to identify the specific chromophore for the process under scrutiny.

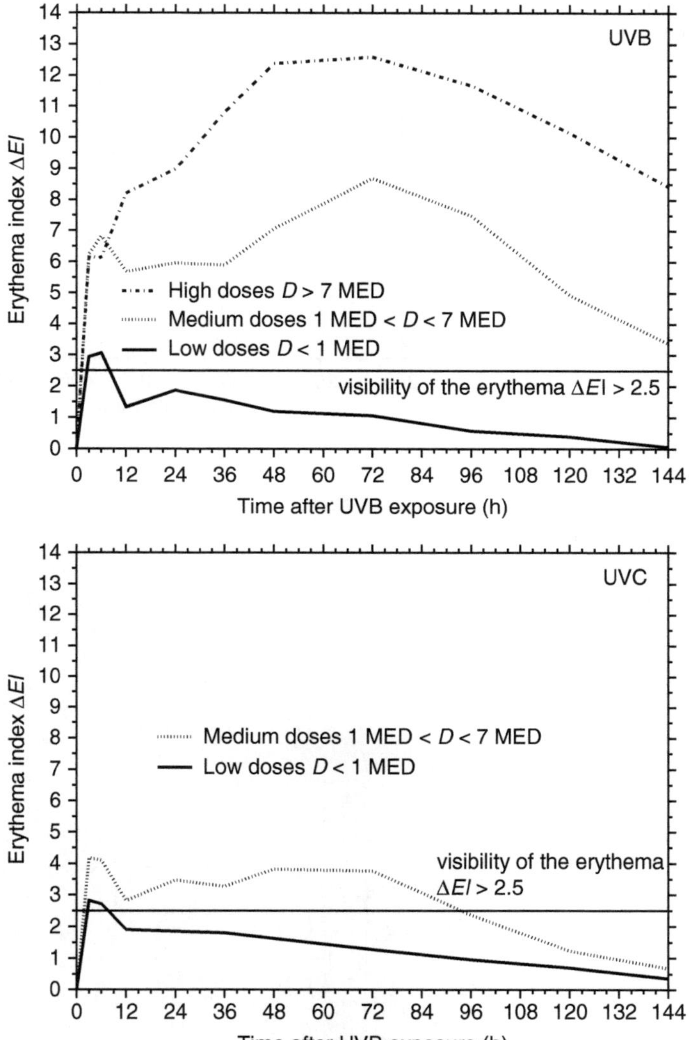

Fig. 22.5 Time course of the erythema intensity caused by UVB and UVC radiation, measured by the erythema index ΔEI for three dose levels [6]

If free monomeric chromophore molecules are present together with bound or aggregated molecules, the situation is more complicated. Aggregated or bound molecules usually have distorted or shifted spectra as compared with monomeric molecules, but are usually less efficient in initiating photochemical processes.

If different chromophores are involved, the AS is a superposition of several absorption spectra.

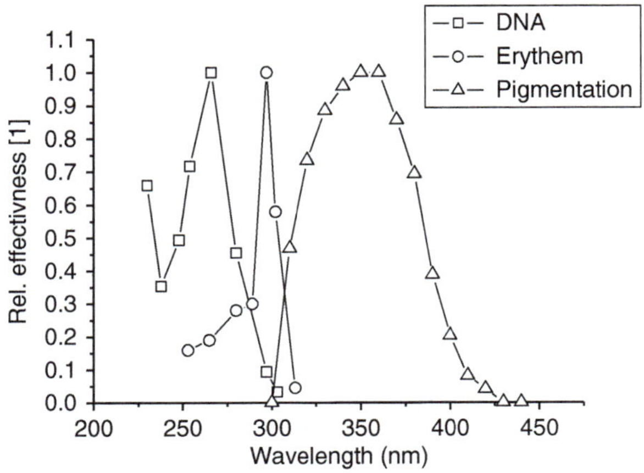

Fig. 22.6 Early known action spectra such as DNA damage [11], erythema in humans [19], and pigmentation [28]

Fig. 22.7 3D model of an animal and its environment

Action spectra are specific to each process. They may obviously differ, as in the cases of DNA damage, Erythema, and Pigmentation see (Figs. 22.6–22.8), depending on the absorbing molecules.

Comparisons between experimentally determined action spectra and absorption spectra of appropriate molecules can sometimes give insight into which molecule is primarily responsible for the effect. Using the bactericidal AS derived by Gates [11] see (Fig. 22.6), it becomes possible to identify DNA as the genetic material [11, 12]. This observation was confirmed with other unicellular organisms and led to the realization that nucleic acids have a fundamental role in ultraviolet photobiology.

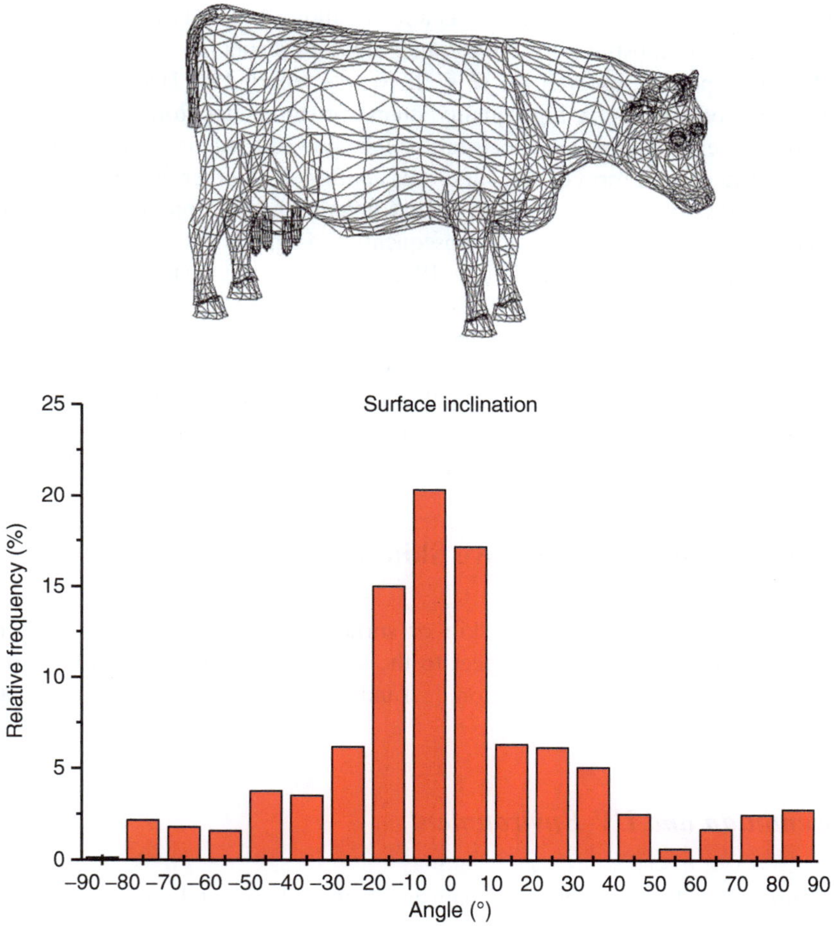

Fig. 22.8 3D model of a cow and frequency distribution of surface tilt (0° denoted vertically, +90° horizontally face up and −90° face down)

Although it is still possible to glean more information about the photoresponses of organisms that do not meet the criteria above, it may be impossible to identify either the molecule(s) or the mechanism(s) involved in the process being studied. It may not be possible to circumvent these limitations if a complex process, such as plant "growth," is to be analyzed. An action spectrum for this latter process may be nothing more than a determination of effects as a function of wavelength, but the exact mechanism causing the effect might remain unknown. Even in this case, such action spectra can be useful in estimating the response to ambient of changing light exposures.

After the discovery of infrared [13, 14] and ultraviolet radiation [15], the crude action spectroscopy was done by dividing whether a certain photobiological effect is caused by UV, or rather visible or infrared radiation. During the nineteenth century action spectroscopy was refined by using colored glass filters to select wavelength ranges. This technique has already allowed identifying chlorophyll as the chromophore most responsible for the growth of plants [16–18].

The first spectral resolved AS (Fig. 22.6) was reported by Hausser and Vahle [19] after constructing a high-quality quartz-prism monochromator. They first reported the erythema action spectrum, which showed a major peak of activity at 297 nm. The general form of this curve was confirmed by other investigators in the following 10 years [20–22]. Then, in 1934, Coblentz and Stair proposed a "standard erythema curve," which was subsequently adopted by the Commission Internationale de l'Eclairage [23] in 1935. This was the first standardized AS, which maintained its status until 1987 [24].

In the 1930s, a variety of other important ASs were derived, like for the killing of bacteria [25] and bacteriophage [26], vitamin D synthesis (healing rickets) [27], pigmentation of human skin [28] (Fig. 22.6), or cell division [29].

Most of the ASs were derived in the 1980s after the ozone depletion issue became known, to the effect that biologically effective radiation became a topic of interest.

Transmission to the Effective Skin Dose

The estimation of the effective skin or surface dose needs knowledge of the orientation of the receiver in respect to the sun, of the nearby UV environment (reflection, scattering, sky obstruction, . . .) and of the protection of the hair coat and behavior (Fig. 22.7).

Orientation and UV Environment

Measured or calculated solar UV data incident on horizontal surfaces are commonly used to assess the risk to a biological system from the local UV environment. Usually, standard solar radiation measurements are performed for horizontal surfaces and include the incoming irradiance from the whole sky. This incident global irradiance on a horizontal surface is the sum of direct beam and scattered diffuse radiation.

In reality, few biological surfaces are horizontal or flat. However, the surface of an organism can be decomposed into many small inclined planes or "facets."

Figure 22.8 depicts such a surface mesh for a cow, along with the corresponding frequency distribution of the tilt of these surface patches. This is convenient for an immobile subject. Obviously, the task of assessing the radiant exposure becomes much more complex when the subject is moving.

The incident irradiance on inclined planes (e.g., body parts of an animal) includes the sum of three radiation components see (Fig. 22.9):

- Direct beam irradiance: The intensity of the incident direct beam on an inclined plane depends not only on the intensity of the direct beam irradiance but also on the incidence angle between the sun and the normal to the plane.

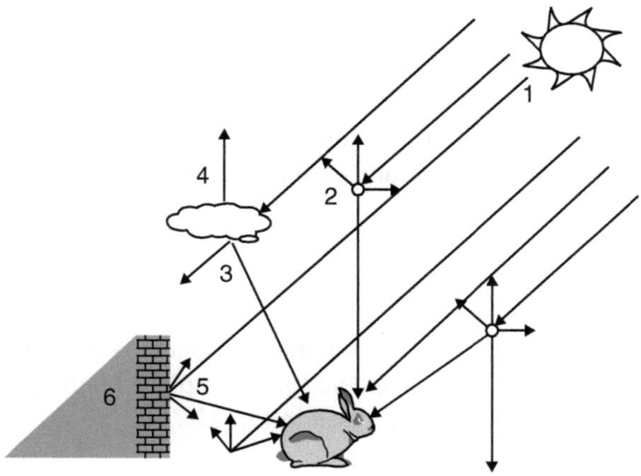

Fig. 22.9 Overview of the various radiation components and phenomena that have an impact on the radiation exposure of an animal. *1* direct beam irradiance, *2* radiation scattered by molecules and aerosols, *3* radiation scattered by clouds. (*2* and *3* make up the diffuse radiation component), *4* radiation reflected by clouds, *5* reflection by the surroundings, *6* shading (here, by a wall)

Table 22.2 Albedo of various surfaces in the UV and visible range

Surface	UV	Visible
Woodland	0.07	0.2
Grass	0.05	0.18
Bare soil	0.1	0.3
New, deep snow	0.9	0.8

Source: Blumthaler and Ambach [30]

- Diffuse sky irradiance: The inclined plane does not "see" the whole hemisphere, and the diffuse irradiance is therefore reduced compared to a horizontal receiver.
- Ground-reflected radiation toward the receiver: This reflected radiation depends on the illumination of the reflecting facet (incident irradiance on ground), on the solid angle of the facet, on the reflectance (albedo) (Table 22.2) of the facet, and of course on the inclination of the receiving plane.

Direct Beam Radiation

The direct beam, I, on a given plane depends on the initial intensity of the direct solar beam irradiance incident on a normal plane (I0), which can be measured or calculated, and on the angle of incidence, β, which is the angle between the solar beam and the surface normal (also called illumination angle)

$$I = \mathrm{I0}\cos\beta \tag{22.7}$$

β is calculated using the following equation:

$$\cos\beta_{\mathrm{i}} = \cos\theta_{\mathrm{s}}\ \cos\theta_{\mathrm{n,i}} + \sin\theta_{\mathrm{s}}\ \sin\theta_{\mathrm{n,i}}\ \cos\left(\varphi_{\mathrm{s}} - \varphi_{\mathrm{n,i}}\right) \tag{22.8}$$

where θ_s is the solar zenith angle, $\theta_{n,i}$ the slope inclination (or tilt angle), $\varphi_{n,i}$ the azimuth (orientation) of the slope, and φ_s the solar azimuth.

Diffuse Irradiance

The diffuse irradiance D incident on a horizontal plane is the integral of the sky radiance I over the whole sky hemisphere (please replace I here by something else to avoid confusion since it is already used in Eq. 22.7).

$$D = \int_0^{\pi/2}\int_0^{2\pi} I \sin\theta \cos\theta \; \mathrm{d}\theta \mathrm{d}\gamma \tag{22.9}$$

where θ is the zenith angle and γ the azimuth angle of a sky element.

For the calculation of the incident radiation on an inclined plane, the sky radiance has to be integrated only over the portion of sky seen from the plane. The diffuse irradiance on a given plane may be approximated, assuming an isotropic hemispherical sky radiance distribution, using the following equation [31, 32]:

$$\mathrm{Dt} = \left(1 + \cos\left(\theta_{\mathrm{n,i}}\right)\right)\mathrm{D}/2 \tag{22.10}$$

where $\theta_{n,i}$ is the tilt angle of the plane, and D is the diffuse horizontal irradiance (Eq. 22.9). Equation 22.10 does, however, not take into account the reflected irradiance emanating from the ground, which should be added to the quantity Dt.

Measurements showed that the ratio of the irradiance on a horizontal plane to the irradiance on the vertical plane has a strong seasonal course, resulting in higher irradiance on the horizontal plane in summer and higher irradiance on the vertical plane in winter at mid- to high-latitudes. The irradiance on vertical planes in winter may be up to 2.5 times higher than in summer [33]. In summer, the ratio of irradiance incident on vertical to irradiance incident on horizontal planes may be as low as 0.2 for solar zenith angles around 45°. Over snow surfaces the incident irradiance can be much higher [34–36]. Figure 22.10 shows the influence of the UV albedo on theUV irradiance incident on an inclined plane.

Obstruction of the Sky

The obstruction of the horizon shields a sensor/subject from a part of the diffuse irradiance. The surroundings, however, also reflect some radiation toward the sensor or subject.

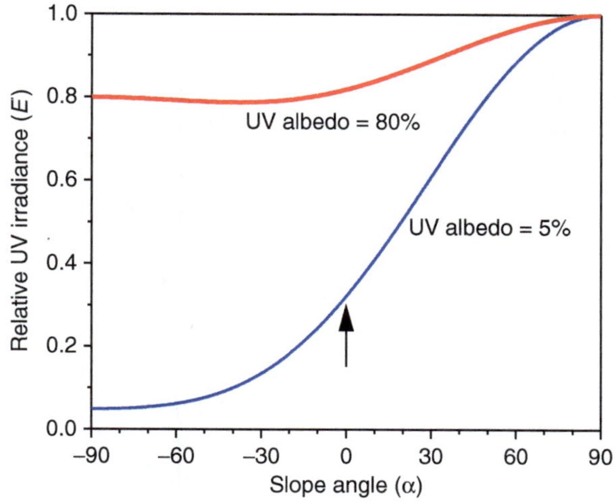

Fig. 22.10 Relative irradiance on a sloped receiving surface in dependence of slope angle and UV albedo

The obstruction of the horizon may be calculated by integrating the shadow line over all azimuth angles [32]:

$$\mathrm{Obstr} = \int \mathrm{d}\alpha\, \mathrm{d}\varphi \tag{22.11}$$

where α is the elevation of the object over the horizon and φ its azimuth.

Obstr is the angular obstruction of the horizon (due to the surrounding obstacles, such as mountains) in addition to the obstruction already occurring due to the inclination of the sensor.

By taking this obstruction of the horizon into account, the effective diffuse irradiance Dt on a given plane may be calculated under the assumption of isotropic sky radiance distribution and no reflected irradiance, by using the following equation [31, 32], which is a modification of Eq. 22.10:

$$\mathrm{Dt} = \left(1 + \cos\left(\theta_{\mathrm{n,i}}\right)\right)\mathrm{D}(1 - \mathrm{Obstr})/2 \tag{22.12}$$

where Obstr is obtained from Eq. 22.11.

Models

Using Eqs. 22.7 and 22.8, the calculation of the incident beam UV radiation on inclined planes is straightforward. Equations 22.10 and 22.12, for the determination of the diffuse irradiance on incident planes, bear some uncertainty, because the sky radiance is not perfectly isotropic. Several geometrical models [37–41] for the calculation of the broadband (rather than UV) diffuse irradiance which account also for the non-isotropic component of the sky radiance are available [42]. (Note that [37, 39–41] deal with

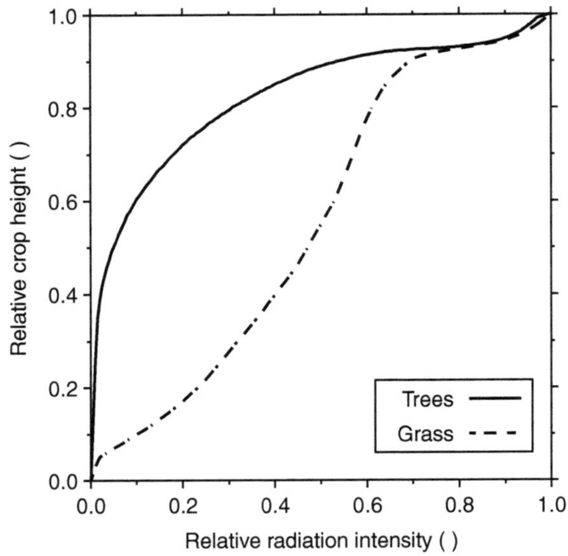

Fig. 22.11 Effect of the relative crop height of trees and grass on the relative radiation intensity normalized to unity by the intensity above the crop, according to Dirmhirn [45]

broadband irradiance, not UV; in the UV, the sky radiance is more isotropic than at longer wavelengths, so that these models should not be used in the present situation ....) Models reproduce adequately the incident UV radiation on planes facing the equator with a mean bias deviation lower than 4.5%. For planes with other orientations, a larger bias may be found, for example, 10% or more. In recent years, radiative transfer models were adapted to calculate the incident UV on inclined planes [43], taking also into account the obstruction of the horizon [44].

Vegetation

In addition to direct shading, vegetation can also be used by animals as protection against UV radiation. The attenuation of radiation results from shadowing by leaves, branches, or stems. Therefore, the geometrical structure of vegetation affects its UV attenuation and protection capacity (Fig. 22.11). Moreover, the protection is dependent on the seasonal changes like height of vegetation, density, and leaf area [45, 46]. Figure 22.11 depicts the influence of the relative crop height of trees and grass on the relative radiation intensity normalized to unity by the intensity above the crop, as modeled by Dirmhirn [45].

Behavior

The biggest problem for estimating the effective skin dose is the consideration of the behavior, because it affects the duration of stay within the vegetation. Many herbivores forage in the open field and afterward go back into higher vegetation for

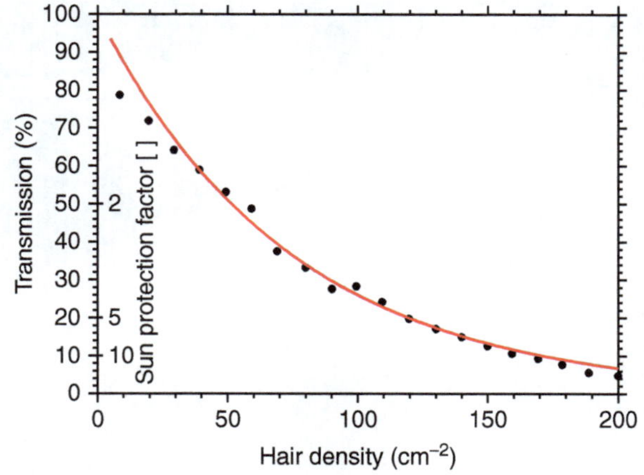

Fig. 22.12 UV Transmission and the resulting UV sun protection factor SPF due to the hair coat of cattle, according to Bianca and Wegmann [54]

cooling, resting, and cowering. For this no investigations are available. The biggest problem for estimating the effective skin dose is the consideration of the behavior, because it affects the duration of exposition. For this only a few investigations are available especially for ecotherms. [please condense these 2 paragraphs]

In 1997, Manning and Grigg [47] found that, for the freshwater turtle, basking (moving and positioning the body to maximize sun exposure) lasts longer than its thermoregulation would require. Ferguson et al. [48, 49] have shown that Panther chameleons can behaviorally regulate UV exposure or regulate absorption by changing skin color, to an absorption level at which the body already responds with vitamin D production at low UV levels. Karsten et al. [50] have shown that in dependence of their dietary Vitamin D_3 intake, Panther chameleons behaviorally regulate their exposure to solar UV to ensure an optimal Vitamin D_3 status. For some species of lizards, it was previously known that they possess a vitamin D_3 receptor in the brain [51]. Their retina is sensitive to the UV [52, 53] and may therefore be helpful for sun seeking and predator avoidance.

Hair Coat

One important barrier that keeps the ambient dose below the effective skin dose is the hair cover of animals. In addition to its important role as the primary thermal protection of the body, the hair cover is an effective UV protection.

Investigations on the hair coat of horses [9, 55, 56] found a biologically effective UV radiation transmission factor to the skin below 2%. This transmission corresponds to a sun protection factor (SPF) of approximately 50. In to the case of cattle, similar results were obtained [8, 54]. Figure 22.12 depicts the relation between transmission and hair density. This leads to the conclusion that, whenever

Fig. 22.13 Erythema of a newborn pig due to outdoor UV exposure

the biological effects of UV radiation are analyzed, the natural protection of the hair coat has to be considered.

The lack of UV protection was described by Chapman et al. [57] for newborn furless sheep and for pigs [58]. In Fig. 22.13, a pig is shown with a whole-body erythema due to outdoor UV exposure.

Transmission Through Water, Snow, and Ice

Water

For aquatic life, the optical properties of the immediate habitat – both atmosphere and water – are of great relevance. At the boundary, the optical properties of the interface between air and water, as well as the spectral reflection properties of the benthic division, are crucial [59]. At the water–air boundary, the index of refraction changes and must be taken into account. Reflection is in the order of 5–7% in the UVB and 6–9% in the UVA (e.g., [60]), where the higher values are measured at lower solar elevations.

The penetration of UV radiation into water depends on the optical properties of water as well as on wavelength. Transmittance in water is much lower than in the atmosphere.

Solutes and salinity decrease transmittance due to increased absorption, but most important is the amount of absorbing and scattering suspended particles in the water. Particulate matter is either of terrigenous (erosion and pulverization) or biological origin (dissolved organic carbon, pigments such as chlorophyll or pheophytin, etc.). Especially in eutrophic fresh water systems and coastal regions, the transparency is affected strongly by these.

Dissolved organic carbon (DOC) is organic material from plants and animals broken down into such a small size that it is "dissolved" into water. DOC is an umbrella term for thousands of different dissolved compounds. The humic or tannin types of DOC are yellow to black in color and can have a great influence on the water's visible color.

DOC attenuates UVB effectively [61, 62]. Moderate levels of DOC may decrease the UVB within 5 cm down to only 2% compared to the level at the surface [63].

DOC is photochemically degraded by UV [64] into dissolved inorganic carbon (CO, CO_2, H_2CO_3, ...) [65]. Through this process, UV indirectly affects its own transmission.

Other important absorbers are photosynthetic pigments like chlorophyll, pheophytin, and others.

UV is also absorbed and scattered by organisms living in water, such as plankton, algae, and sea grasses, which may build large canopies.

In general, transparency is highest in the clear ocean water [61] and high-latitude (dry valleys) [66, 67] or high alpine lakes [68], and lowest in brown (humic) waters (e.g., [69, 70]). Altogether, the depth for a 10% UV transmission can vary from about dozens of meters in the clearest waters to a few centimeters in brown humic waters.

A multicomponent model developed by Smith and Baker [71, 72] describes the total attenuation as the sum of numerous partial attenuation coefficients from water, DOC, pigments, particulate matter, and undefined residuals:

$$\begin{aligned} \mathrm{K_{Total}}(\lambda) &= \mathrm{K_{Water}}(\lambda) + \mathrm{K_{DOC}}(\lambda) + \mathrm{K_{Pigm}}(\lambda) \\ &\quad + \mathrm{K_{PM}}(\lambda) + \mathrm{K_{Residual}}(\lambda) \end{aligned} \tag{22.13}$$

All individual attenuation coefficients on the right-hand side of Eq. 22.13 depend on wavelength (λ). As a result, UVB is attenuated stronger than UVA and the visible radiation (e.g., [70, 73–75]). In most cases, the main contributor to $\mathrm{K_{Total}}(\lambda)$ is either $\mathrm{K_{DOC}}$ (e.g., lowland lakes) or $\mathrm{K_{Pigm}}$ (e.g., high alpine lakes) (e.g., [76]).

Once $\mathrm{K_{Total}}$ is evaluated, the spectral irradiance $E_\lambda{}^z$ at a certain depth z can be approximated [77, 78] by:

$$E_\lambda^z = -E_\lambda^0 \cdot w_\lambda \cdot \left(e^{-\mathrm{K_{Total}}(\lambda)\cdot z} - 1\right) \cdot \frac{1}{\mathrm{K_{Total}}(\lambda)} \tag{22.14}$$

where w_λ is a weighting function for a certain photobiological effect, and E_λ^0 is the spectral irradiance at the surface.

As can be seen from Fig. 22.14, shorter-wavelength UV radiation is absorbed more effectively than longer-wavelength UV [59, 79].

The transparency of water is not constant over time, but shows a temporal (seasonal) variability, which goes hand in hand with the changing amount of solved and dissolved matter in the water (e.g., [64, 69, 80]). As can be seen from

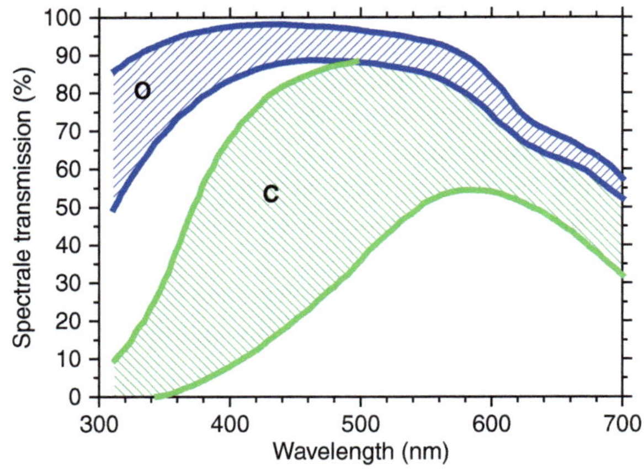

Fig. 22.14 Spectral transmission through 1 m of ocean (o) and coastal waters (c)

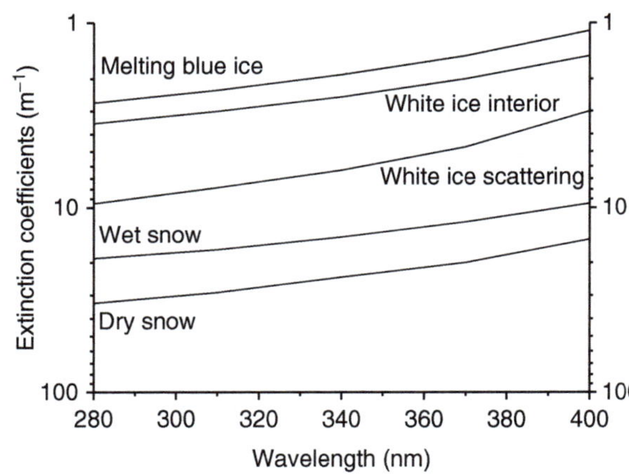

Fig. 22.15 Annual course of transmittance (100 mm) of groundwater (drinking water)

Fig. 22.15, even the transparency of groundwater and well water (drinking water quality) is not constant during the year.

Meteorological factors may also cause variability – snow melting and flood water may transport large amounts of unsolved substances.

In aquatic regions where water becomes occasionally scarce due, for example, to low water flow or, of course, to the tide cycle, UV radiation on sessile aquatic livings may become extreme, weather permitting.

Smith and Baker [78] have weighted the transmitted UV irradiance with the action spectrum for DNA damage [81]. In Fig. 22.16, the spectral biological effective radiation for waters with a small amount of soluble organic material and a small content of chlorophyll (clear oceanic water), as well as offshore water with high biological activity, is depicted. The spectral maximum of the 10-m depth line

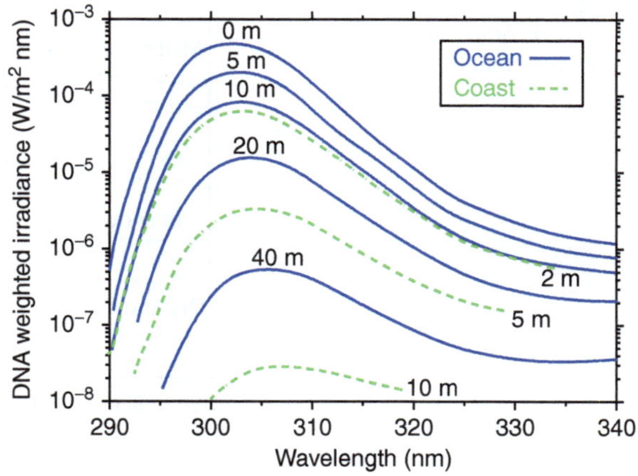

Fig. 22.16 DNA-weighted spectral irradiance in ocean and coastal waters at various depths

is attenuated by less than an order of magnitude, whereas it is four orders of magnitude for offshore waters.

It can also be seen that the ozone-affected UVB range provides the largest amount of irradiance for DNA damage, although the UVB from the sun, after transmittance through water, has a lower magnitude than UVA.

Ice and Snow

Ice and snow reflect a significant part of the incoming UV radiation. New dry snow may reflect over 90% of the incident UVB. New wet snow reflects over 80%. The amounts reflected by old dry snow and old wet snow are 82% and 74%, respectively [82, 83]. The albedo of snow may change within a day by around ± 0.05 with lower values shortly after noon, which may be caused by a change in the optical properties of snow [84], and by an increase in specularity effects at low sun angles.

Organisms living above the snow cover are therefore at serious risk of UV overexposure because they receive high amount of additional UV. Moreover, the intense reflected UV emanates from below, which represents a completely different direction than usual. This upwelling UV irradiance may hit body parts that are seldom exposed to UV, and are therefore less protected by pigmentation or fur, or less shielded by other body parts. A well-known effect is snow blindness, which humans also experience.

In spite of this high reflectance, a certain amount of UV penetrates into the snow cover (Fig. 22.17). Nevertheless, snow cover usually provides high UV protection for the organisms living beneath it (see, e.g., [85]). For instance, a snow layer of 5 cm can reduce the UV radiation down to 15–20%, and a layer of 15 cm can reduce it by nearly two orders of magnitudes [85, 86].

The reflectance of ice is somewhat lower than that of snow. It varies from about 30–80%, as a function of surface structure and temperature [87–89]. The transparency of ice to UV depends on age and clearness of the ice and on wavelength (Fig. 22.17).

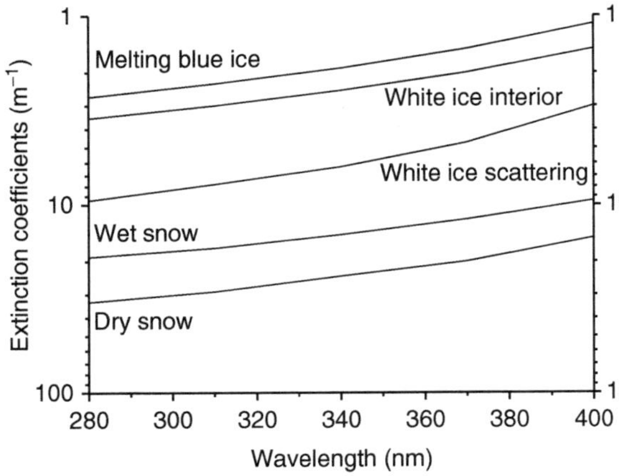

Fig. 22.17 UV extinction coefficients of ice and snow (After data from [89, 92])

UVB is attenuated stronger than UVA or visible radiation. In ice, a layer of 2.5 m can reduce UVA and UVB by up to 2 and 2.5 orders of magnitudes, respectively [90].

One meter of fresh, thin sea ice transmits roughly 20% of the incident UV radiation, whereas cold first-year ice transmits less than 2%. Different layers within the ice, from very white interior ice to almost clear surface ice have different UV extinction coefficients [89, 91, 92].

It was shown by Vincent et al. [67] that the transmitted UV below 3.5 m of very clear ice is still intense enough to inhibit algal growth.

Ice on lakes and rivers is more transparent than that on seawater, because during freeze-up DOC is removed from the ice [93].

Ice is often covered with snow. This combination, of course, affects both albedo and transmittance [87]. A snow cover of ≈5 cm (1–9 cm) on top of ≈0.9 m (0.5–1.3 m) of ice reduces UVB by 2–13% and UVA by 5–19% [94].

Small changes in snow and ice depth can have an extreme influence on UV radiation. This effect is much stronger than that caused by usual changes in atmospheric parameters like cloudiness or total ozone column [95].

Nowadays, special consideration on the transmittance of ice and snow is needed in the context of climate change. During shorter periods of snow and ice cover, the "underlying" aquatic or terrestrial organisms can become exposed, even though they are normally shielded from UV radiation. In addition to changing temperatures, a change in water flow can be responsible. An ecologically very important effect is the thinning of ice cover of the Arctic and Antarctic Seas, which leads to an enhanced exposure of a very large aquatic ecosystem.

Ice and snow also provide habitat for microbial assemblages (microalgae, bacteria, protozoa, . . .). They can be found in melt ponds on the ice surface, in brine channels within the ice, in slush layers between ice and snow, or between black and white ice, attached to the bottom of the ice or in the sub-ice platelet layer (e.g., [96, 97]). In hypersaline melt ponds above the surface of ice, microbial life can be found at temperatures around −10°C.

On the top of ice and snow, the so-called snow algae (cold-tolerant algae and cyanobacteria) can be found during alpine and polar summers. Algal bloom colors ice and snow with red hues, resulting in what is called red snow or watermelon snow.

Biological Effects on Microorganisms

What are called microorganisms or microbes consist of various types of very small life forms. Microorganisms can be found in the air, in waters, in the ground, on surfaces, and inside other organisms. They can also be found in the harshest environments. For instance, bacteria have been found around hydrothermal vents (hot springs, fumaroles, geysers, or "black smoke" at the sea ground), where water effuses at temperatures up to 350°C. These thermophiles grow best at temperatures above 80°C and accept temperature up to 113°C. At the other extreme, microorganisms can grow in hypersaline lakes at temperatures as low as −10°C. Regarding their tolerance to pressure, both extremes can be accommodated: there are barophilic deep-sea bacteria, and there are also microorganisms in the mesosphere.

Microorganisms (such as viruses, bacteria, fungi, and protozoa) are as manifold and different as one can imagine. Their size may be smaller than the wavelength of light (e.g., 40 nm for viruses). Virus and bacteriophage are the most simple microorganisms. They consist only of RNA or DNA surrounded by a protein. Both need a host cell for reproduction. Bacteriophages are viruses that infect bacteria. The cells become lysed, which gave bacteriophages their name: "bacteria eaters."

For humans and animals, a variety of microbes are necessary to stay healthy (e.g., in the gastrointestinal tract, skin, mucous membranes, ...), while other microbes can cause severe diseases (pathogens). The human body consists of approximately 10^{13} cells, but also hosts ten times more microorganisms [98, 99].

Finally, microorganisms play an important role in the biosphere as they build up to 90% of the DNA biomass in marine ecosystems, and play a vital role in the geochemical and global carbon cycle.

UV Environment of Microorganisms

The life span of microorganisms outside a host is defined by temperature, humidity, and solar (UV) radiation. Free moving microorganisms receive UV radiation from all directions. Therefore, the optical properties of the surrounding medium, including its reflectance, are of importance.

On the one hand, solid or organic particulate matter in water or air reduces transmittance of UV, and on the other hand, microorganisms have the ability to adhere on surfaces, so they can make use of particulate matter as radiation protection and vehicles. Clumping of microorganisms is also a strategy of radiation protection in an open environment.

Airborne Microorganisms

Microorganisms can be found at all heights in the atmosphere. In the late 1800s and early 1900s, microorganisms were already collected during balloon flights [100, 101]. In 1936, Rogers and Meier [102] collected microorganisms at a height of 20 km. Imshenetsky et al. [103] could even collect reproductive forms of microorganisms (spores of Cicirnella, Pencicillium, Aspergillus, and Papulaspora, as well as Micrococcus and Mycobacterium) from the mesosphere between 48 and 77 km (−60°C) using meteorological rockets. Five of the six microorganisms had synthesized pigmentation. At such an altitude, UV radiation is several times higher than on the ground. Moreover, the spectral range is broader, extending to the UVC range, since most of ozone is below that level, in the stratosphere at about 20–22 km.

Altitude-dependent UV radiation defines the species composition [104]. Bacteria species in the stratosphere are more resistant than in the troposphere or at the ground [105]. Beside pigmentation, clumping and forming aggregates enhances the chances of survival of microorganisms in the stratosphere [105]. In the boundary layer of the atmosphere (i.e., below about 2 km), only around 65% of microorganisms are pigmented [106]. Predominant species are Staphylococcus, Listeria, and Micrococcus.

Small sizes and low density allow microorganisms to remain airborne for long periods before they sediment to the ground. Long-distance transport of fungal spores was observed already in the 1930s [107]. Desert dust enables intercontinental exchanges by offering a UV-shielding vehicle for microorganisms [108–110]. Biomass fires cause strong convection, which is more effective in bringing up microorganisms that are not burned to heights of 3 km than surface stormy winds. Smoke clouds shield them from UV and act as a vehicle over long distances [111]. This explains why, for example, plant fungi can cause allergic reactions or trigger asthma at locations thousands kilometers away from their origin [112].

Airborne microorganisms are influenced strongly by UV. This can be seen by the fact that nonpigmented microorganisms decline strongly with increasing solar UV radiation during the day and during a year. Pigmented microorganisms exhibit a similar pattern but with lower amplitude [113].

Microorganisms released from humans or animals (coughing, sneezing, and talking) into the air have a rather short range (around 12 m) [114] before adhering on surfaces. From these surfaces, they can be re-aerosoled. Survival may last up to 48–72 h. It is assumed that the survival of some microorganisms like influenza virus in air depends less on humidity and temperature than on UV radiation [115, 116].

Waterborne Microorganisms

In aquatic environments, bacterioplankton may contribute up to 90% of the total cellular DNA [117–121], and up to 40% of the total planktonic carbon biomass [122]. In addition, bacteria may process up to 50% of algal primary productivity in marine systems [123].

The size of most bacterioplankton cells precludes effective cellular shading or protective pigmentation [124]. Their metabolism is affected by solar UVB down to a depth of 5 m, and by UVA down to 15 m in clear ocean waters [125]. It has been shown that bacterial abundance [126] depends on light cycle as well as on metabolism [127], DNA, protein synthesis, and degradative enzyme activities [128]. Various studies have investigated the effect of UV on different aquatic systems, especially in marine environments [129–132]. Studies on the impact of UV radiation on bacterioplankton have also been carried out in other aquatic systems, such as (alpine) lakes [69, 76, 128, 130, 132–134].

Other authors have done research on biodiversity at extreme UV locations, such as the Himalayas [135–138] and the Andes [139–141]. Moreover, viruses were found in ice-covered Antarctic lakes [142]. The effects of UV in rivers are still poorly investigated.

The most resistant waterborne microorganisms are *Bacillus* sp., *Acinetobacter* sp., *Pseudomonas* sp., *Sphyngomonas* sp., *Staphylococcus* sp., and *Stenotrphomonas* sp. These were found in lakes [139, 141] and wetlands [140] up to an altitude of 4,600 m, where UV radiation is approximately 170% that at sea level.

Ecologically and economically important are cyanobacteria (blue-green algae), which are found on wet farmland (rice) all over the world. Cyanobacteria are quite sensitive to UV. Cyanobacteria fix atmospheric nitrogen and are important primary producers in the ocean, but are also found in habitats as diverse as freshwater or hypersaline inland lakes, or arid areas where they are a major component of biological soil crusts.

Effects of UV Radiation on Microorganisms

Damage

Microorganisms can be easily damaged by UV radiation. In 1877, Downes and Blunt [143] established that a beneficial effect of sunlight is bactericidal. The experiments that followed revealed that blue/violet is the most effective color to kill bacteria. Hertel [144, 145] could then definitively show that it is the UV part of the spectrum that is responsible. In parallel, it was also found that the bactericidal effect increases with altitude. A detailed review of the early years of research on the damaging effect by UV is given by Hockberger [146].

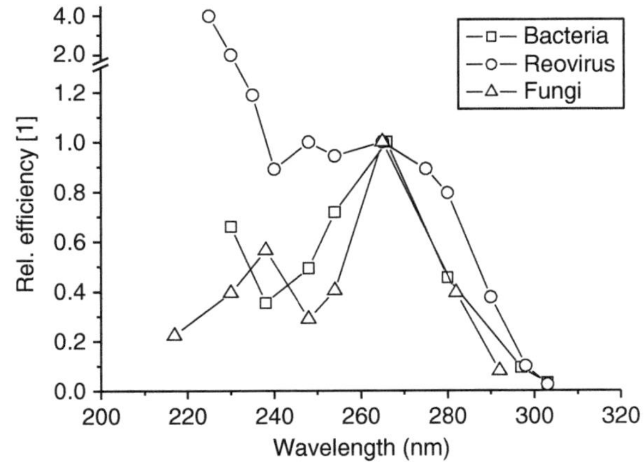

Fig. 22.18 Action spectra for inactivation of Bacteria *Escherichia coli* [11], Reovirus-3 [148], and fungi *Trichophyton mentagrophytes* [12]

Damage can be manifold, and includes growth reduction, decrease of reproduction rate, of metabolism rate, of infectivity, of mobility, or other effects. The main targets for UV damage are RNA, DNA, and proteins.

The wavelength range where UV is most effective in inactivation is quite similar to that where DNA has the highest absorptance (UVC and UVB range) with maximum effectiveness – of course outside the natural UV – i.e., around 265 nm. The range of highest efficiency is also called germicidal range.

Our knowledge of UV-induced damage results mainly from studies in the UVC and UVB, where efficiency is much lower than in the UVA [147]. However, solar UVA is magnitudes higher than solar UVB. It can thus be assumed that solar UVA contributes also to damage, although only a few action spectra have been proposed for the UVA (Fig. 22.18).

Disinfection

Whereas sterilization results in a total absence of microorganisms, disinfection denotes a substantial reduction of the number of microorganism, by at least four orders of magnitude. Hygiene applications like disinfection make use of UV inactivation [149–152]; inactivation means that the existing microorganisms lose their ability to reproduce. UV inactivation results from the direct absorption of UV radiation by the microorganism. This absorption triggers an intracellular photochemical reaction that changes the biochemical structure of nucleic acids. These changes may lead to the inhibition of transcription and replication of nucleic acids, thus making the organism sterile and incapable of spreading infection when entering a host.

At the beginning of the twentieth century, Henry et al. [153] found that UVC's disinfection potential was much stronger than UVB or UVA. After developing a mercury vapor quartz lamp and a quartz tube for protecting the lamp against water, the first application of UV irradiation in disinfection of drinking water was realized in Marseilles, France.

Nowadays, disinfection of air, water, and surfaces is generally done by the application of low- or medium-pressure mercury lamps, which emit up to 90% of their light around 254 nm. However, disinfection of drinking water is also possible with solar UV radiation (e.g., Wegelin et al. [154]). The methods can be as simple as using UV-transmitting PET bottles (e.g., McGuigan et al. [155]) or batch reactors (e.g., Navntoft et al. [156]). Serious pathogens like *Salmonella*, *Shigella dysenteriae*, *Escherichia coli*, *Vibrio cholera*, *Pseudomonas aeuriginosa*, oocysts of *Cryptosporidium parvum*, cysts of *Giardia muris*, *Candida albicans* (yeast), fungus like *Fusarium solani*, or the Poliovirus can be inactivated effectively by solar radiation. With the current booming interest in solar energy applications, various methods of detoxification and disinfection of water by means of concentrated solar radiation are being successfully experimented, both for small-scale (rural) and industrial-scale operations (e.g., [157–159]).

Protection and Repair

The simplest microorganisms have no intrinsic mechanism for UV protection, since their small size does not allow pigmentation or cellular shading. Microorganisms that possess pigments in their outer layer are more resistant. Highly resistant to UV radiation are spores and (oo)cysts. However, UV sensitivity varies a lot, even within a single group, by a factor of up to 10 [160].

Spores and (oo)cysts are reproductive structures that can adapt to dispersal and can survive under unfavorable conditions for extended periods of time. In such a stage, microorganisms have suspended "life," but are able to start again as soon as the environment becomes more favorable.

Mechanical strategies for radiation protection are clumping of microorganisms, forming aggregates, and building films. Dead microorganisms can act as a shield. Additionally, the biological effectiveness of UV is decreased by orders of magnitude if the targeted (pathogenic/parasitic) microorganism has entered a host.

A strategy for UV protection that microorganisms have elaborated is to postpone UV-sensitive processes until the end of the day or nighttime [161]. Microorganisms exhibit a 24-h cell division cycles, whereas DNA replication and cell division occurs during the night, since all division processes are sensitive to solar UV radiation.

Microorganisms – like all other life forms – know how to defend themselves against damage from UV radiation. There are repair mechanisms that may reduce damage to a certain degree. Nucleic acids, for example, can be repaired in a process termed "photoreactivation" in the presence of light, or "dark repair" in the absence of light. Viruses and phages can take advantage of the reactivity capacity of their host.

The action spectra for photorepair exhibit maxima at three different wavelengths: 345, 400, and 430 nm. Figure 22.19 depicts the action spectra for

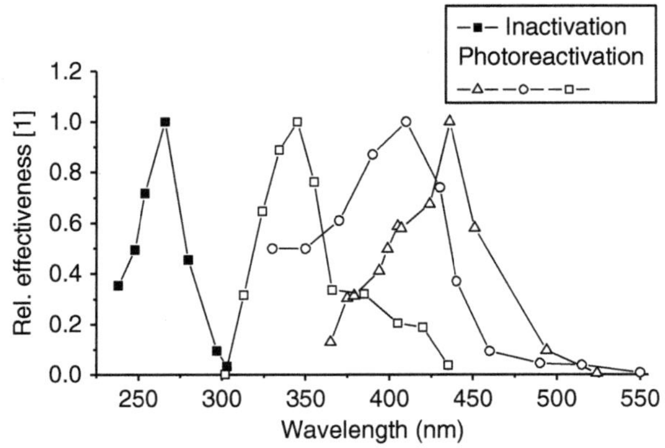

Fig. 22.19 Action spectra for DNA damage and photorepair

damage and photorepair. It can be seen that UVA and blue light are responsible for photorepair.

Since damage is caused by UVB and reactivation by UVA, their ratio is important. This ratio changes during the day, with season, latitude, altitude, total ozone column, and to a certain extent with cloudiness. Low solar radiation, low altitude, high total ozone amount, and high cloudiness tend to favor UVA.

Recent studies have reported an increased UV resistance of environmental bacteria and bacterial spores, compared to lab-grown strains [162]. This means that they could have the ability to adapt to their UV environment.

Vital Effects

Damage is not the only effect of UV radiation on microorganisms. In many species of fungi, for instance, UV may also act by igniting new stages in development, like sporulation. Moreover, a variety of morphogenetics depends on UV radiation. Microorganisms can make use of UV for environmental information too. Metabolism, as well as the circadian clock, can also be triggered by UV.

UV modulates activity of bacteria that are responsible for geochemical transformative reactions in the nitrogen and phosphor cycles [163–165].

Changing UV Environment

As any other organism, microorganisms are well adapted to their environment. However, they have the important advantage to adapt very rapidly to any environmental change, and mutate if necessary.

In general, an increase in UV radiation is detrimental to them, resulting in reduced life span, decreased ability to reproduce, shorter mobility range, reduced habitat, or reduced population.

Still, there are many ways by which microorganisms may profit from environmental or climatic changes. Decreased UV radiation, for example, by enhanced cloudiness, aerosol load or dust, results in decreased damage, which in turn may allow a longer pathway to find a potential host (e.g., [116]). In the case of pathogens, a higher transmission rate and therefore outbreaks of diseases in humans may occur more frequently. For instance, the seasonal outbreak pattern of influenza in many parts of the world is currently believed to result from annual changes in solar UV [166]. Differences in the spring and autumn outbreaks could be explained by different total ozone amounts and cloudiness, which in turn change the relationship between damage and photoreactivation.

If other environmental parameters change (like temperature or the availability of vectors), microorganisms can settle in a new environment where damage through solar radiation is maybe lower and potential hosts are not prepared for the blight. It has been observed that during the past few years, pathogens have already started to migrate toward higher latitudes.

An important factor controlling infections is the immune response of the host which can be suppressed by solar UV [167]. A lowered immune suppression within a human population enables higher infection rates and more frequent outbreaks. How the UV-initiated vitamin D level in humans could influence the outbreak of infection diseases is currently being discussed [168] (Fig. 22.20).

Effects of UV Radiation on Plants

One factor that increases the risk of UVB damage on plants is the fact that they are sessile organisms that are fixed on the ground. UVB damaging effects on plants include the destruction of the cell membranes and all organelles within the cell, including mitochondria, chloroplasts, and deoxyribose nucleic acid (DNA) within the nucleus [169]. These damages to the cell organelles in turn influence the metabolic processes of the plant such as respiration, photosynthesis, growth, and reproduction. These UVB damages eventually impact crop yield and quality. The qualitative effects of UVB radiation exposure on the physiological processes, growth, and yield of plants are shown in Tables 22.3 and 22.4 [170]. Many factors may influence the extent of the UV impact on plants. The intensity and length of exposure, as well as the type of species and cultivar directly influences the sensitivity of plants to UV.

UVB Radiation Damages

Genetic Damage (DNA Damage)

The nucleus of each cell contains deoxyribonucleic acid (DNA), nucleic acid, which contains the genetic instructions used in the development and functioning

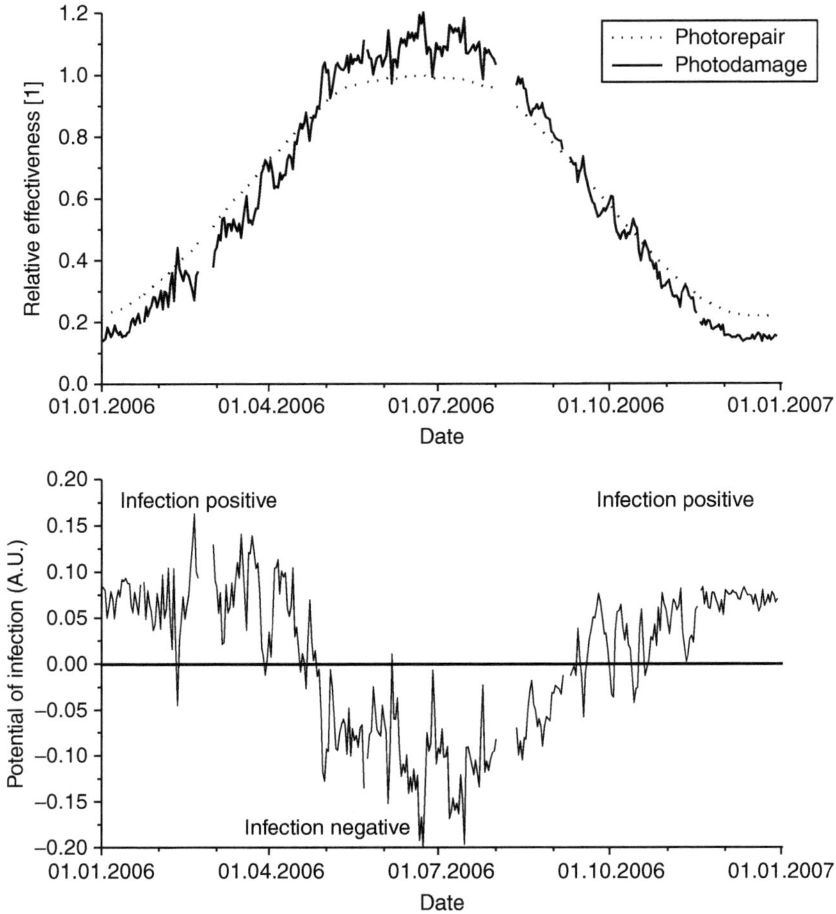

Fig. 22.20 Relative annual course of damaging and reactivating efficiency of solar UV over Vienna, Austria. Calculations are done for clear skies to point out the influence of total ozone

of plants. DNA is very sensitive to UV radiation [81, 171], which is primarily absorbed in the UVB region by the cell. If DNA is exposed to UVB radiation, effects include (1) breakage of bonds in the DNA and DNA-protein crosslinks, (2) chromosomal aberrations, (3) chromosomal breakage, and (4) exchange and production of toxic and mutagenic photoproducts.

Ultrastructural Cell Damage

Ultrastructural changes may occur due to exposure to UV. Among other things, these ultrastructural changes include damage and dilation of the nuclear membrane,

Table 22.3 Effects of UVB exposure on various physiological processes in plants

Trait	Decreases	Increases	No effect
DNA damage		X	
Protein destruction		X	
Fatty acid destruction		X	
Photosynthesis	X		
Photosystem I	X		
Photosystem II	X		
Rubisco	X		
Stomata closure		X	
Chlorophylls	X		
Flavonoids		X	
Waxes		X	
Epidermal hairs		X	
Cuticle thickness		X	
Reproduction			
Pollen viability	X		
Pollen tube growth	X		
Fertilization	X		
Cell division	X		
Cell size			X

Source: Prasad et al. [170]

Table 22.4 Effects of exposure to UVB radiation on various growth and yield parameters in plants

Trait	Decreases	Increases	No effect
Photosynthesis	Y		
Stomatal conductance	Y		
Phenology			Y
Senescence		Y	
Plant height	Y		
Branching		Y	
Leaf area	Y		
Leaf growth and expansion	Y		
Leaf thickness		Y	
Specific leaf weight		Y	
Dry matter production	Y		
Flowering	Y	Y	
Fruit (grain) number	Y		
Fruit (grain) weight	Y		
Yield	Y		
Quality	Y		
Disease incidence			
Powdery mildew	Y		
Rust		Y	
Insect	Y	Y	

Source: Prasad et al. [170]

Fig. 22.21 Example of foliar symptoms (chlorotic patches) on bean leaves strongly enhanced by UV radiation exposure (Taken from [173])

rupture of the chloroplast wall, swelling of chloroplasts, dilation of thylakoids, swollen cisternae in the endoplasmic reticulum, and damage to mitochondria [170].

Inhibition of Photosynthesis

Photosynthesis is the process by which plants convert carbon dioxide and water into organic compounds, especially sugars. The photosynthesis apparatus is one important target site of UVB destruction.

Among the direct effects of UVB that affect photosynthesis are [170]: (1) damage to the ultrastructure of chloroplasts, which are principal sites for photosynthesis; (2) impairment of light energy transfer of photosystem II and to a lower extent photosystem I; (3) reduction of activity of Rubisco (rubilos 1,5-bisphosphate carboxylase-oxygenase); (4) reduction of carbon dioxide fixation and oxygen evolution; and (5) diminution of the starch and chlorophyll content. For field crops [172], a decrease in photosynthesis was observed and was explained both by direct effects (on the photosystem) and indirect effects (decrease in leaf pigments and leaf area).

Plant Morphology and Architecture

UVB leads to strong morphological changes in plants [170]. Leaves that are exposed to enhanced UVB radiation first develop irregular chlorotic patches. With a continuation of exposure to UVB, the chlorotic patches (Fig. 22.21) turn to brown necrotic spots before they later die.

Fig. 22.22 Comparison between leaves (from Patagonian *Jaborosa magellanica* Brisben) from nonirradiated plants (*left*) and leaves from irradiated plants (*right*) (Taken from [174])

Other effects of UVB are: reduction of the plant height, decrease in individual leaf size (Fig. 22.22), decrease of tiller number, and reduction of branch length [172]. These morphological changes also lead to a reduction of the density of the canopy, which in turn intercepts less UVB radiation than a plant grown under favorable conditions.

Phenology and Reproductive Processes

Neither early bud, flower development, nor the time of first flower is influenced by UVB radiation. Shedding of early-formed floral buds is usually the reason for any time delay of first flower observed in crop species. However, in many plant species, UVB radiation affects the size of flowers, anther number, pollen production, pollen germination, and pollen tube growth [170].

In general, the reproductive organs of most plants are highly protected. Sepals, petals, and ovary walls "screen" the reproductive organs from UVB. Pollen is therefore "at risk" when it falls on the stigma. Then pollen germination and rate of pollen tube growth are also affected by UVB radiation. A decrease in pollen tube growth by 10–25% may result. The fertilization process of sensitive plants is affected too, resulting in fewer seeds in these plants. However, the walls of the style and ovary may provide some protection once the pollen tube has penetrated the stigma.

Growth and Dry Matter Production

Increases in length of organs (e.g., roots, leaf, or stem) or in weight are usual indicators of plant growth. For any given plant, several "quantities," such as dry matter increase, increase in cell number, or increase in volume, are generally used to monitor growth [175]. For many plant species, studies in the field [176–179] or in controlled environment [180–182] have shown a decrease in growth of leaves and of stems caused by enhanced UVB radiation (Fig. 22.22). The observed growth reduction is explained by a reduction of the cell division rather than by a reduction in cell dimension. Reduction in plant height in connection with enhanced UVB radiation is explained by decreased growth hormone levels. On one hand, smaller and more compact canopies will better protect the plant by transmitting less solar radiation and UVB. On the other hand, the potential, or total, photosynthetical area, which is essential for growth, will be reduced [170]. Altogether, these factors will lead to a reduction in total dry matter and in biomass production. According to [170] 60% of crop species show a reduction in dry matter production after exposure to UVB radiation, 24% show no change, whereas only 8% show an increase in dry matter production.

Strategies for Protection Against UVB Radiation

During the process of evolution, plants developed repair and defense mechanisms by which UVB damages to the cell are limited and tolerance to solar UV-radiation is increased. Several UV-driven repair and defense mechanisms exist, which modify the optical characteristics of the leaves or other parts of the plant, or which work at the biochemical-molecular level [169, 170].

Repair Mechanisms

Cellular life forms usually possess repair enzymes that can recognize chemically modified bases, including those formed by UV radiation [170, 183]. In addition, a variety of biochemical mechanisms exist to restore the integrity of the genetic material after DNA damage, and thus to maintain its stability. These DNA repair mechanisms include photoreactivation, excision, and postreplication repair. During the process of photoreactivation, an enzyme (photolyase) responsible for the splitting of pyrimidine cyclobutane dimers is involved. Excision repair consists in eliminating the damaged part of DNA by removing the bases in the damaged strand and then by synthesizing the gap. During the process of postreplication repair, the DNA damage is bypassed during DNA replication, and the resulting gaps are later filled in by using the sister duplex information. These kinds or repair mechanisms are observed in chloroplast and nuclear DNA [170, 183]. The ability to repair DNA varies a lot among the different plant species.

Defense Mechanisms

The first kind of defense mechanism is based on a better adaptation of the surface structure, physiology, and composition of the epidermal layer to attenuate the transmission of UVB through the epidermal layers, and eventually to better shield cells from UVB radiation [183]. The protective structures have the capacity to attenuate radiation damage by reflecting, absorbing, and scattering the incident flux. Among other mechanisms, hairs and wax coating have a prominent role [184]. UVB radiation leads to oxidative stress in plant systems, similarly to what is observed for abiotic and biotic stress. As a result, plants increase production of some chemicals such as flavonoids and antioxidant enzymes, which provide defense against UVB radiation [185–187]. Flavonoids, which are produced and mostly deposited in leaf hairs and in epidermal and mesophyll layers, are very efficient in mitigating the effects of UVB radiation. In this way, flavonoids may reduce damage to sensitive cell organs (e.g., DNA, chloroplasts, and mitochondria). Other compounds that could have the potential to protect against UV radiation effects are anthocyanins and carotenoids. They can protect the pollen grains, especially in flowers. They do not directly influence photosynthesis and other physiological processes, since they attenuate incoming radiation only in the UVB spectral range and not in the range of photosynthetic active radiation [170].

Selection or Genetic Improvement

Plant species adapt and protect themselves from UVB in more or less adapted ways. The variable stimulation of protective and repair mechanisms may explain the difference among the various plant species. These variations provide an opportunity for genetic improvement – either through traditional plant breeding techniques (crossing or selection) or through modern molecular biology techniques, such as plant transformation including some genetic modification. As a possible method, it may prove useful to identify tolerant species or cultivars by screening wide germplasm from various locations, and investigate the origin of UV tolerance [170].

Effect of UVB on Aquatic Plants and Macroalgae

In general, UVB radiation constitutes a significant stressor for macroalgae, aquatic mosses, liverworts, and aquatic flowering plants [130].

UVB affects macroalgae on the cellular, molecular, individual, and community levels [188]. UVA and UVB both affect photosynthesis and growth rate, and lead to accumulation of DNA damage. In addition, some studies report about the "impairment" of phototaxis used, e.g., by motile gametes of brown algae to accumulate at the water surface in order to increase the chance of finding a mating (do you mean: mating?) partner [189].

UVB sensitivity varies a lot among species. This results in vertical and in geographical "distributions" of the various species according to their sensitivity to UV. UV-tolerant species populate the tidal zone and are in general closer to the water surface. More sensitive species are found in deeper waters [190, 191]. Seasonal changes in UV and visible radiation also result in a succession of species over the year [192]. Macroalgae show an "ability" to adapt to UV. For instance, young specimens are more sensitive to UV than older specimens, and species collected shortly after the winter are found to be more sensitive than those harvested later in the year [193]. Most macroalgae have efficient repair mechanisms. In many macroalgae, efficient ROS (reactive oxygen species) (has ROS been defined earlier?) scavenging enzymes are found in addition to the DNA repair mechanisms.

Aquatic mosses and liverworts show a UVB-related inhibition of photosynthesis, growth, and pigmentation [194, 195].

For aquatic flowering plants, studies [130] show that the photosynthetic quantum yield dramatically decreases under unfiltered solar radiation, and that removal of UVB or total UV leads to an increase in photosynthetic activity [196]. Experiments show also an efficient adaptation of sea grasses to solar UV [197].

UVB and Agricultural and Natural Terrestrial Plants

A meta-analysis comparing the overall reaction of woody and herbaceous plants under high supplemental UVB levels showed that the changes in "physiological" variables due to UVB observed in woody plants were significantly smaller than those observed in herbaceous plants [198].

Methodologies Used for the UV Experiments

To investigate the effects of UV on plants, two different methodologies are followed:

1. Experiments in climate rooms and glasshouses
2. Experiments in the field

The experiments in climate rooms and glasshouses have the considerable advantage of a controlled environment. The drawbacks of this methodology are the unfavorable PAR to UVB ratio and differences between the PAR and UV spectra of the lamps and that of solar radiation. Results from indoor studies cannot therefore be easily extrapolated to outdoor conditions. Conversely, results from outdoor experiments are more realistic than those from indoor studies [199, 200]. Outdoor studies consist of experiments using some UV lamps to increase, or filters to decrease, the UV levels. In general, the increase or decrease in UV should be comparable to expected realistic changes in UV that may be simulated using a radiative transfer model.

UVB and Woody Plants

Studies showed contradictory results regarding the effect of UVB on growth and photosynthesis in woody plants. Some investigations [201] demonstrated a sensitivity of woody plants to UVB irradiance, and also showed that these kinds of effects may be cumulative [202]. Other studies, however, found that even large increases in UVB radiation would not lead to any harmful effects on growth and photosynthesis in some woody plants [203, 204].

Radiative transfer in tree canopies is very complex: radiation incident on leaves depends on the orientation and inclination of the leaf and on the position of the leaf in the canopy. Whereas the upper foliage may be fully exposed to UVB, the lower foliage is benefits from the shading by branches, leaves and twigs. This results in a very variable radiation environment with sunflecks and gaps. In various forests with closed canopies, as little as 1–2% of the incident UVB may be transmitted to the lower levels [205].

At the individual leaf level, surface reflectance and absorption by pigments are major determining factors for the plant's UVB sensitivity. In general the UVB surface reflectance of leaves is lower than 10% [205]. The UVB surface reflectance depends on leaf surface waxes and hairiness. Leaf hairs are one major protector against UVB radiation [206–208].

In general, the amount of UVB radiation reaching the photosynthetic mesophyll is relatively low in evergreen species and higher in deciduous species [209], due to a higher concentration of absorbing compounds in the epidermis of evergreen plants.

UVB and Herbaceous Plants

Studies showed that herbaceous plants would suffer a negative effect in biomass production under elevated UVB radiation levels. Reductions of 7–14.6% have been found, compared to values under normal ambient UV levels. Elevated UVB leads to decrease in plant height and specific leaf area, as well as increases in herbaceous UVB absorbing compounds [198].

UVB and Field Crops

Indoor or glasshouse experiments conducted with additional UVB lamps to increase UVB, or with filters to reduce UVB, usually show dramatic effects of UVB radiation. Due to the supplemental UVB sources, however, the tested UV levels are usually much higher than what would ever occur under natural outdoor conditions. Several studies also showed that it is much more complex to simulate and investigate the effects of changing UVB that would occur under natural conditions [172].

Table 22.5 Selected results quantifying the influence of increased UV radiation on crop yield

Crop	Simulation of O_3 depletion (%)	Experimental condition	Observed change in yield (%)	Reference
Barley	–	F	−17 to −31	Mazza et al. [176]
Black gram	−15	F	−63	Singh [214]
Corn	−20	F	−22 to −33	Correia et al. [177]
Mung bean	−15	F	−76	Singh [214]
Soybean	−16	F, GH	−41	Teramura and Murali [216]
Wheat	−12, −20, −25	F	−43	Li et al. [215]
	15	F	15	Al-Oudat et al. [217]

F outdoor experiments, *GH* Greenhouse experiment
Source: Kakani et al. [172]

Many of the studies showed that enhanced UVB radiation levels could lead to visual symptoms consisting of chlorotic or necrotic patches on leaves exposed to UVB. UVB radiation alters the vegetative and reproductive morphology. Changes in leaf anatomy consisting of changes in thickness of epidermal, palisade, and mesophyll layers occur. After enhanced exposure to UVB, a decrease in chlorophyll content (10–70%), and an increase in UVB-absorbing compounds (10–300%) are observed in many crops [172]. A decrease in photosynthesis (3–90%) is particularly observed at higher UV doses.

Yield and Yield Components

The decrease in chlorophyll pigments and photosynthesis usually results in lower biomass and yield of crop plants. Some genotypes of crop species show a thicker leaf wax layer, a loss of chlorophyll, and an increase in phenolics (that allow a better UVB protection), ultimately resulting in changes in biomass and yield [172]. The effects of UVB on yield vary very much with crop species, however. In any case, most of the studies show that UVB increases lead to yield loss (Table 22.5).

Some species such as cowpea (*Vigna unguiculata*) [210, 211], millet (*Setaria italica*), and tobacco (*Nicotiana tabacum*) [212] show almost no yield reduction. Other species (e.g., pea [179], barley (*Hordeum vulgare*) [176, 179], mustard (*Brassica nigra*) [213], Black gram, Mung bean [214], and wheat [215]) show a strong reduction (Table 22.6). This yield loss consists of a reduced fruit grain number due to failure in fertilization, destruction of fruiting structures, and reduced fruit size due to decreased supply of assimilates to the growing sink (fruits). UVB also affects the yield quality. For instance, protein content and seed oil are reduced in soybeans that are exposed to enhanced UVB radiation.

Table 22.6 Sensitivity of selected crops to enhanced levels of UVB radiation. Results of experiments performed in controlled environments

Sensitive	Moderately sensitive	Relatively tolerant
Barley (*Hordeum vulgare*)	Common bean (*Phaseolus* spp.)	Corn (*Zea mays*)
Carrot (*Daucus carota*)	Lettuce (*Lactuca sativa*)	Cotton (*Gossyptum hirsutum*)
Cucumber (*Cucurnis sativus*)	Peanut (*Arachis hypogaea*)	Cowpea (*Vigna unguiculata*)
Mustard (*Brassica* spp.)	Pepper (*Piper nigrum*)	Clover (*Trifolium* spp.)
Oats (*Avena sativa*)	Petunia (*Petunia* spp.)	Millet (*Setaria italica*)
Pea (*Pisum sativum*)	Potato (*Solanum tuberosum*)	Radish (*Raphanus sativus*)
Soybean (*Glycine max*)	Rice (*Oryza sativa*)	Sunflower (*Helianthus annuus*)
Sweet corn (*Zea mays* var. *saccharata*)	Rye (*Secale cereale*)	Tobacco (*Nicotiana tabacum*)
Tomato (*Lycopersicon* spp.)	Sorghum (*Sorghum vulgare*)	Wheat (*Triticum aestivum*)

Source: Prasad et al. [170]

Animals

Animals are divided here depending on the utilization of the species in the following four groups: pets (animals kept for companionship and enjoyment, or a household animal), farm animals (including horses for sport reasons), zoo animals, and finally fish and fishery-related animals.

A general finding is that animals show similar UV-induced changes of eye and skin as humans do. For the skin, two distinctive forms can be distinguished: malign melanoma and non-malign skin cancer, predominantly squamous cell carcinoma (SCC), and basal cell carcinoma (BCC).

Pets

UV Hazards

The UV exposure of pets depends strongly on how these animals are kept. Cats and dogs can stay outdoors for longer periods, whereas several other species that are kept as pets live predominantly indoor. For humans, it is well known that the incidence of melanoma and non-melanoma skin cancer depends on the UV exposure, which can be assessed by the geographic latitude, elevation, and/or annual global radiation (itself related to sunshine hours). For dogs, this correlation between environmental factors and the incidence of skin neoplasms could not be found in experiments conducted in Australia (Melbourne and Queensland), USA, and UK [27, 218–227]. Two arguments could explain this discrepancy between humans and dogs: first, the fact that dogs stay predominantly indoors, so that the influence of environmental factors is reduced, and second, the different distribution of breeds in these regions. For farm animals, it is well known that some breeds are not adapted to areas with high UV exposure due to lack of pigmentation (e.g., Herford cattle).

Teifke and Lohr [228] analyzed 106 SSC immunohistochemically for overexpression of p53, which is an indicator of a UV-related incidence. In 9 of 11 (82%) feline SCCs of the ear and in 7 of 14 (50%) feline SCCs of other locations, p53 immunoreactivity was detected. For dogs, 7 of 25 (30%) cutaneous SCCs gave a positive reaction. This is a good indicator that UV exposure is an important external factor for SCC.

The clinical picture of eye cancer, such as cutaneous SCCs in the close vicinity of the eye, could be observed in guinea pigs, hamsters, rats, and mice, which are used as pets as well. Pigmented rats and mice showed a significant reduction in the incidence of development of such cancerous eyes compared with nonpigmented animals [229].

Dogs A study on humans estimated that 0.25–1% of actinic keratosis (AK), which is mainly caused by UV exposure, progress toward SCC. Similar data do not exist for cats or dogs. Nevertheless, it can be observed that AK is common in the vicinity of SCC. In dogs, they are found primarily on the central abdomen; ventral flanks; and medial thighs of short-coated, white-haired, or piebald breeds. Dalmatians, Pit Bull Terriers, Beagles, Basset Hounds, and other dogs with similar coat characteristics have an increased incidence of AK, as well as other solar-induced neoplasms [230]. The occurrence of AK on the nose is often called "collie nose" (nasal solar dermatitis), even though it is not restricted to this breed [231]. As possible therapy, the application of sun screens with a high sun protection factor, or the tattooing of the nonpigmented areas of the nose, is recommended.

A genetically homogeneous group of 991 Beagles was housed in gravel-based, outdoor pens with doghouses in a high-altitude, high-sunshine environment, in Colorado (40.5°N, USA). Solar dermatosis was restricted to the sparsely haired, nonpigmented abdominal skin. Solar dermatosis was diagnosed in 363 of the 991 dogs, representing an incidence of 36.6%. There were 175 cases of hemangiomas, hemangiosarcomas, or SCC of the skin among the 991 dogs. Of these, 129 tumors occurred in dogs with, and only 46 in dogs without, solar dermatosis [232].

Nielsen and Cole [233] found that there is no preferred location for SCC, contrarily to humans, whose preferred sites are the most UV-exposed areas of the body.

The influence of age is mentioned by Nesbitt [234], who observed the occurrence of SCC mainly for dogs older than 5 years. The cumulative incidence for such lesions for Beagles was observed with 47% at the age of 12 years [235]. For a genetically homogeneous group of 991 Beagles, Nikula et al. [232] investigated the role of age on the cumulative incidence of hemangiomas, hemangiosarcomas, and SCC. For SCC, the slope gets steeper for an age of 7 years (Fig. 22.23).

A breed disposition for the incidence of SCC was observed for dogs with insufficient pigmentation on the lips and nose, as well as on toes and ventral

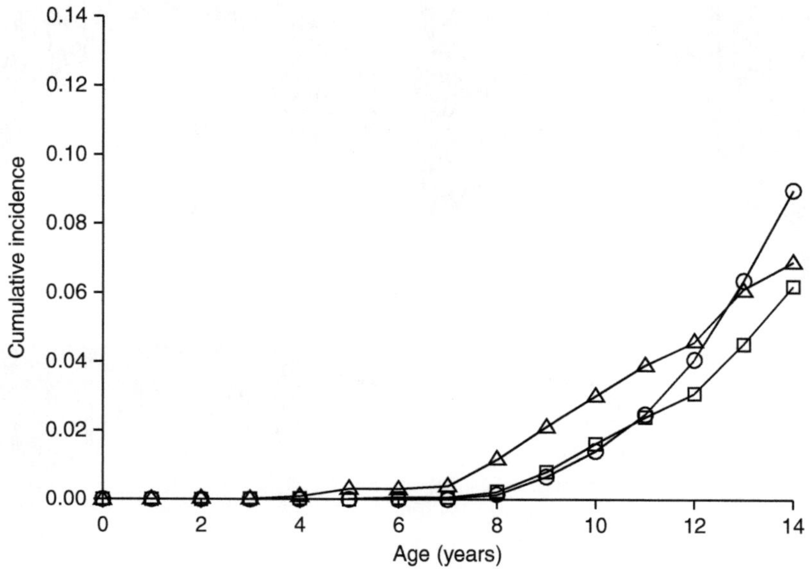

Fig. 22.23 Cumulative incidence of hemangiomas (○), hemangiosarcomas (□), and SCC (Δ) for Beagle dogs (Source: Nikula et al. [232])

abdomen. Worse affected are: White German Shepherd, White Australian Shepherd, and Welsh Corgi [234].

Cats In the case of cats from sunny places, SCC could be found on the head. Risk factors are UV radiation, nonpigmented skin, and thin hair [236]. Actinic keratoses occur most often on the pinnae, nose, and eyelids of white-faced cats [230]. The influence of pigmentation was shown by Dorn et al. [237] for white-faced, blue-eyed cats. The prevalence of SCC in white cats is 11-fold higher than in nonwhite cats (Fig. 22.24).

In cats, melanomas could be hardly observed. After Jörger [236], no race-specific or sex-specific predispositions are known.

Vitamin D in Pets

Cats and Dogs Vitamin D is synthesized by UV in the skin of omnivores (rats, pigs, humans) and herbivores (horses, sheep, cattle). On the contrary, carnivores are solely dependent on oral intake to meet their vitamin D requirement [220–222].

How et al. [223] have shown that there is no photosynthesis of vitamin D_3 in cat skin. Although not strictly carnivores, dogs lack in the ability to photosynthesize

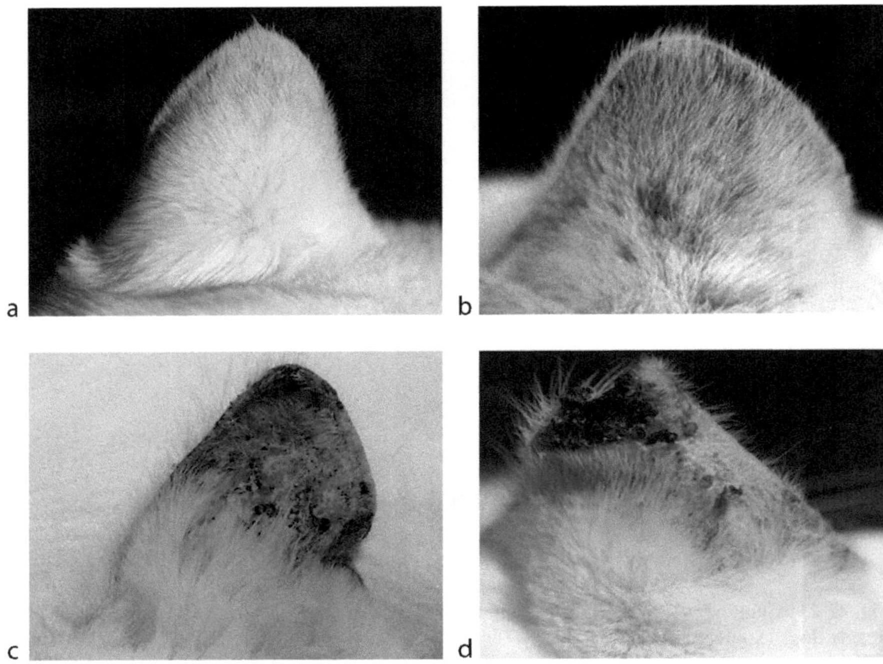

Fig. 22.24 Demonstration of progressive stage of photo damage shown on the ear of cats. (**a**) Normal, (**b**) initial photo damage, with erythema and scaling, (**c**) advanced photo damage, with erythema, scaling, erosion, ulceration, and plaques, and (**d**) advanced photo damage and SCC (Source: Almeida et al. [238])

vitamin D_3 [223]. It could be shown that hypovitaminosis D occurs in dogs irradiated with UV but with no vitamin D in their food [220].

Both natural food and commercially available complete dog food contain sufficient vitamin D_3 to fulfill a dog's requirement. Therefore, in domesticated dogs and cats, rickets is only seen under extreme circumstances, such as a strict vegetarian diet, biliary atresia, and inborn errors of vitamin D_3 metabolism [226]. Hypervitaminosis D and coupled hypercalcemia are rarely reported [225].

Rabbits In rabbits, there is a clear difference in vitamin D concentration between animals housed in hatches and animals kept in free range. A strong seasonal variation in plasma vitamin D_3 concentration has been reported in rabbits [219]. For animals that had the opportunity to bask in the sun, the highest concentration was found between May and September. It takes about 5 months for the vitamin D reserve to become depleted after exposure has stopped, in conjunction with a vitamin D deficient diet [224]. Vitamin D content of forage and vegetables being negligible, only high-quality hay could contribute to the vitamin D level in rabbits.

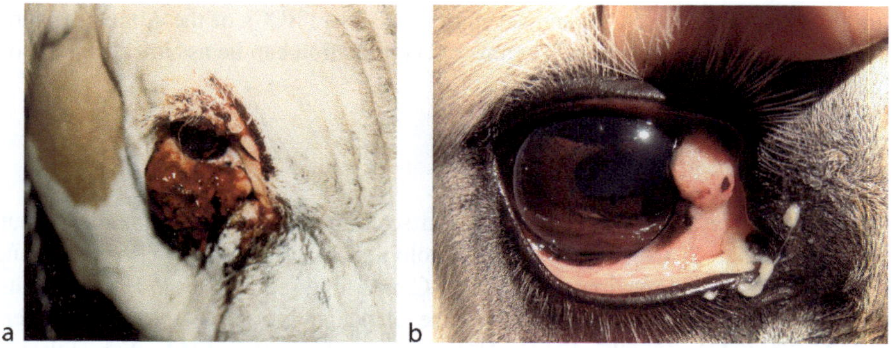

Fig. 22.25 Cancer eye (ocular squamous cell carcinoma) around a sparsely pigmented skin area (**a**) cattle (Source: [9]) and (**b**) horse (Source: [241])

Rats The fact that UV radiation cures rickets in rats was reported by Sonne and Rekling [27]. Beside a dose–response relationship, they were also able to derive an action spectrum. This action spectrum is quite similar to that for humans, with high efficiency in the UVB. How et al. [222] showed directly that after UV exposure an increase in vitamin D_3 in rats could be observed.

Farm Animals

UV Hazards

Beside the known short-term effects like erythema, several long-term effects can be detected, such as elastosis and skin cancer. Farm animals of several species – cattle [9, 239], goats, sheep [240], and horses [241] – can develop skin cancer, mainly squamous cell carcinoma (SCC), in sparsely haired and sparsely pigmented areas of the skin.

Several farm animal species can also develop the so-called cancer eye, which is a squamous epithelial carcinoma near the eye. This disease typically occurs because the eyelid is hairless, and is dependent on both pigmentation and UV exposure. This cancer eye (Fig. 22.25) can be observed predominantly in Herford and Simmentaler cattle, Ayreshire, Shorthorn, black-pied cattle, Indian water buffalo, and their cross breedings [9, 239], but also in horses [241].

Teifke and Lohr [228] analyzed 106 SCC immunohistochemically for overexpression of p53, which is an indicator of a UV-related incidence. All of six

(100%) equine ocular SCCs and seven of nine (78%) SCCs of the equine penis or vulva gave positive reactions. Therefore, UV radiation can be assumed as a major etiological factor for SCC.

Cattle Heeney and Valli [242] reviewed several epidemiological studies about bovine ocular SCC (cancer eye). Etiological factors were UV radiation, circumocular apigmentation, and viruses. Concerning pigmentation, Hereford cattle are commonly affected. The incidence of ocular lesions was found between 8.1–45.7% [243], 35–58% [244], 21% [245], and 25–35% [242]. For Holstein cattle, a much lower incidence was observed, <1% [242], as a result of a denser circumocular pigmentation.

In a study of Gharagozlou et al. [246], 32 cases of ocular neoplasms were diagnosed. The affected animals were female (100%), adult and more than 50% of them aged more than 5 years. In most cases (70%), the lesions were located in the nictitating membrane and palpebral conjunctiva. Intraocular invasion was noted in seven cases (22%). Microscopically, in 12 cases out of 32 (38%) the tumors were noninvasive SCC or carcinoma in situ; 18 cases (56%) were invasive SCC; a single case (3%) was lymphosarcoma, while a further single case (3%) was malignant hemangioendothelioma.

Morris [247] investigated the heritability of predictors that are closely related to skin cancer. The heritability of the eyelid pigmentation was estimated at 0.64–0.83. The Hereford cattle also appear to be highly sensitive to eye cancer. Estimates indicated a moderate heritability: 0.17–0.29 for eye cancer, 0.10 $\pm$ 0.08 for the incidence, and 0.30 $\pm$ 0.09 for the number of tumors.

The influence of UV exposure on the incidence of cancer eye can be related to geographic latitude, duration of sunshine, and altitude. Anderson and Skinner [248] classified 5,000 Hereford cattle into groups, describing the UV exposure based on latitude, sunshine duration, and altitude (Table 22.7). The stratification of the data showed that for all three parameters, the incidence of skin cancer stays proportional to the UV exposure.

Anderson and Badzioch [245] performed epidemiological investigations on 34 Hereford herds in North America (Table 22.8). For some parameters that are correlated with UV intensity (sunshine duration, elevation, cloudiness index, and global radiation) a relationship with ocular SCC was found. The odds ratios for these parameters, as a measure for the risk to develop a SCC, are summarized in Table 22.8. Because of the similar geographical latitudes of the statistical investigated territory, these results can be applied to midlatitude countries and southern Europe as well.

In Zimbabwe, ocular SCC was frequently observed in Simmental cattle, and exposure to intense solar radiation has been proposed as the cause, especially when cattle are kept at high altitude (1,500 masl) in a sunny and warm climate [249]. No cases were observed for fully pigmented cattle breeds.

Table 22.7 Relationship between solar UV exposure and the incidence of cancer eye in Hereford cattle

				Arithmetic mean within each level		
Criterion	UV exposure	# animals	Incidence	Latitude (°)	Sunshine (h)	Altitude (masl)
Latitude (°)	Low	3,445	3.6 ± 0.3	46.4	2,593	809
	Medium	361	7.8 ± 1.4	43.4	2,840	1,143
	High	1,154	9.2 ± 0.8	33.6	3,150	776
Sunshine duration (h)	Low	3,445	3.6 ± 0.3	46.4	2,593	809
	Medium	823	5.6 ± 0.8	38.1	2,800	691
	High	692	11.7 ± 1.2	33.4	3,400	1,068
Altitude (masl)	Low	670	4.8 ± 0.8	36.1	2,880	357
	Medium	3,909	5.1 ± 0.1	44.4	2,718	846
	High	381	9.7 ± 1.5	43.9	2,704	1,433

Source: Anderson and Skinner [248]

Table 22.8 Relationship between solar UV exposure and the incidence of cancer eye in Hereford cattle in North America (After Anderson and Badzioch [245]) (Global radiation is expressed in ly (1 ly = 4.184 J/cm^2), and cloudiness index as the global horizontal irradiation normalized with its extraterrestrial equivalent)

UV exposure described by	No of animals	Affected (%)	Odds ratio
Sunshine duration (h)			
1,800–2,299	273	13.6	1.0
2,300–2,599	335	16.1	1.1
2,600–2,899	1,158	21.2	1.8
2,900–3,499	1,009	25.7	2.5
Global radiation (ly)			
200–299	283	17.3	1.0
300–399	1,455	19.1	2.1
400–499	1,037	25.8	3.1
Cloudiness index (–)			
0.400–0.499	1,067	13.6	1.0
0.500–0.599	690	27.2	3.4
0.600–0.699	1,018	25.7	2.9
Altitude (ft)			
0–1,999	1,308	13.7	1.0
2,000–2,999	564	30.3	2.8
3,000–3,999	381	33.6	4.6
4,000–9,200	522	22.4	3.2

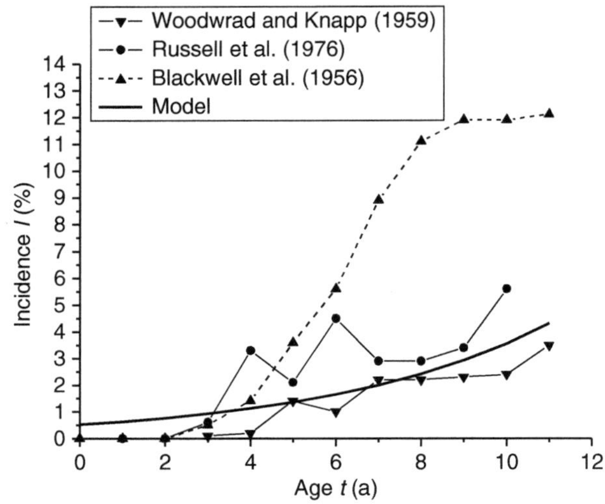

Fig. 22.26 Influence of age (years) of Hereford cattle on the incidence of cancer eye $I = 0.519\ \exp(0.193\ t)$ (%) (Data for the regression model from [244, 250])

The influence of the age t (in years) of Hereford cattle on the incidence I (in %) is depicted in Fig. 22.26, showing an exponential growth according to $I = 0.519\,\exp(0.193\ t)$ (with an adjusted coefficient of determination $r^2 = 0.541$), based on the data from Woodward and Knapp [250] and Russel et al. [244]. The half value period gives 3.6, which means that the incidence I doubles every 3.6 years. The first data set was collected in Montana at a latitude of about 46°N, and the second one in Colorado at a latitude of about 37°N. Anderson and Badzioch [245] found a 2% increase of risk per month of age, compared to 1.6% per month for the exponential model in Fig. 22.26. For New Mexico, Blackwell et al. [251] found an incidence for ocular SSC which was 2.5 times larger than in the two previous data sets. This could be caused by different climatic conditions, combined with a lower geographic latitude of about 35°N. Therefore, these data were not included into the regression model for the incidence (Fig. 22.26).

For cattle, SCC could be observed also on the vulva [232]. Wettmuny [252] found that pigmentation is a main criterion for the incidence of vulva carcinoma in Ayreshire cattle in Ceylon (8°N; 1,800 masl). The same relationship was obtained by Hünermund [253] for Kenya (1°S, 2,200 masl). The incidence of SCC on nonpigmented areas of the vulva reached between 1% in Israel (30°N) [254] and about 10% in Ceylon [255].

Like the skin, the eye is also sensitive to UV exposure. In an epidemiological survey about infectious keratoconjunctivits in Australian cattle, Slatter et al. [239] derived a relationship between season and frequency of disease. Similarly, Kopecky et al. [256] established that solar radiation was an amplifying factor for this disease.

Small Ungulates (Sheep and Goat) For sheep and goat, SCC could be observed on the vulva, eyelid, ear, and other parts poorly covered by wool [232].

Table 22.9 Adjusted odds ratio (OR) of variables that are correlated with the occurrence of ocular SSC in horses (After Dugan et al. [259]). Variables with a nonsignificant OR ($p < 0.05$) are not shown here

Variable		OR	p-value
Breed	Mixed breed	3.1	<0.001
	Paint, pinto	4.5	<0.001
	Apaloosa	7.6	<0.001
	Draft breed (Belgium, Clydesdale, Shire)	21.6	<0.001
Hair color	Chestnut, sorrel	3.8	<0.001
	Buckskin	4.4	<0.01
	Red/white or strawberry/white	4.7	<0.001
	Gray	6.7	<0.001
	Creamello, polomino	13.7	<0.001
	White	26.7	<0.001

The clinical and pathological characteristics of the outbreak of SCC observed in that study suggest that the breed of sheep, its lack of skin pigmentation, the farm's southern latitude (where UV is comparatively stronger than at the same latitude in the northern hemisphere), and factors related to the type of farming played a decisive role in the development of SCC in these sheep [240]. For Marino sheep, Forrest and Fleet [257] could show the lack of tanning capability as well as a distinct erythemal response to UV radiation. Due to a long-term UV exposure of over 28 days, the thickening of the epidermis from 23 to 120 μm was observed.

Horse For horses, a histological examination of 21 suspected neoplasms confirmed the presence of SSC (76% of cases). The remaining cases were diagnosed as lymphoid hyperplasia (14%), mast cell tumor (5%), and sebaceous gland adenocarcinoma (5%) [241]. Beside SCC, melanoma is also a common equine tumor [258]. Approximately 80% of gray horses over 15 years of age show melanotic growths. The neoplasms are generally found at the perineum, vulva, and undersurface of the root of the tail, but also on the male genitalia, neck, and ears. Melanomas remain benign for 10–20 years on the skin; however, the disease is rapidly fatal once vital organs are involved.

Genetic disposition of horses for ocular SCC were found by Dugan et al. [259], who analyzed the databases of 14 veterinarians in the USA. It was found that breed and hair color had a significant effect on the prevalence of ocular SSC. The increasing prevalence with increasing age can be calculated by the odds ratio (OR). The influence of UV radiation on the prevalence of ocular SSC was found as a function of longitude (do you mean: latitude?), altitude, and annual solar exposure [259], win good agreement with epidemiological studies for other species (e.g., cattle [245]).

The adjusted OR results summarized in Table 22.9 provide the relative prevalence $P_{\mathrm{rel}}(x, y)$ describing the change of prevalence of an older horse (age x, in years)

in relation to a young horse (age y, in years) according to $P_{\text{rel}}(x, y) = OR^{x-y}$. The age effect on OR is 1.1 ($p < 0.001$). This means, for instance, that a 12-year-old horse has a 2.6-fold higher risk to develop ocular SSC than a 2-year-old horse (of the same breed and hair color) based on $P_{\text{rel}} = OR^{12-2} = 1.1^{10}$.

Photosensitivity Photosensitivity defines an abnormal reaction of the skin to visible light and UV radiation. Basically, two forms of sensitivity can be distinguished: Phototoxic reactions and photo-allergenic reactions. The former occur after absorption of radiation by molecules, and the membrane and DNA damages that follow. The reaction of the skin is restricted to exposed areas. Photo-allergenic reactions are a more seldom phenomenon, where the agent evokes immunologic reactions. An observed photo-allergenic reaction can persist many years, even when the causing noxa has already been eliminated.

Photosensitive substances can be found in either drugs or feed. For instance, buckwheat (*Fagopyrum esculentum*) is known to cause photosensitivity.

This disease can be observed mainly in sheep and swine, rarely in cattle and goat, and hardly in horse. Saint-John's-wort (*Hypericum perforatum*) causes a disease in sheep, which manifests as a depigmentation by hypericin. Other sensitizers exist, such as furanocoumarin, which can be found in apiaceae or umbelliferae (like hogweed [*Heracleum mantegazzianum*], chervil [*Anthriscus cerefolium*], angelica [*Angelica archangelica*]) and yarrow (*Achillea millefolium*).

The application of veterinary medicines and medicated feed may expose agricultural workers to antimicrobial drugs, tranquilizers, and other chemicals with phototoxic and photoallergic side effects, like phenothiazin, acriflaviniumchlorid, tetrazyklin, mansonil, or sulphonamide. Olaquindox, a growth promoter for swine, is known to cause photosensitization in farmers, probably due to a photoallergic mechanism. Moreover, olaquindox was found to be phototoxic in animal experiments [260–263].

Vitamin D

Vitamin D_3 is synthesized by UV radiation in the skin of omnivores (pigs, humans) and herbivores (horses, cows, sheep, goats). This synthesis works in fur-bearing animals even in the presence of hair.

For animals, both types of vitamin D – namely, cholecalciferol (vitamin D_3) and ergocalciferol (vitamin D_2) – are of importance. Whereas vitamin D_3 is produced in the skin, vitamin D_2 is ingested with feed.

Vitamin D is negligibly low in forage and silage, as well as in corns, roots, seeds, and grains. The only exception is silage made of maturity corn. Contrarily, hey contains a significant amount of vitamin D.

For animals in agricultural use, the antirachitic properties of vitamin D are brought by vitamin D_2 supplementation to food [8]. From animal welfare reasons outdoor areas for farm animals are also quite common. In such a case, supplementation is not necessary [264]. The free-range practice can also be a problem, because most swine have a very low minimal erythema dose (in MED) due to a lack of photoprotection by hair coat or low pigmentation. Exceptions are found in older races, like Mangaliza pigs, which are well adapted to solar UV radiation.

In calves, exposition to UV was shown to enhance weight gain [265] compared to nonexposed calves.

Cattle can receive vitamin D from two sources: either UV radiation or plants exposed to the sun. Vitamin D_2 is produced by UV radiation by ergosterol in forage. However, common forage (grass, grass silage) is often poor in vitamin D [266]. This explains why, in cattle, vitamin D_3 is the major form in cattle blood plasma, whereas vitamin D_2 is minor [267]. When cattle are brought in the sun or specially exposed to UV, vitamin D_3 concentrations increase within the first 5 weeks of exposure up to a plateau level [268]. In pasture animals, the annual cycle of vitamin D_3 is correlated with the annual course of solar UV exposure [269, 270]. A plateau is held from June to August. In animals that have no longer access to UV radiation, the plasma concentration in vitamin D_3 decreases rapidly, whereas the concentrations of calcium and magnesium remain much longer, and even increase for a certain period after UV exposure has stopped [268, 269]. The rather rapid decrease in vitamin D concentration may be caused by tits release into the milk [269]. Milk contains only low vitamin D concentrations – even in summer when cows are at the pasture [269]. The vitamin D content of milk can be enhanced by a factor of 100 just by irradiating it with UV.

In sheep, vitamin D is gained from both dietary sources and photosynthesis in the skin exposed to UV. In the grazing sheep, D_3 plays a dominant role in the vitamin D status [270]. When sheep are kept indoors, they need a diet that delivers vitamin D; otherwise, vitamin D supplementation has to be given. In sheep, D_2 is the dominant form of vitamin D in the blood plasma [264]. However, the presence of vitamin D_3 increases with UV exposure [271, 272], although there seems to be no linear relationship between vitamin D_3 and UV exposure. Experiments have shown that sheep easily reach their optimum vitamin D_3 level through UV exposure. Plasma D_3 and 25-OH D_3 are still detectable 7 days after the UV exposure has stopped (whereas it lasts 14 days in the case of D_2 from diet) increasing for the next 7 weeks up to a plateau. The UV caused vitamin D concentration plateau was two times higher than the plateau caused by vitamin D supplemented diets. After treatment (UV or diet), the vitamin D level stays stable twice as long as that from diet. It is concluded that UV is faster and more effective than diet to improve the vitamin D level.

Sheep kept outdoors show strong seasonal variations in plasma 25-OH D_3 concentration [273]. In winter, its level depends strongly on geographical latitude.

Wool has an important influence on the vitamin D level. As shown by Zintzen and Boyazoglou [274], the concentration of vitamin D in shorn sheep was two to three times greater than in unshorn animals. With increasing Fleece length, the vitamin D level decreases.

Fig. 22.27 Application of a solarium for horses in the Spanish Riding School of Vienna (Source: Weinsberger SW GmbH, Germany)

For goats and pigs, studies on vitamin D with respect to UV radiation are very rare.

Horses exposed to 4–6 h of solar radiation (including cloudy days) can synthesize sufficient quantities of vitamin D [275]. Solaria for horses (Fig. 22.27) have been recommended for animals that were held predominantly in stables, to compensate for the lack of solar UV exposure. Such radiation devices use lamps for visible and infrared radiation, as well as UV lamps, to simulate the entire spectrum of the Sun. From this exposure to simulated solar radiation, several positive physiological effects are expected. The spatial distribution of UV radiation over the entire horse body, as received either from the sun or from solaria, has been investigated by Keck et al. [55], Fig. 22.28.

This study showed that the UV skin dose from outdoor exposure had a totally different spatial distribution than that resulting from exposure beneath a solarium, as depicted in Fig. 22.28. The UV dose distribution was measured by 17 polysulphon dosimeters that were fixed with mastic on the hair coat of the horses. The dose distribution beneath the solarium depends on the geometry of the UV lamps (Fig. 22.27), whereas the solar radiation field is much more homogeneous due to the presence of hemispherical diffuse radiation from the sky. Consequently, in outdoor conditions, hairless or little haired regions of the horse's skin receive

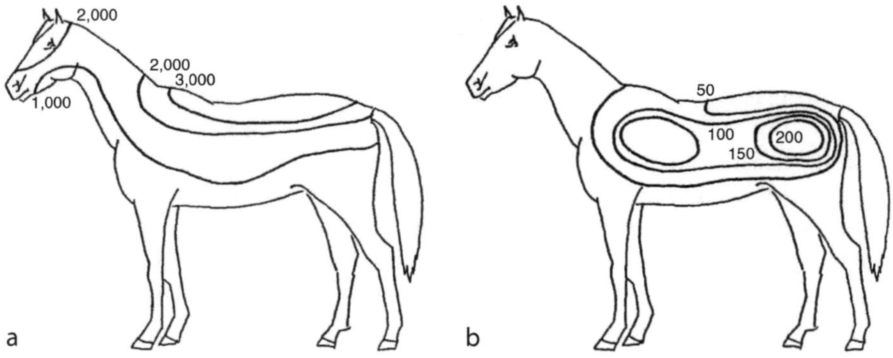

Fig. 22.28 Spatial distribution of the biologically effective UV dose outside of the hair cover of horses (J/m^2) (**a**) typical 1-day outdoor exposure, (**b**) solarium with a recommended exposure time of 22 min (Source: Keck et al. [55])

a considerable portion (up to 60%) of radiation compared to a horizontal receiving surface. This means that UV effects can be assumed especially in these body regions.

The local maximum of the UV dose was ≈3,500 J/m^2 for outdoor exposure and ≈400 J/m^2 for solarium exposure. Considering a UV transmission of the hair coat of horses of less than 2% [55] the effective skin dose is about 70 J/m^2 for outdoor exposure and less than 9 J/m^2 for the solarium. Based on the Grotthus-Draper law, there is only little evidence that the (low) skin dose of horses beneath a solarium can contribute to positive UV effects like vitamin D synthesis. Longer exposure times, or more powerful UV lamps, would have to be used.

Horses exhibit a strong diurnal variation (50%) in vitamin D_3 serum level [276, 277]. This phenomenon seems to be restricted to horses.

Chicken can gain vitamin D from feed and UV exposure. Early studies [278–281] on poultry indicated that 10–30 min of daily sun exposure could prevent all signs of vitamin D deficiency in growing chicken. It has also been known for a long time that a 15-min exposure to a quartz mercury vapor arc lamp produced similar effects [281].

Because of the dense plumage covering the chicken body, photosynthesis occurs in the featherless skin of legs and feet. The chicken's skin of the legs and feet contains eight times as much 7-dehydrocholesterol as their body skin [282]. Bernard et al. [283] pointed out that UV exposure (from lamps) and vitamin supplementation of food had similar efficiency. Higher hardness of egg shells can also be observed in chicken with sufficient vitamin D supply [8].

A known problem with chicken is that viable eggs produced by apparently healthy adults may incubate full term but then fail to hatch. The fully developed, dead embryos appear normal but have poorly mineralized skeletons. This problem can be successfully corrected by providing UV to the adult female prior to oviposition [284–287].

Fish and Fisheries

UV Hazards

Fish and fisheries are part of the aquatic ecosystem, which produces more than 50% of the total biomass of our planet. The primary producers (phytoplankton) in freshwater and marine ecosystems constitute the basis of the food webs, providing energy for the primary (e.g., zooplankton) and secondary (e.g., fish) consumers and are thus important contributors to the production of staple diet. Humans use about 8% of the productivity of the entire oceans, but this amounts to 35% of the productivity of temperate continental shelf systems [130].

Phytoplankton and bacterioplankton are primary producers, since they use solar energy for photosynthetic conversion of CO_2 and nutrients into carbohydrates (in the so-called photic or euphotic zone). They account for about 95% of the primary productivity of oceans, and about half of all primary productivity on earth. Phytoplankton occurs predominantly in cooler, midlatitude zones with sufficient nutrients, especially nitrogen. The two major primary producers are the diatoms, which are located in temperate and polar oceans. They contribute about 60% of the primary productivity of oceans, with a typical size of about 30 μm. In contrast, the coccolithophores, with a typical size of 5–10 μm, dominate in regions of moderate turbulence at midlatitudes in late spring, as well as in subpolar and equatorial regions. They contribute about 15% to the primary productivity of oceans.

Zooplankton is the primary consumer of phytoplankton. They range in size from single-celled organisms to larger multicelled organisms. Small zooplankton is eaten by larger zooplankton. Zooplankton include ciliates (single-celled animals), copepods, shrimp, and larval forms of barnacles, mollusks, fish, and jellyfish.

The UV sensitivity of primary producers (phytoplankton) and consumers (zooplankton) in the food chain is an important predictor of biomass production. The UV protection capability of plankton is inversely related to its size. Bacteria and nano- or pico-plankton are too small to effectively protect themselves against UV radiation by absorption. There is an upper limit for the concentration of absorbing substances due to osmotic restrictions [288]. DNA damage correlates strongly with the penetration of UVB radiation into the water column. As long as the repair mechanisms keep up with the UV-caused damage, the plankton population is not threatened. However, when cyclobutane pyrimidine dimers accumulate under high-UV exposure, the population decreases.

The UV sensitivity of aquatic plankton (bacterioplankton, phytoplankton, and zooplankton) can be expressed by the radiation amplification factor, RAF, which is defined as the relative increase of UV-caused damage for a relative change in total ozone (Table 22.10).

By using a model that describes the UV-induced DNA damage in oceanic bacterioplankton, the sensitivity on varying ozone thickness (which conditions the UV exposure at the sea surface), dissolved organic matter concentration, chlorophyll concentration, wind speed, and mixed layer depth was investigated.

Table 22.10 Radiation amplification factor (RAF) for aquatic species

Species	RAF
Phytoplankton motility (*Evglena gracilis*)	1.5–1.9
Phytoplankton photosynthesis (*Phaeodactylum* sp.)	0.2–0.3
Phytoplankton photosynthesis (*Prorocentrvm micans*)	0.3–0.4
Phytoplankton photosynthesis, in Antarctic community	0.8
Phytoplankton photosynthesis (*Nodularia spumtgena* cyanobacteria)	0.2
Bacterioplankton DNA damage (euphotic zone)	1.7
Bacterioplankton DNA damage (surface water)	2.1–2.2
Green alga *Prasiola stipitata*	0.7
Dinoflagellate *Gyrodinium dorsum*	0.4
Cyanobacterium *Anabaena* sp	1.0
Corals photosynthesis	0.21

Source: References [289–292]

From the model, the total amplification factor (TAF; a relative measure of the increase in UV damage associated with a decrease in ozone thickness) for net DNA damage in the euphotic zone is 1.7, as compared to 2.1–2.2 for UV radiation weighted for damage to DNA at the surface [291]. The link between producers and consumers in the food web has been successfully shown by Kouwenberg and Lantoine [293] for trophic plankton interactions. Diatom (*Skeletonema costatum*) was exposed to UV radiation in doses of 10% higher than the natural UV exposure. UVB-exposed algae showed modifications in cell structure, volume increases, and delay in cell division. These UVB-stressed *S. costatum* cultures were used as food for wild copepods (*Calanus helgolandicus*), which showed an indirect UVB effect on their reproductive output.

The sensitivity of copepoda *Calanus finmarchicus* to UV is an important factor conditioning the planktonic food chain of the St. Lawrence Gulf, Canada [294]. The biological weighting function of the mortality of copepods eggs is similar to that of DNA damage (Fig. 22.29).

Based on the spectral UV sensitivity of copepods eggs, their mortality corresponding to solar UV irradiance at noon was calculated by an exponential law

$$N = N_0 \exp(-E_{biol}\, t) \tag{22.15}$$

where the egg number of live eggs is N_0, and N is the egg number after exposure of the eggs to a biologically effective irradiance E_{biol} during a time t. The weighting of the irradiance was done by a weighting function that was derived from Fig. 22.29. For an egg mortality of 50% at the sea surface, a 2.5 h exposure time was assumed. At a depth of 50 cm the exposure time was rather 4.6 h, due to the UV absorption of the water column [294].

To investigate the natural protection of Antarctic marine organisms against solar UV exposure, 57 species (1 fish, 48 invertebrates, and 8 algae) were collected during austral spring (Anvers Island, Antarctic Peninsula), and were analyzed for the presence of mycosporine-like amino acid – compounds that absorb UV

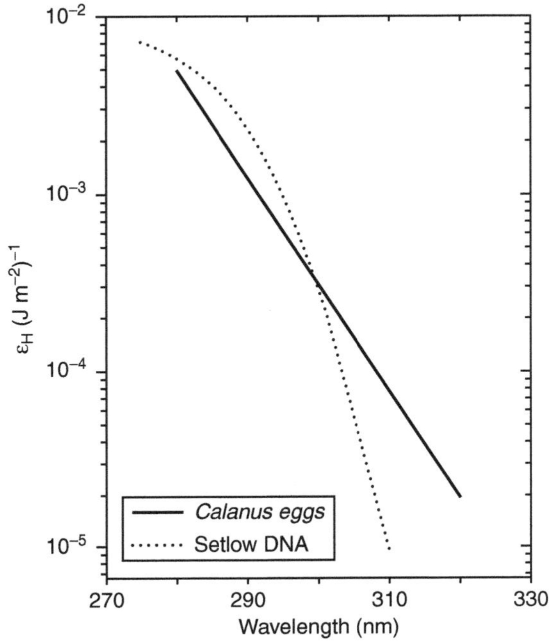

Fig. 22.29 Action spectra of the copepod egg mortality and of the DNA damage [81] (Source: Kouwenberg et al. [294])

radiation and may provide shielding from UV radiation. Nearly 90% of the 57 examined species contained mycosporine-like amino acids [311].

To assess the ecological UV sensitivity of plankton and aquatic animals, the vertical distribution of the species (e.g., Uitz et al. [312], Kesler et al. [313]) and the spectral absorption of the water column have to be taken into account [85, 314, 315].

For fish, several experiments were done to investigate their sensitivity to solar UV radiation. In addition to the mortality of eggs, skin lesions and eye damages were observed and summarized (Table 22.11). Rodgers [305] referred to the UV damage of the skin as the "summer syndrome." Photobiologic responses in fish are also referred to as "sunburn" [308].

The biological weighting function derived for UVB-induced mortality in cod eggs is similar to that reported for DNA [81], which suggests that mortality is a direct result of DNA damage. No evidence was found that UVA had any influence on the UV sensitivity. Calculations based on this biological weighting function indicate that, under current noon surface UV irradiance, 50% of cod eggs located at, or very near below (within 10 cm), the ocean surface would be dead after 42 h of exposure [298]. To counterbalance this high sensitivity to UV exposure (found in the laboratory), many effects actually reduce the solar irradiance on fish eggs: surface water mixing, UV absorption in the water column, cloud cover, etc.

Browman et al. [316] investigated the UV effects on the early life stages of a planktonic Calanoid copepod (*calanus finmarchicus gunnerus*) and of Atlantic cod (*Gadus morhua*). Both are key species for the North Atlantic food webs. The wavelength-specific DNA damage (cyclobutane pyrimidine dimer [CPD] formation per megabase of DNA) is similar to the action spectrum for damage to T7 bacteriophage DNA.

Table 22.11 Biological UV effects in fish

Biological effects	Fish species	Author
Skin lesions, damage on eggs and larvae; damage of the brain, eyes, disorders in growth and development	Anchovies (*Engraulis mordax*)	Hunter et al. [295]
	Pacific mackerels (*Scomber japonicus*)	
Dose-response functions for mortality of eggs	Sockeye salmon (*Oncorhynchus nerka*)	Bell and Hoar [296]
Dose-response functions for mortality	Galaxiids (*Galaxus maculatus*)	cit. by Zagarese et al. [297]
	Odonthestes hatcheri	
Dose-response functions for mortality of eggs	Atlantic cod (*Gadus morhua*)	Kouwenberg et al. [298]
Skin lesions, sunburn	Chinook salmon Arctic char	Brocklebank and Armstrong [299]
Skin lesions, sunburn	Hammerhead shark (*Sphyrna lewini*)	Lowe and Goodman-Lowe [300]
Skin lesions, sunburn	Juvenile rainbow trout (*Salmo gairdneri*)	Dunbar [301], Bullock and Couts [302]
Skin lesions, sunburn	Juvenile paddles fish (*Polyodon spathula*)	cit. by Zagarese et al. [297]
Skin lesions, sunburn	Juvenile plaice (*Pleuronectes platessa*)	cit. by Zagarese et al. [297]
Skin lesions, sunburn	Atlantic salmon (*Salmo salar*)	McArdle and Bullock [303], Kaweewat and Hofer [304], and Rodger [305]
Skin lesions, sunburn	Koi carp (*Cyprinus carpio*)	Bullock et al. [306]
Skin lesions	Rainbow trout (*Oncorhynchus mykiss*)	Markkula et al. [307]
Immune system	Carp (*Cyprinus carpio*)	Markkula et al. [307]
Cataract	Rainbow trout (*Oncorhynchus mykiss*)	Markkula et al. [307]
Cataract	Lake trout (*Salvelinus namayush*)	Roberts [308], Allison [309]
Cataract	Atlantic salmon (*Salmo salar*)	Wall [310]

Cases of skin lesions in fish have been reported long ago, as early as 1930. Skin lesions were described as white-to-gray necrotic areas and erosions [297]. As explained by Roberts [308], the sensitivity of fish skin to UV radiation results from the absence of a keratin layer, the poor melanin content, and the presence of dividing cells in all layers of the skin. Therefore, the skin of fish is more sensitive to UV than that of humans [297]. Lesions are typically found on the head, dorso and tail, because these regions are exposed most. Bullock [317] wrote a detailed review about skin alterations in fish. Such lesions could be found in Koi carps [306], rainbow trouts [301, 302, 318, 319], brown trout [309], Atlantic salmon [303, 305], and plaice [320]. Exposure of Atlantic salmon to enhanced UVB radiation retarded growth, and decreased the hematocrit value and plasma protein concentration. Furthermore, enhanced UVB radiation affected the plasma immunoglobulin concentration. These results also demonstrated that juvenile Atlantic salmons are not able to fully adapt to increased ambient UVB levels over long-term exposures. The interference with the immune system function suggests a negative effect of UVB on disease resistance [321]. Markkula et al. [322] studied the effect of UVB exposure on common carp (*Cyprinus carpio*), based on the immunomodulation in the blood and head kidney. Their results emphasize the potentially harmful impact of increased solar UVB radiation on the fish immune functions, and show that the effects of short- and long-term exposure differ from each other.

Armstrong et al. [323] investigated larvae of Japanese medaka (*Oryzias latipes*) to determine whether pigmentation modifies the UVB-inducible damage. One-day post-hatch medaka were exposed to UVB radiation, with or without photoreactivating light, 7 h per day for 5 days. At the higher UVB dose tested, wild-type melanophore-containing medaka formed significantly more dimers than at least one of the other strains tested. Wild-type medaka also showed significantly less photorepair capability than the white melanophore-lacking medaka. The presence of melanophores in the wild-type medaka may have contributed to an increased level of tissue damage in this strain when compared to the other strains.

Juvenile rainbow trout were also exposed to UVB using a UV-transparent respirometer chamber. A direct relationship between UVB exposure and the oxygen consumption was observed. Increased swimming activity and restless behavior were also noted [324].

Keeping fish in fish farms with high stocking rates results in a much higher UV exposure [325]. A relationship for UV-induced abnormalities was found for fish farms at high elevation in Bolivia [302] and for different seasons [305, 309].

Sunburn is in many cases the first stage before an outbreak of infectious diseases can be observed [326]. For example, fungal infections commonly occur in fish exposed to excessive sunlight. Outbreaks of *flexibacteriosis maritimus* in cultured Atlantic salmon in Tasmania (Fig. 22.30) occurred due to high UV exposure during cloud-free days with water temperatures above 21°C [327]. Markkula et al. [307] studied the effect of UVB on the resistance of rainbow trout (*Oncorhynchus mykiss*) against a bacterium (*Yersinia ruckeri*), the causative agent of enteric red mouth disease, and a trematode parasite (*Diplostomum spathaceum*), which causes cataracts in fish. As a result of increased UVB exposure, a significantly higher

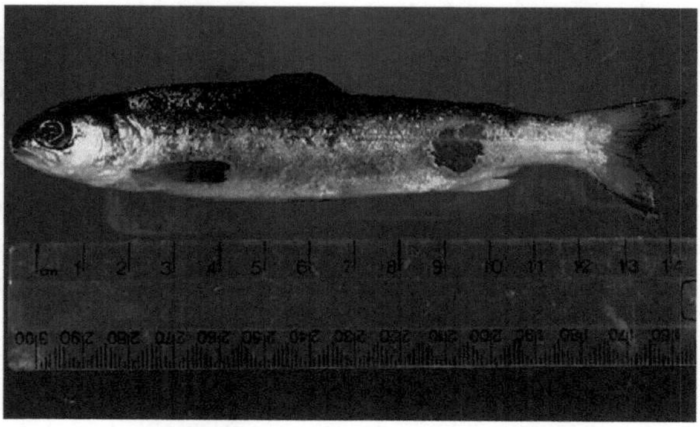

Fig. 22.30 Skin lesions of salmonid cutaneous erosion disease caused by *Flexibacter maritimus* in experimentally infected Atlantic salmon (From Handlinger et al. [327])

number of parasites was detected in the eyes of the irradiated fish, indicating reduced resistance against the pathogen. Furthermore, UVB exposure altered the immunological and hematological parameters of fish, which also verified the immunomodulatory potential of UVB.

In a retrospective study conducted at 13 salmon farms in Western Ireland, Rodgers [305] showed that economic losses resulting from UV-induced skin damages (sunburn) could be reduced by the use of broadband antibiotics. Six out of 13 farms using antibiotics suffered from such losses, mounting to 10–23% of the post-smolt stocks over the sampling period. In 4 out of the 13 farms, where the cages were shaded, no sunburn cells in the histology of the skin could be observed.

In addition to skin lesions, eye damages can be observed in fish. Cullen et al. [328] exposed rainbow trouts (*Onchorynchus mykiss*) to broadband UV exposure over a period of 205 days. The results support the hypothesis that chronic exposure to ambient levels of UV radiation is cataractogenic. For the Atlantic salmon (*Salmo salar*), Wall [310] could show that the incidence of cataracts was very variable between sites (Ireland, Scotland, and Norway). Whereas in some farms 95% of the salmons had cataract, this fraction was less than 5% in other farms. Blind fish could often be seen near the surface. The pattern of these outbreaks suggested that the diet could be involved, beside other factors like UV exposure. For juvenile scalloped hammerhead sharks (*Sphyrny lewini*), the absorption in their corneal tissues, particularly at wavelengths below 310 nm, increases proportionally to UV exposure [329].

Cataract could also be observed in fish [308]. Allison [309] reported on trout kept in open but covered basins. In these sun-protected basins, 6% of the brown trouts developed cataract, whereas it was as much as 22% in uncovered basins.

To study repair mechanisms and the effects of UV radiation in animals, the Zebra fish is becoming an important model system. Their embryos are easily treated with UV radiation. The DNA damage repair pathways appear to be conserved in Zebra fish and mammals [330].

Due to UV exposure, aquatic organisms also developed protection measures and avoidance strategies. For them, a variety of physiological and biochemical mechanisms are available for reducing the damage by UV exposure. Screening mechanisms include both physical and chemical protection measures. Physical methods are UV absorption of mucus, sporopollenin, and multiple cell walls. Chemical screening methods include UV absorption by mycosporines, mycosporine-like amino acids, scytonemin, 3-Hydroxykynurenin, melanin, and fluorescent pigments. When screening does not eliminate penetration of UVR into the cell, there is a variety of quenching and repair processes available to overcome damage to sensitive cellular components (antioxidants, carotenoids, and the xanthophyll cycle). Damaged proteins are usually replaced by de novo synthesis (turnover), whereas damaged DNA is generally repaired. Several different mechanisms exist for repairing DNA, including photoreactivation, nucleotide excision repair, dimer bypass, and recombinational repair [331].

Coral reef fish were recently discovered to have ultraviolet (UV) radiation screening compounds, most commonly known as mycosporine-like amino acids (MAAs), in their external body mucus. However, little is known about the identity and quantity of MAAs in the mucus of reef fish, or what factors affect their abundance and distribution. In seven species of reef fish (*Labroides dimidiatus* and *Thalassoma lunare* [*Labridae*]; *Chlorurus sordidus, Scarus flavipectoralis, S. niger, S. rivulatus, and S. schlegelii* [*Scaridae*]) from Queensland, Australia, the UV absorptance of mucus was measured. The mucus of all fish investigated contained MAAs. Depending on species, different combinations and quantities of the MAAs asterina-330, palythene, and mycosporine-N-methylamine serine were present [332]. Zamzow [333] could show that the UV protective agent in the mucus depends on the level of UV exposure. For shallow-water Caribbean fish (*Scarus iseri* and *Halichoeres bivittatus*), different ways of coping with UV radiation were discovered: *H. bivittatus* shifted the spectral quality of its mucus to increase absorptance of shorter, more damaging wavelengths, whereas *S. iseri* increased the overall UV absorbance of its mucus, with minimal spectral shifting. The influence of the diet and gender on the UV protection by the mucus could be shown for tropical wrasse, *Thalassoma duperrey* [334]. Females had higher rates of skin damage than males. Females sequester UV absorbing compounds in their pelagic eggs as well as their epithelial mucus, whereas males do not sequester these compounds in the testes.

Behavioral avoidance strategies (Table 22.12) are not only protection measures, but also a way to reduce UV exposure. Both laboratory and field experiments have demonstrated that many species are negatively phototactic to UV. In addition, UV photoreceptors were found in a variety of fish and invertebrates, suggesting that UV vision may be prominent in aquatic organisms [335–337]. These UV photoreceptors are thought to be used for navigation, communication, enhanced foraging, and possibly UV radiation avoidance. Given the presence of negative phototactic behaviors as well as UV vision, UV radiation may be an important factor influencing migration and abundance patterns, as well as predator-prey and intraspecific interactions [338].

UV vision has been documented in a variety of terrestrial organisms including insects, birds, amphibians, reptiles, and mammals. It is therefore not surprising that many aquatic organisms also perceive light in the UV range. Most UV

Table 22.12 Behavioral avoidance of aquatic organism caused by UV radiation in laboratory (L) and field (F) experiments

Organism	Behavioral activity in the presence of UV radiation	Laboratory/ field study
Marine cyanobacteria *Microcoleus chthonoplastes*	Migrate to greater depths	L
Ciliate *Btepharisma japonicum*	Backward swimming	L
Cladocerans	Negatively phototactic	L
Daphnia magna	Negatively phototactic	L
Cyclopoid *Cyclops serrulatus*	Avoid UV exposure	L
Marine echinoid larva *Dendraster excentricus*	Avoid UV exposure	L
Oplophorid shrimp *Systellaspis debilis*	Pitching, changing swimming speed, moving the feeding appendages	L
Zebra mussel *Dreissena polymorpha*	Ceased all swimming and crawling motions	L
D. pulicaria	Rapidly descended from the surface waters	F
D. catawba	Negatively phototactic	F
D. catawba	Negatively phototactic	F
Copepod *Diaptomus nevadensis*	Negatively phototactic	F
Marine copepod *Acartia tonsa*	Negatively phototactic	F
Cladoceran *Daphnia magna*	Negatively phototactic	F
Hydromedusan *Polyorchis penicillatus*	Negatively phototactic	F
Juvenile rainbow trout (*Oncorhynchus mykiss*)	Increased swimming activity	F
Juvenile coho salmon (*Oncorhynchus kisutch*)	Negatively phototactic	L/F

Sources: Leech and Johnsen [338], Alemanni et al. [324], Holtby and Bothwell [340]

photoreceptors in aquatic organisms have been described in fish species; however, UV photoreceptors have also been reported in bacteria and algae as well as in some species of protozoan, annelids, cnidarians, and crustaceans. Many UV photoreceptors have a maximum absorptance peak in the UVA range, although UVB photoreceptors have also been documented in some species. One explanation for the rarity of UVB vision is that UVB radiation is potentially more damaging to the eye [338, 339].

Vitamin D

Fish store large quantities of vitamin D in their liver and fat tissues, including the fat associated with muscles, and this makes fish an important dietary source of vitamin D for their predators or for humans. Plankton, the chief food source of fish, was assessed [341] as the possible dietary origin of vitamin D in fish. High amounts of provitamins D and vitamins D (D_2 and D_3) were found in freshwater plankton [342] during summer months. Thus, plankton may be an important contributor to vitamin D in fish.

For some species of fish (rainbow trout and Mozambique tilapia) photosynthesis of vitamin D_3 from 7-DHC was reported [343]. However, it is unlikely that the photosynthesis of vitamin D_3 plays a significant role under natural conditions, where fish are less exposed to UV light.

The role of vitamin D in fish physiology is still enigmatic [344]. Until the 1970s, the consensus was that fish accumulated but not metabolized vitamin D. Now, 4 decades later, there is substantial evidence that fish have a vitamin D endocrine system with similar functions as in mammals.

Zoo Animals

The geographical origins of many exotic animals are areas with high-UV intensity. When brought into zoological gardens, they might suffer under a deficiency of UV exposure, especially at mid- and high-latitudes. On the other hand, increased UV exposure may cause photodamage of the skin and the eyes. To provide these animals with an appropriate environment, their solar UV exposure has to be taken into account.

Lack of UV radiation strongly reduces the availability of vitamin D_3. Therefore, animals can suffer from rachitis (rickets) and osteodystrophy, which causes irreversible bone degeneration of the locomotion system and substantial deformation of the jaw. This handicaps both their movement and food intake. Vitamin D can also be supplied by feed. However, excessive dietary vitamin D can result in toxic effects and death. Conversely, high-UV exposure does not result in excessive vitamin D levels [345].

For some species, the vitamin D supply can be substituted by vitamin D_2 supplementation in the feed. However, Göltenboth [346] illustrated that rachitis in primates could not be healed by such vitamin D_2 supplementation. The symptoms could only be eradicated by exposure to lamps with a high UV output (Ultravitalux-lamps, Osram). Several species of New World primates have an absolute requirement for vitamin D_3, either from the diet or through exposure to UV radiation. They seem to metabolize vitamin D_2 too quickly to maintain adequate serum levels of 1,25-dihydroxyergocalciferol [347]. Hofmann-Parisot [348] described typical differences in diverse primate species concerning the vitamin D metabolism.

Tonge [349] has shown that the application of supplemental UV radiation can prevent the deformation of the tail of lizards. Bosch and Frank [350] pointed out that in terrariums, 11% of turtles and 7% of iguanas display a vitamin D deficiency syndrome.

McCrystal and Behler [351] and Townsend and Cole [352] indicated that for the breeding success in lizards, a sufficient supply in UV is necessary. This can

Fig. 22.31 X-ray image of whiptail lizards (*Cnemidophorus exsanguis*). (**a**) Healthy, wild-caught adult. (**b**) Healthy laboratory-reared adult. (**c**) Laboratory-reared young adult with metabolic bone disease (Source: Townsend and Cole [352])

be attributed to interactions with the calcium metabolism, because calcium is responsible for the mechanical stability of their eggs (Fig. 22.31).

Another fundamental problem for lizards raised in captivity may be that their viable eggs incubate full term but then fail to hatch, even when produced by apparently healthy adults. The fully developed, dead embryos appear normal but have poorly mineralized skeletons. This problem is similar to that of chicken, and can also be corrected by supplementing UV to the adult female prior to oviposition [284, 285].

Turtles receive vitamin D either from dietary or from UV exposure. It was shown by Acierno et al. [353] that UV is much more effective than diet. In 1997, Manning and Grigg [47] found that basking in a freshwater turtle is longer than thermoregulation would require, which indicates that turtles regulate their vitamin D status by basking.

For snakes and turtles, Acierno et al. [354] investigated the vitamin D supply following 28 days of additional UV exposure. As a result of such an additional UV exposure, the plasma concentration of 25-hydroxyvitamin D_3 was about 3.4 times higher than the control group in snakes, and 2.3 times higher in turtles.

For various lizards, snakes, and chameleons [50, 345], it could be shown that the exposure behavior to natural sunlight depends on the vitamin D status. Animals with low dietary vitamin D_3 intake significantly increased their spontaneous exposure to solar radiation, compared to those with a high dietary vitamin D_3 intake. These results demonstrate the importance of basking for nonthermoregulatory purposes and, more specifically, that basking constitutes an integral mechanism for the regulation of a vital hormone, vitamin D_3. Ferguson et al. [49] found some evidence that the vitamin D photosynthesis in the skin may reflect a correlated response associated with the need of vitamin D. However, the possibility that specific adaptations for the photosynthesis of vitamin D and for the protection from skin damage could involve independent mechanisms still needs more investigation.

Ferguson et al. [48, 49] have shown that Panther chameleons (*Furcifer pardalis*) can regulate UV exposure behaviorally or by a change in skin color, so that their vitamin D production is enhanced at low UV levels. As Karsten et al. [50] have shown, Panther chameleons and possibly reptiles, in general, bask also for reasons other than thermoregulation. In dependence of their dietary vitamin D_3 intake, Panther chameleons behaviorally regulate their exposure to solar UV to ensure an optimal vitamin D_3 status with high precision. Their UVA-sensitive retina is helpful in this respect. Vitamin D supplementation in lizards is not easy. It may become toxic, and possibly lethal [284].

Hofmann-Parisot [348] investigated the use of artificial UV sources that were used in keeping zoo animals healthy. It was found that $\approx$60% of the zoo gardens surveyed apply artificial UV radiation to animals. Of that number, 48% use UV only for treating reptiles, and 52% do this also for mammals and avian. Interestingly, this survey also provided a very fuzzy picture of the UV supplementation practice: UV intensity and exposition times, types of lamps, and radiation geometry were highly variable, illustrating that no sufficient information is available concerning the UV needs of specific animal species. As illustrated by the case of chameleons, animal behavior has to be taken into account.

Gehrmann [355] realized the importance of using UV sources in animal husbandries and investigated different types of lamps, as well as the reflection and transmission properties of some materials that are used in zoos [356]. (Note that, in this kind of study, the spectral distribution of these UV sources and the specific animal requirements would have to be considered as well.)

Even though it is well known that most of the observed diseases related to vitamin D deficits are caused by a lack of UV exposure, no systematic investigations have been performed to optimize the UV irradiance or the exposure time for various animals. Most of the applications of artificial UV sources resulted from "educated guess." For snakes and lizards, however, Ferguson et al. [345] estimated the natural UV irradiance by using the UV index of their natural environment. The conversion rate of vitamin D due to UV exposure was investigated by Ferguson et al. [285] for panther chameleons (*Furcifer pardalis*).

Although insufficient natural UV radiation is typical in zoo animals, overexposure to artificial UV can also appear, if inappropriate UV sources are in use. For a ball python (*Python regius*) and a blue tongue skink (*Tiliqua nigrolutea*), the application of an otherwise appropriate UV lamp, but with an extremely high UVB output, showed epidermal erosion and ulceration, with severe epidermal basal cell degeneration and necrosis, and superficial dermatitis as well as bilateral ulcerative keratoconjunctivitis with bacterial colonization [357]. Veterinarians who are consulted for reptiles with ocular and/or cutaneous disease of unapparent cause should fully evaluate the specifics of the vivarium light sources. In general, overexposure to UV radiation can cause eye and skin damage, skin cancer, and poor reproduction in reptiles and amphibians [345].

Future Directions

The so-called ozone hole was detected in ozone measurements of the late 1970s [358]. The reduction of the total ozone content in the stratosphere and the resulting decrease of absorbing capacity for UV radiation [359] cause a hype in research on the biological effects of UV radiation. Up to this moment the solar radiation and therefore also the spectral range of UV radiation was assumed to be constant without remarkable variability. In the following 2 decades, a steep increasing number of investigations were published which were concentrated primarily to the monocausal effects of UV radiation on plants, animals, and humans.

These research activities of biologically effective UV radiation due to ozone depletion were replaced by the climate change discussion starting in the late 1990s. By more sophisticated general circulation models (GCM) also parameters like solar radiation were predicted for the next century. This leads to the opportunity to predict the biologically effective UV radiation as well. These climate change scenarios for a century were the basis for a more holistic approach to describe the interaction between UV radiation and flora as well as fauna. The coupling of the biosphere with GCM can be used to assess effects of the climate change on the biosphere and vice versa (e.g., reduced cloud cover leads to a higher UV exposure of plankton which reduces the plankton reproduction, which causes a reduction of carbon storage and finally an increase of CO_2 release, influencing the radiative forcing as an important input parameter of GCM). For humans the resulting migration scenarios and the adaptation of the behavior to the changing climate can be used to calculate the modification of the UV skin exposure and the health implications.

Beside UV effects, which cause a damage (e.g., eye, skin), also beneficial effects are under discussion, which are correlated with the vitamin D synthesis due to UV radiation. A sufficient supply with vitamin D is necessary for skeletal health. Further it is assumed as protective measure against several kinds of cancer in humans. For these investigations an appropriate animal model would be a great help.

For several UV effects different working hypothesis are in use, which are verified very rarely. First of all the assumption of reciprocity is a simple prerequisite which says that the effect is depending on the UV dose, which is calculated by the intensity of the radiation times the exposure time. Secondly we assume that a polychromatic UV exposure is caused by an unweighted sum of all spectral UV doses. This means that the weighting of the UV radiation and the resulting biologically effective UV radiation can be described as photoadditive, without photoprotection or photoaugmentation due to certain wavelength ranges [360].

Bibliography

Primary Literature

1. Bunsen RW, Roscoe HE (1857) Photochemische Untersuchungen. Poggendorfs Ann 100:51
2. Schindl A, Rosado-Schlosser B, Trautinger F (2001) Die Reziprozitätsregel in der Photobiologie. Eine Übersicht Der Hautarzt 52:779–785
3. Diffey BL, Kochevar IE (2007) Basic principles of photobiology. In: Lim HW, Hönigsmann H, Hawk JL (eds) Photodermatology. Informa Healthcare USA, New York, pp 15–27
4. Schwarzschild K (1900) On the law of reciprocity for bromide of silver gelatin. Astrophys J 11:89
5. Martin JW, Chin JW, Nguyen T (2003) Reciprocity law experiments in polymeric photodegradation: a critical review. Prog Org Coat 47:292–311
6. Reischl S, Henrich J, Schauberger G (2000) Time course and dose-response-relationship of erythema of the pigs induced by UVB- and UVC-radiation. In: 13th international congress on photobiology, San Francisco, 1–6 July 2000, p 136
7. Farr PM, Besag JE, Diffey BL (1988) The time course of UVB and UVC erythema. J Investig Dermatol 91:454–457
8. Mehlhorn G, Steiger A (1974) Künstliche UV-Strahlung in der Tierproduktion. Gustav Fischer Verlag, Jena
9. Kasper I (1986) Abschätzung biologischer UV-Wirkungen durch Bestrahlung von Großtieren mit künstlichen UV-Quellen. Dissertation Veterinärmedizinische Universität Wien, Wien
10. MacKinley AF, Diffey BL (1987) A reference action spectrum for ultraviolet induced erythema in human skin. CIE J 6:17–22
11. Gates FL (1930) A study of the bactericidal action of ultra violet light. III. The absorption of ultra violet light by bacteria. J Gen Physiol 14:31–42
12. Hollaender A, Emmons CW (1941) Wavelength dependence of mutation production in the ultraviolet with special emphasis on fungi. Cold Spring Harb Symp Quant Biol 9:179–185
13. Herschel W (1800) Investigation of the powers of the prismatic colours to heat and illuminate objects; with remarks, that prove the different refrangibility of radiant heat. To which is added, an inquiry into the method of viewing the sun advantageously, with telescopes of large apertures and high magnifying powers. Philos Trans R Soc Lond 90:255–283
14. Herschel W (1800) Experiments on the refrangibility of the invisible rays of the sun. Philos Trans R Soc Lond 90:284–292
15. Ritter JW (1801) Ann Phys 7:527
16. Daubeny C (1836) On the action of light upon plants and plants upon the atmosphere. Philos Trans R Soc Lond 126:149–176
17. Draper JW (1884) Note on decomposition of carbonic acid by the leaves of plants under the influence of yellow light. Philos Mag 25:159–173
18. Engelmann TW (1882) Über Sauerstoffausscheidung von Pflanzenzellen im Microspectrum. Bot Zeit 40:419–426
19. Hausser KW, Vahle W (1921) Die Abhängigkeit des Lichterythems und der Pigmentbildung von der Schwingungszahl (Wellenlänge) der erregenden Strahlung. Strahlentherapie 13:41–71
20. Coblentz WW, Stair R (1934) Data on the spectral erythemic reaction of the untanned human skin to ultraviolet radiation. Bur Stand J Res 12:13–14
21. Luckiesh M, Holladay LL, Taylor AH (1930) Reaction of untannted human skin to ultraviolet radiation. J Opt Soc Am 20:423–426
22. Hausser KW, Vahle W (1927) Sonnenbrand und Sonnenbräunung. Wiss Veröfftl Siemens Konzern 6:101–120
23. CIE (1937) Neuvieme Session Berlin und Karlsruhe Juillet 1935. Cambridge University Press, Cambridge, UK

24. McKinley AF, Diffey BL (1987) A reference action spectrum for ultraviolet induced erythema in human skin. CIE J 6:17–22
25. Duggar BM, Hollaender A (1934) Irradiation of plant viruses and of microorganisms with monochromatic light: I. The virus of typical tobacco mosaic and serratia marcescens as influenced by ultraviolet and visible light. J Bacteriol 27:219–239
26. Gates FL (1934) Results of irradiating staphylococcus aureus bacteriophage with monochromatic ultraviolet light. J Exp Med 60:179–188
27. Sonne C, Rekling E (1927) Behandlung experimenteller Rattenrachitis mit monochromatischem ultravioletten Licht. Strahlentherapie 35:552–558
28. Henschke U, Schulze R (1939) Über Pigmentierung durch langwelliges Ultraviolett. Strahlentherapie 65:14–42
29. Reiter T, Gabor D (1928) Ultraviolette Strahlen und Zellteilung. Strahlentherapie 28:125–132
30. Blumthaler M, Ambach W (1988) Solar UV-B albedo of various surfaces. Photochem Photobiol 54:429–432
31. Monteith JL (1973) Principles of environmental physics. Arnold, London
32. Quaschning V, Hanitsch R (1998) Irradiance calculation on shaded surfaces. Sol Energy 62:369–375
33. Blumthaler M, Ambach W, Ellinger R (1996) UV-Bestrahlung von horizontalen und vertikalen Flächen im Hochgebirge. Wetter Leben 1:25–31
34. Schauberger G (1990) Model for the global irradiance of the solar biologically-effective ultraviolet radiation on inclined surfaces. Photochem Photobiol 52:1029–1032
35. Grant RH (1998) Ultraviolet irradiance of inclined planes at the top of the plant canopies. Agric For Meteorol 89:281–293
36. Webb AR, Weihs P, Blumthaler M (1999) Spectral UV irradiance on vertical surfaces: a case study. Photochem Photobiol 69:464–470
37. Hay JE (1979) Calculation of monthly mean solar radiation for horizontal and inclined surfaces. Sol Energy 23:301–307
38. Schauberger G (1992) Anisotropic model for the diffuse biologically effective irradiance of solar UV radiation on inclined surfaces. Theor Appl Climatol 46:45–51
39. Klucher TM (1979) Evaluation of models to predict insolation on tilted surfaces. Sol Energy 26:179–184
40. Hay JE, McKay DC (1985) Estimating solar irradiance on inclined surfaces: a review and assessment of methodologies. Int J Sol Energy 3:203–240
41. Reindl DT, Beckmann WA, Duffie JA (1990) Diffuse fraction correlations. Sol Energy 45:1–7
42. Utrillas MP, Marin MJ, Esteve AR, Estelles V, Tena F, Canada J et al (2009) Diffuse ultraviolet erythemal irradiance on inclined planes: a comparison of experimental and modeled data. Photochem Photobiol 85:1245–1253
43. Mech M, Koepke P (2004) Model for UV irradiance on arbitrarily oriented surfaces. Theor Appl Climatol 77:151–158
44. Hess M, Koepke P (2008) Modelling UV irradiances on arbitrarily oriented surfaces: effects of sky obstructions. Atmos Chem Phys 8:3583–3591
45. Dirmhirn I (1964) Das Strahlungsfeld im Lebensraum. Akademische Verlagsgesellschaft, Frankfurt
46. Geiger R (1961) Das Klima der bodennahen Luftschicht, 4. Aufl. ed. Viehweg, Braunschweig
47. Manning B, Grigg GC (1997) Basking is not of thermoregulatory significance in the "basking" freshwater turtle Emydura signata. Copeia 1997:579–584
48. Ferguson GW, Gehrmann WH, Karsten KB, Hammack SH, McRae M, Chen TC et al (2003) Do panther chameleons bask to regulate endogenous vitamin D_3 production? Physiol Biochem Zool 76:52–59
49. Ferguson GW, Gehrmann WH, Karsten KB, Landwer AJ, Carman EN, Chen TC et al (2005) Ultraviolet exposure and vitamin D synthesis in a sun-dwelling and a shade-dwelling

species of Anolis: are there adaptations for lower ultraviolet B and dietary vitamin D_3 availability in the shade? Physiol Biochem Zool 78:193–200
50. Karsten KB, Ferguson GW, Chen T, Holick MF (2009) Panther chameleons, furcifer pardalis, behaviorally regulate optimal exposure to UV depending on dietary vitamin D status. Physiol Biochem Zool 82:218–225
51. Bidmon HJ, Stumpf WE (1996) Vitamin D target systems in the brain of the green lizard Anolis carolinensis. Anat Embryol 193:145–160
52. Bowmaker JK, Loew ER, Ott M (2005) The cone photoreceptors and visual pigments of chameleons. J Comp Physiol A Neuroethol Sens Neural Behav Physiol 191:925–932
53. Tovee MJ (1995) Ultra-violet photoreceptors in the animal kingdom: their distribution and function. Trends Ecol Evol 10:455–460
54. Bianca W, Wegmann H (1974) Die Durchlässigkeit von Rinderhaaren für ultravio¬lette Strahlung. Schweiz Arch Tierheilk 116:114–146
55. Keck G, Kasper I, Schauberger G, Cabaj A (1987) Biological effects of UV-irradiation on horses with artificial UV-sources. In: Passchier WF, Bosnjakovic BFM (eds) Human exposure to ultraviolet radiation – risk and regulations. Elsevier, Amsterdam
56. Keck G, Kasper I, Schauberger G, Cabaj A (1988) UV-Wirkungen in Pferdesolarien? Berl Münchner Tierärztl Wochenschr 101:49–52
57. Chapman RS, Cooper KD, De Fabo EC, Frederick JE, Gelatt KN, Hammond SP et al (1995) Solar ultraviolet radiation and the risk of infectious disease: summary of a workshop. Photochem Photobiol 61:223–247
58. Schauberger G (1997) Biological relevance of solar radiation to animals. Wien Tierärztl Mschr 84:2–13
59. Wolfe WL, Zissig GJ (1978) The infrared handbook. Office of Naval Research, Department of the Navy, Washington
60. Chadysiene R, Girgzdys A (2008) Ultraviolet radiation albedo of natural surfaces. J Environ Eng Landscape Manage 16:83–88
61. Morris DP, Zagarese H, Williamson CE, Balseiro EG, Hargreaves BR, Modenutti B et al (1995) The attenuation of solar UV radiation in lakes and the role of dissolved organic carbon. Limnol Oceanogr 40:1381–1391
62. Reitner B, Herndl GJ, Herzig A (1997) Role of ultraviolet-B radiation on photochemical and microbial oxygen consumption in a humic-rich shallow lake. Limnol Oceanogr 42:950–960
63. Booth CR, Morrow JH (1997) The penetration of UV into natural waters. Photochem Photobiol 65:254–257
64. Morris DP, Hargreaves BR (1997) The role of photochemical degradation of dissolved organic carbon in regulating the UV transparency of three lakes on the Pocono Plateau. Limnol Oceanogr 42:239–249
65. Graneli W, Lindell M, Tranvik L (1996) Photo-oxidative production of dissolved inorganic carbon in lakes of different humic content. Limnol Oceanogr 41:698–706
66. Markager S, Vincent WF (2000) Spectral light attenuation and the absorption of UV and blue light in natural waters. Limnol Oceanogr 45:642–650
67. Vincent WF, Rae R, Laurion I, Howard-Williams C, Priscu JC (1998) Transparency of Antarctic ice-covered lakes to solar UV radiation. Limnol Oceanogr 43:618–624
68. Sommaruga R, Psenner R (1997) Ultraviolet radiation in a high mountain lake of the Austrian Alps: air and underwater measurements. Photochem Photobiol 65:957–963
69. Sommaruga R, Augustin G (2006) Seasonality in UV transparency of an alpine lake is associated to changes in phytoplankton biomass. Aquat Sci 68:129–141
70. Huovinen PS, Penttilä H, Soimasuo MR (2003) Spectral attenuation of solar ultraviolet radiation in humic lakes in Central Finland. Chemosphere 51:205–214
71. Smith RC, Baker KS (1981) Optical properties of the clearest natural waters (200–800 nm). Appl Opt 20:177–184
72. Baker KS, Smith RC (1982) Bio-optical classification and model of natural waters. Limnol Oceanogr 27:500–509

73. De Lange HJ (2000) The attenuation of ultraviolet and visible radiation in Dutch inland waters. Aquat Ecol 34:215–226
74. Bukaveckas PA, Robbins-Forbes M (2000) Role of dissolved organic carbon in the attenuation of photosynthetically active and ultraviolet radiation in Adirondack lakes. Freshw Biol 43:339–354
75. Squires MM, Lesack LFW (2003) Spatial and temporal patterns of light attenuation among lakes of the Mackenzie Delta. Freshw Biol 48:1–20
76. Sommaruga R (2001) The role of solar UV radiation in the ecology of alpine lakes. J Photochem Photobiol B 62:35–42
77. Smith RC, Baker KS (1978) Optical classification of natural waters. Limnol Oceanogr 23:260–267
78. Smith RC, Baker KS (1979) Penetration of UV-B and biologically effective close-rates in natural waters. Photochcm Photobiol 29:311–323
79. Jerlov NG (1968) Optical oceanography. Elsevier, Amsterdam
80. Williamson CE, Stemberger RS, Morris DP, Frost TM, Paulsen SG (1996) Ultraviolet radiation in North American lakes: attenuation estimates from DOC measurements and implications for plankton communities. Limnol Oceanogr 41:1024–1034
81. Setlow RB (1974) The wavelengths in sunlight effective in producing skin cancer: a theoretical analysis. Proc Natl Acad Sci USA 71:3363–3366
82. Grenfell TC, Warren SG, Mullen PC (1994) Reflection of solar radiation by the Antarctic snow surface at ultraviolet, visible, and near-infrared wavelengths. J Geophys Res 99: 18669–18684
83. Blumthaler M, Ambach W (1988) Solar UVB-albedo of various surfaces. Photochem Photobiol 48:85–88
84. Meinander O, Kontu A, Lakkala K, Heikkilä A, Ylianttila L, Toikka M (2008) Diurnal variations in the UV albedo of arctic snow. Atmos Chem Phys 8:6551–6563
85. Perovich DK (1993) A theoretical model of ultraviolet light transmission through Antarctic sea ice. J Geophys Res 98:22579–22588
86. Cockell CS, Rettberg P, Horneck G, Wynn-Williams DD, Scherer K, Gugg-Helminger A (2002) Influence of ice and snow covers on the UV exposure of terrestrial microbial communities: dosimetric studies. J Photochem Photobiol B 68:23–32
87. Warren SG, Brandt RE, Grenfell TC (2006) Visible and near-ultraviolet absorption spectrum of ice from transmission of solar radiation into snow. Appl Opt 45:5320–5334
88. King MD, France JL, Fisher FN, Beine HJ (2005) Measurement and modelling of UV radiation penetration and photolysis rates of nitrate and hydrogen peroxide in Antarctic sea ice: an estimate of the production rate of hydroxyl radicals in first-year sea ice. J Photochem Photobiol A 176:39–49
89. Perovich DK (2001) UV radiation and the optical properties of sea ice and snow. In: Hessen D (ed) UV radiation and Arctic ecosystems, pp 73–89
90. Lesser MP, Lamare MD, Barker MF (2004) Transmission of ultraviolet radiation through the Antarctic annual sea ice and its biological effects on sea urchin embryos. Limnol Oceanogr 49:1957–1963
91. Fritsen CH, Priscu JC (1999) Seasonal change in the optical properties of the permanent ice cover on Lake Bonney, Antarctica: consequences for lake productivity and phytoplankton dynamics. Limnol Oceanogr 44:447–454
92. Grenfell TC, Maykutt GA (1977) The optical properties of ice and snow in the Arctic basin. J Glaciol 18:445–463
93. Belzile C, Gibson JAE, Vincent WF (2002) Colored dissolved organic matter and dissolved organic carbon exclusion from lake ice: implications for irradiance transmission and carbon cycling. Limnol Oceanogr 47:1283–1293
94. Belzile C, Johannessen SC, Gosselin M, Demers S, Miller WL (2000) Ultraviolet attenuation by dissolved and particulate constituents of first-year ice during late spring in an Arctic polynya. Limnol Oceanogr 45:1265–1273

95. Kepner JRL, Wharton RA Jr, Collier RD, Cockell CS, Jeffrey WH (2000) UV radiation and potential biological effects beneath the perennial ice cover of an antarctic lake. Hydrobiologia 427:155–165
96. Felip M, Camarero L, Catalan J (1999) Temporal changes of microbial assemblages in the ice and snow cover of a high mountain lake. Limnol Oceanogr 44:973–987
97. Felip M, Wille A, Sattler B, Psenner R (2002) Microbial communities in the winter cover and the water column of an alpine lake: system connectivity and uncoupling. Aquat Microb Ecol 29:123–134
98. Savage DC (1977) Microbial ecology of the gastrointestinal tract. Annu Rev Microbiol 31:107–133
99. Berg RD (1996) The indigenous gastrointestinal microflora. Trends Microbiol 4:430–435
100. Cristiani H (1893) Analyse bacteriologique de l'air des hauteurs, puise un voyage en ballon. I'Inst Pasteur 7:665–671
101. Harz CO (1904) Bakteriologische Untersuchungen der freien Atmosphäre mittels Luftballons nebst Bemerkungen über den atmosphärischen Staub. Fb dtsch Luftsch Verb 1904:293–302
102. Rogers LA, Meier FC (1936) U.S. Army Air Corps stratosphere flight of 1935 in the balloon "Explorer II." Technical papers, The National Geographic Society, Washington, DC, pp 146–151
103. Imshenetsky AA, Lysenko SV, Kazakov GA (1978) Upper boundary of the biosphere. Appl Environ Microbiol 35:1–5
104. Fulton JD (1966) Microorganisms of the upper atmosphere. 3. Relationship between altitude and micropopulation. Appl Microbiol 14:237–240
105. Yang Y, Itahashi S, Yokobori S, Yamagishi A (2008) UV-resistant bacteria isolated from upper troposphere and lower stratosphere. Biol Sci Space 22:18–25
106. Tong Y, Che F, Xu X, Chen M, Ye B (1993) Population study of atmospheric bacteria at the Fengtai district of Beijing on two representative days. Aerobiologia 9:69–74
107. Meier FC, Lindbergh CA (1935) Collecting micro-organisms from the Arctic atmosphere. Sci Mon 40:5–20
108. Griffin DW, Garrison VH, Herman JR, Shinn EA (2001) African desert dust in the Caribbean atmosphere: microbiology and public health. Aerobiologia 17:203–213
109. Griffin DW, Kubilay N, Koçak M, Gray MA, Borden TC, Shinn EA (2007) Airborne desert dust and aeromicrobiology over the Turkish Mediterranean coastline. Atmos Environ 41: 4050–4062
110. Griffin DW, Westphal DL, Gray MA (2006) Airborne microorganisms in the African desert dust corridor over the mid-Atlantic ridge, Ocean Drilling Program, Leg 209. Aerobiologia 22:211–226
111. Mims SA, Mims FM III (2004) Fungal spores are transported long distances in smoke from biomass fires. Atmos Environ 38:651–655
112. Brown JKM, Hovmoll MS (2002) Aerial dispersal of pathogens on the global and continental scales and its impact on plant disease. Science 297:537–541
113. Tong Y, Lighthart B (1997) Solar radiation is shown to select for pigmented bacteria in the ambient outdoor atmosphere. Photochem Photobiol 65:103–106
114. Reiling F (2000) Dissemination of bacteria from the mouth during speaking, coughing, and otherwise. JAMA 284:156
115. Giese AC (1976) Living with our suns's ultraviolet rays. Plenum, New York
116. Mims FM III (2005) Avian influenza and UV-B blocked by biomass smoke. Environ Health Perspect 113(12):A806–A807
117. Coffin RB, Velinsky DJ, Devereux R, Price WA, Cifuentes LA (1990) Stable carbon isotope analysis of nucleic acids to trace sources of dissolved substrates used by estuarine bacteria. Appl Environ Microbiol 56:2012–2020
118. Paul JH, Jeffrey WH, DeFlaun M (1985) Particulate DNA in subtropical oceanic and estuarine planktonic environments. Mar Biol 90:95–101

119. Paul JH, Carlson DJ (1984) Genetic material in the marine environment: implication for bacterial DNA. Limnol Oceanogr 29:1091–1097
120. Fuhrman JA, Sleeter TD, Carlson CA, Proctor LM (1989) Dominance of bacterial biomass in the Sargasso Sea and its ecological implications. Mar Ecol Prog Ser 57:207–217
121. Boehme J, Frischer ME, Jiang SC, Kellogg CA, Pichard S, Rose JB et al (1993) Viruses, bacterioplankton, and phytoplankton in the southeastern Gulf of Mexico: distribution and contribution to oceanic DNA pools. Mar Ecol Prog Ser 97:1–10
122. Cho BC, Azam F (1990) Biogeochemical significance of bacterial biomass in the ocean's euphotic zone. Mar Ecol Prog Ser 63:253–259
123. Azam F, Fenchel T, Field JG, Gray JS, Meyer-Reil LA, Thlngstad F (1983) The ecological role of water-column nilcrobes In the sea. Mar Ecol Prog Ser 10:257–263
124. Garcia-Pichel F (1994) A model for internal self-shading in planktonic organisms and its implications for the usefulness of ultraviolet sunscreens. Limnol Oceanogr 39:1704–1717
125. Aas P, Lyons MM, Pledger R, Mitchell DL, Jeffrey WH (1996) Inhibition of bacterial activities by solar radiation in nearshore waters and the Gulf of Mexico. Aquat Microb Ecol 11:229–238
126. Sieracki ME, Sieburth JM (1986) Sunlight-induced growth delay of planktonic marine bacteria in filtered seawater. Mar Ecol Prog Ser 33:19–27
127. Bailey CA, Neihof RA, Tabor PS (1983) Inhibitory effect of solar radiation on amino acid uptake in Chesapeake Bay bacteria. Appl Environ Microbiol 46:44–49
128. Herndl GJ, Muller-Niklas G, Frick J (1993) Major role of ultraviolet-B in controlling bacterioplankton growth in the surface layer of the ocean. Nature 361:717–719
129. Alonso-Sáez L, Gasol JM (2007) Seasonal variations in the contributions of different bacterial groups to the uptake of low-molecular-weight compounds in Northwestern Mediterranean coastal waters. Appl Environ Microbiol 73:3528–3535
130. Häder DP, Kumar HD, Smith RC, Worrest RC (2007) Effects of solar UV radiation on aquatic ecosystems and interactions with climate change. Photochem Photobiol Sci 6:267–285
131. Joux F, Jeffrey WH, Lebaron P, Mitchell DL (1999) Marine bacterial isolates display diverse responses to UV-B radiation. Appl Environ Microbiol 65:3820–3827
132. Sommaruga R, Obernosterer I, Herndl GJ, Psenner R (1997) Inhibitory effect of solar radiation on thymidine and leucine incorporation by freshwater and marine bacterioplankton. Appl Environ Microbiol 63:4178–4184
133. Halac S, Felip M, Camarero L, Sommaruga-Wögrath S, Psenner R, Catalan J et al (1997) An in situ enclosure experiment to test the solar UVB impact on plankton in a high-altitude mountain lake. I. Lack of effect on phytoplankton species composition and growth. J Plankton Res 19:1671–1686
134. Winter C, Moeseneder MM, Herndl GJ (2001) Impact of UV radiation on bacterioplankton community composition. Appl Environ Microbiol 67:665–672
135. Jiang H, Dong H, Yu B, Liu X, Li Y, Ji S et al (2007) Microbial response to salinity change in Lake Chaka, a hypersaline lake on Tibetan plateau. Environ Microbiol 9:2603–2621
136. Liu Y, Yao T, Jiao N, Kang S, Zeng Y, Huang S (2006) Microbial community structure in moraine lakes and glacial meltwaters, Mount Everest. FEMS Microbiol Lett 265:98–105
137. Zhang G, Niu F, Ma X, Liu W, Dong M, Feng H et al (2007) Phylogenetic diversity of bacteria isolates from the Qinghai-Tibet Plateau permafrost region. Can J Microbiol 53:1000–1010
138. Zhang G, Ma X, Niu F, Dong M, Feng H, An L et al (2007) Diversity and distribution of alkaliphilic psychrotolerant bacteria in the Qinghai-Tibet Plateau permafrost region. Extremophiles 11:415–424
139. Dib J, Motok J, Zenoff VF, Ordonez O, Farias ME (2008) Occurrence of resistance to antibiotics, UV-B, and arsenic in bacteria isolated from extreme environments in high-altitude (above 4400 m) Andean wetlands. Curr Microbiol 56:510–517

140. Fernandez-Zenoff V, Sineriz F, Farias ME (2006) Diverse responses to UV-B radiation and repair mechanisms of bacteria isolated from high-altitude aquatic environments. Appl Environ Microbiol 72:7857–7863
141. Ordonez OF, Flores MR, Dib JR, Paz A, Farias ME (2009) Extremophile culture collection from Andean lakes: extreme pristine environments that host a wide diversity of microorganisms with tolerance to UV radiation. Microb Ecol 58:461–473
142. Kepner RL Jr, Wharton RA Jr, Suttle CA (1998) Viruses in Antarctic lakes. Limnol Oceanogr 43:1754–1761
143. Downes A, Blunt TP (1877) Researches on the effect of light upon bacteria and other organisms. Proc R Soc Med 26:488–500
144. Hertel E (1904) Ueber Beeinflussung des Organismus durch Licht, speziell durch die chemisch wirksamen Strahlen. Z Allg Physiologie 4:1–43
145. Hertel E (1905) Ueber physiologische Wirkung von Strahlen verschiedener Wellenlänge. Z Allg Physiologie 5:95–122
146. Hockberger PE (2002) A history of ultraviolet photobiology for humans, animals and microorganisms. Photochem Photobiol 76:561–579
147. Cabaj A, Sommer R, Pribil W, Haider T (2002) The spectral UV sensitivity of microorganisms used in biodosimetry. Water Sci Technol Water Supply 2(3):175–181
148. Rauth AM (1965) The physical state of viral nucleic acid and the sensitivity of viruses to ultraviolet light. Biophys J 5:257–273
149. EPA (1999) Alternative disinfectants and oxidants guidance manual. EPA 815-R-99-014, United States Environmental Protection Agency, Cincinnati
150. Solsona F, Méndez JP (2003) Water disinfection. PAHO/CEPIS/PUB/03.89
151. WHO (2006) Guidelines for safe recreational water environments, vol 2, Swimming pools and similar environments. WHO, Geneva
152. WHO (2003) Guidelines for safe recreational water environments, vol 1, Coastal and fresh waters. WHO, Geneva
153. Henry V, Helbronner A, Recklinghausen M (1910) Nouvelles recherches sur la sterilization de grandes quantited d'eau par les rayons ultraviolets. Comp Rend Acad Sci 151:677–680
154. Wegelin M, Canonica S, Mechsner K, Fleischmann T, Pesaro F, Metzler A (1994) Solar water disinfection: scope of the process and analysis of radiation experiments. Aqua J Water Supply Res Technol 43:154–169
155. McGuigan KG, Joyce TM, Conroy RM, Gillespie JB, Elmore-Meegan M (1998) Solar disinfection of drinking water contained in transparent plastic bottles: characterizing the bacterial inactivation process. J Appl Microbiol 84:1138–1148
156. Navntoft C, Ubomba-Jaswa E, McGuigan KG, Fernández-Ibáñez P (2008) Effectiveness of solar disinfection using batch reactors with non-imaging aluminium reflectors under real conditions: natural well-water and solar light. J Photochem Photobiol B 93:155–161
157. Sommer B, Mariño A, Solarte Y, Salas ML, Dierolf C, Valiente C et al (1997) SODIS – An emerging water treatment process. J Water Supply Res Technol AQUA 46:127–137
158. Spiewak I, Benmair R, Messalem R, Radchenko O (1998) Water detoxification and disinfection using high solar concentration and homogeneous photocatalysts. J Chem Sci 110:229–238
159. Ubomba-Jaswa E, Fernández-Ibáñez P, Navntoft C, Polo-López MI, McGuigan KG (2010) Investigating the microbial inactivation efficiency of a 25 L batch solar disinfection (SODIS) reactor enhanced with a compound parabolic collector (CPC) for household use. J Chem Technol Biotechnol 85:1028–1037
160. Hijnen WAM, Beerendonk EF, Medema GJ (2006) Inactivation credit of UV radiation for viruses, bacteria and protozoan (oo)cysts in water: a review. Water Res 40:3–22
161. Nikaido SS, Johnson CH (2000) Daily and circadian variation in survival from ultraviolet radiation in Chlamydomonas reinhardtii. Photochem Photobiol 71:758–765
162. Nicholson WL, Munakata N, Horneck G, Melosh HJ, Setlow P (2000) Resistance of Bacillus endospores to extreme terrestrial and extraterrestrial environments. Microbiol Mol Biol Rev 64:548–572

163. Zepp RG, Erickson DJ III, Paul ND, Sulzberger B (2007) Interactive effects of solar UV radiation and climate change on biogeochemical cycling. Photochem Photobiol Sci 6:286–300
164. Zepp RG, Callaghan TV, Erickson DJ (1998) Effects of enhanced solar ultraviolet radiation on biogeochemical cycles. J Photochem Photobiol B 46:69–82
165. Zepp RG, Callaghan TV, Erickson DJ (1995) Effects of increased solar ultraviolet radiation on biogeochemical cycles. Ambio 24:181–187
166. Sagripanti JL, Lytle CD (2007) Inactivation of influenza virus by solar radiation. Photochem Photobiol 83:1278–1282
167. Kripke ML (1984) Immunologic unresponsiveness induced by UV radiation. Immunol Rev 80:87–102
168. Lipsitch M, Viboud C (2009) Influenza seasonality: lifting the fog. Proc Natl Acad Sci USA 106:3645–3646
169. Hollosy F (2002) Effects of ultraviolet radiation on plant cells. Micron 33:179–197
170. Prasad PVV, Kakani VG, Reddy KR (2003) Plants and the environment: ozone depletion. In: Encyclopedia of applied plant sciences. Elsevier Academic, Oxford, pp 749–756
171. Jagger J (1967) Introduction to research in photobiology. Prentice Hall, Englewood Cliffs
172. Kakani VG, Reddy KR, Zhao D, Sailaja K (2003) Field crop responses to ultraviolet-B radiation: a review. Agric For Meteorol 120:191–218
173. Roosien J, van Klaveren P, Van Vloten-Doting L (1983) Competition between the RNA 3 molecules of wildtype alfalfa mosaic virus and the temperature-sensitive mutant Tb ts 7(UV). Plant Mol Biol 2:113–118
174. Cuadra P, Herrera RI, Fajardo V (2004) Effects of UV-B radiation on the Patagonian Jaborosa magellanica Brisben. J Photochem Photobiol B 76:61–68
175. Mohr H, Schopfer P (1992) Pflanzenphysiologie, 4th edn. Springer, Berlin/Heidelberg
176. Mazza CA, Battista D, Zima AM, Szwarcberg BM, Giordano CV, Acevedo A et al (1999) The effects of solar ultraviolet-B radiation on the growth and yield of barley are accompanied by increased DNA damage and antioxidant responses. Plant Cell Environ 22:61–70
177. Correia CM, Coutinho JF, Bjorn LO, Torres PJMG (2000) Ultraviolet-B radiation and nitrogen effects on growth and yield of maize under Mediterranean field conditions. Eur J Agron 12:117–125
178. Pal M, Sengupta UK, Srivastava AC, Jain V, Meena RC (1999) Changes in growth and photosynthesis of mungbean induced by UV-B radiation. Indian J Plant Physiol 4:79–84
179. Mepsted R, Paul N, Stephen J, Nogues S, Corlett JE, Baker NR et al (1996) Effects of enhanced UV-B radiation on pea (Pisum sativum L.) grown under field conditions. Glob Change Biol 2:325–334
180. Tevini M, Mark U, Saile-Mark M (1991) Effects of enhanced solar UV-B radiation on growth and function of crop plant seedlings. Curr Top Plant Biochem Physiol 10:13–31
181. Gonzalez R, Mepsted R, Wellburn AR, Paul ND (1998) Non-photosynthetic mechanisms of growth reduction in pea (Pisum sativum L.) exposed to UV-B radiation. Plant Cell Environ 21:23–32
182. Kumagai T, Hidema J, Kang HS, Sato T (2001) Effects of supplemental UV-B radiation on the growth and yield of two cultivars of Japanese lowland rice (Oryza sativa L.) under the field in a cool rice-growing region of Japan. Agric Ecosyst Environ 83:201–208
183. Taylor RM, Tobin AK, Bray CM (1997) DNA damage and repair in plants. In: Lumsden PJ (ed) Plants and UVB responses to environmental changes. Cambridge University Press, Cambridge
184. Robberecht R, Caldwell MM (1978) Leaf transmittance of ultraviolet radiation and its implications for plant sensitivity to ultraviolet-radiation induced injury. Oceologia 32:277–287
185. Caldwell MM, Robberecht R, Flint SD (1983) Internal filters: prospects for UV-accumulation in higher plants. Physiol Plant 58:445–450

186. Tevini M, Iwanzik W, Teramura AH (1983) Effects of UV-B radiation on plants during mild water stress. II Effects on growth, protein, and flavonoid content. Z Pflanzenphysiol 110:459–467
187. Beggs CJ, Schneider-Ziebert U, Wellmann E (1986) UV-B radiation and Adaptive Mechanisms in Plants. In: Worrest RC, Caldwell MM (eds) Stratospheric ozone reduction, solar ultraviolet radiation and plant life. Springer, Berlin, pp 235–250
188. Bischof K, Wiencke C (2006) Auswirkung der Zunahme der UV-Strahlung. In: Lozan J, Graál H, Hubberten HW, Hupfer P, Piepenburg D (eds) Warnsignale aus den Polarregionen. Druckerei in St. Pauli, Hamburg, pp 259–263
189. Wiencke C, Roleda MY, Gruber A, Clayton MN, Bischof K (2006) Susceptibility of zoospores to UV radiation determines upper depth distribution limit of Arctic kelps: evidence through field experiments. J Ecol 94:455–463
190. Johansson G, Snoeijs P (2002) Macroalgal photosynthetic responses to light in relation to thallus morphology and depth zonation. Mar Ecol Prog Ser 244:63–72
191. Roleda MY, Wiencke C, Hanelt D, van de Poll WH, Gruber A (2005) Sensitivity of Laminariales zoospores from Helgoland (North Sea) to ultraviolet and photosynthetically active radiation: implications for depth distribution and seasonal reproduction. Plant Cell Environ 28:466–479
192. Lotze HK, Worm B, Molis M, Wahl M (2002) Effects of UV radiation and consumers on recruitment and succession of a marine macrobenthic community. Mar Ecol Prog Ser 243:57–66
193. Bischof K, Hanelt D, Aguilera J, Karsten U, Vögele B, Sawall T et al (2002) Seasonal variation in ecophysiological patterns in macroalgae from an artic fjord. I. Sensitivity of photosynthesis to ultraviolet radiation. Mar Biol 140:1097–1106
194. de Bakker NVJ, van Bodegom PM, van de Poll WH, Boelen E, Nat E, Rozema J et al (2005) Is UV-B radiation affecting charophycean algae in shallow freshwater systems? New Phytol 166:957–966
195. Martinez-Abaigar J, Nunez-Olivera E, Beaucourt N, Garcia-Alvaro MA, Tomas R, Arroniz M (2003) Different physiological responses of two aquatic bryophytes to enhanced ultraviolet-B radiation. J Bryol 25:17–30
196. Figueroa FL, Jimenez C, Vinegla B, Perez-Rodriguez E, Aguilera J, Flores-Moya A et al (2002) Effects of solar UV radiation on photosynthesis of the marine angiosperm Posidonia oceanica from southern Spain. Mar Ecol Prog Ser 230:59–70
197. Brandt LA, Koch EW (2003) Periphyton as a UV-B filter on seagrass leaves: a result of different transmittance in the UV-B and PAR ranges. Aquat Bot 76:317–327
198. Li FR, Peng SL, Chen BM, Hou YP (2010) A meta-analysis of the responses of woody and herbaceaous plants to elevated ultraviolet-B radiation. Acta Oecologica 36:1–9
199. Rozema J, Van de Staaij JWM, Tosserams M (1997) Effects of UV-B radiation on plants from agro- and natural ecosystems. In: Lumsden PJ (ed) Plants and UV-B responses to environmental change. Cambridge University Press, Cambridge
200. Caldwell MM, Flint SD (1994) Solar ultraviolet radiation and ozone layer change: implications for crop plants. In: ASo A (ed) Physiology and determination of crop yield. American Society of Agronomy, Madison
201. Sullivan JH, Teramura AH (1992) The effects of UV-B radiation on loblolly pine. 2. Growth of field-grown seedlings. Tree Struct Funct 6:115–120
202. Naidu SL, Sullivan JH, Teramura AH, DeLucia EH (1993) The effects of ultraviolet-B radiation on photosynthesis of different aged needles in field-grown loblolly-pine. Tree Physiol 12:151–162
203. Bassman JH, Edwards GE, Robberecht R (2002) Long-term exposure to enhanced UV-B radiation is not detrimental to growth and photosynthesis in Douglas-fir. New Phytol 154:107–120
204. Trost-Sedej T, Gaberscik A (2008) The effects of enhanced UV-B radiation on physiological activity and growth of Norway spruce planted outdoors over 5 years. Trees 22:423–435

205. McLeod AR, Newsham KK (1997) Impacts of elevated UV-B on forest ecosystems. In: Lumsden PJ (ed) Plants and UV-B responses to environmental change, 64th edn. Cambridge University Press, Cambridge
206. Karabourniotis G, Papadopoulos K, Papamarkou M, Manetas Y (1992) Ultraviolet-B radiation absorbing capacity of leaf hairs. Physiol Plant 86:414–418
207. Karabourniotis G, Kyparissis A, Manetas Y (1993) Leaf hairs of Olea europaea protect underlying tissues agains ultraviolet-B radiation damage. Environ Exp Bot 33:341–345
208. Grammatikopoulos G, Karabourniotis G, Kyparissis A, Petropoulou Y, Manetas Y (1994) Leaf hairs of olive (Olea europaea) prevent stomatal closure by ultraviolet-B radiation. Aust J Plant Physiol 21:293–301
209. Day TA, Howells BW, Rice WJ (1994) Ultraviolet absorption and epidermal-transmittance spectra in foliage. Physiol Plant 92:207–218
210. Nedunchezhian N, Kulandaivelu GL (1991) Effects of UV-B enhanced radiation on ribulose 1, 5-bisphophate carboxylase in leaves of Vigna sinensis. Photosynthetica 25:431–435
211. Premkumar A, Kulandaivelu G (1996) Influence of ultraviolet-B enhanced solar radiation on growth and photosynthesis of potassium deficient cowpea seedlings. Photosynthetica 32:521–528
212. Basiouny FM (1986) Sensitivity of corn, oats, peanuts, rice, rye, sorghum, soybean and tobacco to UV-B radiation under growth chamber conditions. J Agron Crop Sci 157:31–35
213. Pal M, Jain V, Sengupta UK (1998) Influence of enhanced UV-B radiation on mustard: cultivar response. Indian J Plant Physiol 3:188–193
214. Singh A (1995) Influence of enhanced UV-B radiation on tropical legumes. Trop Ecol 36:249–252
215. Li Y, Yue M, Wang XL (1998) Effects of enhanced ultraviolet-B radiation on crop structure, growth and yield components of spring wheat under field conditions. Field Crops Res 57:253–263
216. Teramura AH, Murali NS (1986) Intraspecific differences in growth and yield of soybean exposed to ultraviolet-B radiation under greenhouse and field conditions. Environ Exp Bot 26:89–95
217. Al-Oudat M, Baydoun SA, Mohammad A (1989) Effects of enhanced UV-B on growth and yield of two Syrian crops wheat (Triticum durum var. Horani) and broad beans (Vicia faba) under field conditions. Environ Exp Bot 40(1):11–16
218. Finnie JW, Bostock DE (1979) Skin neoplasia in dogs. Aust Vet J 55:602–604
219. Fairham J, Harcourt-Brown FM (1999) Preliminary investigation of the vitamin D status of pet rabbits. Vet Rec 145:452–454
220. Hazewinkel HAW, Tryfonidou MA (2002) Vitamin D_3 metabolism in dogs. Mol Cell Endocrinol 197:23–33
221. Morris JG (1999) Ineffective vitamin D synthesis in cats is reversed by an inhibitor of 7-dehydrocholestrol-Δ^7-reductase. J Nutr 129:903–908
222. How KL, Hazewinkel HAW, Mol JA (1994) Dietary vitamin D dependence of cat and dog due to inadequate cutaneous synthesis of vitamin D. Gen Comp Endocrinol 96:12–18
223. How KL, Hazewinkel HA, Mol JA (1995) Photosynthesis of vitamin D in the skin of dogs cats and rats. Vet Q 17(Suppl 1):S29
224. Brommage R, Miller SC, Langman CB, Bouillon R, Smith R, Bourdeau JE (1988) The effects of chronic vitamin D deficiency on the skeleton in the adult rabbit. Bone 9:131–139
225. Mellanby RJ, Mee AP, Berry JL, Herrtage ME (2005) Hypercalcaemia in two dogs caused by excessive dietary supplementation of vitamin D. J Small Anim Pract 46:334–338
226. Schulze C, Rothuizen J, Van Sluijs FJ, Hazewinkel HAW, Van Den Ingh TSGAM (2000) Extrahepatic biliary atresia in a border collie. J Small Anim Pract 41:27–30
227. Ladds PW, Kraft H, Sokale A, Trueman KF (1983) Neoplasms of the skin of dogs in tropical Queensland. Aust Vet J 60:87–88

228. Teifke JP, Lohr CV (1996) Immunohistochemical detection of P53 overexpression in paraffin wax-embedded squamous cell carcinomas of cattle, horses, cats and dogs. J Comp Pathol 114:205–210
229. UNEP (1991) Environmental effects of ozone depletion: 1991 update. United Nations Environment Programme (UNEP), Nairobi
230. Gross TL, Ihrke PJ, Walder EJ, Affolter VK (2005) Skin diseases of the dog and cat: clinical and histopathologic diagnosis. Blackwell, Oxford
231. Moss LC, Severin GA (1963) Tattooing as therapy for chronic solar dermatitis (Collie nose) in the dog. J Am Vet Med Assoc 142:609–610
232. Nikula KJ, Benjamin SA, Angleton GM, Saunders WJ, Lee AC (1992) Ultraviolet radiation, solar dermatosis, and cutaneous neoplasia in beagle dogs. Radiat Res 129:11–18
233. Nielsen SW, Cole CR (1960) Cutaneous epithelial neoplasms of the dog-a report of 153 cases. Am J Vet Res 21:931
234. Nesbitt GH (1983) Canine and feline dermatology: a systematic approach. Lea & Febiger, Philadelphia
235. Hargis AM, Thomassen RW, Phemister RD (1977) Chronic dermatosis and cutaneous squamous cell carcinoma in the beagle dog. Vet Pathol 14:218–228
236. Jörger K (1988) Skin tumors in cats. Occurrence and frequency in the research material (biopsies from 1984 to 1987) of the Institute for Veterinary Pathology, Zurich. Schweiz Arch Tierheilk 130:559–569
237. Dorn CR, Taylor DO, Schneider R (1971) Sunlight exposure and risk of developing cutaneous and oral squamous cell carcinomas in white cats. J Natl Cancer Inst 46:1073–1078
238. Almeida EMP, Caraca RA, Adam RL, Souza EM, Metze K, Cintra ML (2008) Photodamage in feline skin: clinical and histomorphometric analysis. Vet Pathol 45:327–335
239. Slatter DH, Edwards ME, Hawkins CD, Wilcox GE (1982) A national survey of the occurrence of infectious bovine keratoconjunctivitis. Aust Vet J 59:65–68
240. Méndez A, Pérez J, Ruiz-Villamor E, García R, Martín MP, Mozos E (1997) Clinicopathological study of an outbreak of squamous cell carcinoma in sheep. Vet Rec 141:597–600
241. Payne RJ, Lean MS, Greet TRC (2009) Third eyelid resection as a treatment for suspected squamous cell carcinoma in 24 horses. Vet Rec 165:740–743
242. Heeney JL, Valli VEO (1985) Bovine ocular squamous cell carcinoma: an epidemiological perspective. Can J Comp Med 49:21–26
243. Sloss V, Smith TJ, De Yi G (1986) Controlling ocular squamous cell carcinoma in Hereford cattle. Aust Vet J 63:248–251
244. Russell WC, Brinks JS, Kainer RA (1976) Incidence and heritability of ocular squamous cell tumors in Hereford cattle. J Anim Sci 43:1156–1162
245. Anderson DE, Badzioch M (1991) Association between solar radiation and ocular squamous cell carcinoma in cattle. Am J Vet Res 52:784–788
246. Gharagozlou MJ, Hekmati P, Ashrafihelan J (2007) A clinical and histopathological study of ocular neoplasms in dairy cattle. Veterinarski Arh 77:409–426
247. Morris CA (2007) A review of genetic resistance to disease in Bos Taurus cattle. Vet J 174:481–491
248. Anderson DE, Skinner PE (1961) Studies on bovine ocular squamous carcinoma ("Cancer Eye") XI. Effects of sunlight. J Anim Sci 20:474–477
249. Radostits OM, Arundel JH, Gay CC, Blood DC, Hinchcliff KW (2000) Veterinary medicine: a textbook of the diseases of cattle, sheep, pigs, goats and horses. W.B. Saunders, London
250. Woodward RR, Knapp B Jr (1950) The hereditary aspect of eye cancer in hereford cattle. J Anim Sci 9:578–581
251. Blackwell RL, Anderson DE, Knox JH (1956) Age incidence and heritability of cancer eye in hereford cattle. J Anim Sci 15:943–951
252. SGdS W (1974) Das Vulvacarcinom beim Rind auf Ceylon. Zbl Vet Med A21:834–843

253. Hühnermund G (1973) Disposition der Ayreshirerasse zu karzinomatösen Veränderungen an Vulva und Membrana nicitans bei extremen Sonneneinwirkungen in den Tropen. Berl Münchn Tierärztl Wsch 21:414–415
254. Yeruham I, Perl S, Orgad U, Yakobson B (1999) Tumours of the vulva and vagina in cattle – a 10-year survey. Vet J 158:237–239
255. Wettimuny SG (1974) Bovine vulval carcinoma in Ceylon. Zentralbl Veterinärmed A 21:834
256. Kopecky KE, Pugh GW Jr, Hughes DE (1980) Wavelength of ultraviolet radiation that enhances onset of clinical infectious bovine keratoconjunctivitis. Am J Vet Res 41:1412–1415
257. Forrest JW, Fleet MR (1985) Lack of tanning in white Merino sheep following exposure to ultraviolet light. Aust Vet J 62:244–246
258. Turner AS, Mansmann RA, McAllister ES, Pratt PW (1982) Equine medicine and surgery, 3rd edn. American Veterinary, Santa Barbara
259. Dugan SJ, Curtis CR, Roberts SM, Severin GA (1991) Epidemiologic study of ocular/adnexal squamous cell carcinoma in horses. J Am Vet Med Assoc 198:251–256
260. Eberlein B, Bergner T, Przybilla B (1992) Demonstration of olaquindox phototoxicity in vitro. Photodermatol Photoimmunol Photomed 9:63–66
261. Bovenschen HJ, Peters B, Koetsier MI, Van Der Valk PG (2009) Occupational contact dermatitis due to multiple sensitizations in a pig farmer. Contact Dermat 61:127–128
262. Schauder S, Schröder W, Geier J (1996) Olaquindox-induced airborne photoallergic contact dermatitis followed by transient or persistent light reactions in 15 pig breeders. Contact Dermat 35:344–354
263. Belhadjali H, Marguery MC, Journé F, Giordano-Labadie F, Lefebvre H, Bazex J (2002) Allergic and photoallergic contact dermatitis to Olaquindox in a pig breeder with prolonged photosensitivity. Photodermatol Photoimmunol Photomed 18:52–53
264. Bartussek H (1988) Haltung. In: Haiger A, Storhas R, Bartussek H (eds) Artgemäße Nutztierhaltung. Ulmer, Stuttgart
265. Sozonov JI, Inesin VA (1975) Ergebnisse der UV-Bestrahlung von Kälbern. Landwirtschaftliches Zentralbl 4:460
266. Hidiroglou M, Proulx JG, Roubos D (1979) 25-hydroxyvitamin D in plasma of cattle. J Dairy Sci 62:1076–1080
267. Horst RL, Littledike ET, Riley JL, Napoli JL (1981) Quantitation of vitamin D and its metabolites and their plasma concentrations in five species of animals. Anal Biochem 116:189–203
268. Hidiroglou M, Williams CJ, Proulx JG (1985) Plasma vitamin D3 response in cattle and sheep exposed to ultraviolet radiation. Int J Vitam Nutr Res 55:41–46
269. Flachowsky E, Graf H, Matthey M, Ochrimenko WI, Beyersdorfer S, Dorn W et al (1993) Einfluß von Jahreszeit, Haltungsform und einer Vit.D3 Zulage auf den Vit.A, D und E Gehalt von Kuhmilch sowie die 25-OH Vit-D3-Konzentration im Blutplasma von Milchkühen. Mh Vetmed 48:197–202
270. Richter GH, Flachowsky G, Matthey M, Ochrimenko K, Wolfram D, Schade T (1990) Einfluß der Stall- bzw. Auslaufhaltung auf die 25-OH-Vitamin-D3-Blutplasmakonzentration bei Mastbullen. Mh Vet Med 45:227
271. Hidiroglou M (1987) Kinetics of intravenously administered 25-hydroxyvitamin D3 in sheep and the effect of exposure to ultraviolet radiation. J Anim Sci 65:808–814
272. Hidiroglou M, Karpinski K (1989) Providing vitamin D to confined sheep by oral supplementation vs ultraviolet irradiation. J Anim Sci 67:794–802
273. Smith BS, Wright H (1984) Relative contributions of diet and sunshine to the overall vitamin D status of the grazing ewe. Vet Rec 115:537–538
274. Zintzen H, Boyazoglu PA (1973) Vitamin D and its relation to calcium phosphorus metabolism in ruminants. J S Afr Vet Med Assoc 44:25–36
275. El Shorafa WM, Feaster JP, Ott EA, Asquith RL (1979) Effect of vitamin D and sunlight on growth and bone development of young ponies. J Anim Sci 48:882–886

276. Piccione G, Assenza A, Fazio F, Bergero D, Caola G (2008) Daily rhythms of serum vitamin D-metabolites, calcium and phosphorus in horses. Acta Vet Brno 77:151–157
277. Piccione G, Assenza A, Grasso F, Caola G (2004) Daily rhythm of circulating fat soluble vitamin concentration (A, D, E and K) in the horse. J Circadian Rhythms 2(1):3
278. Hart EB, Steenbock H, Lepkovsky S, Halpin JG (1923) The nutritional requirements of baby chicks: III. The relation of light to the growth of the chicken. J Biol Chem 58:33–42
279. Bethke RM, Kennard DC, Kik MC (1925) Nutritional studies of the growing chick: I. The relation of sunlight and green clover to leg weakness in chicks. J Biol Chem 63:377–390
280. Mussehl FE, Bancroft PM (1925) Nutrient requirements of growing chicks. Poult Sci 4:118–122
281. Heuser GF, Norris LC (1929) Rickets in chicks: III. The effectiveness of mid-summer sunshine and irradiation from quartz mercury vapour arc in preventing rickets. Poult Sci 8:89–98
282. Scott ML, Nesheim MC, Young RJ (1982) Nutrition of the chicken. M. L. Scott and Associates, Ithaca
283. Bernard JB, Watkins BE, Ullrey DE (1989) Manifestations of vitamin D deficiency in chicks reared under different artificial lighting regimes. Zoo Biol 8:313–395
284. Ferguson GW, Jones JR, Gehrmann WH, Hammack SH, Talent LG, Hudson RD et al (1996) Indoor husbandry of the panther chameleon Chamaeleo [Furcifer] pardalis: effects of dietary vitamins A and D and ultraviolet irradiation on pathology and life-history traits. Zoo Biol 15:279–299
285. Ferguson GW, Gehrmann WH, Chen TC, Dierenfeld ES, Holick MF (2002) Effects of artificial ultraviolet light exposure on reproductive success of the female panther chameleon (Furcifer pardalis) in captivity. Zoo Biol 21:525–537
286. Narbaitz R, Tsang CPW (1989) Vitamin D deficiency in the chick embryo: effects on prehatching motility and on the growth and differentiation of bones, muscles, and parathyroid glands. Calcif Tissue Int 44:348–355
287. Packard MJ, Clark NB (1996) Aspects of calcium regulation in embryonic lepidosaurians and chelonians and a review of calcium regulation in embryonic archosaurians. Physiol Zool 69:435–466
288. Häder DP, Sinha RP (2005) Solar ultraviolet radiation-induced DNA damage in aquatic organisms: Potential environmental impact. Mutat Res Fundam Mol Mech Mutagen 571:221–233
289. Madronich S, Flocke S (1995) Theoretical estimation of biologically effective UV radiation at the Earth's surface. In: Zerefos CS, Bais AF (eds) Solar ultraviolet radiation: modelling, measurements, and effects, pp 23–48
290. Rozema J, Björn LO, Bornman JF, Gaberscik A, Häder DP, Trost T et al (2002) The role of UV-B radiation in aquatic and terrestrial ecosystems-An experimental and functional analysis of the evolution of UV-absorbing compounds. J Photochem Photobiol B 66:2–12
291. Huot Y, Jeffrey WH, Davis RF, Cullen JJ (2000) Damage to DNA in bacterioplankton: a model of damage by ultraviolet radiation and its repair as influenced by vertical mixing. Photochem Photobiol 72:62–74
292. Zepp RG (2003) UV exposure of coral assemblages in the Florida Keys. Environmental Protection Agency Report No. 600/R-03/095. U.S. Environmental Protection Agency, Washington, DC
293. Kouwenberg JHM, Lantoine F (2007) Effects of ultraviolet-B stressed diatom food on the reproductive output in Mediterranean Calanus helgolandicus (Crustacea; Copepoda). J Exp Mar Biol Ecol 341:239–253
294. Kouwenberg JHM, Browman HI, Runge JA, Cullen JJ, Davis RF, St-Pierre JF (1999) Biological weighting of ultraviolet (280–400 nm) induced mortality in marine zooplankton and fish. II. Calanus finmarchicus (Copepoda) eggs. Mar Biol 134:285–293

295. Hunter JR, Taylor JH, Moser HG (1979) Effect of ultraviolet irradiation on eggs and larvae on the northern anchovy, Engraulis mordax, and the Pacific mackerel, Scomber japonicus, during the embryonic stage. Photochem Photobiol 29:325–338
296. Bell G, Hoar W (1950) Some effects of ultraviolet radiation on sockeye salmon eggs and alevins. Can J Res 28:35–43
297. Zagarese HE, Williamson CE (2001) The implications of solar UV radiation exposure for fish and fisheries. Fish Fish 2:250–260
298. Kouwenberg JHM, Browman HI, St-Pierre JF, Runge JA, Cullen JJ, Davis RF (1999) Biological weighting of ultraviolet (280–400 nm) induced mortality in marine zooplankton and fish. I. Atlantic cod (Gadus morhua) eggs. Mar Biol 134:269–284
299. Brocklebank JR, Armstrong RD (1994) Solar dermatitis in hatchery-reared salmonids in British Columbia. Can Vet J 35:651–652
300. Lowe C, Goodman-Lowe G (1996) Suntanning in hammerhead sharks. Nature 383:677
301. Dunbar C (1959) Sunburn in fingerling rainbow trout. Prog Fish Cult 21:74
302. Bullock AM, Coutts RR (1985) The impact of solar ultraviolet radiation upon the skin of rainbow trout, Salmo gairdneri Richardson, farmed at high altitude in Bolivia. J Fish Dis 8:263–272
303. McArdle J, Bullock AM (1987) Solar ultraviolet radiation as a causal factor of "summer syndrome" in cage-reared Atlantic salmon, *Salmo solar* L.: a clinical and histopathological study. J Fish Dis 10:255–264
304. Kaweewat K, Hofer R (1997) Effect of UV-B radiation on goblet cells in the skin of different fish species. J Photochem Photobiol B 41:222–226
305. Rodger HD (1991) Summer lesion syndrome in salmon: a retrospective study. Vet Rec 129:237–239
306. Bullock AM, Roberts RJ, Waddington P, Bookless WD (1983) Sunburn lesions in koi carp. Vet Rec 112:551
307. Markkula SE, Karvonen A, Salo H, Valtonen ET, Jokinen EI (2007) Ultraviolet B irradiation affects resistance of rainbow trout (Oncorhynchus mykiss) against bacterium Yersinia ruckeri and trematode Diplostomum spathaceum. Photochem Photobiol 83:1263–1269
308. Roberts R (2001) Fish pathology. Churchill Livingstone, Edinburgh
309. Allison L (1960) "Sunburning" fingerling lake trout with ultraviolet light and the effect off a niacin-fortified diet. Prog Fish Cult 22:114–116
310. Wall AE (1998) Cataracts in farmed Atlantic salmon (Salmo salar) in Ireland, Norway and Scotland from 1995 to 1997. Vet Rec 142:626–630
311. Karentz D, McEuen FS, Land MC, Dunlap WC (1991) Survey of mycosporine-like amino acid compounds in Antarctic marine organisms: Potential protection from ultraviolet exposure. Mar Biol 108:157–166
312. Uitz J, Claustre H, Morel A, Hooker SB (2006) Vertical distribution of phytoplankton communities in open ocean: an assessment based on surface chlorophyll. J Geophys Res 111:C08005
313. Kessler K, Lockwood RS, Williamson CE, Saros JE (2008) Vertical distribution of zooplankton in subalpine and alpine lakes: Ultraviolet radiation, fish predation, and the transparency-gradient hypothesis. Limnol Oceanogr 53:2374–2382
314. Yamashita Y, Tanoue E (2009) Basin scale distribution of chromophoric dissolved organic matter in the Pacific Ocean. Limnol Oceanogr 54:598–609
315. Loiselle SA, Bracchini L, Dattilo AM, Ricci M, Tognazzi A, Cózar A et al (2009) Optical characterization of chromophoric dissolved organic matter using wavelength distribution of absorption spectral slopes. Limnol Oceanogr 54:590–597
316. Browman HI, Vetter RD, Rodriguez CA, Cullen JJ, Davis RF, Lynn E et al (2003) Ultraviolet (280–400 nm)-induced DNA damage in the eggs and larvae of Calanus finmarchicus G. (Copepoda) and Atlantic cod (Gadus morhua). Photochem Photobiol 77:397–404

317. Bullock AM (1982) The pathological effects of ultraviolet radiation on the epidermis of teleost fish with reference to the solar radiation effect in higher animals. Proc R Soc Edinb 81B:199–210
318. Bullock AM, Roberts RJ (1979) Skin photosensitisation by phenothiazine in cultured rainbow trout (Salmo gairdneri Richardson). Vet Rec 104:55
319. Bullock AM, Roberts RJ (1981) Sunburn lesions in salmonid fry: a clinical and histological report. J Fish Dis 4:271–275
320. Bullock AM (1985) The effect of ultraviolet-B radiation upon the skin of the plaice, Pleuronectes platessa L., infected with the bodonid ectoparasite Ichthyobodo necator (Henneguy, 1883). J Fish Dis 8:547–550
321. Jokinen IE, Markkula ES, Salo HM, Kuhn P, Nikoskelainen S, Arts MT et al (2008) Exposure to increased ambient ultraviolet B radiation has negative effects on growth, condition and immune function of juvenile Atlantic salmon (Salmo salar). Photochem Photobiol 84:1265–1271
322. Markkula SE, Salo HM, Immonen AK, Jokinen EI (2005) Effects of short- and long-term ultraviolet B irradiation on the immune system of the common carp (Cyprinus carpio). Photochem Photobiol 81:595–602
323. Armstrong TN, Reimschuessel R, Bradley BP (2002) DNA damage, histologial changes and DNA repair in larval Japanese medaka (Oryzias latipes) exposed to ultraviolet-B radiation. Aquat Toxicol 58:1–14
324. Alemanni ME, Lozada M, Zagarese HE (2003) Assessing sublethal effects of ultraviolet radiation in juvenile rainbow trout (Oncorhynchus mykiss). Photochem Photobiol Sci 2:867–870
325. Bullock AM (1982) The pathological effects of ultraviolet radiation on the epidermis of teleost fish with reference to the solar radiation effect in higher animals. Proc R Soc Edinb B Biol Sci 81:199–210
326. Nowak BF (1999) Significance of environmental factors in aetiology of skin diseases of teleost fish. Bull Eur Assoc Fish Pathol 19:290–292
327. Handlinger J, Soltani M, Percival S (1997) The pathology of Flexibacter maritimus in aquaculture species in Tasmania, Australia. J Fish Dis 20:159–168
328. Cullen AP, Monteith-McMaster CA, Sivak JG (1994) Lenticular changes in rainbow trout following chronic exposure to UV radiation. Curr Eye Res 13:731–737
329. Nelson PA, Kajiura SM, Losey GS (2003) Exposure to solar radiation may increase ocular UV-filtering in the juvenile scalloped hammerhead shark, Sphyrna lewini. Mar Biol 142:53–56
330. Zeng Z, Richardson J, Verduzco D, Mitchell DL, Patton EE (2009) Zebrafish have a competent p53-dependent nucleotide excision repair pathway to resolve ultraviolet B-induced DNA damage in the skin. Zebrafish 6:405–415
331. Banaszak AT (2003) Photoprotective physiological and biochemical responses of aquatic organisms. In: Helbling EW, Zagarese HE, Zagarese H (eds) UV effects in aquatic organisms and ecosystems. Royal Society of Chemistry, Cambridge, UK, pp 329–356
332. Eckes MJ, Siebeck UE, Dove S, Grutter AS (2008) Ultraviolet sunscreens in reef fish mucus. Mar Ecol Prog Ser 353:203–211
333. Zamzow JP (2007) Ultraviolet-absorbing compounds in the mucus of shallow-dwelling tropical reef fishes correlate with environmental water clarity. Mar Ecol Prog Ser 343:263–271
334. Zamzow JP (2004) Effects of diet, ultraviolet exposure, and gender on the ultraviolet absorbance of fish mucus and ocular structures. Mar Biol 144:1057–1064
335. Rick IP, Bakker TCM (2008) Color signaling in conspicuous red sticklebacks: do ultraviolet signals surpass others? BMC Evol Biol 8:189
336. Gacic Z, Damjanovic I, Mickovic B, Hegedis A, Nikcevic M (2007) Spectral sensitivity of the dogfish shark (Scyliorhinus canicula). Fish Physiol Biochem 33:21–27

337. Matsumoto T, Ihara H, Ishida Y, Yamamoto S, Murata O, Ishibashi Y (2010) Spectral sensitivity of juvenile chub mackerel (Scomber japonicus) in visible and ultraviolet light. Fish Physiol Biochem 36(1):63–70
338. Leech DM, Johnsen S (2003) Behavioral response – UVR avoidance and vison. In: Helbling EW, Zagarese HE, Zagarese H (eds) UV effects in aquatic organisms and ecosystems. Royal Society of Chemistry, London, pp 329–356
339. Leech DM, Johnsen S (2009) Light, biological receptors. In: Likens GE (ed) Encyclopedia of inland waters. Elsevier, Oxford, pp 671–681
340. Holtby LB, Bothwell ML (2008) Effects of solar ultraviolet radiation on the behaviour of juvenile coho salmon (Oncorhynchus kisutch): avoidance, feeding, and agonistic interactions. Can J Fish Aquat Sci 65:701–711
341. Rao DS, Raghuramulu N (1996) Food chain as origin of vitamin D in fish. Comp Biochem Physiol A Physiol 114:15–19
342. Copping AM (1934) Origin of vitamin D in cod-liver oil: vitamin D content of zooplankton. Biochem J 28:1516–1520
343. Rao DS, Raghuramulu N (1997) Vitamin D_3 in Tilapia mossambica: relevance of photochemical synthesis. J Nutr Sci Vitaminol 43:425–433
344. Lock EJ, Waagbo R, Wendelaar Bonga S, Flik G (2010) The significance of vitamin D for fish: a review. Aquac Nutr 16:100–116
345. Ferguson GW, Brinker AM, Gehrmann WH, Bucklin SE, Baines FM, Mackin SJ (2010) Voluntary exposure of some western-hemisphere snake and lizard species to ultraviolet-B radiation in the field: how much ultraviolet-B should a lizard or snake receive in captivity? Zoo Biol 29(3):317–334
346. Gölthenboth R (1982) Höhensonne für Zootiere im Zoo Berlin. Berliner Münchner Tierärztliche Wochenschrift 95(11):220
347. Morrisey JK, Reichard T, Lloyd M, Bernard J (1995) Vitamin-D-deficiency rickets in three colobus monkeys (Colobus guereza kikuyuensis) at the Toledo Zoo. J Zoo Wildl Med 26(4):564–568
348. Hofmann-Parisot M (1995) Result of a European-wide inquiry about the use of artificial UV radiation sources in zoo animal husbandry with special regard to the mammals (Primates/Simiae/monkeys). Berl Münch Tierärztl Wschr 108:185–191
349. Tonge S (1985) The management of juvenile telfair's skinks leiolopisma telefairii with particular reference to the role of ultra violet light. In: Reptiles: breeding, behaviour and veterinary aspects. British Herpetological Society, London, pp 61–72
350. Bosch H, Frank W (1983) Häufige Erkrankungen bei im Terrarium gehaltenen Amphibien und Reptilien. Salamandra 19:29–54
351. McCrystal HG, Behler JL (1982) Husbandry and reproduction of captive Giant ameiva lizards. Int Zoo Yearb 22:159–163
352. Townsend CR, Cole CJ (1985) Additional notes on requirements of Captive whiptail lizards (cnemidophorus) with emphasis on ultraviolet radiation. Zoo Biol 4:49–55
353. Acierno MJ, Mitchell MA, Roundtree MK, Zachariah TT (2006) Effects of ultraviolet radiation on 25-hydroxyvitamin D_3 synthesis in red-eared slider turtles (Trachemys scripta elegans). Am J Vet Res 67:2046–2049
354. Acierno MJ, Mitchell MA, Zachariah TT, Roundtree MK, Kirchgessner MS, Guzman DSM (2008) Effects of ultraviolet radiation on plasma 25-hydroxyvitamin D_3 concentrations in corn snakes (Elaphe guttata). Am J Vet Res 69:294–297
355. Gehrmann WH (1987) Ultraviolet irradiances of various lamps used in animal husbandry. Zoo Biol 6:117–127
356. Burger RM, Gehrmann WH, Ferguson GW (2007) Evaluation of UVB reduction by materials commonly used in reptile husbandry. Zoo Biol 26:417–423
357. Gardiner DW, Baines FM, Pandher K (2009) Photodermatitis and photokeratoconjunctivitis in a ball python (Python regius) and a blue-tongue skink (Tiliqua spp.). J Zoo Wildl Med 40:757–766

358. Farman JC, Gardiner BG, Shanklin JD (1985) Large losses of total ozone in Antarctica reveal seasonal ClO_x/NO_x interaction. Nature 315:207–210
359. Blumthaler M, Ambach W (1990) Indication of increasing solar ultraviolet-B radiation flux in alpine regions. Science 248:206–208
360. Rünger TM (2007) How different wavelengths of the ultraviolet spectrum contribute to skin carcinogenesis: The role of cellular damage responses. J Investig Dermatol 127:2103–2105

Books and Reviews

Björn LO (2008) Photobiology. The science of life and light. Springer, New York
Cockell C, Blaustein AR (2001) Ecosystems, evolution, and ultraviolet radiation. Springer, New York
Gao W, Schmoldt DL, Slusser JR (2010) UV radiation in global climate change. Measurements, modeling and effects on ecosystems. Springer, Heidelberg
Ghetti F, Checcucci G, Bornman JF (2006) Environmental UV radiation: impact on ecosystems and human health, NATO science series, IV earth and environment sciences. Springer, Dordrecht
Helbling EW, Zagarese HE, Zagarese H (2003) UV effects in aquatic organisms and ecosystems, Comprehensive series in photochemical and photobiological sciences. Royal Society of Chemistry, Cambridge
Lumsden PJ (1997) Plants and UV-B: responses to environmental change, Society for experimental biology seminar series. Cambridge Univeristy Press, Cambridge
Mora SJD, Demers S, Vernet M (2000) The effects of UV radiation in the marine environment, Cambridge environment chemistry series. Camridge University Press, Cambridge
Reichrath J (2008) Sunlight, vitamin D and skin cancer, Advances in experimental medicine and biology. Springer, New York

Chapter 23
Xenobiotic Protection/Resistance Mechanisms in Organisms

Christopher J. Kennedy and Keith B. Tierney

Glossary

Absorption	In living organisms, the process by which environmental molecules move across biological membranes.
Adaptation	The intergenerational genetically based selection of individuals for specific traits at the population level.
Avoidance	The act of keeping away from or withdrawing from an undesirable environmental chemical.
Biotransformation	The enzymatic conversion of a foreign or endogenous chemical to another form (metabolite) in living organisms.
Contaminant	Any substance that enters a system where it is not normally found.
Defense	A strategy or mechanism by which an organism protects itself or maintains its function in the face of a xenobiotic challenge.
Hydrophilicity	Hydrophilicity refers to a physical property of a molecule that can transiently bond with water (H_2O) (dissolve in it) through hydrogen bonding.
Hydrophobicity	Hydrophobicity (Gr. *Water fearing*) is the physical property of a molecule that is repelled from a mass of water.
K_{ow}	Octanol water partition coefficient, which generally describes the degree of hydrophobicity of a chemical. High Log K_{ow} values >4 are typical of hydrophobic chemicals.

This chapter was originally published as part of the Encyclopedia of Sustainability Science and Technology edited by Robert A. Meyers. DOI:10.1007/978-1-4419-0851-3

C.J. Kennedy (✉)
Department of Biological Sciences, Simon Fraser University, Burnaby, BC V5A 1S6, Canada
e-mail: ckennedy@sfu.ca

K.B. Tierney
Department of Biological Sciences, University of Alberta, Edmonton, AB T6G 2E9, Canada

E.A. Laws (ed.), *Environmental Toxicology: Selected Entries from the Encyclopedia of Sustainability Science and Technology*, DOI 10.1007/978-1-4614-5764-0_23,

Lipophilicity	Lipophilicity (Gr. *fat-liking*) refers to the ability of a chemical compound to dissolve in fats, oils, lipids, and nonpolar solvents.
Phase I reactions	Also called asynthetic or functionalization reactions and include oxidation, reduction, and hydrolysis reactions. In these reactions, a hydrophobic xenobiotic is usually converted to more water soluble metabolite.
Phase II reactions	Also called synthetic or conjugation reactions in which a xenobiotic is covalently bound to an endogenous molecule forming a water soluble metabolite.
Phase III transport	The products of Phase I or II biotransformation reactions and some parent xenobiotics are transported out of cells by various protein transport families collectively termed Phase III transport. In some cases, these transporters prevent the cellular entry of xenobiotics.
Pollutant	A contaminant that adversely alters properties of the environment.
Resistance	The capacity of a population of organisms to withstand the effects of a harmful environmental agent through genetic selection of less susceptible individuals.
Tolerance	Reduced sensitivity of an individual organism, tissue, or cell to a foreign chemical and its toxic effects through acclimatory processes.
Toxicodynamics	The process of the interaction of chemical substances with target sites (molecules, cells, tissues, or organs) leading to adverse effects.
Toxicokinetics	Processes of the uptake of potentially toxic substances, their biotransformation, distribution, and elimination.
Xenobiotic	Chemicals that are foreign to biological systems.

Definition of Subject

Organisms have always been exposed to chemicals that are foreign to them. Evolution has produced protective mechanisms against natural chemicals, mechanisms that are currently used against exposures to anthropogenically sourced chemical contaminants. Protection can occur at the individual level, often called tolerance, or at the population level, which is called resistance. The earliest studies on tolerance and resistance mechanisms come from the insect/pesticide literature; however recent environmental and health research is providing a wealth of information on these mechanisms in other organisms including humans. There are two major categories of protective mechanisms: (1) toxicokinetically derived mechanisms, which alter the way in which organisms absorb, biotransform, and excrete chemicals; and (2) toxicodynamically derived mechanisms, in which target sites are modified to reduce sensitivity. In both of these categories, protection can occur at the molecular and genetic level, through the cellular/tissue and organ levels, and up to and including whole organism responses including changes in behavior.

Introduction

Organisms have been exposed to foreign chemicals, or xenobiotics, through evolutionary history. The challenge of this chemical affront is to maintain appropriate concentrations of particular molecules to function, while excluding xenobiotics that will interact with cellular constituents. This is problematic, as organisms must remain open to a chemically dynamic environment for essential life processes. Chemical homeostasis employs cellular, physiological, and behavioral regulatory processes in response to the accumulation of xenobiotics that may cause departures from a state of dynamic equilibrium with the environment and reduce performance (e.g., reduced growth, reproduction) or cause death. The early evolution of systems capable of reducing overall xenobiotic accumulation and their susceptibility to them, point to the evolutionary significance of xenobiotic protection mechanisms and their survival advantages.

Organisms have always been exposed to xenobiotics of natural origin. For example, host-plant resistance strategies often employ defensive secondary metabolites, known as allelochemicals (e.g., alkaloids such as nicotine, caffeine, morphine, colchicines, strychnine, and phenolics such as the cannabinoids), to influence the behavior, growth, or survival of herbivores. Herbivorous insects and mammals have evolved common protective defenses against the ingestion of allelochemicals. For example, Australian marsupials are constantly exposed to a diverse array of toxic xenobiotics that come from a wide variety of plant sources, such as Eucalyptus terpenes [1]. Marsupials display several significant xenobiotic processing mechanisms at the biochemical level for enhanced terpene detoxification, which is critically important for surviving on their unique Eucalyptus diet.

Organisms have adapted (to various degrees) to xenobiotic exposure of natural origin by evolving chemical defense/protective/resistance mechanisms. The relatively recent phenomenon of industrial pollution and the release of novel synthetic compounds into the environment at high concentrations and relatively short durations has not allowed for much evolutionary adaptation (although some) to synthetics. Most organisms rely on constituent chemical defense mechanisms for protection against this new chemical arsenal. The dramatic and catastrophic effects of pollution on many species points to the inadequacy of such defenses; however, in some cases the protection afforded by such mechanisms works equally well for anthropogenic chemicals. For example, the Great Lakes and Puget Sound, WA, have accumulated tons of PCBs, PCDDs, and PCDFs; yet despite chronic exposures to such compounds at concentrations that elicit toxicity in mammals and birds, populations of fish can flourish at very polluted locations. Although species diversity at these contaminated sites is reduced, signs of overt toxicity, with the exception of carcinogenesis, are rare. It is believed that populations chronically exposed to high or even moderate levels of these contaminants have evolved or developed resistance mechanisms to these compounds [2].

There are two main possible means by which organisms can reduce their sensitivity to xenobiotics; those that occur at the level of the individual through acclimatory processes (often termed tolerance) and those that occur on populations of organisms in which the genetic constitution of a population is changed in response to the chemical, so that a greater number of individuals are able to resist it compared to the unexposed population (often termed resistance). Acclimation occurs at the level of the individual, where organisms become more tolerant as a consequence of a prior exposure through genetic or epigenetic mechanisms. Tolerance brought about through acclimation is not transmitted across multiple generations, therefore it is believed to wane if exposure is removed, for example, in remediated environments.

Adaptation is the intergenerational genetically based selection of more resistant individuals at the population level that results in individuals contributing relatively more offspring to subsequent generations. In order for adaptation to occur, resistant phenotypes must be present in sensitive populations before exposure, albeit at low frequencies. After remediation, adaptations may also disappear from populations because the selective pressure has been removed, but this occurs slower than for the acclimation process. Examples of heavy metal tolerance in plants are common and demonstrate dramatic changes in susceptibilities over a few generations. The evolution of industrial melanism in moths is an example of resistance that is widely cited as an example of contamination as a selective pressure for evolution.

Insect resistance to pesticides has been a well-studied area in this regard due to its scientific, economic, ecological, and health significance; pesticide resistance is a significant cause of failures in agricultural and pest management programs. Resistance to insecticides was first documented in 1914 by A. L. Melander for scale insects to an inorganic insecticide. Between 1914 and 1946, 11 additional cases of resistance to inorganic insecticides were recorded. Since 1945, it has been estimated that between 500 and 1,000 species of pests have developed a resistance to a pesticide. It was believed that the development of organic insecticides such as DDT would prevent insecticide resistance; however, by 1947 DDT resistance was exhibited by houseflies. With the introduction of every new insecticide class – cyclodienes, carbamates, formamidines, organophosphates, pyrethroids, even *Bacillus thuringiensis* – cases of resistance surfaced within 2–20 years [3].

Xenobiotic resistance is also significant in the area of human health. It is a major cause of failure of human therapies and has implications in antibacterial, anticancer, antipaludic, and antihuman immunodeficiency virus-1 (HIV-1) therapies. The mechanisms behind this resistance are conserved among bacteria, eukaryotic cells, parasites, and viruses [4].

Mechanisms of Resistance

Organisms use a variety of mechanisms to avoid the effects of toxic substances. These strategies range from the molecular to the behavioral level. Several different mechanisms have been proposed whereby resistance to chemical contaminants can develop in animal populations. The literature cites several classification schemes; however, here they are organized according to:

1. Mechanisms of decreased exposure through toxicokinetically derived resistance. Specific mechanisms in this category include reduced uptake or increased elimination of toxicants, thereby decreasing body burdens. Increased metabolic transformations of xenobiotics to less toxic or persistent forms can contribute to resistant phenotypes. Also, the sequestration of contaminants within cellular organelles or binding to proteins are means by which toxicants are unable to exert their toxic effects despite their accumulation.
2. Mechanisms of decreased sensitivity through toxicodynamically derived resistance. Mechanisms here include: increases in target receptor insensitivity (target site resistance) through changes in receptor structure, up- or down-regulation of target receptors, or decreases in the responsiveness of signal transduction pathways. Circumvention is the bypassing of an inhibited pathway through an alternate route; limited opportunities for this mechanism exist, however it can occur.

Toxicokinetically Derived Resistance Mechanisms

Reductions in Xenobiotic Uptake

Limits to Exposure: Avoidable Exposures Ideally, organisms act to limit their exposures to harmful substances by sensing the substance and moving into a cleaner environment (Fig. 23.1). For this to occur, four conditions are necessary: (1) the substance must exist in the environment in a gradient, (2) the organism must be able to perceive the substance's presence, (3) the organism must identify the substance as harmful, and (4) the organism must be able to move itself away from exposure. When met, these conditions provide a means of behavioral avoidance.

Environmental gradients – In aerial and aquatic environments of the past and present world, the first condition (a substance varying in concentration over space) is satisfied over small to large scales for innumerable toxic substances of natural and human origin. In the past, substances that organisms sought to avoid included hydrocarbons, toxins from plants and algae, and some arising from volcanic activity. Today, there is an ever increasing array of synthetics including

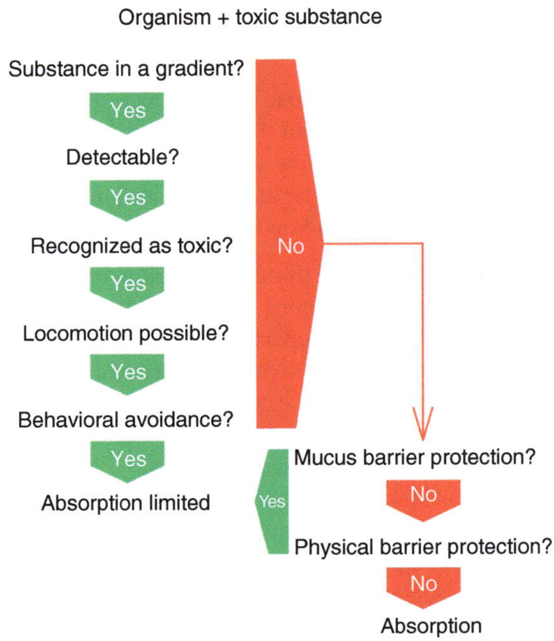

Fig. 23.1 There are many factors that determine whether an organism existing in a contaminated environment will end up absorbing any given contaminant. These factors include sensory and behavioral responses (e.g., the ability to detect a contaminant and move away from it), as well as the characteristics of the outermost structures (e.g., the permeability of keratin of skin and the chitin of arthropod cuticle)

pesticides, flame retardants, surfactants, and pharmaceuticals. Many of these synthetics are based on substances of natural origin, for example, pyrethroid insecticides are structurally similar to compounds of the chrysanthemum flower, and ethinylestradiol used in birth control pills is a more potent form of estrogen. Overall, there are few organisms that do not experience gradients of substances of both natural and synthetic origin.

Perception of environmental gradients – To meet the second condition of behavioral avoidance, organisms possess exteroceptive sensory mechanisms to gather environmental chemical and/or conditional information. Such mechanisms utilize intensity receptors, regardless of species. Intensity receptors can be relatively simple and exist at the molecular level in single-celled organisms such as in *Euglena* spp. [5], or more complex, such as those relying on multisensory neuron integration. The majority of our knowledge comes from the second category, for organisms with few sensory neurons (e.g., ~32 in *Caenorhabditis elegans*) [6] to millions (e.g., humans have ~six million olfactory sensory neurons alone) [7]. These sensory neurons rely on a diverse array of G-protein coupled receptors (GPCRs) that transduce molecular signals from the outside of the organism to the inside. The array of receptors in any given organism can be enormous – mice, for example, have 1,000–1,200 distinct olfactory receptors [8]. All intensity receptors have evolved to detect "natural" substances, and not novel compounds of human origin. Nevertheless, given the fact that many synthetic toxic compounds are based on naturally produced compounds, it is not unreasonable to expect "historic" sensory machinery to be able to function to some degree in the current environment.

Perceiving a substance as harmful – There are two routes by which a substance may be identified as harmful: through innately (genetically) recognizing it as so, or through acquiring (learning) its harmful nature. The former constitutes the most basic form of avoidance, that is, instinctive and not learned responses, which include reflexes and fixed action patterns (FAPs; a series of contiguous responses that continue until completion, e.g., sneezing). The latter involves memory and cognition, and so is restricted to organisms capable of these functions.

With innate avoidance responses, organisms respond aversively to unconditioned stimuli – stimuli that need not have been previously encountered. Reflexes, for example, do not even involve the brain (e.g., nociceptive withdrawal reflex to painful stimuli). Other innate responses may involve higher order neurological signal integration, such as in fishes that instinctively avoid sulfur dioxide and chlorine, which presumably helps maintain acidic and ionic balance [9]. Similarly, *Drosophila* instinctively avoids CO_2, apparently because CO_2 is associated with stressed conspecifics [10]. While these example substances can be increased in the environment as a result of human activities, their avoidance response is based on receptors and neural circuits. With synthetic, perceptible substances, there is no guarantee that an organism will identify them as harmful and so move down a concentration gradient. In fact, the opposite can occur – fishes are known to swim up concentration gradients of various pesticides, including bentazone, dalapon, and prochloraz [11]. Furthermore, innate responses may change over time, becoming reduced or enhanced in magnitude. Reduced responses can occur following successive exposures through neurological habituation (also referred to as "non-associative learning"). Decreased gill withdrawal response to painful stimuli is well documented in sea slugs (*Aplysia* spp.) [12]. Conversely, enhanced responses can occur with neurological sensitization. Drug seeking behaviors to amphetamine and cocaine are good examples [13]. What these innate response modifying mechanisms suggest is that over time, desirable avoidance responses may be lost or attenuated, and undesirable attraction responses may be enhanced.

In the absence of an innate response, organisms may limit future exposures to toxic substances by learning to avoid them. A mechanism by which this occurs is through associative learning, which occurs when an organism learns to associate a conditioned stimulus (e.g., certain tastes or smells) with an unconditioned unpleasant or otherwise undesirable stimulus (e.g., pain). Through their pairing, the organism may even learn to respond aversively to the conditioned stimulus alone. This process is widely employed by organisms, using not only exteroceptive sensors, but also interoceptive sensors (i.e., those responding to internal stimuli). A well-known natural example of exteroceptive-based associative learning is the avoidance animals will show to skunks following an unfortunate interaction.

Presumably this could occur for a toxic substance if an organism learned to pair it and its effects with a location or odor, although experimental examples are absent. A more tested form of learned avoidance is based on interoceptive responses. Ingestional aversive learning, for example, occurs when an organism ingests a specific food item, later experiences internally based pain or distress, and subsequently chooses to avoid that food in the future. Interoceptive avoidance learning is used by many organisms with naturally occurring toxins, from grasshoppers [14] to humans [15]. Furthermore, while associative learning was previously thought only to occur in organisms that posses neurological systems, recent evidence suggests the contrary: gene-regulatory networks appear to impart this ability in single-celled organisms [16]. Interoception is also the basis for the well-used conditioned taste aversion (CTA) assay. For CTA, a novel taste (conditioned stimulus) is typically paired with an injected toxin, and the rate at which the organism develops CTA is determined. This rate can be rapid in mice, just one pairing of the insecticide metabolite paraoxon with a novel taste was sufficient for development of CTA [17]. Clearly, should an animal survive an initial exposure to a toxic substance and associate a stimulus with that substance, future exposures have the potential to be avoided.

Innate and acquired avoidance responses may be adversely affected when an exposure is unavoidable and toxic substances are taken up (e.g., ingested) by the organism, especially if the substances are neurotoxic. Most of the studies of this type of impairment have focused on how natural avoidance responses are lost following contaminant exposure (e.g., the common toad (*Bufo bufo*)) failed to behaviorally respond to chemical cues released by Turkish crayfish (*Astacus leptodactylus*) following exposure to the herbicide amitrole [18]. However, at least one study noted that contaminant exposure will modify future contaminant avoidance. Specifically, vendace (*Coregonus albula*) given weeklong exposures to bleached kraft mill effluent (BKME) and then subsequently given the choice to select clean versus contaminated water chose (preferred) BKME-contaminated water [19]. Predicting whether an organism will make the right choice when given the option to avoid exposure almost certainly requires empirical evidence.

Aversive locomotion – Those organisms with physical means to move in response to stimuli can do so by means of one of two sensory routes, and these are based on the complexity of their sensory mechanism(s). At the most simple extreme, the input from a single sensor can be used to guide locomotion. The resulting movement, referred to as klinotaxis, is evident in a behavioral phenotype characterized by "side-to-side" searching. This oscillatory motion is necessary for the organism to establish a gradient over the receptor, and therefore over the environment itself. Klinotaxis is generally understood to refer to positive motion, that is, up a gradient; however, the reverse is also possible. For more complex organisms, that is, those with paired intensity receptors, tropotaxis can occur. This motion is behaviorally evident in more linear movement, since the paired receptors can provide information of a concentration gradient across the location of the receptors, not just the environment. As for structures that enable movement, even single-celled spp. have the means, for example, *Euglena* spp. and its flagellum. A consideration is that locomotory ability is known to be highly sensitive to impairment by toxic substances, such as to the numerous anticholinesterase (anti-AChE) organophosphorus insecticides [20, 21].

Limits to Absorption: Unavoidable Exposures Arguably all organisms today receive exposure to synthetic substances. Toxic effects may ensue if any one or more of the four necessary avoidance conditions are not met and the substance passes into or through the external and internal coverings, that is, protective barriers, and into the tissues of the organism. Barriers include secretions, secretion clearance mechanisms, the epithelium, and other physical barriers to uptake.

Across phyla, organisms produce mucous to cover external and internal surfaces and cavities with an environmental interface. With vertebrates such as fish and mammals, goblet cells within mucous membranes are responsible for the production of mucus and are found in places such as the mouth, stomach, nose, and genitals. With invertebrates, including those in marine (e.g., *Polychaeta* worms) and terrestrial (e.g., slugs) [22, 23] environments, mucous is produced in mucous cells (mucocytes), which are in similarly sensitive areas of environmental exposure such as the malpighian tubules of arthropods [24]. Mucus fulfills multiple roles and contains constituents such as mineral salts, antibacterial enzymes, agglutinating glycoproteins (mucins), immunoglobulins, and other specialized proteins. These substances together form a mucous matrix, which serves as a potential "trap" for environmental substances, such as with tetracycline [25]. For vertebrates, proteins can be secreted into the mucous to achieve an additional function, namely, detoxification (i.e., biotransformation). Specifically, Phase II (conjugation) enzymes (e.g., glutathione *S*-transferases [GSTs]) can be found in mucus [26]. In sensory neurons, the presence of defensive biotransformation enzymes is critical not only for protection, but also to "inactivate" odorants and facilitate their removal [27]. Currently the presence of biotransformation enzymes in insect mucus remains to be explored, although biotransformation enzymes are known to be found in sensory organs (e.g., in the antennae of the moth *Mamestra brassicae*) [28]. Whether biotransformation occurs or not, for organisms with cilia, mucus (and any trapped substances) may be removed by mucociliary clearance, that is, propulsion of mucous away from cavities.

The outermost physical barrier of many organisms is covered with skin or epithelium, and fibrous protein or sugar-based coverings. Epithelium functions as a permeability barrier to toxic substances, limiting percutaneous absorption (cutaneous permeation), such as can occur with organophosphorus insecticides [29]. The resistance of skin is largely due to the stratum corneum, which contains dead cells rich in fibrous proteins known as keratin [30]. These dead cells are continuously sloughed off and replaced, sending any trapped substances with them. In many organisms, keratin is also found in more rigid and resistant structures such as beaks, hooves, (reptilian) shells, and scales. In invertebrates and fungi, a barrier of polysaccharide termed chitin exists, and it has a similar strength to keratin. Chitin can be so formidable a barrier that numerous synthetic agents exist to inhibit its formation [31]. Nevertheless, outer barriers are not always uniform in strength; in fact there can be features that limit protective function. In mammals, for example, the pilosebaceous unit, which is composed of hair, its follicle, and the sebaceous gland, can provide an avenue of uptake [32]. Ultimately, the ability of outer physical barriers to prevent the uptake of foreign substances is a function of exposure concentration, duration, and the physicochemical properties of the substance.

Increases in Xenobiotic Elimination and Reductions in Accumulation

Biotransformation Biotransformation is defined as the conversion of a substance from one form to another by the actions of enzymes within an organism and is distinguished from physical-chemical processes that can also effect chemical conversions (e.g., photolysis). Biotransformation usually refers to the combined transformations, reflecting the reality that a particular chemical may undergo a number of chemical changes. A molecule or its biotransformational product, its metabolite, may be modified at a number of different positions with a variety of modifications, possibly resulting in several metabolites.

Xenobiotic metabolism is basically a process that introduces hydrophilic functionalities onto the foreign molecule to facilitate excretion. The availability of xenobiotics for biotransformation reactions and the rates of metabolism therefore have important implications on the tissue levels of toxicants, metabolite patterns, and xenobiotic half-life, all of which affect the bioaccumulation, persistence, and the severity and duration of a toxic response. For example, it has been shown that the four order of magnitude difference in the bioaccumulation of two structurally related compounds, DDT (2-bis(*p*-chlorophenyl)-1,1,1-trichloroethane) and 2-bis(*p*-methylthiophenyl)-1,1,1-trichloroethane, with similar lipophilic characteristics, was due to the latter compound's ability to be biotransformed by the mosquitofish (*Gambusia affinis*) [33].

Hydrophobic (high log K_{ow} [octanol:water partition coefficients] value) and lipophilic chemicals (e.g., PCBs, cyclosporine A) are not easily excreted, as they partition back into cells and tissues from excretory media (e.g., urine, bile). Water-soluble chemicals are not reabsorbed because lipid membranes of cells lining excretory routes act as barriers to their reuptake. This is why many lipophilic compounds tend to accumulate: they are easily absorbed and poorly excreted. For example, southern resident Killer Whales in the Pacific Northwest contain PCB concentrations as high as 146 mg/kg lipid weight, placing them among the most PCB-contaminated marine mammals in the world [34]. Therefore, the general scheme of biotransformation processes is the conversion of lipophilic compounds to more polar hydrophilic metabolites.

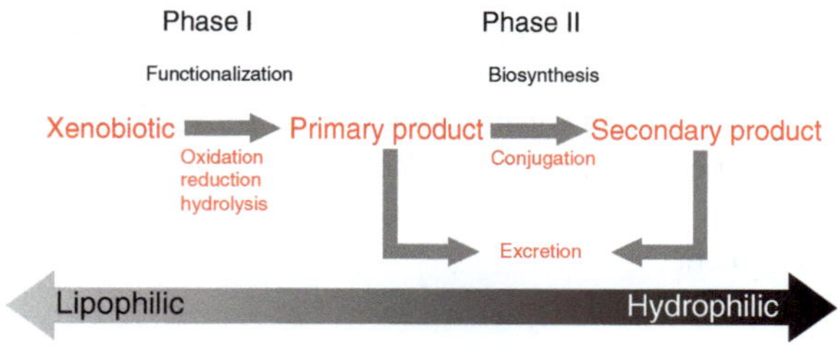

Fig. 23.2 The general biotransformation scheme for xenobiotics in organisms through phase I and phase II reactions. The general trend is to convert lipophilic compounds to more hydrophilic metabolites, which are more easily excreted

All organisms have the ability to metabolize foreign compounds that are taken up; however, their evolved abilities in this regard vary widely. The highest biotransformation ability is usually found in mammals and birds, followed by fish and reptiles, and invertebrates. Major sites of biotransformation include respiratory tissues, nasal mucosa, and the skin and gastrointestinal tract (all sites of potential entry of xenobiotics). The major site of biotransformation in organisms is the liver or functional equivalent. This organ has a high lipid content (to enhance lipophilic chemical accumulation), a high blood flow (increased delivery), and high titers of biotransformation enzymes.

Biotransformation generally occurs through a sequence of reactions termed phase I and phase II reactions (Fig. 23.2). In phase I reactions a nucleophilic functional group (e.g., –OH, –SH, $-NH_2$, –COOH) is introduced or exposed in the parent molecule. In phase II reactions, the phase I metabolite or parent molecule already containing a functional group is conjugated to an endogenous molecule (e.g., sugar, amino acid, sulfate). Phase I metabolites are more polar than the parent molecule, increasing excretion potential, as well as rendering them suitable for phase II enzymatic reactions. Phase II reactions form exceptionally water-soluble products that undergo significant ionization at physiological pH. The endogenous conjugating moieties are also substrates for secretion proteins in transport epithelia. Phase I reaction products can be less toxic than the parent compound or may be bioactivated to more toxic metabolites (see below in oxidation reactions), a significant process in the toxicity of many environmental contaminants (e.g., dibrominated diphenyls) and drugs (e.g., acetominophen).

Phase I Reactions (Asynthetic, Functionalization Reactions) Phase I reactions are the predominant biotransformation pathway for most xenobiotics and include microsomal monooxygenations, cytosolic and mitochondrial oxidations, co-oxidations in the prostaglandin sythetase reaction hydrolysis reactions, epoxide hydration, and reduction reactions. The common definition of Phase I reactions includes oxidation, reduction, and hydrolysis reactions, a grouping that lacks any mechanistic coherence. The modern systematic classification of enzymes (Enzyme Commission numbering system) groups together oxidations and reductions (catalyzed by oxidoreductases, EC class 1) but places hydrolysis reactions in a completely separate class (class 3). Not uncommonly, the term "phase I metabolism" continues to be used, mainly as an outdated synonym for P450-dependent oxidations. Regardless of nomenclature, the common theme to these reactions is to either unmask or introduce a polar functional group in or onto the parent xenobiotic.

Oxidation

The cytochrome P-450 (CYP)-dependent mixed function oxidase (MFO) system of eukaryotes consisting of membrane lipids (e.g., mostly localized to the endoplasmic reticulum [ER], but also present in mitochondrial inner membranes), the enzyme cytochrome P450 (P450s), and an electron transfer chain (NADPH-cytochrome P450 reductase in the ER, and redoxin reductase/redoxin in mitochondria) is crucial for oxidative, peroxidative, and reductive metabolism. P450s have two functions in general: (1) a protective role in the degradation or provision of polar handles for the solubilization of xenobiotics in preparation for excretion, and (2) a broad functional role in the biosynthesis of critical signaling molecules used for the control of development and homeostasis.

P450s constitute a superfamily of heme enzymes present in every living species on Earth from archaebacteria to humans. Despite their occasionally minimal sequence similarity, all CYPs have a similar structural fold with a highly conserved core. Due to its high diversity, a systematic classification of individual CYP forms into families and subfamilies has emerged. The protein sequences within a given gene family are at least 40% identical (e.g., CYP2A6 and CYP2B6), and the sequences within a given subfamily are >55% identical (e.g., CYP2A6 and CYP2A7). Individual members of a family or subfamily are labeled again by Arabic numerals (e.g., CYP3A4, CYP3A7). There are 17 different families and 59 functional P450 genes currently known in humans. In this respect, hepatic drug oxidation is a major source of interindividual variation in pharmacokinetics and therapeutic response. The expression of individual P450 proteins in the liver is influenced by a number of factors, such as genetic make-up, disease, ageing, and environmental factors (e.g., smoking, alcohol, nutrition, pollutants) [35].

Early P450 research, suggested that only a few types of organic compounds served as P450 substrates, but it is now known that P450 substrates are an enormous and diverse group of compounds, including endobiotics and xenobiotics, and the list continues to grow. P450s do not conform to the typical textbook definition of an enzyme as a highly specific biological catalyst, and it has been predicted that the substrate base could be a million or more compounds [35]. Endobiotic substrates include such compounds as steroids, bile acids, fatty acids, prostaglandins, fat-soluble vitamins, amino acids, eicosanoids and retinoids, lipid hydroperoxides,

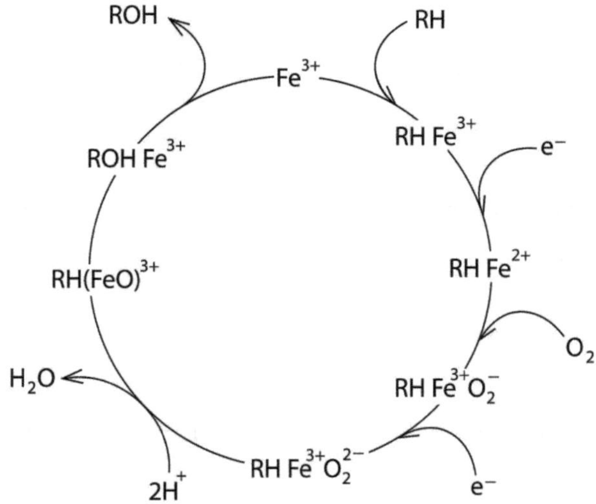

Fig. 23.3 The cytochrome P450 catalytic cycle

and leukotrienes. Xenobiotic substrates include most of the therapeutic drugs and environmental contaminants, antioxidants, dyes, and plant products such as flavorants and odorants. The enzymes in the families 1–3 are mostly active in the metabolism of xenobiotics, whereas the other families have important endogenous functions. Interestingly, mutations that inactivate CYP enzymes with physiological functions often lead to serious diseases, whereas similar mutations in xenobiotic-metabolizing CYPs rarely do.

It should also be noted that P450 cytochromes occur in a variety of tissues other than liver. For example, P450 has been found in olfactory and respiratory nasal mucosa of animals, including fish.

Nature has found many ways to utilize atmospheric dioxygen to functionalize molecules for their degradation or synthesis through the use of a diverse set of cofactors. P450 is a heme (Fe at the active site) containing oxygenase, conscripted to metabolize atmospheric dioxygen in an oxygenase catalytic cycle (Fig. 23.3), resulting in the incorporation of one oxygen atom into a substrate. This hydroxylation reaction has the following overall stoichiometry:

$$RH + O_2 + NADPH + H^+ \rightarrow ROH + H_2O + NADP^+,$$

where RH represents a xenobiotic or endogenous substrate and ROH is the product. The most important property of all known P450s is their ability to bind and activate two atoms of oxygen. A second characteristic property of P450s is that their active site is a heme macrocycle with the central atom, the heme iron, being bound to the protein structure through the anionic, thiolate sulfur of a cysteine residue. The mode by which the heme is bound to apoprotein enables the ability of all P450s to activate the Fe–O–O moiety. The third common property of all P450 enzymes is their similar overall structure and shape, in which most of the secondary-structure elements, as well as the overall folding pattern, are conserved. NADPH-cytochrome

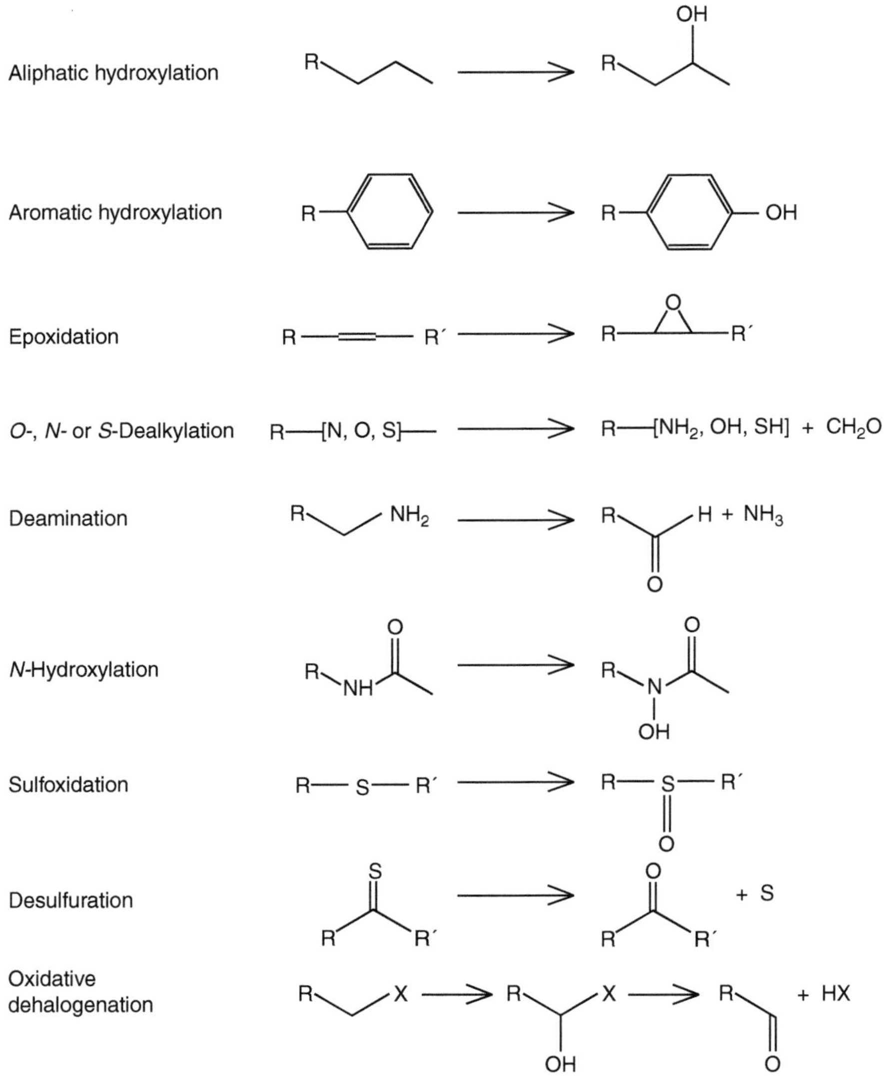

Fig. 23.4 Typical oxidation reactions catalyzed by cytochrome P450 monooxygenases

P450 reductase uses NADPH as a cofactor that, after binding of the substrate to oxidized CYP P450, transfers electrons to CYP P450, thereby reducing it, allowing for several steps whereby one atom of oxygen from O_2 is transferred onto the xenobiotic and the other oxygen is used to form water. The oxygenation and other alterations of the very diverse number of substrates by P450s may seem indiscriminate, but in many instances the modification is positionally and even stereochemically specific. Some examples of reactions catalyzed by the microsomal cytochrome P450 system are given in Fig. 23.4.

Biotransformation generally results in reduced toxicity through the genesis of less toxic metabolites, although bioactivation to more toxic products (bioactivation) is a significant process in the toxicity of many compounds. Reactive intermediates can be formed directly through either phase I (oxidative or reductive reactions) or phase II reactions (less common), or by a rearrangement of a metabolite to an unstable form. Reactive intermediates are usually electrophilic and can react with nucleophilic sites on molecules such as the sulfhydryl group of glutathione and cysteine, or the guanine base in DNA. For example, the addition of oxygen to PAH C=C can produce an epoxide that can nonenzymatically react with macromolecular nucleophiles, resulting in covalent bond formation. Benzo[*a*]pyrene is a procarcinogen, and via a series of oxidation and hydrolysis reactions, a highly reactive metabolite (+)-benzo[*a*]pyrene-7,8-diol-9,10-epoxide is formed, which can react with cellular DNA and initiate the carcinogenic process [36].

Other important Phase I oxidative enzymes include flavin-containing monooxygenase, which catalyzes the oxygenation of nitrogen and sulfur compounds, and the epoxide hydrolases, which hydrolyze various epoxides to diols. The cytochrome P-450 and flavin-containing monooxygenase enzymes are localized in the microsomes, while epoxide hydrolases occur in both the microsomal and cytosolic fractions of the cell.

Reduction

Reduction reactions are also catalyzed by P450, owing to the transfer of reducing equivalents of P450 reactions to the xenobiotic substrate, rather than molecular oxygen. Reductions are most likely to occur with xenobiotics in which oxygen tension is low. Reductions can occur across nitrogen–nitrogen double bonds (*azo reduction*) or on nitro groups (NO_2). Frequently, the resulting amino compounds are oxidized forming toxic metabolites. Some chemicals can be reduced to free radicals, which are reactive with biological tissues.

Hydrolysis

Hydrolysis reactions are typically not carried out by the P450 system. Grouping hydrolysis together with oxidation (as in phase I reaction) means separating hydrolysis from glutathione conjugation (phase II reaction), however, both hydrolysis and glutathione conjugation are reactions of an abundant cellular nucleophile (water or GSH) with a xenobiotic electrophile. In most cases, hydrolysis and glutathione conjugation are competing pathways for the disposition of xenobiotics (e.g., PAH dihydrodiol epoxides, alkylating agents, *O*-acetoxy esters of aromatic amines, etc.) but they are certainly not sequential. The most significant hydrolyzing enzymes are carboxylesterases, arylesterases, cholinesterases, and serine endopeptidases. The hydrolysis of xenobiotic esters and amides generate carboxylic acids, alcohols, and amines. Such metabolites are all susceptible to phase II conjugation and excretion.

Phase II Reactions (Biosynthetic, Conjugation Reactions) Phase II reactions involve the biosynthetic union or coupling (conjugation) of xenobiotics or functionalized intermediates from phase I reactions with endogenous molecules generating hydrophilic conjugated metabolites that are devoid of toxicological activity and easily eliminated. Various endogenous molecules that are ionizable at physiological pH such as carbohydrates, proteins or amino acids, sulfur or phosphate components, or intermediates in lipid metabolism, are used in these reactions. Conjugated metabolites are substrates for specific transport proteins in epithelial tissues and are secreted with great efficiency. Conjugation reactions have a dual purpose; they have a role in intermediary metabolism of endogenous compounds as well as in the biotransformation of xenobiotics.

These reactions can be categorized by the mechanism of interaction between the conjugating agent and the xenobiotic. In type I reactions (e.g., glycosides, sulfates, methylated, and acetylated conjugates), ATP is used to form an activated "high-energy" intermediate. This activated coenzyme is conjugated to the substrate at the site of one of the functional groups. In type II conjugation reactions, the substrate itself is activated and then combines with an unactivated coenzyme (e.g., an amino acid), forming the phase II metabolite. In type III reactions, the xenobiotic or its metabolite already possesses the intrinsic reactivity to enzymatically react with a conjugating agent without the preformation of a high-energy intermediate (e.g., formation of glutathione conjugates). The enzymes that catalyze the reaction between the substrate and activated coenzyme are collectively called transferases. These reactions are usually considered as detoxication reactions, but in certain cases, bioactivation occurs (e.g., N-glucuronides of N-hydroxyarylamines are transported to the bladder, where β-glucuronidase hydrolysis in the acidic urine produces hydroxylarylamines, which spontaneously form electrophilic arylnitrenium ions and cause bladder cancer [37]). Major phase II reactions include glucuronidation, sulfation, acetylation, and conjugation with glutathione or amino acids.

Glucuronidation

Glucuronidation represents the major route of sugar conjugation, and qualitatively and quantitatively, glucuronide formation is the most important form of conjugation of both xenobiotics and endogenous compounds in most organisms. Four general categories of O-, N-, S-, and C-glycosides are formed and increase the compounds' polarity and ultimately their excretion in urine or bile. In glucuronidation, the high-energy nucleotide uridine diphosphate (UDP)-glucuronic acid (UDPGA; Fig. 23.5) interacts with the functional group on the acceptor molecule (the substrate or aglycone) forming a conjugate with D-glucuronic acid. The transfer of glucuronic acid from UDPGA to the aglycone is catalyzed by a family of enzymes called UDP-glucuronosyltransferases (UGTs). There are multiple forms of UGTs that have different substrate specificities in the same species and in different species. For example, the domestic cat, *Felis catus*, can form glucuronide conjugates of thyroxin and bilirubin, but is unable to form other glucuronide conjugates. UGTs are localized to the ER of all tissues, but the majority of activity is found in the liver. Location in the ER is a strategy whereby it has direct access to products (including reactive intermediates) generated by microsomal P450, an excellent example of the integration of the sequestration of lipophilic xenobiotics in the lipids of the ER, the addition or unmasking of a functional group, and the conjugation of this functional group with the highly polar D-glucuronic acid.

Sulfation

As with glucuronidation, these conjugation reactions are found in most organisms and are important to both endogenous (e.g., thyroid and steroid hormones, catecholamine neurotransmitters, the tyrosinyl group of peptides and proteins, and bile acids) and xenobiotic compounds, particularly those chemicals with hydroxyl groups, but also contributes to the biotransformation of 1°, 2°, and 3° alcohols, amines, and to a lesser extent, thiols. The resulting compounds are generally less toxicologically active and more polar, thus more readily excreted in the urine, with little in bile (this however, is species dependent). However, a number of compounds are converted to highly labile sulfate conjugates that form reactive intermediates that have been implicated in carcinogenesis and other toxicities. Sulfate conjugation is a multistep process, comprising activation of inorganic sulfate, first, by converting it via ATP to adenosine-5′-phosphosulfate (APS), and further to the activated form, 3′-phosphoadenosine-5′-phosphosulfate (PAPS; Fig. 23.5) by ATP sulfurylase and APS kinase. PAPS is synthesized in all cells, but the highest concentrations are found in the liver or functional homologue in organisms without livers. Interestingly, the cat has very high levels of this conjugating enzyme system, perhaps compensating for the lack of glucuronidation as a primary conjugation mechanism. Sulfotransferases catalyze the reaction, whereby the SO_3^- group of PAPS is readily transferred in a reaction involving nucleophilic attack of the phenolic oxygen or the amine nitrogen on the sulfur atom with the subsequent displacement of adensine-3′,5′-diphosphate. Interestingly, sulfation is a high affinity, low capacity conjugation reaction, which is opposite to glucuronidation. At low phenol concentrations, sulfation predominates; at high concentrations, glucuronidation is the most important pathway. The major classes of sulfotransferases that metabolize xenobiotics are aryl sulfotransferases, alcohol sulfotransferases, tyrosine-ester sulfotransferases, and amine N-sulfotransferases.

Methylation

Methylation involves the transfer of methyl groups from one of two methyl donor substrates, *S*-adenosylmethione (SAM (Fig. 23.5)) or N^5-methyltetrahydrofolic acid. Methylation is a common biochemical reaction of endogenous compounds (e.g., proteins, nucleic acids, phospholipids), but is not usually of quantitative

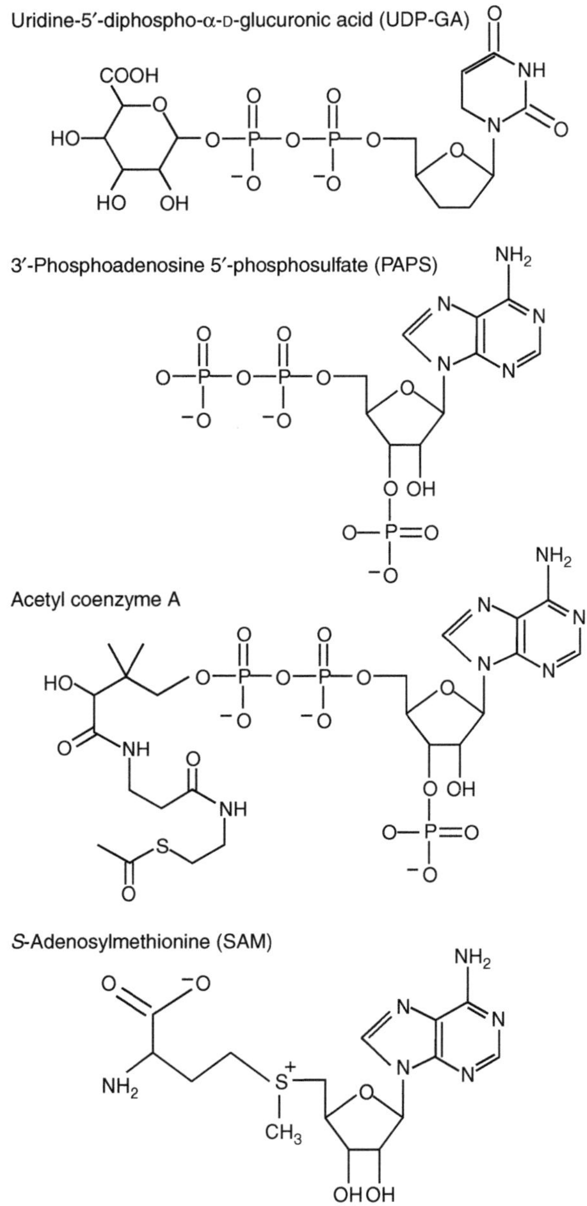

Fig. 23.5 Phase II high energy cofactors

significance in the biotransformation of xenobiotics. Methylation differs from most other conjugation reactions in that it masks functional groups, which may reduce water solubility and/or impair its ability to be excreted or participate in other conjugation reactions. N-methylation is an important reaction in metabolizing 1°, 2°, and 3° amines. Methyl groups from SAM are also transferred to the thiol-containing xenobiotics (e.g., thiopyrimidine drugs).

Acylation

There are two types of acylation reactions based on the activated component. Acylation reactions involve foreign carboxylic acids and amines to form amide conjugates. In the first type, an activated conjugating intermediate acetyl CoA (Fig. 23.5) reacts with a xenobiotic through acetyl CoA: amine N-acetyl transferases (mostly cytosolic enzymes) in a reaction called acetylation. Acetylation reactions require a specific cofactor, acetyl-CoA, which is obtained mainly from the glycolysis pathway via direct interaction of acetate and coenzyme A, or from the catabolism of fatty acids or amino acids. The primary site of acetylation is the liver, although extrahepatic sites have been identified as well (e.g., spleen, lung, and gut). This reaction has been studied extensively in humans because of its important consequences in drug therapy and in the carcinogenicity of certain xenobiotics.

In the second type of acylation reaction, amino acids are conjugated to carboxylic acid containing xenobiotics following activation of the foreign compound to form an acyl CoA derivative catalyzed by the enzyme acid:CoA ligase (a mostly cytosolic enzyme). The acyl CoA subsequently reacts with an amino acid (most frequently glycine, glutamine, alanine and histidine, and ornithine in birds, and taurine in fish) to form an acylated amino acid conjugate, catalyzed by N-acyltransferase. The reaction is categorized as a high affinity, low-medium capacity system, and given that it has only a relatively small substrate base such as aromatic, heteroaromatic and arylacetic acids, it is not considered as effective as the main conjugation reactions (glucuronidation, sulfate, GSH-conjugation).

Glutathione Conjugation and Mercapturic Acid Formation

Glutathione (GSH) is a tripeptide that contains cysteine, glutamic acid, and glycine. In its reduced form, it acts as an antioxidant due to its nucleophilic nature at the sulfur group, and as such, it is essential in protecting cells from reactive oxygen species (ROS). GSH is found almost exclusively in its reduced form since the enzyme that reverts it from its oxidized form, GSH reductase, is constitutively active, and is present in high concentrations in all tissues, particularly the liver.

The glutathione *S*-transferases (GSTs) are a family of mainly cytosolic enzymes (microsomal and mitochondrial forms do exist) that play an important role in xenobiotic resistance by several distinct mechanisms: increased detoxication, sequestration of the xenobiotic, and decreased xenobiotic activation. Arguably the most important function is to catalyze the initial reaction in the formation of N-acetylcysteine (mercapturic acid) derivatives of a diverse group of xenobiotics with the following characteristics in common: some degree of hydrophobicity, presence of an electrophilic carbon atom, and some tendency to react with glutathione nonenzymatically (e.g., electrophilic substrates including herbicides, insecticides, pharmaceuticals and anesthetics, anticancer agents, carcinogens, antibiotics, etc.) GSTs are found ubiquitously in all tissues, with the highest levels in the testes, liver, intestine, kidney, and adrenal glands of humans. They have also been found in every organism examined.

The initial step in mercapturic acid formation (Fig. 23.6, with example) involves the formation of a glutathione conjugate. GSH conjugation involves the formation of a thioether link between the GSH and electrophilic compounds. The reaction can be considered as the result of nucleophilic attack by GSH on electrophilic carbon atoms; thus conjugation with glutathione usually results in detoxication of electrophilic compounds by preventing their reaction with nucleophilic centers in other biomolecules. The enzyme γ-glutamyltranspeptidase, a membrane-associated enzyme found in high concentrations in cells that provide absorptive or excretory functions, removes glutamic acid from the GSH-conjugate, forming a cysteinyl glycine conjugate. Glycine is then removed from this metabolite by the enzyme cysteinyl glycine dipeptidase or aminopeptidase M, to form the cysteine S-conjugate (pre-mercapturic acid), which is acylated by N-acetyltransferases, resulting in the formation of the N-acetyl derivative, mercapturic acid.

Modifiers of Biotransformation The dramatic significance of biotransformation and the mechanisms to achieve the protection and resistance of organisms from the effects of xenobiotic exposure have been discussed. There are a number of endogenous (physiological) and environmental factors that affect the abilities of different organisms to use biotransformation in a protective role, factors which can modify their susceptibility to toxicants. Some of the more important intrinsic factors include the following:

- Species, strain, and other genetic variations (particularly interesting with individual sensitivities to either drugs or environmental contaminants from a health perspective)
- Development (under complex ontogenic control)
- Sex (males typically > females)
- Age (low in fetus and infants, increasing in age and declines with old age)
- Hormones (complex interrelationship: antagonistic, additive, and synergistic relationships)
- Pregnancy (decrease in maternal enzymes)
- Disease (tissue and organ-specific as to magnitude, but generally decreases with hepatic disease, infections, cancer)
- Diurnal and seasonal cycles (direction with circadian rhythms depends mostly on the endogenous function of enzymes; season in wildlife usually reflects hormonal fluctuations)
- Nutritional effects (deficiencies generally decrease enzymes) [38]

Extrinsic factors that can modify biotransformation include: pre-exposure to xenobiotics, resulting in either inhibition of biotransformation enzyme systems or induction (increases) in enzyme concentrations following exposure. In ectothermic

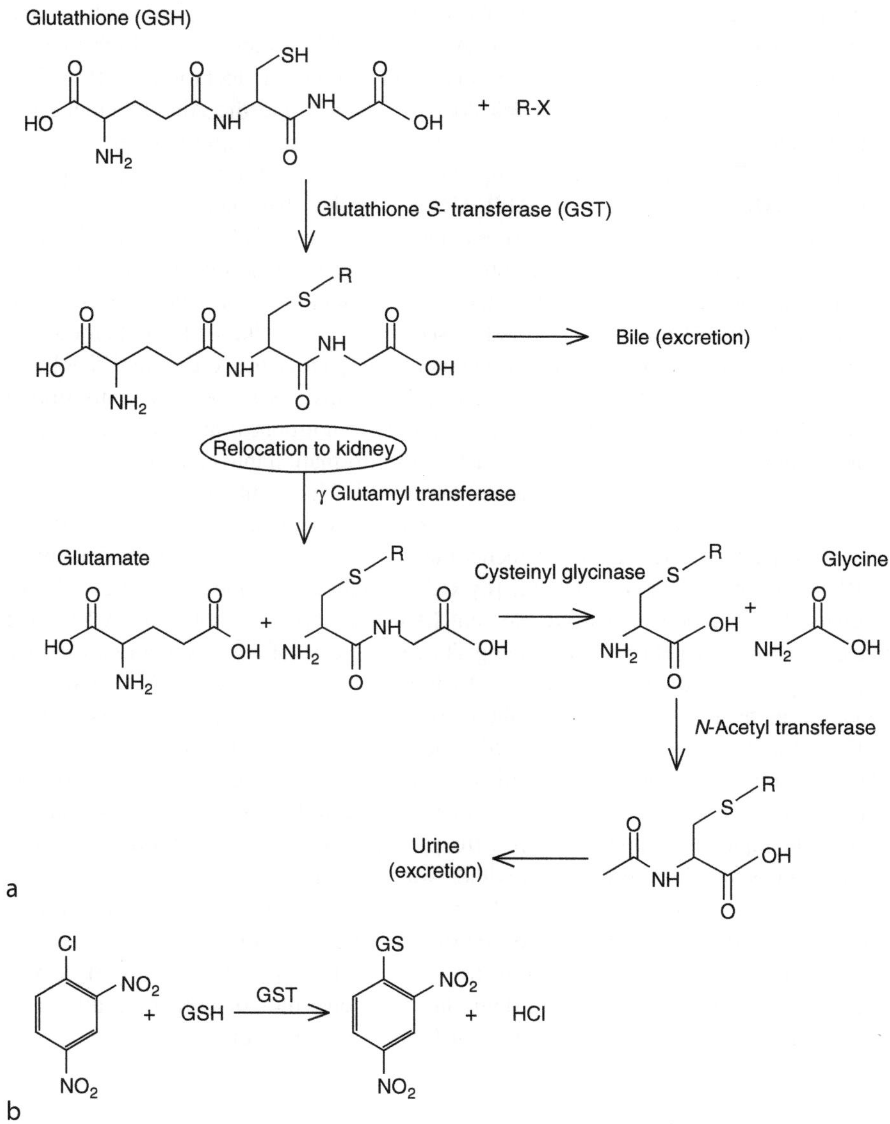

Fig. 23.6 (**a**) Glutathione conjugation and mercapturic acid formation. (**b**) An example of glutathione conjugation

organisms, environmental temperature and the rate of temperature change can play a large role in altering xenobiotic metabolism.

Prior exposure to some contaminants can increase the biotransformation capability of organisms by a process called *induction*. For example, P450 isoforms in the 1A (CYP1A) subfamily are inducible by planar PAH and HAH, including chlorinated biphenyls, dibenzo-*p*-dioxins, and dibenzofurans, as well as several drugs including the classic (well studied pharmacologically) inducers phenobarbital and 3-methylcolanthrene. The Ah receptor (AhR), a soluble protein with a high binding affinity for planar PAH and HAH, controls the induction of cytochrome CYP1A. The aryl hydrocarbon receptor nuclear translocator (ARNT) and the AhR together form a transcription factor complex through which altered gene expression and toxicity occurs [39]. Their dimerization is dependent upon high affinity binding of planar PAH and HAH to the AhR. The AhR:ARNT complex then binds specific genomic recognition sites (xenobiotic response elements, XRE) on DNA. This alters the transcription of target genes that include both phase I and phase II biotransforming enzymes. Other nuclear receptors are involved in the induction of biotransformation enzymes, and considerable current research effort is focused in this area. Like AhR, orphan nuclear receptors (e.g., constitutive androstane receptor [CAR], pregnane X receptor [PXR], and retinoid X receptor [RXR]) comprise a gene superfamily that encodes the transcription factors that sense endogenous (e.g., small lipophilic hormones) and exogenous (e.g., drugs, xenobiotics) compounds, and transfer this information into cellular processes by regulating the expression of their target genes [40]. The regulation and induction of phase II enzymes is less well understood, and recent research has revealed the existence of several *cis*-acting regulatory elements including the antioxidant response element (ARE/electrophile response element [EpRE]), xenobiotic responsive element (XRE/aromatic hydrocarbon responsive element [AhRE]), activator protein-1 (AP-1), and nuclear factor-kappa B (NF-kB) binding sites. In the case of ARE, xenobiotics induce stress responses leading to potential sulfhydryl modification of Keap1-Nrf2 and other signaling pathways, which leads to the activation of the transcription factors such as Nrf2/Maf and increases ARE-mediated gene expression including phase II enzymes [41]. Maximum induction with chemicals is dependent on species, inducer (variation exists in inducer strength), route of administration, and dose; however, induction can occur quite rapidly (within hours in in vitro assays, days in vivo), and enzyme activities can remain elevated for weeks upon withdrawal of the inducer.

Unlike induction (a physiologically regulated and evolved process for protection), the inhibition of biotransformation enzymes can include a multitude of mechanisms including competition for active sites or cofactors of enzymes, decreased biosynthesis, increased degradation of enzymes or cofactors, allosteric changes in enzyme conformation, loss of functional tissue, etc.

Phase III Transport In many cases, these transporters act as a first line of defense, preventing toxic chemicals from entering the cell. However, if a compound is not recognized by these transporters and enters the cytoplasm, detoxifying enzymes in the cell may modify the chemical to a more hydrophilic form. In this case, related cellular transporters can again come into play, effluxing the modified products out of the cell. There are two main types of these: *ATP-binding cassette* (ABC) and *solute carrier* (SLC) proteins (Fig. 23.7).

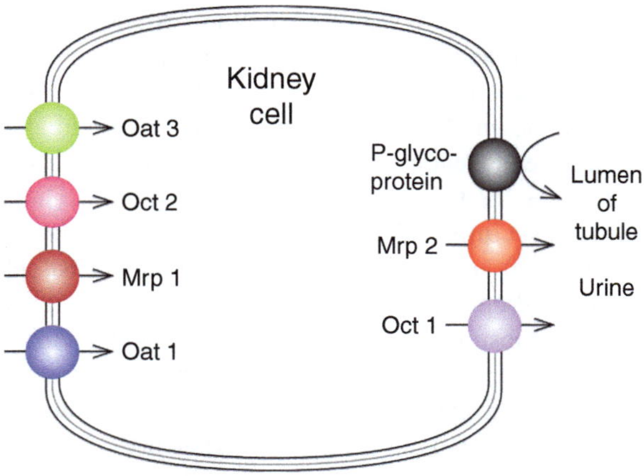

Fig. 23.7 Vectorial transport of xenobiotics and their metabolites is accomplished by efflux transporters as pictured in this kidney cell. *Oat* Organic anion transporter, *Oct* Organic cation transporter, *Mrp* multidrug resistance-associated protein

ABC Proteins

ABC transporters are evolutionarily ancient, and three ABC subfamilies of efflux (multidrug) transporters involved in the biochemical defense against toxicants have both health and environmental significance [42]. ABC drug efflux transporters have a wide phylogenetic distribution and are found in vertebrates as well as in deuterostome invertebrates, protostome invertebrates, fruitflies, protozoans, and yeast. Homologous proteins are also present in plants, though their roles in multidrug efflux have not been firmly established. ABC transporters include members of the P-glycoprotein (P-gp [ABCB]) family, the multidrug resistance protein (MRP [ABCC]) family and the breast cancer resistance protein (BCRP or MXR [ABCG]) family. These ABC proteins are membrane-bound transporters coupling ATP hydrolysis to the translocation of substrates across biological membranes. These efflux pumps provide important protection to internal tissues (e.g., blood–brain barrier, liver, kidney, and placenta) and to sites of entry of environmental contaminants (e.g., gills, gastrointestinal tract).

P-Glycoprotein (ABCB Family) The multidrug resistance (MDR due to its remarkably nonspecific substrate base) protein, P-glycoprotein (P-gp for "permeability glycoprotein," [ABCB1]), transports both endogenous and exogenous substrates and is found in a broad range of taxa [43]. It is a transmembrane energy-dependent pump that mediates the efflux of a large number of moderately hydrophobic compounds. P-gps confer MDR to tumor cell lines [44] and tumors of human patients [45], and have been described in many aquatic wildlife species [46]. It appears that, in addition to normal cell function, P-gp activity contributes to the relative hardiness of fish living in contaminated environments, providing multixenobiotic resistance (MXR).

Although the high resolution structures of ABC proteins and the structure–function relationships of mammalian P-gp have been characterized in recent years, how it works as an efflux pump or its role in normal physiology is still unknown. A role for P-gp certainly exists in altering the pharmacokinetics of natural endogenous compounds and xenobiotics, although the full range of substrates for P-gp has not yet been explored. The factors that regulate P-gp are also not yet fully understood, and understanding the regulation of these efflux pumps is essential in determining their role in normal physiology. Limited evidence suggests P-gp is up-regulated (induced) in response to a number of signals, including specific substrates, toxicants, heat stress, and nuclear factors [47].

P-glycoproteins are expressed in many of the same tissues as biotransformation enzymes (e.g., cytochromes P450); however, biotransformation is not a prerequisite for efflux. Compounds are transported as parent molecules, and among P-gp substrates, no consensus structure has been defined, although within each class of chemicals certain chemical features have been defined that seem to be essential for functional interaction (moderate hydrophobicity, small size, and positively charged domains). Their interaction with biotransformation systems, or other transporters (e.g., multidrug resistance protein [MRP]) are unclear. Although it is now clear that the induction of CYP enzymes and P-gp are not co-regulated, they may interact through CYP-mediated production of P-gp substrates or CYP-mediated- or glutathione (GSH) depletion-mediated increases in intracellular reactive oxygen species (iROS) [48]. As a first line of cellular defense against xenobiotic influx, P-gp may be overwhelmed by high concentrations, or may not recognize certain classes of compound, which enter cells and are acted on by biotransformation enzymes such as P450, followed by the conjugation of the metabolite. The conjugated metabolite can then be eliminated by another ABCC family of efflux transporters, which recognize the conjugated moiety (e.g., MRP).

Multidrug Resistance-Associated Protein (ABCC Family) Conjugates require export across cell membranes, and conjugated metabolites may be subject to further metabolism (e.g., the pathway of mercapturic acid formation) or transported without further modification out of cells. The use of the term "Phase III metabolism" to describe membrane transport of drug conjugates [49] is also potentially misleading, since proteins such as MRP1 effect transport of some drugs in their parent form, without any prior biotransformation, and actually prevent cellular uptake. Some members of the multidrug resistance-associated protein (MRP, ABCC) family also contribute to multidrug efflux by acting like P-gp to efflux unmodified xenobiotics. However, members of this family also act on endogenous substrates that are normal products of metabolism, or act on xenobiotics that have been modified as a part of the above mentioned phase I and phase II biotransformation processes. The resulting conjugated, negatively charged molecules are recognized by various ABCC transporters and then exported from the cell [48].

Multixenobiotic Resistance Protein (ABCG Family) ABCG2, a member of the ABCG family commonly referred to as breast cancer resistance protein (BCRP), also causes significant resistance to a limited group of chemotherapeutic treatments (substrate specificity is less broad than for ABCB and ABCC [50]) as well as effluxing dietary toxicants [51]. A difference from the aforementioned transporter types, which are so-called full transporters (the gene product constitutes one structural unit), is that ABCG2 is a half transporter, in which two protein molecules are assembled to form a structural unit active as a homodimer. ABCG2 also plays a role in the translocation of essential compounds, such as riboflavin, into breast milk and may also transport certain toxicants into milk. In fact, in an environmental sense, transfer of milk to offspring of top predators such as whales, puts calves at a higher trophic level than the mothers, making young the highest risk group [52].

Multixenobiotic Resistance (MXR) MXR is a phenomenon that increases the potential for survival of aquatic organisms in polluted environments through enhanced expression of one or more MDR-like transport mechanisms (e.g., P-gp or MRP) that restrict xenobiotic uptake [53]. As mentioned earlier, these transporters act as first lines of defense against xenobiotics and are found in tissues where exchange with the environment can ensue. Such transport systems are present at some basal level in animals maintained in unpolluted environments, and they appear to be up-regulated on exposure to pollutants.

Mixtures of xenobiotics can compromise this defense because competitive binding of different chemicals to the transporters can sabotage the activity of transporter binding sites or saturate them. In addition, some of the chemicals in the mixture might be direct inhibitors of the transporter activity. The resultant competition or inhibition, termed chemosensitization, can decrease transporter activity of chemicals in a mixture such that toxic substances normally excluded from the cell can now enter.

Solute Carrier Proteins

Elimination in most organisms occurs via the kidneys and hepatobiliary route or functional equivalent. In kidney tubules and in hepatocytes, as well as other tissues, there exist a number of other xenobiotic (or metabolite) export pumps that mediate the vectorial transport of substrates from the basolateral extracellular fluid into the cells on the basolateral sides of cells and export them to excretory media on the luminal side of cells (e.g., facing canaliculi or lumen) (Fig. 23.7). Some of these transporters include the aforementioned ABC proteins. At the basolateral membrane are members of the SLC superfamily of proteins, the organic anion transporters (Oat1, Oat3), and organic cation transporters (Oct2). Oat1 exhibits a broad substrate specificity and is responsible for the cellular uptake of most amphiphilic organic anions. High levels of expression of organic cation/carnitine transporter (OCTN1) are seen in barrier and excretory tissues as well [54].

Regulation of Efflux Transporters Pregnane X receptor (PXR), constitutive androstane receptor (CAR), peroxisome proliferators activated receptors (PPAR), liver X receptor (LXR), and farnesoid X receptor (FXR) are all nuclear receptors that bind to specific DNA response elements that produce effects on phase III efflux transporters, although which specific transporters and the mechanisms are still unclear [41]. PXR appears to regulate several efflux transporters including P-gp, MRPs, and Oats. Common mechanisms exist between phase I and phase II xenobiotic metabolizing enzymes through AhR/ARNT, and binding to AhR-xenobiotic response elements and subsequent induction of mRNA suggesting coordinate regulation. Moreover, several types of phase II enzyme inducers have been empirically shown to increase the excretion of phase II metabolites. Since phase II metabolites are transported out of cells by P-gp, MRPs, and OatP2, phase III transport processes may be coordinately regulated; however, the precise mechanisms are not clear.

Storage

The distribution of organic xenobiotics to sites of storage are in most cases not directed, but are the results of simple chemical partitioning. Storage sites may be the target sites (e.g., paraquat in lung), or sites of noninteraction. Therefore, storage can only be considered a protective mechanism in the latter case. Major storage sites can include the following: plasma proteins (e.g., Cu is stored by ceruloplasmin [55]), liver and kidney (e.g., organic acids bid to Y protein/ligandin in liver), fat (e.g., chlordane, DDT, PCBs, dioxins), and bone (e.g., Pb). Xenobiotic sequestration inside the cell appears to be an uncommon protective mechanism. Doxorubicin (chemotherapy drug) was shown to be sequestrated into small cellular vesicles whose concentration was correlated to the cellular resistance level to the drug [56]. This storage mechanism could participate in exocytosis and elimination. Glutathione *S*-transferase and other transferases contain lipophilic domains that can serve as binding sites for a number of exogenous compounds that are not substrates for reactions [57]. Organisms have few mechanisms for protection against metal contaminants. Metallothioneins (MTs) are ubiquitous low molecular weight proteins rich in cysteines with a high affinity for specific metals. They function in metal homeostasis by binding essential metals (e.g., Zn) and in defense by sequestering nonessential metals (e.g., cadmium). MTs have been identified in most tissues, and can be greatly induced in tissues that are active in metal uptake, storage, and excretion [58].

Toxicodynamically Derived Resistance Mechanisms

Toxicodynamically derived resistance refers to alterations in xenobiotic-receptor interactions. This can be manifest by alterations in target site, increased or decreased concentrations of target molecules, or circumvention of target function. Whether any of these resistance mechanisms arises and is retained in a population depends on a number of factors, most importantly its presence in the population and the physiological cost of the mechanism to the organism. Variable resistance mechanisms will come into play on different time scales; for example, a single major resistance mechanism should arise first with some associated cost, a cost that can potentially be corrected by epistasis with subsequent less costly mechanisms.

Cellular Detoxification

Many xenobiotics have similar modes of action, and the evolutionary chemical history of organisms has equipped them with various mechanisms in dealing with toxic effects they are likely to encounter. Many xenobiotics or their metabolites are electrophilic and highly reactive. For example, alkylating agents and drugs have lethal activity that is linked to electrophilic (quinines, cations) or radical species. The structure of GSH and its chemical characteristics have been described previously. The GSH thiol group performs a nucleophilic attack on the target electrophile and forms a stable conjugate that is nonreactive.

Target Site Alterations

Structural Modification In contrast to the above example, many xenobiotics (particularly therapeutic drugs) are targeted to a specific molecule or cellular component. An organism may resist the action of a xenobiotic due to target site modification that affords it an efficient resistance mechanism. Non-silent point mutations within structural genes are the most common cause of target-site resistance [59]. These changes can be brought about by alterations in amino acid sequences, the distances between critical amino acid residues, and protein topography. These mechanisms can be rapidly acquired when the xenobiotic has a unique cellular receptor or when the xenobiotic resembles the target's endogenous substrate. Structural mutations (non-silent) that give rise to this type of resistance to the effects of a given xenobiotic can incur costs to the organism; altered binding/catalysis by structural changes can lead to biological effects from reduced functionality of the target in its normal physiological roles. For selection of the mutations to occur, the resultant amino acid change must reduce the binding of the xenobiotic without causing a significant loss of the primary function of the target site. The number of possible amino acid substitutions, therefore, can be very limited. It is not surprising that identical resistance-associated mutations are commonly found across highly diverged taxa [59]. The degree to which function is impaired by the resistance mutation is reflected in the fitness of resistant individuals in the absence of xenobiotic selection and the persistence of resistance in the field.

There are several classic examples of structural target site modification as a mechanism in insecticide resistance; resistance to organophosphates (OPs) and carbamates, legacy organochlorine insecticides, and the naturally derived insecticides (pyrethrins) are discussed as targets to components of the insect nervous system.

Acetylcholinesterase-based resistance has been well documented in numerous strains of mites, ticks, houseflies, and mosquitoes. The OPs and carbamates target acetylcholinesterase (AChE), an enzyme localized in the synapse that hydrolyzes the neurotransmitter acetylcholine once it has excited postsynaptic nerve membranes. The OP insecticides are converted to their oxon analogues via the action of monooxygenases before acting as AChE inhibitors. Alterations in AChE in OP-and carbamate-resistant insects result in a decreased sensitivity to inhibition of the enzyme by these insecticides [60]. The determination of kinetic constants of inhibition of AChE indicated that the binding affinities between AChE and the oxon metabolites of OPs or carbamates were lower in more resistant insect strains.

GABA receptors belong to a superfamily of neurotransmitter receptors that include nicotinic acetylcholine receptors. These receptors are formed by the oligomerization of five subunits around a central transmitter-gated ion channel. The GABA receptor in insects is a heteromultimeric gated chloride-ion channel, a widespread inhibitory neurotransmission channel in the insect's central nervous system and in neuromuscular junctions. The $GABA_A$ receptor-chloride channel complex is known to be the target of cyclodiene insecticides. Dieldrin first enhances and then suppresses γ-aminobutyric acid (GABA)-induced chloride currents. An alanine-to-serine substitution in the putative channel-lining domain of the GABA receptor confers resistance to cyclodienes such as dieldrin in a broad range of dieldrin-resistant insects [61]. The only variation in resistant insects is that glycine rather than serine can sometimes be the substituted amino acid residue. Despite the widespread switch away from the use of cyclodiene insecticides for agricultural and public health use the resistance allele is still found at relatively high frequencies in insect field populations. The insect GABA receptor is implicated as a site of action for some pyrethroids and avermectins as well as cyclodienes.

DDT and pyrethroids cause persistent activation of Na^+ channels by delaying the normal voltage-dependent mechanism of inactivation. Insensitivity of Na^+ channels to insecticide inhibition, also called knockdown or *kdr-like resistance*, was inferred from cross-resistance between DDT and pyrethroids, which act on the same site within the

Na^+ channel. Several mechanisms of resistance have been proposed and involve a reduction in the number of Na^+ channels, changes in the fluidity of nerve membranes, or alterations in the binding characteristics between Na^+ channels and insecticides. Changes associated with resistance are more variable than those seen in with GABA receptors but are still limited to a small number of regions on this large channel protein. The first mutation to be characterized in *kdr* insects was in the housefly. The Na^+ channel of housefly contains 2,108 amino acids, which fold into 4 hydrophobic repeat domains. A leucine to phenylalanine point mutation in the S6 transmembrane segment of domain II produced 10- to 20-fold resistance to DDT and pyrethroids [62]. More than 500-fold resistance is found in "*super-kdr*" houseflies, which possess a mutation in a second methionine to threonine substitution further upstream in the same domain.

Overexpression of Target The up-regulation of target is also be a resistance mechanism by which the target is allowed to fulfill its biological function even in the presence of the xenobiotic at physiological concentrations and resultant toxicity. This is a compensatory mechanism for the portion of the target that is occupied. Thymidilate synthase (TS) is the major cellular target of 5-fluorouracyl (5-FU), a chemotherapy agent. 5-FU is activated to 5′-monophosphate-5-fluorouridine, which interrupts the action of TS, blocking the synthesis of the pyrimidine thymidine, a nucleotide required for DNA replication, resulting in a thymidine deficiency and cell death in replicating cells. Resistant tumor cells show increased TS activity through increases in the amount of target per cell [63].

Down-regulation (Including Gene Disruption or Silencing) Finally, decreasing target site sensitivities can be achieved by down-regulation of receptor levels or responsiveness of signal transduction pathways. This can be problematic, as down-regulation of an important physiological process can lead to reduced function. Down-regulation can also occur for repressors; for example, AhR repressor mutations in regulatory factors causing disruption in silencer elements can result in overexpression of AhR and may carry a high cost due to dominant expression in the wrong tissues; like structural mutations, these types of mutations may arise early in resistance, but disappear when less deleterious alternatives come about. One example of down-regulation occurs in resistance to cancer chemotherapy. Topoisomerase II is a homodimeric enzyme that is essential for DNA replication since it modifies DNA topology by transient breakage of DNA and religation. During this process, an intermediate complex between the enzyme and DNA forms and is susceptible to several cytotoxic chemotherapy drugs (e.g., mitoxanthrone) that stabilize the complex leading to DNA fragmentation and chromosomal aberrations and rearrangements. In resistant tumor cells, topoisomerase II is down-regulated to levels just required for cell division, thereby reducing the amount of DNA damage caused by these drugs [63, 64].

Circumvention of Target

This is a rare phenomenon of resistance but has been documented very early in organisms that have alternative mechanisms to bypass the affected site of action. An example of circumvention occurs in citrus-scale insects resistant to hydrogen cyanide. Widespread application of hydrocyanic acid (HCN formulation) was initially limited to the fumigation of valuable tree crops such as citrus fruits, which led to the development of other HCN-based insect and rodent control products for the fumigation of ships, stores, factories, and even residential buildings. The target molecule is cytochrome oxidase, in the fourth complex of the electron transport chain. HCN binds to Fe within this protein, preventing it from transferring electrons to oxygen, resulting in an inability to aerobically produce ATP. Citrus-scale mites utilize an HCN-insensitive flavoprotein as an alternate electron carrier in the place of cytochrome oxidase, enabling electron transport [65].

Conclusions and Future Directions

Understanding the role of protection (tolerance and resistance) has numerous significant implications for the quality of life and the environment. It impacts the area of human health in disease treatment (e.g., chemotherapy), pharmacology (e.g., new drugs, individual susceptibilities, safety), and public health protection programs (disease and vector control). It is also crucial for many areas of the environment in determining sensitive and sentinal species, biomarker development, establishment of water quality guidelines, new chemical development, and risk assessment. As well, it has considerable economic ramifications (e.g., modern agriculture and pest resistance, cross-resistance, genetically modified food, biological control, integrated pest management programs). Few real tools for delaying or preventing the onset of resistance to chemicals such as pesticides and chemotherapeutic agents have yet to be put into practical use. Furthering basic scientific understanding of biological relationships (e.g., herbivory and allelochemicals), biological toxins and antidotes, development theories for resistance evolution, etc. are also of value.

Imagining future directions for a subject area as broad and multidisciplinary is fraught with potential omissions. Directions for more research are as diverse as the applications that would be supported. One thing is clear, however: With the enormous advances being made in molecular biology (e.g., toxicogenomics, bioinformatics, metabolomics) and its sophisticated tools, insights into the mechanisms and manipulations of protection against xenobiotics are fast approaching. These new tools and information will be transforming, and if researchers make a conscious effort to link molecular data to population or ecological levels, gains in this sphere of study will be vast.

Bibliography

1. Ngo NT, Stupans I, McKinnon RA (2009) Xenobiotic detoxification in Australian marsupials – an overview. Comp Biochem Physiol Part A 154:S5
2. Wirgin I, Waldman JR (2004) Resistance to contaminants in North American fish populations. Mut Res 552:73–100
3. Mallet J (1989) The evolution of insecticide resistance: have the insects won? Tree 4:336–340
4. Saves I, Masson J-M (1998) Mechanisms of resistance to xenobiotics in human therapy. Cell Mol Life Sci 54:405–426
5. Ray K, Levandowsky M (2001) Effects of ionic blocking agents on shape, motility and chemosensory responses in Euglena. Abstracts of the General Meeting of the American Society for Microbiology 101:734
6. Bargmann CI (2006) Chemosensation in *C. elegans*. In: WormBook (ed) The *C. elegans* Research Community, WormBook, 29 pp. doi:10.1895/wormbook.1.123.1, http://www.wormbook.org
7. Doty RL (2001) Olfaction. Ann Rev Psychol 52:423–452
8. Rouquier S, Giorgi D (2007) Olfactory receptor gene repertoires in mammals. Mut Res/Fund Mol Mech Mutagenesis 616:95–102
9. Hall LW, Burton DT, Graves WC, Margrey SL (1984) Behavioral modification of estuarine fish exposed to sulfur dioxide. J Toxicol Environ Health 13:969–978
10. Suh GSB, Wong AM, Hergarden AC, Wang JW, Simon AF, Benzer S, Axel R, Anderson DJ (2004) A single population of olfactory sensory neurons mediates an innate avoidance behaviour in Drosophila. Nature 431:854–859
11. Tierney KB, Baldwin DB, Hara TJ, Ross PS, NScholz NL, Kennedy CJ (2009) Review: olfactory toxicity in fishes. Aquat Toxicol 96:2–26
12. Kandel ER (2009) The biology of memory: a forty-year perspective. J Neurosci 29:12748–12756
13. Pierce RC, Kalivas PW (1997) A circuitry model of the expression of behavioral sensitization to amphetamine-like psychostimulants. Brain Res Rev 25:192–216
14. Lee JC, Bernays EA (1990) Food tastes and toxic effects: associative learning by the polyphagous grasshopper *Schistocerca americana* (Drury) (Orthoptera: Acrididae). Animal Behav 39:163–173
15. Craig AD (2002) How do you feel? Interoception: the sense of the physiological condition of the body. Nat Rev Neurosci 3:655–666
16. Fernando CT, Liekens AML, Bingle LEH, Beck C, Lenser T, Stekel DJ, Rowe JE (2009) Molecular circuits for associative learning in single-celled organisms. J R Soc Interface 6:463–469
17. Lopez-Grancha M, Sanchez-Amate C, Navarro M, Carvajal F, Sanchez-Santed F, Cubero I (2006) Lateral parabrachial lesions disrupt paraoxon-induced conditioned flavor avoidance. Toxicol Sci 91:210–217
18. Mandrillon A-L, Saglio P (2007) Herbicide exposure affects the chemical recognition of a non native predator in common toad tadpoles (*Bufo bufo*). Chemoecology 17:31–36
19. Myllyvirta TP, Vuorinen PJ (1989) Avoidance of bleached kraft mill effluent by pre-exposed *Coregonus albula* L. Water Res 23:1219–1227
20. Patil VK, David M (2010) Behavioral and morphological endpoints: as an early response to sublethal malathion intoxication in the freshwater fish, *Labeo rohita*. Drug Chem Toxicol 33:160–165
21. Tierney KB, Casselman M, Takeda S, Farrell AP, Kennedy CJ (2007) The relationship between cholinesterase inhibition and two types of swimming performance in chlorpyrifos-exposed coho salmon (*Oncorhynchus kisutch*). Environ Toxicol Chem 26:998–1004
22. Martin GG (1978) Ciliary gliding in lower invertebrates. Zoomorphology 91:249–261

23. Storch V, Welsch U (1972) The ultrastructure of epidermal mucous cells in marine invertebrates (Nemertini, Polychaeta, Prosobranchia, Opisthobranchia). Mar Biol 13: 167–175
24. Taylor HH (1971) The fine structure of the type 2 cells in the Malpighian tubules of the stick insect, Carausius morosus. Cell Tissue Res 122:411–424
25. Khanvilkar K, Donovan MD, Flanagan DR (2001) Drug transfer through mucus. Adv Drug Delivery Rev 48:173–193
26. Debat H, Eloit C, Blon F, Sarazin B, Henry C, Huet J-C, Trotier D, Pernollet J-C (2007) Identification of human olfactory cleft mucus proteins using proteomic analysis. J Proteome Res 6:1985–1996
27. Starcevic SL, Zielinski BS (1995) Immunohistochemical localization of glutathione S-transferase pi in rainbow trout olfactory receptor neurons. Neurosci Lett 183:175–178
28. Maïbèche-Coisne M, Jacquin-Joly E, François MC, Nagnan-Le Meillour P (2002) cDNA cloning of biotransformation enzymes belonging to the cytochrome P450 family in the antennae of the noctuid moth *Mamestra brassicae*. Insect Mol Biol 11:273–281
29. Boutsiouki P, Thompson J, Clough G (2001) Effects of local blood flow on the percutaneous absorption of the organophosphorus compound malathion: a microdialysis study in man. Arch Toxicol 75:321–328
30. Moon KC, Maibach HI (1991) Percutaneous absorption in diseased skin: relationship to the exogenous. In: Menne T, Maibach HI (eds) Exogenous dermatoses: environmental dermatitis. CRC Press, Boca Raton, pp 217–226
31. Matsumura F (2010) Studies on the action mechanism of benzoylurea insecticides to inhibit the process of chitin synthesis in insects: a review on the status of research activities in the past, the present and the future prospects. Pesticide Biochem Physiol 97:133–139
32. Singh P, Sihorkar V, Jaitely V, Kanaujia P, Vyas SP (2000) Pilosebaceous unit: anatomical considerations and drug delivery opportunities. Indian J Pharmacol 32:269–281
33. Kapoor IP, Metcalf RL, Herwe AS, Coats JR, Khalsa MS (1973) Structure-activity correlations of biodegradability of DDT analogs. J Agric Food Chem 21:310–315
34. Ross PS, Ellis GM, Ikonomu MG, Barrett-Lennard LG, Addison RF (2000) High PCB concentrations in free-ranging Pacific killer whales, *Orcinus orca*: effects of age, sex and dietary preference. Mar Poll Bull 40:504–515
35. Coon MJ (2005) Cytochrome P450: nature's most versatile biological catalyst. Ann Rev Pharmacol Toxicol 45:1–25
36. Kennedy CJ, Tierney KB (2008) Energy intake affects the biotransformation rate, scope for induction, and metabolite profile of benzo[a]pyrene in rainbow trout. Aquat Toxicol 90:172–181
37. Cui L, Sun H, Wishnok JS, Tannenbaum SR, Skipper PL (2007) Identification of adducts formed by reaction of N-acetoxy-3, 5-dimethylaniline with DNA. Chem Res Toxicol 20:1730–1736
38. Hodgson E, Levi PE (1994) Biochemical toxicology, 2nd edn. Appleton & Lange, Norwalk
39. Denison MS, Pandini A, Nagy SR, Baldwin EP, Bonati L (2002) Ligand binding and activation of the Ah receptor. Chem Biol Interact 141:3–24
40. Levine SL, Perdew GH (2001) Aryl hydrocarbon receptor (AhR)/AhR nuclear translocator (ARNT) activity is unaltered by phosphorylation of a periodicity/ARNT/single-minded (PAS)-region serine residue. Mol Pharmacol 59:557–566
41. Xu C, Li CY-T, Kong A-NT (2005) Induction of Phase I, II and III drug metabolism/transport by xenobiotics. Arch Pharm Res 28:249–268
42. Epel D, Luckenbach T, Stevenson CN, MacManus-Spencer LA, Hamdoun A, Smital T (2008) Efflux transporters: newly appreciated roles in protection against pollutants. Environ Sci Technol 42:3914–3920
43. Cooper P, Vogelbein W, Van Veld W (1996) Immunohistochemical and immunoblot detection of P-glycoprotein in normal and neoplastic fish liver. In: Ostrander G (ed) Techniques in aquatic toxicology. CRC Press, Boca Raton, pp 307–325

44. Juliano RL, Ling V (1976) A surface glycoprotein modulating drug permeability in Chinese hamster ovary cell mutants. Biochim Biophys Acta 455:152–162
45. Gerlach JH, Kartner N, Bell DR, Ling V (1986) Multidrug resistance. Cancer Surv 5:25–46
46. Kurelec B, Krca S, Lucic D (1996) Expression of multixenobiotic resistance mechanism in a marine mussel *Mytilus galloprovincialis* as a biomarker of exposure to polluted environments. Comp Biochem Physiol 113:283–289
47. Bard S (2000) Multixenobiotic resistance as a cellular defense mechanism in aquatic organisms. Aquat Toxicol 48:357–389
48. Felix R, Barrand A (2002) P-glycoprotein expression in rat brain endothelial cells: evidence for regulation by transient oxidative stress. J Neurochem 80:64–72
49. Ishikawa T (1992) The ATP-dependent glutathione S-conjugate export pump. Trends Biochem Sci 17:463–468
50. Litman T, Druley TE, Stein WD, Bates SE (2001) From MDR to MXR: new understanding of multidrug resistance systems, their properties and clinical significance. Cell Mol Life Sci 58:931–959
51. van Herwaarden AE, Jonker JW, Wagenaar E, Brinkhuis RF, Schellens JHM, Beijnen JH, Schinkel AH (2003) The breast cancer resistance protein (BCRP1/ABCG2) restricts exposure to the dietary carcinogen 2-amino-1-methyl-6-phenylimidazo[4, 5-b]pyridine. Cancer Res 63:6447–6452
52. Ross PS (2006) Fireproof killer whales (*Orcinus orca*): flame- retardant chemicals and the conservation imperative in the charismatic icon of British Columbia, Canada. Can J Fish Aquat Sci 63:224–234
53. Kurelec B (2002) The multixenobiotic resistance mechanism in aquatic organisms. Crit Rev Toxicol 22:23–43
54. Miller DS (2002) Xenobiotic export pumps, endothelin signaling, and tubular nephrotoxicants – a case of molecular hijacking. J Biochem Mol Toxicol 16:122–127
55. Twomey PJ, Viljoen A, House IM, Reynolds TM, Wierzbicki AS (2005) Relationship between serum copper, ceruloplasmin, and non–ceruloplasmin-bound copper in routine clinical practice. Clin Chem 51:1558–1559
56. Cornwell MM, Safa AR, Felsted RL, Gottesman MM, Pastan I (1986) Membrane vesicles from multidrug resistant human cancer cells contain a specific 150- to 170-kDa protein detected by photoaffinty labelling. Proc Natl Acad Sci USA 83:3847–3850
57. Kaplowitz N (1980) Physiological significance of glutathione S-transferases. Am J Physiol 239 (Gastrointest Liver Physiol 2): G439–G444
58. Sigel A, Sigel H, Sigel RKO (2009) Metallothioneins and related chelators. Metal ions in life sciences, vol 5. RSC Publishing, Cambridge
59. Taylor M, Feyereisen R (1996) Molecular biology and evolution of resistance to toxicants. Mol Biol Evol 13:719–734
60. Skouras PJ, Margaritopoulos JT, Seraphides NA, Ioannides IM, Kakani EG, Mathiopoulos KD, Tsitsipis JA (2007) Organophosphate resistance in olive fruit fly, *Bactrocera oleae*, populations in Greece and Cyprus. Pest Manage Sci 63:42–48
61. Gao J, Kozaki T, Leichter CA, Rinkevich FD, Shono T, Scott JG (2007) The A302S mutation in Rdl that confers resistance to cyclodienes and limited cross-resistance to fipronil is undetectable in field populations of house flies from the USA. Pesticide Biochem Physiol 88:66–70
62. Soderlund DM (2008) Pyrethroids, knockdown resistance and sodium channels. Pest Manage Sci 64:610–616
63. Skovsgaard T, Nielson D, Maare C, Wassermann K (2004) Cellular resistance to cancer chemotherapy. Int Rev Cytol 156:77–157
64. Vassetzky YS, Alghisi GC, Gasser SM (1995) DNA topoisomerase II mutations and resistance to anti-tumour drugs. Bioessays 17:767–774
65. Hoskins WM, Gordon HT (1956) Arthropod resistance to chemicals. Ann Rev Entomol 1:89–122

Index

E.A. Laws (ed.), *Environmental Toxicology: Selected Entries from the Encyclopedia of Sustainability Science and Technology*, DOI 10.1007/978-1-4614-5764-0,

Z

MIX
Papier aus verantwortungsvollen Quellen
Paper from responsible sources
FSC® C105338

If you have any concerns about our products,
you can contact us on
ProductSafety@springernature.com

In case Publisher is established outside the EU,
the EU authorized representative is:
Springer Nature Customer Service Center GmbH
Europaplatz 3, 69115 Heidelberg, Germany

Printed by Libri Plureos GmbH
in Hamburg, Germany